MATHEMATICAL SOCIOLOGY

Mathematical Sociology

AN INTRODUCTION TO FUNDAMENTALS

Thomas J. Fararo

A Wiley-Interscience Publication

JOHN WILEY & SONS, New York • London • Sydney • Toronto

Copyright © 1973, by John Wiley & Sons, Inc.

All rights reserved. Published simultaneously in Canada.

No part of this book may be reproduced by any means, nor transmitted, nor translated into a machine language without the written permission of the publisher.

Library of Congress Cataloging in Publication Data:

Fararo, T   J
  Mathematical sociology: an introduction to fundamentals.

  Bibliography: p.
  1. Sociology—Methodology.   2. Sociology—Mathematical models.   3. Game theory.   I. Title.

HM24.F35   301'.01'51   73-3204
ISBN 0-471-25460-6

Printed in the United States of America

10 9 8 7 6 5 4 3 2 1

The paradox is now fully established
that the utmost abstractions are the
true weapons with which to control
our thought of concrete fact.

ALFRED NORTH WHITEHEAD

PREFACE

This book constitutes a comprehensive introduction to the techniques, methodology, and achievements of mathematical sociology. It is intended primarily for graduate students and advanced undergraduates in sociology and related fields. All the mathematics is introduced along with the applications, so as to make the ideas accessible to a wide audience. The book is the outcome of an aim toward expository adequacy with respect to a wide variety of modes of formal representation. The main device used to achieve this aim is to balance formal definition with an abundance of concrete examples at every stage. A second device is the use of certain key ideas concerning model-building that are repeated and amplified in a variety of contexts, each new topic providing a fresh perspective. These and other didactic aspects of the book account for its length, despite its elementary character: it is only a gateway to the field.

In the introductory chapter, there is an overview of each of the following four parts of the book. As an approximate image of the overall structure, one might think of the four parts as forming a sequence starting with formal prerequisites, passing to methodology, and then to theoretical models followed by an in-depth study of one particular abstract framework. In terms of content emphasis, an image of the sequence is: relational system, dynamic system, relational structure and process, and rule-governed relational structure.

The book can be used in self-study, seminar, or lecture contexts. For the study of special topics—mathematical, methodological, or sociological—it is recommended that one consult the index: it must be remembered that in no one place is the whole story told about any such topic. For example, the discussion of measurement in Chapter 7 is conducted with reference to the concept of an empirical system. But this concept is not fully understood apart from the extensive example of measurement provided in Chapter 20. Indeed, the entirety of Part 4 is in one aspect a study of the deep interplay between a theoretical framework and an associated measurement basis. As a further aid in the study of special topics, the following diagram shows lines of logical dependence among the various chapters. Only Chapter 15 is not shown; it merely summarizes some principles of estimation, making reference to Chapters 13 and 14.

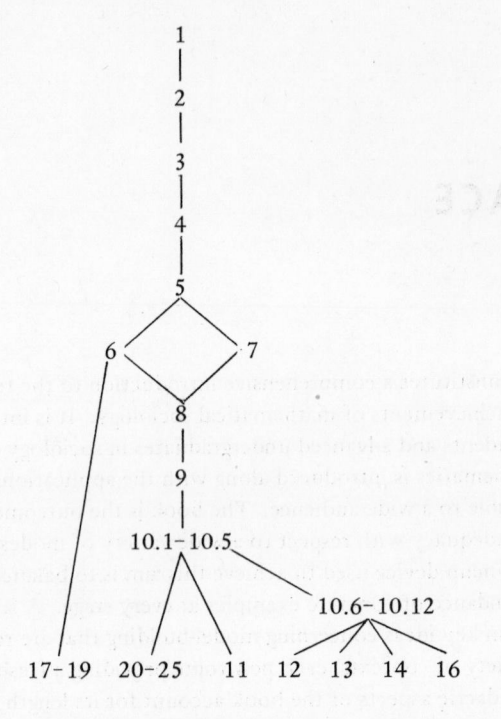

In terms of expository style, in Parts 1 and 2 (Chapters 2-11) the topics are presented in small chunks with frequent pauses for examples; in Parts 3 and 4 (Chapters 12-25) the sections are longer, reflecting longer discussions and longer chains of concatenated formal reasoning processes.

Some limitations of the book should be mentioned. First, it does not aim to produce a sociological synthesis. Second, it is not a study in the foundations (e.g., the metaphysical basis) of the field. These topics are clearly of the utmost importance, but they are excluded from the scope of the present task. Third, there are the obvious limitations arising from selectivity in the sociological topics treated. Indeed, space constraints induced certain omissions (for example, n-person game theory and random net theory) and would have required still more cuts, in the usual publication format. In the spirit of producing a book that was self-contained and yet comprehensive of the field, a method of composition was chosen that would reduce costs and thereby permit more text material.

In terms of the mathematics in the book there are two additional limitations. First, there are no explicitly labeled exercises. I had hoped to produce a separate booklet with problems and solutions, but for various reasons I have not yet been able to carry out this plan. However, there is a list of supplementary readings at the end of each part of the book; these readings often contain exercises one can draw upon in studying these materials. In addition, I frequently follow the practice of taking a development up to a certain point with the next steps left to the reader. In proof contexts, I usually prove only some of a larger set of listed results at the same level of difficulty. Second, a decision was made not to teach differential and integral calculus and to keep its use at a very minimal level. In certain places this means the reader need only see that a certain notation refers to a rate of change: this is purely "notational usage." For reference, the sections using some calculus are: some examples in 8.3 and 8.4; only notational usage in 8.6 and elementary use in 8.7; in estimation contexts in 11.14, 15.2, and 15.4; extensively in one mobility model in 16.4 and only notational usage in 16.5; in game theory in 25.11 and 25.12. Also, Chapter 13 is an attempt to explain Coleman's stochastic process methodology with only notational usage of the calculus. Hence there are only rare exceptions to the general rule: the book is addressed to the reader with no calculus background.

It is a pleasure to acknowledge the influence of Linton Freeman in shaping my sociological inclinations. As my mentor during my graduate student days, he was an important source of my commitment to exact methods in social theory. My ideas about mathematical social science were formed and transformed under diverse influences. To name but a few that come to mind: the disciplined imagination of Anatol Rapoport; the rigorous and wide-ranging set-theoretic style of Patrick Suppes; the philosophy of theoretical science framed by Stephen Toulmin, which coordinates so well with the experiences of model-building; the abstract sociological theories advocated and formulated by Joseph Berger, Bernard P. Cohen, and Morris Zelditch; and the achievements of James Coleman and Harrison White, who made mathematical sociology empirically meaningful and thereby established its place in modern sociological research.

Certain formal organizations deserve mention in terms of how I acquired the background for the writing of this book. The National Institute of Mental Health granted me a three-year postdoctoral fellowship for the study of pure and applied mathematics. Stanford University provided the rich intellectual nexus in which I spent those three years. Both in my graduate work and subsequently, I received financial support from the National Science Foundation. Several of the chapters of this book contain material reporting results of research supported by NSF under GS-2538.

To my colleagues in the Department of Sociology at the University of Pittsburgh I feel a special sense of gratitude. Their spirit of diversified intellectual adventure, articulated so well by Burkart Holzner and Roland Robertson, was the basis for a teaching program including formal analysis in a significant role.

Mario Bunge was directly responsible for my embarking upon the project of writing this book. My hope is that he and his colleagues in philosophy will find some value in the outcome, in spite of its obvious shortcómings from a philosophical standpoint.

In writing this book I was fortunate to have working with me a dedicated graduate student, Klaus Teuter, who studied the entire first draft and provided a page-by-page commentary, including detection of errors. In a technical work written for students, this was an essential element in the process of rewriting to strike the right balance between abstraction and concreteness. I am also grateful to three of my colleagues who read and commented on selected chapters of the early draft: Richard Conviser, Patrick Doreian, and Norman Hummon. Professor Doreian adopted Part 4 as a basis for a game theory seminar and provided a detailed commentary on the teaching of the materials, which led to improvements in the text. Kenneth Land read most of the first draft. He provided a general critical evaluation that led to the inclusion of certain topics I had omitted in the sheer exhaustion that set in eventually. I was not able to incorporate all his suggestions, but I thank him warmly for his counsel.

For the massive job of typing and proofreading the manuscript, I am indebted to Josephine Stagno and Klaus Teuter. More than technical efficiency on their part was involved; there was pride in a task well done, an autonomous outpouring of energy, and amid the welter of manuscript pages in various stages of revision, a sure organizational grip leading to an orderly outcome.

There is the final acknowledgment that seems pitifully inadequate when put into words: to my wife Irene, whose good sense is marred only by her incredible faith in me—which, by the way, made this book possible.

Thomas J. Fararo

University of Pittsburgh

June 1973

CONTENTS

Chapter Five   Mappings and Algebraic Systems 97

Chapter Six   Vectors and Matrices 125

Suggestions for Further Reading (Part One) 150

Part Two   Measurement, Process, and Probability

Chapter Seven   Basic Measurement Concepts 153

SYMBOLS AND ABBREVIATIONS

Topic	Symbol or Abbreviation	Meaning	Sec. Ref.
Logic	$\exists x, \ldots$	There exists an x such that . . .	2.3
	$\forall x, \ldots$	For every x, . . .	2.3
	$P \Rightarrow Q$	If P, then Q	2.8
	$P \Leftrightarrow Q$	P if and only if Q	2.8
	P iff Q	P if and only if Q	2.8
	$P \wedge Q$	P and Q	2.17
	$P \vee Q$	P or Q	2.17
	$\neg P$	Not the case that P	2.17
	Px, Pxy, Pxyz	Sentential function	2.12–2.15
Sets	$\{x : Px\}$	The set of all x such that Px	3.1
	$x \in A$	x is an element of set A	3.1
	$x \notin A$	x is not an element of A	3.1
	$\emptyset$	The null set	3.3
	$A \subseteq B$	A is a subset of B	3.3
	$A \subset B$	A is a subset of B and $A \neq B$	3.3
	$A \cup B$	Union of A and B	3.4
	$A \cap B$	Intersection of A and B	3.4
	$A - B$	Difference of A and B	3.4
	$\overline{A}$	Complement of A	3.4
	$A \triangle B$	Symmetric difference of A and B	11.3

Topic	Symbol or Abbreviation	Meaning	Sec. Ref.
	2^A	Power set of A	3.5
	$A \times B$	Cartesian product of A and B	3.7
	(x, y)	Ordered pair	3.7
Relations	xRy	Relationship between x and y	4.2
	R_A	All pairs from domain A which satisfy xRy: binary relation on A	4.6
	A/E	Set A modulo equivalence E	4.12
	$\bar{x}$	Equivalence class of element x	4.12
	$R \circ S$	Composition of relations R, S	4.16
	R^{-1}	Inverse of relation R	4.16
	R^n	nth power of relation R	4.16
	I	Identity relation	4.16
	$(A, R, S, \ldots)$	Relational system over set A	4.11
Mappings and Operations	$f: A \to B$	f is a mapping from A to B	5.2
	$x \mapsto y$	Element x is mapped into y = f(x)	5.2
	f^{-1}	Inverse of function f	5.4
	$g \circ f$	Composition, f followed by g	5.6
	$x \circ y$	Binary operation	5.11
	$(A, \circ, *, \ldots)$	Algebraic system	5.12
	$St(A, \ldots)$	Structure of system $(A, \ldots)$	5.16
	$\varepsilon = (E, R, \circ, \ldots)$	Empirical system	7.3
Vectors and Matrices	$\mathbf{x}$	Vector $\mathbf{x} = (x_1, x_2, \ldots, x_n)$	6.1
	$\mathbf{x} \cdot \mathbf{y}$	Dot product of vectors x, y	6.1
	$\mathbf{A} \atop (a_{ij})$	Matrix of elements a_{ij}, $\mathbf{A} = (a_{ij})$	6.2
	$\mathbf{AB}$	Row-by-column product of $\mathbf{A}$ and $\mathbf{B}$	6.2
	δ_{ij}	Kronecker delta	6.2

Topic	Symbol or Abbreviation	Meaning	Sec. Ref.
	$\mathbf{I}$	Identity matrix, $\mathbf{I} = (\delta_{ij})$	6.2
	$\mathbf{A}^{-1}$	Inverse of matrix $\mathbf{A}$	6.2
	$\mathbf{A}^T$	Transpose of matrix $\mathbf{A}$	6.2
	Det $\Big\}$ Δ	Determinant of $\mathbf{A}$, det $\mathbf{A} = \Delta$	6.3
	A_{ij}	Cofactor of element a_{ij} in $\mathbf{A}$	6.3
	$\mathbf{A}_R$	Adjacency matrix of relation R	6.4
	$\mathbf{A}_R \oplus \mathbf{A}_S$	Boolean sum of adjacency matrices	6.4
	$\mathbf{A}_R \times \mathbf{A}_S$	Element-wise product of matrices	6.4
	$\mathbf{A}_R \mathbf{A}_S (\oplus)$	Row-by-column product using Boolean addition	6.4
	R_*	Reachability relation based on R	6.4
	$|\mathbf{x}|$	Length of vector $\mathbf{x}$	24.6
	$d(\mathbf{x}, \mathbf{y})$	Distance from $\mathbf{x}$ to $\mathbf{y}$	24.6
	$\dot{\mathbf{X}}$	Rate of change of state vector $\mathbf{X}$	8.8
Probability and Statistics	$P(E)$	Probability of event E	9.6
	$P(E|F)$	Probability of E, given F	9.9
	$P(X = x)$	Probability that random variable X takes value x	10.4
	$E(X)$	Expectation of X	10.5
	$VAR(X)$	Variance of X	10.5
	$\overline{X}$	Observed mean of X	10.5
	$\hat{\theta}$	Estimator of parameter θ	15.1
	$L(\theta)$	Likelihood function	15.2
Groups	I	Identity element of group	17.2
	x^{-1}	Inverse of element x	17.2
	$(g_1, g_2 : \dots)$	Defining relations in terms of generators g_1, g_2	17.5

Topic	Symbol or Abbreviation	Meaning	Sec. Ref.
	$G_1 \cong G_2$	G_1 and G_2 are isomorphic	17.5
	C_N	Cyclic group of order N	17.5
	C_∞	The infinite cyclic group	17.5
	$C_M \times C_N$	Product of cyclic groups C_M, C_N	17.6
	D_N	Dihedral group of index N	17.6
	G/N	Factor group of G modulo N	19.2
Categories	$Mor(A, B)$	Set of morphisms for pair A, B	19.5
	$Ob(\mathcal{C})$	Objects of category $\mathcal{C}$	19.5
	$Ar(\mathcal{C})$	Arrows (morphisms) of category $\mathcal{C}$	19.5
Utility	NM	von Neumann and Morgenstern	20.1
	$(p_1 A_1, p_2 A_2, \ldots, p_r A_r)$	Probability distribution over $A_1, A_2, \ldots, A_r$	20.2
	$x \succsim_\alpha y$	α weakly prefers x to y	20.2
	$x \succ_\alpha y$	α strictly prefers x to y	20.3
	$x \sim_\alpha y$	α is indifferent between x and y	20.3
	$\tilde{A}_i$	A_i corresponds to alternative A_i via expression (20.3.5)	20.3
Games	$E(\mathbf{x}, \mathbf{y})$	Expected payoff, zero-sum game	23.8
	$E_i(\mathbf{x}, \mathbf{y})$	Expected payoffs, general game	24.1
	$E_i(\mathbf{z})$	Expected payoffs, under joint mixed strategy z	25.3
	$\mathbf{z} = (z_{ij})$	Joint mixed strategy	25.3
	(a_{ij})	Payoff matrix, zero-sum game	23.1
	$((a_{ij}, b_{ij}))$	Payoff matrix, general game	24.2
	$\left. \begin{array}{l} (P_\sigma(x)A_x) \\ (x_i y_j{}^o{}_{ij}) \\ (z_{ij}{}^o{}_{ij}) \end{array} \right\}$	Special forms of $(p_1 A_1, p_2 A_2, \ldots, p_r A_r)$	23.8
	v	Value of game, zero-sum	23.5

Topic	Symbol or Abbreviation	Meaning	Sec. Ref.
	(v_1, v_2)	Pairs of values, general game	25.9
	(u_0, v_0)	Shapley solution	25.9
	o_{ij}	Normal form outcome of game	22.3
	R^*	Payoff region	25.3
	P	Pareto optimal set	25.4
	$\mathfrak{N}$	Negotiation set	25.5
	$\mathfrak{N}^-$	Set of joint strategies mapped into $\mathfrak{N}$	25.5
	$\left.\begin{array}{cc} \max & \min \\ x & y \\ \min & \max \\ y & x \end{array}\right\}$	Operators applied to expectation	23.10
	BSL	Best security level	23.5
	BMSSL	Best mixed strategy security level	23.9
Point-sets	$L(u, v)$	Line segment from u to v	24.4
	$H(S)$	Convex hull of set S	24.5
	$N(x_0, r)$	Neighborhood of x_0 with radius r	24.6
	$b(S)$	Boundary of set S	24.6
	$i(S)$	Interior of set S	24.6
Miscellaneous	$\mathfrak{R}$	The real number system	5.2; 5.18
	$d(x, y)$	Distance from x to y in metric space	11.3
	$m(A)$	Measure of set A	11.3
	$\mathfrak{M}(\theta)$	Model $\mathfrak{M}$ as function of parameter θ	15.1
	BKM	Blumen-Kogan-McCarthy	16.3
	pms	Prescribed marriage system	18.2
	$i(x)$	Indegree of point x	21.2
	$o(x)$	Outdegree of point x	21.2

Topic	Symbol or Abbreviation	Meaning	Sec. Ref.
	$\left.\begin{array}{l}\max_{i} \\ \min_{j}\end{array}\right\}$	Operators on vector	23.4
	$\left.\begin{array}{l}\max_{i}\min_{j} \\ \min_{j}\max_{i}\end{array}\right\}$	Operators on matrix	23.4
	$\dfrac{dx}{dt} = x'(t)$	Derivative of $x(t)$; also $\dot{x}(t)$	8.3

MATHEMATICAL SOCIOLOGY

CHAPTER ONE INTRODUCTION

1.1. Aims of the Chapter. A central aspect of theoretical science involves the representation of phenomena in a systematic manner. One aim of this brief chapter is to give the reader some intuitive feel for the concrete meaning of the idea of representation. This is best attained by discussing the nature of model-building and providing a concrete example of a simple character that nevertheless exemplifies the general methodology. A second aim of this chapter is to introduce the remainder of the book by way of showing how its structure emerges out of a conception of the relation of mathematics to theory in general and to contemporary sociology in particular.

1.2. Phases of Model-Building. Three basic phases may be specified in model-building. First, there is the setup of the model. In this first phase the aim is to set down the primitive (formally undefined) notions and to embed these concepts in a set of assumptions. The notions and the assumptions, of course, depend in part on the concrete system under analysis and in part on the general framework within which such systems are analyzed. The second phase of model-building is the analysis phase. To analyze a model means to ferret out the consequences of the assumptions given in the setup of the model. Thus, in this second phase the procedures are those of conjecture and proof, with the aim of understanding the properties of the model. In the third phase the model is applied. In theoretical science, this last phase may be considered the testing phase: that is, the function of the application phase is that of testing the conception of the phenomenon proposed by the assumptions of the model. Bush and Mosteller (1955) demarcate three features in this phase: identification, estimation, and goodness-of-fit. An elaboration of these features yields the sequential subphases: identification of abstractions; estimation of parameters; calculation of numerical predictions; checking goodness-of-fit; and returning to an earlier phase or subphase based on a negative evaluation (model reconstruction) or based on a positive evaluation (model generalization).

1

These are only rough guidelines to the considerations involved in passing from the abstract, theoretical phases of setting up and analyzing a model to the more empirical phase of applying the model. An example will help to clarify these aspects of the process.

A simple example can be given in a probability model for group problem-solving (see Lorge and Solomon, 1960). The basic question is this: Given a puzzle that a person may or may not be able to solve, when several persons are brought together and asked to produce a solution, how does the solution probability depend upon the size of the group and the individual probability of finding a solution?

Concretely, an experiment is conducted in which a probability, say P_I, that an individual of a certain class of individuals can solve the specific puzzle is estimated by finding the relative frequency of individuals obtaining a solution. Then, a certain group size, n, is decided upon and a number of groups of that size are formed from the class of individuals. If the same puzzle is used in this second stage, then the groups must contain individuals who did not participate in the earlier part of the experiment. In any case, this provides a way of estimating $P_{G(n)}$, the probability that a group of size n can solve the puzzle, using the relative frequency of groups of size n that solve the puzzle.

We have yet to formulate any assumption about the way in which the group probability $P_{G(n)}$ depends upon n and P_I. One assumption—which may turn out to require modification but which does provide a definite conception—is that the group solves the puzzle if and only if at least one individual in the group has the ability to solve it. This is based on the simple idea that the phenomenon "group solves the puzzle" really amounts to no more than a shorthand way of saying that some individual solves the puzzle. For example, it neglects any interaction effects. The analysis of the model is now conducted with the purpose of ferreting out a formula relating $P_{G(n)}$ to n and P_I, based upon this assumption.

Note first that if P_I is the chance of solution by an individual, then $1 - P_I$ is the chance that he cannot solve the puzzle. The event that the group solves the puzzle has unknown probability $P_{G(n)}$. Therefore, the probability that the group fails to solve the puzzle is, in terms of this unknown, $1 - P_{G(n)}$. Now the group fails to solve the puzzle if and only if each individual cannot solve it; this is a direct use of the assumption in setting up the model. Thus, if we let $(1 - P_I)$ apply to each of n individuals, the system of n individuals fails to solve with probability $(1 - P_I)^n$, provided we make an assumption that the probabilities multiply. This latter assumption is a new assumption, not anticipated in setting up the model. The assumption is known as independence. Therefore, the basic assumption is augmented to read as follows: The group solves the puzzle if and only if at least one individual in the group can solve it and the individuals arrive at solutions independently. The word "independently" is used in the exact sense of probability theory. It will be explained later in the book. With this definite model, the result is

$$1 - P_{G(n)} = (1 - P_I)^n$$

and, therefore,

(1.3.1) $$P_{G(n)} = 1 - (1 - P_I)^n$$

Formula (1.3.1) is the desired formula expressing the group probability in terms of the individual parameter P_I and the controlled group size n. Since n is controlled, it is known directly by the experimenter, say n = 3 in a particular experiment. On the other hand, P_I is estimated from the performances of the first wave of individuals attempting solutions. Thus, P_I is estimated from the data. When this is done, a definite numerical equation results. For instance, in one experiment reported by Lorge and Solomon, of 91 individuals, 13 solved the puzzle. Then, the estimate of P_I is .15, and expression (1.3.1) becomes,

$$P_{G(3)} = 1 - (1 - .15)^3$$

since n = 3. Thus, $P_{G(3)}$ = .39, by numerical computation. This is a numerical prediction for the second part of the experiment, in which groups attempt to solve the problem. It was mentioned earlier that $P_{G(3)}$ itself is estimated by the relative frequency of groups arriving at a solution. In one set of data, of 23 groups of size 3, only 3 arrived at a solution. Then, the estimation of $P_{G(3)}$ is .13. Although Lorge and Solomon conducted many experiments and their model evaluation is therefore based on a large body of data, let us abstract from the real context to the sheer logic of the situation based on this one experiment.

At this point then the numerical prediction is

$$P_{G(3)} = .39$$

and the direct empirical estimate

$$P_{G(3)} = .13$$

The latter may be called the observed value of the variable, while the former is the predicted value based on the model, as identified in the case at hand. Thus, the prediction considerably overestimates the chance that the group can solve the puzzle.* The goodness-of-fit of the model and the experimental phenomenon is not adequate and we have a negative evaluation of the model. According to our general phase description of the process of model-building, we should return to the earlier phases for reconstruction. It is important at this point to realize that any aspect of the application process could possibly account for the discrepancy

---

*The reverse is usual: the group does better than predicted. It is interesting to note that both types of error occur, because a revised model needs to cope with both possibilities. In this discussion, it is assumed also that the difference between the observed and the theoretically expected values cannot be attributed to chance.

between predicted and observed values, and so, one must scrutinize the experimental identification of the concepts, the method of estimation of parameters, and even the numerical calculations. If all these aspects of the application seem in order, then evaluation loops back to the theoretical phase of work: the setting up and the analysis. The analysis phase can be checked as to whether abstract calculations and derivations were correct. If so, the negative evaluation rests firmly on the assumptions of the model. In this case, with a relatively simple model structure, we conclude either that the fundamental "individualist" assumption is wrong or that the specific technical assumption of independence is wrong. (Of course, possibly both are incorrect.) Precisely by modifying these assumptions, further work has proceeded with this model.

To summarize this brief introduction to the logic of model-building, one should recall what the three phases were in the concrete example. The first phase involved setting up a model for group problem-solving in terms of individual parameters and group size. The variables were introduced and related to each other in a specific assumption that led, in the analysis phase, to a definite formula of the desired type—that is, one obtained a formula expressing the group probability as a function of the two parameters, group size and individual probability of solution. The third phase, application, consisted in identifying the formally undefined ideas "solving the puzzle," "individual," and "group" in a specific experimental setting. This allowed an estimation of the parameters of the model and so a calculation of a prediction for an observable quantity. The comparison of the prediction and the observed value led to a process of evaluation of the various phases leading to the discrepancy and settled on the assumptions of the model as in need of some modification. Further work begins by a new setting up phase based on somewhat altered assumptions.

Perhaps one point should be made about an atypical aspect of the above example. It was chosen because it exemplifies the model-building process in a very simple form. However, actual models differ rather dramatically from our example in that they ordinarily make a large number of predictions about the phenomenon rather than one prediction. The evaluation phase involves not one quantity, nor even one array of quantities, but a whole series of arrays of quantities in many cases. In such models the evaluation process becomes more complex, especially since the number of assumptions is also large. In short, a good deal of intuition is required to make good guesses about which aspects of the model are in need of modification.

A second point about a possibly misleading aspect of this example is worth noting: its emphasis on numerical calculation. It is not at all an essential aspect of model-building that numbers enter into the construction. Instead, the assumptions may define a relational system of a non-numerical type and the subsequent phases of analysis and application employ mathematical techniques which are not numerical. (See section 4.18 for a relevant discussion in more conceptual detail.)

Indeed the formulation of a non-numerical, relational-system is the more fundamental manner of theoretical model-building: it provides an essential component of the entire enterprise as outlined and exemplified in this book.

1.3. Mathematics and Theory. In one aspect, theoretical science is characterized by its search for novel and systematic ways of representing phenomena (emphasized in Toulmin, 1960). For example, in the following list, the phrases on the left refer to some class of phenomena and the associated phrases on the right refer to a systematic way of representing such phenomena:

Phenomenon	Representation
orbital path	ellipse
light	straight line
learning	stimulus sampling process
waiting-line length	queueing process
social mobility	Markov process

We may say that on the left the existence of the phenomenon is merely noted, while on the right an exact characterization of its form is provided. Thus, an orbital path might have been rectangular, but it is not. The adequacy of the representation is always an issue of importance. Once the representation is accepted, certain questions are meaningful and others are virtually ruled out. For example, to ask for the length of the side of an orbital path is nonsense; to ask for the focal distance is meaningful. Thus, modes of representation specify modes of raising questions about phenomena. Conversely, in the context of discovery of the representation, the modes of asking questions may suggest the appropriate representation. For example, if one finds a question of chance frequently raised by scientists, a probabilistic representation is suggested. In social mobility studies, for instance, sociologists talk in intuitive ways about the chances of a son of a rich father being downwardly mobile. This suggests a probabilistic conceptualization of the mobility phenomenon.

Another aspect of theoretical science is the attempt to provide a rationale for why certain representations are empirically adequate. For example, one can construe Newton's theoretical work as providing a rationale for Kepler's discovery that orbital paths could be represented as ellipses. Typically, this rationalization occurs in the process of construing the empirical phenomenon as merely one of a large class of phenomena and providing a generic method of representing any of these phenomena. Then, the generic method, when instantiated to the particular phenomenon, provides a way of logically deriving the accepted representation. Thus, the dynamical model for the motion of a planet relative to its sun, which is obtained by the Newtonian method of representing motion, yields the result that the motion forms an ellipse. This rationalization process is an explanation of the representational law—that is, it is explanatory of the fact that a certain way of representing the phenomenon is adequate.

The generic method of representation which provides the set of analytical concepts and techniques for deployment in explanations of laws may be said to form the framework of a branch of a science. Framework-construction, then, is part of the activity of theoretical science. Hence, the outline of the phases of model-building can be embedded in a wider scientific context of framework construction somewhat as follows:

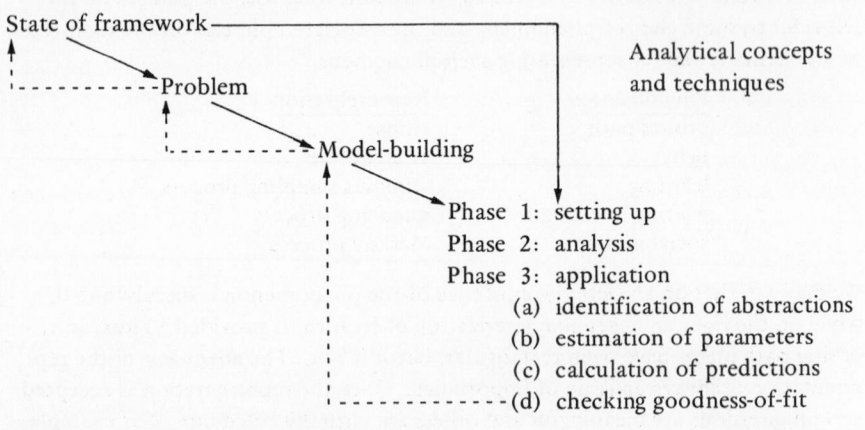

The "state of the framework" determines, in conjunction with real-world phenomena, some scientific problem. This problem leads to model-building efforts, and such an effort, in its phases, may draw upon the current state of the framework for concepts and techniques, especially in the setting up and analysis phases. In an advanced science, the very nature of what counts as a model of a phenomenon within its scope may be defined by the framework. For example, a model for physical motion "must" employ the conceptual categories of space, time, and mass, along with principles that "govern" the setting up without determining it completely, as in the use of certain general forms of equations.

When the stage of evaluation is reached, following the application of the model, a "feedback loop" exists that has the effect of model-reconstruction or model-generalization to a wider problem. In this case, the problem has been broadened or otherwise respecified, so that there is a feedback loop from model-building to concrete scientific problems. In turn, the development over time of new problems and models either reaffirms the framework, leaving its state unaltered, or exerts pressure for framework alterations, formalizations, reconstructions, and the like. This feedback loop, then, has the effect of redefining the state of the framework, and a new cycle of problems and model-building efforts starts.

It is useful to note that the entire feedback system has both theoretical and empirical components in terms of the scientific activity involved. The upper part of the system, including framework and model-building phases of setting up and analysis, may be termed theoretical activity; the lower part of the system, involving model-application, may be termed empirical activity. We see that model-building itself overlaps theoretical and empirical components of the system.

This book attempts to present ideas, principles, and tools useful in the theoretical activity of framework construction, model setup, and model analysis. It also covers at various points the techniques and special problems arising from the application phase of model-building, with strongest emphasis on (a) identification of abstractions and (b) estimation of parameters, with relatively minor examples of the activities of (c) calculation of predictions and (d) checking goodness-of-fit. In regard to (a) and (b), in fact, the procedures involve a conceptual and mathematical component that marks them as especially crucial as linkages from the model, as set up and analyzed, to observational materials.

In these terms, the representational focus of mathematical sociology "extends upward" to cover problems of clarifying conceptual schemes, arriving at acceptable formal definitions, inventing new formalisms in which to express the intuitions of a currently informal framework, and so forth. These conceptual emphases, typical at the framework level, require the use of logicomathematical skills not based in calculation, whether abstract or concrete, but in formal conceptualization: in the use of formalisms of qualitative and relational character rather than familiar numerical formalisms (algebra and calculus).

With these remarks kept in mind, we may roughly describe two roles (not kinds of people) in a theoretical science, say X: mathematical X, in which the role specialization is built around the representation of the form of phenomena, including framework-level activity; and theoretical X, in which the role specialization is built around the explanation of phenomena.

Previous discussion indicated that there is a very close connection between these two tasks and their corresponding specialities. To explain the form of a phenomenon, the theoretician needs a representation of that form and hence an input from the mathematical side. But, if no representation of a formal kind is available, the theoretician employs descriptions in an ordinary language framework, however inadequate, to "tune" his audience to the intuition he has as to the form of the phenomenon. This is an input to the mathematical side of the field: it provides a felt problem requiring solution in terms of the achievement of a representation in formal terms.

In a very mature field, which has solved a good many representation problems, the theoretician also must employ a mathematical apparatus in order to explain the mathematical representations or laws accepted in that field. There appears to be a dynamic convergence in a developing science in which the equilibrium condition is one in which the two roles are filled by one person. At the earlier stages of representational efforts, on the other hand, it appears that the two roles are segregated. For instance, sociological theory is ordinarily taught without using any mathematics and the people called sociological theorists tend not be identified simultaneously as mathematical sociologists. Yet, if our sketch of the relation

between representation and theory is at all correct, both mathematical and theoretical sociologists will benefit from adoption of a common set of commitments
centered around formal methods.

1.4. Overview of Part 1. This book is written as a contribution to the accelera-
tion of the movement of sociology toward a state in which theoretical frameworks
are formal and model-building is closely linked to theory, that is, explanation.

Part 1 introduces the formal methods essential for framework construction and
for model-building. Formal reasoning (Chapter 2) is the common element in all
particular contexts of model-building as well as in efforts to formalize theoretical
frameworks. The logic of the application, or testing, phase of model-building is
governed by the logic of falsifiability studied in Chapter 2. On the other hand,
any attempt to formalize intuitive sociological frameworks requires knowledge of
the concepts of tautology, tautological implication, sentential function, and the
like, presented in that chapter.

Yet, if framework-construction stays at the level of mere symbolization of
ordinary-language sentences, experience teaches that the formalization remains
unsatisfactory. The missing element in such sentence-symbolizing efforts is the
generic method for representation of the phenomena within the scope of the
theory. It is one thing to symbolize "mobility" by "M" but a totally different
thing to represent adequately the processes of mobility. The key to representation
of virtually any kind is the preliminary step of locating a domain of objects and subsets of objects standing in specified relations. This is not the whole of the task, but
again experience teaches that merely trying to specify the various sets of objects
under discussion shows the paucity of theoretical knowledge disguised by freely
flowing verbal discourse. It spurs theoretical activity to greater detail of specification. Chapter 3 presents an introduction to the basic notions of set theory, and
Chapter 4 immediately applies and extends them to relations. It almost goes
without saying that the proper study of any phenomenon inheres in a sharp focus
on the relations involved. Relation theory, as an abstract device, is embedded
within set theory, allowing application of readily understood set-theoretic operations to various relations. Graphs represent relations, and in Chapter 4 they are
used to conveniently depict the structure of relations. Chapter 4 itself is a mere
starting point for manifold ideas depending upon it: for instance, algebraic methods
for kinship analysis (Chapter 18) and measurement theory (Chapter 7).

Chapter 5 deals with the manner of representing rules for associating things, or
"mapping." This is one of the essential elementary ideas of modern mathematics.
After illustrating its use in formalizing sociological concepts, it is shown how
algebraic ideas are built up: operation, algebraic system, isomorphism, and so
forth. The focus is on the parallel with relational systems and is preparatory to
the work of the following chapters, which frequently use algebraic ideas. In
Chapter 6 the special algebraic system of vectors and matrices is outlined, both

to exemplify algebraic ideas and to prepare for its use later. Once again, the material is related to relations and graphs.

The reader should not expect to see many elaborate substantive illustrations in these early, introductory chapters. Part 1 is merely a first exposure to a tool kit. Later parts, however, will go into some topics in depth, utilizing formal methods to construct frameworks and to set up and analyze definite models for particular classes of social phenomena.

1.5. Overview of Part 2. The bridge from elementary formal methods to sociological model constructions dealing with social process and social structure (Part 3) is made up of an array of ideas: measurement concepts (Chapter 7), the general language of processes (Chapter 8), and probabilistic methods (Chapters 9, 10, and 11).

Measurement is one form of representation. It has a logical structure that can be exhibited in terms of the concept of a relational system. Empirical systems form a species of relational systems. So do numerical systems. The important notion of homomorphism of systems, defined and illustrated in Part 1, is now applied to show the logical structure of fundamental measurement: a homomorphic mapping of an empirical system into a representing numerical system. We are able to exploit this idea to define in a rigorous way the notion of empirical meaningfulness of a numerical relation and to apply the definition to statistics, thus codifying some well-known ideas in applied statistics. This first chapter, in which formal methods are applied, then, has the advantage of being devoted to a topic familiar to most sociologists and to students who have studied statistics and sociological methodology at the introductory level.

The thrust of the book is not completely understood until one studies Chapter 8, where for the first time we talk in a systematic manner about process. Once again, formal methods, as well as the measurement concepts of the prior chapter, are used to develop a subject systematically: the concept of an analytical system— and, above all, of a dynamic system—is treated in formal terms and instantiated with sociological examples. Such topics as equilibrium and stability are formally defined. Dynamic-system behavior is conceptualized to include probabilistic systems, as a prelude to the subsequent heavy focus on probabilistic methods in the remainder of Part 2. Finally, two important methodological rules for the construction of process models, rules actively applied and exemplified in the remaining chapters of the book, are noted. The important notion of a Markov process is first mentioned in this chapter, but a detailed treatment is reserved until an adequate probability basis has been presented.

In Chapters 2-8 the structure-process interplay so characteristic of the remainder of the book receives its first exposition, at the introductory level. At this level, no particular model is explored in any depth: this is reserved for later chapters.

The third basic concept of Part 2 is that of a probability model. In Chapter 9 probability theory is defined as the study of abstract probability models. Topics in elementary probability modelling are introduced along with the basic axiomatic structure of any probability model. Both applied techniques and rigorous axiomatic notions are introduced and explored together: the tree diagram is a tool used to considerable advantage in this regard. In Chapter 10 additional concepts of probability theory are introduced in the context of the setup and analysis of models for processes. The chapter constitutes an introduction to the theory of Markov chains. It includes an extended example of the setup and analysis of a process model, starting from axioms about the process and culminating in detailed derivations about its properties. Also included in this chapter is a complete classification of all finite-state, discrete-time Markov chains, making use of relation theory. Two special types of chains, regular and absorbing, are given detailed treatments.

We conclude our introduction to probabilistic methods with a chapter in psychology: the theory of individual choice based on a probabilistic conceptualization. The ideas of Frank Restle (1961), based on a set-theoretic representation of choice situations, are used in Part 1 to illustrate a set-theoretic conceptualization of a phenomenon. They are extended here to provide the conceptual basis for supposing that choice is best understood in probabilistic terms. But then we go to our real topic: the statement of Luce's (1959) theory of individual choice behavior. This theory exemplifies the massive power of deduction conferred upon a theory framed in a mathematical manner: one axiom, many significant logical consequences. But our interest in Luce's theory is not merely pedagogic. His theory includes a measurement model for individual valuations scaled at a ratio level. This result could be of great importance in the measurement of status variables, where not the concrete characteristic (occupation, education) alone counts but its differential evaluation. Hence, Chapter 11 concludes with an application of this measurement model to the scaling of occupational prestige. This provides a context for explaining some detailed formal procedures of estimation and goodness-of-fit, the book's first extended example of the logic of the application phase of model-building.

1.6. Overview of Part 3. In Part 3 we are able to apply the entire formal and conceptual basis of the earlier chapters. The focus is sociological, with emphasis on the logic of the representational effort in a particular domain. The first three chapters of this part (Chapters 12-14) present three examples of sociological efforts based on the idea that some process of interest is a Markov process. In Chapter 12 the aim is to exhibit an explanation of empirical generalizations about differential imagery of stratification, this explanation being achieved by showing how the stable states of a Markov process implied by the axioms of a theory of image formation and transformation correspond to the structure of images actually observed. In Chapter 13 the context is interpersonal influence. Individuals are linked into small networks within which processes of influence occur. The flow over time of the state of such a small network forms a Markov process. The whole methodology is part of the "continuous-time stochastic process" approach, introduced into sociology by James Coleman (1964). In Chapter 14 the focus shifts to the study of how expectations, general and specific, influence behavior in concrete social situations. The methodology exemplifies the "Stanford approach" of Joseph Berger, Bernard Cohen, and Morris Zelditch (1966). The aim is explanatory; the tools are balance theory and the notion of a Markov process.

In Chapter 15 we pause in our analysis of sociological models to collate the basic ideas of estimation: the concept of estimation of parameters from the decision-theoretic perspective, and the logic of three methods of estimation used in this book (maximum likelihood, method of moments, and least squares).

Then, Chapter 16 returns to sociology with a selective review of models of social mobility. This is an area of sociology that has received much attention by model-builders. Here we cover the simple Markov model for intergenerational mobility, the mover-stayer model, and its "semi-Markov" formulation by Blumen, Kogan, and McCarthy (1955). Some continuous-time models suggested by Thomas Mayer (1972) are then outlined. Sections 16.6 and 16.7 present two quite different approaches to social mobility. In section 16.6 the path analytic model of Blau and Duncan (1967) is presented. This is a context for explaining the general ideas of path analysis. In section 16.7, the more recent "vacancy chain" model developed by Harrison White (1970) is outlined. The chapter concludes with a review of what appears to be the basic representational problem of mobility theory, including ideas drawn from the work of Robert McGinnis (1968) and others.

The remainder of Part 3 constitutes an illustration of the use of abstract algebra in the analysis of social structure. Chapter 17 is pure mathematics: an introduction to the ideas of group theory. However, rather than a conventional introduction it presents the material organized around the group-graph correspondence, which not only has pedagogical value but is significant for applications. In Chapter 18 the concept of a prescribed marriage system is formally defined and then group theory is applied to the problem of classifying and counting the various structurally distinct systems conforming to a set of axioms. This chapter, then, forms an introduction to the algebra of kinship. In the following chapter, the ideas are extended. The central idea is homomorphism. The focus begins with kinship but moves to the treatment of an arbitrary social relational system in which there are many relations connecting many actors. Category theory, a relatively new idea in modern mathematics, is explicated and then applied to the problem of social structural analysis. All the material in this third part relies heavily on the work of Kemeny, Snell, and Thompson (1966) and Harrison White (1963).

1.7. Overview of Part 4. In this final part of the book, the aim is to exhibit a strong framework of conceptual and deductive developments. To achieve this aim, the theory of games is introduced, starting at the foundations: the formulation of the basis theory of utility (Chapter 20); the exposition of the basic formal concept of a game in the context of institutional analysis and with a strong focus on the interpretation of various mathematical ideas (Chapter 21); and the formulation of the passage to normal or standard form for the purpose of strategic analysis (Chapter 22). Following these foundation chapters are three chapters on two-person game theory: the theory of zero-sum two-person games (Chapter 23) and the extension of the theory to general two-person games (Chapters 24 and 25). The central concept traced out in these chapters is the extension of the idea of rationality from its intuitive origins to its "cooperative form" in the general two-person game. An attempt is also made to apply the phases conceptualization within the game-theory framework; to this end, at various points we discuss the setup of a game model of a social situation or institution, the strategic analysis of the model, and the application and evaluation of the model.

1 INTRODUCTION TO FORMAL METHODS

CHAPTER TWO   FORMAL REASONING

2.1. Aim of the Chapter. Formal reasoning involves the explicit use of logico-mathematical techniques. Apart from such specific techniques, however, there are some rudimentary and invariant aspects of such formal reasoning. These pervasive and presupposed aspects will be explicitly discussed in this chapter.

2.2. The Concept of a Logical Variable. A fundamental aspect of formal reasoning is the explicit use of logical variables. By a logical variable, one means a place holder for the name of an object in a specified domain of objects.

Example. (a) The domain of objects is the set of all numbers. The logical variable is x. The use of the variable is exemplified by the expression

(2.2.1)   x is even.

Expression (2.2.1) is not yet a sentence because it is neither true nor false. Two methods can be used to convert (2.2.1) into a sentence. The first method involves replacing the place holder x by a definite numeral. For example, replacing x by 5 we obtain

5 is even,

which is false; replacing x by 6, on the other hand,

6 is even

is obtained, which is true. The second method of converting (2.2.1) into a sentence is to assert something general about numbers. For example, we can say,

(2.2.2)   For all x, x is even.

This means that whatever x may be, when replaced by a numeral, the definite sentence obtained "x is even" will be true. Thus, (2.2.2) means, among other things, that

5 is even

is true. Since this latter sentence is false, we conclude that (2.2.2) is a false sentence. By contrast, the general statement

(2.2.3)   There is some x such that x is even

is true, since it only requires that for some one numeral replacing x the resulting definite sentence be true. Since

6 is even

is such a sentence, it follows that (2.2.3) is true.

   Example.   (b) For a second domain of objects take a specified set of persons. For a logical variable choose the Greek letter alpha, α. One use of this place holder is given by the expression,

(2.2.4) α is a sociologist.

We note that expression (2.2.4) is neither true nor false. The two methods of converting (2.2.4) into a sentence are exemplified by, first,

Merton is a sociologist,

assuming Merton to be a member of the chosen domain of persons and that the name "Merton" refers to Robert Merton, the author of *Social Theory and Social Structure*. Then, with this replacement for the logical variable α, expression (2.2.4) becomes a true sentence. Second, if we choose to make a general statement, we can say,

(2.2.5)   For all α, α is a sociologist.

This means that the domain contains only sociologists, because (2.2.5) is the same as

Every α is a sociologist

(with α in the domain). We cannot say this is true or false until the domain is actually specified, because the expression "a set of persons" itself is a logical variable, with domain the collection of all possible sets of persons.

   Example.   (c) More than one variable may be needed for asserting something about objects in a given domain. Thus, let the domain be a set of persons. Let two logical variables be α and β (beta). For a use of these place holders, consider the expression

(2.2.6)   α is a friend of β.

In conjunction with (2.2.6) we can consider that one might make the assumption about the domain that everybody has a friend in the domain. This is the conversion of (2.2.6) into the statement

(2.2.7)   For all β, there exists an α such that α is a friend of β.

Here, in sentence (2.2.7), the phrases "for all β" and "there exists an α" occur in a definite order to indicate that if you choose any person at all (β), then holding that person (β) fixed, the expression

                          α is a friend of β

will be true for at least one replacement of α by the name of a person in the domain. For example, if it is assumed that everybody is a friend of himself,

(2.2.8)   For all β, β is a friend of β,

then when β is fixed, the expression (2.2.6) can be converted into a true sentence by replacing α by β. For example, let β be replaced by John, so that (2.2.6) becomes

                          α is a friend of John.

Then replacing α by β means replacing α by "John" to obtain

                          John is a friend of John,

which is true if (2.2.8) is the working assumption about this domain and this relationship. If we reverse the order of the expressions "for all β" and "there exists an α" in (2.2.7) we obtain

(2.2.9)   There exists an α such that, for all β, α is a friend of β.

This means that if one considers the expression

(2.2.10)   For all β, α is a friend of β,

then there is at least one domain member such that when his name replaces α, we obtain a true sentence in (2.2.10), which by itself is neither true nor false. For example, suppose that "Harris" is the name of a person who is everybody's friend. Then,

(2.2.11)   For all β, Harris is a friend of β

is true. If (2.2.11) is true, then there does exist a person α such that (2.2.10) is true. Therefore, (2.2.9) is true. The meaning of (2.2.9), therefore, is radically different from the meaning of (2.2.7): the order of logical expressions of the sort "for all α" and "for some β" does indicate different propositions. In the case at hand (2.2.7) says that everybody has a friend, while (2.2.9) says that there is somebody who is everybody's friend.

2.3. Quantifiers. The expressions "for some x" and "for all x" used in conjunction with logical variables are called logical quantifiers. The quantifier "for some x" means the same as "there exists an x such that," which is equivalent to "there is at least one x such that." Similarly, "for every x" is the same as "for all x" and "for any x." The former quantifier, using "some" or "at least one," is called existential; the latter, using "all" or "every," is called universal.

The logical symbol for the existential quantifier is $\exists x$, when the logical variable is x, $\exists \alpha$ when the logical variable is α, and so forth. Thus, the expression

$$\exists x, x \text{ is even}$$

is the same as

$$\text{For some x, x is even,}$$

a true statement if x has domain the set of all numbers.

The logical symbol for the universal quantifier is $\forall x$, when the logical variable is x, $\forall \alpha$ when the logical variable is α, and so forth. Thus, the expression

$$\forall x, x \text{ is even}$$

means

$$\text{For all x, x is even,}$$

a false statement if x has domain the set of all numbers.

2.4. Names and Things. A basic aspect of conceptually clear thinking is the distinction between name and thing named. Symbols are employed to talk about objects. A symbol used to designate an object is called a name for that object. Thus, "5," "five," and "fünf" are all names of the number that is one greater than four. "The number one greater than four" is also a name for this object.

The name is not the thing. To test the distinction, compare

(2.4.1)   "Five" has four letters

with

(2.4.2)   Five has four letters.

Of these two sentences, (2.4.1) is meaningful and true; but (2.4.2) is not even meaningful, since numbers are not entities that are comprised of letters.

Since names are only symbols standing for objects, we can invent new names to facilitate discourse. For instance, we can let a = Chicago and b = New York, so that "a" now names Chicago, just as "Chicago" names that city. This naming is contextual in that in some context "a" might name something else, even New York. "Chicago," on the other hand, is a folk symbol that designates Chicago in many interactional contexts of the modern world. (Of course, one may find it important in some analyses to deal with sets of such folk symbols, employing names designating symbols.)

To talk about any pair of cities one cannot use the names "Chicago" or "New York," since their meanings are established in folk usage. On the other hand, a and b are free for general use. Thus, we can say,

Let $S = \{a,b\}$ be a set of two cities,

and not be committed to a domain of particular cities. But this means that S is here a logical variable with domain the collection of all possible pairs of cities.

2.5. Exact Concepts and Mathematical Theories.

However, what is a city? Certainly it is not a mere domain or class of persons nor even necessarily such persons in specified relationships, for, to be general, why constrain cities to humankind? Perhaps, then, we need organisms, with suitable relationships. But why organisms? The answer appears to be that organisms are the sort of entities that can enter into the sort of relationships that make up a city. For example, organisms can present stimuli to each other. Also, some of them can build things.

We see, then, that a city is an arbitrary collection of objects forming a certain structure of relationships. Passing from the intuitive idea of the city to the precise specification of this structure—in the abstract, relative to an arbitrary set of objects—is an example of the creation of an exact concept. Precisely by an abstract relational focus did the game notion pass from its intuitive usage to an exact concept as the foundation of game theory, as shown later in section 21.4. With respect to the city notion, of course, one does not have such a formal conceptualization, but the form of the exact concept would be

Let A be an arbitrary set. Let $R_1, R_2, \ldots, R_N$ be (certain types of) relations among objects in A. Then the relational system $(A, R_1, R_2, \ldots, R_N)$ is a city if and only if . . . (followed by axiomatic conditions).

In this statement, A is a logical variable with domain the class of all sets. This is the meaning of the expression "Let A be an arbitrary set," just as a numerical logical variable, say x, might be introduced by the expression "Let x be an arbitrary number." One also sees the expression "Let A be an abstract set." This means that the objects in A are not specified. Thus, A is a "variable set," an entity like x in referring to numbers.

A concept defined with respect to an abstract set is called an axiomatic, or exact, concept (cf. Carnap, 1958). With an exact concept we make a connection to phenomenal, concrete entities by asking if the concrete set of things satisfies the concept. The method of answering such a question would exemplify the model-building methodology: construct a "city-model" for the referent domain and evaluate its "fit." This idea will be concretely exemplified in the fullest sense in the discussion of probability theory and the notion of a probability model.

At this point it bears noting that such a concept is not simply a given intuitive idea; it must be constructed. In general, mathematics has two essential component processes: construction and deduction. Construction often appears under the guise of definition (cf. Carnap, 1969). A mathematical theory is the deductive elaboration of a constructed exact concept. Having explored in a preliminary intuitive manner the nature of the constructive aspect of formal reasoning, we turn now to a first look at logic, the tool for deductive elaboration. We will return to the idea of an exact concept in section 4.18.

2.6. Atomic and Molecular Sentences. Logic may be defined as the discipline that deals with theories of inference. In application to science, logic provides a basic system of rules of inference. These rules are justified by appeal to the general theory of validity in regard to ordinary declarative sentences. The theory begins by noting that all such sentences can be partitioned into two classes: in one class, the sentences make no use of certain basic logical connectives; in the second class, the sentences are comprised of sentences of the first kind combined by the use of connectives. The connectives are "and," "or," "not," "if . . . then," and "if and only if." Sentences not using these connectives are called atomic; others are called molecular, or compound. Thus, in the following list only sentences (1) and (5) are atomic:

(1) Society s is industrial.
(2) If society s is industrial, then it is complex.
(3) Society s is not industrial.
(4) Society s is urban and it is industrial.
(5) Society s is urban.

2.7. Truth Assignments. The logical connectives are used to create compound sentences from atomic constituents. Given any finite list of atomic sentences, say,

$$A = \{a_1, a_2, \ldots, a_n\}$$

where each a_i is atomic, the class of all sentences obtained by compounding sentences in A is denoted A^*. This "sentential system" has the following interesting property: once a definite "truth-value" (i.e., "true" or "false") is assigned to each atomic sentence in A, then each sentence in A^* has a unique truth-value.

Example. Consider the set A of three atomic sentences given by

Society s is industrial.   (a_1)
Society s is complex.   (a_2)
Society s is urban.   (a_3).

Then, among the sentences in A^* are those given by (1) through (5) of section 2.6. Assume that all three atomic sentences are true. Then sentences (2) and (4) are true, but sentence (3) is false since it denies a true sentence.

The number of possible truth-value combinations for n atomic sentences is 2^n, which is seen by appeal to the tree diagram in Figure 2.7.1.

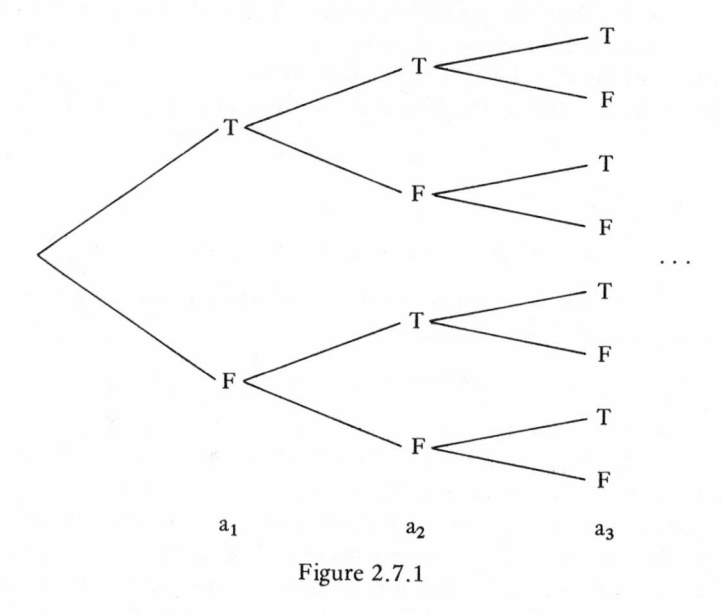

a_1 a_2 a_3

Figure 2.7.1

Thus, with three atomic sentences, there are eight possible paths on the tree, corresponding to eight possible truth-value combinations for a_1, a_2, and a_3. The above illustrative assignment was the path TTT. For two atomic sentences there are only four possible truth-assignments.

2.8. Basic and Derived Truth-Tables. The above argument about the dependency of the truth-value of a molecular sentence following directly from the truth-assignment for its component atomic sentences assumes that one sees the logical connectives as establishing unique conventions about truth-values. In particular, the following truth-tables give these conventions for three of the connectives:

Negation			Conjunction				Disjunction		
P	not-P		P	Q	P and Q		P	Q	P or Q
T	F		T	T	T		T	T	T
F	T		T	F	F		T	F	T
			F	T	F		F	T	T
			F	F	F		F	F	F

Here P and Q have domain the class A* of all sentences: given any two sentences, whether atomic or molecular, these three truth-tables provide a mapping of their truth-values into a unique truth-value for the constructed molecular sentences (not-P, P and Q, and, P or Q).

Example. To continue the example from sections 2.6 and 2.7, the expression "not-a_1" means "It is not the case that a_1 holds." This is the meaning of sentence (3), for to say that society s is not industrial is to claim that it is not the case that it is industrial. We say that not-a_1 is false. Now let P be not-a_1 and let Q be a_3, which is assumed true. Hence, P is false and Q is true. Thus, the sentence

$$\text{Not-}a_1 \text{ and } a_3,$$

or

$$\text{Society s is not industrial and it is urban,}$$

has a truth-value of "false," as shown in the third row of the conjunction truth-table. On the other hand, the sentence

$$(\text{not-}a_1 \text{ and } a_3) \text{ or } a_2$$

is true: let P be (not-a_1 and a_3) and Q be a_2. Then, P is false and Q is true, so that the third row of the disjunction table yields a "true" value. In words, this sentence says: Either society s is not industrial and it is urban, or it is complex.

To introduce tables for "if . . . then" and "if and only if" one argues as follows. By "If P, then Q" we mean "It is not the case that P and yet not Q." For example, to say that when copper is heated (P), it expands (Q) is to say that it is not the case that in fact copper is heated and (but) it does not expand. Thus, one can compute the truth-values for "If P, then Q" by appeal to this meaning. To find the truth-value of not-(P and not-Q) one proceeds from the inside-out, making repeated use of the above truth-tables. Thus:

Step 1.   Truth-values of not-Q from truth-values of Q

Step 2.   Truth-values of (P and not-Q) from those of P and those of not-Q

Step 3.   Negation applied to the result of Step 2.

This is shown in the following computation:

P	Q	(1) not-Q	(2) P and not-Q	(3) not-(P and not-Q)
T	T	F	F	T
T	F	T	T	F
F	T	F	F	T
F	F	T	F	T

Finally, record the last column in the induced truth-table for "If P, then Q":

Conditional

P	Q	If P, then Q
T	T	T
T	F	F
F	T	T
F	F	T

As was to be expected, a conditional sentence is false when its condition P holds, but its consequent Q does not (line two). However, it is otherwise to be considered true. The interpretation of lines three and four is that if P is false, we are not wrong when we claim that if P holds, then Q holds. The assignment of "false" to the statements would overlook the conditional character of the claim.

By "P if and only if Q" one means the conjunction "If P, then Q, and also, if Q, then P." Thus, by computation from the prior tables:

Step 1. Record truth-values of "If P, then Q."

Step 2. Record truth-values of "If Q, then P."

Step 3. Apply conjunction to (1) and (2).

P	Q	(1) If P, then Q	(2) If Q, then P	(3) (1) and (2)
T	T	T	T	T
T	F	F	T	F
F	T	T	F	F
F	F	T	T	T

The last column gives us the induced truth-table:

Biconditional

P	Q	P if and only if Q
T	T	T
T	F	F
F	T	F
F	F	T

We will occasionally write "iff" for the biconditional "if and only if." Also, we sometimes use "$P \Rightarrow Q$" to mean "If P, then Q," and "$P \Leftrightarrow Q$" to mean "P if and only if Q."

2.9. Tautologies. Further computations may be illustrated with the objective of showing the following: given $A = \{a_1, a_2, \ldots, a_n\}$ containing n atomic sentences, there are sentences of the molecular type in A^* that have the truth value T no matter which of the 2^n possible truth-assignments holds for the atomic basis. For example, consider the sentence

$$\text{If } a_1 \text{ and } a_2, \text{ then } a_1.$$

It is intuitively clear that we need not inquire into the truth of the condition to affirm that the entire sentence cannot be false. By computation we have the following:

		(1)	(2)
a_1	a_2	a_1 and a_2	If a_1 and a_2, then a_1
T	T	T	T
T	F	F	T
F	T	F	T
F	F	F	T

Column 1 is obtained by applying the basic conjunction table with P replaced by a_1 and Q replaced by a_2. Then, for column 2 we go to the basic conditional table, replacing P by (a_1 and a_2) and replacing Q by a_1. From the invariant value T in column 2, we conclude that this sentence is true no matter what the states of affairs described by the atomic sentences a_1 and a_2. Such a sentence is called a tautology.

(2.9.1) Definition. A sentence is a tautology if and only if it is true for all possible truth-assignments to its atomic constituents. This leads to:

(2.9.2) Definition. A conditional sentence is a tautological implication if and only if it is a tautology.

From the definitions we see that the truth-values of the atomic constituents in a tautological implication can be varied at will without affecting the truth of the implication itself. A tautological implication is thereby useful in logical inference. For example, consider the inference:

$$\frac{a_1 \text{ and } a_2}{\text{Hence, } a_1,}$$

where the line separates the premises from the conclusion. Clearly this inference is warranted because we know that the conditional sentence "If a_1 and a_2 then a_1" is true in all possible actualities which establish the actual truth-values of the atomic constituents a_1, a_2. For another example, consider:

$$\frac{\begin{array}{l} a_1 \\ a_1 \Rightarrow a_2 \end{array}}{\text{Hence, } a_2}$$

This inference is warranted by the fact that the conditional sentence

$$\text{If} \quad a_1 \quad \text{and} \quad a_1 \Rightarrow a_2, \quad \text{then } a_2$$

is a tautological implication. The computation is as follows:

		(1)		(2)		(3) If a_1 and $a_1 \Rightarrow a_2$,
a_1	a_2	$a_1 \Rightarrow a_2$	a_1	and	$a_1 \Rightarrow a_2$	then a_2
T	T	T		T		T
T	F	F		F		T
F	T	T		F		T
F	F	T		F		T

We can see that each tautological implication warrants an inference from the condition (the set of premises) to the conclusion.

Example. A concrete illustration is

If society s is industrial, then it is complex.	$a_1 \Rightarrow a_2$
Society s is industrial.	a_1
Hence, it is complex.	Hence, a_2.

In this illustration the premises may be assumed to be true on empirical grounds, but the argument itself exemplifies a tautological implication in which the conclusion must be accepted as true once the premises are taken to be true.

(2.9.3) Definition. Two sentences are tautologically equivalent if and only if the biconditional formed from them is a tautology.

Thus, P and Q are tautologically equivalent when "P if and only if Q" is a tautology. This amounts to "P and Q are tautologically equivalent if and only if P tautologically implies Q and Q tautologically implies P." For example, "P and Q" is tautologically equivalent to "Q and P."

A contradiction is shown by truth-table computation leading to the assignment of F in each row. For example, for any P, the sentence "P and not-P" is a contradiction; the corresponding truth-table is:

P	(1) not-P	(2) P and not-P
T	F	F
F	T	F

2.10. Valid Arguments. Note the following examples of valid arguments (i.e., in which the inference is warranted), with varying truth-values for the constituent atomic sentences:

Argument 1.

If wood is a metal, then wood melts.	(T) ⎫
Wood is a metal.	(F) ⎬ (F)
Hence, wood melts.	(F)

Here we have a false set of premises, because wood is not a metal, and also a false conclusion, since wood does not melt. Nevertheless, inference is warranted by the fact that the premises logically imply the conclusion.

Argument 2.

If iron is a metal, then iron melts.	(T) ⎫
Iron is a metal.	(T) ⎬ (T)
Hence, iron melts.	(T)

Here we have true premises as well as a true conclusion, so that not only do the premises logically imply the conclusion, but we can accept all constituent sentences as true.

Argument 3.

If wood is a metal, then wood burns.	(T) ⎫
Wood is a metal.	(F) ⎬ (F)
Hence, wood burns.	(T)

In this case, the conclusion is true but one of the premises is false. Note that the first premise is true because we first consider the constituent atomic sentences:

Wood is a metal.	(F)
Wood burns.	(T)

Then, the conditional table, with P replaced by "Wood is a metal" and Q replaced by "Wood burns," gives a "true" value for the conditional.

A fourth argument could be constructed to exemplify the combination of a set of premises whose conjunction is true and whose conclusion is false. Such an argument could never exemplify a valid inference because from a true condition, a tautological implication guarantees a true conclusion. Thus, a valid argument with a true condition and false conclusion is impossible. If an argument exemplifies a tautological implication, then it is valid.

The above three valid arguments make it clear that from the fact of valid reasoning one cannot conclude that the conclusion is true; however, one sees also that to do valid reasoning does not require that the premises be in fact true. Thus, hypothetical reasoning, even with known contrary-to-fact premises, is entirely possible and, in many cases, worthwhile. The general point is that validity is related to truth only in that a valid form of inference assures that from true premises one can never acquire false conclusions.

2.11. Contrapositive and Falsifiability. A basic implication of the ideas in section 2.10 is that from a theoretical conclusion established as true by appeal to experience, one cannot conclude that one's theoretical system forms a true condition.

However, as Popper (1959) has emphasized, one can reject theories if their logical consequences are shown to be false. This is based on the following idea. Any sentence of the form "If P, then Q" has an associated sentence of the form "If not-Q, then not-P" called its contrapositive. The following theorem is basic to Popper's argument:

(2.11.1) Theorem. Let P and Q be any two sentences. Then, the conditional "If P, then Q" and its contrapositive "If not-Q, then not-P" are tautologically equivalent.

A truth-table proof of this theorem is as follows:

	(1)	(2)	(3)	(4)	(5)
P Q	If P, then Q	not-Q	not-P	If not-Q, then not-P	(1) iff (4)
T T	T	F	F	T	T
T F	F	T	F	F	T
F T	T	F	T	T	T
F F	T	T	T	T	T

When this theorem is applied to scientific philosophy, one wants "If P, then Q" to mean that P is the molecular sentence consisting of the assumptions of the theory and Q is the conclusion that logically follows from P. Thus, we require that P tautologically imply Q. If this is so, then because of the equivalence of the contrapositive, we conclude that not-Q tautologically implies not-P. Thus, from observations that negate Q, we conclude that the theory is "wrong."

In detail, the "experimental inference" takes the form, initially,

If P, then Q	(L-T)
Not-Q	(E-T)
Hence, not-P	(?)

In this argument the first premise can be replaced by its equivalent contrapositive; its truth-value is not dependent on the experiment but on the fact (we assume) that Q is tautologically implied by P. Thus, the theorist correctly reasoned from P to Q. We label this L-T for "logically true." Also, not-Q is true, based on the experiment; we label this E-T. The question mark refers to an initially unknown truth-value of the denial of the theory. By using the contrapositive one gets the form

If not-Q, then not-P
Not-Q
Hence, not-P

This makes it clear that the argument is valid. Turning now to the truth-value to be assigned to the denial of the theory, we reason as follows: the argument is valid and each premise is true, and so, the conclusion is true. To deny this would be to challenge the truth of the premises. But of the two premises, only the "(E-T) premise" can be challenged once we check the logical argument by which Q was derived from P. Hence, if one wishes to preserve the theory, one challenges the observations leading to the result not-Q. If these are incontravertible, the theory has been refuted.

In sociology it is often the case that verbal theories are too informal to generate any definite logical conclusions. Thus the statement "If P, then Q" is a "psychological conviction" and not a tautological implication. Q is not generated by valid argument from the theory P; it is merely suggested by it. Then, the "experimental inference" has the form

If not-Q, then not-P   (C-T)

Not-Q   (E-T)

Hence, not-P   (?)

Here, "(C-T)" means "assigned a truth-value T by conviction but not by logical inference." Then, once again, the experimental argument is seen to be valid. Again we wish to examine what is the truth-status of the conclusion, the denial of the theory. If we grant T to both premises, not-P is also true because of the valid form of argument. Then the theory requires revision. But suppose we wish to save the theory: then we examine the data leading to assigning not-Q a value of T. Let this be convincing. Then, unlike the earlier situation when Q was a logical inference from P, we cannot claim the theory is refuted, because one could simply claim that "If P, then Q" was wrong. The conviction-true (C-T) status differs from the logically true (L-T) status in precisely this way: the community of agreement as to the truth-value T is enormously greater for L-T than for C-T. A rational man would probably never doubt an L-T sentence, while he might well quarrel with a C-T sentence. From this one concludes that when a theory does not produce logical conclusions (that can be tested), it is substantially less refutable on the basis of evidence.

The model-builder's emphasis on deriving consequences can be seen as a firm commitment to the rationality of logical argument not only at the level of theory but also at this level of "experimental argument" to disprove theories. (For a further discussion, see section 4.18).

2.12. Inference with Quantifiers; Symbolization of Ordinary Language Sentences. The domain of logical inference is not exhausted by tautological implications. We can see this from the following argument:

If society x is industrial, then it is complex.

France is industrial.

Hence, France is complex.

This argument is without doubt valid, but it contains four distinct atomic sentences, as we see by introducing names:

S = Society x is industrial.
C = Society x is complex.
F = France is industrial.
f = France is complex.

Then the argument is

If S, then C
F
Hence, f

In this form, the argument reveals clearly that the premises do not tautologically imply the conclusion. The difficulty is that the validity of the argument depends upon the internal sentence structure and the presence of logical variables. To see this, let η (eta) be a domain of societies. Then let x be a logical variable with domain η. Note that France is in η. Thus, "France" is one possible replacement for the variable x. The idea behind the first premise is that for any x, if x is industrial, then x is complex:

$\forall$x,   if x is industrial, then x is complex.

Since France is in the domain of the variable, we obtain

If France is industrial, then France is complex.

Then, the argument becomes

If France is industrial, then France is complex.
France is industrial.
Hence, France is complex.

This is now warranted by tautological implication.

To make the inference more transparent it is customary to call an expression like "is industrial" a predicate and to symbolize it. The meaning of the symbol is introduced in the following form:

Ix   means   x is industrial.

Similarly,

Cx   means   x is complex.

Then, introducing a contextual name for France, a = France, we can write "Ia" and say that a was substituted for x in Ix. This is sometimes termed specification or, as here, instantiation. The argument now has the form

$\forall$x,   if Ix, then Cx   (premise)
If Ia, then Ca   (instantiation)
Ia   (premise)
Hence, Ca

Here the step from the first line to the second involves instantiation of a general assumption about the objects in a specified domain to some particular member of the domain. The argument reaches its ultimate in symbolization if one makes the conditional connective symbolic:

$$\forall x, \quad Ix \Rightarrow Cx$$
$$Ia \Rightarrow Ca$$
$$\underline{Ia}$$
$$\text{Hence, } Ca$$

Similarly, the existential quantifier "for some x" plays a role in inference. Consider:

> No person can fly.
> James is a person.
> ――――――――――――
> Hence, James cannot fly.

Again the argument is valid but its validity is not that of tautological implication, which gives the clearly invalid argument

$$P$$
$$\underline{Q}$$
$$R$$

The validity is made formal by noting that to say "No person can fly" is to say "It is not the case that there exists a person who can fly":

$$\text{not-(Some person who can fly)}$$

Then, let A be a domain of persons and let α be a logical variable with domain A. We have

$$\text{not-}(\exists \alpha, \alpha \text{ can fly})$$

and, symbolizing the predicate,

$$F\alpha \quad \text{means} \quad \alpha \text{ can fly,}$$

we have

$$\text{not-}(\exists \alpha, F\alpha)$$

for the general premise. But this assertion reveals itself to be logically equivalent to the expression

$$\forall \alpha, \text{not-}F\alpha$$

that is, "For any α, it is not the case that $F\alpha$," meaning "If α is a person, then α cannot fly." Now let a = James so that the argument runs

$$\frac{\forall\alpha,\ \text{not-}F\alpha}{\text{Hence, not-}Fa}\qquad\text{(premise)}\\ \text{(instantiation)}$$

because a is in the domain of α. Practice in sentence symbolization follows.

Example. (a) Every metal conducts heat.

Variable: x   (domain: physical objects)

> Mx   means   x is a metal
> Cx   means   x conducts heat

Then, (a) becomes,

$$\forall x,\qquad Mx \Rightarrow Cx$$

(For every x, if x is a metal, then x conducts heat.)

Example. (b) Every planet travels in an ellipse.

Variable: z   (domain: physical objects)

> Pz   means   z is a planet
> Tz   means   z travels in an ellipse

Then (b) becomes

$$\forall z,\qquad Pz \Rightarrow Tz$$

(For all z, if z is a planet, then z travels in an ellipse.)

Example. (c) Some people cannot read. (version 1)

Variable: α (domain: organisms)

> Pα means α is a person
> Rα means α can read

Then (c) becomes

$$\exists\alpha,\qquad P\alpha\ \text{and not-}R\alpha$$

(There exists an α such that α is a person and it is not the case that α can read.)

Example. (d) Some people cannot read. (version 2)

Variable: α (domain: all people)

> Rα means α can read

Then (d) becomes

$$\exists\alpha,\ \text{not-}R\alpha$$

(There exists an α such that it is not the case that α can read.)

Example. (e) All people can read.

Variable: β   (domain: organisms)

Pβ   means   β is a person
Rβ   means   β can read

Then (e) becomes

$$\forall\beta, \quad P\beta \Rightarrow R\beta$$

(For any β, if β is a person, then β can read.)

Example. (f) Some persons cannot read and cannot write.

Symbolization as in (e),
plus Wβ   means   β can write.

Then (f) becomes

$$\exists\beta, \quad P\beta \text{ and not-}R\beta \text{ and not-}W\beta$$

(There exists a β such that β is a person and [but] it is not the case that β can read and it is not the case that β can write.)

Example. (g) Every man loves some woman.

Variables: x, y   (domain: all people)

Mx (My)   means   x (y) is a man
Wx (Wy)   means   x (y) is a woman
Lxy   means   x loves y

Then (g) becomes

$$\forall x, \quad Mx \Rightarrow (\exists y, Wy \text{ and } Lxy)$$

(For every x, if x is a man, then there exists a y such that y is a woman and x loves y.)

Note here that the y variable takes a value that depends on x. Also, there is a possibility x loves more than one woman, because ∃y merely means "at least one."

Example. (h) If two people are friends, then they interact frequently.

Variables: α,β   (domain: all people)

Fαβ   means   α and β are friends
Iαβ   means   α and β interact frequently

Then (h) becomes

$$\forall\alpha, \ \forall\beta, \qquad F\alpha\beta \Rightarrow I\alpha\beta$$

Alternative notations for the "two-termed predicates" F and I are "$\alpha F\beta$" and "$\alpha I\beta$." Also, one often writes "$\forall\alpha, \beta$" to mean "$\forall\alpha, \forall\beta$."

Then (h) becomes

$$\forall\alpha,\beta, \qquad \alpha F\beta \Rightarrow \alpha I\beta$$

Example. (i) If an instrumental leader emerges, then a socioemotional leader also emerges.

Variable: g   (domain: all small groups)

> Ig   means   g has an instrumental leader
> Sg   means   g has a socioemotional leader

Then (i) may be roughly rendered

$$\forall g, \qquad Ig \Rightarrow Sg$$

The loss of time implicit in the emergence concept in (i) is unfortunate. But logical notation to cope with time is no problem once we dig a little deeper into mathematics (see section 8.6). For the present we note only that the basic point for mathematical analysis is to represent the process from which (i) was developed rather than to represent the sentence (i). This is basically the distinction between mathematics and logic: the former is employed to represent more-or-less complex aspects of patterns in the world; the latter is employed to represent sentences for the sake of making inference more transparent. (In mathematical logic these two enterprises meet: the mathematical logician studies patterns of inference by mathematical methods.)

Finally, we should note that in mathematical work it is useful to place quantifiers off to the right in parentheses:

$$x^2 + 1 = 0 \qquad\qquad (\exists x)$$
$$x + y = y + x \qquad\qquad (\forall x, y)$$

Also, we often use quantifying expressions without the quantifier symbols. Thus, the first five examples above would be written

(a) $Mx \Rightarrow Cx$ (for every x)
(b) $Pz \Rightarrow Tz$ (for every z)
(c) $P\alpha$ and not-$R\alpha$ (for some α)
(d) not-$R\alpha$ (for some α)
(e) $P\beta \Rightarrow R\beta$ (for every β)

2.13.   Rules of Inference and Proof.   The use of tautologies in science is not usually explicit; they appear under the guise of tacit rules of inference. As we saw in the prior sections, the two basic types of tautologies are implications and equivalences. Thus, if $P \Rightarrow Q$ is a tautological implication, we obtain the rule of inference: from P, one may conclude: Q. For example, for any P and Q,

$$P \text{ and } Q \quad \Rightarrow \quad Q$$

is a tautology; thus,

From P and Q, one may conclude: Q

is a rule of inference. The argument

$$\underline{2 \text{ is even and 2 is less than 3.}}$$
Hence, 2 is less than 3

exemplifies the use of this rule. Similarly,

$$\text{not-(P and Q)} \quad \Leftrightarrow \quad \text{not-P or not-Q}$$

is a tautological equivalence. Thus, the corresponding two rules of inference hold:

From not-(P and Q), one may conclude: not-P or not-Q.
From not-P or not-Q, one may conclude: not-(P and Q).

For example, illustrating the second of these two rules,

$$\underline{\text{The equation is not solvable or the solution is not a real number.}}$$
Hence, it is not the case that the equation is solvable and the solution is a real number, that is, it has no real solution.

In addition to the rules of inference obtained from tautological implications and equivalences, there are the rules of inference associated with the use of logical variables and quantifiers. Four of these are very frequently, if tacitly, used:

(1) universal specification (U.S.): From $\forall x, Px$, one may conclude: Pa.
(2) existential specification (E.S.): From $\exists x, Px$, one may conclude: Pa_0.
(3) existential generalization (E.G.): From Pa_0, one may conclude: $\exists x, Px$.
(4) universal generalization (U.G.): From Px, one may conclude: $\forall x, Px$.

In these rules (cf. Suppes, 1957), x is a logical variable with a domain, in which a and a_0 are located. The name "a" is a constant: a definite entity is designated by it. The name "a_0" is termed ambiguous because one need not know the object referred to—it names something because it is only used by specification of the meaning of $\exists x$, but one may be able to say no more. For example, we may know that an equation has some solution, possibly not unique. We can call such a solution a_0 in further discourse. Rule 4 formulates the idea that if one has shown that Px holds by reasoning with the logical variable, this amounts to saying that it holds for any replacement, that is, $\forall x$, Px.

An example will show the use of rules of inference to produce a deduced conclusion.

Example.   Some men are sociologists.   (Premise 1)

Every man is an animal.   (Premise 2)

Hence, some animals are sociologists.

We formalize this argument as follows:

Variable: x   (domain: set of living things)

Mx	means	x is a man
Sx	means	x is a sociologist
Ax	means	x is an animal.

Then we have the argument with justifying rules to the right as follows:

(1)	$\exists x$,	Mx and Sx	(Premise 1)
(2)	$\forall x$,	Mx $\Rightarrow$ Ax	(Premise 2)
(3)	Ma_0 and Sa_0		From (1), using E.S.
(4)	$Ma_0 \Rightarrow Aa_0$		From (2), using U.S.
(5)	Ma_0		From (3), using:
			From P and Q, conclude: P
(6)	Aa_0		From (4) and (5) using:
			From (P $\Rightarrow$ Q) and P, conclude: Q
(7)	Sa_0		From (3), as in step (5)
(8)	Aa_0 and Sa_0		From (6) and (7) using:
			From P, Q, conclude: P and Q
(9)	$\exists x$,	Ax and Sx	From (8), using E.G.

And line 9 is the desired conclusion, since "Some animals are sociologists" means "There is an x, such that x is an animal and x is a sociologist."

2.14. Open Formulas and Existence Theorems.   A formula of the form P_x where P is a predicate and x is a logical variable, is called open, because it is not really a definite assertion. As discussed in section 2.2, it may be closed in three possible ways:

$$(1) \ \forall x, Px$$
$$(2) \ \exists x, Px$$
$$(3) \ Pa$$

where a is in the domain of x. The first two ways produce general statements that are closed in the sense that they are true or false. The third way produces a particular, or singular, sentence about object a and, as such, is closed in the sense that it is either true or false. By contrast, Px is simply awaiting closure. It is open.

The same idea applies to predicates associating two or more terms: Pxy (or xPy), Pxyz, and so forth. The third method replaces all variables with names of objects. For example,

$$Pab \ (or \ aPb)$$
$$Pabc$$

The first and second ways become more ramified. One can write

$$\forall x, xPy$$

but the formula is still open because y is a logical variable not yet either quantified (by $\exists y$ or $\forall y$) or replaced by a constant. Thus, to make open formula xPy closed in the sense of a general rather than singular sentence requires that both variables be quantified.

Considering all logically possible double quantification means that we must choose one variable to appear first (two ways), quantify this first variable (two ways), and then quantify the second variable (also two ways). This makes a total of eight forms (picture a tree diagram of successive independent binary choices). However, the logical meaning of "for every x and for every y" is the same as that of "for every y and for every x." Also, the meaning of "for some x and for some y" is the same as that of "for some y and for some x." This leaves six forms of double quantification. The following list gives these six forms, together with an interpretation in terms of

$$xAy \qquad iff \qquad x \text{ is acquainted with } y$$

Form	Interpretation
(1) $\forall x, \forall y, xAy$	Everyone is acquainted with everybody.
(2) $\forall x, \exists y, xAy$	Everyone is acquainted with somebody.
(3) $\exists x, \forall y, xAy$	Someone is acquainted with everybody.
(4) $\exists x, \exists y, xAy$	Someone is acquainted with somebody.
(5) $\forall y, \exists x, xAy$	Everyone is an acquaintance of somebody.
(6) $\exists y, \forall x, xAy$	Someone is an acquaintance of everybody.

In a different context, form (2) is exemplified by

For any number, x, there exists some number y such that $x + y = 0$.

We know from experience that y is actually unique. But the logic of $\exists y$ is "at least one." Thus, to prove that a certain condition is satisfied uniquely we need two steps:

Step 1. Existence: prove or assume that the condition is satisfied by some (at least one) entity in the domain.

Step 2.   Uniqueness: prove that if "two" objects in the domain satisfy the condition, then they are in fact equal.

For example, let us imagine that we did not know that for a given x the number y is unique. We assume only that there exists at least one such number. Suppose there were two such numbers, say y_0 and y_1. Then, because y_0 is such a number,

$$x + y_0 = 0$$

Adding y_1 to both sides,

$$(x + y_1) + y_0 = y_1$$

and since y_1 satisfies

$$x + y_1 = 0$$

we conclude that

$$y_0 = y_1$$

That is, if at least one such number exists (our assumption), then that number is unique.

2.15. Sentential Functions in One and Several Variables.

In the previous section it was noted that formulas containing variables may be open or closed, depending upon whether each of the variables is quantified. A variable that is not associated with a quantifier is termed free and otherwise termed bound. Thus,

$$\forall x, Pxy$$

has x as a bound variable but y as a free variable. A formula containing variables is a sentence if and only if each of the variables is bound. A formula with at least one free variable (and so, an open formula) is also called a sentential function: it is a form awaiting quantification (or specification) to become a sentence. If n is the number of free variables in a sentential function, then it is called a sentential function in n free variables. For example, the following are sentential functions in two free variables:

(1) 3 is between x and y.
(2) x is married to y.
(3) x gave the ball to y.

These might be symbolized

Bx3y

Mxy

Gxby

with their more general forms given by

$$Bxzy$$
$$Mxy$$
$$Gxzy$$

B is said to be a three-place predicate. Similarly, M is a two-place predicate and G is a three-place predicate. But in (1) and (3), the z variable had already been (tacitly) specified to the definite entities "3" and "the ball," respectively. Thus, as found in (1) and (3), the sentential functions had only two free variables.

2.16. Horatio and Discrimination. From a logical standpoint, experimentation involves the identifications of abstract notions and the production of premises in more-or-less complex arguments. We give an example of the logic of "experimental proof" as an exercise in formal reasoning. "Horatio" is the name of a hypothetical entity that seems to exhibit voluntaristic behavior (Galanter, 1966). Horatio is a smooth spherical creature whose sensitive reactions to its environment are novel and strange. Because Horatio has no obvious sense receptors, to learn what it is that Horatio knows and is capable of doing entails being quite explicit about otherwise tacit assumptions about organisms. To exhibit this explicit analysis, suppose that one wishes to learn whether or not Horatio can discriminate brightness as a factor in its environment. For this purpose, a box with two regions is constructed, with a view from above, as shown in Figure 2.16.1.

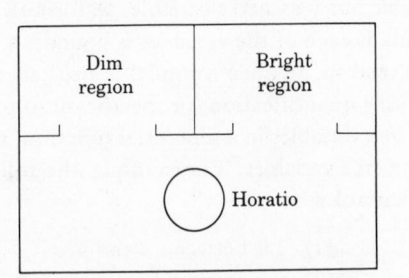

Figure 2.16.1

Intuitively, suppose we give Horatio a reason for action by putting something it likes in the bright region. If it can tell one region from the other, we should observe it move into the bright region with the object it likes. However, this may only mean Horatio has a bias to move toward the right. Thus, we should need to do a second experiment, shifting the location of the brightness. Also, we should be sure to pair the "reward" with the dim region in one type of experiment for otherwise we have only shown that Horatio can detect the bright-reward combination but not necessarily the bright-no-reward combination.

To formalize the experimental argument, we let x, y be arbitrary stimuli. Further,

xDy	means	Horatio can discriminate x from y;
xPy	means	Horatio prefers x to y.

Then, assume

(2.16.1)   If xPy, then xDy   ($\forall$x, y)

Thus, for any stimuli, we assume that if Horatio exhibits a preference, then it surely can tell the difference between the two objects.

To follow the plan described above, we need to know some type of object that Horatio likes. Now, suppose that natural observation of Horatio shows that it seems to orient to rough surfaces rather than smooth surfaces. It seems to enjoy the rough surface. A little tinkering with some sandpaper shows what appears to be a positive reaction by Horatio: perhaps, it joggles about a bit, and when moved off the sandpaper, it rolls back on to it. We conclude, with s = sandpaper and $\bar{s}$ = smooth surface,

(2.16.2) $sP\bar{s}$

Clearly, from (2.16.1) and (2.16.2), replacing x by s and y by $\bar{s}$ in the former, we have

$$\begin{array}{l} \text{If } sP\bar{s}, \text{ then } sD\bar{s} \\ \underline{sP\bar{s}} \\ \text{Hence, } sD\bar{s} \end{array}$$

We want to conclude either $bD\bar{b}$ or not-$bD\bar{b}$, where b = brightness, $\bar{b}$ = dimness. We will assume that Horatio does not evidence the same sort of obvious preference based on brightness, so that the problem is more complex. We introduce the notation,

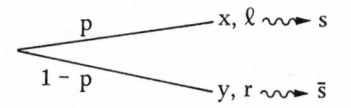

to mean that with stimulus condition x on the left yielding the rewarding sandpaper, Horatio goes to this left region with probability p. With probability 1 - p Horatio goes to the right, which has stimulus y, yielding no reward.

The rule for identification of D can be formally stated as follows:

(2.16.3) D-identification Rule.

If   (1)

$$\overbrace{\qquad}^{\displaystyle p \nearrow \quad x,\ \ell \rightsquigarrow s}_{\displaystyle 1-p \searrow \quad y,\ r \rightsquigarrow \bar{s}} \qquad\qquad \text{and } p > \tfrac{1}{2}$$

and (2)

$$\overbrace{\qquad}^{\displaystyle 1-q \nearrow \quad y,\ \ell \rightsquigarrow \bar{s}}_{\displaystyle q \searrow \quad x,\ r \rightsquigarrow s} \qquad\qquad \text{and } q > \tfrac{1}{2}$$

and (3)

$$\overbrace{\qquad}^{\displaystyle 1-p' \nearrow \quad x,\ \ell \rightsquigarrow \bar{s}}_{\displaystyle p' \searrow \quad y,\ r \rightsquigarrow s} \qquad\qquad \text{and } p' > \tfrac{1}{2}$$

and (4)

$$\overbrace{\qquad}^{\displaystyle q' \nearrow \quad y,\ \ell \rightsquigarrow s}_{\displaystyle 1-q' \searrow \quad x,\ r \rightsquigarrow \bar{s}} \qquad\qquad \text{and } q' > \tfrac{1}{2}$$

then xDy (for any x, y in a stimulus domain).

Note that the reward object s is placed in all possible conditions: (1) with x on the left, (2) with x on the right, (3) with y on the right, and (4) with y on the left.

Now, if not-xDy, then conditions (1) and (2) are indiscriminable for Horatio so far as the stimulus conditions in the two regions are concerned. In showing differential choice behavior [in (1) left orientation and (2) right orientation], Horatio betrays its ability to discriminate x from y, at least when x is coupled with the reward. The conditions (3) and (4) show that this ability is sustained when the reward object is coupled with y rather than x. We are willing to conclude that this constitutes determinate evidence that in fact xDy rather than not-xDy.

Now, suppose four experiments are done with the specifications that arbitrary stimulus x is in fact b, brightness, and y is $\bar{b}$, dimness. Then, suppose one observes

(1)

$$\overbrace{\qquad}^{\displaystyle .9 \nearrow \quad b,\ \ell \rightsquigarrow s}_{\displaystyle .1 \searrow \quad \bar{b},\ r \rightsquigarrow \bar{s}}$$

(2)

$$\overbrace{\qquad}^{\displaystyle .1 \nearrow \quad \bar{b},\ \ell \rightsquigarrow \bar{s}}_{\displaystyle .9 \searrow \quad b,\ r \rightsquigarrow s}$$

(2.16.4)

(3)

$$\overbrace{\qquad}^{\displaystyle .2 \nearrow \quad b,\ \ell \rightsquigarrow \bar{s}}_{\displaystyle .8 \searrow \quad \bar{b},\ r \rightsquigarrow s}$$

(4)

$$\overbrace{\qquad}^{\displaystyle .8 \nearrow \quad \bar{b},\ \ell \rightsquigarrow s}_{\displaystyle .2 \searrow \quad b,\ r \rightsquigarrow \bar{s}}$$

Then, from (2.16.3), with x replaced by b and y replaced by $\bar{b}$, we conclude that $bD\bar{b}$.

Finally, we note that the experimental conclusion was based on three premises: first, a general rule for identifying xDy relative to a known reward object; second, a premise asserting that s (sandpaper), is a reward object for Horatio; and third, the premise (2.16.4) constituting the experimental evidence cast in the "canonical form," required for the application of the first premise. If the empirical meaning postulate (2.16.3) is accepted, say on grounds of conforming to our intuitions about discrimination, if we are correct in assuming from natural observations that sP$\bar{s}$ (so that s is a reward object) and if we believe that (2.16.4) is a true rendition of the experimental results, then we are impelled to conclude that bD$\bar{b}$. The argument is a tautological implication, of course, but the premises betray their foundations in theoretical and empirical intuition [namely, premise (2.16.3)], in nonexperimental observation [namely, (2.16.2)], and, finally, in "proper" experimentation [namely, (2.16.4)].

In a sense the remainder of this book constitutes a series of applications of logic. For more "direct" applications, one can consult Suppes (1957), Tarksi (1946), or Carnap (1958). An interesting application of logic to the problem of providing an exact concept corresponding to the fuzzy notion of culture is provided by Anderson and Moore (1962). Briefly, the culture "definitions" proposed in textbooks and in conceptual reviews are surveyed by Anderson and Moore. They then pull out the common underlying meaning and show that a formal concept of culture as "the set of all learnables" of a given actor or set of actors satisfies the intuitive requirements for the concept. Details are omitted here.

2.17.   A Table of Tautologies.   It is useful to have at hand a list of frequently used tautologies. These are best displayed in fully symbolic form, letting

$\wedge$ mean "and"
$\vee$ mean "or"
$\neg$ mean "not"
$\Rightarrow$ mean "if, then"
$\Leftrightarrow$ mean "if and only if"

Table 2.17.1   Some Useful Tautologies

Implications			Equivalences		
$P \wedge (P \Rightarrow Q)$	$\Rightarrow$	Q	$P \Rightarrow Q$	$\Leftrightarrow$	$\neg Q \Rightarrow \neg P$
$\neg Q \wedge (P \Rightarrow Q)$	$\Rightarrow$	$\neg P$	$\neg (P \wedge Q)$	$\Leftrightarrow$	$\neg P \vee \neg Q$
$(P \Rightarrow Q) \wedge (Q \Rightarrow R)$	$\Rightarrow$	$(P \Rightarrow R)$	$\neg (P \vee Q)$	$\Leftrightarrow$	$\neg P \wedge \neg Q$
$P \wedge Q$	$\Rightarrow$	P	$P \wedge Q$	$\Leftrightarrow$	$Q \wedge P$
$\neg P \wedge (P \vee Q)$	$\Rightarrow$	Q	$P \vee Q$	$\Leftrightarrow$	$Q \vee P$
$(P \wedge Q) \Rightarrow R$	$\Rightarrow$	$P \Rightarrow (Q \Rightarrow R)$	$P \Rightarrow Q$	$\Leftrightarrow$	$\neg P \vee Q$
$P \Rightarrow (Q \wedge \neg Q)$	$\Rightarrow$	$\neg P$	P	$\Leftrightarrow$	$\neg \neg P$

In reference to Table 2.17.1, each tautological implication gives rise to a valid rule of inference, in which there is the form,

From . . . , conclude: . . .

For instance, the last implication yields,

From P implies both Q and not-Q, conclude: not-P.

Each tautological equivalence is a two-way implication and, so, in an analogous manner, gives rise to two rules of inference.

In the next chapter, we use the context of set theory to show how the elementary principles of formal reasoning, outlined in this chapter, are applied in mathematics.

CHAPTER THREE ELEMENTARY SET THEORY

3.1 Specification of Sets. To each sentential function in one free variable we associate the concept of the class or collection of all objects in the domain of the variable satisfying the sentential function—that is, making it true when their names replace the variable. We call this concept the set notion. If x is the variable and Px is the sentential function, then the class defined is denoted

$$\{x: Px\}$$

which is read, "the set of all x such that Px." Once we know the set is well-defined in this manner, we can give it a new contextual name, say S:

$$S = \{x: Px\}$$

Example. (a) Variable: α (domain: all persons)
Sentential function: α is a sociologist.
Set defined: $S = \{\alpha: \alpha$ is a sociologist$\}$
Here S is the set of all sociologists.

Example. (b) Variable: x   (domain: all numbers)
Sentential function: x is even.
Set defined: $E = \{x: x$ is even$\}$
Thus, E is the set of all even numbers.

If S is a well-defined set, the objects in the domain that comprise the set are called its elements or members. If x is an element of set S, we write

$$x \in S$$

which is read, "x is in S." Sometimes we also write

$$x \text{ in } S$$

For instance, in example (a) above,

$$\text{Robert Merton} \in S,$$

and in example (b),

$$2 \in E$$

If an object in the domain is not an element of a set S, we write

$$x \notin S$$

Thus, in example (a),

$$\text{Bismarck} \notin S$$

and in example (b),

$$3 \notin E$$

The symbol $\in$ is said to refer to the membership relation holding between elements and their corresponding sets. Clearly, if definite object $a \in S$ where $S = \{x: Px\}$, then Pa holds—that is, a replacing x makes the sentential function into a true sentence. Conversely, if Pa holds, then $a \in S$. Thus, being a member of a certain set is the same as satisfying a certain sentential function.

Sets can also be explicitly defined by listing the elements, a procedure that only works for finite sets. For example, we can let a set A be given by

$$A = \{\text{George Washington, 3, Dartmouth College}\}$$

if (for some reason) in some context we need to talk about this peculiar collection. This is a special case of defining a set of objects by the use of a sentential function, because we can specify

$$Px \quad \text{means} \quad x = \text{George Washington}$$
$$\text{or } x = 3$$
$$\text{or } x = \text{Dartmouth College}$$

and then,

$$A = \{x: Px\}$$

letting x range over any convenient domain including the three entities under discussion.

3.2. Axiom of Extension. Two sets are said to be identical when and only when they have the same elements. This is the basic axiom of the theory of sets and, so, of all mathematics. Explicitly,

Axiom of Extension. For any two sets A, B,

$$(3.2.1) \qquad A = B \quad \text{iff} \quad \forall x, \quad x \in A \Leftrightarrow x \in B$$

From this axiom it is seen that two sets could arise from totally different sentential functions and yet be equal on some specified domain.

Example. Variables: x, y, z   (domain: $\{1, 2, 3\}$)
Sentential function 1: x is even.

Sentential function 2: $\exists y, \exists z, \quad x = \dfrac{y+z}{2}, (y \neq z)$

From the first sentential function, noting the domain of the variables,

$$S_1 = \{x: \ x \text{ is even}\} = \{2\}$$

From the second sentential function we want the "solutions" of the equation in one free variable, since y and z are bound,

$$x = \frac{y+z}{2}$$

where all three variables range from one to three. Here we are looking for those x that make this expression true, for some y, and for some z, with $y \neq z$. In the expression (y+z)/2 we replace the variables by numerals as follows:

$$\left.\frac{y+z}{2}\right|_{\substack{y=1 \\ z=2}} \quad = \quad 1.5 \quad = \quad \left.\frac{y+z}{2}\right|_{\substack{y=2 \\ z=1}}$$

$$\left.\frac{y+z}{2}\right|_{\substack{y=1 \\ z=3}} \quad = \quad 2 \quad = \quad \left.\frac{y+z}{2}\right|_{\substack{y=3 \\ z=1}}$$

$$\left.\frac{y+z}{2}\right|_{\substack{y=2 \\ z=3}} \quad = \quad 2.5 \quad = \quad \left.\frac{y+z}{2}\right|_{\substack{y=3 \\ z=2}}$$

This exhausts all possible patterns of replacement of variables by numerals. We see that the computed values are 1.5, 2, and 2.5. Returning to the defined set,

$$S_2 = \left\{x: \ x = \frac{y+z}{2}, \quad \exists y, \quad \exists z, \quad y \neq z\right\} = \{2\}$$

because only 2 is in the domain and satisfies this rather complicated sentential function in one free variable x. Comparing S_1 and S_2, using (3.2.1), we conclude,

$$S_1 = S_2$$

although quite different predicates led to the sets. Note, however, that on the domain of all integers the two predicates lead to distinct sets because one set will contain all and only even numbers, while the other will also contain odd numbers.

Examples. Note the following concrete examples of set identity:

$$\text{(a) } \{1, 2, 3\} = \{3, 2, 1\}$$
$$\text{(b) } \{1, 1, 1\} = \{1\}$$

In (a) we show that order of writing down names of elements is irrelevant because the set consists of objects in a domain and not their names. For the same reason, (b) is true. Also, note that sets can be comprised of sets as elements. For example, let

$$N = \{1, 3, 4\}$$
$$P = \{2, 1, 10, 11\}$$

and then let

$$S = \{N, P\}$$
$$= \{\{1, 3, 4\}, \{2, 1, 10, 11\}\}$$

so that

$$N \in S$$

that is,

$$\{1, 3, 4\} \in S$$

On the other hand,

$$3 \notin S$$

because S has just two elements, N and P. N itself is a set with three elements. In other words, from

$$x \in N \quad \text{and} \quad N \in S$$

we do not infer

$$x \in S$$

For example, each person "belonging" to a certain nation is not thereby a member of the United Nations, although the nation may indeed be a member.

3.3. Containment; Proofs in Set Theory.
A basic relationship between sets is containment. We say that set A is contained in set B if and only if the conditional

$$x \in A \quad \Rightarrow \quad x \in B$$

is true for all values of x. Then we write $A \subseteq B$.

Example. (a) $$\{1, 2\} \subseteq \{1, 2, 3\}$$

because we have

$$\underbrace{\underbrace{1 \in \{1, 2\}}_{T} \Rightarrow \underbrace{1 \in \{1, 2, 3\}}_{T}}_{T}$$

$$\underbrace{\underbrace{2 \in \{1, 2\}}_{T} \Rightarrow \underbrace{2 \in \{1, 2, 3\}}_{T}}_{T}$$

$$\underbrace{\underbrace{3 \in \{1, 2\}}_{F} \Rightarrow \underbrace{3 \in \{1, 2, 3\}}_{T}}_{T}$$

assuming x ranges over $\{1, 2, 3\}$. Here we indicate how the truth-table for the conditional is employed. Because a conditional with a false condition is true ("it doesn't apply"), as shown in the third truth-computation, we really need examine only the membership of A and ask of each the elements: Is it also in B? That is, is "x $\in$ B" true, as well?

Example. (b) $$\text{not-}\big(\{1, 2\} \subseteq \{1, 3, 7\}\big)$$

To see this, we look at the membership of the first set, A = $\{1, 2\}$, and ask for each x $\in$ A: Is x $\in \{1, 3, 7\}$? We have

$$x = 1: \quad \underbrace{\underbrace{1 \in \{1, 2\}}_{T} \Rightarrow \underbrace{1 \in \{1, 3, 7\}}_{T}}_{T}$$

$$x = 2: \quad \underbrace{\underbrace{2 \in \{1, 2\}}_{T} \Rightarrow \underbrace{2 \in \{1, 3, 7\}}_{F}}_{F}$$

Here we see clearly that the conditional sentence is false for 2, and so (because it takes only one object to negate a universal quantifier), it follows that the sentence

$$\text{For all x if x} \in \{1, 2\}, \text{ then x} \in \{1, 3, 7\}$$

is false. Thus, it is not the case that $\{1, 2\} \subseteq \{1, 3, 7\}$.

(c) $$\{1, 2\} \subseteq \{1, 2\}$$

$$1 \in \{1, 2\} \;\Rightarrow\; 1 \in \{1, 2\}$$
$$2 \in \{1, 2\} \;\Rightarrow\; 2 \in \{1, 2\}$$

$$\underbrace{\underbrace{}_{T} \qquad \underbrace{}_{T}}_{T}$$

Here we see that we are "forced" to the conclusion that $\{1, 2\}$ is contained in itself. This is a special case of the general result

(3.3.1) $A \subseteq A$

It is a good exercise in formal reasoning to show this. What does it mean? It means, by definition,

$$\forall x,\, x \in A \Rightarrow x \in A$$

But the formula

$$x \in A \Rightarrow x \in A$$

instantiates, for each value of x,

$$P \Rightarrow P$$

a tautological implication. Thus, the formula is true of each x. Hence, the universal statement is true. Hence, $A \subseteq A$.

Consider now the special set

$$\{x \colon x \neq x\}$$

which, of course, cannot contain any elements. A contradiction, in other words, leads to an empty set. How many empty sets are there? Suppose A and B are empty. Then,

$$A = B \quad \text{iff} \quad \forall x,\, x \in A \Leftrightarrow x \in B$$

The conjecture is that A = B. To prove it examine the formula,

$$x \in A \Leftrightarrow x \in B$$

and compute truth-values:

$$\underbrace{\underbrace{x \in A}_{F} \;\Leftrightarrow\; \underbrace{x \in B}_{F}}_{T}$$

This is so because A and B are both empty; hence, for any x, "$x \in A$" is false and "$x \in B$" is false. The final truth-value T comes from the biconditional table. We see that we are forced to conclude A = B: there is only one empty set. Thus, we can give it a unique symbolic name: $\emptyset$. We say that $\emptyset$ is the empty or null set. We write $\{\ \}$ for $\emptyset$ in some contexts.

Consider, also, the expression

(3.3.2) $$\emptyset \subseteq A$$

What does it mean? It means

$$\forall x, x \in \emptyset \Rightarrow x \in A.$$

But "$x \in \emptyset$" is false for each x. Thus, "$x \in \emptyset \Rightarrow x \in A$" is true for each x, whether or not that x is in A. Thus, the defining condition is true, and we see that by definition, $\emptyset \subseteq A$. This is shorthand for

For any set A, $\emptyset \subseteq A$.

By universal specification on the domain of all sets, this yields

$$\emptyset \subseteq \emptyset$$

by replacing "A" by "$\emptyset$". And this agrees with (3.3.1).

The following proposition is very useful:

(3.3.3) *Proposition*

$$A = B \quad \text{iff} \quad A \subseteq B \text{ and } B \subseteq A$$

To prove this will further illustrate the use of logical techniques. Two things need to be shown:

(a) $A = B \quad \Rightarrow \quad A \subseteq B \text{ and } B \subseteq A$
(b) $A \subseteq B \text{ and } B \subseteq A \quad \Rightarrow \quad A = B$

To prove a proposition of the form $P \Rightarrow Q$, we assume P and produce a string of logically derived consequences terminating in Q.

Proof of (a). Suppose that A = B. What does it mean? By (3.2.1) it means that for any x,

(1) $x \in A \quad \text{iff} \quad x \in B.$

But then if $x \in A$, then $x \in B$, for any x. By the definition of containment, we conclude $A \subseteq B$. Also (1) implies that if $x \in B$, then $x \in A$, any x. Thus, $B \subseteq A$. Thus (1) implies $A \subseteq B$ and $B \subseteq A$. Therefore, if A = B, then $A \subseteq B$ and $B \subseteq A$.

Proof of (b). Suppose that $A \subseteq B$ and $B \subseteq A$. Let $x \in A$ (i.e., let x be a logical variable with domain A.) Then $x \in B$, because $A \subseteq B$. Thus, $x \in A \Rightarrow x \in B$, any x. (Universal generalization applied here.) Now suppose $x \in B$. Then, $x \in A$ because this is what $B \subseteq A$ means. Thus, $x \in B \Rightarrow x \in A$, any x. Hence, $A \subseteq B$ and $B \subseteq A$ implies both: $x \in A \Rightarrow x \in B$, and $x \in B \Rightarrow x \in A$ (any x). Thus, by the meaning of the biconditional, $A \subseteq B$ and $B \subseteq A$ implies that $x \in A$ iff $x \in B$, any x. The axiom of extension says this is the meaning of A = B. Thus, if A ⊆ B and B ⊆ A, then A = B.

The refrain "What does it mean?" is central to the carry-through of a proof. The important thing to realize is that in mathematics the single most important elements of a proof are the prior definitions. These establish meanings that function in grinding out consequences. In a sense this is the meaning of saying that "mathematics is analytical"—everything follows from meanings (using rules of inference based in logic).

Here is another example:

(3.3.4)   A ⊆ B and B ⊆ C   ⇒   A ⊆ C   (any sets A, B, C)

Proof of (3.3.4). Suppose x ∈ A. (Introduce logical variable.) Then, since A ⊆ B, this means x ∈ B as well. Then, since B ⊆ C, this means x ∈ C as well. Thus if x ∈ A, then x ∈ C. But this holds for arbitrary x (universal generalization now), so that for all x, x ∈ A ⇒ x ∈ C. Then, by definition of containment, A ⊆ C.

We note that one can write A ⊂ B to mean A ⊆ B and A ≠ B.

3.4. Operations; Venn Diagrams; More on Proofs. In the preceding section the concept of containment as a relationship between sets was introduced. Perhaps the reader noticed the analogy between A ⊆ B for sets and x ≤ y for numbers. This suggests analogies with addition and multiplication: define operations on sets producing new sets.

The following are the basic operations with so-called Venn diagrams illustrating their meaning:

(3.4.1) Definitions of Set Operations.

(a) Union

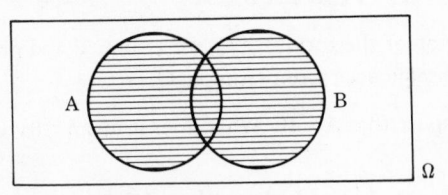

A ∪ B = {x: x ∈ A or x ∈ B}
For example: {1, 2, 3} ∪ {2, 4} = {1, 2, 3, 4}

(b) Intersection

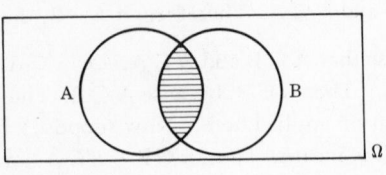

A ∩ B = {x: x ∈ A and x ∈ B}
For example: {1, 2, 3} ∩ {2, 4} = {2}

(c) Difference

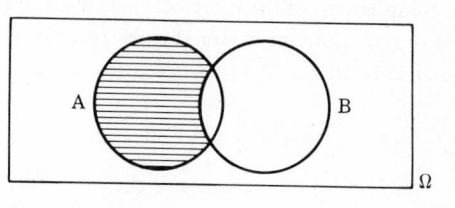

$A - B = \{x: x \in A \text{ and } x \notin B\}$
For example: $\{1, 2, 3\} - \{2, 4\} = \{1, 3\}$

(d) Complement

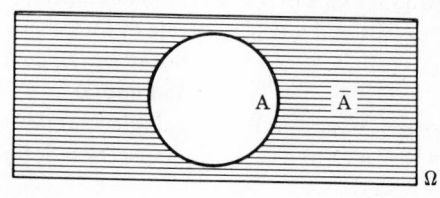

$\overline{A} = \{x: x \notin A\} = \Omega - A$
For example: Let $\Omega = \{1, 2, 3, 4, 5, 6\}$, $A = \{1, 2, 3\}$.
Then $\overline{A} = \{4, 5, 6\}$.

In this definition, the rectangle represents the domain Ω (omega) associated with the logical variable x and the regions represent specified sets.

Let us prove some so-called identities to practice the idea of asking "What does it mean?" in the grinding out of a deduction.

(3.4.2) $A - B = A \cap \overline{B}$

Proof of (3.4.2). To establish that these two sets are equal we need to show two things, according to the meaning of set identity:

(1) $A - B \subseteq A \cap \overline{B}$
(2) $A \cap \overline{B} \subseteq A - B$

To show that (1) holds, let $x \in A - B$. What does this mean? By (c) under (3.4.1) this means that $x \in A$ and $x \notin B$. But $x \notin B$ means that $x \in \overline{B}$, according to (d) under (3.4.1). Thus, $x \in A$ and $x \in \overline{B}$. Therefore, according to the meaning of intersection, we conclude that $x \in A \cap \overline{B}$. Thus, $x \in A - B \Rightarrow x \in A \cap \overline{B}$ for arbitrary x. This means that $A - B \subseteq A \cap \overline{B}$. This proves (1). To prove (2), let $x \in A \cap \overline{B}$. Then $x \in A$ and $x \in \overline{B}$, by meaning of intersection. But $x \in \overline{B}$ means, according to the definition of complement, that $x \notin B$. Thus, if $x \in A \cap \overline{B}$, then $x \in A$ and $x \notin B$. Thus, by (c) under (3.4.1), this means $x \in A - B$. Therefore, for any x, if $x \in A \cap \overline{B}$, then $x \in A - B$. Thus, by the meaning of containment, $A \cap \overline{B} \subseteq A - B$. This proves (2).

(3.4.3) $\overline{A \cap B} = \overline{A} \cup \overline{B}$

This is read: The complement of the intersection is the union of the complements. The proof by two-way containment is straightforward. For example, if $x \notin A \cap B$, then either $x \notin A$ or $x \notin B$. Hence, $\overline{A \cap B} \subseteq \overline{A} \cup \overline{B}$.

(3.4.4) $$\overline{A \cup B} = \overline{A} \cap \overline{B}$$

This is read: The complement of the union is the intersection of the complements. Here is a different style of proof, letting x be a variable:

$$x \in \overline{A \cup B} \Leftrightarrow x \notin A \cup B \qquad \text{(complement definition)}$$
$$\Leftrightarrow \text{not-}(x \in A \text{ or } x \in B) \text{ (union definition)}$$
$$\Leftrightarrow x \notin A \text{ and } x \notin B \qquad \text{(rule of inference: From}$$
$$\text{not-(P or Q), conclude: not-P}$$
$$\text{and not-Q)}$$
$$\Leftrightarrow x \in \overline{A} \text{ and } x \in \overline{B} \qquad \text{(complement definition)}$$
$$\Leftrightarrow x \in \overline{A} \cap \overline{B} \qquad \text{(intersection definition)}$$

Therefore,

$$\forall x, \quad x \in \overline{A \cup B} \Leftrightarrow x \in \overline{A} \cap \overline{B}$$

Hence, by the axiom of extension,

$$\overline{A \cup B} = \overline{A} \cap \overline{B}$$

Just as in the case of tautologies, it is useful to have at hand those properties of sets most frequently needed in applications. Table 3.4.1 provides some of the key properties. We give these properties names often associated with them. A general discussion of algebraic properties is given later in Chapter 5. In the table, it is assumed that all sets are subsets of some domain Ω.

3.5. Power Set. Another concept of some importance is the idea of the power set based on a set A. This is defined as the set of all subsets of A. For example, if $A = \{1, 2\}$, then the power set has as its elements

$$\{1\}$$
$$\{2\}$$
$$\{1, 2\}$$
$$\emptyset$$

since each of these four sets is contained in A and no set not in this list is contained in A. To construct explicitly the power set of a small set one can use a tree diagram. At each juncture the logical question is, Do we put element x in the subset? The symbol "$\in$" denotes a yes answer; "$\notin$" denotes a no. For instance, for two elements as above, we have the tree diagram in Figure 3.5.1.

Table 3.4.1   Properties of Sets

Commutative laws:
$$A \cup B = B \cup A \qquad\qquad A \cap B = B \cap A$$

Associative laws:
$$A \cup (B \cup C) = (A \cup B) \cup C \qquad\qquad A \cap (B \cap C) = (A \cap B) \cap C$$

Identities:
$$A \cup \emptyset = A \qquad\qquad A \cap \Omega = A$$

Special identity properties:
$$A \cup \Omega = \Omega \qquad\qquad A \cap \emptyset = \emptyset$$

Idempotent laws:
$$A \cup A = A \qquad\qquad A \cap A = A$$

Distributive laws:
$$A \cap (B \cup C) = (A \cap B) \cup (A \cap C) \qquad\qquad A \cup (B \cap C) = (A \cup B) \cap (A \cup C)$$

Self-distributive laws:
$$A \cup (B \cup C) = (A \cup B) \cup (A \cup C) \qquad\qquad A \cap (B \cap C) = (A \cap B) \cap (A \cap C)$$

Containment relations:
$$A \cap B \subseteq A \subseteq A \cup B$$
$$A \subseteq \Omega$$
$$\emptyset \subseteq A$$
$$A \subseteq B \quad \Rightarrow \quad A \cup B = B$$
$$A \subseteq B \quad \Rightarrow \quad A \cap B = A$$

Complementation laws:
$$A \cup \overline{A} = \Omega \qquad\qquad A \cap \overline{A} = \emptyset$$
$$\left. \begin{array}{l} \overline{A \cup B} = \overline{A} \cap \overline{B} \\ \overline{A \cap B} = \overline{A} \cup \overline{B} \end{array} \right\} \quad \text{De Morgan's Laws}$$
$$\overline{\overline{A}} = A$$

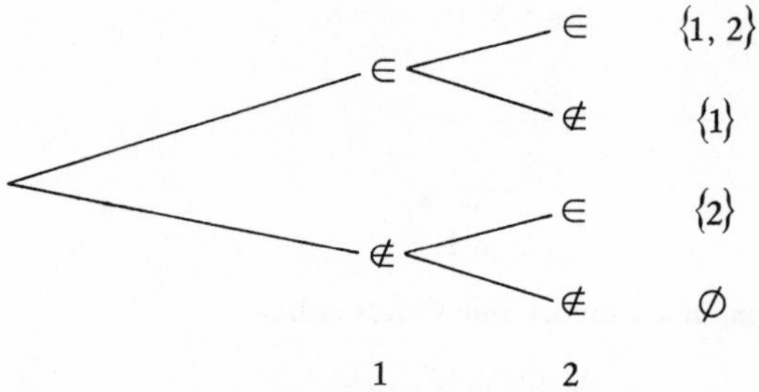

Figure 3.5.1

For $A = \{1, 2, 3\}$, we have the diagram in Figure 3.5.2.

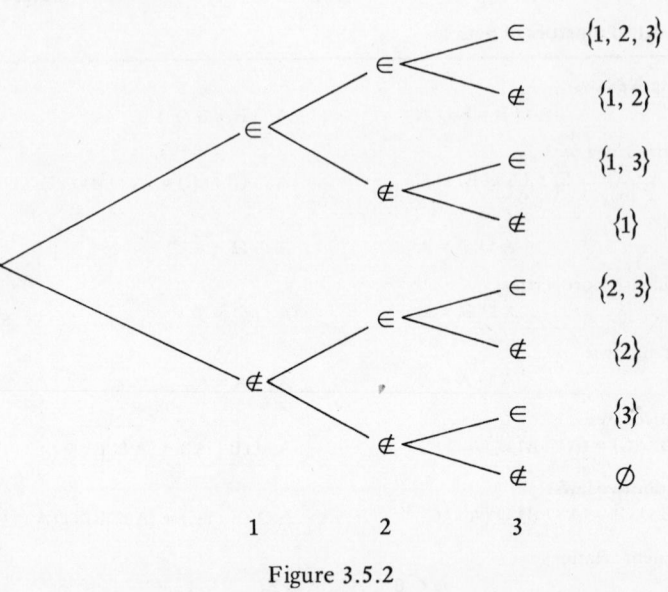

Figure 3.5.2

We see that for each of n elements we ask a binary, two-valued question. Thus, there are 2^n paths in the tree to represent all possible patterns of answers and, hence, all possible subsets. This is stated in the following proposition.

(3.5.1) The number of elements in the power set of a finite set of n elements is 2^n.

For this reason, mathematicians often denote the power set of A by the symbol

$$2^A$$

3.6. Special Notations. Just as one writes either

$$x_1 + x_2 + \ldots + x_n$$

or

$$\sum_{i=1}^{n} x_i$$

for a sum of numbers, in set theory one writes either

$$A_1 \cup A_2 \cup \ldots \cup A_n$$

or

$$\bigcup_{i=1}^{n} A_i$$

for a union of sets. By definition,

$$x \in \bigcup_{i=1}^{n} A_i \text{ if and only if } x \in A_i, \text{ for some } i = 1, 2, \ldots, n.$$

Also, for the intersection of a series of sets one writes either

$$A_1 \cap A_2 \cap \ldots \cap A_n$$

or

$$\bigcap_{i=1}^{n} A_i$$

Here

$$x \in \bigcap_{i=1}^{n} A_i \text{ if and only if } x \in A_i, \text{ all } i = 1, 2, \ldots, n.$$

It is often useful to write $\bigcup_i A_i$ and $\bigcap_i A_i$ for the more explicit forms $\bigcup_{i=1}^{n} A_i$ and $\bigcap_{i=1}^{n} A_i$, respectively.

3.7. Cartesian Product. Given two sets, say A and B, we can form a new set, denoted A x B, whose members are all pairs in which the first component is from A and the second from B. A pair is denoted (x, y), where x is in A and y is in B.

(3.7.1) Definition. Cartesian product:

$$A \times B = \{(x, y): x \in A \text{ and } y \in B\}$$

Examples. (a) Let $A = \{a, b\}$, $B = \{c, d\}$. Then

$$A \times B = \{(a, c), (a, d), (b, c), (b, d)\}$$

(b) Let $A = \{1, 2\}$, $B = \{a, b, c\}$. Then

$$A \times B = \{(1, a), (1, b), (1, c), (2, a), (2, b), (2, c)\}$$

(c) As in (b), let $A = \{1, 2\}$, $B = \{a, b, c\}$. Then

$$B \times A = \{(a, 1), (a, 2), (b, 1), (b, 2), (c, 1), (c, 2)\}$$

As example (c) shows, in general we cannot assume that A x B = B x A. Note that we can let B = A and form A x A, as a special case.

Given two "dimensions," say ability and race, one may form the Cartesian product to obtain a "compound" dimension, whose levels are based on pairs drawn from the dimensional characteristics.

Example. (d) Let

$$C_1 = \{\text{high, low}\} \text{ refer to some ability}$$
$$C_2 = \{\text{black, white}\} \text{ refer to race}$$

Then,

$$C_1 \times C_2 = \{(\text{high, black}), (\text{high, white}), (\text{low, black}), (\text{low, white})\}$$

If the dimensional characteristics are themselves ordered in some way—for example, a particular social context makes ability more important than race—then the order of the product $C_1 \times C_2$ has some social significance. In other cases, however, the order of the product may be of no substantive significance; then the choice of order is like the arbitrary scale unit in measurement, in that some such choice is needed to attain a representation but the choice is empirically arbitrary.

The product may be extended to several sets. Then we write, as in the case of unions and intersections,

$$\mathop{\times}_{i=1}^{n} A_i = A_1 \times A_2 \times \ldots \times A_n = \{(x_1, x_2, \ldots, x_n): x_i \in A_i\}$$

3.8. Restle Foundations. In the next few sections, the aim is to apply immediately the set-theoretic basis and to further extend it in a substantive context. Restle, in his study in psychological theory (1961), presents a mathematical foundation for the study of the choice behavior of higher organisms in which the set-theoretic concept of partition plays a central role. Hence, we will need a definition of the concept. Let A be a set. Then a collection of subsets of A forms a partition of A if, and only if,

(1) the subsets have a union that is A
(2) the subsets are pairwise disjoint
(3) each subset is nonempty.

By "disjoint" we mean the intersection is empty, as in Figure 3.8.1.

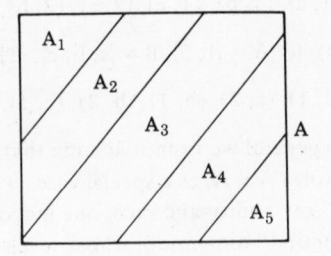

Figure 3.8.1

Formally, we have the following:

(3.8.1) Definition. A collection $\{A_i\}$ of sets forms a partition of a set A if and only if

(1) $\bigcup_i A_i = A$

(2) $A_i \cap A_j = \emptyset$ (for all i, j with i ≠ j)

(3) $A_i \neq \emptyset$ (all i)

Concretely, consider the set

$$A = \{1, 2, 3, 4\}$$

Then one partition of A is given by

$$A_1 = \{1, 2\}, \quad A_2 = \{3, 4\},$$

because $A_1 \neq \emptyset, A_2 \neq \emptyset, A_1 \cap A_2 = \emptyset, A_1 \cup A_2 = A$. Another partition of A is given by

$$B_1 = \{3\}, \quad B_2 = \{1, 2, 4\}$$

In general, choose any nonempty subset of A, say X. Then $\{X, \overline{X}\}$ forms a partition. Each such partition has just two parts. But one could have more parts. For example, another partition of A is

$$C_1 = \{1\}, \quad C_2 = \{2, 4\}, \quad C_3 = \{3\}$$

In general, let A be a set. Then there exists another associated set such that this set contains all the partitions of A. Thus, in this set, for the given A above, one would find the elements, P_i,

$$P_1 = \{A_1, A_2\}$$
$$P_2 = \{B_1, B_2\}$$
$$P_3 = \{C_1, C_2, C_3\}$$

among others; that is, the associated set is:

$$\{P_1, P_2, P_3, \ldots\}$$

We begin our introduction to the Restle foundations of choice theory with two primitive (that is, undefined) notions: situations and stimulus variables. The latter are treated as classifications of situations, as follows.

(3.8.2) Definition. A pair of sets $(\mathcal{S}, \mathcal{P})$ is termed a situation complex if and only if

(1) $\mathcal{S}$ is a nonempty set of situations;

(2) $\wp$ is the associated collection of all possible partitions* of $\mathcal{S}$. The elements of $\wp$ are called stimulus variables.

Organisms respond in situations according to aspects that become salient. But aspects can be defined in terms of the above notions.

(3.8.3) Definition. Let v be a stimulus variable. To each situation s in $\mathcal{S}$ there exists a corresponding class in v. Then this class is called an aspect (under v) of s.

(3.8.4) Definition. $A^* = \{a:$ a is an aspect of some $s \in \mathcal{S}\}$

(3.8.5) Definition. $A_s = \{a \in A^*:\ s \in a\}$

Example. For illustrative clarity, let $\mathcal{S}$ contain the following three figures:

$$\diamondsuit \quad \circ \quad \bigcirc$$

Then $\wp$ contains

$$P_1 = \{\{\diamondsuit, \circ\}, \{\bigcirc\}\}$$

$$P_2 = \{\{\diamondsuit, \bigcirc\}, \{\circ\}\}$$

$$P_3 = \{\{\diamondsuit\}, \{\circ, \bigcirc\}\}$$

$$P_4 = \{\{\diamondsuit\}, \{\circ\}, \{\bigcirc\}\}$$

$$P_5 = \{\{\diamondsuit, \circ, \bigcirc\}\} = \{\mathcal{S}\}$$

Thus,

$$\wp = \{P_1, P_2, P_3, P_4, P_5\}$$

According to definition (3.8.2), each P_i is a stimulus variable. According to definition (3.8.3), we generate aspects as follows:

First consider P_1. Then, we have

Situation	Aspect under P_1

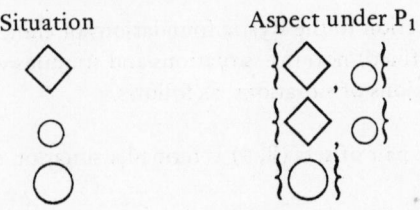

*The set $\wp$ is induced as soon as we know $\mathcal{S}$; hence, it is not really a new primitive notion. A general discussion of this distinction is given in section 4.18.

Similarly,

Situation	Aspect under P_2	Aspect under P_3	Aspect under P_4	Aspect under P_5
◇	{◇, ○}	{◇}	{◇}	{◇, ○, ○}
○	{○}	{○, ○}	{○}	{◇, ○, ○}
○	{◇, ○}	{○, ○}	{○}	{◇, ○, ○}

Thus, an aspect of a situation is a manner in which it is classed with other situations. We form A^* according to (3.8.4):

$$A^* = \{\{◇, ○\}, \{◇\}, \{◇, ○\}, \{○\}, \{◇\}, \{○, ○\}, \{◇, ○, ○\}\}$$

Also, according to 3.8.5),

$$A_◇ = \{\{◇, ○\}, \{◇, ○\}, \{◇\}, \{◇, ○, ○\}\}$$

(3.8.6) $\quad A_○ = \{\{◇, ○\}, \{○\}, \{○, ○\}, \{◇, ○, ○\}\}$

$$A_○ = \{\{○\}, \{◇, ○\}, \{○, ○\}, \{◇, ○, ○\}\}$$

3.9. Representation of Situations.

Later on we shall be in a position to prove the following directly from the definitions:

(3.9.1) Theorem. Each situation uniquely corresponds to a set of aspects.

For the present we note that in our example

$$A_◇ \neq A_○ \neq A_○$$

Thus, although a situation is not just a set of aspects—that is, $s \neq A_s$—there is this one-to-one (1-1) correspondence that enables a mathematical representation of a situation by a set of aspects.

Not all aspects are salient or relevant in a given occasion. Thus, we are led to introduce an additional notion. Namely, for each $s \in \mathcal{S}$, we may select a specific subset of A_s, denoted A_s', and call A_s' the set of relevant aspects of s.

For example, in (3.8.6), we may select,

$$A_◇' = \{\{◇, ○\}\}$$

(3.9.2) $\quad A_○' = \{\{○, ○\}\}$

$$A_○' = \{\{○\}, \{◇, ○\}, \{○, ○\}\}$$

According to this selection we can say that while $\{\Diamond, \bigcirc\}$ is an aspect of $\Diamond$, it is not a relevant aspect. There is only one relevant aspect of $\Diamond$, namely $\{\Diamond, \bigcirc\}$.

3.10. Ideal Schema and Response Probability. Restle introduces the primitive notion of ideal situations, in terms of a set I of such situations. I is primitive because it is a special subset of $\mathcal{S}$ and cannot be "induced" or defined in terms of the earlier notions. Also, a primitive set R is introduced called the set of responses. The purpose of the theory can now be framed: to assign a probability to each element of R, using relevant aspects and ideal situations. The intuitive idea is that the higher organism holds a schema assigning a response to each ideal situation. The response in the actual situation is dependent upon similarity of the situation to the various ideals. Thus, to each $s \in I$, there exists a response under the schema.

To determine a probabilistic structure over R, Restle deals with various measures assigned to the sets of aspects. For a simple illustration, let us imagine that the measure of a set of aspects is simply the number of aspects in it. Suppose the organism confronts situation s and holds two ideals with associated responses:

$$s_1, R_1$$
$$s_2, R_2$$

so that we want to compute the probability of R_1 and the probability of R_2. Let the relevant aspects be in the sets,

$$A_s', A_1', A_2',$$

for the three situations s (the actual), s_1 (ideal situation 1) and s_2 (ideal situation 2). Then if m(X) denotes the measure assigned to any set X of aspects, Restle assumes that

$$(3.10.1) \qquad \Pr(R_1) = \frac{m(A_s' \cap A_1' \cap \bar{A}_2')}{m(A_s' \cap A_1' \cap \bar{A}_2') + m(A_s' \cap \bar{A}_1' \cap A_2')}$$

The intuition behind this formula can be captured through a Venn diagram and a concrete example. The following Venn diagram shows the various sets whose measures are being considered in (3.10.1):

$(1) = A_s' \cap A_1' \cap \bar{A}_2'$
$(2) = A_s' \cap \bar{A}_1' \cap A_2'$
$(3) = A_s' \cap A_1' \cap A_2'$
$(4) = \bar{A}_s' \cap A_1' \cap A_2'$

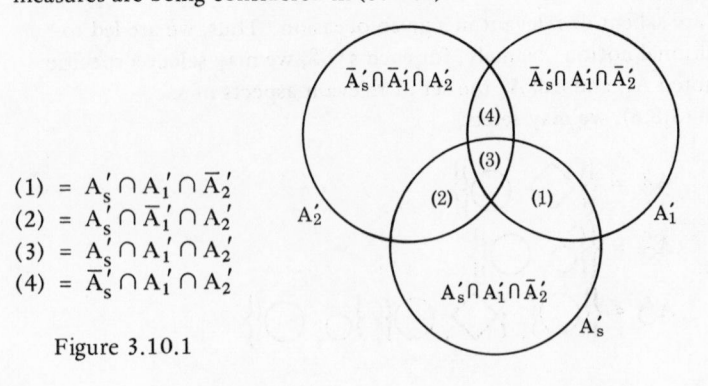

Figure 3.10.1

In these terms,

$$(3.10.2) \qquad \Pr(R_1) = \frac{m(1)}{m(1) + m(2)}$$

Example. If region (1) is empty, then s shares no aspect with s_1 and we can take $m(1) = 0$. Then, if $m(2) \neq 0$, we conclude $\Pr(R_1) = 0$. On the other hand, if (2) is empty, take $m(2) = 0$. Then, provided (1) is such that $m(1) \neq 0$, we conclude $\Pr(R_1) = 1$. In general, $\Pr(R_1)$ depends on the salience measure of the ideals and their aspect similarity to the actual situation.

3.11. Concrete Example. The following example reviews all the developments given above, but in the context of a fresh illustration.

Example. For $\mathcal{S}$ take the three elements

Then for the family of partitions $\mathcal{P}$ we obtain the collections

We assume the relevant aspect sets take the form for each "situation":

$$(3.11.1)$$

(For later reference, call these A_1', A_2' and A_s' respectively.)

For example, under v_2 the black flag-waver has the aspect given by grouping with the black non-flag-waver; under v_3 and under v_4, he has an "individuality," while under v_1 he is classed together with the white flag-waver. Here we have three situations, each identified with a person-figure; we have five stimulus variables v_1, v_2, v_3, v_4, and v_5; and each person-figure has relevant aspects.

We let the person-figure, call him s,

be the actual situation confronting our psychological subject, and let the black

be ideal situation s_1, and the white flag-waver

be the ideal situation s_2, and we assume schema of the form,

$\sim\!\!\!\rightarrow$ "black"

$\sim\!\!\!\rightarrow$ "patriot"

Then we want to compute the chances of the two defining responses when the black flag-waver is "the situation" confronting our subject. Initially, let us assume that m(X)—the measure of set X of aspects—is just the number of elements in the set. To find the various sets needed to apply formula (3.10.2) we calculate

$$(1) \; = \; A_s' \cap A_1' \cap \overline{A}_2' \; = \; \{a\text{: } a \text{ is an aspect of s, and of } s_1, \text{ but not of } s_2\}$$

Note that

(3.11.2)
$$A_s' \cap A_1' \; = \; \left\{ \text{}, \text{} \right\}$$

and that the only element in this set is not in A_2' and hence is in $\overline{A}_2'$. Thus, this set is actually $A_s' \cap A_1' \cap \overline{A}_2'$. This gives us region (1) of Figure 3.10.1.

Also, to find region (2), note that

(3.11.3)
$$A_s' \cap A_2' \; = \; \left\{ \text{}, \text{} \right\}$$

and that since this single element (aspect) is not in set A_1' it is in $\overline{A}_1'$. Thus (3.11.3) is actually $A_s' \cap A_2' \cap \overline{A}_1' = A_s' \cap \overline{A}_1' \cap A_2'$, which is region (2).

Taking the size measures, we obtain

$$m(1) = 1, \; m(2) = 1,$$

and so, by (3.10.2),

$$Pr(R_1) = Pr(\text{``black''}) = \frac{1}{2}$$

For an alternative, assume each aspect has a measure based on a salience weight. Suppose: weight of black = 2, weight of flag-waving = 1. Then,

$$m(1) = 2$$
$$m(2) = 1$$

and so,

$$Pr(\text{``black''}) = \frac{2}{3} > \frac{1}{2}$$

because "black" counts more than "flag-waving" in the defining response.

3.12. Concluding Remarks. Restle's conceptualization is quite general. The connection with labeling in deviance theory is clear from the example. In addition, one can think of reference-group phenomena in these terms: The situation confronting a person is one thing; the background of ideals with response schema can be interpreted in terms of possibly competing norms orienting the person to one or another mode of response; and, of course, there are the relevant aspects and their salience weights, which determine the probability of one or another response. The main problem is that Restle's formulation, while attractive in conceptualization, does not have a powerful underlying principle that governs the choice process. Even from a conceptual viewpoint, however, there is the obvious difficulty that novelty of response cannot be generated by reference to preexisting schema: hence there is a serious scope restriction on the applicability of the framework.* Also, a good deal of information must be fed into the Restle scheme; yet it provides a sound foundation for more deductively rich formulations in probabilistic choice theory, as we shall see subsequently, in Chapter 11.

---

*There are levels of novelty. Verbal novelty is least fundamental. A man may describe a woman as a "female surgeon" and operate with schemas which require her to reduce the situational salience of her sex to generate his defining her as a surgeon, i.e., behaving toward her in appropriate ways. In this case, the male aspect is tied to "surgeon" in the way that the white aspect is tied to the "patriot" defining response in section 3.11.

CHAPTER FOUR REPRESENTATION OF PSYCHOLOGICAL AND SOCIAL RELATIONS

4.1. Aim of the Chapter. Knowledge of abstract relation theory provides a precise way to think about social relations. In this chapter, we introduce this theory by showing how it emerges naturally from the logical and set-theoretical basis; examples of how one thinks about actual relationships with this theory are given.

4.2. Specification of Relations. To each sentential function in two free variables we associate the concept of the class of pairs of objects satisfying the sentential function—that is, making it true when the pair-names replace the variables. We call such a class a binary relation.

Example. (a) Variables: x, y   (domain: set of persons)
Sentential function (in 2 free variables): x is a friend of y.
Binary relation defined: F $= \{(x, y): x \text{ is a friend of } y\}$

Example. (b) Variables: α, β (domain: set of persons)
Sentential function: α is an ancestor of β.
Binary relation defined: A $= \{(\alpha, \beta): \alpha \text{ is an ancestor of } \beta\}$

Binary relations can also be specified merely by listing ordered pairs of objects in some domain. For instance,

$$R = \{(1, 3), (10, 2), (7, 9)\}$$

This relation R can be thought of as containing three points in the usual Cartesian coordinate plane.

By defining relations as we have, one thing is accomplished that is of considerable importance: the relation concept is imbedded in set theory. That is, we can carry over all the relationships, operations, and notations from sets to relations. Thus, if in Example (a) we have the defined relation

$$F = \{(\text{Jones, Smith}), (\text{Harris, Jones}), (\text{Smith, Jones}), (\text{Jones, Harris})\}$$

then we can say,

$$(\text{Jones, Smith}) \in F.$$

We would read this, "Jones and Smith are in the friendship relation," or, "The pair (Jones, Smith) is in the friendship relation." Clearly, we have

$$(x, y) \in F \qquad \text{iff} \qquad x \text{ is a friend of } y$$

That is, the pair is in the relation if and only if the sentential function is satisfied.

Further examples of the carryover of set-theoretic ideas to relation theory follow.

Example. (c) If

$$R = \{(1, 3), (4, 5), (2, 1)\}$$
$$S = \{(2, 6), (1, 3), (1, 1)\}$$

then,

$$R \cup S = \{(1, 3), (4, 5), (2, 1), (2, 6), (1, 1)\}$$
$$= \{(x, y): (x, y) \in R \text{ or } (x, y) \in S\}$$
$$R \cap S = \{(1, 3)\}$$
$$= \{(x, y): (x, y) \in R \text{ and } (x, y) \in S\}$$
$$R - S = \{(4, 5), (2, 1)\}$$
$$= \{(x, y): (x, y) \in R \text{ and } (x, y) \notin S\}$$

Example. (d) If

$$N = \{(a, b), (a, c), (a, d)\}$$
$$M = \{(b, a), (a, b), (c, a), (a, c), (d, a), (a, d)\}$$

then,

$$N \subseteq M$$

because for every $(x, y) \in N$, we have such $(x, y) \in M$.

4.3. Relations and the Axiom of Extension.

Just as we noted in section 3.2 that different sentential functions can give rise to the identical sets, we must note the same here in regard to relations. Indeed, in regarding relations as sets of ordered pairs (specified by sentential functions in two free variables), we must accept the consequences of the axiom of extension (3.2.1). How this affects our thinking is shown as follows:

Let α, β be variables with domain a specified finite set G of persons—for instance, all persons in a given classroom. Let

$$F_G = \left\{(\alpha, \beta): \alpha \text{ thinks of } \beta \text{ as a friend}\right\}$$
$$F_G' = \left\{(\alpha, \beta): \beta \text{ thinks that } \alpha \text{ thinks of } \beta \text{ as a friend}\right\}$$

The axiom of extension is brought forward to relation theory by letting the logical variable x of (3.2.1) be a variable with domain a set of ordered pairs, and, in fact, replacing variable x with variable-pair (α, β). Thus, the axiom now states,

$$\text{(Relation) A = (Relation) B} \quad \text{iff} \quad \text{for } \forall \, (\alpha, \beta),$$
$$(\alpha, \beta) \in A \quad \text{iff} \quad (\alpha, \beta) \in B$$

For the previous example,

$$F_G = F_G' \quad \text{iff} \quad \forall \, (\alpha, \beta),$$
$$(\alpha, \beta) \in F_G \quad \text{iff} \quad (\alpha, \beta) \in F_G'$$

Clearly, the sentential functions determining F_G and F_G' are quite distinct in meaning, but one can readily imagine a small cohesive group in which whenever a person (α) thinks of another (β) as a friend, that other (β) perceives this to be the case and, conversely, such perceptions are always accurate. In such a group,

$$(\alpha, \beta) \in F_G \quad \text{iff} \quad (\alpha, \beta) \in F_G'$$

Hence, applying the axiom of extension, we conclude $F_G = F_G'$. The two binary relations are identical, although the two sentential functions are distinct in meaning.

In semantics (see Carnap, 1956), one would say that F_G and F_G' are identical in the extensional sense although they differ in the "intensional," or meaning, sense. However, the intension of a relation concept can be formalized, as we shall see subsequently (see, especially, sections 4.17 and 4.18) by using the device of setting down axioms on an abstract (unspecified) set.

One last point on the consequence of the axiom of extension is that it permits us to see an analogy: property is to class as relationship is to binary relation. A property determines a class of objects, given some domain, and likewise, a relationship (or relation in the intensional sense) determines a binary relation, given some domain.

4.4. Meaning Postulates. We see that it could happen that on a particular domain Ω, a relationship has a property it "ordinarily" does not have; that is, in other domains, the same relationship produces extensions with different relational character. For example, the sociometric relation

$$\alpha \text{ chooses } \beta \text{ (on criterion } \ldots)$$

is ordinarily not one characterized by total mutuality (α names β and β names α, all α and β), but for a small group on some occasion, it is not impossible that the induced relation is totally mutual. Thus, if G is this group and if we let

$$\alpha C\beta \qquad \text{iff} \qquad \alpha \text{ chooses } \beta \; .$$

then the relation C on group G, defined by

$$C_G = \{(\alpha, \beta): \alpha C\beta\}$$

will be said to be mutual, although in all generality one would not want to say that C is mutual. On the other hand, consider the property "α chooses α, for some α." If C_G has this property, we would perhaps regard it as an error because we think of C carrying the convention, "no person may choose himself."

As a second example, consider the friendship relationship. Let

$$\alpha F\beta \qquad \text{iff} \qquad \alpha \text{ and } \beta \text{ are friends}$$

and consider a group G so that

$$F_G = \{(\alpha, \beta): \alpha F\beta\}$$

A postulate for friendship might be that it is symmetric:

(4.4.1) $\qquad\qquad$ If $\alpha F\beta$, $\quad$ then $\beta F\alpha$, $\quad$ all $\alpha, \beta \; (\alpha \neq \beta)$

from which two possibilities arise:

(1) For any domain G, binary relation F_G is determined such that (4.4.1) can be false. For example, one asks α, "Is β your friend?" and one asks β, "Is α your friend?" We say that (4.4.1) is then contingent, based on the choice relationship C.

(2) For any domain G, binary relation F_G is determined by procedures in conformity with (4.4.1); that is, the latter is a meaning postulate for F. Thus, one might observe interactions between α and β, collect historical information on their pattern of past interactions, and collect psychological information on their planned activities in the future, all in order to decide

$$(\alpha, \beta) \in F_G?$$

and to make this decision such that (4.4.1) must be satisfied.

To satisfy (4.4.1) we could easily consume enormous amounts of time and energy by procedure (2). Yet, our only purpose is to determine the current state of affairs; this state of affairs is essentially a reflection of the past history and a disposition for present and future observable activities: it is a state of a system. Thus, instead of extensive observations, we could revert to simply discovering the present state directly by applying method (1) to each person. Then, because not all pairs may choose each other, we may adopt a rule, in conformity with (4.4.1):

(4.4.2) $\qquad\qquad (\alpha,\beta) \in F_G \qquad \text{iff} \qquad \alpha C_G \beta \text{ and } \beta C_G \alpha$

This rule says that we will determine F to exist between a pair in G when, and only when, each answers yes to the question tapping their present dispositions toward each other. This makes F_G symmetric, because

(1) $(\alpha,\beta) \in F_G \Rightarrow \alpha C_G \beta$ and $\beta C_G \alpha$ (4.4.2)

(2) $\Rightarrow \beta C_G \alpha$ and $\alpha C_G \beta$ (From P and Q, one may conclude: Q and P)

(3) $\Rightarrow (\beta,\alpha) \in F_G$ (4.4.2)

The entire chain shows

(4.4.3) $(\alpha,\beta) \in F_G \Rightarrow (\beta,\alpha) \in F_G$

When a postulate like (4.4.1) functions in this way it is called a meaning postulate, as mentioned previously. Once we know that (4.4.1) is such a postulate, we know that it specifies an analytic property of the relationship and not merely a contingent property of the extension in a particular group. A fundamental job of a formal approach to a subject-matter area is to try to collect together these analytic properties and to show their logical consequences. This, too, is knowledge, and it is significant in reducing the degree of semantic squabbling that frequently characterizes nonmathematized fields. (For a closely related discussion, see section 4.18.)

4.5. Human Relations. We now turn to further examples in applied relation theory.

Example. (a) We introduce symbols for certain social relationships:

xDy   iff   x dominates y,

xSy   iff   x has the right to dominate y,

yOx   iff   y has the duty to obey x.

These three relations may now be used to generate defined relations. The procedure involves setting up a formalism that reflects one's intuitions fairly well. For example, we may agree that x has authority over y and write xAy when we have both xSy and yOx. If, in addition, we have this same x and y standing in the relation D—that is, x dominates y (as a matter of fact)—then we may agree to say that x has authoritative power over y and write xPy. This whole definitional procedure may then be written formally as follows:

Definition. xPy.

xPy   iff   xSy, yOx and xDy   (all x,y)

Alternatively, we may proceed in two stages:

Definition. xAy.

$$xAy \quad iff \quad xSy \text{ and } yOx \quad (all\ x,y)$$

Definition. xPy.

$$xPy \quad iff \quad xAy \text{ and } xDy \quad (all\ x,y)$$

Example. (b) A relation may hold between a person and a group. Just as we use different letters for different relations, it is helpful to use different letters for different kinds of objects. For example, to note that person x would like to belong to a group g, we may write xLg. If x actually belongs to g, we may write xBg.

Example. (c) A relation may hold between two groups. For example, gCh if and only if group g is in conflict with group h.

Example. (d) A relation may hold between two collections of groups. For example, two collections of groups (say G, H) may be said to be disjoint if it is not the case that there is some person x such that both xBg and xBh, with g in G and h in H; that is, $g \cap h = \emptyset$ (all $g \in G$, $h \in H$).

Example. (e) A relation may vary in time. In this case we may amplify our notation and write xRy at t to mean that x and y stand in the relation R at the moment or period t.

4.6. Properties of Relations. Relations have properties. For example, the relation may hold between all members of a population, in which case it has the property of completeness. There are several such types of properties to be dealt with. The point is that in specifying any kind of social relation one ought to consider these properties.

In discussing such properties of a concrete relation, as discussed in section 4.3, one must always beware of the fact that what one counts as the same relationship on intuitive grounds may have markedly different properties in different populations. One need only recall that the relation of interpersonal communication may be complete in some small groups and very "sparse" in large, natural populations. In keeping with this variation, in introducing the ideas of relation theory that describe properties of relations, we shall use an explicit notation. We shall write

$$xR_A y \text{ or } (x,y) \in R_A$$

and mean that x and y are arbitrary members of the population A and that x and y stand in the relation R as it obtains as a matter of fact on population A. (This notation was used in section 4.4.)

We now discuss some basic properties of such concrete relations. These are concepts one may use to classify various kinds of relations. We begin with those that have to do with symmetry.

4.7. Symmetry. We shall say that R_A is symmetric if whenever we have $xR_A y$ we also have $yR_A x$. In other words, whenever x is in the relation with y, then y is also in the relation with x, for all x, y in A. This does not exclude the possibility that in the given population some members may not be related by R_A at all. For example, the marriage relation is symmetric but of course does not hold between any arbitrary pair of persons.

Some relations are asymmetric, meaning that whenever we have $xR_A y$ we do not have $yR_A x$. In other words, if one party is in the relation with the other, the second party is not in the relation with the first. For example, we have an established pecking order among a flock of hens F, then whenever hen x dominates hen y, the reverse does not obtain: $xD_F y$ implies $\text{not-}yD_F x$.

A relation that is asymmetric fails to have symmetry in a very strong sense. But many relations are not symmetric and yet have a kind of partial symmetry, in the sense that some couples may be mutual and others not. A familiar example comes from sociometry, where the choice relation usually involves mutuality in some cases and nonmutuality in others. If we want to say that a relation fails to be symmetric without committing ourselves to the strong property of asymmetry, we say that the relation is nonsymmetric. Thus, an asymmetric relation is a special kind of nonsymmetric relation, where the absence of symmetry holds throughout the population.

Whether or not a relation is symmetric will often depend on methodological decisions. Consider the love relation. This relation will be nonsymmetric in most populations (i.e, there will be some asymmetric couples). If it is claimed this is not really love, then it is intended that symmetry be taken as a meaning postulate for the love relation.

A final notion in regard to symmetry is the property exemplified by the containment relation on a domain of sets. Namely,

$$A \subseteq B \text{ and } B \subseteq A \;\Rightarrow\; A = B \qquad (\text{all } A, B)$$

The general property is

$$xR_A y \text{ and } yR_A x \;\Rightarrow\; x = y \qquad (\text{all } x, y \in A)$$

This is termed antisymmetry. We will encounter its significant role later when we discuss orderings.

The various definitions related to symmetry can be summarized formally as follows:

(4.7.1) Definition. R_A is symmetric   iff $xR_A y \;\Rightarrow\; yR_A x$ (all x, y $\in$ A).

(4.7.2) Definition. R_A is asymmetric   iff $xR_A y \Rightarrow \text{not-}yR_A x$
(all $x, y \in A$).

(4.7.3) Definition. R_A is nonsymmetric   iff   not-(R_A is symmetric).

(4.7.4) Definition. R_A is antisymmetric   iff $xR_A y$ and $yR_A x \Rightarrow x = y$
(all $x, y \in A$).

The tree diagram in Figure 4.7.1 may help the reader to relate these concepts to each other.

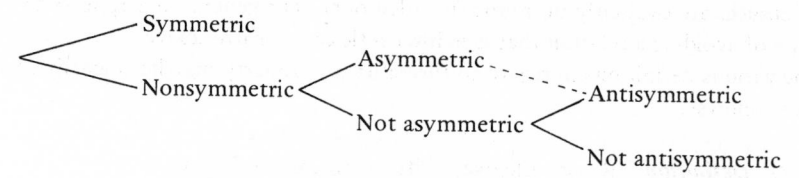

Figure 4.7.1

The paths in the tree generate exactly four mutually exclusive and exhaustive classes of relations: (a) symmetric, (b) asymmetric, (c) neither symmetric nor asymmetric, but antisymmetric, and (d) none of the above. The usual sociometric relation, for instance, falls into the residual category (d). The reason for specifying type (c) by the three properties instead of antisymmetry alone relates to the dotted line in the tree. If a relation is asymmetric, then it is vacuously antisymmetric. To see what this means, we first write definition (4.7.4),

$$xR_A y \text{ and } yR_A x \Rightarrow x = y$$

Now, if R_A is asymmetric, the condition on the left cannot be true. We have a case of a conditional sentence with a false antecedent. Hence, the conditional is vacuously true and R_A is said to be vacuously antisymmetric.

4.8. Reflexivity. The admonition Know Thyself would indicate that it is possible for a man not to know himself. This concept appears to mean "having an accurate model of." If this model relation is defined on a set of persons A, then it is likely that we will have persons x for whom $xM_A x$ obtains (x has an accurate model of himself) and others for whom we have not-$xM_A x$ (x does not have an accurate model of himself). On the other hand, no one is married to himself, but in most populations everyone is (at least mildly) acquainted with himself.

These variations lead to the following distinctions: A relation R_A is said to be reflexive if for all x we have $xR_A x$. For example, every person in a given population is acquainted with himself. A relation is said to be nonreflexive if there is some x such that not-$xR_A x$ obtains. So, a relation is nonreflexive if the reflexive property fails for at least one object. An example is the relation of "having an accurate model of" when considered on the conceptual population of all persons. We also may have a property akin to asymmetry: here the reflexive property fails to hold in every instance. We say that a relation R_A is irreflexive if not-xR_Ax obtains for every x. For example, no one is married to himself.

An operationally defined relation will often be arbitrarily declared to have a certain reflexive property quite apart from the intuitive connotations of the concept involved. For example, the liking relation is nonreflexive in ordinary experience, but a sociometric liking relation is often stipulated to be irreflexive (self-choices are explicitly or implicitly ruled out). The general rule appears to be one of avoiding a relation that is neither reflexive nor irreflexive.

The various definitions in regard to the reflexive property may be formally set out as follows:

(4.8.1) Definition. R_A is reflexive   iff xR_Ax (all $x \in A$).

(4.8.2) Definition. R_A is irreflexive   iff   not-xR_Ax (all $x \in A$).

(4.8.3) Definition. R_A is nonreflexive   iff   not-(R_A is reflexive).
The corresponding tree diagram is shown in Figure 4.8.1.

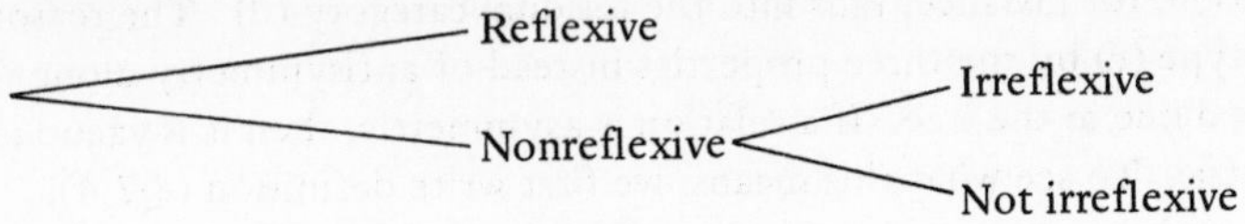

Figure 4.8.1

The diagram makes clear that the three distinct paths yield a typology with three categories: reflexive, irreflexive, and neither reflexive nor irreflexive.

We note that if R_A is not irreflexive, then it is not asymmetric, because if x_0 satisfies $x_0R_Ax_0$, then the conjunction with the asymmetric condition that $x_0R_Ax_0 \Rightarrow$ not-$(x_0R_Ax_0)$ yields a contradiction. This may be shown explicitly by constructing a truth-table computation terminating in a final column of all F. Let P be the sentence $x_0R_Ax_0$. Then:

	(1)	(2)	(3)
P	not-P	P $\Rightarrow$ not-P	(2) and P
T	F	F	F
F	T	T	F

Note that $x_0R_Ax_0 \Rightarrow$ not-$(x_0R_Ax_0)$, column (2), is not itself contradictory.

Finally, we note that the tautologically equivalent contrapositive statement corresponding to "If R_A is not irreflexive, then R_A is not asymmetric," is the more intuitively understandable statement, "If R_A is asymmetric, then R_A is irreflexive."

4.9. Transitivity. A relation R_A is transitive if whenever we have both xR_Ay and yR_Az, then we also have xR_Az. For example, the ancestor relation on any collection of humans is transitive: if x is an ancestor of y and y is an ancestor of z, then x is also an ancestor of z. Again, if x may directly or indirectly communicate with y and y may directly or indirectly communicate with z, then also x may directly or indirectly communicate with z. In these examples, the transitivity is an analytical property of the relationship.

We saw in the case of symmetry and again with reflexivity that we have two additional concepts in each case: the failure of the property for at least one test case and its failure throughout a population. The same considerations apply in regard to transitivity. If a relation fails to be transitive, we say that it is nontransitive, meaning that there exists in the set A under discussion at least one "triple" of objects, x, y, and z, such that we have

$$xR_Ay \text{ and } y R_Az$$

but we also have

$$\text{not-}xR_Az$$

For example, in a typical sociometric test, we arrive at such triples: x names y and y names z, but x does not name z. The overwhelming majority of social relations defined on the set of all persons appear to be nontransitive. For another example, the relation of dominance among hens is nontransitive: there are hens in the usual barnyard such that x pecks y and y pecks z, but x does not peck z.

Some relations are nontransitive in the strong sense that no transitive triples exist. The relation R_A is said to be intransitive if whenever we have

$$xR_Ay \text{ and } yR_Az$$

then we also have

$$\text{not-}xR_Az$$

for all x, y, z in A. The biological fatherhood relation on the set of all persons is of this type, for if x is the father of y and y is the father of z, then x is the grandfather of z and hence not the father of z. Thus, the fatherhood relation is intransitive.

Summing up these ideas formally, we have the following definitions:

(4.9.1) Definition. R_A is transitive iff

$$xR_Ay \text{ and } yR_Az \;\Rightarrow\; xR_Az \qquad \text{(all } x, y, z \in A\text{)}$$

(4.9.2) Definition. R_A is intransitive iff

$$xR_Ay \text{ and } yR_Az \;\Rightarrow\; \text{not-}(xR_Az) \qquad \text{(all } x, y, z \in A\text{)}$$

(4.9.3) Definition. R_A is nontransitive iff not-(R_A is transitive)
The tree diagram is shown in Figure 4.9.1.

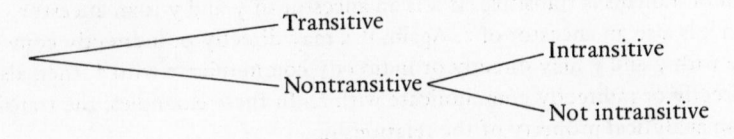

Figure 4.9.1

This yields a typology with three categories: transitive, intransitive, and neither transitive nor intransitive. Once again the usual sociometric relations fall in the last, residual category.

4.10. Completeness. For a given population a relation may or may not be quite "dense" in that population. For example, we mentioned earlier that interpersonal communication is quite dense in small groups and comparatively sparse in large populations. The relation R_A is said to be complete if whenever x and y are distinct objects, either xR_Ay or yR_Ax obtains. We purposefully let the relation be complete regardless of the reflexive property, for in most cases of applications we want to say that the relation is complete even when it is irreflexive. This is the reason for using the phrase "distinct objects." If we drop this stipulation, then we say that R_A is strongly complete if for any x, y, either xR_Ay or yR_Ax holds. If a relation fails to be complete, then we simply call it incomplete. It should be clear that the majority of social relations on the usual populations are incomplete. It should be noted in passing that in the literature of logic the words "connectedness" and "connectivity" are often used in place of "completeness." We reserve these words for a different notion.

One should note that R_A is complete if both xR_Ay and yR_Ax hold for all distinct pairs, because working within the standard logical basis given in chapter 2, the disjunctive molecular sentence

$$xR_Ay \text{ or } yR_Ax$$

is true when both atomic components are true, as well as when only one of them is true.

Finally, it might be mentioned that completeness is sometimes referred to as a matter of pairs being comparable. For example, any two real numbers are comparable in regard to the ordering given by the relation $\leqslant$. That is, $x \leqslant y$ or $y \leqslant x$ for all real numbers x, y.

The formal definitions for completeness, then, are as follows:

(4.10.1) Definition. R_A is complete iff xR_Ay or yR_Ax (all x, y $\in$ A; x $\neq$ y).

(4.10.2) Definition. R_A is incomplete iff not-(R_A is complete).

(4.10.3) Definition. R_A is strongly complete iff $xR_A y$ or $yR_A x$ (all x, y). A trivial deduction yields

(4.10.4) Proposition. R_A is strongly complete $\Rightarrow$ R_A is complete.

4.11. Relational Systems. If we consider a population of objects, A, then a relation R_A is said to structure the set or population. The set, together with the relation structure of that set, we term a binary relational system of the simplest type—or more simply, a binary relational system. The binary relational system is denoted by the ordered couple (A, R_A).

Examples. (a) A collection of persons, together with the authoritative power relation defined in section 4.5, example (a), constitutes a binary relational system.

(b) A collection of social groups, together with a conflict relation, is a binary relational system.

(c) A collection of persons in an interaction process analysis setting, together with the relation defined by noting whether or not x originated an act toward y in a time interval T, is a binary relational system.

(d) A collection of persons, together with their sociometric choices, is a binary relational system.

(e) A collection of occupations, together with the relation of having the same prestige, is a binary relational system.

(f) A collection of hens, together with the dominance or pecking-right relation among the hens, constitutes a binary relational system.

4.12. Equivalence Relations and Quotient Sets. Relations may be classified according to the particular combination of properties they possess. Two generic types of relations are of enormous significance: the relations that refer to an aspect of "sameness," or equivalence, of entities and hence create unity out of diversity; and the relations that refer to an aspect of ordering the objects in the domain.*

(4.12.1) Definition. A relation R_A is termed an equivalence relation if and only if (1) R_A is reflexive, (2) R_A is symmetric, and (3) R_A is transitive.

Example. (a) The relationship of "having the same parents as" is an equivalence relation on any set of human beings: for all x, x has the same parents as x, so that (1) holds; if x has the same parents as y, then y has the same parents as x, so that (2) holds; and if x has the same parents as y and y has the same parents as z, then x has the same parents as z, so that (3) holds. Thus,

---

*Definition (4.12.1) illustrates the idea of exact concept. Also, each condition is a meaning postulate for equivalence.

$$S_A = \{(x, y): x \text{ has the same parents as } y\}$$

is the full sibling relation on A and sorts persons in A into categories within which the persons are full siblings [i.e., $(x, y) \in S_A$] and across which the persons are not full siblings [i.e., $(x, y) \notin S_A$].

Example. (b) Occupations in the United States may be related to each other via the relationship "has the same prestige as" for a particular group or for the entire adult population. Once again the "sameness" means that (1), (2), and (3) will hold. On the other hand, not the use of the word "same" but the referential aspect determined by the empirical procedure is what must be shown to satisfy the conditions. (See section 7.5).

Example. (c) Trivially, on any set, the identity relation,

$$(x, y) \in R_A \qquad \text{iff} \qquad x = y$$

is an equivalence relation, since

(1) $x = x$ (all x in A)
(2) $x = y \quad \Rightarrow \quad y = x$ (all x, y in A)
(3) $x = y$ and $y = z \quad \Rightarrow \quad x = z$ (all x, y, and z in A)

Example. (d) Consider a collection of behaviors of a person in a social role. Call this set B. Persons in the role system, including the role-player, may be indifferent to certain behaviors. For example, suppose that a teacher may or may not use a fountain pen in correcting assignments. We want to say, then, that there exists a relation, say I_B, on B (for the "actors" in a set A) such that

(1) $xI_B x$ (all x in B)
(2) if $xI_B y$, then $yI_B x$ (x, y in B)
(3) if $xI_B y$ and $yI_B z$, then $xI_B z$ (all x, y, z in B)

so that I_B is an equivalence relation on B for these actors and for this role.

Associated with every equivalence relation there is the induced division or partition of the domain of the relation. This partition is defined as follows: Let x be in the domain of equivalence relation E. We specify that

(4.12.2) $\bar{x} = \{y: xEy\}$

so that $\bar{x}$ is the class of all objects equivalent to x (and so, by the reflexive property, including x). Then $\bar{x}$ is termed the equivalence class of x, under E.

We indicate proof-technique a little more by showing that

(4.12.3) $\bar{x} = \bar{y} \qquad \text{iff} \qquad xEy$ (all x, y in domain)

To prove this biconditional we must prove two conditionals hold:

(1) $\bar{x} = \bar{y} \quad \Rightarrow \quad xEy$
(2) $xEy \quad \Rightarrow \quad \bar{x} = \bar{y}$

To prove a conditional, we suppose that the condition holds and then show that by prior meanings and assumptions (and tacit use of tautologies) the conclusion holds. Thus, to prove (1), suppose that $\bar{x} = \bar{y}$. Then, for any object (say z),

$$z \in \bar{x} \quad \text{iff} \quad z \in \bar{y}$$

by the axiom of extension. For "z," put "y":

$$y \in \bar{x} \quad \text{iff} \quad y \in \bar{y}$$

But, $y \in \bar{y}$ holds because every equivalence class of an object contains that object. Thus, also $y \in \bar{x}$. But, then, by (4.12.2) we conclude that xEy. To prove (2), suppose xEy. Then we want to show that $\bar{x} = \bar{y}$. This means, by the axiom of extension, we want to show that

$$z \in \bar{x} \quad \text{iff} \quad z \in \bar{y} \quad \text{(any z)}$$

Now this desired conclusion itself is a biconditional. Thus, we employ the logic of proving two conditions: $z \in \bar{x} \Rightarrow z \in \bar{y}$, and, $z \in \bar{y} \Rightarrow z \in \bar{x}$. First, suppose that $z \in \bar{x}$. Then xEz, by (4.12.2). But, by symmetry, then zEx. Thus, both zEx and xEy hold. By transitivity, we conclude that zEy, and so, by (4.12.2), $z \in \bar{y}$. Thus, $z \in \bar{x} \Rightarrow z \in \bar{y}$. Second, suppose that $z \in \bar{y}$. Then yEz, and since we are supposing that xEy, we conclude from both xEy and yEz that xEz holds. Thus, $z \in \bar{x}$, by (4.12.2). Hence, $z \in \bar{y} \Rightarrow z \in \bar{x}$.

The set of all equivalence classes induced by E on a domain A is denoted A/E; it is A divided into classes by E. In fact, A/E is called a quotient set. We say "A modulo E" or "A mod E." Thus, in example (a), people are divided into classes, modulo "have the same parents." The siblings are the whole population modulo same parentage. In example (b), occupations are divided into classes by equal prestige. The classes so induced constitute the domain of jobs "mod equal prestige." In example (c), the objects are divided modulo identity; then the equivalence class of object x contains only x. Thus, essentially this division recapitulates the identity of the original objects. It is the most discriminating equivalence relation on any set.

In example (d), behaviors are grouped mod indifference. The set B/I_B is the collection of equivalence classes of behaviors. These behavior classes are essentially groups of substitutible ways of accomplishing a task without violating any norms. To indicate that there may not be one relation I_B for all actors, we might let α be a logical variable with domain A, the set of actors, and form the class,

$$\{I_B^\alpha, \alpha \in A\}$$

that is, the set of all indifference relations on B, indexed by the actor holding that relation. Then we also have

$$\{B/I_B^\alpha, \alpha \in A\}$$

so that each α is associated with a possibly distinctive system of behaviors mod indifference.

4.13.   Order Relations.   A second type of relation, as determined by the configuration of its properties, is a partial order relation.

(4.13.1)   Definition.   By a partial order relation on a set A we mean a relation R_A such that (1) R_A is reflexive, (2) R_A is antisymmetric, and (3) R_A is transitive.

Example.   (a)   Consider the relation $\leqslant$ on the domain of all real numbers. Then the following properties hold:

$$
\begin{array}{lll}
(1) & x \leqslant x & \text{(all x)} \\
(2) & x \leqslant y \text{ and } y \leqslant x \quad \Rightarrow \quad x = y & \text{(all x, y)} \\
(3) & x \leqslant y \text{ and } y \leqslant z \quad \Rightarrow \quad x \leqslant z & \text{(all x, y, z)}
\end{array}
$$

Thus, $\leqslant$ is a partial order.

Example.   (b)   In section 3.3, it was shown that the containment relationship between sets satisfies

$$
\begin{array}{ll}
(1) & A \subseteq A, \\
(2) & A \subseteq B \text{ and } B \subseteq A \quad \text{iff} \quad A = B, \\
(3) & \text{if } A \subseteq B \text{ and } B \subseteq C, \text{ then } A \subseteq C,
\end{array}
$$

for any sets A, B, and C. Thus, containment has all three properties. Now consider a domain Ω. Let 2^Ω be the family of all subsets of Ω. For instance,

$$
\Omega = \{1, 2\}
$$
$$
2^\Omega = \{\emptyset, \{1\}, \{2\}, \{1, 2\}\}
$$

Then containment is a partial order relation on the set 2^Ω. Thus, containment partially orders the subsets of Ω.

Mathematicians sometimes use "Hasse diagrams" to represent partial orders. The diagram for the order of subsets of $\Omega = \{1, 2\}$ is Figure 4.13.1.

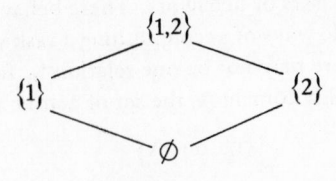

Figure 4.13.1

The diagram is based on the same principle as an organization chart showing authority lines. If a path can be traced upwards in the diagram this indicates the object represented by the starting point stands in the partial order relation to the object represented by the terminal point of the path.

The fundamental property required for any ordering of a domain of entities is transitivity. Additional properties generate special types of order. We have examined one important type: partial order. A partial order that is also complete is called a simple order. Example (a) provides an instance of a simple order because $x \leqslant y$ or $y \leqslant x$, for all numbers x, y. But example (b) exemplifies a partial order that is not simple; for instance, subsets $\{1\}$ and $\{2\}$ are "not comparable" under the containment relation. This difference shows up in the Hasse diagram. For positive integers, for instance, the Hasse diagram of relation $\leqslant$ is sketched in Figure 4.13.2.

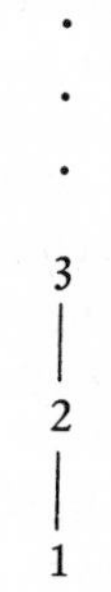

.

.

.

3
|
2
|
1

Figure 4.13.2

A different family of order relations is obtained if the antisymmetry and reflexivity properties of partial order are replaced by asymmetry. (The asymmetry, in turn, logically implies that the relation is irreflexive, as we noted earlier.) The prototype of all such orders is $<$ on the numbers. Note that this relation is asymmetric, transitive, and, in addition, complete. A relation that is transitive and asymmetric may be termed strict. If a strict order is also complete, as in the case of the numerical inequality relations, it is called linear. Hence, a linear relation satisfies (a) transitive, (b) asymmetric, and (c) complete. The terminology in the field of order relations is very unsettled, however, and the reader should always look for explicit defining conditions in terms of the fundmental relational properties based on symmetry, reflexivity, transitivity, and completeness.

Example. (c) A rational preference relation over a set of commodities provides an example of an incomplete strict ordering. A rational person does not prefer y to x if he prefers x to y (asymmetry); if he prefers x to y and y to z, then he prefers x to z (transitivity); therefore, a rational preference relation constitutes a strict ordering of the domain of objects of preference. But a rational preference relation is not a complete ordering because of the phenomenon of indifference: given two commodities, the rational man may prefer neither one to the other.

Example. (d) A complete strict ordering (i.e., a linear order) is provided by an idealized authority relation whose closest realization is likely to be an army in which clear rules exist for deciding, for persons in the same ostensible rank, who has precedence over whom. In this case, for any pair of distinct personnel, one of them "has rank over" the other, so the relation is complete. The ranking is asymmetric. Further, if x has rank over y and y has rank over z, then x has rank over z. Thus, it is transitive. Therefore, the ranking constitutes a linear ordering. The resulting series is formally identical to an array of distinct numbers ordered by the "less than" relation.

One additional type of relation deserves mention because it is substantively similar to an order relation but does not possess the transitivity property. A relation that is asymmetric and complete is called a dominance relation. As an example of this type of relation, consider pecking rights among a flock of hens. Suppose that the hens are segregated at some initial time into two subgroups, with no encounters between subgroups. Within each subgroup a pecking right will develop such that if x pecks y, then y does not peck x. The relation will ordinarily be nontransitive in the sense that hens exist such that although x pecks y and y pecks z, x does not peck z. Further, within each subgroup, given any pair of hens, one will have the pecking right over the other. Therefore, within the subgroups, we will have a dominance relation. On the other hand, the failure of hens in the different subgroups to encounter each other means we have no established pecking right between pairs of hens drawn from different subgroups. Therefore, the total relation over the entire flock is not a dominance relation. It may be presumed that if the hens were desegregated then after a transient period of encounters, a new pecking-right system would develop and be complete over the entire flock.

From a methodological viewpoint, note that we did not state anything about reflexivity or transitivity in the definition of a dominance relation. From asymmetry, we conclude irreflexivity. The absence of an axiom as to transitivity means that we allow concrete relations to be termed "dominance" even if there are triples which are counterexamples to transitivity. Hence, the absence of an axiom implies nontransitivity. Also, since counterexamples to intransitivity are allowable, there will be concrete dominance relations that are of the residual type in which neither transitivity nor intransitivity obtains. The general rule is: if no axiom is stated about a transitive property, then the type is residual in the sense of our earlier tree diagram.

4.14. Temporal Relational Systems. In example (e) of section 4.5, it was mentioned that when we wish to consider a relation varying in time, we use the notation

$$xRy \quad \text{at} \quad t$$

to denote the fact that x and y stand in R at the moment or period t. When x and y are relativized to some set A, we write, first,

$$xR_A y \qquad \text{at} \qquad t$$

as an extension of this notation, but then, for the sake of compactness, we change to

$$xR_{At}y$$

to mean that x and y of population A stand in relation R at time t, where R is defined on A.

If we now extend this notation to binary relational systems, we have (A, R_{At}) for a binary relational system at time t. It remains only to note that we are interested in the entire set of such systems varying in time, and so, we define a temporal binary relational system as a family of systems at time t, as t varies in the index set T. T may be a set of integers $\{0, 1, 2, \ldots\}$ or some interval of the real numbers. The notation we use for the temporal binary relational system is $(R_{At}, t \in T)$.

A nontemporal system may be regarded as a special kind of temporal system; namely, T shrinks to contain a single number, and so, t takes on just one value. A nontemporal system is then a temporal system considered at some fixed time.

As an example of a temporal binary relational system, let us look at interaction process analysis. Let a 50-minute session be broken up into 2-minute intervals, and number these intervals 1, 2, ..., 25. This set of numbers will be T. If x addresses a verbal act toward y in an interval t of T, we write $xR_{At}y$, where A is the collection of persons in the group. Each interval gives us a binary relational system that will usually be nonsymmetric and nontransitive. The entire sequence of such systems defines a temporal binary relational system associated with the situation.

An instance of the continuous time case may be mentioned in passing. Let A be some large natural population. Let $xR_{At}y$ mean that at moment t, x is interacting with y. For each moment, we have a binary relational system (A, R_{At}), and for some time interval T, we have the temporal binary relational system $(R_{At}, t \in T)$. By meaning postulate for the term "interaction" as here used, the relation is always symmetric, but its transitivity and completeness may vary in time.

4.15. Graph Representation of Relational Systems. It will be convenient at this time to discard the notation R_A for a relation on A in favor of the simpler notation R, with the domain understood from the context. Let R be a relation on A. Represent objects in A by points in the plane of the paper and indicate that $(x, y) \in R$ by drawing a directed line segment from x to y. The resulting graph represents the relational system. For example,

$$A = \{1, 2, 3\}$$
$$R = \{(1, 1), (2, 3)\}$$

has the representation shown in Figure 4.15.1.

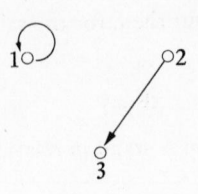

Figure 4.15.1

Points are labeled by the numbers they represent.

Corresponding to a relation on A that is symmetric, all lines appearing on the graph must be mutually directed; corresponding to asymmetry, any line on the graph cannot appear with its "reciprocal" line; the residual category of neither symmetry nor asymmetry will show up on the graph with a mixture of symmetric and asymmetric pairs. These cases are shown in the graphs in Figure 4.15.2.

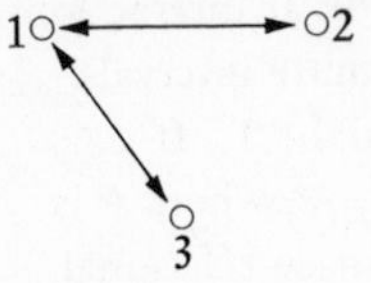

$\Big\{$ represents symmetric relation on $\{1, 2, 3\}$; namely, $\{(1, 3), (3, 1), (1, 2), (2, 1)\}$.

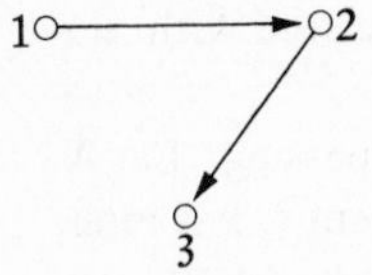

$\Big\{$ represents an asymmetric relation on $\{1, 2, 3\}$; namely, $\{(1, 2), (2, 3)\}$.

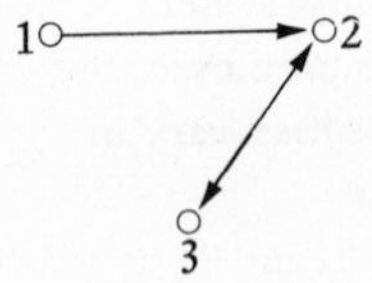

$\Big\{$ represents a nonsymmetric relation on $\{1, 2, 3\}$, which is also not asymmetric; namely, $\{(1, 2), (2, 3), (3, 2)\}$.

Figure 4.15.2

Corresponding to a reflexive relation, we must see a "loop" on each point of the graph; corresponding to an irreflexive relation, no loop must appear on the graph. This is shown in Figure 4.15.3.

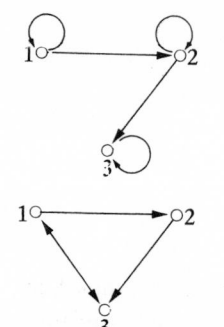

$\Big\{$ represents a reflexive relation on $\{1, 2, 3\}$; namely, $\{(1, 1), (2, 2), (3, 3), (1, 2), (2, 3)\}$

$\Big\{$ represents an irreflexive relation on $\{1, 2, 3\}$; namely, $\{(1, 2), (2, 3), (3, 1), (1, 3)\}$

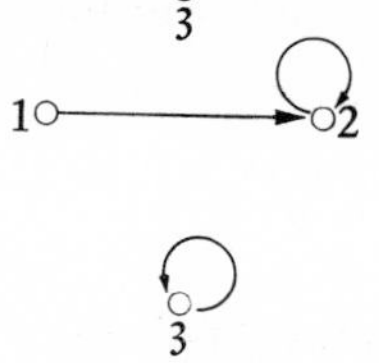

$\Big\{$ represents a nonreflexive relation on $\{1, 2, 3\}$, which is also not irreflexive; namely, $\{(1, 2), (2, 2), (3, 3)\}$.

Figure 4.15.3

Corresponding to transitivity we will see patterns of the form

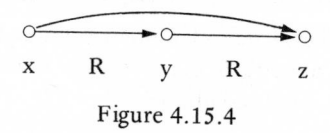

x   R   y   R   z

Figure 4.15.4

shown in Figure 4.15.4, in which xRz appears as the connecting link verifying the conclusion appropriate to the hypothesis xRy and yRz.

Corresponding to intransitivity, we will see a realization of the hypothesis, namely, xRy and yRz and not the link xRz, as indicated in Figure 4.15.5.

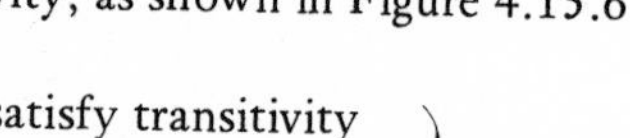

x   R   y   R   z

Figure 4.15.5

For the residual category we will find some points satisfying transitivity and some satisfying intransitivity, as shown in Figure 4.15.6.

$\}$ satisfy transitivity

$\}$ satisfy intransitivity

$\Big\}$ relation neither transitive nor intransitive

Figure 4.15.6

Suppose we have a graph of the form shown in Figure 4.15.7.

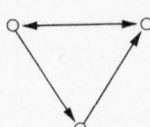

Figure 4.15.7

There is completeness in the sense that every pair of distinct points are joined, in one direction or the other (or both). Such a graph represents a complete relation. For example,

$$A = \{1, 2, 3\}$$
$$R = \{(1, 2), (1, 3), (2, 1), (3, 2)\}$$

is represented by the graph in Figure 4.15.7.

Corresponding to an equivalence relation, we will have a graph that "falls apart" into disjoint subgraphs. Within each subgraph the relation holds, but across subgraphs it does not hold. For example, let

$$A = \{1, 2, 3\}$$
$$E = \{(1, 1), (2, 2), (3, 3), (1, 2), (2, 1)\}$$

Let us verify that E is an equivalence relation. It is reflexive because for all $x \in A$, we have (x, x) in E. It is symmetric because $(1, 2)$ in E and also $(2, 1)$ in E. Finally it is transitive: for instance, the condition

$$(1, 2) \in E \text{ and } (2, 1) \in E$$

is satisfied and so is the conditional conclusion $(1, 1) \in E$. Also,

$$(2, 1) \in E \text{ and } (1, 2) \in E$$

and, in addition, to verify the conclusion of the conditional

$$(2, 2) \in E.$$

The graph of the binary relational system (A, E) is shown in Figure 4.15.8.

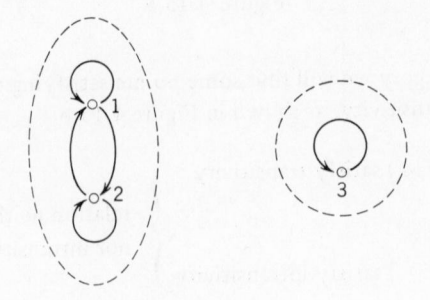

Figure 4.15.8

Note that, first, the number of subgraphs is two and they are complete within themselves but not cross-related. Second, the "algebraic division" of A modulo E gives

$$A/E = \{\{1, 2\}, \{3\}\}$$

having two equivalence classes: this corresponds to the number of complete subgraphs.

4.16. Composition of Relations. Given two relations R and S on a set A, we can form a new relation denoted R o S as follows:

(4.16.1) Definition. R o S = $\{(x, y): xRz \text{ and } zSy, \text{ for some } z\}$

Example. Where A = $\{1, 2, 4\}$

$$R = \{(4, 2), (1, 1)\}$$
$$S = \{(2, 2), (2, 1)\}$$

Then we conclude:

$$(4, 1) \in R \text{ o } S, \text{ since } 4R2 \text{ and } 2S1$$
$$(4, 2) \in R \text{ o } S, \text{ since } 4R2 \text{ and } 2S2$$

But,

$$(1, 2) \notin R \text{ o } S, \text{ since there is no } z \text{ such that } 1Rz \text{ and } zS2.$$
$$(1, 1) \notin R \text{ o } S, \text{ since there is no } z \text{ such that } 1Rz \text{ and } zS1.$$

Thus, in this particular case,

$$R \text{ o } S = \{(4, 1), (4, 2)\}$$

A pair (A, R) where R is a relation on set A was termed a simple binary relational system. More generally, by a relational system over a set A, one means an ordered object

$$(A, R, S, \ldots)$$

consisting of A together with a sequence of relations on A. Thus, in the above, we had a system (A, R, S).

We can represent (A, R, S) graphically as indicated in Figure 4.16.1.

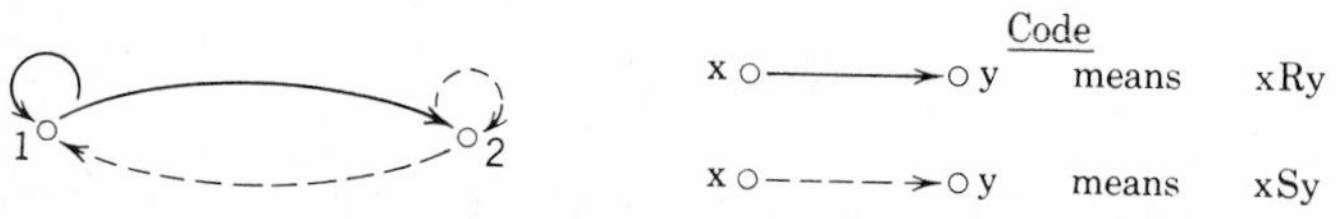

	Code		
x o———→o y	means		xRy
x o– – – –→o y	means		xSy

Figure 4.16.1

Figure 4.16.1 represents the system

$$\underbrace{\langle 1, 2 \rangle}_{A}, \underbrace{\{(1, 2), (1, 1)\}}_{R}, \underbrace{\{(2, 2), (2, 1)\}}_{S}\rangle$$

For more than two relations, Figure 4.16.2 provides an example.

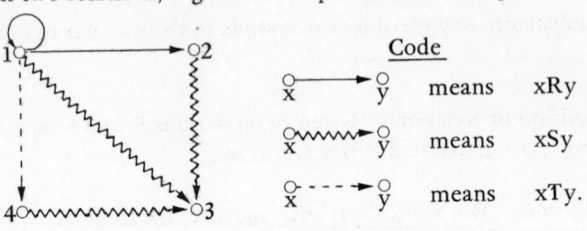

Figure 4.16.2

Relational composition now is apparent on the graph; for example,

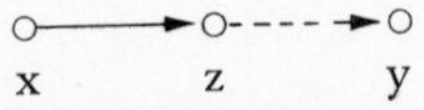

shows a z such that x and y are linked through z.

In sociometry one may ask for choices based on two distinct criteria. Write

xRy   iff   x names y on criterion 1
xSy   iff   x names y on criterion 2.

Then the outcome is the system (A, R, S) where A is the population questioned. Persons x and y are in R o S if and only if some person z exists such that x names z on criterion 1 and z names y on criterion 2. If R is based on liking and S is based on disliking, then x(R o S)y means that x likes somebody who dislikes y. The two findings

$$(x, y) \in R$$
$$(x, y) \in R \circ S$$

together show what has been called imbalance (cf. Heider, 1958): x likes y but likes someone else who dislikes y (Figure 4.16.3).

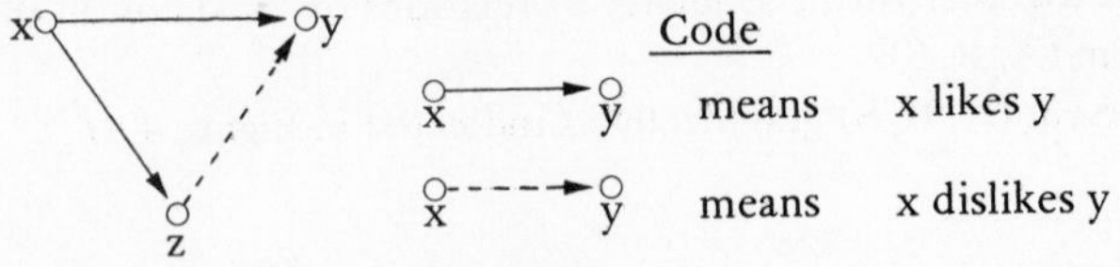

Figure 4.16.3

Since for any relations R and S defined a common domain A the relation R o S exists, then R o S may be composed with a third relation T to yield (R o S) o T.

Also, since S and T are relations, we obtain S o T. Then R can be composed with relation S o T to yield R o (S o T). It is an analytic property of this concept of relation composition that

(4.16.2) $(R \circ S) \circ T = R \circ (S \circ T)$

that is, it is "associative." The proof runs as follows, juxtaposing symbols to indicate composition so that RS means R o S.

$x(RS)Ty$	iff	$xRSz_1$ and z_1Ty (some z_1)	(composition definition)
	iff	xRz_2 and z_2Sz_1 and z_1Ty (some z_2)	(composition definition)
	iff	xRz_2 and z_2STy	(composition definition)
	iff	$xR(ST)y$	(composition definition)

The axiom of extension now yields (4.16.2). The justification of each step, shown on the right, involves appeal only to the definition, as was indicated in saying that the property is "analytic" of the concept of relation composition.

The relation composition notion may be applied to one relation repeatedly, as in taking powers of a number. We define "powers" of R by

(4.16.3)   (1) $R^1 = R$
                  (2) $R^{n+1} = R \circ R^n$ (n = 1, 2, . . .)

Then from (2), $R^2 = R \circ R^1 = R \circ R$. Using this result, we apply (2) again to obtain $R^3 = R \circ R^2 = R \circ R \circ R$, and so forth. This is called an "inductive definition" of the idea of the nth power of a relation R. For instance, if R is "father of" then R^2 is "grandfather of," R^3 is "great-grandfather of," and so forth.

We also define, for any relation R, its inverse, denoted R^{-1}:

(4.16.4) $xR^{-1}y$ iff yRx

For example, $>$ is the inverse of $<$, on a domain of numbers. Similarly, if R is "father of," then R^{-1} is "child of."

Another useful relation is given by

$$I = \{(x, y): x = y\}$$

the identity relation (on a given domain). It is represented by a graph containing only loops, one on each node.

The following proposition is now to be proved to illustrate proof techniques; the concepts will be applied later in the book.

(4.16.5) Proposition.

                  (1) R is transitive   iff $R^2 \subseteq R$
                  (2) R is symmetric   iff $R \subseteq R^{-1}$
                  (3) $I \circ R = R$
                  (4) $R \circ I = R$

Proof of (1).   R is transitive

$$\begin{array}{ll} \text{iff} & x R y \text{ and } y R z \Rightarrow x R z \qquad (\text{all } x, y, z) \\ \text{iff} & x(R \circ R)z \Rightarrow x R z \\ \text{iff} & x R^2 z \Rightarrow x R z \\ \text{iff} & R^2 \subseteq R \end{array}$$

(In passing to the second line of this proof, we used y for the intermediate link between x and z. If no such y actually exists, then $(x, z) \notin R^2$ and this is an irrelevant pair. If for no pair (x, z) is there such a y, then R is vacuously transitive, in the sense that the hypothetical condition—xRy and yRx—is never true. In this case $R^2 = \emptyset$. Since $\emptyset \subseteq R$, because the empty set is formally contained in any set, we still obtain a valid result.)

Proof of (2).   R is symmetric

$$\begin{array}{ll} \text{iff} & x R y \Rightarrow y R x \qquad (\text{all } x, y) \\ \text{iff} & x R y \Rightarrow x R^{-1} y \\ \text{iff} & R \subseteq R^{-1} \end{array}$$

Proof of (3). $x(I \circ R)y$ iff   xIz and zRy   (some z)

   iff   xIx and xRy $(z = x)$

Now xIx is true for all x, and so $x(I \circ R)y$ will be true if, and only if, xRy. Hence, $I \circ R = R$.

Proof of (4).   Prove this as in (3).

The following proposition also holds and allows us further practice in deduction.

(4.16.6) Proposition.

(1) $(R \circ S)^{-1} = S^{-1} \circ R^{-1}$

(2) $(R \cup S) \circ T = (R \circ T) \cup (S \circ T)$

(3) $(R \cup S)^{-1} = R^{-1} \cup S^{-1}$

(4) $(R \cap S)^{-1} = R^{-1} \cap S^{-1}$

(5) $\bar{R}^{-1} = \overline{R^{-1}}$ ($\bar{R}$ is the complement of R)

For proofs, we can simplify notation and write RS for $R \circ S$, $(R \cup S)T$ for $(R \cup S) \circ T$, and so forth. We will prove (1) and (2), leaving the remainder for the reader.

Proof of (1). $x(RS)^{-1}y$ iff   yRSx

   iff   yRz and zSx   (some z)

   iff $zR^{-1}y$ and $xS^{-1}z$

   iff $xS^{-1}z$ and $zR^{-1}y$

   iff $xS^{-1}R^{-1}y$

Proof of (2). x(R $\cup$ S)Ty   iff   x(R $\cup$ S)z and zTy   (some z)

                         iff   (xRz or xSz) and zTy

                         iff   (xRz and zTy) or (xSz and zTy)

                         iff   xRTy or xSTy

                         iff   x(RT $\cup$ ST)y

In the proof of (2), the first line is a matter of definition of composition, the second follows from the definition of union. The next step of the proof relies on a tautological equivalence, which may be verified by a truth-table. Finally, we "recombine" to obtain a union of products.

4.17. Preference and Indifference. In this section we illustrate further how one may deduce sentences pertaining to relations from definitions and axioms, using the rules of logic. The axioms are interpretable as those pertaining to idealized or "rational" preferences. The theorems simply state various things about a rational man's system of preferences.

We begin with the weak preference axioms stated by Suppes (1957). We have a collection of objects C and a relation Q defined on C that satisfies for all x, y, z in C:

Axiom.

(4.17.1)   xQy and yQz implies xQz   (transitivity)

Axiom.

(4.17.2)   xQy or yQx   (strong completeness)

The axioms are construed as applying to a single individual, and the set C represents the collection of objects he is evaluating. Axiom (4.17.1) says that his weak preference Q is transitive and axiom (4.17.2) says that the collection C is such that the person can express some weak preference for every pair of objects in C. In the formalism no letter represents the individual; the letters x and y stand for evaluated objects (e.g., commodities, ideas, people).

Given the system described by these two axioms, the two theoretically important notions appear as definitions. We have

(4.17.3) Definition. Strict preference relation P:

                xPy   iff   not-yQx

(4.17.4) Definition. Indifference relation I:

             xIy   iff   xQy and yQx

The first definition says that if and only if y is not (at least) weakly preferred to x, then x is strictly preferred to y. The second definition says that in the event that the person weakly prefers x to y and vice versa, this means he is indifferent to this pair.

It is evident that (4.17.3) may be written in the form

$$P = \bar{Q}^{-1}$$

so that we see that P is the inverse of the complement of Q. (Equivalently, using (5) under (4.16.6), $P = \overline{Q^{-1}}$. That is, P is the complement of the inverse of Q.)

The precise meaning of these definitions and axioms comes now from the deductions. These deductions play the same role in the formalism as the activity in intuitive discourse of examining an alleged definition, wondering what it implies. Here we calculate explicit consequences of these stipulations.

We prove a series of theorems in much greater detail than would ordinarily be the case, in order to show the simplicity of each step in the chain of thinking.

(4.17.5) Theorem. I is an equivalence relation.

We note by axiom (4.17.2), with y replaced by x, that xQx for all x. So by definition (4.17.4), with y replaced by x, we conclude that xIx. This shows that I is reflexive. To prove that I is symmetric, we have to show that if xIy, then yIx. Suppose xIy. Then, by definition (4.17.4), xQy and yQx. But this is tautologically equivalent to yQx and xQy, which, by definition (4.17.4), means yIx, the desired conclusion. To show that I is transitive assume that xIy and yIz, which by definition means that xQy, yQx, yQz, and zQy. But by axiom (4.17.1), Q is transitive, and therefore, we conclude from xQy and yQz that xQz, and from zQy and yQx, we conclude that zQx. But by applying definition (4.17.4) from xQz and zQx we conclude xIz, so that I is transitive. Since I is reflexive, symmetric, and transitive, I is an equivalence relation.

(4.17.6) Theorem. P is a strict ordering of C—that is, P is asymmetric and transitive.

Assume xPy. Then, by definition (4.17.3), not-yQx. By axiom (4.17.2), therefore, we must have xQy, which by appeal again to definition (4.17.3) (in the equivalent form, not-xPy if and only if yQx), shows not-yPx. So P is asymmetric. To show that P is transitive, suppose that xPy and yPz. Note that yPz implies not-zPy, by asymmetry, and hence, yQz, by definition of P. On the other hand, suppose that in contradiction to our desired result, we have not-xPz, and so also zQx. Then yQz and zQx imply yQx, by axiom (4.17.1), and so not-xPy, using (4.17.3); but xPy was one of our assumptions. Hence, we obtain a contradiction. So P is transitive.

As a final deduction, we state an important theorem on substitutability of equivalent objects.

(4.17.7) Theorem. A substitution rule:

$$xIy \text{ and } yPz \text{ implies } xPz$$

To prove the theorem by contradiction, suppose we have xIy and yPz but not-xPz. The latter implies zQx, while yPz implies not-zQy. On the other hand, xIy implies xQy, so that we have zQx, xQy, and not-zQy. But zQx and xQy imply zQy, and so we have a contradiction. Hence, we must have xPz.

The following theorem may readily be proved, but we omit the details:

(4.17.8) Theorem. Trichotomy condition: Exactly one of the following holds for all x, y:

$$xPy, yPx, xIy$$

Since the relation I of indifference is an equivalence relation, it partitions C into a collection of disjoint and exhaustive subsets (recall section 4.12). Let $\bar{x}$ be the equivalence class containing x. We define a relation on these classes, namely, the relation P^* such that

$$\bar{x}P^*\bar{y} \quad \text{iff} \quad xPy$$

The resulting binary relational system $(C/I, P^*)$ is such that P^* is asymmetric, transitive, and complete. Thus, P^* is a linear ordering of the equivalence classes of C.

Example. (a) Consider the set

$$C = \{\text{lawyer, doctor, carpenter}\}$$

Let an individual be asked to express an occupational prestige rating of each job against each other. Thus, he is given three subsets,

$$\{\text{lawyer, doctor}\}, \{\text{lawyer, carpenter}\}, \{\text{doctor, carpenter}\}$$

and in each instance he must choose one job or claim they are the same in prestige. We may wish to determine, among other things, whether his choices satisfy our axioms. For this purpose, we identify the abstract relations Q, P, and I as follows on the domain C:

(1) P is identified by pairs (x, y) such that x is chosen when x and y comprise the choice set. This defines a concrete relation P_C.

(2) I is identified by pairs (x, y) such that the claim of "sameness" in prestige is made when x and y comprise the choice set. This defines a concrete relation I_C.

(3) Q is identified by placing (x, y) in Q_C if (x, y) in P_C or (x, y) in I_C.

It is clear that P and I are the important relations, so that we would check the deduced properties of these relations as compared with the observed relations. Suppose that the empirical outcomes are

$$P_C = \{(\text{doctor, lawyer}), (\text{lawyer, carpenter}), (\text{carpenter, doctor})\}$$
$$I_C = \emptyset$$

According to (4.17.6), P is transitive, but the observed relation P_C is not: the doctor "should" have been chosen over the carpenter. Hence, contrary to our assumption, our chooser is not "rational" in the specific sense of the axioms. On the other hand, suppose we obtained the outcomes

$$P_C = \{(\text{doctor, carpenter}), (\text{lawyer, carpenter})\}$$
$$I_C = \{(\text{doctor, lawyer})\}$$

We see that P_C is asymmetric and transitive (in the vacuous sense), and this means (4.17.6) is satisfied. However, I_C is not yet an equivalence relation because it is not symmetric and not reflexive. By a convention without empirical import, we "add in" the reflexive pairs and also put (lawyer, doctor) in I_C. Transitivity is the only empirical dubious property of I, but we do not have enough pairs to check it. All in all, these last outcomes do not refute the rationality assumption.

Note that the method of identifying abstract concepts in this application involves what Braithwaite (1956) might call the "zipper upward" mode: the defined notions are identified first, and they in turn confer empirical meaning on the primitive notion Q. The alternative approach, of course, would involve direct empirical identification of Q.

Example. (b) Consider the same set of jobs and ask the respondent to rank the set. Identify the relations as follows:

(1) P is identified by placing (x, y) in P_C if x is ranked above y.

(2) I is identified by tied ranks.

(3) Q is identified by again combining P_C and I_C:

$$xQ_C y \quad \text{iff} \quad xP_C y \quad \text{or} \quad xI_C y.$$

Under this empirical method, the relation P_C has the properties of asymmetry and transitivity as a sheer matter of the definition of the task. It is not possible to falsify transitivity as it was in the pair comparison case. Hence, theorem (4.17.6) is analytically true. Similarly, identifying I via ties yields (4.17.5) with no possibility of falsification. Hence, Q_C in turn has properties (4.17.1) and (4.17.2) as a sheer matter of the imposed ranking task. If our objective is to learn if our chooser is rational, the task is not well chosen, because it imposes rationality on the chooser. A different viewpoint toward rankings is possible in which hypotheses about preferences may be tested on ranking data. The approach is probabilistic. It will be discussed in Chapter 11.

Several points are being made here: first, in some applications, because of empirical operations, axioms can be made to be true analytically, even though in other applications in the same field this is not the case. The generic point is that any statement functions differently in different contexts. (Also, for example, definitions are sometimes given nonanalytic interpretations.) Second, definitions, of course, function as premises: the label applied to a statement does not affect its role in the logic of deductive inference. Third, abstract theories when applied need not be applied by direct identification of the undefined notions entering the axioms. Instead, ideas formally appearing as definitions may be identified, in which case they provide empirical meaning for the undefined concepts. In this respect, the P and I identification leading back to the empirical identification of Q was not necessarily typical: in many interesting cases, an undefined notion remains observationally unidentified, but carrying indirect empirical meaning (for instance, the "stimulus elements" in stimulus sampling theory, as shown in Suppes and Atkinson, 1960). Examples of this kind will be encountered in the subsequent sections of this book.

4.18. Axiomatics. To construct an axiomatic system, one provides two related clusters of ideas. The first cluster we will term the primitive basis; the second will be termed the induced set.

The primitive basis, in turn, has two components. First, there is a collection of undefined notions. We say that each of these refers to a primitive entity. The intuition here is that such entities are not displayed as constructed or induced in the system, but taken as givens. Second, there is a collection of postulates such that (a) aside from logical connectives and exact concepts from mathematical theories forming a background for the system, the only terms appearing are primitive notions, and (b) these postulates are interpretable as conditions on the structure or pattern determined by the system of primitive entities. Such postulates are termed axioms. Collectively, the primitive basis serves to define a new exact concept: namely, just such an entity whose abstract composition is given by the primitive entities in such-and-such relatedness according to the axioms.

The induced set also has two components. First, there are the definitions. Each definition names an object already given in the primitive basis by virtue of the pattern it determines. For instance, the term "circle of radius r" is a defined term in axiomatic Euclidean geometry because the primitive basis of points (and other objects) yields the existence of sets of points related to each other as being equidistant (with measure r) from a given point. Second, there are the propositions that are provable by taking the axioms as premises, using any prior definitions, and showing a chain of reasoning. In this chain, the rules of inference are based on the tautologies of logic and on accepted mathematical theories functioning as systems of inference rules. For example, the axioms of arithmetic function in this way in most scientific systems. Each provable proposition may be regarded as describing an aspect of the internal composition of the complex entity defined by the primitive basis.

Axiomatic systems vary enormously in complexity. For a simple example, consider the ideas studied in the prior section concerning preference and indifference. A partial sketch of the axiomatic aspect may be instructive:

Primitive basis
  1. Primitive entities: C, Q
  2. Axioms: Q transitive and strongly complete on domain C

Induced set
  1. Definitions: P, I
  2. Propositions: for instance Theorems (4.17.7) and (4.17.8)

To coordinate to the abstract discussion, note that C and Q are simply given sets, with Q a relation on C. Together they form a pattern which is a binary relational system (C, Q). The axioms are structural conditions as to this pattern. The primitive basis can be taken to define the exact concept of "a rational preference system." As to the induced set, the terms P and I are well defined as names of certain sets of pairs induced by the given basis. They are introduced as equivalences with right-hand sides containing the primitive term Q. As we say in the prior section, certain propositions are provable statements, using prior definitions and tautologies.

By an identification of such an axiomatic system one means an instantiation of the ideas to a concrete system. For instance, C was identified with a set of three definite job titles and Q was constructed from choices. Whether or not the resulting identified pattern (corresponding to the complex entity given by the system of primitive entities apart from the axiomatic conditions) satisfies the axioms depends upon the context of identification. In the most interesting cases, the question is empirical: the axioms, or some subset of them, function as falsifiable claims about an empirical system. In section 4.17, we have treated one example of this contextual matter.

As we have noted above, an exact concept is one which is defined by an axiomatic system. If the system formalizes some intuitive ideas in a given scientific community, then the definition is not merely nominal so far as this community is concerned. In other words, a concept has received an essential definition via axiomatization. For example, this was the great achievement of Kolmogorov (1933) in regard to the working concept of probability within pure and applied mathematics. However, the axiomatics need not be comprehensive of an entire field in order to be of scientific significance. The modest system outlined in section 4.17 is an example; many others will appear in the remainder of the book.

Because in many cases the transition from intuition to exact concept is of scientific consequence, one finds disputes as to the adequacy of the primitive basis. In this context we can distinguish two modes of semantic orientation. First, in a pre-axiomatic semantic quarrel, discourse about the intuitive idea is caught in a nexus of dissensus so that one term has many meanings and the relation of the meanings to each other is obscure. (Consider, for instance, the term "anomie.") Second, when a formal working tradition begins to be established, these several meanings each can be the source of an exact concept. However, with an explicit primitive basis for each such meaning, it becomes possible to ferret out the analytical consequences of each definition and to relate the various definitions. From a semantic viewpoint, the one intuitive term, say t, is replaced by its versions $t_1, t_2, \ldots$, and the formal relations among the t_i are provable. In this way, otherwise endless and unproductive debates about meanings are transformed into a process of systematic inquiry. Note that under each t_i there are axioms which in various contexts may function as empirical claims. Hence, systematic inquiry includes empirical research relevant to such claims, as well as formal work. For an excellent example of this kind of orientation in practice, see Luce and Suppes (1965) in regard to preference systems.

Given an exact concept, and so an axiomatic system, the axioms and proved propositions of the system may be said to form the theory concerning that concept. For example, it is meaningful in this sense to talk of the theory of probability or the theory of games. Let T denote this theory. Then the process of identification of the axiomatic system, discussed above, may be termed "constructing a T-model." For instance, one constructs probability models or game models. If analytical mechanics is put in axiomatic form (see Suppes, 1957), then the idea of constructing a mechanical model is in line with this terminology. But there is no model construction of any consequence without some problem to which the model-building effort is germane. In Part 3, this point of view governs our presentation of various model-building efforts in sociology.

We return briefly at this point to the relevance of the notion of falsifiability in the context of axiomatics. In section 2.11, we treated falsifiability in the form,

$$P \Rightarrow Q \quad \text{(L-true)}$$
$$\frac{\text{not-}Q}{\text{not-}P} \quad \text{(E-true)}$$

In this logical representation, P refers to the conjunction of the premises of a theory, Q is a proposition proved from these premises, and empirical inquiry yields the result that Q is false.

To relate the notion of a theory T concerning an exact concept to this formulation of falsifiability logic, we note that P cannot literally be the abstract theory T. This is because Q makes an empirical claim about some definite circumstances involving some definite concrete systems. Hence P in the above formulation is in identified form, that is, we can interpret P as the specification of a T-model. It follows that the conclusion not-P is a rejection not of T but of a specific T-model. There is no logical entailment that T is rejected, per se, in this argument.

At this point, two policies concerning abstract theory T in relation to such rejected T-models are possible. For reference, let us call them Policy 1 and Policy 2 as follows.

Policy 1. If a T-model is rejected, construct a different T-model for the phenomenon. For example, if a weak preference system fails to be satisfied by the choices made by a person, one might argue that more alternatives were entertained by the chooser (thus changing the identification of C). As another example, if an equally likely probability model for a phenomenon is rejected, the usual procedure is to construct another probability model with altered probabilities. Note that under this policy, there is no suggestion that the primitive basis of theory T be changed.

Policy 2. If a T-model is rejected, reconstruct the primitive basis of T itself. For example, upon rejecting a particular preference model because the observed indifference relation is not transitive, one might try to alter the assumptions about the Q relation. This might lead to a distinct exact concept. The previous concept would "survive," but if its models proved generally inadequate it might function only in purely theoretical treatments of the phenomenon. The new axiom system, assuming reconstruction to fit the phenomenon, will play a stronger role in science. Hence, under this second policy there is an evolutionary process in the direction of greater conformity with wide experience obtained in model-building efforts. The theory is under empirical control.

These are two ideal types of policy, which implies that the real situation surrounding axiomatic theories is more complex than this discussion indicates. For instance, one may suggest the hypothesis that scientific ideas display a developmental trend in the direction of Policy 1, starting from Policy 2, with various mixed phases. In other words, theoretical systems are constructed under empirical control but eventually become accepted frameworks under which scientists implement Policy 1.

These remarks indicate that the relation between abstract theory and the construction and evaluation of models is not deductive but cybernetic. The theory, via the model-building efforts, controls the form of data (as in realizations of the notions of the primitive basis). But it does not control the goodness-of-fit. Under Policy 1, there is no reciprocal control over the theory by feedback of evaluative considerations based on goodness-of-fit. The feedback loop goes to the model construction or setup, but not to the dominant abstract theory. Under Policy 2, the loop does return to the abstract theory so that the control system is reciprocal.

A scientific theory is one for which a norm exists in a particular scientific community to the effect that Policy 2 applies. But norms form only one class of element determinative of scientific process; hence it is possible to have scientific theories which are virtually released from actual empirical control.

For related discussions in more substantive contexts, see especially Chapters 11 and 14. For further remarks on the mathematical aspects of axiomatics, see section 9.7.

CHAPTER FIVE MAPPINGS
AND ALGEBRAIC SYSTEMS

5.1. Aim of the Chapter. In Chapter 4 we applied the logic of formal reasoning and the set-theoretic basis to the generic problem of representation of psychological and social relations. In this chapter, we introduce some algebraic ideas that will provide a foundation for our subsequent treatment of such topics as measurement theory, probabilistic choice theory, expectation-states theory, formal kinship analysis, and the theory of games. We begin with the concept of mapping (or function), illustrating its use in formalizing some sociological ideas related to status; in the later sections of this chapter some basic abstract algebraic ideas are introduced.

5.2. Concept of Mapping. Figure 5.2.1 shows that we associate 2 with 1, 4 with 2, 6 with 3.

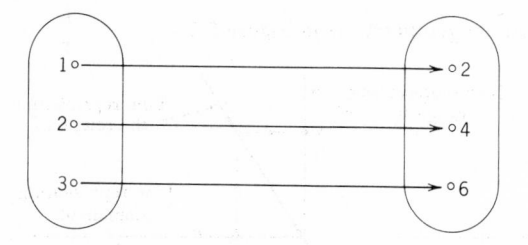

Figure 5.2.1

The table

x	f(x)
1	2
2	4
3	6

provides the same information. The two sets

$$A = \{1, 2, 3\}$$
$$B = \{2, 4, 6\}$$

together with the set of ordered pairs

$$\{(1, 2), (2, 4), (3, 6)\}$$

also provide the same information. We use the latter to define the concept of a mapping.

(5.2.1) **Definition.** An ordered triple (A, B, f) is a mapping if and only if A and B are sets and f, as a relation, has a unique pair (x, y) $\in$ f for each x $\in$ A.

Because of the uniqueness, as is apparent from the table and picture of the mapping, we can replace the notation (x, y) $\in$ f with

$$f(x) = y$$

and we say "f sends x into y." We really do not think of the triple (A, B, f) as such, but of some conceptual connection, some imagery, that associates y with x. The triple serves to make this imagery a definite object of analysis. We say that two mappings (A, B, f) and (A', B', f') are the same, or identical, when A = A', B = B', and f = f'. Also, the term "function" may be used for "mapping," especially in contexts where A and B are sets of numbers.

Example. (a) For each x $\in$ $\Re$ (set of all real numbers), let

$$y = 2x.$$

Then this defines a mapping ($\Re$, $\Re$, f) such that f(x) = 2x; that is,

$$f = \{(x, y) : x \in \Re, y \in \Re, \text{ and } y = 2x\}$$

In this case we can use geometry, as in Figure 5.2.2.

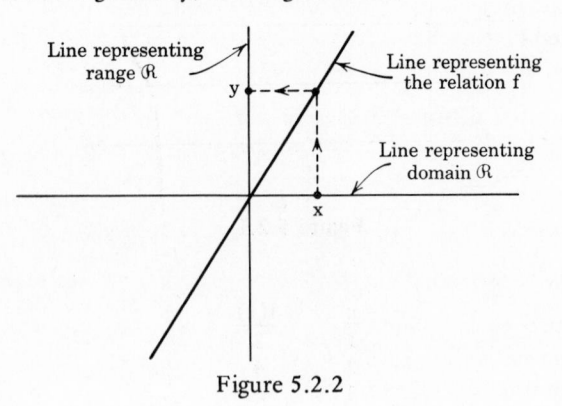

Figure 5.2.2

Example. (b) Let P be a set of persons. Let C be a set with two "states":

$$C = \{\text{conforms to } \eta, \text{does not conform to } \eta\}$$

where η (eta) is some norm relevant among the members of P. Then a conceptualized mapping is useful for theoretical purposes: associate with each $x \in P$, a definite state in C (for a given time). This specifies a mapping.

The best notation for abstract work is as follows. If (A, B, f) is a mapping, write

$$f: A \rightarrow B \quad \text{or} \quad A \xrightarrow{f} B$$

to think "f maps A into B." Or, write in terms of the typical element x in A,

$$x \mapsto f(x)$$

to think "x is mapped into f(x)," or write

$$x \xmapsto{f} y$$

to think "x is mapped into y, under f."

Example. (c) Let c: $P \rightarrow C$ map persons into norm-conformity states as in example (b); then $x \mapsto c(x)$ means person x is associated with a state of conformity, $c(x)$ in C. Concretely, a display by a table for a definite mapping (i.e., an identification of the abstract mapping c: $P \rightarrow C$) might be as follows:

x	y
Tom	conforms to η
Dick	conforms to η
Harry	does not conform to η

Here x is a logical variable with domain the set of persons P identified in the concrete case as $\{$Tom, Dick, Harry$\}$. Also, y is a logical variable with domain the set C containing states of conformity to a specified norm. The mapping consists in the two sets together with the association we understand tacitly in looking at the table. Explicitly, however, the concrete identification of c: $P \rightarrow C$ in this case is

$$\begin{aligned}
c(\text{Tom}) &= \text{conforms to } \eta \\
c(\text{Dick}) &= \text{conforms to } \eta \\
c(\text{Harry}) &= \text{does not conform to } \eta
\end{aligned}$$

or, in relation form,

$$c = \{(\text{Tom, conforms}), (\text{Dick, conforms}), (\text{Harry, does not conform})\}$$

Note that we satisfy definition (5.2.1) in that for each element in the first set (now P), there exists a unique pair in the relation c: each person is in a determinate state of conformity to norm η. Thus, as a table, a mapping consists in the first column, the second column, and the association established by the rows of the table. These are, respectively, the elements of the defining triple (A, B, f).

5.3. Properties of Mappings. Let f: A → B be a mapping. A is termed the domain of the mapping and B is called the codomain or range. If x ∈ A, the object f(x) ∈ B is termed the image of x under f. The set

$$\left\{f(x): x \in A\right\}$$

is the set of all images under f and is called the image set or the image of A under f. We denote it by f(A). Thus, f(A) ⊆ B. If f(A) = B, we call f onto or surjective; if x ≠ y implies f(x) ≠ f(y)—"distinct objects have distinct images"—we call f either 1-1 or injective.   If f is both injective and surjective (1-1 onto) it is termed bijective ("1-1 correspondence").

A minor point to note is that we have two distinct concepts, both denoted by the word "domain." First, the concept of a logical variable (cf. Chapter 2) requires a specified class of objects over which it draws its values. This is the logical concept of domain. Second, a mapping associates an object (in some set) to each entity in a given set; this given set is called its domain. This is the mathematical concept of domain. Thus, each mapping "has a domain," which is A in the mapping (A, B, f). But, also, each of the sets A, B, and f comprising the mapping has members; to discuss these members in generality, we are required to introduce logical variables: one variable ranging over A, one variable ranging over B, and one pair-variable ranging over f. Thus, we have

$$x \quad \text{with domain A}$$
$$y \quad \text{with domain B}$$
$$(x, y) \quad \text{with domain f}$$

And, in this case, only A is the domain of the mapping (A, B, f).

Example. (a)
$$A = \left\{1, 2, 3, 4\right\}$$
$$B = \left\{1, 2, 3, 4, 5\right\}$$
$$f(x) = x \qquad (\text{all } x \in A)$$

The image of 2 is 2. The set of all images is A itself. f is injective but not surjective, because no x ∈ A is mapped into 5.

Example. (b)
$$A = \left\{1, 2, 3, 4\right\}$$
$$B = \left\{1, 2, 3, 4, 5, \ldots\right\}$$
$$f(x) = 2x \qquad (\text{all } x \in A)$$

Then, x ↦ 2x; hence, 3 ↦ 6; f(A) = $\left\{2, 4, 6, 8\right\}$; f is injective; f is not surjective.

Example. (c)
$$A = \left\{\text{Dick, Charles, Harry, Bill}\right\}$$
$$B = \left\{\text{conforms, does not conform}\right\}$$

$$c(\text{Dick}) = \text{conforms}$$
$$c(\text{Charles}) = \text{conforms}$$
$$c(\text{Harry}) = \text{conforms}$$
$$c(\text{Bill}) = \text{does not conform.}$$

Then c: $A \to C$ is surjective; Dick $\mapsto$ conforms; $c(A) = C$; c is not injective because, for example,

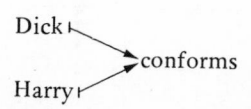

so that distinct persons do not go into distinct states.

5.4. Types of Mappings. The following examples further illustrate the notions introduced in section 5.3.

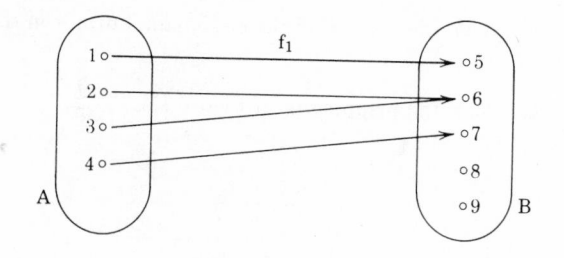

Figure 5.4.1

Example. (a) We see from Figure 5.4.1 that $f_1: A \to B$ is given by

$$f_1(1) = 5 \quad (\text{or } 1 \overset{f_1}{\mapsto} 5)$$
$$f_1(2) = 6 \quad (\text{or } 2 \overset{f_1}{\mapsto} 6)$$
$$f_1(3) = 6 \quad (\text{or } 3 \overset{f_1}{\mapsto} 6)$$
$$f_1(4) = 7 \quad (\text{or } 4 \overset{f_1}{\mapsto} 7)$$

Therefore, the image set is

$$f_1(A) = \{f_1(1), f_1(2), f_1(3), f_1(4)\} = \{5, 6, 7\}$$

This mapping is not injective because

$$f_1(2) = f_1(3)$$

This mapping is not surjective, because $8 \in B$ but there is no $x \in A$ with $f_1(x) = 8$. In other words, $f_1(A) \neq B$ shows that the mapping is not surjective.

Example. (b) Next, we consider Figure 5.4.2.

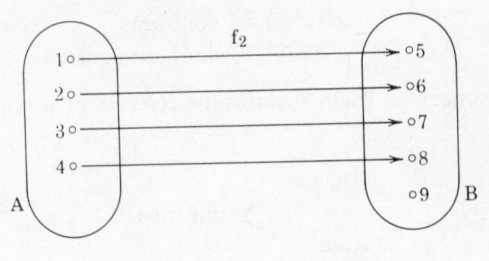

Figure 5.4.2

We see that f_2 is injective because distinct points in A are mapped by f_2 into distinct points in B. But f_2 is not surjective, because

$$f_2(A) = \{5, 6, 7, 8\}$$

and $9 \in B$, but $9 \notin f_2(A)$: that is, 9 is in the codomain, but not in the image set, of f_2.

Example. (c) Consider the mapping f_3 of Figure 5.4.3.

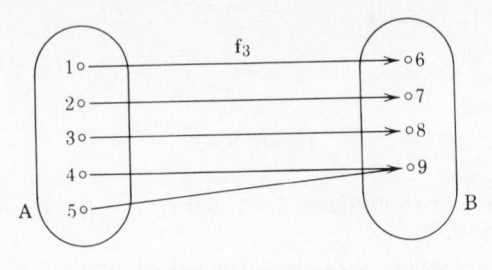

Figure 5.4.3

This mapping is surjective because

$$B = \{6, 7, 8, 9\}$$

and

$$f_3(A) = \{f_3(1), f_3(2), f_3(3), f_3(4), f_3(5)\} = B$$

But f_3 is not injective, since $f(4) = f(5)$.

Example. (d) On the other hand, consider Figure 5.4.4. This mapping f_4: $A \rightarrow B$ is both injective and surjective, because distinct objects in the domain have distinct images and the image set is the range. Thus, by definition, f_4 is bijective.

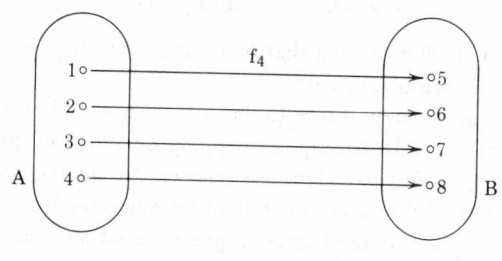

Figure 5.4.4

Summarizing this information we present Table 5.4.1 for convenience.

Table 5.4.1 Types of Mappings

Type	Example
Neither injective nor surjective	Figure 5.4.1
Injective and not surjective	Figure 5.4.2
Not injective but surjective	Figure 5.4.3
Both injective and surjective ("bijective")	Figure 5.4.4

Given any mapping f: A → B, we see that if the mapping f is injective, then to each point y in the image of A under f there corresponds a unique point x in A. Hence, we can define for any such y,

$$f^{-1}(y) = x \quad \text{iff} \quad f(x) = y$$

and call f^{-1} the inverse of f. The domain of the inverse is the image under f of the domain of f. The codomain is A. If f is also surjective, then f(A) = B and so f^{-1} maps B onto A. For example, in Figure 5.4.4 the inverse of f_4, namely f_4^{-1}, is given by reversing the arrows.

5.5. Intension, Extension, and Rules. We remind the reader that set is the extensional meaning of a property concept, while relation is the extensional meaning of a relationship concept. In connection with mappings one often reads that "a mapping is a rule associating to each object in a domain a unique object from some set of objects." The notion here comes from classical mathematics, in which "mapping" meant "function," which in turn meant "formula relating numbers, y = f(x)." The modern concept distinguishes between formula and function. A formula generates a function. That is, it is a certain type of linguistic entity; a formula is no more a function (mapping) than an expression of a language is a man. "Tom is a man" is simply a sentence; clearly it is not a man. Similarly, a formula specifies a function in the usual set-theoretic manner. Thus, the formula

$$y = 2x \qquad (\text{all } x \in \Re)$$

specifies the function $f: \Re \to \Re$ such that $f(x) = 2x$. This mapping is the extensional meaning of the formula, or rule.

Thus, we extend our analogy of intensional and extensional meanings: set is to property, as relation is to relationship, as mapping is to rule, in the sense that the left side of the analogy refers to extensions (classes of objects), while the right side refers to "meanings" that may generate or be generated by such classes. For example, plot number-pairs in the Cartesian plane, based on readings of time and distance traveled under certain conditions. The plot shows the mapping. The rule relating distance traveled to time is conceptualized out of the mapping. The rule is satisfactory if it regenerates the relation—the pairs of numbers—as its extension. Thus, to a certain extent science is an effort to replace extensions by regenerative ("fitting") intensions: formulas or rules that sum up the mapping (or relation) in compact form. But, this characterizes more than science: man, indeed any organism to some extent, develops rules for coding his extensional surroundings. People map each other, often by use of compact formulas—"If he drives a Rolls-Royce, he is rich"—and so, sociology can employ mapping concepts to describe the conceptual activity of the actors. This will be a central theme in section 5.7.

5.6. Composition of Mappings. Let

$$f: A \to B$$
$$g: B \to C$$

be two mappings. Then,

$$x \mapsto f(x)$$

and since $f(x) \in B$, by use of the second mapping,

$$f(x) \mapsto g[f(x)]$$

so that taking both mappings into account,

$$x \mapsto g[f(x)]$$

Thus, if we define g o f, termed the composition of the maps, by

(5.6.1) $\qquad (g \circ f)(x) = g[f(x)] \qquad (\text{all } x \in A)$

we have a new "induced" mapping from A to C:

$$g \circ f: A \to C.$$

Example. Let

$$A = \{1, 2, 3\}$$
$$B = \{a, b, c\}$$
$$C = \{*, **, ***\}$$

and let

$$f(1) = a, f(2) = b, f(3) = b$$
$$g(a) = **, g(b) = *, g(c) = *$$

Then,

$$(g \circ f)(1) = g[f(1)] = g[a] = **$$
$$(g \circ f)(2) = g[f(2)] = g[b] = *$$
$$(g \circ f)(3) = g[f(3)] = g[b] = *$$

so that under g o f we have

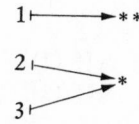

For practice in formal computation of compositions, as well as for later reference, let us consider how to construct a "multiplication table"—where "multiplication" refers to composition of functions—whose row and column headings are functions and whose entries record the function obtained by composing the row function with the column function. It will be convenient to take a small set and consider all possible mappings with domain and codomain given by this set.

Hence, let

$$A = \{1, 2\}$$

and let

$$F[A] = \{f: (A, A, f) \text{ is a mapping}\} = \{f: A \xrightarrow{f} A\}$$

Concretely, element 1 can be mapped into either of two elements, and element 2 can be mapped into either of the two elements. Hence, there are a total of 2 x 2 = 4 mappings given by

f_1	f_2	f_3	f_4
$1 \mapsto 1$	$1 \mapsto 2$	$1 \mapsto 1$	$1 \mapsto 2$
$2 \mapsto 1$	$2 \mapsto 2$	$2 \mapsto 2$	$2 \mapsto 1$

Here all possible mappings of A to A are exhausted by f_1, f_2, f_3, and f_4; thus,

$$F\left[\{1, 2\}\right] = \{f_1, f_2, f_3, f_4\}$$

Combining:

$$(f_1 \circ f_1)(1) = f_1[f_1(1)] = f_1(1) = 1 \qquad \text{[using definition (5.6.1)]}$$
$$(f_1 \circ f_1)(2) = f_1[f_1(2)] = f_1(1) = 1$$

Hence, $f_1 \circ f_1 = f_1$. Continuing,

$$(f_1 \ o \ f_2)(1) = f_1[f_2(1)] = f_1[2] = 1$$
$$(f_1 \ o \ f_2)(2) = f_1[f_2(2)] = f_1[2] = 1$$

We are able, by this point, to see the general pattern; that is, f_1 applied to anything gives back f_1:

$$f_1 \ o \ f_1 = f_1; \ f_1 \ o \ f_2 = f_1; \ f_1 \ o \ f_3 = f_1; \ f_1 \ o \ f_4 = f_1$$

Similarly, f_2 always gives back f_2:

$$f_2 \ o \ f_1 = f_2; \ f_2 \ o \ f_2 = f_2; \ f_2 \ o \ f_3 = f_2; \ f_2 \ o \ f_4 = f_2$$

But f_3 has no effect because it simply preserves the "action" of the prior map, for example

$$(f_3 \ o \ f_4)(1) = f_3[f_4(1)] = f_4(1)$$
$$(f_3 \ o \ f_4)(2) = f_3[f_4(2)] = f_4(2)$$

Thus,

$$f_3 \ o \ f_1 = f_1; \ f_3 \ o \ f_2 = f_2; \ f_3 \ o \ f_3 = f_3; \ f_3 \ o \ f_4 = f_4$$

To compute f_4 combinations:

$$(f_4 \ o \ f_1)(1) = f_4[f_1(1)] = f_4(1) = 2 \cdot \qquad \text{(hence, } 1 \mapsto 2)$$
$$(f_4 \ o \ f_1)(2) = f_4[f_1(2)] = f_4(1) = 2 \qquad \text{(hence, } 2 \mapsto 2)$$

so $f_4 \ o \ f_1 = f_2$. Further,

$$(f_4 \ o \ f_2)(1) = f_4[f_2(1)] = f_4(2) = 1 \qquad \text{(hence, } 1 \mapsto 1)$$
$$(f_4 \ o \ f_2)(2) = f_4[f_2(2)] = f_4(2) = 1 \qquad \text{(hence, } 2 \mapsto 1)$$

and so, $f_4 \ o \ f_2 = f_1$

$$(f_4 \ o \ f_3)(1) = f_4[f_3(1)] = f_4(1) = 2 \qquad \text{(hence, } 1 \mapsto 2)$$
$$(f_4 \ o \ f_3)(2) = f_4[f_3(2)] = f_4(2) = 1 \qquad \text{(hence, } 2 \mapsto 1)$$

so that $f_4 \ o \ f_3 = f_4$. Finally,

$$(f_4 \ o \ f_4)(1) = f_4[f_4(1)] = f_4(2) = 1 \qquad \text{(hence, } 1 \mapsto 1)$$
$$(f_4 \ o \ f_4)(2) = f_4[f_4(2)] = f_4(1) = 2 \qquad \text{(hence, } 2 \mapsto 2)$$

Hence, $f_4 \ o \ f_4 = f_3$.

The multiplication table below summarizes the results in which we have simply sequentially produced each row.

(5.6.2)

o	f_1	f_2	f_3	f_4
f_1	f_1	f_1	f_1	f_1
f_2	f_2	f_2	f_2	f_2
f_3	f_1	f_2	f_3	f_4
f_4	f_2	f_1	f_4	f_3

In concluding this introduction to the notion of composition of mappings, we note that since f is a relation and we showed in section 4.16 that relation composition is associative, it follows that composition of mappings is associative. That is,

$$h \circ (g \circ f) = (h \circ g) \circ f$$

whenever the mappings can be composed

5.7. Characterizations and Evaluations.

Since people classify and evaluate each other, we can represent this activity by mathematical concepts referring to mappings. Further, we can see certain relations thereby induced over the set of people. For this type of work, we use the term "actor" to mean an element of an abstract set, say A, such that the intended interpretation is that A be identified with a set of entities with a capacity for cognitive activity, in any specific identification. Detailed treatments of a variety of phenomena in terms of the framework outlined here may be found in Fararo (1968, 1970b, 1970c, 1972b).

When actors classify each other, two types of mental activity are involved: (1) noting aspects of each other and (2) evaluating such aspects. These are the social ramifications of the elementary psychological relations of discrimination and preference. Let A be an arbitrary set of actors. Let $\alpha \in A$. Let C be an arbitrary characteristic, and let $x \in C$. Then, also, let $\beta \in A$. When α maps A into C, he "sends" each $\beta \in A$ into some "state" of C. A display shows this clearly:

Members of A (β)	States of C (x)
Tom	North Hills
Dick	South Hills
Harry	Inner City

In this display we are showing what some $\alpha \in A$—for instance, α = Tom—believes to be the residential locations of each $\beta \in A$, including $\beta = \alpha$ (his own residential location). To name this type of mapping, we denote it by c_α: $A \to C$, and call c_α a characterization (by α, of actors in A, in terms of C). Thus, if the above display represents Tom's mapping, we have

$$c_{Tom}(Tom) = \text{North Hills}$$
$$c_{Tom}(Dick) = \text{South Hills}$$
$$c_{Tom}(Harry) = \text{Inner City}$$

and in the alternate notation, for example,

$$\text{Harry} \xmapsto{c_{Tom}} \text{Inner City}$$

"Harry is mapped into the Inner City residential location by Tom's characterization." More plainly: Tom characterizes Harry, in terms of residential location, as living in the Inner City.

As α varies over A, we have the class or family of characterizations:

$$\{c_\alpha, \alpha \in A\}$$

which in the case at hand consists of

$$\{c_{Tom}, c_{Dick}, c_{Harry}\}$$

Note that only c_{Tom} has been given explicitly.

Now clearly two mappings are the same if and only if they have the same domain and range and associate the same image to each object in the domain. Hence, if the above display were also Harry's, it would be indistinguishable from Tom's. Thus,

(5.7.1) $c_\alpha = c_\beta$

if and only if

$$c_\alpha(\gamma) = c_\beta(\gamma) \qquad \text{(all } \gamma)$$

where γ (gamma) is an arbitrary member of A.

We say that the family of characterizations $\{c_\alpha, \alpha \in A\}$ is uniform when (5.7.1) holds for all α and β. Then, of course, this set contains just one object: the mapping common to all $\alpha \in A$, and we can write $c: A \to C$ without ambiguity.

Now this kind of mapping is only one ingredient in status mappings. The other central ingredient is some evaluation of the actors, by the actors, in terms of C. For this evaluative structure, we can use the axioms of section 4.17 dealing with idealized (rational) preference-indifference. We do this as follows. Let $\alpha \in A$. Let Q_α be a relation on C associated with α: it is α's "weak preference" over C. Assume that Q_α satisfies axioms (4.17.1) and (4.17.2) of section 4.17: Q_α is transitive and strongly complete on C. Then P_α and I_α are induced by definitions (4.17.3) and (4.17.4) and satisfy theorems (4.17.5), (4.17.6), and (4.17.8). As shown in section 4.17, this means we have a complete ordered system of the form

(5.7.2) $(C/I_\alpha, P_\alpha^*)$

for each $\alpha \in A$.

The form (5.7.2) consists in (1) a set of evaluative equivalence classes over the characteristic C, according to α's evaluation, and (2) an ordering of these classes, according to α's evaluation.

Suppose, for example, that Tom regards North Hills and South Hills as about equally good places to live and better than the Inner City. Then

$$C/I_{Tom} = \{\{\text{North Hills, South Hills}\}, \{\text{Inner City}\}\}$$

Now the residential locations in C have been mapped by Tom's evaluation into equivalence classes

$$x \mapsto \bar{x}$$

namely,

North Hills $\longmapsto$ {North Hills, South Hills}

South Hills

Inner City $\longmapsto$ {Inner City}

We shall denote this type of mapping of x into $\bar{x}$ by σ (sigma) appropriately subscripted:

$$\sigma_{Tom}: C \to C/I_{Tom}$$

where

$$\sigma_{Tom}(x) = \bar{x} \qquad (\text{all } x \in C)$$

Finally, since

$$\{\text{North Hills, South Hills}\} P^*_{Tom} \{\text{Inner City}\}$$

we let

$$\{\text{North Hills, South Hills}\} \mapsto 2$$
$$\{\text{Inner City}\} \mapsto 1$$

for a defined mapping v_{Tom} of classes into numbers to preserve evaluative order P^*_{Tom}. The higher number represents the more favorably evaluated class of residential locations. Thus,

β	x	$\bar{x}$	$v_{Tom}(\bar{x})$
Tom	North Hills	{North Hills, South Hills}	2
Dick	South Hills	{North Hills, South Hills}	2
Harry	Inner City	{Inner City}	1

From this display we see that the composition of three mappings induces an evaluative order over A (by Tom):

c_{Tom}: Tom's characterization of A in terms of C
σ_{Tom}: Tom's classifying of C in terms of indifference
v_{Tom}: Our numerical ordering to represent Tom's preference relation

This can be written

$$A \xrightarrow{c_{Tom}} C \xrightarrow{\sigma_{Tom}} C/I_{Tom} \xrightarrow{v_{Tom}} \Re$$

In all generality, let

(5.7.3) $$A \xrightarrow{c_\alpha} C \xrightarrow{\sigma_\alpha} C/I_\alpha \xrightarrow{v_\alpha} \Re \qquad (\alpha \in A)$$

be a sequence of mappings satisfying

(1) A is a set of actors with $\alpha \in A$

(2) c_α is a mapping from A into C

(3) I_α and P_α^* exist satisfying

  (a) I_α is an equivalence relation on C

  (b) P_α^* is a linear order relation on C/I_α

(4) σ_α maps state x into class $\bar{x}$ in C/I_α

(5) v_α represents P_α^*

An object (5.7.3) satisfying these axioms is called a status chain. Then the composition

$$(5.7.4) \qquad s_\alpha(\beta) = v_\alpha \overline{[c_\alpha(\beta)]} \qquad \text{(all } \beta \in A\text{)}$$

is called a status mapping. We also write

$$s_\alpha = v_\alpha \sigma_\alpha c_\alpha$$

using juxtaposition to represent composition of mappings. Expression (5.7.4) says that the status α accords to β is the valuation that α has of the class to which α assigns β.

5.8. Preservation of Status Relations.

Relations are induced over A by the existence of I_α and P_α^* and the characterization c_α. Namely, define

$$(5.8.1) \qquad \begin{array}{llll} \beta H^\alpha \gamma & \text{iff} & \overline{[c_\alpha(\beta)]} \, P_\alpha^* \, \overline{[c_\alpha(\gamma)]} & \text{(all } \beta, \gamma \text{ in A)} \\ \beta E^\alpha \gamma & \text{iff} & c_\alpha(\beta) = c_\alpha(\gamma) & \text{(all } \beta, \gamma \text{ in A)} \end{array}$$

This says, first, that (β, γ) stand in H^α if and only if the characterizing class to which β is assigned by α is "better"—we should say "α-better"—than the characterizing class to which γ is assigned by α. Thus, "β has higher α-status than γ." Then, (β, γ) in E^α if and only if these characterizing classes are one and the same. Thus, "β and γ are α-status equals."

To practice abstract calculations, we show that the ordering of (5.8.1) is precisely the ordering given by the status mapping, because

$$\begin{array}{llll} \overline{[c_\alpha(\beta)]} \, P_\alpha^* \, \overline{[c_\alpha(\gamma)]} & \text{iff} & v_\alpha \overline{[c_\alpha(\beta)]} > v_\alpha \overline{[c_\alpha(\gamma)]} & \text{(definition of } v_\alpha) \\ & \text{iff} & s_\alpha(\beta) > s_\alpha(\gamma) & \text{(definition of } s_\alpha) \end{array}$$

And, also,

$$\begin{array}{llll} \overline{c_\alpha(\beta)} = \overline{c_\alpha(\gamma)} & \text{iff} & v_\alpha \overline{[c_\alpha(\beta)]} = v_\alpha \overline{[c_\alpha(\gamma)]} & \text{(definition of } v_\alpha) \\ & \text{iff} & s_\alpha(\beta) = s_\alpha(\gamma) & \text{(definition of } s_\alpha) \end{array}$$

Hence, for any $\alpha \in A$,

$$(5.8.2) \qquad \begin{array}{llll} \beta H^\alpha \gamma & \text{iff} & s_\alpha(\beta) > s_\alpha(\gamma) \\ \beta E^\alpha \gamma & \text{iff} & s_\alpha(\beta) = s_\alpha(\gamma) \end{array}$$

for all β, γ in A. This indicates that a status mapping is, after all, just a representation of an ordering of actors. But it is not an arbitrary ordering, in the sense that this order emerges out of the conceptual and evaluative activity of actors themselves.

5.9. Special Cases. The fact may be that for some C, for the given α, each member of A gets mapped to one state of C. For instance,

$$c_\alpha(\beta) = \text{American} \qquad (\text{all } \beta \in A)$$

in a certain group where C = {American, not-American}. Then,

$$c_\alpha(\beta) = c_\alpha(\gamma) \qquad (\text{all } \beta, \gamma \in A)$$

Hence, of course,

$$\overline{c_\alpha(\beta)} = \overline{c_\alpha(\gamma)}$$

and so, also,

$$v_\alpha \overline{[c_\alpha(\beta)]} = v_\alpha \overline{[c_\alpha(\gamma)]}$$

that is,

$$s_\alpha(\beta) = s_\alpha(\gamma) \qquad (\text{all } \beta, \gamma \in A)$$

This shows that if no cognitive discrimination is made between the actors in A, by α, as to how they are characterized for a given C, then they are all α-status equals in terms of this C. Also, suppose

$$c_\alpha: A \to C$$

is not constant, but for all x, y in C

$$\bar{x} = \bar{y}$$

That is, while α can tell the difference between actors—for instance he knows where they live—he regards this variation with evaluative indifference: they all live in a certain evaluatively uniform neighborhood, for example. Then, even though $c_\alpha(\beta) \neq c_\alpha(\gamma)$, we have

$$\overline{c_\alpha(\beta)} = \overline{c_\alpha(\gamma)} \qquad (\text{for all } \beta, \gamma)$$

and

$$s_\alpha(\beta) = s_\alpha(\gamma) \qquad (\text{all } \beta, \gamma \in A)$$

Thus, if all actors are seen in terms such that no evaluative differentiation occurs, then they are α-status equals.

Sometimes we find it useful to treat theoretical models in which for each α, $\beta \in A$

$$v_\alpha = v_\beta$$

Then we are saying that

$$(C/I_\alpha, P_\alpha^*) = (C/I_\beta, P_\beta^*)$$

since the former merely represents the latter. We say that A is v-homogeneous (with respect to C) or is evaluatively uniform.

By considering all combinations of uniformity with respect to both ingredients of a status mapping (i.e., characterization and valuation), we see that for any characteristic C and any set of actors A, there are four uniformity types:

(1) fully uniform: v-homogeneous and c-uniform
(2) v-homogeneous but not c-uniform
(3) not v-homogeneous but c-uniform
(4) neither v-homogeneous nor c-uniform.

Also, one may develop different meanings for "v-homogeneity" aside from the agreement at the level of I and P^* relations. For example, one may let the observed disagreements among the various P_α, $\alpha \in A$, suggest a probability approach in which "v-homogeneity" assumes a different empirical meaning. This we shall do later in the book (section 11.13).

5.10. Universalism-Particularism. We may apply this set of concepts dealing with status to formalize the "operational" version of the universalism-particularism distinction introduced by Blau (1964). The distinction appears in intuitive form in much of the work of Talcott Parsons. (See, for example, Parsons, Bales, and Shils, 1953.) It applies to expectations as well as evaluations, but here we restrict ourselves to the latter context. According to Blau, particularistic values are reflected in evaluations of the form: the actors prefer their own characterizing state. Universalistic values are reflected in evaluations of the form: the actor's preferences do not depend on the characterizing state of the actor. Particularism implies group identity and potential for action in concert; universalism implies stratification.

We state this idea in our formalism as follows. Let A be a set of actors and C a characteristic. For the particularistic case, we may keep in mind the identification

$$C = \{Jew, Protestant, Catholic\}$$

For the universalistic case, we may keep in mind the identification,

$$C = \{high\ school\ diploma, college\ degree\}$$

We assume a characterization, made by the sociologist,

$$c: A \to C$$

assigning states of C to actors in A.

For each α in A, let us assume a common I relation on C such that I is the identity relation: every state of C is evaluatively distinguishable from every other. (For any C for which this seems unreasonable, we would first pass to C/I and then the following discussion holds with C/I in place of C.)

For each α in A, let v_α represent α-preference over C. Thus, $v_\alpha(x)$ is the valuation of state x of C, from the α standpoint.

(5.10.1) Definition. We say that v_α is α-particularistic if and only if

$$c(\alpha) \neq c(\beta) \Rightarrow v_\alpha[c(\alpha)] > v_\alpha[c(\beta)] \qquad (\text{all } \beta \in A)$$

This says that the valuation v_α over C is α-particularistic if whenever another actor has a state differing from that of α, then α prefers his own state. For example, α has a "religious preference" in terms of "being Catholic."

Suppose, now, that for each α in A the corresponding v_α is α-particularistic. In this case, provided the characterization c: A $\rightarrow$ C is not constant, so that people differ in terms of C, it follows that the valuations are not homogeneous. More than this is true, however, since we have the special regularity amid the inhomogeneity, given by α-particularism for each α. In this instance, we can say that the collection $\{v_\alpha, \alpha \in A\}$ of valuations is particularistic. With the set of actors understood contextually, we can also say that C is particularistic. For example, religion is particularistic, family is particularistic, nationality is particularistic. In each case, an appropriate set of actors is required to obtain the nonconstant characterization. * In the case of nationality, for instance, the set A must contain actors drawn from distinct countries.

The universalistic case is best interpreted simply as v-homogeneity. This implies that we cannot have α-particularism for all actors. For instance, a high school graduate who evaluates a college degree as higher than a high school diploma will fail to exemplify α-particularism. Hence, universalism-particularism defines a pair of special cases in the general conceptualization:

(1) universalism: v-homogeneity
(2) particularism: no v-homogeneity, but α-particularism, for all α.

5.11. Binary Operations. Just as a relational system is a set together with one or more relations defined on that set, (A, R, S, . . .), so an algebraic system is a set with one or more operations defined on that set. By an "operation" we mean some mode of combination of objects that produces an object of the same class. For example, addition of numbers is an operation, which we could show in particular cases as

$$(2, 3) \mapsto 5$$
$$(3, 4) \mapsto 7$$

---

* Otherwise we have "vacuous particularism" because of the conditional in (5.10.1).

and so forth. In other words, we can represent an operation as a mapping of pairs of elements in some set into some element in that same set.

(5.11.1) Definition. By a binary operation we mean a mapping (A x A, A, o). That is, o maps A x A into A.

Rather than write o(x, y), we write x o y. Hence,

$$(x, y) \mapsto x \text{ o } y$$

is the generic idea of a binary operation.

Examples. (a) Take o to be addition of numbers:

$$(x, y) \mapsto x+y$$

(b) Take o to be multiplication of numbers:

$$(x, y) \mapsto xy$$

(c) Take o to be union of subsets:

$$(A, B) \mapsto A \cup B$$

(d) Take o to be intersection of subsets:

$$(A, B) \mapsto A \cap B$$

(e) Take o to be function composition:

$$(f, g) \mapsto f \text{ o } g$$

(f) Take o to be relation composition:

$$(R, S) \mapsto R \text{ o } S$$

5.12. Groupoids. The simplest algebraic system consists of two objects: a set and a binary operation defined on that set. In other words, if A x A $\overset{o}{\to}$ A is a mapping, then (A, o) is an algebraic system in which A is the domain of some mode of combination of objects in A given by o. We term (A, o) a groupoid.

Examples. (a) Addition of numbers:
  $\Re$ is the set of real numbers;
  + is addition;
  $(\Re, +)$ is a groupoid.

(b) Multiplication of numbers:
  $\Re$ is the set of real numbers;
  · is multiplication of numbers;
  $(\Re, \cdot)$ is a groupoid.

(c)   Union of subsets:

Ω is some set;

2^{Ω} is the power set of Ω;

$\cup$ is union of subsets of Ω;

$(2^{\Omega}, \cup)$ is a groupoid.

(d)   Intersection of subsets:

Ω is some set;

2^{Ω} is the power set of Ω;

$\cap$ is intersection of subsets of Ω;

$(2^{\Omega}, \cap)$ is a groupoid.

(e)   Function composition:

A is some set;

F[A] is the set of all functions on A into A;

o is function composition;

(F[A], o) is a groupoid.

(f)   Relation composition:

A is some set;

R[A] is the set of all relations on A;

o is relation composition;

(R[A], o) is a groupoid.

The concept of a groupoid (A, o), then, is analogous to the concept of a binary relational system (A, R). Whereas o combines elements of A, in the latter system R relates elements of A. Also, just as the relational system concept generalizes to (A, R, S, . . .) so does the groupoid concept generalize to (A, o, *, . . .) where several modes of combination are defined on the same domain. This latter system is the generic concept of an algebra. Combining the two types of entities yields more complex systems, some of which will appear in our subsequent work.

5.13. Properties of Operations and Types of Groupoids.

Just as there are four basic typological properties for relations (referent to symmetry, reflexivity, transitivity, and completeness), so in the case of operations we have some basic properties. Just as in confronting a relation we may ask for its "value" on each of the four properties, so in operations we may investigate them by determining whether or not key properties are exemplified. These properties are now defined.

(5.13.1) Definition. o is commutative   iff   x o y = y o x   (all x, y).

(5.13.2) Definition. o is associative   iff   x o (y o z) = (x o y) o z   (all x, y, z).

(5.13.3) Definition. o has an identity   iff   for some element e

$$x o e = x \qquad \text{(all x)}$$
$$e o x = x \qquad \text{(all x)}$$

(5.13.4) *Definition.* o has the inversion property   iff   it has an identity e
and for any element x, there is an element (say, x^{-1}) such that

$$x \text{ o } x^{-1} = e$$
$$x^{-1} \text{ o } x = e$$

Examples. (a) Addition of numbers: $(x, y) \mapsto x + y$
  (1) $x + y = y + x \Rightarrow +$ is commutative
  (2) $x + (y + z) = (x + y) + z \Rightarrow +$ is associative
  (3) $x + 0 = x, 0 + x = x \Rightarrow +$ has an identity, 0
  (4) $x + (-x) = 0, (-x) + x = 0 \Rightarrow +$ has the inversion property

(b) Multiplication of numbers: $(x, y) \mapsto xy = x \cdot y$
  (1) $xy = yx \Rightarrow \cdot$ is commutative
  (2) $x(yz) = (xy)z \Rightarrow \cdot$ is associative
  (3) $x1 = x, 1x = x \Rightarrow \cdot$ has an identity, 1
  (4) 0 is a number, but there is no number such that 0 times the
  number yields $1 \Rightarrow \cdot$ does not have the inversion property.

(c) Union of subsets: $(A, B) \mapsto A \cup B$
  (1) $A \cup B = B \cup A \Rightarrow \cup$ is commutative
  (2) $A \cup (B \cup C) = (A \cup B) \cup C \Rightarrow \cup$ is associative
  (3) $A \cup \emptyset = A, \emptyset \cup A = A \Rightarrow \cup$ has an identity element, $\emptyset$
  (4) If A is nonempty there is no set, say X, such that $A \cup X$
  $= \emptyset \Rightarrow \cup$ does not have the inversion property.

(d) Intersection of subsets: $(A, B) \mapsto A \cap B$
  (1) $A \cap B = B \cap A \Rightarrow \cap$ is commutative
  (2) $A \cap (B \cap C) = (A \cap B) \cap C \Rightarrow \cap$ is associative
  (3) $A \cap \Omega = \Omega \cap A = A \Rightarrow \cap$ has an identity, Ω
  (4) If $A \subset \Omega$, there is no set X such that $A \cap X = \Omega \Rightarrow \cap$ does
  not have the inversion property.

(e) Function composition: $(f, g) \mapsto f \text{ o } g$
  (1) $f \text{ o } g \neq g \text{ o } f$, in general $\Rightarrow$ o not commutative
  (2) $f \text{ o } (g \text{ o } h) = (f \text{ o } g) \text{ o } h \Rightarrow$ o is associative
  (3) Let I be the identity function $x \overset{I}{\mapsto} x$. Then $f \text{ o } I = I \text{ o } f$
  $= f \Rightarrow$ o has an identity element, I.
  (4) Define $f(x) = x_0$, where x_0 is fixed as x varies, so f is a constant function. Then, $f \text{ o } g = f$, for any g, and so there is no g such
  that $f \text{ o } g = I \Rightarrow$ o does not have the inversion property.

(f) Relation composition: $(R, S) \mapsto R \text{ o } S$
  (1) $R \text{ o } S \neq S \text{ o } R$, in general $\Rightarrow$ o not commutative
  (2) $R \text{ o } (S \text{ o } T) = (R \text{ o } S) \text{ o } T \Rightarrow$ o is associative
  (3) Let I be the identity relation, containing only pairs (x, x).
  Then, $R \text{ o } I = I \text{ o } R = R \Rightarrow$ o has an identity element, I.
  (4) Earlier (section 4.16) we defined

$$R^{-1} = \{(y, x): xRy\}$$

But, if we form $R \circ R^{-1}$, we obtain (x, x) only for those x such that $(x, y) \in R$. Moreover, suppose we have xRz and yRz, for some z. Then, xRz and $zR^{-1}y$ imply that $(x, y) \in R \circ R^{-1}$, even if $x \neq y$. Hence, $R \circ R^{-1} \neq I$, in general $\Rightarrow \circ$ does not have the inversion property.

In the above, we have investigated six different operations to determine their configuration of properties. Just as combinations of relation properties yield important types of relations (equivalence, order), so combinations of operation properties yield important types of algebraic operations. These are used to define special types of groupoids, as follows.

(5.13.5) Definition.   By a semigroup we mean a groupoid whose operation is associative.

(5.13.6) Definition.   By a monoid we mean a semigroup whose operation has an identity element.

(5.13.7) Definition.   By a group we mean a monoid whose operation has the inversion property.

(5.13.8) Definition.   A groupoid is called commutative (or abelian) if its operation is commutative.

Based on the prior work on our six operations, we have the following:

(1) addition of numbers: $(\Re, +)$ is a commutative group
(2) multiplication of numbers: $(\Re, \cdot)$ is a commutative monoid
(3) union of subsets: $(2^\Omega, \cup)$ is a commutative monoid
(4) intersection of subsets: $(2^\Omega, \cap)$ is a commutative monoid
(5) function composition: $(F[A], \circ)$ is a noncommutative monoid
(6) relation composition: $(R[A], \circ)$ is a noncommutative monoid.

5.14. Homomorphisms.   Let $(X, \circ)$ and $(Y, \circ)$ be two groupoids where $\circ$ on X may have a totally different meaning than the $\circ$ on Y. A map $f: X \to Y$ is termed a homomorphism (of groupoids) if and only if the image of a combination is the combination of images:

(5.14.1) $f(x \circ y) = f(x) \circ f(y)$

A homomorphism that is bijective is called an isomorphism.

Two artificial groupoids are provided by the two tables:

o	A	B
A	A	A
B	A	B

o	C	D
C	C	C
D	C	D

It is clear that "abstractly" these are two different nomenclatures for the same abstract table, whose pattern is clear. The mapping $f(A) = C$, $f(B) = D$ shows this explicitly, since

$$f(A \circ A) = f(A) \qquad \text{(from the first table)}$$
$$= C \qquad \text{(by definition of f)}$$

Also

$$f(A) \circ f(A) = C \circ C \qquad \text{(by definition of f)}$$
$$= C \qquad \text{(by the second table)}$$

Thus, $f(A \circ A) = f(A) \circ f(A)$. The other combinations verify the general result: f is a homomorphism since it satisfies (5.14.1). But f is also a one-to-one correspondence (a bijection); thus, f is an isomorphism. Two groupoids that are linked by some isomorphism are called isomorphic: the isomorphism provides the evidence that the groupoids have the same abstract structure.

The two tables

o	A	B
A	A	A
B	A	B

o	E	F
E	F	F
F	F	F

show nonisomorphic groupoids: no isomorphism could possibly be found. This is obvious, but let us show the logic of the impossibility proof. To deny the systems are isomorphic means that it is not the case that there exists a mapping $f: \{A, B\} \rightarrow \{E, F\}$ that is an isomorphism. To show this we use the logic of contradiction: assume such an f exists and deduce a contradiction. Thus, let f be an isomorphism. Then either $f(B) = E$ or $f(B) = F$. Suppose $f(B) = E$. Then, since f is an isomorphism,

$$f(B \circ B) = f(B) \circ f(B) \qquad \text{[f an isomorphism, so (5.14.1) applies]}$$
$$= E \circ E \qquad \text{[assuming } f(B) = E]$$
$$= F \qquad \text{(second table).}$$

But also,

$$B \circ B = B \qquad \text{(first table)}$$

so applying f to both sides,

$$f(B \circ B) = f(B)$$
$$= E$$

Comparing the two computations of $f(B \circ B)$, we conclude $E = F$, which contradicts the given groupoid containing two distinct elements, $E \neq F$. Now suppose $f(B) = F$. Then, since f is an isomorphism, it is an injection and $f(A) \neq f(B)$. Thus, $f(A) = E$. From this we obtain

$$
\begin{aligned}
f(A \circ A) &= f(A) \circ f(A) && \text{(f an isomorphism)} \\
&= E \circ E && [\text{since } f(A) = E] \\
&= F && \text{(first table)}
\end{aligned}
$$

Also,

$$
\begin{aligned}
A \circ A &= A && \text{(second table)} \\
f(A \circ A) &= f(A) && \text{(applying f)} \\
&= E && [\text{since } f(A) = E]
\end{aligned}
$$

again, a contradiction, $E = F$. Since either possible case leads to a contradiction, the assumption of an isomorphism leads to a contradiction. Hence, there exists no isomorphism.

Here is a more complex example of a homomorphism, using numerical systems. Let the first groupoid be $(\Re^+, \cdot)$, the positive real numbers and the multiplication operation. Let the second groupoid be $(\Re, +)$, the real numbers with the addition operation. We want to show that these two algebraic systems are homomorphic. We need a mapping $f: \Re^+ \to \Re$ such that

$$f(xy) = f(x) + f(y) \qquad (\text{all } x, y \in \Re^+)$$

The reader may recall that

$$\log xy = \log x + \log y$$

provided the arguments x and y are positive. Thus, the log function is a homomorphism. It replaces one computation (xy) by another ($\log x + \log y$), but the real reason it works is that we can return to the "answer" by "antilog." This could not be done if log mapped two distinct numbers into the same image. It does not: it is injective. Moreover any number y has an associated positive number x such that $\log x = y$. Thus, log is also surjective. From this it follows that log establishes an isomorphism between the positive real numbers and the entire set of real numbers. Working with logs is working with an isomorphic representation of one's original problem.

Homomorphisms preserve structure but not with total "fidelity" unless they are bijective.

Consider the two groupoids

o	A	B
A	A	A
B	A	B

o	E	F	G
E	E	E	G
F	E	F	E
G	F	G	G

Here the first table can be "imbedded" in the second table: $A \mapsto E$, $B \mapsto F$. But, clearly, no isomorphism can exist, because the systems cannot be put in one-to-one correspondence. Thus, we have homomorphism between the two systems, although their structures are not exactly the same. Within the second groupoid, we have an isomorphic image of the entire first groupoid, but the entire second groupoid is only homomorphically related to the first. Another example is shown by the following two groupoids.

o	A	B	C	D
A	A	A	B	B
B	A	A	B	B
C	C	C	D	D
D	C	C	D	D

o	E	F
E	E	E
F	F	F

Here E corresponds to class $\{A, B\}$, F to $\{C, D\}$ so that the homomorphism is defined by

$$f(A) = f(B) = E$$
$$f(C) = f(D) = F$$

The second table reduces the first to an essential structural feature. In later work (see Chapters 12 and 19), we will see sociological examples of this kind, featuring "homomorphic reduction."

A second significant point about homomorphism is that we sometimes have an equivalence relation in one groupoid and all objects that are equivalent have to be represented by the same object in the second system. For example, in measurement of mass, if two objects balance each other they are assigned the same number. The mathematical function used is m: $E \to \Re$ where E is the empirical domain and $\Re$ the real number system, with m the homomorphism called the mass function. Thus, in some cases, as in measurement, we do not desire isomorphism but only homomorphism. This topic will be treated in detail in Chapter 7.

5.15. Relational Homomorphisms. The concept of homomorphism of groupoids suggests an analogous concept for relational systems.

(5.15.1) Definition. Let (A, R) and (B, S) be binary relational systems. Thus, R is a relation on A, and S is a relation on B. A relational homomorphism between the systems is a mapping f: $A \to B$ such that

$$(x, y) \in R \quad \text{iff} \quad [f(x), f(y)] \in S \qquad \text{(all x, y in A)}$$

Thus, relationships are preserved. If f is also bijective, it is called a relational isomorphism.

Example. (a) Consider Figure 5.15.1, where $A = \{1, 2, 3\}$, $B = \{a, b, c\}$.

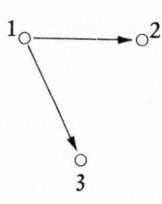 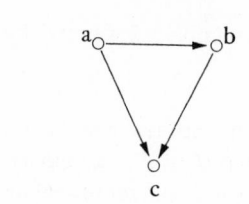

Figure 5.15.1

The mapping $1 \mapsto a$, $2 \mapsto b$, $3 \mapsto b$ (and call it f) is a relational homomorphism, for

$(1, 2) \in R$	and	$(a, b) \in S$
$(1, 3) \in R$	and	$(a, b) \in S$
$(2, 3) \notin R$	and	$(b, b) \notin S$

illustrate that in general, for all x and y in A

$$(x, y) \in R \quad \text{iff} \quad [f(x), f(y)] \in S$$

It is intuitively clear that no isomorphism can be found. Logically, let f be any one-to-one mapping; then for f to be an isomorphism $(2, 3) \notin R$ and $(3, 2) \notin R$ would have to correspond to $[f(2), f(3)] \notin S$ and $[f(3), f(2)] \notin S$; but with $f(2) \neq f(3)$, this implies a pair of distinct points linked in neither direction. Such a pair does not exist in the second graph, and so, no isomorphism is possible. Thus, the two systems have a different structure, although the homomorphism shows their similarity.

Example. (b) On the other hand, the two graphs in Figure 5.15.2 represent isomorphic systems.

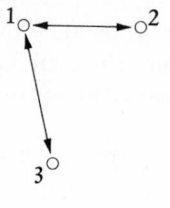 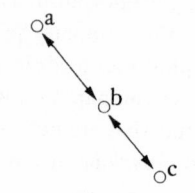

Figure 5.15.2

To see this rotate the (1, 2) linkage. Algebraically, define the mapping

$$1 \mapsto b$$
$$2 \mapsto a$$
$$3 \mapsto c$$

which is injective and surjective and a relational homomorphism. Thus, the systems represented by these two graphs are isomorphic.

Example. (c) In section 5.7 we introduced status mappings

$$A \overset{c}{\to} C \overset{\sigma}{\to} C/I \overset{v}{\to} \mathfrak{R}$$

We can now make the intuitive connection between preferences and the mapping v clear. Consider that $(C/I, P^*)$ is a binary relational system. Also, $(\mathfrak{R}, >)$ is a binary relational system: the set of real numbers with the usual ordering relation $>$. The valuation v is a homomorphism into $\mathfrak{R}$; also, the mapping v homomorphically imbeds the empirical relational system in the numerical relational system. The homomorphism is the job we want v to do: to faithfully represent P^*. Thus, we defined v so that

$$\bar{x}P^*\bar{y} \quad \text{iff} \quad v(\bar{x}) > v(\bar{y}) \qquad (\text{all, } \bar{x}, \bar{y} \in C/I)$$

which specifies a homomorphism v. It is not an isomorphism, because although it is injective (distinct classes are mapped into distinct numbers), it is not surjective (ordinarily, since C/I will usually be finite, and so, its image cannot exhaust the set of real numbers).

5.16. Structure. This section illustrates an abstractive technique based on quotient sets (cf. section 4.12).

Let REL be the class of all relational systems of the form (A, R) where A is a set and R is a relation on A. The following propositions are provable:

(1) (A, R) is isomorphic to itself;

(2) If (A, R) is isomorphic to (B, S), then (B, S) is isomorphic to (A, R);

(3) If (A, R) is isomorphic to (B, S) and (B, S) is isomorphic to (C, T), then (A, R) is isomorphic to (C, T).

We see that, regarding isomorphism as a relation* on REL, we have (1) the reflexive property; (2) the symmetry property; and (3) the transitive property. It follows that isomorphism is an equivalence relation. Thus, abbreviating isomorphism as "iso," we can partition REL into classes REL/iso such that distinct classes contain systems that are not isomorphic.

By a relational structure one means an element of REL/iso. Thus, each concrete system can be said to represent the structure in which it is contained. If we let $St(A, R)$ be the class containing (A, R), then

$$St(A, R) = St(B, S)$$

if and only if (A, R) is isomorphic to (B, S), and $St(A, R)$ is "the structure of (A, R)."

---

*Logically, we could then have a paradox because (REL, iso) is a binary relational system in its own right; a rigorous distinction between types of sets avoids the paradox. See MacLane and Birkhoff (1967).

Form the class $\mathcal{G}$ of all groupoids, that is, systems of the form (A, o) where A is a set and o is a mode of combination of objects in A. Groupoid isomorphism is an equivalence relation on $\mathcal{G}$, and so, we can partition $\mathcal{G}$ into classes. If St(A, o) is the class containing (A, o), then

$$St(A, o) = St(B, o)$$

if and only if (A, o) is isomorphic to (B, o), and St(A, o) is the "structure of (A, o)."

5.17. Representation of a Relational System. Consider the class of finite (simple) relational systems: systems (A, R) such that A is finite. We may consider graphs in the plane as geometric relational systems. It is intuitively clear that no matter what finite relational system one chooses, that system can be represented by a graph in the plane.

Figure 5.17.1 illustrates the conceptual situation. For each (A, R) in the class of all finite relational systems there exists some isomorphic (G, L), a graph, in the subclass of systems that are "geometric." This is an example of one of the most fundamental ideas in mathematics: a representation theorem. We need also an associated "uniqueness" theorem: any two graphs isomorphically representing a given relational system are themselves "equivalent"—a one-to-one correspondence exists between points such that oriented lines are preserved.

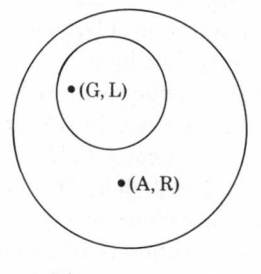

Figure 5.17.1

The notions of representation and uniqueness apply in the same way to measurement. The larger class includes "empirical" systems—the objects are observable things—and the smaller class of interest is a class of numerical systems of the same type (for example, numerical groupoids). The uniqueness theorem provides the theoretical foundation for the idea of "scale-type." This will be treated subsequently, in Chapter 7.

5.18. Other Algebras. If a groupoid is the simplest algebraic system, a step upward in complexity is obtained by dealing with a system with two operations. We treat this subject very briefly at this point. (For extensive details on this and many other topics, see MacLane and Birkoff, 1967).

(5.18.1) Definition. By a ring we mean a system (A, o, *) satisfying the three conditions

(1) (A, o) is a commutative group,
(2) (A, *) is a monoid,
(3) A "distributive law" connects algebras (1) and (2), meaning

$$x * (y \circ z) = (x * y) \circ (x * z)$$
$$(x \circ y) * z = (x * z) \circ (y * z)$$

Example. Since $(\Re, +)$ is a commutative group, since $(\Re, \cdot)$ is a monoid, and since

$$x(y + z) = xy + xz$$
$$(x + y)z = xz + yz$$

we conclude that $(\Re, +, \cdot)$ forms a ring. It is a commutative ring, since $(\Re, \cdot)$ is a commutative monoid.

(5.18.2) Definition. By a field we mean a commutative ring (A, o, *) in which the only element of (A, *) that does not exhibit the inversion property is the identity element of (A, o).

Example. The commutative ring $(\Re, +, \cdot)$ is a field because only 0, which is the identity element for addition, is not invertible; that is, there is no x such that $0x = 1$, which means "you cannot divide by 0."

Other fields include the rational numbers with addition and multiplication and the complex numbers (of the form $x \pm iy$). In algebra it is shown that there are many non-numerical rings and, in particular, fields.

The two commutative monoids based on union and intersection, namely $(2^\Omega, \cup)$ and $(2^\Omega, \cap)$, together comprise an instance of a type of system called a Boolean algebra. Essentially, the defining conditions for a Boolean algebra are those exemplified in set algebra, as shown in Table (3.4.1). A related algebraic system, built upon the partial order provided by the containment relation, is a semi-lattice. These two types of algebras have been applied to sociological problems by Friedell (1967, 1969).

Finally, we note that the real number system is formally defined as a field satisfying certain conditions (see Bartle, 1964). In the category of all such fields, it is a fact that any pair of such fields are isomorphic. Hence, $\Re$ is a unique structure.

CHAPTER SIX   VECTORS
AND MATRICES

6.1. Vectors. Mathematical sociology makes extensive use of the mathematics introduced in the preceding sections. But, in addition, certain special topics are crucial. Among these is the topic of vectors and matrices, which will now be reviewed.

As usual, let $\Re$ be the set of real numbers. By the Cartesian product $\Re \times \Re$ we mean, as indicated earlier (section 3.7), the set of all (ordered) pairs of numbers: these are "points" in $\Re \times \Re$. We think of this set in terms of a coordinate plane; this suggests drawing an arrow from the origin to the point of interest as in Figure 6.1.1.

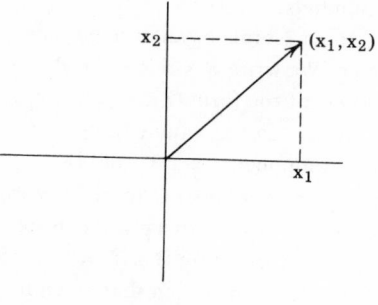

Figure 6.1.1

Given an origin point and units on both axes, there is a bijective (1–1 onto) correspondence between the points and the pairs of numbers in $\Re \times \Re$. The arrows also correspond to the number pairs. Thus, it is a matter of intuitive appeal and contextual usage as to precisely which of the three entities is under discussion when one says "vector in 2-space." The number pair (x_1, x_2) may be thought of as a numerical name for the point or for the vector to which it corresponds.

In regard to the arrows, however, we take the view that such an arrow is "the same" when it is "moved" if length and direction remain the same. This corresponds to the usual analytical geometry computations, as shown in Figure 6.1.2.

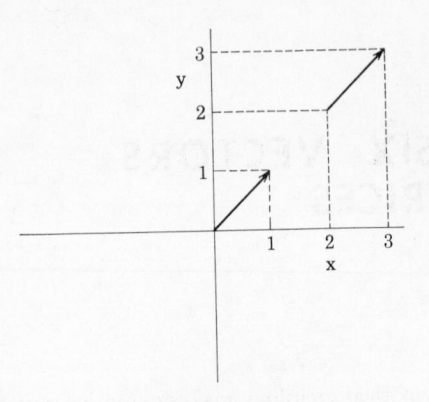

Figure 6.1.2

In Figure 6.1.2, the two arrows are regarded as the same vector. The vector (as arrow) corresponds to the pair $(1, 1)$. Hence, we are indifferent to its location in the plane if its length and direction are kept constant. This turns out to be quite useful in the geometric visualization of vector operations.

We write $\Re \times \Re \times \Re$ for 3-space: the set of all ordered triplets (x_1, x_2, x_3) such that $x_i \in \Re$, $i = 1, 2, 3$. This corresponds to coordinate 3-space. A vector in 3-space is thought of in three useful ways: a point, an arrow from the origin to the point, or the triplet of numbers.

In general, we call (x_1, x_2) a 2-tuple or an ordered pair. We call (x_1, x_2, x_3) a 3-tuple or an ordered triplet. We write $\Re \times \Re \times \ldots \times \Re$ (n times) to mean the set of all "n-tuples" of numbers of the form $(x_1, x_2, \ldots, x_n)$. We call any n-tuple a vector. Two vectors, $(x_1, x_2, \ldots, x_n)$ and $(y_1, y_2, \ldots, y_n)$, are identical if and only if they agree in each component: $x_i = y_i$, (all i).

To have a simple symbol for a vector so as to make work with them easier, we introduce x to mean $(x_1, x_2, \ldots, x_n)$. In a given context, vectors are always in some one particular "n-space" formed by $\Re \times \Re \times \ldots \times \Re$ (n times).

The following operations are defined such that when $n = 1$, we retrieve all the ordinary arithmetical operations as special cases. Let $x = (x_1, x_2, \ldots, x_n)$ and $y = (y_1, y_2, \ldots, y_n)$ be arbitrary n-tuples or vectors. Then we want to define $x + y$, $-x$, and so forth, for any $n = 1, 2, \ldots$.

(6.1.1) Definitions of Operations on Vectors.

    (1) Addition:

$$x + y = (x_1 + y_1, x_2 + y_2, \ldots, x_n + y_n)$$

For example:

$$x = (1, 2), y = (10, -4)$$
$$x + y = (1 + 10, 2 - 4) = (11, -2)$$

(2) Negative:

$$-x = (-x_1, -x_2, \ldots, -x_n)$$

For example:

$$x = (3, -7, 17)$$
$$-x = (-3, 7, -17)$$

(3) Zero:

$$\mathbf{0} = (0, 0, \ldots, 0) \qquad \text{(n-tuple if in n-space)}$$

(4) Subtraction:

$$x - y = (x_1 - y_1, x_2 - y_2, \ldots, x_n - y_n)$$

For example:

$$x = (7, -4, 0, 14), y = (12, 2, 0, 0)$$
$$x - y = (7 - 12, -4 - 2, 0 - 0, 14 - 0)$$
$$= (-5, -6, 0, 14)$$

(5) Scalar multiplication:

$$cx = (cx_1, cx_2, \ldots, cx_n) \qquad (c \in \Re)$$

For example:

$$c = 5, x = (3, 10, -1)$$
$$cx = (15, 50, -5)$$

(6) Dot (inner) product:

$$x \cdot y = \sum_{i=1}^{n} x_i y_i$$

For example:

$$x = (1, 3, -2, 0), y = \left(\frac{1}{3}, \frac{1}{3}, 4, -1\right)$$
$$x \cdot y = 1 \cdot \frac{1}{3} + 3 \cdot \frac{1}{3} + (-2) \cdot 4 + 0 \cdot (-1)$$
$$= \frac{1}{3} + 1 - 8 = -6 \frac{2}{3}$$

We can use the correspondence with arrows in the plane to illustrate some of these defined concepts geometrically. Addition can be thought of as follows: To add y to x (both now thought of as arrows in the plane), move the y arrow—preserving length and direction—to place its starting point at the terminal point of x. Then the terminal point of y, as placed, is the point corresponding to x + y. An example is given in Figure 6.1.3.

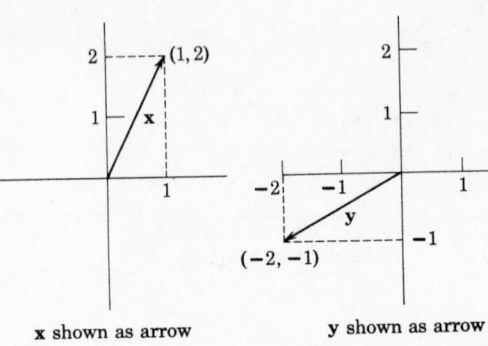
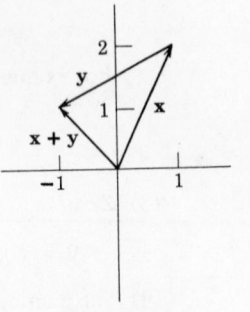

x shown as arrow   **y** shown as arrow   **x + y** shown

Figure 6.1.3

Note that in Figure 6.1.3,

$$x = (1, 2), y = (-2, -1)$$

Hence,

$$x + y = (-1, 1)$$

Consider, then, that if y = -x, we have the situation in Figure 6.1.4.

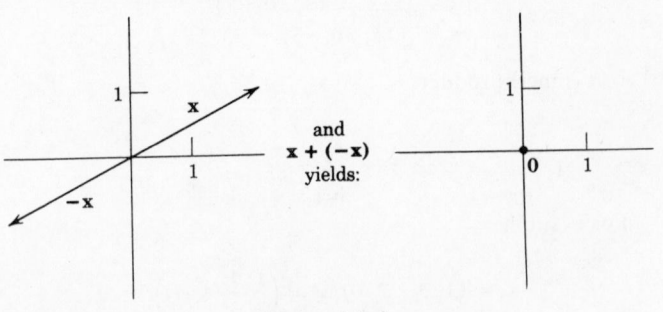

Figure 6.1.4

That is, -x is just a reflection of x through the origin; adding -x to x, arrow-wise, produces the zero point (vector).

Subtraction can be thought of as x + (-y). Arrow interpretation rests upon a succession of operations: flip y through the origin to obtain -y, then add in arrow-sense. An example is shown in Figure 6.1.5.

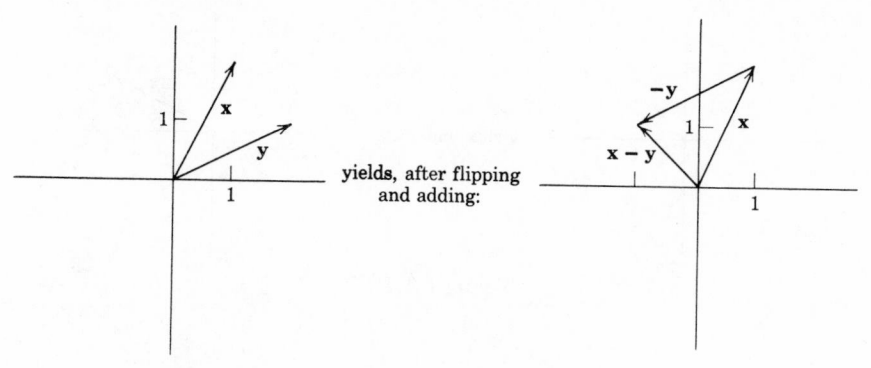

yields, after flipping
and adding:

Figure 6.1.5

Keeping in mind that the arrow representing $x - y$ need not start at the origin, we see in Figure 6.1.6 that we could obtain the difference directly instead of flipping and then adding.

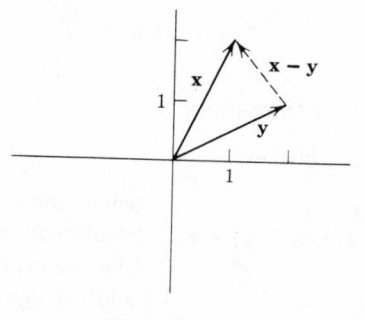

Figure 6.1.6

In general, the rule is that the arrow $x - y$ must be oriented such that by arrow addition we have $y + (x - y) = x$. This will be so if the start of $x - y$ is at the tip of y.

Scalar multiplication is interpretable as stretching or shrinking the arrow representation as indicated in Figure 6.1.7. In corresponding numerical terms, we have $x = (1, 1/2)$, $2x = (2, 1)$. Thus, $-2x$ would be a succession of operations: stretch by a factor of two, then flip through the origin. Or; flip through the origin, then stretch by a factor of two. Also, $(1/2)x$ would shrink by factor of two; that is, cut x down to half its length. If $x = (1, 1/2)$, for example, then $(1/2)x = (1/2, 1/4)$ as shown in Figure 6.1.8.

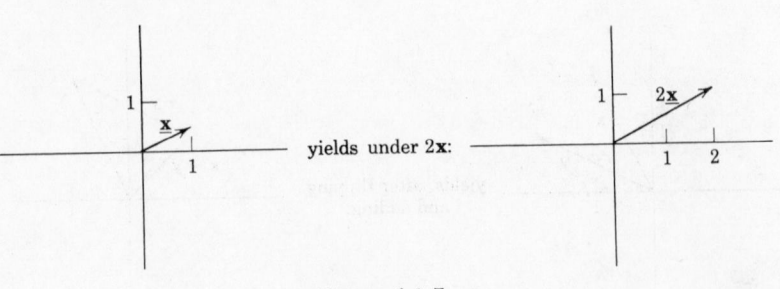

Figure 6.1.7

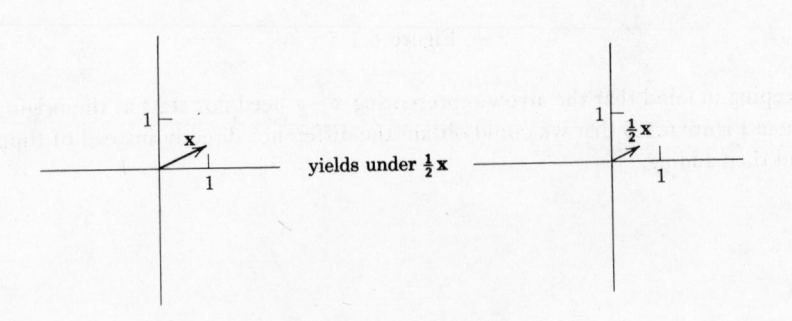

Figure 6.1.8

(6.1.2) Properties of Vector Operations.

(1) Properties of addition:

$$x + y = y + x \qquad \text{(commutative law)}$$
$$x + (y + z) = (x + y) + z \qquad \text{(associative law)}$$
$$x + 0 = x \qquad \text{(zero element)}$$
$$x + (-x) = 0 \qquad \text{(additive inverse)}$$

These properties show that $\Re \times \Re \times \ldots \times \Re$ (n times) is a commutative group under addition of n-tuples as the binary operation.

(2) Properties of scalar multiplication:

$$c(x + y) = cx + cy$$
$$(c_1 + c_2)x = c_1 x + c_2 x$$
$$c_1(c_2 x) = (c_1 c_2)x$$

(3) Properties of dot product:

$$x \cdot y = y \cdot x \qquad \text{(commutative law)}$$
$$x \cdot (y + z) = x \cdot y + x \cdot z \qquad \text{(distributive law)}$$

Note, however, that the dot product is not a binary operation in the sense of section 5.11, since it maps a pair of n-tuples into a number, not another n-tuple. A dot product gives a number, not a vector.

All the properties listed in (6.1.2) follow from the properties of numerical operations and the definitions in (6.1.1). For example, that addition is commutative is proved by

$$x + y = (x_1 + y_1, x_2 + y_2, \ldots, x_n + y_n) \qquad \text{(definition)}$$
$$= (y_1 + x_1, y_2 + x_2, \ldots, y_n + x_n) \qquad \text{(numerical + is commutative)}$$
$$= y + x \qquad \text{(definition)}$$

The properties may also be verified geometrically. For example, Figure 6.1.9 shows the first identity of part (2) of (6.1.2), with c = 1/2.

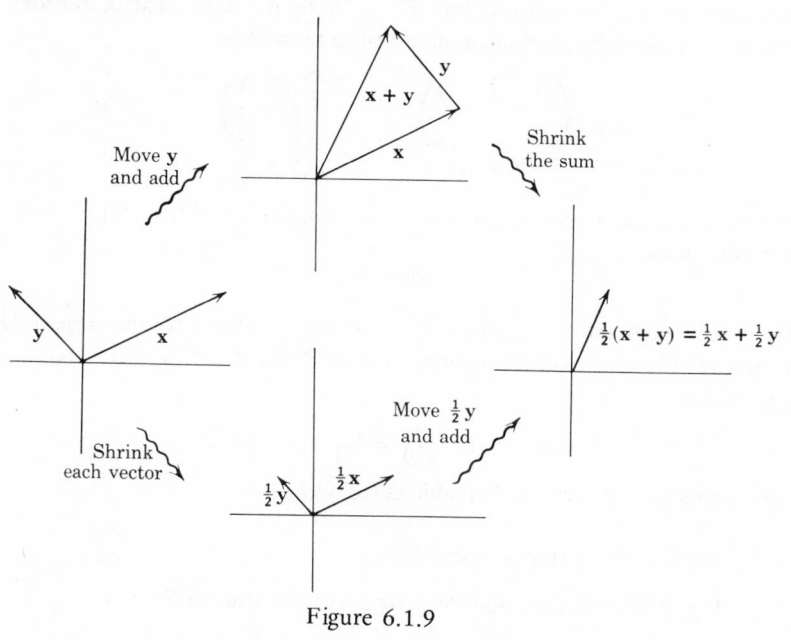

Figure 6.1.9

6.2. Matrices.

A matrix is a rectangular array of numbers that, for our purposes, are drawn from $\mathfrak{R}$, the set of all real numbers. In the matrix,

$$A = \begin{pmatrix} 7 & 3 & 1 \\ 4 & 2 & 8 \end{pmatrix},$$

we say that there are two rows and three columns. In fact, the rows are 3-tuples: (7, 3, 1) and (4, 2, 8). Also, the columns are 2-tuples, (7, 4), (3, 2), (1, 8). We call a typical element or entry a_{ij} if it is at the intersection of row i and column j. Thus, a_{12}, the entry in row 1, column 2, is equal to 3.

The generic notation for an arbitrary matrix is

$$\mathbf{A} = (a_{ij})$$

The letters i and j are mere logical variables. The first letter is called the row index: its domain is $\{1, 2, \ldots, m\}$, where $\mathbf{A}$ has m rows. The second letter is called the column index and has domain $\{1, 2, \ldots, n\}$ where $\mathbf{A}$ has n columns. Indifferently, one might write

$$\mathbf{A} = (a_{ij}), \mathbf{B} = (b_{jk}), \mathbf{C} = (c_{kl})$$

provided that in each case the first letter is the row index and the second is the column index for the given matrix. The order of a matrix means the number pair (m, n) of rows and columns, respectively. Since the number of entries is m x n, the order is often written "m x n" or "m by n." But a matrix of order n x m is generally different from a matrix of order m x n:

$$\begin{pmatrix} 7 & 3 & 1 \\ 4 & 2 & 8 \end{pmatrix} \qquad \begin{pmatrix} 7 & 4 \\ 3 & 2 \\ 1 & 8 \end{pmatrix}$$
$$(2 \times 3) \qquad\qquad (3 \times 2)$$

When we deal with an abstract matrix $\mathbf{A}$, we sometimes find it useful to show its order as follows:

$$\mathbf{A}^{m \times n}$$

Two matrices, $\mathbf{A}$ and $\mathbf{B}$, are identical, if and only if, they have the same order and they are identical in corresponding entries. Thus, if $\mathbf{A} = (a_{ij})$, $\mathbf{B} = (b_{ij})$ and both are m x n, then

$$\mathbf{A} = \mathbf{B} \quad \text{iff} \quad a_{ij} = b_{ij} \quad \text{(all i, j)}$$

Operations on matrices are definable as follows.

(6.2.1) *Definitions of Matrix Operations.*

(1) Addition is defined for matrices of the same order by

$$(a_{ij}) + (b_{ij}) = (a_{ij} + b_{ij})$$

This notation means "the typical (i, j) entry of the sum is the sum of the corresponding entries in the summands."

For example:

$$\begin{pmatrix} 1 & 6 \\ 0 & -1 \end{pmatrix} + \begin{pmatrix} 0 & 1 \\ -1 & 2 \end{pmatrix} = \begin{pmatrix} 1 & 7 \\ -1 & 1 \end{pmatrix}$$

(2) Negative:

$$-(a_{ij}) = (-a_{ij})$$

For example:

$$-\begin{pmatrix} 7 & 1 & -2 \\ -1 & 0 & 1 \end{pmatrix} = \begin{pmatrix} -7 & -1 & 2 \\ 1 & 0 & -1 \end{pmatrix}$$

(3) Zero:

$$\mathbf{0} = (0)$$

That is, the zero matrix of any order contains zero in each entry.

(4) Subtraction is defined if **A** and **B** are the same order by

$$(a_{ij}) - (b_{ij}) = (a_{ij} - b_{ij})$$

For example:

$$\begin{pmatrix} 6 & 1 \\ 7 & 0 \\ 8 & -1 \end{pmatrix} - \begin{pmatrix} -1 & 0 \\ 2 & 1 \\ -3 & 7 \end{pmatrix} = \begin{pmatrix} 7 & 1 \\ 5 & -1 \\ 11 & -8 \end{pmatrix}$$

(5) Scalar multiplication:

$$c(a_{ij}) = (ca_{ij}) \qquad (c \in \Re)$$

For example:

$$3 \begin{pmatrix} 1 & 7 \\ 2 & 0 \\ 7 & -3 \end{pmatrix} = \begin{pmatrix} 3 & 21 \\ 6 & 0 \\ 21 & -9 \end{pmatrix}$$

(6) Multiplication of matrices **A** and **B** requires that they first be conformable for multiplication, meaning that given

$$\mathbf{A}^{m \times n}$$

then we must have

$$\mathbf{B}^{n \times p}$$

that is, the number (n) of columns of **A** is identical to the number (n) of rows of **B**. This is so because we use the fact that rows of **A** can be taken to be n-tuples and that columns of **B** can be taken to be n-tuples. Then dot product (section 6.1) can be exploited:

$$\mathbf{AB} = \left(\sum_{k=1}^{n} a_{ik} b_{kj} \right)$$

The typical (i, j) entry of **AB** is the dot product of row i of **A**, which is an n-tuple since **A** has order $m \times n$, and column j of **B**, which is an n-tuple since **B** has order $n \times p$.

For example:

$$A = \begin{pmatrix} 7 & 1 \\ 0 & 4 \\ 2 & 0 \end{pmatrix} \quad B = \begin{pmatrix} 0 & -1 & 3 \\ 1 & 2 & 1 \end{pmatrix}$$

Here we have $A^{3 \times 2}$ and $B^{2 \times 3}$ —so conformable for the **AB** product:

$$AB = \begin{pmatrix} 7 \cdot 0 + 1 \cdot 1 & 7 \cdot (-1) + 1 \cdot 2 & 7 \cdot 3 + 1 \cdot 1 \\ 0 \cdot 0 + 4 \cdot 1 & 0 \cdot (-1) + 4 \cdot 2 & 0 \cdot 3 + 4 \cdot 1 \\ 2 \cdot 0 + 0 \cdot 1 & 2 \cdot (-1) + 0 \cdot 2 & 2 \cdot 3 + 0 \cdot 1 \end{pmatrix}$$

$$= \begin{pmatrix} 1 & -5 & 22 \\ 4 & 8 & 4 \\ 0 & -2 & 6 \end{pmatrix}$$

We note that **AB** is of order 3 × 3. In general,

$$A^{m \times n} B^{n \times p} = (AB)^{m \times p}$$

Multiplication is not commutative: order counts. Thus, let

$$A = \begin{pmatrix} 7 & 1 \\ 0 & 8 \end{pmatrix}$$

$$B = \begin{pmatrix} 2 & -1 \\ 1 & 3 \end{pmatrix}$$

$$AB = \begin{pmatrix} 14 + 1 & -7 + 3 \\ 0 + 8 & 0 + 24 \end{pmatrix} = \begin{pmatrix} 15 & -4 \\ 8 & 24 \end{pmatrix}$$

$$BA = \begin{pmatrix} 14 + 0 & 2 - 8 \\ 7 + 0 & 1 + 24 \end{pmatrix} = \begin{pmatrix} 14 & -6 \\ 7 & 25 \end{pmatrix}$$

and $AB \neq BA$.

(7) Identity matrix (always a square matrix, m = n):

$$I = (\delta_{ij})$$

where

$$\delta_{ij} = \begin{cases} 1 & \text{if} \quad i = j \\ 0 & \text{if} \quad i \neq j \end{cases}$$

and δ_{ij} is termed the Kronecker delta.
For example:

$$\begin{pmatrix} 1 & 0 \\ 0 & 1 \end{pmatrix} \quad \begin{pmatrix} 1 & 0 & 0 \\ 0 & 1 & 0 \\ 0 & 0 & 1 \end{pmatrix}$$

Thus, for each n, there is an identity matrix of that order: n × n.

(8) Inverse: Let A be a square matrix of order n x n. If there exists a matrix B of order n x n such that

$$AB = BA = I$$

we call B an (the) inverse of A. If an inverse exists, it is unique, so we can say "the" inverse. The inverse of A, if it exists, is denoted A^{-1}. Thus, by definition,

$$AA^{-1} = I = A^{-1}A$$

(6.2.2) *Properties of Matrix Operations.*

(1) Properties of addition:

$A + B = B + A$	(commutative law)
$A + (B + C) = (A + B) + C$	(associative law)
$A + 0 = A$	(zero)
$A + (-A) = 0$	(additive inverse)

Thus, the matrices of order n x m form a group with respect to the binary operation of addition. The operation is, moreover, commutative.

(2) Properties of scalar multiplication:

$$c(A + B) = cA + cB$$
$$(c_1 + c_2)A = c_1A + c_2A$$
$$c_1(c_2A) = (c_1c_2)A$$

(3) Properties of matrix multiplication:
   (a) For any matrices conformable for multiplication:

$A(BC) = (AB)C$	(associative law)
$AI = A$	(identity)
$A(B + C) = AB + AC$	(distributive law)
$(A + B)C = AC + BC$	

   (b) If the matrices are square and if inverses exist,

$$(A^{-1})^{-1} = A$$
$$(AB)^{-1} = B^{-1}A^{-1}$$

These properties follow from the properties of real numbers and the definitions (6.2.1). For example, the commutative property of matrix addition is shown by

$A + B = (a_{ij}) + (b_{ij})$	(notation convention)
$= (a_{ij} + b_{ij})$	(definition of matrix sum)
$= (b_{ij} + a_{ij})$	(commutative property of numerical +)

$$= (b_{ij}) + (a_{ij}) \qquad \text{(definition of matrix sum)}$$
$$= B + A \qquad \text{(notation convention)}$$

Similarly, to show that multiplication is associative, we write

$$A(BC) = (a_{ij})(p_{jk}) \qquad \text{(where } p_{jk} \text{ is the typical entry in the } BC \text{ product)}$$
$$= \left(\sum_j a_{ij} p_{jk}\right) \qquad \text{(definition of matrix product)}$$
$$= \left(\sum_j a_{ij} \sum_\ell b_{j\ell} c_{\ell k}\right) \qquad \left(\text{since } p_{jk} = \sum_\ell b_{j\ell} c_{\ell k}\right)$$
$$= \left(\sum_j \sum_\ell a_{ij} b_{j\ell} c_{\ell k}\right) \qquad \text{(distributive law for numbers)}$$
$$= \left[\sum_\ell \left(\sum_j a_{ij} b_{j\ell}\right) c_{\ell k}\right] \qquad \text{(commutative and distributive laws for numbers)}$$
$$= \left(\sum_\ell q_{i\ell} c_{\ell k}\right) \qquad \left(\text{where } q_{i\ell} = \sum_j a_{ij} b_{j\ell}, \text{ typical entry in } AB\right)$$
$$\overset{?}{=} (AB)C \qquad \text{(definition of matrix product)}$$

The second distributive law is shown by

$$(A + B)C = (a_{ij} + b_{ij})(c_{ij}) \qquad \text{(definition of matrix sum)}$$
$$= (s_{ij})(c_{ij}) \qquad \text{(letting } s_{ij} = a_{ij} + b_{ij})$$
$$= \left(\sum_k s_{ik} c_{kj}\right) \qquad \text{(definition of matrix product)}$$
$$= \left[\sum_k (a_{ik} + b_{ik}) c_{kj}\right] \qquad \text{(substituting for } s_{ik})$$
$$= \left[\sum_k (a_{ik} c_{kj} + b_{ik} c_{kj})\right] \qquad \text{(distributive law for numbers)}$$
$$= \left(\sum_k a_{ik} c_{kj} + \sum_k b_{ik} c_{kj}\right) \qquad \text{(distributive law for numbers)}$$
$$= \left(\sum_k a_{ik} c_{kj}\right) + \left(\sum_k b_{ik} c_{kj}\right) \qquad \text{(definition of matrix sum)}$$
$$= AC + BC \qquad \text{(definition of matrix product)}$$

From these properties of the two matrix operations, we see that if we consider addition only, we have a commutative group. Multiplication is not defined for an arbitrary pair of matrices. However, if we consider only square matrices of order n x n, then it follows that all such matrices are conformable for multiplication. Hence, multiplication of square matrices is a binary operation. Since it is associative, we have a semigroup, and the identity matrix I implies that in fact we have a monoid. Also, the distributive law connecting addition and multiplication holds. Hence, the three defining properties for a ring, given in definition (5.18.1), hold for square matrices. Hence, mathematicians speak of a "matrix ring." Because many square matrices do not have an inverse, it follows that this ring fails to be a field and in this way is distinguishable from the algebra of real numbers, in which every number except zero has an inverse (reciprocal).

A matrix of order m x 1 is called a column vector and a matrix of order 1 x n is called a row vector. The vectors we previously dealt with were n-tuples. For them the concepts of row and column are irrelevant. A "column vector" is defined as a matrix of a certain type; but it is a matrix. The correspondence between the n-tuples and these matrices is that to each n-tuple in $\Re \times \Re \times \ldots \times \Re$ (n times) there corresponds both a row vector and a column vector. The fact that the same n-tuple has two representations in matrix algebra is no cause for worry: it means that from a scientific standpoint (in which the n-tuple represents the possible results of n distinct measurement mappings) the distinction between column vector and row vector is trivial. We choose the one we happen to want for some calculating purpose.

The two representations of the n-tuple are called transposes of each other. We write

$$(x_1, x_2, \ldots, x_n)^T = \begin{pmatrix} x_1 \\ x_2 \\ \cdot \\ \cdot \\ \cdot \\ x_n \end{pmatrix}$$

Since row and column vectors are matrices, definitions (6.2.1) and properties (6.2.2) apply to them in operations with other matrices. In particular, a row vector of order 1 x n is conformable for multiplication with a matrix of order n x p, and the product is a row vector of order 1 x p. For instance,

$$(3, 2, -1) \begin{pmatrix} 1 & 0 \\ 1 & 1 \\ 2 & -1 \end{pmatrix} = (3, 3)$$

In general,

$$x^{1 \times n} A^{n \times p} = (xA)^{1 \times p}$$

Similarly, a matrix of order m x n is conformable for multiplication with a column vector x of order n x 1, producing a column vector of order m x 1. For instance,

$$\begin{pmatrix} 1 & 1 & 2 \\ 0 & 1 & -1 \end{pmatrix} \begin{pmatrix} 3 \\ 2 \\ -1 \end{pmatrix} = \begin{pmatrix} 3 \\ 3 \end{pmatrix}$$

In general,

$$B^{m \times n} (x)^{n \times 1} = (Bx)^{m \times 1}$$

A basic use of matrices is to operate by multiplication upon vectors to produce new vectors. Conceptually, this is a mapping of one n-tuple into another n-tuple, a generalization of the concept of function in the classical numerical sense. We shall see many examples later, but here is a brief example now.

Example. Let $x_t = (.30, .70)$ be a 2-tuple representing the fraction of people in two social classes at a given time t in a given two-class society. From social mobility analysis, it is known that of those in class I at a given time t, 90 percent remain in class I during a time interval Δt, while 10 percent move to class II. On the other hand, 20 percent of those in class II move to class I, while only 80 percent stay in class II in this time interval. We can represent this information in the form

$$
A = \begin{array}{c} \text{Class I} \\ \text{at t} \\[1em] \text{Class II} \\ \text{at t} \end{array}
\begin{pmatrix} .90 & .10 \\[1em] .20 & .80 \end{pmatrix}
$$

with column headings

$$
\begin{array}{cc} \text{Class I} & \text{Class II} \\ \text{at t+}\Delta t & \text{at t+}\Delta t \end{array}
$$

To find the fraction in class I at t+Δt, we reckon intuitively that we need to take 90 percent of the 30 percent and add 20 percent of the 70 percent:

$$.90(.30) + .20(.70) = .41$$

But this is just the dot product of $x = (.30, .70)$ with the first column of A. Similarly, to find the fraction in class II at t+Δt, we have to take 10 percent of the 30 percent and add 80 percent of the 70 percent, which is the dot product of $x = (.30, .70)$ with the second column of A. Hence,

$$x_{t+1} = x_t A = (.30, .70) \begin{pmatrix} .90 & .10 \\ .20 & .80 \end{pmatrix} = (.41, .59)$$

Hence, A has the effect of mapping the vector $x_t = (.30, .70)$ into the vector $x_{t+1} = (.41, .59)$. A matrix that operates upon proportions or probabilities in this manner is termed a transition matrix: it yields a transition from a "system state" at time t to a system state at t+1, where the n-tuples represent the state of the system. We shall deal in detail with such transition matrices in Chapter 10, after discussing processes and probability theory in Chapters 8 and 9, respectively.

The operation of transposition can be applied to each row (or column) vector in a matrix. Then, in an n x n matrix, the n row vectors, under transposition, become the n column vectors. For example,

$$
\begin{pmatrix} 1 & 3 & 0 \\ 4 & 0 & 7 \\ 2 & 1 & -1 \end{pmatrix} \mapsto \begin{pmatrix} 1 & 4 & 2 \\ 3 & 0 & 1 \\ 0 & 7 & -1 \end{pmatrix}
$$

Thus, we could call the image matrix the transpose of the given matrix. If $\mathbf{A}$ is what we begin with, then $\mathbf{A}^T$ is the transpose: the rows of $\mathbf{A}$ have become the columns of $\mathbf{A}^T$. This makes sense even if the number of rows and columns is not equal; for instance,

$$\begin{pmatrix} 6 & 1 \\ 4 & -3 \\ 2 & 0 \end{pmatrix} \mapsto \begin{pmatrix} 6 & 4 & 2 \\ 1 & -3 & 0 \end{pmatrix}$$

simply making rows into columns. Thus, if $\mathbf{A}$ is m x n, then $\mathbf{A}^T$ is n x m. Now we can regard the transposition action as defined over all possible matrices of numbers, with row vectors and column vectors regarded as 1 x n and n x 1 matrices, respectively. Also, it is clear that $(\mathbf{A}^T)^T = \mathbf{A}$.

Also, we note that

$$(\mathbf{x}\mathbf{A})^T = \mathbf{A}^T\mathbf{x}^T$$

which says that the transpose of the product of row vector $\mathbf{x}$ and matrix $\mathbf{A}$ is the product of the transposes $\mathbf{A}^T$ and $\mathbf{x}^T$. For example, letting

$$\mathbf{x} = (3, 2, -1) \qquad \mathbf{A} = \begin{pmatrix} 1 & 0 \\ 1 & 1 \\ 2 & -1 \end{pmatrix}$$

we saw above that $\mathbf{x}\mathbf{A} = (3, 3)$. On the other hand, we computed the product of the transposes and found $(3, 3)^T$.

6.3. Matrix Inversion. Computation of the inverse $\mathbf{A}^{-1}$ of matrix $\mathbf{A}$ is a very frequently encountered problem in applied mathematics. In this section, one procedure is outlined. A different procedure will be shown later. The abstract general statement of the formulas involved is given first, followed by explanations and examples.

(6.3.1) Inversion Procedure. Let $\mathbf{A}$ be a square matrix, $\mathbf{A} = (a_{ij})$. Define

(6.3.2) $$A_{ij} = (-1)^{i+j}M_{ij}$$

where

(6.3.3) M_{ij} = the determinant of the matrix remaining from $\mathbf{A}$ when row i and column j are deleted.

And let,

(6.3.4) $$\Delta = \text{the determinant of } \mathbf{A} \text{ (or "det } \mathbf{A}")$$

Then the inverse of $\mathbf{A}$ is given by,

(6.3.5) $$A^{-1} = \tfrac{1}{\Delta}(A_{ji})$$

Note that in formula (6.3.5), the notation A_{ji} is correct; in other words,

$$A^{-1} = \tfrac{1}{\Delta} (A_{ij})^T$$

Determinants are the main headache of the procedure. For a 2 x 2 matrix (a_{ij}) the determinant is

(6.3.6) $$\Delta = a_{11}a_{22} - a_{12}a_{21}$$

while, formally, a "1 x 1 matrix" or number, is its own determinant, and one can compute higher-order determinants by the relationship,

(6.3.7) $$\Delta = \sum_{i=1}^{n} a_{ir}A_{ir} \qquad \text{(r a fixed column)}$$

The A_{ij} are called cofactors, so that one calls this way of obtaining Δ the method of cofactors.

Examples of Determinant in the 2 x 2 Case.

(a)
$$A = \begin{pmatrix} 2 & 4 \\ 1 & 0 \end{pmatrix}$$

$$\det A = \Delta = a_{11}a_{22} - a_{12}a_{21} = 2 \cdot 0 - 4 \cdot 1 = {}^-4$$

(b)
$$A = \begin{pmatrix} 6 & 5 \\ 6 & 5 \end{pmatrix}$$

$$\det A = \Delta = 30 - 30 = 0$$

(c)
$$A = \begin{pmatrix} 4 & 8 \\ 3 & 6 \end{pmatrix}$$

$$\det A = \Delta = 24 - 24 = 0$$

(d)
$$A = \begin{pmatrix} a & ca \\ b & cb \end{pmatrix}$$

$$\det A = acb - cab = 0 \qquad \text{(all a, b, c)}$$

Example of Determinant in the 3 x 3 Case.

$$A = \begin{pmatrix} 1 & 3 & 0 \\ 4 & 1 & 2 \\ 0 & 2 & 1 \end{pmatrix}$$

We want to use (6.3.7), choosing the fixed column to minimize the number of nonzero summands. We take column 1 (r = 1). Thus,

$$\Delta = \sum_{i=1}^{3} a_{i1}A_{i1} = a_{11}A_{11} + a_{21}A_{21} + a_{31}A_{31}$$

or since $a_{11} = 1$, $a_{21} = 4$, and $a_{31} = 0$,

(6.3.8) $$\Delta = A_{11} + 4A_{21}$$

Now we need particular cofactors A_{11} and A_{21}, so we need to appeal to the definition (6.3.2).

We find that this means computing M_{11} and M_{21} as follows, since each of these is based on a 2 x 2 submatrix:

$$M_{11} = \det \begin{pmatrix} 1 & 2 \\ 2 & 1 \end{pmatrix} \qquad \text{(deleting row 1 and column 1 from } \mathbf{A})$$

$$= 1 - 4 = -3 \qquad \text{(definition of 2 x 2 determinant)}$$

$$M_{21} = \det \begin{pmatrix} 3 & 0 \\ 2 & 1 \end{pmatrix} \qquad \text{(deleting row 2 and column 1 from A)}$$

$$= 3 - 0 = 3 \qquad \text{(2 x 2 determinant)}$$

Thus,

$$A_{11} = (-1)^{1+1} M_{11} = -3$$
$$A_{21} = (-1)^{2+1} M_{21} = -3$$

Substituting into (6.3.8),

(6.3.9) $$\Delta = -3 + 4(-3) = -15$$

Examples of Inversion in the 2 x 2 Case.

(a) $$A = \begin{pmatrix} 2 & 4 \\ 1 & 0 \end{pmatrix}$$

$$\Delta = -4$$

                    (example (a) of 2 x 2 determinants)

$M_{11} = 0$, $M_{12} = 1$, $M_{21} = 4$, $M_{22} = 2$ (using 6.3.3)

$A_{11} = 0$, $A_{12} = -1$, $A_{21} = -4$, $A_{22} = 2$ (using 6.3.2)

$$A^{-1} = -\tfrac{1}{4} \begin{pmatrix} A_{11} & A_{21} \\ A_{12} & A_{22} \end{pmatrix} \qquad \text{(using 6.3.5)}$$

$$= -\tfrac{1}{4} \begin{pmatrix} 0 & -4 \\ -1 & 2 \end{pmatrix}$$

$$= \begin{pmatrix} 0 & 1 \\ \tfrac{1}{4} & -\tfrac{1}{2} \end{pmatrix}$$

(Check: compute AA^{-1} and look for product I)

(b) $$A = \begin{pmatrix} 6 & 5 \\ 6 & 5 \end{pmatrix}$$

$$\Delta = 0 \Rightarrow A^{-1} \text{ does not exist} \qquad \text{(check 6.3.5: no division by 0)}$$

(c) $$A = \begin{pmatrix} 4 & 8 \\ 3 & 6 \end{pmatrix}$$

$$\Delta = 0 \Rightarrow A^{-1} \text{ does not exist}$$

(d) $A = \begin{pmatrix} a & ca \\ b & cb \end{pmatrix}$

$\Delta = 0$, for all a, b, and c in $\Re \Rightarrow$
A^{-1} does not exist, for all a, b, and c in $\Re$

The work in the 2 x 2 case can be simplified by direct appeal to the following result.

(6.3.10) In general in the 2 x 2 matrix

$$A = \begin{pmatrix} a & b \\ c & d \end{pmatrix}$$

$$\Delta = ad - bc$$

$$A^{-1} = \frac{1}{\Delta} \begin{pmatrix} d & -b \\ -c & a \end{pmatrix}$$

This can be shown by appeal to (6.3.1).

Example of Inversion in the 3 x 3 Case.

$$A = \begin{pmatrix} 1 & 3 & 0 \\ 4 & 1 & 2 \\ 0 & 2 & 1 \end{pmatrix}$$

$$\Delta = -15 \qquad\qquad \text{[given by (6.3.9)]}$$

$$\begin{pmatrix} M_{11} & M_{12} & M_{13} \\ M_{21} & M_{22} & M_{23} \\ M_{31} & M_{32} & M_{33} \end{pmatrix} = \begin{pmatrix} -3 & 4 & 8 \\ 3 & 1 & 2 \\ 6 & 2 & -11 \end{pmatrix}$$

Here each M_{ij} is the determinant of a 2 x 2 submatrix of A. For example,

$$M_{12} = \det \begin{pmatrix} 4 & 2 \\ 0 & 1 \end{pmatrix} = 4 - 0 = 4$$

Then matrix (A_{ij}) differs from (M_{ij}) only in change of sign when $i + j$ is odd. Hence, after transposing (A_{ij}), we obtain

$$A^{-1} = -\frac{1}{15} \begin{pmatrix} -3 & -3 & 6 \\ -4 & 1 & -2 \\ 8 & -2 & -11 \end{pmatrix}$$

(One can check this result by matrix multiplication: we obtain $AA^{-1} = I$.)

6.4. Matrix Representation of Relations. It is sometimes convenient to use not only a graph representation of a binary relation but a matrix. This is especially so for computational considerations. The basic ideas of the representation follow (for extensive details, see Harary et al., 1965; for a briefer treatment, Bishir and Drewes, 1970).

Let (A, R) be a relational system. Label the objects in A with integers 1, 2, ..., n. Then define

$$a_{ij} = \begin{cases} 1 \text{ if object i in relation R to object j} \\ 0 \text{ if not} \end{cases}$$

Hence, $A_R = (a_{ij})$ is a matrix of zeros and ones corresponding to R. It is termed the adjacency matrix of R.

Example. (a) Let the relation be such that we have the graph in Figure 6.4.1.

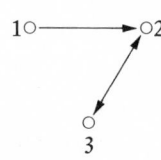

Figure 6.4.1

Then the adjacency matrix is given by

$$A_R = \begin{pmatrix} 0 & 1 & 0 \\ 0 & 0 & 1 \\ 0 & 1 & 0 \end{pmatrix}$$

Example. (b) The adjacency matrix of identity relation I is identity matrix **I**.

Given two relations on domain A, say R and S, we have two matrices A_R and A_S. The matrix representation will be useless unless we know how to calculate results corresponding to $R \cup S$ and $R \cap S$, for instance. In fact we want to find matrices corresponding to $R \cup S$, $R \cap S$, R^{-1}, and R o S.

Note that we want $A_{R \cup S}$ to have 1 in the (i, j) entry if, and only if, 1 appears in A_R or 1 appears in A_S in the (i, j) entry. If we merely add, then we could get for some entries $1 + 1 = 2$, which is not a proper value to enter into an adjacency matrix. Hence, we define a binary operation on $\{1, 0\}$ by the table:

$\oplus$	1	0
1	1	1
0	1	0

This is called Boolean addition: the sum is 1 if either term is 1.

Next we define a Boolean sum of two adjacency matrices; the sum is across corresponding entries, as in the usual matrix sum, but it involves the Boolean operation. This is denoted $A_R \oplus A_S$.

Example. Given the two relations in Figure 6.4.2,

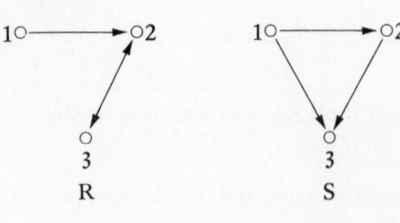

Figure 6.4.2

we obtain

$$A_R \oplus A_S = \begin{pmatrix} 0 & 1 & 0 \\ 0 & 0 & 1 \\ 0 & 1 & 0 \end{pmatrix} \oplus \begin{pmatrix} 0 & 1 & 1 \\ 0 & 0 & 1 \\ 0 & 0 & 0 \end{pmatrix} = \begin{pmatrix} 0 & 1 & 1 \\ 0 & 0 & 1 \\ 0 & 1 & 0 \end{pmatrix}$$

From this calculation we know we have $A_{R \cup S}$. Hence, the graph of $R \cup S$ may be drawn from the matrix and is given in Figure 6.4.3.

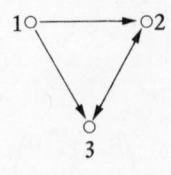

Figure 6.4.3

We see that, in general,

(6.4.1) $$A_{R \cup S} = A_R \oplus A_S$$

Next we consider the intersection matrix $A_{R \cap S}$. We want this to show a 1 in the (i, j) entry when, and only when, the (i, j) entries of both A_R and A_S are 1. Hence, we can use ordinary multiplication of 0 and 1:

·	1	0
1	1	0
0	0	0

Then, to obtain the proper product in matrix terms, however, we need to multiply corresponding entries, which is not, of course, ordinary matrix multiplication. Hence, we denote this operation, called "elementwise product," on matrices by x. Then the rule is: to obtain $A_{R \cap S}$, compute $A_R \times A_S$.

Example. Continuing the preceding example, we find

$$A_R \times A_S = \begin{pmatrix} 0 & 1 & 0 \\ 0 & 0 & 1 \\ 0 & 1 & 0 \end{pmatrix} \times \begin{pmatrix} 0 & 1 & 1 \\ 0 & 0 & 1 \\ 0 & 0 & 0 \end{pmatrix} = \begin{pmatrix} 0 & 1 & 0 \\ 0 & 0 & 1 \\ 0 & 0 & 0 \end{pmatrix}$$

Hence, the graph of $A_{R \cap S}$ is as shown in Figure 6.4.4.

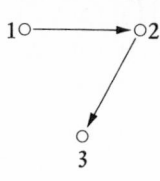

Figure 6.4.4

We see that, in general,

(6.4.2) $$A_{R \cap S} = A_R \times A_S$$

Now consider the inverse relation R^{-1} corresponding to relation R. We want the (i, j) entry of $A_{R^{-1}}$ to be 1 if, and only if, the (j, i) entry of A_R is 1. Hence, the adjacency matrix of the inverse relation is simply the transpose of the matrix of R:

(6.4.3) $$A_{R^{-1}} = A_R^T$$

Example.   Given R, as above, we have

$$A_{R^{-1}} = A_R^T = \begin{pmatrix} 0 & 1 & 0 \\ 0 & 0 & 1 \\ 0 & 1 & 0 \end{pmatrix}^T = \begin{pmatrix} 0 & 0 & 0 \\ 1 & 0 & 1 \\ 0 & 1 & 0 \end{pmatrix}$$

The graph of R^{-1}, constructed from this matrix, is shown in Figure 6.4.5.

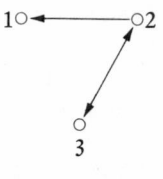

Figure 6.4.5

Consider the adjacency matrix $A_{R \circ S}$ of the composition of R and S. Recall that

$$i(R \circ S)j \quad \text{iff} \quad iRk \text{ and } kSj \qquad \text{(some } k)$$

Hence, we want a 1 in the (i, j) entry of $A_{R \circ S}$ if and only if there is a number k such that the (i, k) entry of A_R is 1 and the (k, j) entry of A_S is 1. This can be accomplished by matrix multiplication in the row-by-column sense but using Boolean addition.

Example.   Given A_R and A_S as above, we find

$$A_{RoS} = \begin{pmatrix} 0 & 1 & 0 \\ 0 & 0 & 1 \\ 0 & 1 & 0 \end{pmatrix} \begin{pmatrix} 0 & 1 & 1 \\ 0 & 0 & 1 \\ 0 & 0 & 0 \end{pmatrix} = \begin{pmatrix} 0 & 0 & 1 \\ 0 & 0 & 0 \\ 0 & 0 & 1 \end{pmatrix}$$

The graph of R o S is shown in Figure 6.4.6.

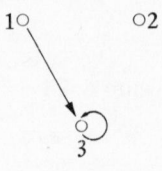

Figure 6.4.6

We introduce the notation $A_R A_S$ ($\oplus$) to mean matrix multiplication (row-by-column) with Boolean addition of the terms. Hence,

(6.4.4) $A_{RoS} = A_R A_S$ ($\oplus$)

We have shown the following correspondences in (6.4.1)-(6.4.4):

Relation	Adjacency Matrix	
$R \cup S$	$A_R \oplus A_S$	
$R \cap S$	$A_R \times A_S$	
R^{-1}	A_R^T	
$R \circ S$	$A_R A_S$	($\oplus$)

A point j is said to be reachable in one step from i if a line is directed from i to j or, equivalently, if iRj.

Point j is reachable in two steps from i if, for some point k, we have k reachable in one step from i and if j is reachable from k in one step or, equivalently, iRk and kRj. That is, iR^2j.

Point j is reachable from point i in three steps if there is some point k reachable from i in two steps and if j is reachable from k in one step or, equivalently, iR^2k and kRj. That is, iR^3j.

In general, we say that j is reachable from i in m steps if and only if iR^mj (m = 1, 2, . . .).

Finally, we introduce the convention that j is reachable from i in 0 steps if and only if i = j. Hence, 0-reachability is the identity relation I.

We say that j is reachable from i if, and only if, j is reachable from i in m steps, for some m $\geqslant$ 0: some path goes from i to j.

For an n-point graph, we can take n-1 as the maximum for m. Hence, if R_* denotes this reachability relation,

(6.4.5) $R_* = I \cup R \cup R^2 \cup R^3 \cup \ldots \cup R^{n-1}$

since iR_*j (meaning a path goes from i to j) is given by

$$iR_*j \quad \text{iff} \quad i = j \text{ or } iRj \text{ or } iR^2j \text{ or} \ldots \text{ or } iR^{n-1}j.$$

Hence, using (6.4.1)

(6.4.6) $\qquad A_{R_*} = I \oplus A_R \oplus A_{R^2} \oplus A_{R^3} \oplus \ldots \oplus A_{R^{n-1}}$

Now, from the fact that $A_{R \circ S} = A_R A_S$ $(\oplus)$, and taking S = R,

$$A_{R^2} = A_{R \circ R} = A_R A_R \ (\oplus) = A_R^2 \ (\oplus)$$
$$A_{R^3} = A_{R^2 \circ R} = A_{R^2} A_R \ (\oplus) = A_R^2 A_R \ (\oplus) = A_R^3 \ (\oplus)$$

and, in general, for any positive integer m,

(6.4.7) $\qquad A_{R^m} = A_R^m \ (\oplus)$

Hence, substituting (6.4.7) into (6.4.6), with Boolean addition understood,

(6.4.8) $\qquad A_{R_*} = I \oplus A_R \oplus A_R^2 \oplus A_R^3 \oplus \ldots \oplus A_R^{n-1}$

This matrix A_{R_*} is called the reachability matrix. If $i \neq j$, it shows a nonzero entry in the (i, j) position if and only if a sequence of directed lines can be traced from i to j on the graph of the relation R.

We define a communications relation based on R by saying that two points communicate if and only if they are mutually reachable:

(6.4.9) $\qquad iCj \quad \text{iff} \quad iR_*j \text{ and } jR_*i$

and since $iR_*^{-1}j$ is the same as jR_*i,

$$iCj \quad \text{iff} \quad iR_*j \text{ and } iR_*^{-1}j$$

or

(6.4.10) $\qquad C = R_* \cap R_*^{-1}$

Hence,

$$A_C = A_{R_* \cap R_*^{-1}}$$

and using (6.4.2)

$$A_C = A_{R_*} \times A_{R_*^{-1}}$$

Hence, from (6.4.3)

(6.4.11) $\qquad A_C = A_{R_*} \times A_{R_*}^T$

That is, the communications matrix is given by the elementwise product of the reachability matrix and its transpose.

Example. Given

$$A_R = \begin{pmatrix} 0 & 1 & 0 \\ 0 & 0 & 1 \\ 0 & 1 & 0 \end{pmatrix}$$

we find

$$A_R^2 \;(\oplus) = \begin{pmatrix} 0 & 1 & 0 \\ 0 & 0 & 1 \\ 0 & 1 & 0 \end{pmatrix} \begin{pmatrix} 0 & 1 & 0 \\ 0 & 0 & 1 \\ 0 & 1 & 0 \end{pmatrix} = \begin{pmatrix} 0 & 0 & 1 \\ 0 & 1 & 0 \\ 0 & 0 & 1 \end{pmatrix}$$

Hence, since n = 3, (6.4.8) becomes

$$A_{R_*} = I \oplus A_R \oplus A_R^2 = \begin{pmatrix} 1 & 0 & 0 \\ 0 & 1 & 0 \\ 0 & 0 & 1 \end{pmatrix} \oplus \begin{pmatrix} 0 & 1 & 0 \\ 0 & 0 & 1 \\ 0 & 1 & 0 \end{pmatrix} \oplus \begin{pmatrix} 0 & 0 & 1 \\ 0 & 1 & 0 \\ 0 & 0 & 1 \end{pmatrix}$$

$$= \begin{pmatrix} 1 & 1 & 1 \\ 0 & 1 & 1 \\ 0 & 1 & 1 \end{pmatrix}$$

This shows point 1 is reachable from no other point, while every point can reach points 2 and 3. The reader should check this computation by tracing sequences of lines in the graph of R in Figure 6.4.2. The communications matrix is next obtained:

$$A_C = A_{R_*} \times A_{R_*}^T = \begin{pmatrix} 1 & 1 & 1 \\ 0 & 1 & 1 \\ 0 & 1 & 1 \end{pmatrix} \times \begin{pmatrix} 1 & 0 & 0 \\ 1 & 1 & 1 \\ 1 & 1 & 1 \end{pmatrix} = \left(\begin{array}{c|cc} 1 & 0 & 0 \\ \hline 0 & 1 & 1 \\ 0 & 1 & 1 \end{array}\right)$$

We see that the set of points is partitioned. This is always true and follows from the properties of the C relation, as we now show.

It is clear that the communications relation is symmetric:

iCj	iff	iR_*j and jR_*i	(definition of C)
	iff	jR_*i and iR_*j	(tautological equivalence)
	iff	jCi	(definition of C)

Also, C is transitive: suppose that iCj and jCk; then,

$$iR_*j \text{ and } jR_*i \text{ and } jR_*k \text{ and } kR_*j$$

Since iR_*j and jR_*k there is a sequence of directed lines from i to j and from j to k. Hence, the two sequences together yield the fact that k is reachable from i, or iR_*k. Similarly, from kR_*j and jR_*i we conclude that i is reachable from k, via point j. Hence, kR_*i. Since both iR_*k and kR_*i, we conclude from the definition of C that iCk. Hence, C is transitive.

Finally, the 0-reachability convention makes C reflexive. From our previous work (section 4.12), we conclude that C is an equivalence relation. Thus, C partitions the set of points of the graph or, equivalently, objects in the domain of relation R. Each class contains mutually reachable points.

If there is only one such class, so that all points are mutually reachable, the relation R on set A is said to be strongly connected. In terms of the graph, this means we can trace a path of directed lines from any point to any other point. In the matrix representation, if matrix A_C contains no zero entry then the relation is strongly connected. In the example above, the relation is not strongly connected.

The graph representation suggests a weaker notion of connectivity: in tracing lines, ignore the particular direction on each link. Formally, we define a relation S by

$$iSj \quad iff \quad iRj \quad or \quad jRi$$

so that

$$S = R \cup R^{-1}$$

In all the formulas above, we now replace R by S. If the final communications matrix contains no zero entry this means that S is strongly connected. By definition, we then say that R is weakly connected.

In the example,

$$A_S = A_R \oplus A_R^T$$

$$= \begin{pmatrix} 0 & 1 & 0 \\ 0 & 0 & 1 \\ 0 & 1 & 0 \end{pmatrix} \oplus \begin{pmatrix} 0 & 0 & 0 \\ 1 & 0 & 1 \\ 0 & 1 & 0 \end{pmatrix} = \begin{pmatrix} 0 & 1 & 0 \\ 1 & 0 & 1 \\ 0 & 1 & 0 \end{pmatrix}$$

The computation of

$$A_{S_*} = I \oplus A_S \oplus A_S^2$$

yields a matrix with no zero entry. In turn, this implies a communications matrix with no zero entry. Hence, S is strongly connected. We conclude that R is weakly connected, even though it is not strongly connected.

An extensive treatment of connectivity is given by Harary, Norman, and Cartwright (1965). The matrix representation of relations will be used subsequently in Markov chain theory (Chapter 10) and in social structural analysis (Chapter 19).

SUGGESTIONS FOR FURTHER READING (PART ONE)

Chapter 2. A text in logic that is filled with examples from empirical science is by Suppes (1957). Also recommended is Tarksi (1946). At a very technical level, there is Carnap (1958).

Chapter 3. A good text in set theory is that by Zehna and Johnson (1962). The second part of Suppes (1957) introduces the idea of axiomatization within set theory, a procedure followed informally in this book but most formally exemplified in game theory (section 21.4).

Chapter 4. Relation theory is dealt with in Kurosh (1963), as well as in Suppes (1957) and Tarksi (1946). The graph theoretic representation is comprehensively covered in Harary, Norman, and Cartwright (1965).

Chapter 5. A compact introduction to modern algebra may be found in Hu (1965), including mappings. An elegant survey of the field from a very modern standpoint is given by MacLane and Birkhoff (1967). Again, Kurosh (1963) can also be recommended.

Chapter 6. For vectors and matrices one might study the relevant chapters of Kemeny, Snell, and Thompson (1966). The matrix representation of graphs is developed in Harary, Norman, and Cartwright (1965). The presentation of matrix algebra and the study of adjacency matrices is carried out in Bishirs and Drewes (1970).

2 MEASUREMENT, PROCESS, AND PROBABILITY

CHAPTER SEVEN BASIC MEASUREMENT CONCEPTS

7.1. Introduction. The purpose of this chapter is to outline, with examples, the basic ideas of the modern (formalized) theory of measurement. The basic work by Scott and Suppes (1958), subsequently extended and clarified in Suppes and Zinnes (1963), is the basis for this chapter, but the summary monograph by Pfanzagl (1968) is also of importance. As opposed to "scaling theory" the "theory of basic measurement" aims to present the conceptual basis for measurement.

For subsequent reference we note that scientific concepts may be classified in a preliminary manner as follows, using the familiar scheme suggested by Hempel (1952):

(1) qualitative
    (a) classificatory
    (b) comparative
(2) quantitative

Some physical examples are types of metals (classificatory), hardness (comparative), and length (quantitative). Some social examples are types of jobs (classificatory), power (comparative), and group size (quantitative).

7.2. Numerical Assignments. Measurement means the assignment of numbers to represent objects of experience. Measurement, that is, involves numerical assignment.

Some numerical assignments arise by using previously given numerical ideas. For example, population density may be defined as the number of population units per unit area. Two given quantitative concepts, number of units and area, are presumed in this definition. Area itself can be defined in terms of length. But length is definable by appeal to experience with objects: empirical relations and operations are supplied to explicate "length." Thus, some numerical assignments are derived (e.g., population density) and others are fundamental (e.g., length). The term "fundamental" does not imply "highly significant." It is a mere technical term for the category of numerical assignment in which the justification for the use of the numbers as they are to be used is given by appeal to non-numerical relations and operations.

The classificatory schema "derived-fundamental" is independent of the scheme in terms of qualitative-quantitative. Thus, if each person in a group is assigned a code number for convenience, a fundamental numerical assignment has been made. This illustrates not only that "fundamental" does not imply "highly significant" but also that it is not necessarily referent to "quantitative" in the sense of the first classification. To some extent, the theory of measurement provides a conceptually more precise way of using the schema: "classificatory-comparative-quantitative." Put another way, scientific intuition and common sense begin with this schema; measurement theory builds upon and refines this intuition. In the following we first attend to fundamental measurement.

7.3. The Scale Concept. We say that we know a scale when we know (1) a specified empirical system, (2) a system of numbers, and (3) a numerical assignment to represent the system (1) by the system (2). To make this clear, we use the formalisms of the earlier chapters.

An empirical system is a set E of concrete systems, or empirical objects, together with a list of relations or operations defined over E. This can be made even more precise if we recall (see section 5.11) that a binary operation is a mapping of a set of the form A x A into A, so that under the operation,

$$(x, y) \mapsto x \circ y \in A$$

This was the representation of the intuitive idea: x and y are combined, in the order (x, y), via operation o, to produce another object of the same kind, denoted x o y. Thus, an empirical operation is exemplified by any of the following:

(1) x o y   iff   physical objects x and y are placed together on a pan of a balance, the result being regarded as a composite physical object.

(2) x o y   iff   alternatives x and y are made the two possible outcomes of a gamble, itself to be taken as an alternative in some choice situation.

(3) x o y   iff   entities x and y form a unit in Heider's (1958) theory, so that the unit is an entity with two parts x and y.

Empirical relations were exemplified extensively in chapter 4. A typical empirical system for a comparative concept has the form

$$(E, R)$$

where R is an order relation on E. For example,

E = a set of metallic objects
xRy   iff   x scratches y

will serve to provide an empirical basis for the comparative concept of hardness.
A typical empirical system for a classificatory concept has the form

$$(E, \equiv)$$

where $\equiv$ is an equivalence relation on E. For example,

E = a set of people
x ≡ y   iff   x and y have the same job

provides a job classification of people.
A typical empirical system for a quantitative concept has the form

$$(E, R, \equiv, o)$$

where R relations and $\equiv$ are on E and o is a binary operation on E. For example,
using a balance to measure mass,

E = set of stones
xRy   iff   the pan with x falls below the pan with y
x ≡ y   iff   the pan with x balances the pan with y
x o y   iff   x and y are together on the same pan

which serves as an empirical system for the determination of mass (for stones).
These three types of empirical systems are themselves a bit oversimplified for
expository purposes; moreover, the concept of empirical system is far wider than
these examples indicate. Finally, the determination of whether a scale provides a
"quantitative" determination of a concept has much more flexibility to it than
indicated by an association of only certain forms [e.g., $(E, R, \equiv, o)$] with certain
types of concepts. We will return to a more exact treatment of this matter in a
later portion of this chapter. For the moment, the aim is to concretize the idea
of an empirical system in connection with its set-theoretical formulation.

Returning now to the item (2) of a scale, the numerical system, we mean here
that the basic domain is $\Re$, the set of real numbers or one of its subsets and there
are relations and operations defined over $\Re$ that are explicitly appealed to. For
example, the following are numerical relational systems:

$$(\Re, >)$$
$$(\Re, =)$$
$$(\{1, 2, 3\}, <)$$

The following are numerical algebras:

$$(\Re, +)$$
$$(\Re, \cdot)$$
$$(\{x: x \text{ is a positive integer}\}, +)$$

$$(\{x: x \text{ is an integer}\}, \cdot)$$

where "$\cdot$" refers to multiplication of numbers. The following are numerical systems:

$$(\Re, >, +)$$
$$(\Re, <, \cdot)$$
$$(\Re, +)$$

so that a numerical system is simply the "real line," $\Re$, together with defined numerical relations or operations.

Finally, item (3) in the notion of a scale says that we need a numerical assignment from the empirical to the numerical system.

But an "assignment" is just an association of a particular number to each object in E. For example,

(a) Assign a number n(p) to each person p in a set E of persons.
(b) Assign a number m(s) to each object s in a set E of physical objects.
(c) Assign a number u(A) to each alternative A in a set E of alternatives.

Thus, an assignment is a mapping. The domain is the set E of concrete systems in the empirical relational system, and the range is the set of numbers of the numerical relational system. However, we recall that we want the assignment to fulfill a basic purpose: "to represent system (1) by system (2)." That is, we want the relations and operations on E to be represented by the relations and operations on the real number system $\Re$. This means that we want to use numerical relations for empirical relations, and so, the mapping must "preserve" the structure of the empirical relation. Similarly, it must preserve the structure of the empirical operation. Recalling the ideas of sections 5.14–5.16, we can say that we want the mapping to be a homomorphism for any empirical relation and a homomorphism for any empirical operation.

For example, we want $x > y$, where x and y are numbers assigned to persons A and B, respectively, to represent the empirical relation ARB, where R is a suitable (social) order relation. Similarly, we want $z = x + y$, where x, y, and z are in $\Re$, to represent the fact that when A (assigned x) and B (assigned y) are placed on one pan of a balance, they just balance an object C assigned the number z. In other words, the numerical identity relation represents the empirical equivalence relation.

We sum up as follows:

(7.3.1) *Definition.* By a scale we mean

$$X: \varepsilon \to \Re$$

where (1) ε is an empirical system over a domain E of concrete systems, (2) $\Re$ is a numerical system over the real line $\Re$, and (3) X is a homomorphism for any relation or operation in $\mathcal{E}$. If $X: \mathcal{E} \rightarrow \mathfrak{N}$ is a scale, then we call X a scientific variable. Thus, a scientific variable is a homomorphic mapping from an empirical system to a numerical system.

Example. (a) Let

$$\mathcal{E} = (E, \equiv)$$

where

$$E = \{\text{architect, plumber, lawyer}\}$$

and let $\equiv$ be an equivalence relation on E defined by

$$a \equiv b \quad \text{iff} \quad a \text{ and } b \text{ are "professions"} \quad (\text{all } a, b \in E)$$

Then, define $X: \mathcal{E} \rightarrow \mathfrak{N}$ by defining a mapping such that it has domain E, has range the real line, and preserves or represents the relation $\equiv$ on E. This can be done in many ways. One way is

$$
\begin{aligned}
X(\text{architect}) &= 1 \\
X(\text{plumber}) &= 0 \\
X(\text{lawyer}) &= 1
\end{aligned}
$$

so that the system $\mathfrak{N}$ is given by

$$\mathfrak{N} = (\mathfrak{R}, =)$$

in the sense that numerical identity represents empirical equivalence. The statement

$$X(a) = X(b) \quad \text{iff} \quad a \equiv b$$

where a, b vary in E, shows that X is a relational homomorphism. Briefly, identity of numbers is made to represent empirical equivalence (in the specified sense).

Example. (b) Let

$$\mathcal{E} = (E, R)$$

where for all $a, b \in E$,

$$E = \{\text{doctor, nurse, orderly}\}$$

and

$$aRb \quad \text{iff} \quad \text{from the perspective of any occupants, position } a \text{ is more important than position } b \quad (\text{all } a, b \in E)$$

Suppose that

$$R = \{(\text{doctor, nurse}), (\text{doctor, orderly}), (\text{nurse, orderly})\}$$

and define X by

$$X(doctor) = 1$$
$$X(nurse) = 2$$
$$X(orderly) = 3$$

Then,

$$\mathfrak{N} = (\mathfrak{R}, <)$$

and $<$ is representing R on the real line. X is a relational homomorphism, since

$$X(a) < X(b) \quad \text{iff} \quad aRb$$

holds for all a, b in E.

Example. (c) Let

$$\varepsilon = (E, I, P)$$

where

$$E = \{\text{North Hills, South Hills, Inner City}\}$$

so that E is the residential characteristic C of section 5.7, as the basis for a status differentiation. Let A be the set of actors,

$$A = \{\text{Tom, Dick, Harry}\}$$

We now review the evaluative aspect of the status mapping in the light of measurement theory. Assume that over A evaluative uniformity holds in that the identical I and P are valid for each actor, where for each a, b $\in$ E,

$$aIb \quad \text{iff} \quad \text{actor } \alpha \text{ is indifferent as between a and b}$$
$$aPb \quad \text{iff} \quad \text{actor } \alpha \text{ prefers a to b}$$

and suppose that, in obvious abbreviations,

$$I = \{(N, N), (S, S), (C, C), (N, S), (S, N)\}$$
$$P = \{(N, C), (S, C)\}$$

Now define X by

$$X(N) = 1$$
$$X(S) = 1$$
$$X(C) = 0$$

and observe the following statements are true for all a, b in E,

$$(1) \quad X(a) = X(b) \quad \text{iff} \quad aIb$$
$$(2) \quad X(a) > X(b) \quad \text{iff} \quad aPb$$

Here statement (1) indicates that X is a relational homomorphism for the I relation on E, representing it in terms of = on the real line; statement (2) shows that X is a relational homomorphism for P, representing it in terms of $>$ on the real line. Thus,

$$\mathfrak{N} = (\mathfrak{R}, =, >)$$

is the numerical system for the scientific variable X: $\varepsilon \to \mathfrak{N}$, whose value for any $a \in E$ represents various relational facts about a and other elements of E.

Now, if we divide E by I, as in section 4.12, we obtain the empirical relational system

$$\varepsilon^* = (E/I, P^*)$$

where

$$E/I = \Big\{ \{N, S\}, \{C\} \Big\}$$
$$P^* = \Big\{ \langle \{N, S\}, \{C\} \rangle \Big\}$$

so that E/I contains only the evaluatively significant distinction shown by P^*. Taking ε^* as the empirical system, we could let X^* be defined by

$$\{N, S\} \mapsto 1$$
$$\{C\} \mapsto 0$$

and so, letting $\bar{N} = \bar{S} = \{N, S\}$ and $\bar{C} = \{C\}$,

$$X^*(\bar{a}) > X^*(\bar{b}) \quad \text{iff} \quad \bar{a}\, P^* \bar{b} \quad \text{(all } \bar{a}, \bar{b} \in E/I)$$

shows that X^* is a relational homomorphism into the numerical system

$$(\mathfrak{R}, >)$$

but now X^* is injective (i.e., distinct objects map into distinct numbers), while the original X was not injective. The empirical difference is null; it is merely that the division creating ε^* from ε permits us to assert that the empirical system is represented by an isomorphic image in a subset of the real line. The isomorphism emerges out of the purely formal fact that X^* is injective and X is not. The isomorphic image of the system

$$\varepsilon^* = (E/I, P^*)$$

is the system

$$\mathfrak{N}^* = (\{1, 0\}, >)$$

that is, the subset $\{1, 0\}$ of the real numbers, and $>$ as restricted in extension to $\{1, 0\}$. We may say that $\mathfrak{N}^*$ is a subsystem of $\mathfrak{N} = (\mathfrak{R}, >)$. Thus, by the method of quotient sets applied to empirical domains using empirical equivalence relations, we can restore to formal validity the intuitive idea that fundamental measurement means establishing an isomorphic correspondence with a numerical system.

Example. (d) For a final example, let us consider an empirical system with a component binary operation. Let

$$\mathcal{E} = (E, \equiv, R, o)$$

where E is a set of sticks of varying lengths, and for all sticks, a, b $\in$ E, define

a $\equiv$ b	iff	a coincides with b
aRb	iff	a extends over b
a o b	iff	a is extended by adjoining b at its right endpoint

Suppose that an experiment on E is performed realizing these relational and operational concepts and that as a result we determine that

(1) $\equiv$ is an equivalence relation on E
(2) R is irreflexive, asymmetric, transitive
(3) if not-(a $\equiv$ b), then aRb or bRa
(4) o is such that on E/$\equiv$ (so that we deal with classes $\bar{a}$, $\bar{b}$, and $\bar{c}$)
    (a) it is commutative: $\bar{a} \, o \, \bar{b} = \bar{b} \, o \, \bar{a}$
    (b) it is associative: $(\bar{a} \, o \, \bar{b}) \, o \, \bar{c} = \bar{a} \, o \, (\bar{b} \, o \, \bar{c})$
    (c) if $\bar{a}R\bar{b}$ and $\bar{c}R\bar{d}$, then $(\bar{a} \, o \, \bar{c})R(\bar{b} \, o \, \bar{d})$

and so forth (exact conditions may be found in the cited references on measurement theory). As a result we see that the empirical system $\mathcal{E}$ has a very rich and lawful structure. For instance, condition (4)(a) means that if you take any one of the sticks coincident with (having the "same length as") a and extend it by any b-equivalent stick, the resulting stick is coincident with the result obtained by first taking any stick equivalent to b and extending it by a stick that is equivalent to a. Thus, the equivalence classes exhibit highly homogeneous behavior vis-à-vis the operation.

From $\mathcal{E} = (E, \equiv, R, o)$ we can construct the quotient system[†]

$$\mathcal{E}^* = (E/\equiv, R, o)$$

and represent this by

$$\mathfrak{N} = (\mathfrak{R}, >, +)$$

using a mapping X such that,

$$X(\bar{a}) > X(\bar{b}) \quad \text{iff} \quad \bar{a} \, R \bar{b}$$
$$X(\bar{a} \, o \, \bar{b}) = X(\bar{a}) + X(\bar{b})$$

so that X is both a relational homomorphism and a homomorphism for the operation on E/$\equiv$. Note that condition (4)(c) above is represented by the true numerical statement

---

[†]We omit the asterisks on R and o defined on classes, in conformity with (4) above.

$$\text{if } x > y \text{ and } x' > y', \text{ then } x + x' > y + y'$$

Finally, in this example, we need to choose a unit. This can be done by purely practical considerations. Let u be in E, and then

$$X(\bar{u}) = 1$$

is required. Give the standard unit a name. For novelty, call it the new-unit. Then, for any stick u' in $\bar{u}$ (so that $u' \equiv u$, meaning u' coincides with u), we say that u' is a new-unit. (For example, any stick that would coincide with the standard inch stick is an inch.) Then for any stick-class $\bar{a}$ in E/$\equiv$, we have

$X(\bar{a}) = x$	(the image of $\bar{a}$ is some number x)
$= x \cdot 1$	(numerical law $x = x \cdot 1$)
$= x \cdot X(\bar{u})$	(1 is the value assigned to the unit)

Suppose, for simplicity that $x = 3$. Then,

$X(\bar{a}) = 3X(\bar{u})$	(putting $x = 3$)
$= X(\bar{u}) + X(\bar{u}) + X(\bar{u})$	[meaning of 3 times a number $X(\bar{u})$]
$= X(\bar{u} \circ \bar{u} \circ \bar{u})$	(using homomorphism property of mapping X)

and now since identity on the real line corresponds to identity in E/$\equiv$, we conclude

$$\bar{a} = \bar{u} \circ \bar{u} \circ \bar{u}$$

Thus,

$$a \equiv u_1 \circ u_2 \circ u_3 \qquad (u_i \in \bar{u})$$

which means that a coincides with the juxtaposition of three new-units. Thus, this is the empirical meaning of the numerical shorthand statement "a is three new-units" or "The length of a is 3 new-units."

A numerical property, even at the ratio level of measurement, then, is not referent to something inherent in the empirical object but instead is referent to a whole unlimited class of relational possibilities for this object vis-à-vis other objects. These relational possibilities (in the example above, of coincidence and extension over) are realized in our experience only occasionally, as we need them, but they help define the extensive character of the experienced world.

In the above examples, we indicated in concrete detail in what way scales are structure-preserving mappings. The question arises, however, as to how far the empirical structure goes in the determination of the numerical assignment X. If X can be virtually any mapping that merely preserves empirical equivalence, then the numbers obtained via X have no special priority over other numerical names for the classes. On the other hand, X may be determined uniquely. Or, it may be determined except for the practical choice of a standard unit. Thus, we need to discuss the uniqueness of X as well as its existence.

7.4. Scale-Types. We define now the concept of a scale-type. If X is a homomorphism out of ε into $\mathfrak{N}$, it need not be the only possible such structure-preserving mapping. Let X' be another such mapping. For example, X might be centigrade, X' Fahrenheit. This means that X and X' are representing the same empirical system. Any differences between them are empirically irrelevant. Thus, we have

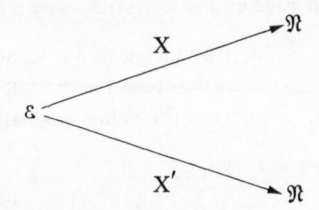

and the question is, How are any two such assignments related? Obviously they must be related by a mapping from $\mathfrak{N}$ to $\mathfrak{N}$, that is, by a real-valued function of a real variable.

Scales are classified by the character of such functions, which are called admissible transformations.

(7.4.1) Definitions. To say that, a scale is

(1)	absolute	means	X' = X	
(2)	ratio	means	X' = cX	($\exists c > 0$)
(3)	interval	means	X' = cX + b	($\exists c > 0$)
(4)	ordinal	means	X' preserves the order on X	
(5)	nominal	means	X' preserves the identity on X	

for any X, X' that are scales for a given empirical system, ε.

We are essentially talking about the type of mapping f that can be employed in the diagram

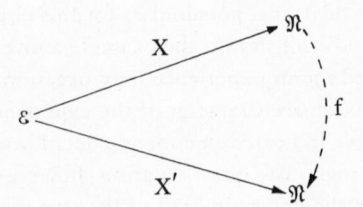

such that

(7.4.2) $f(X) = X'$

which, since X and X′ are functions, is shorthand for composition of functions (see section 5.6) and means that for any object $A \in E$,

$$(f \text{ o } X)(A) = X'(A)$$

that is,

$$A \mapsto (f \text{ o } X)(A)$$

and

$$A \mapsto X'(A)$$

are the identical mapping. We say that the function f "makes the diagram commutative" when it satisfies (7.4.2). Thus, given X, applying f will take us into another perfectly acceptable way of talking about ε.

Definition (7.4.1) shows that the only admissible transformation f for an absolute scale is "leave it alone." Here $f(x) = x$, all x in $\Re$. For example, count the number of people in a room. This is an absolute scale: there is no other numerical system for talking about this number; it is unique.

A ratio scale has the admissible transformations

$$f(x) = cx, \qquad (x \in \Re, c > 0)$$

which, geometrically, are straight lines through the origin. For example, if we write

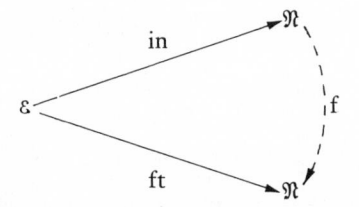

we mean that the "inch" mapping and the "foot" mapping are both just equivalent ways of talking about a certain empirical system, ε, and have

$$f(x) = \frac{x}{12} \qquad (\text{all } x \in \Re)$$

For example, let a certain object A be in E, the domain of ε. Then suppose that

$$in(A) = 24$$
$$ft(A) = 2$$

then,

$$(f \text{ o } in)(A) = f[in(A)] = f[24] = \frac{24}{12} = 2 = ft(A)$$

where the atrocious notation is used to make the point that scaling in inches and scaling in feet are two different scales defined on the same empirical system. Note that we might be given

$$\varepsilon \xrightarrow{\text{in}} \Re$$

and want to use smaller numbers; therefore, we might invent a new unit. Call this unit "un" for the moment. Let us say that what we want is that something that measures 36 under "in" comes out as measuring 1 under "un." So we write

$$un(A) = 1 \quad \text{iff} \quad in(A) = 36$$

Thus, the diagram for an object A^* which is 36 inches is

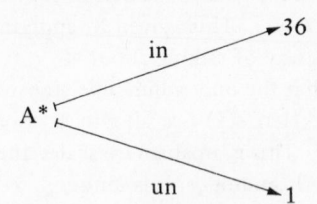

which yields the correspondence, to hold under our transformation,

$$36 \mapsto 1$$

from which it follows

$$f(x) = \frac{1}{36} x \qquad (\text{all } x \in \Re)$$

and we have the new scale completely defined by

$$un(A) = \frac{1}{36} in(A)$$

which makes it clear that "un" is "yard." To find the number of yards assigned to object A, divide the number of inches assigned to A by 36. Here the object A^* that maps to 1 under "un" or "yard" can be regarded as "a yard." The object, say A^+, that maps to 1 under "in" can be regarded as "an inch." Then the new unit is essentially established, with the corresponding empirical meaning:

$$\underbrace{A^+ \text{o } A^+ \text{o} \ldots \text{o } A^+}_{36 \text{ times}} \equiv A^*$$

where o designates empirically placing copies of A^+ end to end and $\equiv$ designates empirically the fact that the resulting object neither extends over A^* nor is it extended over by A^*. Thus, o and $\equiv$ refer to relations and operations in the empirical world. Numerical statements, such as

$$36 \text{ in.} = 1 \text{ yd.}$$

are shorthand formulas for the complex of potentialities in the empirical world as determined by the empirical relations and operations in the empirical system.

Interval scales differ from ratio scales in that the function f can have the form

$$f(x) = cx + b \qquad \text{(all } x; c > 0)$$

and so (when b = 0) includes all the admissible transformations of a ratio scale as a subset: the class of admissible transformations is widened and thereby, any particular scale becomes more arbitrary.

For an ordinal scale, the admissible transformations are order-preserving. That is, they are order-relation homomorphisms from $\Re$ to $\Re$:

$$f(x) > f(y) \qquad \text{iff} \qquad x > y$$

Thus, they include all the functions for the interval class (because c > 0 means a straight line with positive slope).

For a nominal scale the function f need only preserve identity; this is the widest class of admissible transformations: $x \neq y \Rightarrow f(x) \neq f(y)$.

Finally, if the scale is not even nominal, it is totally arbitrary in that any function f is admissible. These "scales" are of no interest to anyone. (Such a scale would not arise, then, by fundamental measurement, but deductions from various defined numerical variables could well lead to a totally arbitrary scale.)

Example. Let us illustrate this situation in sociology. Let five categories be employed in rating occupations according to prestige, as in the well-known N.O.R.C. scaling of prestige (see Hodge, Siegel, and Rossi, 1966). Assume these categories are

$$E_1 = \{\text{excellent, very good, average, below average, poor}\}$$

Let us assume, for a typical person, the categories are related in the form

excellent $\blacktriangleright$ very good $\blacktriangleright$ average $\blacktriangleright$ below average $\blacktriangleright$ poor

where $\blacktriangleright$ means "better." Hence, $(E_1, \blacktriangleright)$ is an empirical relational system. Introducing variable X, a scale of the ordinal type is given by

excellent $\mapsto$ 5
very good $\mapsto$ 4
average $\mapsto$ 3
below average $\mapsto$ 2
poor $\mapsto$ 1

Assume that jobs A and B are rated by each person in a specified set of persons. The proportions rating A and B in various categories are as follows:

	excellent	very good	average	below average	poor
A	.50	.50	0	0	0
B	.60	.30	.10	0	0

It should be noted that proportions constitute an absolute scale. We define a job's prestige score by its weighted sum of X values. Hence, calling this score s,

$$s(A) = (.50)5 + (.50)4 = 4.5$$
$$s(B) = (.60)5 + (.30)4 + (.10)3 = 4.5$$

we conclude that A and B have the same prestige. This is an "induced" empirical equivalence relation. If, now, there exists a scale X', obtained by an admissible transformation of X, such that A and B are not equivalent, then the "prestige score" s is not even nominal. To show this, let f be the transformation $X' = f(X)$ such that

$$f(5) = 10$$
$$f(4) = 5$$
$$f(3) = 3$$
$$f(2) = 2$$
$$f(1) = 1$$

Let s be the prestige score in terms of X':

$$s'(A) = (.50)10 + (.50)5 = 7.5$$
$$s'(B) = (.60)10 + (.30)5 + (.10)3 = 7.8$$

Hence, under the admissible transformation, B has higher prestige than A. It follows, since not even an identity on s is preserved under an admissible transformation of X, that s is not even a nominal scale.

One final remark on scale-types. The general definition of "scale" shows that in logical potentiality many other types of scales are conceivable. Some have been explored in the important work by Suppes and Zinnes (1963), but for our purposes, the above "standard" types are sufficient.

7.5. Empirical Meaningfulness. The concept of empirical meaningfulness of a numerical statement is of great importance for an appreciation of the sorts of restrictions that apply to various mathematical operations, not only in statistics but in other contexts. The following is paraphrased from Suppes and Zinnes (1963).

(7.5.1) Definition. A statement S about a numerical concept is empirically meaningful if and only if its truth-value is invariant under any admissible transformation.

Example. (a) Consider the numerical statement

S_1:   Today it is twice as cold as yesterday, that is, today's minimum temperature is one-half of yesterday's

This statement rests upon two data. For example, if (in Fahrenheit),

$$F^{\circ}(\text{today}) = 14$$
$$F^{\circ}(\text{yesterday}) = 28$$

the S_1 is true. Definition (7.5.1) says that this statement is not empirically meaningful if we can find at least one admissible transformation making it false. Now we know that

$$C^{\circ} = \frac{5}{9}(F^{\circ} - 32^{\circ})$$

and so,

$$C^{\circ}(\text{today}) = -10$$
$$C^{\circ}(\text{yesterday}) = -\frac{20}{9} \cong -2.2$$

whereby we see that under the admissible transformation

$$F^{\circ} \mapsto C^{\circ}$$

the numerical statement becomes false. This shows that S_1 is not empirically meaningful.

Example. (b) Consider the data, say the maximum at a given point in space:

$$F^{\circ}(\text{Sunday}) = 14,$$
$$F^{\circ}(\text{Monday}) = 28,$$
$$F^{\circ}(\text{Tuesday}) = 35,$$

which leads to the statement

S_2: The rise in temperature from Sunday to Monday was twice what it was from Monday to Tuesday.

This statement is true on the Fahrenheit scale, on which it is based. The transformation

$$F^{\circ} \mapsto C^{\circ}$$

as before, shows

$$C^{\circ}(\text{Sunday}) = -10$$
$$C^{\circ}(\text{Monday}) = -20/9$$
$$C^{\circ}(\text{Tuesday}) = 15/9$$

The rise from Sunday to Monday is 70/9; from Monday to Tuesday, it is 35/9. Thus, S_2 is still true. But this is not sufficient to show that S_2 is empirically meaningful, for we have to show that for any admissible transformation it is true. Now since temperature is a scientific variable scaled on an interval scale-type, we know that any admissible transformation has the form

$$f(x) = cx + b \qquad (c > 0)$$

Thus, we set up a little table:

Empirical Object	$F°$	$cF° + b$	Differences
Sunday	14	$14c + b$	
Monday	28	$28c + b$	$(28c + b) - (14c + b) = 14c$
Tuesday	35	$35c + b$	$(35c + b) - (28c + b) = 7c$

This table shows in the column of differences that for all values of c and b and, so, for all admissible transformations, we have the ratio

$$\frac{14c}{7c} = 2$$

so that S_2 is true no matter which transformation is applied. This shows that S_2 really refers to a property of the empirical system and is not a mere artifact of a particular numerical assignment among all such assignments that are equivalently representing the empirical system.

Example. (c) An example from sociology can be given if it is assumed for the sake of illustration that the N.O.R.C. occupational prestige scale is an interval scale. For simplicity of computation, furthermore, suppose the scores for three jobs are as follows:

	(N.O.R.C.)
Object	X
Supreme Court justice (J)	100
Machinist (M)	75
Dockworker (D)	50

Consider the statement

S_3: J has twice as much prestige as does D.

Of course, we make this statement by examining the table and noticing the two scores. Thus, S_3 is true in the given scale. By assumption, the scale is of the interval type. Hence, any admissible transformation has the form

$$f(x) = cx + b \qquad (c > 0)$$

Now we suspect, primarily (let us assume) from a course in statistics, that this statement "does not make sense" on an interval scale. This is our conjecture. To support this conjecture we need the proof. This will be given if we can find a single counterexample to the universal "for any admissible transformation." Thus, we seek constants c_o, b_o such that when we transform the N.O.R.C. scores by the rule,

$$X' = c_o X + b_o$$

the statement S_3 becomes false. A little exploration shows that the constants $c_o = 1, b_o = -50$ do the job, as shown by

$$X'(J) = X(J) - 50 = 100 - 50 = 50$$
$$X'(D) = X(D) - 50 = 50 - 50 = 0$$

whereby the statement S_3 is false. Thus, S_3 is not empirically meaningful.

Consider, relative to the same data,

S_4: The prestige difference between J and M is the same as the prestige difference between M and D.

This is true for the N.O.R.C. scores, X. By conjecture, we believe it true for the empirical system: the jobs and their empirical prestige differences. Thus, we investigate what happens under any arbitrary linear transformation $X' = cX + b$ $(c > 0)$:

Object	X	$X' = cX + b$	Differences
J	100	$100c + b$	$(100c + b) - (75c + b) = 25c$
M	75	$75c + b$	
D	50	$50c + b$	$(75c + b) - (50c + b) = 25c$

Thus, for any c and b, since $25c = 25c$, we conclude that S_4 is true under any admissible transformation, and so, by our definition, S_4 is empirically meaningful.

Let the integers 1, 2, 3, 4, and 5 correspond to the five scale-types of definition (7.4.1) in the order, nominal is 1, . . . , absolute is 5.

Let a_i $(i = 1, \ldots, 5)$ be the set of all admissible transformations for a scale of type i. Then we have the important relationship

$$(7.5.2) \qquad a_5 \subset a_4 \subset a_3 \subset a_2 \subset a_1$$

For example, let $f \in a_4$. Then f is a positive-sloped linear function through the origin; since it is linear, $f \in a_3$. Since it is positive-sloped, it is order-preserving. Thus, $f \in a_2$. But, then, it is also injective, that is, it preserves identity. Thus, $f \in a_1$.

Generally speaking, we can see what kinds of numerical statements will be of interest:

(1) statements of identity: $X(A) = X(B)$, as in

$$\text{length of A} = \text{length of B}$$
$$\text{prestige of A} = \text{prestige of B}$$

(2) statements of inequality: $X(A) > X(B)$, as in

$$\text{length of A} > \text{length of B}$$
$$\text{prestige of A} > \text{prestige of B}$$

(3) statements of equality of differences: $X(A) - X(B) = X(C) - X(D)$, as in

the prestige difference between architect and dentist is equal
to the prestige difference between machinist and carpenter.

(4) statements of ratio: $X(A) = cX(B)$, as in

$$\text{length of A} = 2 \text{ (length of B)}$$

(5) statements of numerical value: $X(A) = 5$, as in

$$\text{size of A} = 5$$

Let us label these five types of statements as follows:

(1) equivalence (E) statements: $X(A) = X(B)$
(2) order (0) statements: $X(A) > X(B)$
(3) distance (D) statements: $X(A) - X(B) = X(C) - X(D)$
(4) ratio (R) statements: $X(A) = cX(B)$
(5) cardinal (C) statements: $X(A) = k$

Using the definition of empirical meaningfulness, one can verify the following
table, where $\sqrt{}$ means that for any scale of the given type, that type of statement
is empirically meaningful.

Table 7.5.1 Empirically Meaningful
Statements by Scale-Type

Scale-Type	Type of Statement				
	E	O	D	R	C
5. absolute	√	√	√	√	√
4. ratio	√	√	√	√	
3. interval	√	√	√		
2. ordinal	√	√			
1. nominal	√				

In fact, what Table 7.5.1 makes clear is that the numerical assignment

$$\text{absolute} \mapsto 5$$

.

.

.

$$\text{nominal} \mapsto 1$$

is a homomorphism into the real line for the relation between scale-types implicitly defined by the cumulation in the table; namely, "supports more types
of numerical statements as empirically meaningful." In this sense, then, the
table defines a scale of scale-types (as restricted to the five basic types).

Moreover, it is clear this scale is ordinal. Scale-types literally higher in the table are "higher" in the usual usage of being supportive of a wider range of empirically meaningful numerical statements.

7.6. Applications to Statistics. Derived Measures. Having come this far, we explore empirical meaningfulness in the context of statistics. We proceed rather rapidly, depending upon intuitive familiarity with statistical concepts.

By X-data one means a set of observed values of the numerical variable X over some identified empirical domain. We write $X_1, X_2, \ldots, X_N$ for N such data, or often $\{X_i\}$. Each X_i may be termed a value for case i, i = 1, . . . , N, where the "case" is the empirical object. By a statistic one means any real-valued function of X-data, where X is an arbitrary numerical variable. Thus,

$$S_1 = f_1\left(\langle X_i \rangle\right) = \Sigma X_i$$
$$S_2 = f_2\left(\langle X_i \rangle\right) = \frac{\Sigma X_i}{N}$$
$$S_3 = f_3\left(\langle X_i \rangle\right) = X_1/X_2 + 10$$

are statistics. It is clear that not all statistics are equally interesting; also, for some X, some statistics may not be meaningful. The following is paraphrased from Pfanzagl (1968).

(7.6.1) Definition. A statistic $S = f\left(\langle X_i \rangle\right)$ is said to be meaningful for X if and only if E-statements involving S are empirically meaningful; that is, if and only if the statement

$$S^{(1)} = S^{(2)}$$

is empirically meaningful, where

$$S^{(1)} = f\left(\langle X_i^{(1)} \rangle\right) \qquad \text{(data-set 1)}$$
$$S^{(2)} = f\left(\langle X_i^{(2)} \rangle\right) \qquad \text{(data-set 2)}$$

Example. (a) Let the statistic be the average of two data. That is,

$$S = f(X_1, X_2) = \frac{X_1 + X_2}{2}$$

For two sets of such data, we have

$$S^{(1)} = \frac{X_1^{(1)} + X_2^{(1)}}{2}, \quad S^{(2)} = \frac{X_1^{(2)} + X_2^{(2)}}{2}$$

and the statement to be tested for empirical meaningfulness is:

$$S: \quad \frac{X_1^{(1)} + X_2^{(1)}}{2} = \frac{X_1^{(2)} + X_2^{(2)}}{2}$$

The basic results (Pfanzagl, 1968) are

(7.6.2)   (a) If X is not at least interval in the scale of scale-types of
                Table 7.5.1, then S is not empirically meaningful.
            (b) If X is at least interval, then S is empirically meaningful.

It follows that arithmetic averaging is not meaningful for X such that X is nominal or ordinal.

To give a concrete case of the absurdity of averaging in certain instances, let X be an ordinal power scale with possible values $1, 2, \ldots, 5$. Then, suppose we have the data:

Object i	X_i
1	4
2	3
3	5
4	2

Thus,

$$\frac{X_1 + X_2}{2} = \frac{7}{2}, \quad \frac{X_3 + X_4}{2} = \frac{7}{2}$$

This leads to the assertion of identity of averages in two empirical systems, one with domain {Object 1, Object 2} and the other with domain {Object 3, Object 4}. However, let X' be obtained by the admissible transformation

X	X'
1	1
2	3
3	5
4	9
5	10

Then the table is augmented:

Object i	X_i	X_i'
1	4	9
2	3	5
3	5	10
4	2	3

From the new values X_i', we obtain

$$\frac{X_1' + X_2'}{2} = \frac{14}{2} = 7$$

$$\frac{X_3' + X_4'}{2} = \frac{13}{2} \neq 7$$

and so the identity statement becomes false in the new scale, arrived at by an admissible (in this case arbitrary order-preserving) transformation. It follows that in no way does the identity of averages for X-data where X is ordinal say anything about the empirical system: it is an artifact of using X when one could just as well have used X'.

Proposition (7.6.2)(b) is not difficult to show and is instructive in showing how we think about such proofs. First, note that it suffices to demonstrate the meaningfulness of sums, simply because in a statement like S above we can multiply both sides by two. Now suppose that

$$X_1^{(1)} + X_2^{(1)} = X_1^{(2)} + X_2^{(2)}$$

As in the last concrete example it is less cumbersome to write

$$X_1 + X_2 = X_3 + X_4$$

Let

$$X' = cX + b \qquad (c > 0)$$

be an admissible transformation; when we put b = 0, we obtain the smaller class for ratio scales, and finally putting c = 1, we obtain the identity transformation for absolute scales. Then,

$$X_i' = cX_i + b \qquad (i = 1, 2, 3, 4)$$

Hence, we want to show that $X_1 + X_2 = X_3 + X_4$ implies

$$X_1' + X_2' = X_3' + X_4'$$

Now,

$$
\begin{aligned}
X_1' + X_2' &= cX_1 + b + cX_2 + b \\
&= c(X_1 + X_2) + b + b \\
&= c(X_3 + X_4) + b + b \\
&= cX_3 + b + cX_4 + b \\
&= X_3' + X_4'
\end{aligned}
$$

Thus, if $X_1 + X_2 = X_3 + X_4$, then $X_1' + X_2' = X_3' + X_4'$. The summation identity is therefore meaningful: addition is meaningful if the scale-type is at least interval.

Now definition (7.6.1) establishes a usage in good conformity to plain common sense. A statistic specifies, on the face of it, a numerical property of a body of data. Each body of data is mapped out of a set or aggregate of empirical objects. Thus, the statistic either implicitly defines an empirical property for the aggregate or is meaningless.

Example. (b) Given

$$
\begin{array}{rcc}
 & \text{Job i} & X_i \\
\text{Nonmanual} & \left\{\begin{array}{l} 1 \\ 2 \end{array}\right. & \begin{array}{l} 100 \\ 80 \end{array} \\
\text{Manual} & \left\{\begin{array}{l} 3 \\ 4 \end{array}\right. & \begin{array}{l} 60 \\ 20 \end{array}
\end{array}
$$

and suppose X is an interval-type scale. Then the fact that

$$\overline{X}^{(1)} = \frac{X_1 + X_2}{2} = 90$$

$$\overline{X}^{(2)} = \frac{X_3 + X_4}{2} = 40$$

shows that the aggregate of jobs $\{$Job 1, Job 2$\}$ differs empirically from the aggregate of jobs $\{$Job 3, Job 4$\}$.

Further, in the case of averaging, it is clear that higher-order statements (of order, etc.) can be meaningful. In this case,

$$\overline{X}^{(1)} > \overline{X}^{(2)}$$

is invariant under any admissible transformation. Thus, a relation over the domain

$$E^* = \{\text{Nonmanual, Manual}\}$$

is induced—namely, that which reflects the computed numerical relation.

The conclusion is that, in defining statistics, one can employ them in a manner that discovers empirical relations over aggregates, beginning with fundamental or derived measurement over the elements in these aggregates.

By exploring the properties of various common statistics, for each type of common scale-type of data, we can learn something about the kinds of meaningful statements about sets of aggregates (in relation to each other) that are supported. Put another way, one can determine the induced scale-type of the statistic when the X-data are a given scale-type. The statistic is then regarded as a measurement—that is, as a derived scale with a location in the scale of scale-types. Table 7.6.1 reports some results understood in the folklore of behavioral science but exactly provable by the methods indicated here.

Some remarks on this table may be useful. First note that the mode is a point on the numerical series given by the X-data. Thus, the mode is an X-datum. It therefore carries the same scale as the data. For example, the modal religion in Community A may be that which was assigned "2," while in Community B it may be that which was assigned "1." Thus, although $2 > 1$, this is no empirical fact because we know that religion is a nominal scale, that therefore the mode for religion is a nominal scale, and that at most E-statements (Table 7.5.1) are meaningful. Hence, $2 \neq 1$ shows that these communities are inequivalent in

Table 7.6.1   Scale-Type of Common Statistics
by Scale-Type of Data

Statistic	Data	Derived Scale-Type of Statistic
Mode	nominal	nominal
	ordinal	ordinal
	interval	interval
Median	nominal	not meaningful
	ordinal	ordinal
	interval	interval
Mean	nominal	not meaningful
	ordinal	not meaningful
	interval	interval
Standard deviation	nominal	not meaningful
	ordinal	not meaningful
	interval	ratio

the specified sense. It shows no more insofar as any relational structure linking the two communities is concerned.

Again, the median is a point on the numerical series that forms the X-data. However, the median employs the numerical inequality relation; thus, its very definition is not meaningful for nominal data. Otherwise the median carries the scale-type of the data. For instance, the median income for one class of jobs is $10,000; for another class of jobs it is $20,000. Since income is a ratio scale, the median itself assumes that scale-type; since $20,000 is twice $10,000 and this R-type statement (Table 7.5.1) is meaningful for ratio scales, it follows that to say that median for the first class of jobs is double that of the other is empirically meaningful.

The standard deviation is computed by using differences of the form $\overline{X} - X_i$ where $\overline{X}$ is the mean of the X-data. Thus, at least an interval scale X is required. But note that the standard deviation itself supports R-statements. This can be shown as follows:

By definition,

$$s = \sqrt{\frac{\Sigma x_i^2}{N}}$$

where

$$x_i = \overline{X} - X_i \qquad (i = 1, \ldots, N)$$

Apply the arbitrary admissible transformation,

$$X' = cX + b$$

Then,

$$s' = \sqrt{\frac{\Sigma(x_i')^2}{N}}$$

where

$$x_i' = \overline{X'} - X_i' \qquad (i = 1, \ldots, N)$$

is the standard deviation in the new numerical assignment. But,

$$\overline{X'} = \overline{cX + b} = c\overline{X} + b$$

and so,

$$
\begin{aligned}
x_i' &= \overline{X'} - X_i' \\
&= (c\overline{X} + b) - (cX_i + b) \\
&= c\overline{X} - cX_i \\
&= c(\overline{X} - X_i) \\
&= cx_i
\end{aligned}
$$

and substitution in the formula for s' yields

$$s' = \sqrt{\frac{\Sigma(cx_i)^2}{N}} = c\sqrt{\frac{\Sigma x_i^2}{N}} = cs$$

so that $s \mapsto s' = cs$ under an arbitrary admissible transformation. Thus, s, regarded as a new scale (over aggregates of empirical systems of the same kind), is a ratio scale. For instance, it is meaningful to say that the standard deviation of a set of temperature measurements made today is three-quarters of what it was yesterday, even though the data are only interval-level. Like many other derived measures, the standard deviation—once we know it is meaningful—requires empirical interpretation. This process is essentially that of asking for a way to think about the bare mathematical relation (in extension) induced by the statistic.

7.7. Measurement Models. What was given in example (d) in section 7.3 as a list of conditions on an empirical relational system would ordinarily be called a measurement model. Given a measurement model, two problems immediately arise from a formal standpoint: to prove that a scale exists and to prove that it is of a certain scale-type.

To show that a scale exists, for an abstract measurement model, means proving a theorem. In relating systems of the same abstract type in this manner, mathematicians call such a theorem a representation theorem. For instance, for the algebraic systems termed groups, it is known that any group can be represented as a so-called "permutation group" (see Chapter 17). This means that a theorem has been proved demonstrating that if G is any group, there exists a special kind of group—a group of permutations—that is isomorphic to it. The mapping from G into the representing group may be called the representation mapping. In measurement models, much the same situation exists. To see this, let us recall the discussion of section 5.17: there the set of graphs was taken as the class of representing systems. We do not consider this measurement, for if we do, there is no way of distinguishing measurement from any other representation mapping.

The appropriate picture for the generic representation situation is given in Figure 7.7.1. The class of all relational-algebraic systems of the same type—for

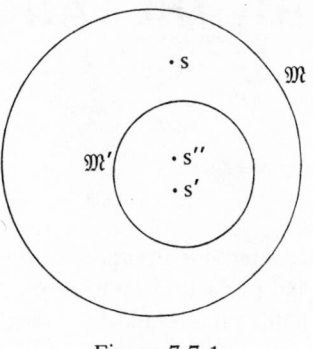

Figure 7.7.1

example, (A, R, o)—is the interior of the region $\mathfrak{M}$. The representing subclass of such systems is denoted by the interior of the region $\mathfrak{M}'$. Note that $\mathfrak{M}' \subset \mathfrak{M}$. Now the two theorems to be shown by a measurement model are as follows:

(7.7.1) Theorem. For any system, say s, in $\mathfrak{M}$, there exists a system, say s' in class $\mathfrak{M}'$ such that a homomorphic mapping carries s into s'.

(7.7.2) Theorem. Suppose that both s' and s'' in $\mathfrak{M}'$ are homomorphic images of s. Then s' and s'' are themselves related by a mapping f of the form . . . .

As applied to measurement, take $s = \varepsilon$, let $\mathfrak{M}'$ be the class of numerical systems, and let s' and s'' be particular numerical representations of ε. The indifference, from an empirical standpoint, is between s' and s'', and thereby, we need the second type of theorem.

The bulk of scientific measurement is derived rather than fundamental. Indirect procedures, lawlike relations, parameter estimation, and so on, provide means of attaining numerical assignments. But in Chapters 11 and 20 we shall study theories which exemplify the most abstract ideas of this section in real scientific contexts. Also, knowledge of the abstract theory plays a strong role in our thinking about representing social concepts and processes, as was seen in the case of the status-mapping notion and will be seen repeatedly in this book in virtually every chapter.

CHAPTER EIGHT THE LANGUAGE OF PROCESS

8.1. Introduction. In this chapter, our attention turns to the outline of conceptual aspects of processes and to the development of a generic way of thinking about processes, which in suitable cases is linked to a mathematical apparatus. The ideas are subsequently applied in Chapter 10 and in Part 3.

8.2. Analytical Systems. Central to a proper appreciation of the way in which we approach processes in mathematical terms is the idea of system.

By a "system" we mean (see Hall and Fagen, 1968) a set of objects of any kind together with relationships between the objects and/or between their qualitative and quantitative attributes. We have noted that the notion of an empirical system—a set of empirical objects together with relationships between the objects—is a basis for explicating measurement theory. Each such empirical system gives rise to a scientific variable. Hence, considering a variety of relationships based on the same set E of empirical objects, we have empirical systems $\mathcal{E}_1, \mathcal{E}_2, \ldots, \mathcal{E}_n$ (for instance, giving rise to mass X_1, temperature $X_2, \ldots$, length X_n). The scientific variables when examined in relationship to each other exemplify the second part of the system idea, involving relationships between the (measured) attributes of the objects. We shall use the term "analytical system" for a set of scientific variables $\{X_i\}$ and relationships between such variables. If the variables are indexed in time so that we can conceive of the values changing through time, we write $\{X_{it}\}$ and called the analytical system a dynamic system.[*]

It will be convenient to distinguish between two levels of analytical systems: frameworks and models.

---

[*]Just as the general concept of mapping embraces the "degenerate" special case in which all objects in the domain go into a single image point, here the general concept embraces the static systems as special cases.

By a "framework" one means a domain E of empirical objects with abstract relational and operational structure yielding variables $X_1, X_2, \ldots, X_n$, together with any general principles satisfied by these variables.

By a "model" one means a specification of one or more objects in E and a corresponding addition, to the framework, of hypotheses relating the variables as defined over these particular objects. Thus, a model is a construction within a framework, using the generic variables and principles of the framework.

Given a framework, the set of measurement models that yield the scales $X_1, X_2, \ldots, X_n$ over the empirical systems based on set E of concrete systems (empirical objects) may be said to form the measurement basis.

Hence, a scientific analytical system has the components:

(1) measurement basis yielding $X_1, X_2, \ldots, X_n$ over domain E
(2) analytical system over $X_1, X_2, \ldots, X_n$, consisting of
    (a) framework
    (b) models for particular concrete systems in E.

Example. The following columns exemplify some of these components of Newtonian mechanics:

Measurement Basis	Framework	Model of x	Concrete Systems x in E
Example: mass	(Partial sketch):		
Primitives:	*Primitives:*	Use the abstract	a pendulum
xHy means x is heavier	mass (m), length,	form	a planet-sun system
than y (x, y in E).	time, force (F)	$F = ma$	a vibrating string
x o y means x combined	*Definitions:*		
with y (x, y in E).	position, velocity,		
Axioms (example):	acceleration (a),	to set up a particular	
If xHy and x'Hy',	center of gravity	system of equations	
then (x o x') H(y o y').	*Axioms* (example):	exemplifying the ab-	
Theorems:		stract form so as to	
(1) There exists a numeri-	$F = ma$	represent the form of	
cal mapping m preserving	*Theorems:*	the motion of x in	
(E, H, o).	If . . . , then	given circumstances.	
(2) The mapping m is			
unique up to multiplica-			
tion by a positive constant.			

The idea of framework is exemplified by the second column: this is mechanics viewed as pure theory. It is a mathematical system of a very complex character in which it is known by everyone that a certain type of application is intended: that dealing with the motion of physical systems. The physical systems themselves exemplify the idea of concrete system: a particular pendulum located in a particular place is a complex entity that in one aspect has a description in terms of its motion. Between the pure mechanics and the pendulum investigated stands the activity of model-building, which culminates in a well-fitting model for the mechanical aspect of a pendulum in its given circumstances. This is the mechanical model of the pendulum. This model is seen to be, in this case, described or set up as a set of differential equations. In these equations, forces and masses and times and distances are presupposed. This is when the measurement basis makes itself concretely felt. In the abstract (columns 1 and 2), the measurement basis for a primitive mechanical concept (e.g., mass) deprimitivizes the concept, as is seen by the basic measurement theorem. Its primitives (H, o) are non-numerical concepts: an order relation and an empirical binary operation, both over an arbitrary set of bodies.

As components of an analytical system, then, we have the exemplifications:

(1) measurement basis: yielding the mass, length, and time variables, and other variables by derived measurements

(2) analytical system:

(a) framework: pure mechanics

(b) model: mechanical model of the pendulum

We could say, using the language of Parsons (1937), that mass is an analytical element. An analytical element is nothing but one of the variables of an analytical system. As such, it is a mapping with domain some empirical system.

Three distinct variants of the system concept have been employed in the above discussion. We summarize these in the following:

(1) Concrete system: a particular thing or entity felt as a unity (for example, a particular rock; a specified man; a particular pendulum).

(2) Empirical system: a set of concrete systems, together with relationships between them (for example, a set of physical bodies together with relations of extension and combination determining a length variable; essentially, the domain of a scientific variable).

(3) Analytical system: a set of scientific variables which have a common domain, together with relationships between them (for example, temperature, volume, and pressure variables based on common physical bodies).

We see a clear order of abstraction. Type (1) is the experienced entity, type (2) is obtained by conceptualization and operations with a class of such entities in terms of aspects giving rise to variables, and type (3) relates the variables.

Corresponding to the temporal changes in a concrete system, values of variables change. Hence, analytical systems of the dynamic type "return" to the concrete to analyze the concrete flux by formal methods. In this sense, the term "dynamic system" may be used to refer to the concrete system in flux as well as to the analytical system whose variables are time-dependent. For instance, a social group is a dynamic system—concrete system in flux—described by an analytical system with time-dependent changes in the variables. This is what process is all about.

8.3. Dynamic System Behavior: Basic Elements. In this section we outline the main elements of an arbitrary dynamic system, with illustrations.

The variables of a dynamic system may be grouped into two classes: (a) dynamical variables, time-varying in the duration of the system analysis, and (b) parametric variables, time-invariant in the duration of the system analysis.

(8.3.1) Definition. The vector of dynamical variables is called the state of the dynamic system. We write

$$X = (X_1, X_2, \ldots, X_n)$$

when no time index is needed; otherwise,

$$X(t) = [X_1(t), X_2(t), \ldots, X_n(t)]$$

gives the state of the system at time t.

Example. (a) Consider an analysis of waiting lines in post offices. We can take as variables

(1) number of counters for service,
(2) number of people at each counter,
(3) average service time per customer.

Here the post office is the concrete system. The dynamical variables are the waiting-line lengths at each counter, while we can assume a time duration such that structural features of the post office, including the number of counters and the average service time, are not time-varying. If there are n counters and if

$$X_i(t) = \text{number of people at counter i at time t}$$

for i = 1, 2, . . . , n and with t ranging over some time domain, then the state of the system is given by

$$X = (X_1, X_2, \ldots, X_n)$$

Example. (b) To analyze the spread of a rumor in a group, we might consider the variables

(1) number of people who have heard the rumor,
(2) number of people who believe the rumor,
(3) average number of contacts per unit time per person, called the contact density.

For this analysis, we can take

$$X_1 = \text{variable (1)}$$
$$X_2 = \text{variable (2)}$$

and let the state be

$$\mathbf{X} = (X_1, X_2)$$

Here variable (3), the density, may be assumed constant during the flow of the rumor. (This is a setup assumption, defining a postulated model property; whether a given group in fact satisfies this condition may be determined in the application phase of the model-building process.) Then the contact density is a parameter.

Generality of process description is achieved in part by the use of unspecified parameter values (i.e., parametric variables). Each instantiation of a numerical value to a parameter specializes the dynamic system model to particular conditions under which the process occurs.

(8.3.2) Definition.   Let T be a set of values of a time variable, with t = 0 in T representing a time at which the system is to be initially observed. Then by the "initial condition" of the system we mean the value of the state vector at t = 0.

Example. (c) In example (b), above, suppose we observe the group at eight A.M. on a certain day, when we know a rumor is first started. Call this time, t = 0. Suppose we also know that the rumor starts with two group members who both believe it. Then the initial condition of the group (or, more exactly, of the rumor-spread process in the group) is given by

$$\mathbf{X}_0 = (2, 2)$$

where

$$\mathbf{X}_0 = \mathbf{X}(0) = \text{the state at time } t = 0$$

(8.3.3) Definition.   If $T = \{0, 1, 2, \ldots\}$, the system is said to be given in discrete time; if T is the nonnegative real line, so that the system state is defined at any arbitrary moment, we say that the system is given in continuous time.

(8.3.4) Definition.   Let $\mathbf{X}$ be the state vector of a discrete-time system. Then a rule of the form

$$\mathbf{X}(t) \mapsto \mathbf{X}(t+1) \qquad (t = 0, 1, \ldots)$$

is termed a dynamic law or a transition rule.

(8.3.5) Definition.   Let $\mathbf{X}$ be the state vector of a continuous-time system. Then a rule of the form, with Δt small,

$$\mathbf{X}(t) \mapsto \mathbf{X}(t+\Delta t) \qquad (t \text{ in } T; \Delta t > 0)$$

is termed a dynamic law or a transition rule.

Example. (d) In the rumor-spread process of example (b), let time be counted by the "distance" the rumor has traveled in terms of the length of the paths in the evolving network of transmissions as in Figure 8.3.1.

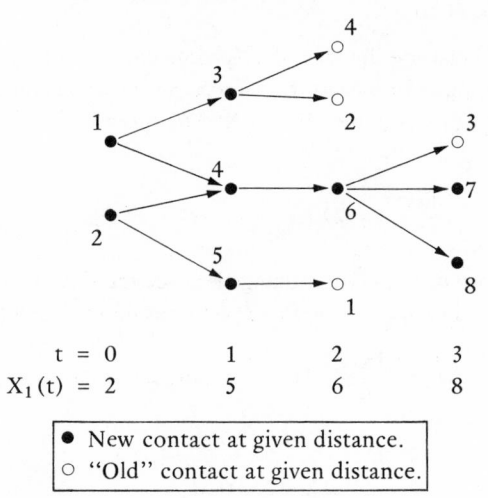

t = 0	1	2	3
$X_1(t) = 2$	5	6	8

●	New contact at given distance.
○	"Old" contact at given distance.

Figure 8.3.1

In the figure, we make the assumption that the variable X_1 is measured by cumulation, in which people who hear the rumor after their first exposure no longer spread it. In any case, this way of marking off time yields a discrete-time process.

Example. (e) At time t in the post office of example (a) above, the state $\mathbf{X}(t)$ is well defined for any $t \geqslant 0$, even if no one is on any line, since then $\mathbf{X}(t) = (0, 0, \ldots, 0)$.

Example. (f) To illustrate the concept of a dynamic law in the discrete-time case, suppose that we write X(t) for the population of a certain group or society at time t. Suppose that the time domain is discrete, so that we consider X(0), X(1), X(2), . . . , a discrete sequence of population sizes. One dynamic law is

$$(8.3.6) \qquad X(t+1) = 2X(t) \qquad (t = 0, 1, 2, \ldots)$$

which says that the size doubles in every period. An initial condition $X_0 = X(0)$ is given, and then, in terms of this condition,

$$X(1) = 2X_0 \qquad [t = 0 \text{ in } (8.3.6)]$$
$$X(2) = 2X(1) = 2^2 X_0 \qquad [t = 1 \text{ in } (8.3.6)]$$
$$X(3) = 2X(2) = 2^3 X_0 \qquad [t = 2 \text{ in } (8.3.6)]$$
$$\vdots \qquad \vdots$$

$$(8.3.7) \qquad X(t) = 2^t X_0 \qquad (t = 0, 1, 2, \ldots)$$

Hence, we can say that the dynamic law (8.3.6) and the initial condition together imply the formula (8.3.7), which gives the system state at every time t in T in terms of the initial condition: we say it describes a trajectory. The population grows geometrically, of course.

Example. (g) To illustrate the idea of a dynamic law in a continuous-time case, we can treat population size as if it were a continuous variable and use calculus methods for this example. We let $X(t)$ be the population size and assume

$$(8.3.8) \qquad \frac{dX(t)}{dt} = 2X(t) \qquad (\text{all } t \geqslant 0)$$

This says that the rate of population change is twice the size at any time. To see how this compares with definition (8.3.5), let us define the differential,

$$\Delta X = \frac{dX(t)}{dt} \Delta t \qquad (\text{any } \Delta t \geqslant 0)$$

Hence,

$$\Delta X = 2X(t)\Delta t$$

and where

$$X(t+\Delta t) \cong X(t) + \Delta X$$

so that

$$X(t+\Delta t) \cong X(t) + 2X(t)\Delta t$$

This shows that (8.3.8) maps $X(t)$ into $X(t+\Delta t)$, the mapping becoming exact "in the limit" as $\Delta t \to 0$. The solution of (8.3.8) is obtained by "separation of variables." We first write (8.3.8) in the form

$$\frac{dX}{dt} = 2X$$

then divide by X,

$$\frac{dX}{dt} \frac{1}{X} = 2$$

On the left we recognize the derivative of $\ln X(t)$. Hence, integrating both sides,

$$\int_0^t \frac{dX}{dt} \frac{1}{X} \, dt' = \int_0^t 2 \, dt'$$

$$\ln X(t') \Big|_0^t = 2t \qquad \text{(fundamental theorem of calculus)}$$

$$\ln X(t) - \ln X(0) = 2t$$

$$\ln \frac{X(t)}{X(0)} = 2t \qquad \text{(property of logarithm)}$$

$$\exp\left[\ln \frac{X(t)}{X(0)}\right] = \exp(2t) \qquad \text{(exponential function of both sides)}$$

$$\frac{X(t)}{X(0)} = e^{2t} \qquad \text{(exp and ln are inverses)}$$

(8.3.9) $\qquad\qquad X(t) = X(0)e^{2t} \qquad \text{(all } t \geqslant 0)$

Hence, (8.3.9) is the rule that gives the trajectory: the state of the system at each time $t \geqslant 0$ in terms of the initial condition. Note the parallel to (8.3.7), where the continuous exponential curve now replaces the discrete geometric increase.

A fundamental idea to be kept in mind with regard to dynamic laws is that the generality given by parametric variables will allow considerable divergence of temporal behavior of a system under different values of the parameters. We illustrate this important point in the context of population processes.

Example. (h) Let $X(t)$ be the population size of a collectivity with birthrate b, death rate d, and net migration zero. Then the dynamic law (8.3.8) may be written in the general case

(8.3.10) $\qquad\qquad \dfrac{dX(t)}{dt} = (b - d)X(t) \qquad (t \geqslant 0)$

or, more compactly,

$$\frac{dX}{dt} = (b - d)X$$

Accordingly, as $b = d$, $b > d$, or $b < d$, we obtain totally distinct behaviors as shown in Figure 8.3.2.

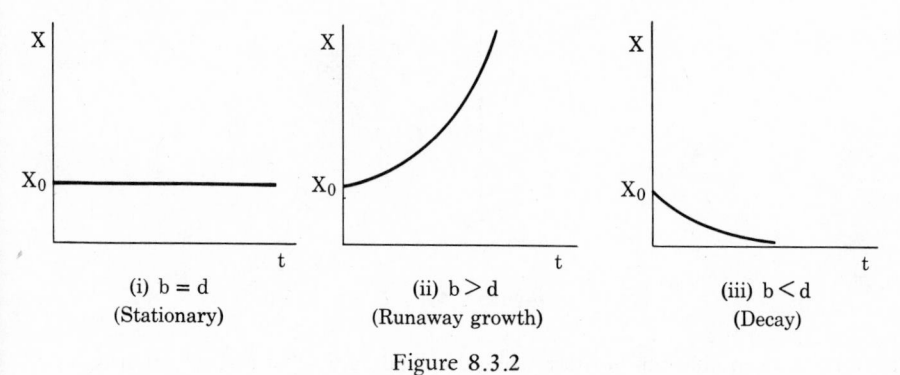

$\quad$(i) b = d $\qquad\qquad\qquad$ (ii) b > d $\qquad\qquad\qquad$ (iii) b < d
$\,$(Stationary) $\qquad\qquad$ (Runaway growth) $\qquad\qquad\quad$ (Decay)

Figure 8.3.2

These totally different observable temporal developments are all subsumed under one general dynamic law, namely (8.3.10). The generality arises precisely from the unspecified character of the parameters in the formula.

8.4. Dynamic System Behavior: Equilibrium and Stability. We turn now to related concepts that arise in dynamic systems analysis. Most important is the following simple idea:

(8.4.1) *Definition.* Let X be the state vector of a process described by a specified dynamic law. Then by an equilibrium state, denoted X_e, we mean a value of X such that under the dynamic law

$$\text{if } X(t) \ = \ X_e, \text{ then } X(t+1) \ = \ X_e \qquad \text{(discrete-time case)}$$
$$\text{if } X(t) \ = \ X_e, \text{ then } X(t+\Delta t) \ = \ X_e \qquad \text{(continuous-time case)}$$

Put another way, the law maps this state of the system into itself: it is a "fixed point" of the process.

It is often the case that writers in model-building contexts combine other notions with equilibrium to blur the distinctive definitive property

$$X_e \ \mapsto \ X_e \qquad \text{(under the law)}$$

We urge the reader to conceptually fix on this definition as analytically distinct from questions of stability, to be discussed next. An equilibrium need not be "something one gravitates toward," since the definition is conditional on being in that state: if in X_e, then again in X_e (under given parametric conditions and dynamic law).

Example. (a) Consider a pendulum, as in Figure 8.4.1.

Figure 8.4.1

Then E_1 is an equilibrium position when the pendulum is "unforced"—that is, is not made to swing back and forth by some device. But, also, E_2 is an equilibrium position if you can get the bob in that position precisely without imparting any velocity to it.

Example. (b) Consider a population size X(t) and the dynamic law

(8.4.2) $$\frac{dX}{dt} = cX(X_M - X) \qquad (c > 0)$$

In this law, X_M is a parametric variable: it is a size of population that can be supported in a given environment. Excess of X_M creates a downturn in the population process, while the population beginning below X_M will grow toward it. To find the equilibrium sizes of X(t) we consider that what is required is a value X_e such that it is constant in time, so that X_e satisfies

(8.4.3) $$\frac{dX_e}{dt} = 0$$

Combining (8.4.2) and (8.4.3), we have the equation

$$0 = cX_e(X_M - X_e)$$

to be solved for X_e. It is clear that there are two solutions

(8.4.4) $$X_e = 0, \qquad X_e = X_M$$

and these two values are equilibrium population sizes under dynamic law (8.4.2). Note how the whole idea of an equilibrium state depends upon the assumed dynamic law in the setup of the dynamic model. Note, also, that more than one equilibrium may exist.

(8.4.5) Definition. An equilibrium state $\mathbf{X}_e$ is said to be (asymptotically) stable if, and only if, when the system state initially is not $\mathbf{X}_e$, then the behavior of the system implied by the dynamic law consists in moving toward the state $\mathbf{X}_e$:

$$\mathbf{X}(t) \to \mathbf{X}_e \qquad (\text{as } t \to \infty)$$

For instance, in example (b) with law (8.4.2), we can test both of the values in (8.4.4) for stability. If $X(0) \neq 0 = X_e$, then (since $c > 0$) the population will increase [if $X(0) < X_M$] until it reaches X_M or decrease [if $X(0) > X_M$] until it reaches X_M. In either case, it is not the case that X(t) approaches 0. Hence, $X_e = 0$ is not a stable equilibrium. On the other hand, if we consider $X_e = X_M$, our argument shows that when $X(0) \neq X_M$, then the system approaches X_M. Hence, this is a stable equilibrium state.

8.5. Probabilistic Systems. These ideas can be extended to the probabilistic case, which shall be done in terms of an example, the Asch situation (see Asch,

1958). Asch was concerned about the view of man prevalent, he thought, in systematic psychology of the stimulus-response or Freudian varieties to the effect that a human individual could be made to do almost anything under appropriate social-stimulus conditions. To this "irrational" view of man, Asch opposed a "cognitive" view. This led him to construct experimental situations with the following properties:

(a) a single individual faced a unanimous group opposed to his opinion;

(b) the opposition was based on an apparent difference in perception of a physical stimulus: a line was to be judged as most similar in length to one of the three standard lines of differing lengths; but

(c) the group, apart from the focal individual, consisted of paid confederates of the experimenter, who responded counter to the sometimes obvious correct answer.

The individuals gave their judgments in order, going around a table at which they were seated and things were arranged so that the focal individual was last to announce his judgment. Hence, at this time he had two apparently irreconcilable inputs: the one based on the stimulus, the perceptual judgment, and the other, on the other individuals, the unanimous counterperceptual judgment. This procedure was repeated in a series of trials to observe temporal trends. Cohen (1963; see also, Berger, et al., 1962) both extended the number of trials and worked out a mathematical model for the trial-to-trial behavior. He found that most individuals, after a transient period, adopted a stable mode of response: some adopted the response of going along with the group, while others stayed with the correct perceptual judgment in announcing their opinion to the group.

A subject in such a situation is an example of a concrete system. We are interested in the dynamics of response of this system. Let C be the conforming response. This varies with the trial of exposure to the stimulus, under fixed boundary conditions (i.e., confederates sitting around a table, a fixed stimulus, a single naive subject). A model for the dynamics in terms of C is bypassed in favor of a model for the dynamics of P(C), the probability of a conforming response. This is still a function of time.

However, a basic distinction is introduced by this use of probability. There is never mere probability; there is always the probability of something. Thus, if a system is analyzed probabilistically, this implies a prior analysis of the "somethings" that have the probability. Thus, to have P(C), you must first have C. Every probabilistic system has an analytical ground system, or grounding. For example, one can treat the pendulum probabilistically; then, the mechanical concepts provide the grounding. At a minimum, a grounding is a coding of the concrete system into relevant aspects; in more developed sciences, the grounding itself establishes lawful relations between such aspects.

Thus, to return to the Asch situation, Cohen first treated the problem probabilistically in terms of the ground system:

$$\{C, \overline{C}\} = \{conforms, does not conform\}$$

on any trial. The probabilistic system over this grounding is given by taking (with "p" now instead of "X," for probabilistic systems)

$$p = P(C)$$
$$t = trial$$

Cohen first tried the "degenerate" dynamic law

$$p_t \mapsto p_{t+1} = p_t$$

This means p is actually constant in this model:

$$p(t) = p_0 \qquad (\text{all } t \geqslant 0)$$

Further, he assumed independent trials (see Chapter 9), which means that the system should behave as a coin repeatedly tossed, with fixed probability p_0 of heads. In particular, there should be no "zeroing in" on one mode of response, say nonconformity, by a given subject. But, on the contrary, concrete systems (persons) do zero in—some into the repeatedly nonconforming behavior, others into the repeatedly conforming behavior. No more than a coin thrown 40 times eventually zeros in on heads (tails) is this expected under this first model. Thus, the model is inadequate. Moreover, even if one takes only the responses before the stable mode of response is reached, the coinlike behavior fails to be exemplified. Thus, revision is required.

To revise a probabilistic system, two modes are possible: revision in the ground system or revision in the probabilistic system only, corresponding to the fact that any probabilistic system has a selected analytical grounding. The former means a reconceptualization of the analytical grounding, the latter means a change in the dynamic law involving the probabilities but no change in the analytical grounding.

Cohen eventually took the important step of revision in the grounding. He did this by proposing a simple dichotomy between the mental state of the subject and the response of the subject. In other words, he abandoned a purely behavioristic analysis in the grounding of his probabilistic system.

Cohen proposed, instead, a set of four mental states:

M_1: absorbed nonconformity
M_2: temporary nonconformity
M_3: temporary conformity
M_4: absorbed conformity

Thus, he conceptualized two "endpoints" for the mental "state-space" and two transient states. He made these designations mathematically meaningful by proposing transition probabilities between them as shown in Figure 8.5.1. For example, if a person is in M_2, then with chance α he goes into M_1 on a given trial.

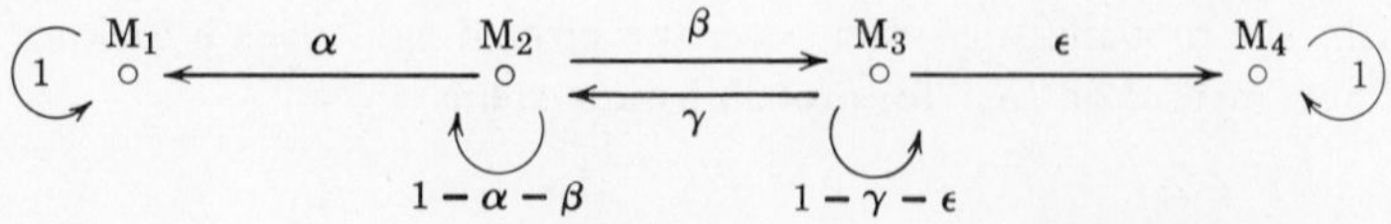

Figure 8.5.1

States are represented by nodes; possible transitions, by directed arrows; the unknown parameters, which are the magnitudes of the transition probabilities, by letters. For example, if we take

$$\alpha = 1, \beta = 0, \gamma = 0, \epsilon = 1$$

then we have the system of Figure 8.5.2.

Figure 8.5.2

For the extreme probabilities of Figure 8.5.2, two things can happen; if the person begins in a transient state, he immediately moves into the associated absorbed state; if he begins at an absorbed state, he stays there.

The model is further specified by a rule linking mental states, which are after all not directly observable, to responses. As part of the meaning of the analytical grounding this rule is given as follows:

(8.5.1)   The subject responds with $\overline{C}$ iff   he is in state M_1 or state M_2.

The analytical grounding is given by $\{M_1, M_2, M_3, M_4\}$; $\{C, \overline{C}\}$; and rule (8.5.1). Mathematically this rule specifies a mapping from the mental domain to the response codomain as follows:

$$R(M_i) = \overline{C} \quad \text{if} \quad i = 1, 2$$
$$R(M_i) = C \quad \text{if} \quad i = 3, 4$$

or, equivalently, in a form often used in model-building,

(8.5.2) $$R(M_i) = \begin{cases} \overline{C} & \text{if} \quad i = 1, 2 \\ C & \text{if} \quad i = 3, 4 \end{cases}$$

The response rule (8.5.1) is the empirical meaning of the mapping given in (8.5.2).

Given this grounding, we already have constructed the essential part of the probabilistic system: namely, the transition diagram, since this diagram provides the transition rule for the process. In probability theory, it is postulated (see Chapter 9) that the probability of an event that can occur in two mutually exclusive ways is the sum of the separate probabilities. By (8.5.2) it follows that $P(\overline{C})$ is obtained by adding the probability that the subject is in M_1 and the probability that he is in state M_2, on any given trial.

Finally, we set up the probabilistic system to exemplify general systems concepts. For the state vector p, we take,

$$p = (p_1, p_2, p_3, p_4)$$

where

$$p_i = P(M_i) \qquad (i = 1, 2, 3, 4)$$

so that the state of the (probabilistic) system is the probability distribution over the mental states. The dynamic law will be treated later in this book in Markov chain theory (see Chapter 10), but for the moment we can let it be denoted by the abstract notation

$$p(t) \mapsto p(t + 1) \qquad \text{(via Figure 8.5.1)}$$

meaning that the transition probabilities of Figure 8.5.1 have the effect of carrying a given probability distribution into a new probability distribution. Markov chain theory yields a solution for p as a function of t, depending upon (1) the initial condition p_0, which is a given or postulated probability distribution over mental states at the start of the experiment, and (2) the parametric conditions in terms of the parameter list:

$$(\alpha, \beta, \gamma, \epsilon)$$

Thus, given a subject who in this situation has time-invariant transition probabilities and given his original chance of being in each mental state, we derive the chance that at time t he is in state M_i, for all t and all i. In particular, Figure 8.5.1 implies the result: with probability 1, he eventually ends up in one of the two absorbed states. This result is necessary, but not sufficient, to show that the model has a good fit to the observable behavior. More significantly, the model predicts the proportion of subjects who eventually adopt each of the two stable modes of behavior.

8.6. **Linear Systems.** Consider a continuous-time dynamic system with dynamic law in the form

(8.6.1)
$$\frac{dX}{dt} = f(X, t)$$

This says: at each point (X, t), which consists of a system state and a time, there is a well-defined rate of change of X. This is a very general form. A somewhat less general form lets the rate depend only on the current state and not on the time. Then,

(8.6.2)
$$\frac{dX}{dt} = f(X)$$

We term (8.6.2) autonomous.

We say that (8.6.2) is linear if f is a linear form in vector X, which means that, with X now thought of as a column vector,

$$f(X) = AX + B$$

where A and B are matrices of suitable orders (for instance, B is a column vector). When $B = 0$, we call the system homogeneous.

If we let either of matrices A and B vary in time, then we have a special case of (8.6.1), say

$$f(X, t) = A(t)X + B(t)$$

Example. (a) Adopting the heuristic approach to functionalism suggested by Stinchcombe (1968), we can treat Malinowski's functional argument concerning magic in terms of a linear system. The diagram in Figure 8.6.1 shows three variables, amount of magic used (M), amount of collective anxiety (A), and amount of uncertainty in relation to the environment (U). As U increases, so does A and as A increases, so does M, which, however, has a compensatory feedback effect on A, reducing it.

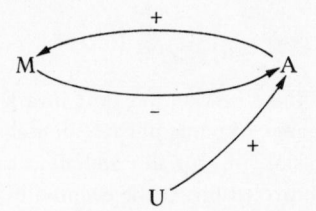

Figure 8.6.1

We interpret the typical linkage in Figure 8.6.1

$$X \rightarrow Y$$

in dynamic terms:

$$\frac{dY}{dt} = f(X) \qquad (f\ \text{linear})$$

because then

$$\Delta Y = f(X)\Delta t$$

yields a change in Y due to X in time interval Δt, a good rendition of a causal transmission. Hence, for Figure 8.6.1, one exact representation in linear form is obtained by adding causal influences into a given variable to obtain

$$\frac{dM}{dt} = c_1 A$$

$$\frac{dA}{dt} = c_3 U - c_2 M$$

where $c_i > 0$ ($i = 1, 2, 3$). The variable U can be taken as parametric (for example, U has one value in the lagoon, another in the open sea). Combining c_3 with U and writing U' for $c_3 U$

$$\frac{dM}{dt} = c_1 A$$

$$\frac{dA}{dt} = U' - c_2 M$$

Forming column vectors,

$$\begin{pmatrix} \dfrac{dM}{dt} \\ \dfrac{dA}{dt} \end{pmatrix} = \begin{pmatrix} c_1 A \\ U' - c_2 M \end{pmatrix} = \begin{pmatrix} c_1 A \\ -c_2 M \end{pmatrix} + \begin{pmatrix} 0 \\ U' \end{pmatrix}$$

Hence, we can write

$$\frac{d}{dt} \begin{pmatrix} M \\ A \end{pmatrix} = \begin{pmatrix} 0 & c_1 \\ -c_2 & 0 \end{pmatrix} \begin{pmatrix} M \\ A \end{pmatrix} + \begin{pmatrix} 0 \\ U' \end{pmatrix}$$

and so

(8.6.3) $$\frac{dX}{dt} = AX + B$$

where

$$X = \begin{pmatrix} M \\ A \end{pmatrix}$$

$$A = \begin{pmatrix} 0 & c_1 \\ -c_2 & 0 \end{pmatrix}$$

$$B = \begin{pmatrix} 0 \\ U' \end{pmatrix}$$

It follows that the functionalist system, so explicated, forms a linear dynamic system such that:

(1) the state is the amount of magic and the amount of collective anxiety, these being the two dynamical variables;

(2) the parameters include not only the uncertainty U', but the coefficients c_1 and c_2 indicating the amount of effect each variable has on the other;

(3) the dynamic law is the linear system (8.6.3);

(4) the equilibrium state X_e satisfies

$$
\begin{aligned}
AX_e + B &= 0 \\
AX_e &= -B \\
X_e &= -A^{-1}B
\end{aligned}
$$

and since, by (6.3.10),

$$
A^{-1} = \begin{pmatrix} 0 & \dfrac{-1}{c_2} \\ \dfrac{1}{c_1} & 0 \end{pmatrix}
$$

we have, using $-A^{-1}$,

$$
X_e = \begin{pmatrix} 0 & \dfrac{1}{c_2} \\ \dfrac{-1}{c_1} & 0 \end{pmatrix} \begin{pmatrix} 0 \\ U' \end{pmatrix} = \begin{pmatrix} \dfrac{U'}{c_2} \\ 0 \end{pmatrix}
$$

Hence, in equilibrium,

$$
M_e = \frac{c_3}{c_2}U, \quad A_e = 0
$$

That is, in equilibrium, the amount of magic is a linear function of the uncertainty, with coefficient c_3/c_2, while the collective anxiety is reduced to zero. Magic "works" in this model, although perhaps too well to fit the phenomenon. Note that without a measurement basis for M and A, the work does not have the serious import it would otherwise have.

In the preceding example, we did not attempt to compute the formula for $X(t)$ in terms of initial conditions and parameters. Techniques for doing so rely heavily on matrix and calculus techniques that we are not presupposing or introducing in this book (for an example using matrix methods, see Land 1970). Instead, we give a second example, with emphasis on the setup phase.

Example. (b) Homans (1961) has state five general propositions dealing with the dynamics of interaction. A simple rendition of part of this theoretical work is as follows. Imagine an interactive exchange between person A and person B, involving help and thanks (see Figure 8.6.2). Four aspects are central: the value A places on social approval (that is, on thanks from B), denoted v_T; the value B places on help from A, denoted v_H; the quantity of help which A supplies to B, denoted q_H; and the quantity of thanks which B gives to A, denoted q_T. Three of Homans' propositions can be stated in terms of these variables:

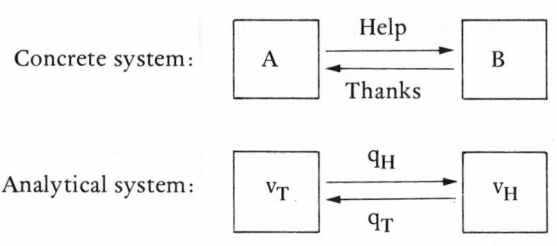

Figure 8.6.2

1. Reinforcement proposition

$$q_H \xrightarrow{+} q_T \quad \text{and} \quad q_T \xrightarrow{+} q_H$$

2. Value proposition

$$v_H \xrightarrow{+} q_T \quad \text{and} \quad v_T \xrightarrow{+} q_H$$

3. Satiation proposition

$$q_H \xrightarrow{-} v_H \quad \text{and} \quad q_T \xrightarrow{-} v_T$$

For example, the first part of the reinforcement proposition states that help induces thanks: q_H produces an increase in q_T in a small time interval. Equivalently, attending to rates, this means a change in q_H produces a change in the rate of q_T: if q_H rises, then the rate of q_T rises; if q_H falls, then the rate of q_T drops. The important point is that there is a representation of a mechanism acting through time.

To continue our example, each activity has an associated cost satisfying

$$c_H \xrightarrow{-} q_H \quad \text{and} \quad c_T \xrightarrow{-} q_T$$

This means: the greater the cost, the lower the quantity emitted (interpreted dynamically). Hence, taking into account the forms under each of the three propositions and the cost terms, we can combine causal mechanisms in the network shown in Figure 8.6.3.

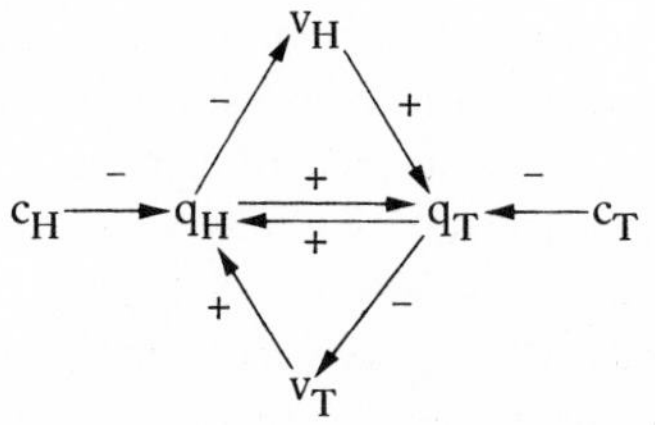

Figure 8.6.3

The corresponding differential equations for the dynamic system are given by

$$\frac{dv_H}{dt} = -c_1 q_H$$

$$\frac{dv_T}{dt} = -c_2 q_T$$

$$\frac{dq_H}{dt} = c_3 v_T + c_4 q_T - c_5 c_H$$

$$\frac{dq_T}{dt} = c_6 v_H + c_7 q_H - c_8 c_T$$

where $c_i > 0$ $(i = 1, 2, \ldots, 8)$. In matrix form,

$$\frac{d\mathbf{X}}{dt} = \mathbf{AX} - \mathbf{B}$$

where,

$$\mathbf{X} = \begin{pmatrix} v_H \\ v_T \\ q_H \\ q_T \end{pmatrix}$$

$$\mathbf{A} = \begin{pmatrix} 0 & 0 & -c_1 & 0 \\ 0 & 0 & 0 & -c_2 \\ 0 & c_3 & 0 & c_4 \\ c_6 & 0 & c_7 & 0 \end{pmatrix}$$

$$\mathbf{B} = \begin{pmatrix} 0 \\ 0 \\ c_5 c_H \\ c_8 c_T \end{pmatrix}$$

If the costs are fixed, then $\mathbf{B}$ is time-invariant. The equilibrium state $\mathbf{X}_e$ of the dynamic exchange system is the vector satisfying

$$\frac{d\mathbf{X}_e}{dt} = \mathbf{A}^{-1}\mathbf{B}$$

At this point, we will terminate our formalization of Homans' two-person exchange system. Two difficulties militate against too serious an interpretation of this particular formalization. First, the rule of distributive justice must be represented so that normative considerations enter as a causal mechanism partially determinative of the dynamics of the system. Secondly, an explicit measurement model is required for each of the two types of variables, value and quantity. Moreover, for empirical meaningfulness of the mechanisms, the value measurement must be higher than ordinal.

One final point may be made in reference to Figure 8.6.3. Namely, the dynamics involved here are quite complex since some feedback loops are positive and others negative. For instance, considering the reinforcement effect alone we are dealing with a positive feedback loop. This implies runaway growth (in frequencies of activities) or possibly zero activity. But the potentiality for such extremes is moderated by the negative feedback loops based on satiation. Also, the cost input moderates the reinforcement effect. The extent of complexity is considerably amplified if, in addition, the distributive justice rule is taken into account.

A general point about formalization of intuitive social psychological or sociological frameworks emerges from the prior examples. Namely, such frameworks very often constitute highly sophisticated descriptions of complex dynamic patterns. Formalization along the lines of sentence symbolization as in section 2.12, as opposed to dynamic interpretations of the kind just studied, prove remarkably uninformative. For further examples of dynamic interpretations, see the papers in Part II of Simon (1957).

8.7. Stability Analysis. It is often useful to study the stability of a linear system—that is, of its equilibrium state—using a matrix analysis. The procedure is as follows:

Step 1. Make the equilibrium state the zero state in a new coordinate system; define

$$x = X - X_e$$

so that when $X = X_e$, then $x = 0$. Then,

$$
\begin{aligned}
\frac{dx}{dt} &= \frac{d}{dt}(X - X_e) \\
&= \frac{dX}{dt} - \frac{dX_e}{dt} \\
&= \frac{dX}{dt} \qquad \left(\text{since } \frac{dX_e}{dt} = 0\right) \\
&= AX + B \\
&= A(x + X_e) + B \\
&= Ax + AX_e + B \\
&= Ax + \frac{dX_e}{dt} \\
&= Ax
\end{aligned}
$$

Hence, the system written in terms of x is homogeneous.

Step 2. A basic idea from matrix theory is now used, the concept of eigenvalue. If A is a matrix, then think of the product of A with a vector x as a mapping of that vector into a vector. Visualize the two-dimensional plane, for example. A number λ such that, for some nonzero vector x, the image of x under matrix A is merely a shrinking or stretching by factor λ is called an eigenvalue of A. Formally, λ is an eigenvalue of **A** if there exists a nonzero x such that

(8.7.1) $Ax = \lambda x$

The proposition that links eigenvalues and stability is as follows:

(8.7.2) *Proposition.* Let

$$\frac{dx}{dt} = Ax$$

be the dynamic law of a system. Then,

(1) there is a unique equilibrium state, namely, zero;
(2) this equilibrium is stable in the sense of definition (8.4.5) if and only if all eigenvalues of **A** have negative real parts;
(3) if the eigenvalues are pure imaginaries, then the state stays in the vicinity of the equilibrium but does not approach it.

To utilize this proposition requires explicit calculation of eigenvalues. We first note that (8.7.1) can be written in the form

$$Ax - \lambda x = 0$$

so that, since $Ix = x$,

$$Ax - \lambda Ix = 0$$

and now, using the distributive law for matrices,

(8.7.3) $(A - \lambda I)x = 0$

We again think of the matrix operating on vector x as a mapping. Then x goes into $(A - \lambda I)x$. We note that for any matrix, say **M**, the product **M0** is **0**. Hence, if $x = 0$, then (8.7.3) is satisfied. Put another way, $A - \lambda I$ maps **0** into **0**. However, we want nonzero x to go into **0** for a given λ. Hence, mapping $A - \lambda I$ must not be injective (1-1). Hence, it cannot have an inverse, in which image points return to their unique origins by reversing the mapping. That is, if λ is an eigenvalue, then $(A - \lambda I)^{-1}$ does not exist. Hence, if λ is an eigenvalue then $\det(A - \lambda I) = 0$. It follows that to find the eigenvalues of matrix **A** we must solve

(8.7.4) $\det(A - \lambda I) = 0$

for λ.

   Example. (a) Let

$$A = \begin{pmatrix} 0 & b \\ c & 0 \end{pmatrix}$$

Hence, for **I**, take

$$I = \begin{pmatrix} 1 & 0 \\ 0 & 1 \end{pmatrix}$$

and then λI is

$$\lambda I = \begin{pmatrix} \lambda & 0 \\ 0 & \lambda \end{pmatrix}$$

Hence,

$$A - \lambda I = \begin{pmatrix} 0 & b \\ c & 0 \end{pmatrix} - \begin{pmatrix} \lambda & 0 \\ 0 & \lambda \end{pmatrix} = \begin{pmatrix} -\lambda & b \\ c & -\lambda \end{pmatrix}$$

From formula (6.3.6), we see that

$$\det(A - \lambda I) = \lambda^2 - bc$$

Therefore, (8.7.4) becomes

(8.7.5) $$\lambda^2 - bc = 0$$

to be solved for λ. It is clear that

$$\lambda_1 = \sqrt{bc}, \quad \lambda_2 = -\sqrt{bc}$$

constitute the two solutions of (8.7.5). Hence:

(1) if b and c have the same sign, so that bc is positive, then both eigenvalues are real, but λ_1 is positive and so, by proposition (8.7.2), the equilibrium is not stable;

(2) if b and c have opposite signs, then bc is negative and $\sqrt{bc}$ is imaginary. Hence, the eigenvalues are pure imaginary and part (3) of proposition (8.7.2) applies. This says that the (two-dimensional) state of the system will not move away from the equilibrium, although it will not approach it; the "motion" of the system consists in oscillation around equilibrium.

Example. (b) To study the stability of the Malinowski system of example (a) in section 8.6, we note that

$$A - \lambda I = \begin{pmatrix} -\lambda & c_1 \\ -c_2 & -\lambda \end{pmatrix}$$

Hence, in terms of (8.7.5), $b = c_1$, $c = -c_2$ and conclusion (2) applies. Hence, the behavior of the system consists in oscillation around the equilibrium, which is the origin of the homogeneous system. Hence, the behavior of a concrete tribe using magic in environmental circumstances measured by U involves temporal cycles of magical activity that we interpret as allaying collective anxiety A, itself oscillating around zero, while the amount of magic oscillates around $\frac{c_3}{c_2}$ U.

For examples of nonlinear system stability analysis, see Kemeny and Snell (1962) and Simon (1957). For an example of stability analysis following upon a framework construction for status equilibration analysis, see Fararo (1972a). Another good example of stability analysis is given in Rapoport (1960) treating the Richardson (1960) process models for arms races. The same topic is treated by Saaty (1968). For further mathematical study, see Sanchez (1968) or Brauer and Nohel (1967). A related approach to dynamics via difference equations is outlined by Goldberg (1961) and applied, for instance, by McPhee (1966).

8.8. State-Space Approach. Let us return to the basic ideas of dynamic systems analysis by noting an important conceptual aspect of the idea of state. This can be done in the context of explication of the state-space approach to processes. This approach was developed by electrical engineers, but it is conceptually quite general (see, for example, Schwartz and Friedland, 1965).

Let us consider a concrete system that is described by a model with observable vector **R**. Here, **R** may be a whole list of records; it is a vector called the response variable. Consider two systems, both described in terms of **R**, both time-varying, and both characterized by the same parameters. Also, now, let us agree both are subjected to inputs—as in the stimuli facing humans.

If under identical input conditions, we have

$$\mathbf{R}_1(t) \; = \; \mathbf{R}_2(t) \qquad \text{(all t)}$$

we say that these two systems are **R**-same. Thus, observationally—in terms of the aspect under which they are observed—the two systems behave identically.

Now let us consider two systems such that under identical input conditions, they are not **R**-same. We say that they are **R**-different. If now all pairs of **R**-different systems never exhibited the same behavior (**R**) at the same time, we would have the situation exemplified in Figure 8.8.1 for a one-dimensional response variable. Such a class of systems may be said to be such that their paths never cross. In particular, from distinct "presents" at t_0 we predict distinct "pasts" $(t < t_0)$ and distinct futures $(t > t_0)$. In contrast to this situation is the pair of systems of Figure 8.8.2. Here the observable records cross paths at t_0. A model built on the principle of a dynamic equation with the observed response as the state variable **X**, as outlined earlier, would fail to discriminate the distinct futures of these two systems. (Given the dynamic law and parameters, identity of initial condition implies identical trajectory.) In other words, the model would fail at the application phase. The problem here is to preserve the elegance of the formulation of

$$\dot{\mathbf{X}} \; = \; \text{some function of } \mathbf{X} \text{ and inputs}$$

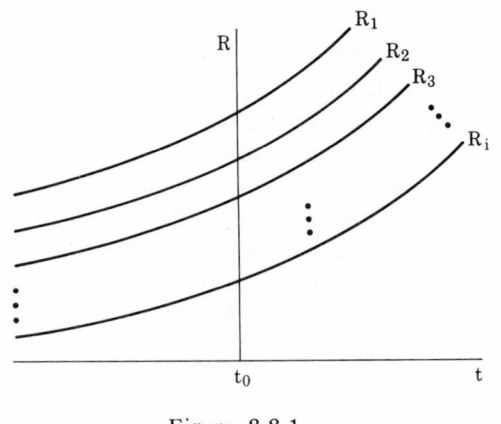

Figure 8.8.1

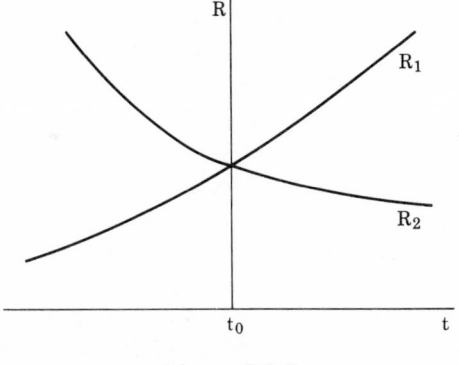

Figure 8.8.2

where $\dot{\mathbf{X}}$ is a generic notation for a time rate of change of $\mathbf{X}$, covering both the discrete-time and continuous-time cases. We can do this if we somehow invent an analytical state concept, $\mathbf{X}$, such that $\mathbf{X}$ is constructively employed with the $\dot{\mathbf{X}}$ dynamic equation but then also

$$\mathbf{R} = \text{some function of } \mathbf{X} \text{ and inputs}$$

and we do this such that the systems of Figure 8.8.2 show

$$\mathbf{X}_1(t_0) \neq \mathbf{X}_2(t_0)$$

even though $\mathbf{R}_1(t_0) = \mathbf{R}_2(t_0)$.

Example. In the case of the pendulum in example (a) of section 8.4, when it swings back and forth after an initial perturbation, it seems to swing through the observable equilibrium position. Yet we said that if a state is an equilibrium state, then what is there stays there. How then could the pendulum move through the E_1 position? The answer is that if "position" means location in space, this is like our R above: it is not sufficient to distinguish different future positions. This is clear since the very same bob could be passing through E_1 or be merely stationary there at t_0. The resolution of this apparent difficulty is the Galilean representation: the state of a mechanical system has to include both the velocity and the position if predictability of position itself is to be achieved.

Thus, what the state-space concept amounts to is a search for a concept of state, X, relative to an observable variable, R, such that

(8.8.1)
$$\dot{X} = \text{some function of } X \text{ and inputs}$$
$$R = \text{some function of } X \text{ and inputs}$$

so that R is seen as an "output" of the system with state X.

Corresponding to this idea is the simple diagram of Figure 8.8.3.

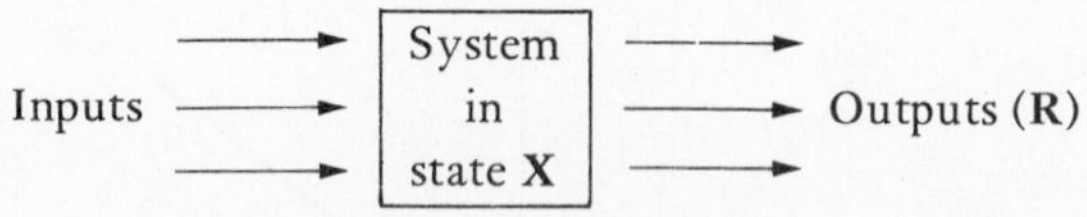

Inputs → System in state X → Outputs (R)

Figure 8.8.3

By definition, outputs are observable, but X is an analytical construction that may not be a simple list of observables. We term the flow of R over time the observable response process. Then, if (8.8.1) is satisfied, we term the flow of X over time the generating process (see Fararo, 1969a, b). "Inputs" are time-varying variables that we conceive as external causes of changes in state and as stimulus conditions for response, given the state. Such variables are called exogenous in the economic literature (see Tintner, 1968). For instance, if we take R to be group effectiveness, a likely conceptual candidate for X is the group's cohesion. Then, as a group faces ever-changing input problems, characterized by such variables as complexity and novelty, it changes cohesion under the impact of these inputs and responds with an effectiveness that depends on both the current cohesion and the particular problem input. If we take R to refer to observable deference relations, then a likely candidate for X involves expectations. If we take R to refer to actual trial-by-trial behavior in a learning experiment, then a good candidate for X is the subject's knowledge state. We shall focus on this approach later in the study of Markov social processes (see Part 3). For an example of the state-space approach in the context of the algebraic representation of language phenomena, see Chomsky and Miller (1963).

8.9. Two Methodological Rules for Process Analysis. We suggest, on the basis of the above ideas, the following two rules for the construction of dynamic models:

(8.9.1) Rule 1. State-space formulation: analyze the dynamics of the observable behavior of a concrete system by the use of an analytical system that is constructed by appeal to the state-space approach, so that (8.8.1) is the generic form of the model.

(8.9.2) Rule 2. Construction in the small: analyze the dynamics of the observable behavior of a concrete system by thinking and constructing in the temporally small [in other words, attempt to state dynamic laws in the sense of definitions (8.3.4) and (8.3.5)] and deducing in the temporally large.

We have discussed the meaning of Rule 1 in the preceding section. Rule 2 refers to the sort of constructive and deductive work shown in example (f) of section 8.3. The setup is in terms of a transition law for time t to t + 1—hence, "in the small." But the deduction (8.3.7) yields the state of the system at any time t, however distant from the initial time, and, so, is "in the large." This deduction is carried out relative to a clear assumption: the same dynamic law applies in transitions of state from t = 0 to t = 1 as applies from t = 1 to t = 2, from t = 2 to t = 3, and so forth. This is a special time-homogeneity assumption to the effect that the parameters of the dynamic law, as well as its form, remain time-invariant. A picture to have in mind is shown in Figure 8.9.1.

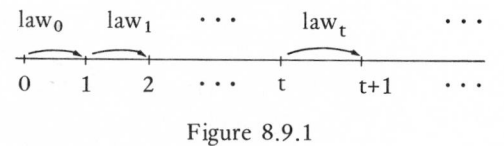

Figure 8.9.1

This picture assumes that the parameters change over time, so that law_t is itself a dynamical element, so to speak. For example, the transition chances in Cohen's model for the Asch situation, as discussed in section 8.5, may not be the same on trial 10 as they are initially. Needless to say, the analysis of such systems, given their setup, is not simple (for a discussion of this problem in the context of social mobility, see Chapter 16). Rule 2 is designed to alert the model-builder to the fact that a time-homogeneity assumption is made—or else a nonhomogeneous model is set up by appeal to the rule but with the particular law itself time-varying. In the small, however, there is but one law: the transition rule for changing state valid at that time.

8.10. General Process Classification. An analytical system of the dynamic type that is constructed in the mode of the state-space approach is called state-determined (cf. Hall and Fagen, 1968). A classification of such systems rests upon the possibilities for the state variable and for the time domain. A state variable can be discrete or continuous. Time may be discrete (trials, periods) or continuous (changes can occur at arbitrary times). The system may be purely analytical or probabilistic. (Traditionally, the systems without a probability model superimposed, the purely analytical type, are called deterministic.) These possibilities are as follows:

Component	Possible values
state	discrete or continuous
time	discrete or continuous
law	deterministic or probabilistic

A state-determined system of the probabilistic variety is called Markov. An exact definition will be given in Chapter 10 (for a similar approach, see Bellman, 1961). We shall see many examples of these processes later, in Part 3.

We may classify processes as shown in the tree diagram in Figure 8.10.1.

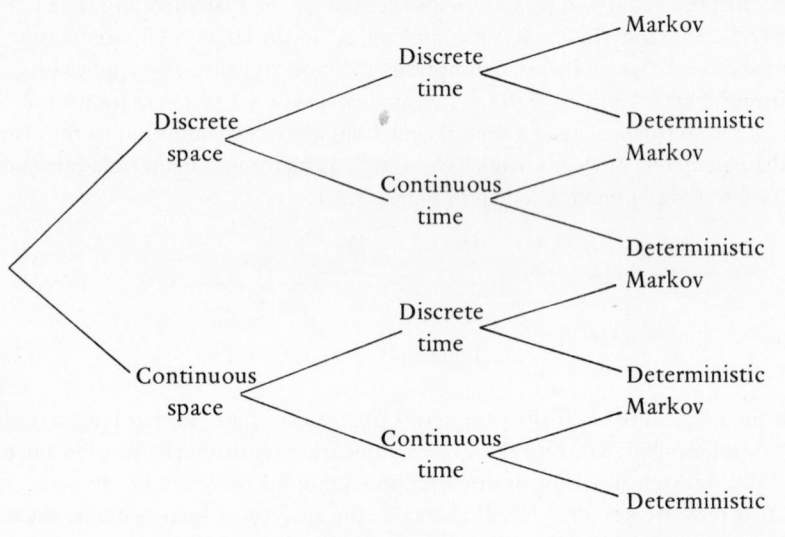

Figure 8.10.1

The various categories of processes tend to be described by rather specialized formalisms. Wherever "continuous space" appears it means either (a) classical deterministic difference or differential equation systems, as in Fararo (1970) and Land (1970), or (b) stochastic diffusion or jump processes, each rather difficult to handle (see Feller, 1966). Discrete space processes, on the other hand, are more manageable; the Markov type are called Markov chains in discrete or in continuous time. The deterministic type in discrete time has a special subclass in which the state is nominal and the state-space approach becomes very algebraic, involving semigroup techniques, for example (see Langer et al., 1968). Another example of this type occurs in the analysis of individual problem solving using information processing representations (see Newell and Simon, 1972).

With reference to this classification, then, the main point is that in Part 3 we will focus on Markov social processes: state-determined systems of the probabilistic variety. The reader can see that this is a selection from a larger set of variants. This selection is made to keep the mathematics elementary and to reflect the dominant role the Markov setup plays in modern mathematical sociology. Roughly speaking, the Markov formalism stands to mathematical sociology as the deterministic differential equation formalism stands to classical physics. In neither case does the particular type of dynamic system exhaust the working methods, but each represents a distinctive dominance in the respective fields.

CHAPTER NINE ELEMENTARY PROBABILISTIC METHODS

9.1. Introduction. The function of this chapter and the next is to introduce the reader to the probabilistic ideas and techniques that play an essential role in the construction of finite probability models. When these ideas are conjoined with the process approach of Chapter 8, we generate the Markov chain models that play a role in current social science modeling analogous to differential equations in mathematical physics. In sociology proper, also, a substantial role is played by continuous-time processes. These processes continue to employ the fundamental techniques of probabilistic modeling, but the mathematical setup changes in the direction of greater analytic complexity. Therefore, we shall introduce the fundamentals of probability modeling unencumbered by techniques or concepts drawn from these continuous-time applications, but at every point we shall build toward these applications by deepening the probabilistic skills and intuitions as we go along. In addition to the role of probability in the subsequent setup and analysis of processes, another major role of probability relates to basic measurement and concept formation. For the latter reason, especially, the reader needs to see not only the "applied" techniques that "work" in setting up and analyzing probability models, but also he needs to become appreciative of the axiomatic structure of probability theory. Thus, in this chapter, the aim is to develop probabilistic techniques along with an introduction to the axiomatic structure. Perhaps in this probability area above all others, the social scientist has a chance to see the close interplay between exact (axiomatic) concept, informal background, deduction, calculation, and application.

9.2. Structure of Any Finite Probability Model. Probability theory may be defined as the analytical study of probability models in abstraction from the particularities of scientific context in which these models are used. Our first task, then, is to specify the structure of any probability model. As the title of this section indicates, we remain within the confines of the finite case, although the ideas do generalize to the infinite case.

Three fundamental notions or objects comprise any probability model. To know how to construct a probability model means to know how to see phenomena in terms of these three notions. They are (1) possible outcome space, (2) events, and (3) probability (of events).

9.3. Outcome Space and Trees. These three notions are primitive in that they appear in any probability model as the entities constructed from identifications made in the application phase of model-building. But we can indicate what they mean in terms of the sorts of identifications that are made. Think about any process of inquiry, real or imagined; it may be seen as realizing one among a collection or set of possible outcomes. In mathematics, when a set of objects is under discussion and these objects are possibilities, the term "space" tends to be used. Thus, a set of possible outcomes of an experiment or of any process may be constructed (i.e., in the finite case listed or otherwise specified by some property). Then this set is called the outcome space of the model (some authors say "sample" space). The following are some simple situations and each case an associated outcome space:

 Examples. (a)   Toss a coin: $\{H, T\}$

      (b)   Roll a die: $\{1, 2, 3, 4, 5, 6\}$

      (c)   Observe a subject about to press either of two response buttons, one to his left, one to his right: $\{left, right\}$

      (d)   Observe two individuals during a given day, with the question of whether they ever interact: $\{interact, not\text{-}interact\}$

      (e)   Think of the sociometric protocol of an individual, say c, who is asked to choose one other person in the population $\{a, b, c\}$: $\{a, b\}$.

Thus, as these examples indicate, the basic idea of an outcome space is that of listing, or otherwise specifying, all possible conceptualized outcomes so that no empirical result fails to show the occurrence of an outcome and so that it shows only one outcome, and the space has no elements that are irrelevant.

Here are some "bad" outcome spaces:

(1) Toss a coin: $\{T\}$
(2) Roll a die: $\{1, 2, 3, 4, 5, 6, Charles\}$
(3) Roll a die: $\{1, 2, 3, 4, 5, 6, odd\}$

The problem with (1) is that the empirical result of heads fails to realize any of the outcomes in the proposed space; the problem with (2) is that although each possible empirical result will realize one, and only one, of the outcomes shown, the outcome "Charles" is irrelevant. The problem with (3) is that the empirical result may show the occurrence of, say, "3" and "odd." This violates the principle that it must show the occurrence of only one outcome. One can code the empirical world in many ways, but the coding possibilities do not form a true outcome space if they do not classify the empirical phenomena: each result must realize one, and only one, of the notions called possible outcomes.

Any finite outcome space may be represented as a tree whose branches have labels at their endpoints corresponding to the possible outcomes. For example, a tree appropriate for the observation of a subject as in example (c) above is shown in Figure 9.3.1.

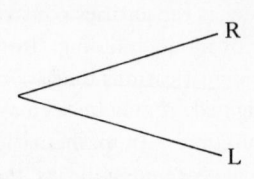

Figure 9.3.1

The tree diagram is a basic tool for finite probability model construction, because when we begin thinking about some process a natural thing to do is to write down a tree. In fact, one might "play" with several trees, seeking the most intuitively appealing conceptualization.

9.4. Representation-of-Events Principle. Given a tree, or outcome space, for a phenomenon, the next problem is to translate events into the tree or space representation. The particular phenomenon—this subject about to press this button on trial 10—is of interest because it realizes some event of interest. But what is an event? Think of the die and the space,

$$\Omega = \{1, 2, 3, 4, 5, 6\}$$

If we think about the event "the die comes up odd," we see that we will want to talk about the chance of this event. Yet, what does it mean to say "it comes up odd"? Clearly, it means one observes the possible outcome 1, the possible outcome 3, or the possible outcome 5. Thus, the event occurs if and only if some one of a specified subset of possible outcomes occurs. This is true of any problem in probability. Any event, when analyzed in terms of the conceptualization given by the outcome space—in other words, analyzed in terms of the basic coding of the phenomenon—will be said to occur if and only if some particular outcomes are observed. Thus the fundamental:

(9.4.1) Representation-of-Events Principle. Represent any event by a set of possible outcomes such that if one observed any of these outcomes, one would say the event occurred.

Example.   (a) $\Omega = \{1, 2, 3, 4, 5, 6\}$

(1) Event E_1 occurs   iff   the die comes up less than 4.

$$\text{Representation: } E_1 \ = \ \{1, 2, 3\}$$

(2) Event E_2 occurs   iff   the die comes up with a number divisible by 3.

$$\text{Representation: } E_2 \ = \ \{3, 6\}$$

(3) Event E_3 occurs   iff   either E_1 or E_2 occurs.

$$\text{Representation: } E_3 \ = \ \{1, 2, 3\} \cup \{3, 6\} \ = \ \{1, 2, 3, 6\}$$

(4) Event E_4 occurs   iff   the die comes up seven.

$$\text{Representation: } E_4 \ = \ \{ \ \} = \varnothing$$

In terms of the tree diagram, Ω of example (a) is shown in Figure 9.4.1,

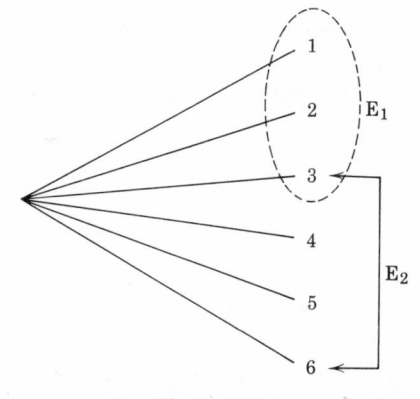

Figure 9.4.1

with events shown by various devices as indicated.

Example.   (b)   A man can shift from his present class position, say lower class (L) to any of three classes: upper (U), middle (M) and lower. Let $\Omega = \{U, M, L\}$.

(1) Event E_1 occurs   iff   he shifts to a different class.

$$\text{Representation: } E_1 \ = \ \{U, M\}$$

(2) Event E_2 occurs   iff   he does not shift (i.e., "shifts" to L).

$$\text{Representation: } E_2 \ = \ \{L\}$$

(3) Event E_3 occurs   iff   he shifts to the middle class.

$$\text{Representation: } E_3 \ = \ \{M\}$$

(4) Event E_4 occurs   iff   he does not shift to the upper class.

Representation: $E_4 = \{M, L\} = \{M\} \cup \{L\} = E_3 \cup E_2$

An appropriate tree diagram is shown in Figure 9.4.2.

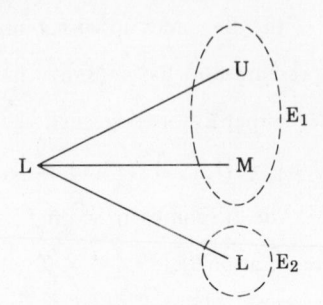

Figure 9.4.2

Example. (c) A more complex example is as follows: A learning process is conceptualized such that, at any time, a person is in either a state of having learned (called the solution state S) or he is not ($\bar{S}$). See Figure 9.4.3.

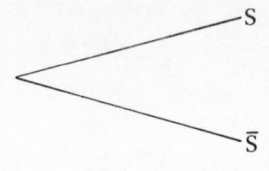

Figure 9.4.3

But, then, if he is in S, he responds correctly (C) to the stimulus input, while if he is in $\bar{S}$, he may or may not ($\bar{C}$) respond correctly. Thus, we obtain Figure 9.4.4.

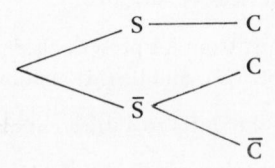

Figure 9.4.4

Then, if he responds correctly in state S, he is going to retain his solution, while if he responds correctly in state $\bar{S}$, he is not going to learn (e.g., his false hypothesis about the situation is verified, so he holds on to it). On the other hand, if he responds incorrectly ($\bar{C}$) in state $\bar{S}$, then he may learn. Thus, Figure 9.4.5 shows

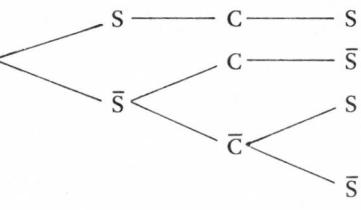

Figure 9.4.5

the complete tree diagram carrying us from the onset of the learning trial at time t to its conclusion at time t+1, the latter, by definition, culminating in a possible change in state. The reader should see the connection to the process language:

$$S, \bar{S}: \text{ state-space of process}$$
$$C, \bar{C}: \text{ response-space of process}$$

so that variables with these domains have the "canonical" relation of the state-space approach: the tree shows that response (C or $\bar{C}$) depends on the state (of confronting the input stimulus, for which no notation was used), while the change in state (recall the dynamic equation) depends upon both the state and, because this is a feedback system, the input as modified by the response. That is, $\bar{C}$ is really a conjunction of pure response and its situational evaluation (input). A probability model for this episode of process has the ground system shown by the tree diagram. The corresponding possible outcome space is

(9.4.2) $\Omega = \{SCS, \bar{S}C\bar{S}, \bar{S}\bar{C}S, \bar{S}\bar{C}\bar{S}\}$

One event of obvious interest is

E_1 occurs   iff   the person learns in this trial

The representation is

$$E_1 = \{\bar{S}\bar{C}S\}$$

assuming that "learning in this trial" means a transition from the $\bar{S}$ to S state. Thus, this event occurs only if an error is made. Note that this conceptualization is "probability-free," in the sense that we have the conceptualization of possible outcomes and events made independently of the problem of the chances of various outcomes.

9.5. Probability Assignments. The last two sections dealt with the outcome space and with events, as represented in a probability model. These are the first two of the three fundamental parts of a probability model. We treat now the notion of probability itself. We can begin our treatment of probability by simply saying it is a weighting of the branches of a tree diagram so that the weights are nonnegative and sum to 1. Let us consider, in turn, each of the outcome spaces of examples (a)-(e) in section 9.3.

Example. (a) Assume the coin is fair. Then we have Figure 9.5.1.

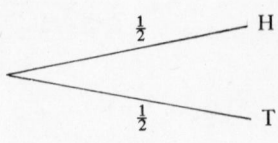

Figure 9.5.1

Example. (b) Assume the die is fair; see Figure 9.5.2.

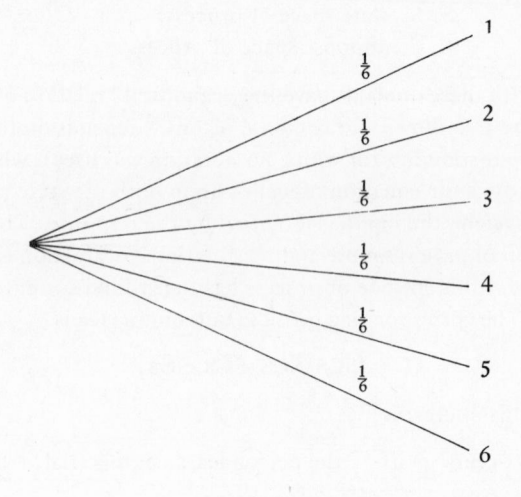

Figure 9.5.2

Example. (c) Assume that the subject presses the button to the left with probability p and that with probability $1 - p = q$ he presses the other button; see Figure 9.5.3.

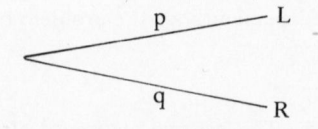

Figure 9.5.3

Example. (d) Assume they interact at least once during the day with chance p and with chance $q = 1 - p$, they do not interact; see Figure 9.5.4.

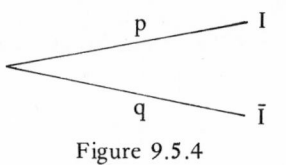

Figure 9.5.4

Example. (e) Suppose c favors a over b to the extent that the chance of his choosing a is twice the chance of his choosing b. Since the numbers must sum to 1: $2x + x = 1$, and $x = 1/3$, $2x = 2/3$. See Figure 9.5.5.

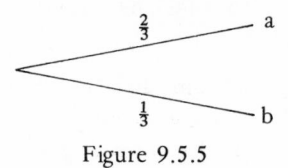

Figure 9.5.5

Using the probability assignment to the outcomes in the outcome space, we now supplement the representation principle (9.4.1) with the following:

(9.5.1) Addition Rule for Events. The probability of an event, as represented as a set of possible outcomes, is the sum over the probabilities assigned to the possible outcomes in that representing set.

Example. (f) Consider the tree with probabilities given in Figure 9.5.2 and the events defined in example (a), section 9.4. Clearly,

(1) The probability of $E_1 = \{1, 2, 3\}$ is $3/6 = 1/2$.
(2) The probability of $E_2 = \{3, 6\}$ is $2/6 = 1/3$.
(3) The probability of $E_3 = \{1, 2, 3, 6\}$ is $4/6 = 2/3$.
(4) The probability of $E_4 = \emptyset$, is 0.

Let us return to the learning process of example (c), in section 9.4. We could think about how we constructed the tree in a sequence from Figure 9.4.3 to Figure 9.4.5. With a probability assignment our new problem, it would be much more natural to arrive at the assignment on the final outcome space (9.4.2) by employing the intuition of the tree construction: first this, then that, and then such-and-such. Let us try it by putting in a probability assignment at each sequential step and try to arrive at the final assignment in the manner shown in Figure 9.5.6.

The problem now is how we can use tree (3) of Figure 9.5.6 to arrive at a probability assignment for the outcome space $\{SCS, \overline{S}C\overline{S}, \overline{S}CS, \overline{S}C\overline{S}\}$. Think about these numbers and parameters as relative frequencies for a moment: start with N people, then, from tree (1) of Figure 9.5.6 we see that

$$bN \text{ go into } \overline{S}$$

(1)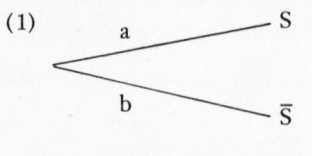

(The probability is a that he already is in state S, and b = 1 − a.)

(2)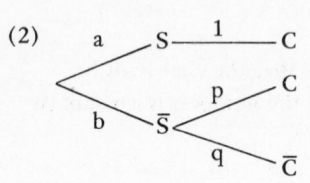

(He is in the learned state, so he is correct with chance 1.)

(He might just happen to be right, with chance p.)

(But then he is wrong with chance q = 1 − p.)

(3)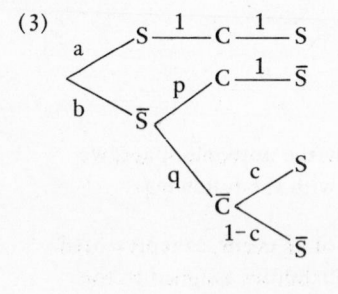

(He already is in S; he is reinforced, so he stays there with chance 1.)

(He has the wrong idea, but it was substantiated, so he keeps it with chance 1.)

(He has the wrong idea and discovers this because of $\overline{C}$, so he has a chance, say c, to learn.)

(Otherwise, he still hits on a wrong idea with chance 1 − c.)

Figure 9.5.6

and of these bN, according to tree (2),

$$q(bN) \text{ go into } \overline{C}$$

and of these qbN, according to tree (3),

$$c(qbN) \text{ go into } S$$

Thus, of the original N,

$$\frac{cqbN}{N}$$

is the relative frequency of people going through the transition: $\overline{S}\overline{C}S$. This is cqb. This suggests that the probability of the outcome $\overline{S}\overline{C}S$ is cqb.

Drawing upon a relative-frequency interpretation of the assigned probabilities, we are led to formulate the following fundamental:

(9.5.2) *Product Rule for Path Probabilities.* The probability of a path in a tree is the product of the probabilities of its component branches.

With this rule we obtain from tree (3) of Figure 9.5.6 the situation of Figure 9.5.7.

Path probabilities

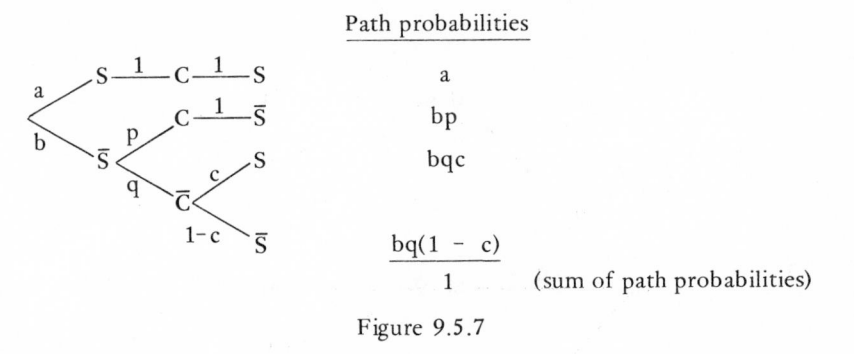

	a
	bp
	bqc
	bq(1 - c)
	1 (sum of path probabilities)

Figure 9.5.7

Note that the sum of the path probabilities is 1. Also, the product of the non-negative terms along any path is again a nonnegative number. Thus, we conclude that by the use of the product rule for path probabilities we are led to an assignment of numbers to paths such that each of the numbers is nonnegative and their sum is 1. Hence, we have a valid probability assignment. Each path in the tree corresponds to a possible outcome.

Example. (g) Here is another example. A man born in class L, who may be mobile or not to classes M or U or may stay in L has a son who in turn may be mobile or not in the same class structure. Thus, we have a two-generational system.

For the first generation we have Figure 9.5.8.

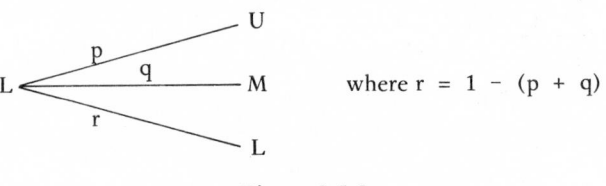

where r = 1 - (p + q)

Figure 9.5.8

And now, for the second generation, we require first the augmentation of the tree, as shown in Figure 9.5.9. For example, the path LUU means: the man in L is mobile to U and his son stays in U. Between the constant starting point L and the terminal points U, M, or L for the son we have a transition structure mediated by the mobility, if any, of father and son. At this point, following the tactic of assignments in the small, we need to ask for a decision between two possibilities: (a) son's mobility chances are the same as father's (p, q, r) or (b) son's mobility chances are altered in some way.

Let us assume a system whose form is possibility (a). Then the tree can be completed and the product rule employed to obtain Figure 9.5.10, although we shall criticize this assignment in a moment.

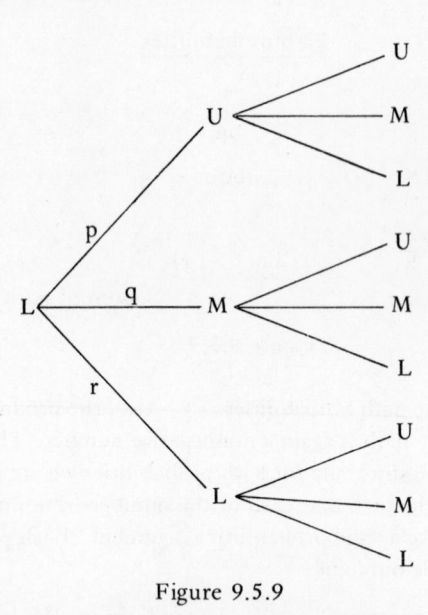

Figure 9.5.9

Path probabilities

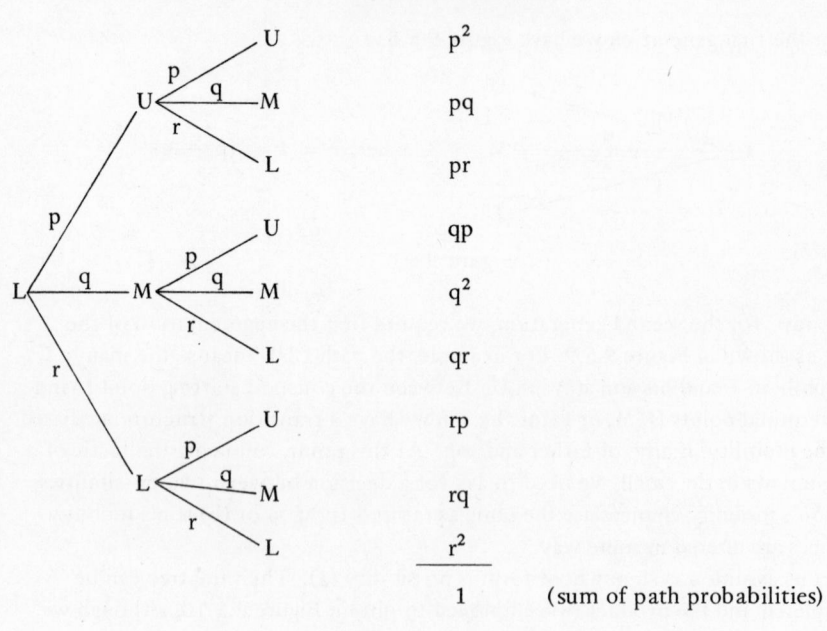

p^2	
pq	
pr	
qp	
q^2	
qr	
rp	
rq	
r^2	
1	(sum of path probabilities)

Figure 9.5.10

The possible outcomes for this process, as conceptualized, correspond one-to-one to the paths in the tree:

$$\Omega = \{LUU, LUM, LUL, LMU, LMM, LML, LLU, LLM, LLL\}$$

One critical remark on this example: when we decided on letting son's mobility chances be the same as father's, we took it very literally. Figure 9.5.10 above shows that even if the son is starting from his father's achieved U-class, his chance of "shifting to U"—which now means merely staying in the upper class—is the same as his father's chance of moving out of L into U, namely, p. This makes no good intuitive sense. Thus, an exploration of an alternative might be based on the assumptions: (a) if son is in L, then his mobility chances are the same as his father's; (b) if son is in M, then the chances are p', q', r'; and (c) if son is in U, the chances are p'', q'', and r''. Here we anticipate $p' \neq p$, $p'' \neq p'$, $p'' \neq p$, and so forth.

9.6. Axiomatic Probability: First Treatment. The axiomatic structure of a finite probability model is as follows: We have a triple of objects $(\Omega, \mathfrak{B}, P)$ such that Ω is a finite set; $\mathfrak{B}$ is the power set of Ω; and P is a mapping from $\mathfrak{B}$ to the unit interval $\{x: 0 \leqslant x \leqslant 1\}$, which satisfies certain axioms. Here Ω is the outcome space; $\mathfrak{B}$ is the set of all subsets of Ω, each of which "may be needed" as a representation of an event in accordance with representation principle (9.4.1). Finally, P is an assignment of a number between zero and one to each of the event representations. We will discuss the "certain axioms" about this P shortly. A triple $(\Omega, \mathfrak{B}, P)$, as above, is termed a finite probability model. We saw above, really, that drawing a tree and weighting it appropriately in effect constructs a finite probability model. Symbol $\mathfrak{B}$ is used to suggest "Boolean algebra of subsets."

Example. (a) Consider Figure 9.5.1. Then, for this model,

$$\Omega = \{H, T\}$$
$$\mathfrak{B} = \}\Omega, \varnothing, \{H\}, \{T\}\{$$
$$P(\Omega) = 1$$
$$P(\varnothing) = 0$$
$$P(\{H\}) = 1/2$$
$$P(\{T\}) = 1/2$$

We read "$P(\{H\})$" as "the probability of the event $\{H\}$," or "the probability of heads," in this case. We read "$P(\Omega)$" as "the probability of the logical certainty." To see this, define

E occurs   iff   H or T occurs,

so that, by the representation principle,

$$E = \{H, T\} = \Omega$$

and clearly, if Ω is really an outcome space (it exhausts the logical possibilities), $P(\Omega) = 1$.

Example. (b) Consider Figure 9.5.2. Here, we have

$$\Omega = \{1, 2, 3, 4, 5, 6\}$$
$$\mathfrak{B} = \{X : X \text{ is a subset of } \Omega\}$$
$$= 2^{\Omega}$$

(so that $\mathfrak{B}$ has 2^N subsets, where N = number of elements in Ω)

$$P(\{i\}) = \tfrac{1}{6} \qquad (i = 1, 2, \ldots, 6)$$
$$P(X) = \sum_{i \in X} P(\{i\})$$
$$= \frac{N_X}{6}$$

where N_X is the number of elements in set X. For instance, if $X = \{3, 4\}$, then

$$P(X) = P(\{3, 4\})$$
$$= P(\{3\}) + P(\{4\})$$
$$= \tfrac{1}{6} + \tfrac{1}{6}$$
$$= \tfrac{1}{3}$$

Example. (c) Consider Figure 9.5.3. Here we have

$$\Omega = \{L, R\}$$
$$\mathfrak{B} = \{\Omega, \varnothing, \{L\}, \{R\}\}$$
$$P(\Omega) = 1$$
$$P(\varnothing) = 0$$
$$P(\{L\}) = p$$
$$P(\{R\}) = q$$

In this example, the model is the triple

$$[\Omega, \mathfrak{B}, P(p)] \qquad (0 \leqslant p \leqslant 1)$$

where p is a parameter. When p is specified, a definite triple is determined. The family of models, as p varies between zero and one, may be thought of as a model-family. It is perhaps what a mathematical social scientist always means intuitively by "model," because ordinarily we use parameters that are not numerically specified but are estimated in the application phase. After estimation a definite model of the model-family is determined.

Example. (d) Consider the mobility of father, in Figure 9.5.8. This is the model

$$\Omega = \{U, M, L\}$$
$$\mathcal{B} = 2^{\Omega}$$
$$= \{\Omega, \emptyset, \{U\}, \{M\}, \{L\}, \{U, M\}, \{U, L\}, \{M, L\}\}$$

(check 2^{Ω} has $2^N = 2^3 = 8$ subsets)

$$P(\Omega) = 1,$$
$$P(\emptyset) = 0,$$
$$P(\{U\}) = p, P(\{M\}) = q, P(\{L\}) = 1 - (p + q) = r$$
$$P(\{U, M\}) = p + q$$
$$P(\{U, L\}) = p + r = 1 - q$$
$$P(\{M, L\}) = q + r = 1 - p$$

Thus, here we again do not have a definite model until parameters are specified numerically. But there are two parameters: p, q (not three because r is determined as soon as we know p and q). Thus, the model-family is given by

$$[\Omega, \mathcal{B}, P(p, q)] \qquad (0 \leqslant p, q \leqslant 1)$$

This is a two-parameter family of models. Normally, p and q are not specified until the application phase; then their estimates specify a definite model.

Example. (e) Finally, let us treat the learning model of Figure 9.5.7. The path probabilities are sufficient to determine P, because of the addition rule. Here the probability component of the usual triple is $P(a, p, c)$, where $0 \leqslant a, p, c \leqslant 1$, because b, q, and $1 - c$ are determined once a, p, and c are known; on the other hand, these three parameters are analytically independent. Thus, we have a three-parameter model-family. This is for the single trial or learning episode, whereas the usual learning theory—done in mathematical form—deals with multitrial processes. These will be illustrated subsequently in section 10.7.

The two-generational mobility model with the identical mobility chances for son and father, no matter where son begins relative to father's achievement, is a two-parameter family. Thus in passing from one to two generations [as shown in Figure 9.5.10] no new parameters were introduced; but note that we regard the model as violating intuition. When p', q', p'', q'' are introduced to build in the intuition that mobility chances depend upon the location, we end up with a probability component of the form

$$P(p, q, p', q', p'', q'')$$

a six-parameter family. Thus, estimation may well make it difficult to test. One can always increase model complexity by dropping explicit or implicit homogeneity or uniformity assumptions; but, then, apart from any augmentation of difficulty in the model analysis phase, there may be considerable loss of ability to test the model or even to estimate the parameters.

To sum up, a finite probability model is constructed by the heuristic and visualizable method of tree diagrams or similar aids to the imagination, and what it is that is constructed from a mathematical viewpoint is a triple of entities forming a finite probability model as a definite entity of the type encountered in the study of mappings and algebras. The former heuristic, applied method of setting up such a model—especially as exemplified in the learning model—is important in the context of discovery. Recall that we aim to discover novel ways of representing the phenomena. The latter, axiomatic methods are important in other contexts: in particular, in chapter 11, in framing a choice theory, the axiomatic approach provides a definite aid to the imagination.

9.7. Axiomatic Probability: Axioms and Theorems.   At this point we present the axiomatic treatment of an arbitrary finite probability model. The point is to build in the promised axioms omitted in section 9.6, with the more general aim of illustrating the idea of an exact concept.

(9.7.1) Definition.   If $(\Omega, \mathcal{B}, P)$ is a triple such that

Axiom 1. Ω is a finite set

Axiom 2. $\mathcal{B}$ is the power set of Ω

Axiom 3. P is a real-valued mapping with domain $\mathcal{B}$ such that, for any E, F $\in \mathcal{B}$:

$$P_1: P(E) \geqslant 0$$
$$P_2: P(E \cup F) = P(E) + P(F), \text{ if } E \cap F = \emptyset$$
$$P_3: P(\Omega) = 1$$

then one calls $(\Omega, \mathcal{B}, P)$ a finite probability model. The objects in Ω are termed outcomes; the subsets of Ω, as contained in $\mathcal{B}$ as elements, are termed events; and P(E) is termed the probability of E.

The following remarks as to axiomatics are in order: the triple $(\Omega, \mathcal{B}, P)$ is a logical variable; any replacement of it that makes the axioms true is a definite entity forming a finite probability model. In other words, the pure mathematics viewpoint is that a new sentential function is introduced, namely,

$(\Omega, \mathcal{B}, P)$ is a finite probability model.

The class of all triples satisfying this (complex, axiomatic) sentential function is the subject matter of the mathematical theory of probability (when it is artificially restricted to the finite outcome space, as in the case here for pedagogical reasons).

If one now asks where these axioms "come from," it is clear they come from the probabilistic intuition exemplified in the earlier paragraphs. Note that the mathematician insists on calling any member of $\mathcal{B}$ an event, whereas we pointed to the far-reaching representation principle (9.4.1) to guide us in passing from empirical events to subsets—now axiomatically endowed with the event nomenclature.

But no harm is done, presumably, for as the reader can see, the terminology pre-supposes the principle, while the mathematician ignores the principle in favor of its pragmatic consequence: scientists will be using subsets to represent events, so call such subsets events.

What about the probability assignment on the tree? The axiomatic concept mentions no such object. From the standpoint of the definite entity $(\Omega, \mathcal{B}, P)$, we might arrive at P by any means we like, so long as the conditions of axiom 3 are satisfied. But scientifically we arrive at P via the assignment of probabilities to outcomes, and thence, by the addition principle, (9.5.1) we obtain P.

The conditions P_i of axiom 3 then are natural. P_1 says let any probability be nonnegative. Anticipating estimates in terms of relative frequencies, this is what we want. P_2 says something a little more complicated. Note that once we use the representation rule for events, the latter are associated with particular sub-sets. Then Venn diagrams apply, as in Figure 9.7.1.

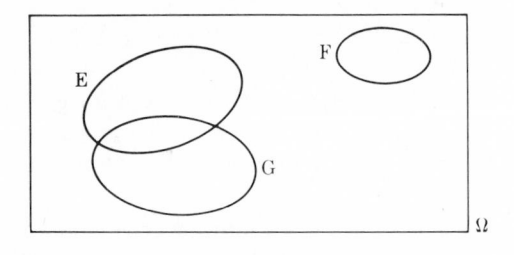

Figure 9.7.1

Regions show various events; here the correspondence is as follows: an outcome is a point in the region Ω, and a subset of points, representing an event, is a sub-region of Ω. Two regions that fail to overlap (E and F) have no common out-come: there does not exist an outcome such that if we observed it, we would agree that E occurred and also that F occurred. Then the region $E \cap F$ is clearly empty. Condition P_2 says when this intersection region is empty—the two regions fail to overlap—the probabilities add. For instance, let

$$\Omega = \{1, 2, 3, 4, 5, 6\}$$
$$E = \{1, 2\}$$
$$F = \{5, 6\}$$
$$G = \{1, 2, 3\}$$

and assume each outcome is assigned 1/6. Then, note that $E \cap F = \emptyset$. We have

$$P(E) = \tfrac{2}{6} = \tfrac{1}{3}$$
$$P(F) = \tfrac{2}{6} = \tfrac{1}{3}$$
$$P(E \cup F) = P(\{1, 2\} \cup \{5, 6\}) = P(\{1, 2, 5, 6\}) = \tfrac{4}{6} = \tfrac{2}{3}$$

Thus,

$$P(E \cup F) = P(E) + P(F)$$

On the other hand, note that $E \cap G \neq \emptyset$. And

$$P(G) = \tfrac{3}{6} = \tfrac{1}{2}$$
$$P(E \cup G) = P(\{1, 2\} \cup \{1, 2, 3\}) = P(\{1, 2, 3\}) = \tfrac{1}{2}$$

Thus,

$$\tfrac{1}{2} = P(E \cup G) \neq P(E) + P(G) = \tfrac{1}{3} + \tfrac{1}{2} = \tfrac{5}{6}$$

Thus, addition holds if the events do not overlap (i.e., are "disjoint"); if they do overlap, then it may not hold, as in the case of E and G in this example. Addition will hold even in these overlapping cases if the region of overlap turns out to have probability zero, but this is infrequent in applied work.

Condition P_2 implies the more general principle

$$(9.7.2) \qquad P(\overset{k}{\underset{i=1}{\cup}} E_i) = \overset{k}{\underset{i=1}{\Sigma}} P(E_i) \qquad \text{(for any } k = 1, 2, \ldots)$$

provided all the E_i are mutually disjoint (i.e., not overlapping).

Finally, condition P_3 of axiom 3 merely assures that probabilities will never exceed one and that the "sure event" represented by Ω has chance one of occurring.

The Venn diagram (9.7.1) suggests an intuitive way of conjecturing probabilistic properties: namely, if an event corresponds to a region, let the probability correspond to the area. One need not worry about proportionate regions and pretty drawings: the point is to "see intuitively" why probability should have certain properties. Note that if we let $\alpha(E)$ be the area of region E and let area be scaled so that Ω has area of unity, then,

$$\alpha(E) \geqslant 0$$
$$\alpha(E \cup F) = \alpha(E) + \alpha(F) \qquad \text{(if } E \cap F = \emptyset)$$
$$\alpha(\Omega) = 1$$

and so α and P satisfy the same "measure" axioms. Thus, area properties "seen" in a Venn diagram are probably true α properties and, so, true P properties. Here is an example (see Figure 9.7.2). The ruled region may be denoted F − E. It is the set of points in the larger circular region F that are not in the smaller circular region E. Now we are sure that $\alpha(F - E) = \alpha(F) - \alpha(E)$, and so $P(F - E) = P(F) - P(E)$. Thus, we propose the following:

(9.7.3) Theorem. If $E \subseteq F$, then

$$P(F - E) = P(F) - P(E)$$

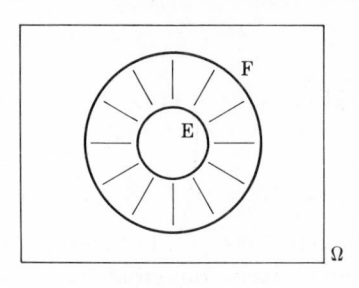

Figure 9.7.2

Here is a formal proof from the axioms: Since $E \subseteq F$,

$$F = E \cup (F - E)$$

Hence, because E and F − E are disjoint,

$$P(F) = P(E) + P(F - E)$$

Then we solve for $P(F - E)$. Note that if $E = F$,

$$P(F - E) = P(\emptyset) = 0$$

Also, by letting $F = \Omega$ in theorem (9.7.3) and by using $P(\Omega) = 1$, we conclude the following:

(9.7.4) Theorem. $P(\overline{E}) = 1 - P(E)$

This is because $\Omega - E = \overline{E}$ by the definition of the complement of a set.

Further, we have the following idea, suggested by the area interpretation (see Figure 9.7.3):

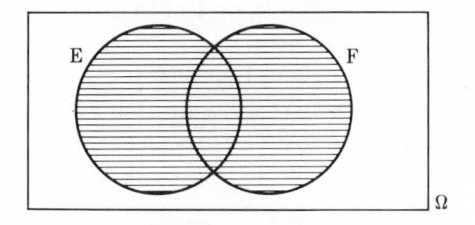

Figure 9.7.3

Suppose E and F overlap. We want the area of the whole region $E \cup F$ shown by horizontal lines. If we add $\alpha(E)$ to $\alpha(F)$, the area in the overlap, denoted $\alpha(E \cap F)$, has been counted twice. Thus, we should subtract it to obtain

$$\alpha(E \cup F) = \alpha(E) + \alpha(F) - \alpha(E \cap F)$$

This suggests:

(9.7.5) Theorem. $P(E \cup F) = P(E) + P(F) - P(E \cap F)$

The formal proof is

$$E \cup F = (E - F) \cup F,$$

the right being a "disjoint union " (i.e., the two parts $E - F$ and F are disjoint). The Venn diagram is crucial for seeing this intuitively. But then,

(9.7.6) $P(E \cup F) = P(E - F) + P(F)$

and since we have another disjoint union,

$$E = (E - F) \cup (E \cap F) \qquad \text{(check Venn diagram)}$$

we have,

(9.7.7) $P(E) = P(E - F) + P(E \cap F)$

Solving for $P(E - F)$ in both (9.7.6) and (9.7.7) and then equating the resulting terms,

$$P(E \cup F) - P(F) = P(E) - P(E \cap F)$$

which yields the desired conclusion (9.7.5).

Now suppose region E is contained in region F. Then $\alpha(E) < \alpha(F)$. In fact, allowing even $E = F$, then $\alpha(E) \leqslant \alpha(F)$. This suggests:

(9.7.8) Theorem.   If $E \subseteq F$, then $P(E) \leqslant P(F)$

A proof from the axioms is as follows: If $E \subseteq F$, then by (9.7.3),

$$P(F - E) = P(F) - P(E)$$

But $F - E$ is an event, so $P(F - E) \geqslant 0$. Thus, $P(F) - P(E) \geqslant 0$, and (9.7.8) follows.

Here is a problem that is suggested by intuition but with no obvious conjectured theorem. What is the probability of exactly one of two events occurring? We draw a Venn diagram as in Figure 9.7.4.

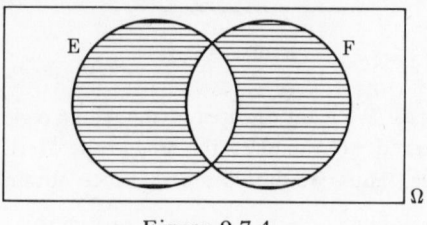

Figure 9.7.4

The event of interest will be said to occur if and only if one of the possible outcomes in E − F occurs or one of the possible outcomes in F − E occurs. The representation rule for events says to represent this event ("exactly 1") by the set of all possible outcomes that realize it if they occur. Thus, since E − F and F − E do the job partially, their union, (E − F) ∪ (F − E), is the proper representation. Now we see that this event is really just

$$(E \cup F) \quad - \quad (E \cap F)$$

At least one   Both

occurs   occur

Thus, this way of looking at the problem is to picture the region and then use prior results. We have a result on E ∪ F. Call the shaded region E^*. Then

$$P(E^*) = P[(E \cup F) - (E \cap F)]$$
$$= P(E \cup F) - P(E \cap F)$$

which holds because (E ∪ F) − (E ∩ F) exemplifies a difference with the second set contained in the first, as in (9.7.3). Thus, using (9.7.5),

(9.7.9) Theorem $P(E^*) = P(E) + P(F) - 2P(E \cap F)$

Thus, from the axioms, additional properties of any finite probability model have been derived: these properties are stated in the theorems, say T for a typical theorem. This means that the statement

If $(\Omega, \mathcal{B}, P)$ is a finite probability model, then T

is a tautological implication.* Hence, one may conclude from the fact that one has constructed such a model (which is usually obvious from the tree diagram or the like) that all the properties given by theorems hold for the model. A related point is that these can be used as rules of inference in contexts of application:

From $(\Omega, \mathcal{B}, P)$ is a finite probability model, conclude: T.

9.8. Applied Probability Theory. The basic principle connecting probability models to experience is the following:

(9.8.1) Principle of Applied Probability. The probability P(E) of any event in a given finite probability model is the expected relative frequency of event E in independent repetitions of the phenomenon being modeled.

---

*Strictly, a universal quantifier, "for any $(\Omega, \mathcal{B}, P)$," should occur prior to the conditional sentential form.

Example. (a) A group of size N has probability

$$P_{G(N)} = 1 - (1 - P_I)^N$$

of solving a problem under one particular finite probability model, as discussed in chapter 1. Here,

E occurs   iff   group solves problem,

and there are 23 repetitions recorded in which three of the groups actually solved the problem. This is the actual relative frequency in the experiment, as replicated 23 times: $3/23 \cong .13$. Its expectation is $P_{G(N)}$, which we recall was in this case given by .39.

Example. (b) A coin is tossed 100 times. This repeats the basic experiment 100 times, in which for each time we have the model in Figure 9.5.1. Thus, if the coin turns up H on 33 percent of the repetitions, this number is the actual or realized relative frequency; its expectation is 50 percent.

The principle of applied probability has certain difficulties associated with it. In our two examples, we came to the data with numerical values for P(E) to compare with the relative frequencies. But we may wonder just how much difference between expectation and relative frequency can be tolerated, and we may not have the numerical value of P(E) to begin with. The first problem is the technical problem of assessing goodness of fit of a probability model. It is treated in terms of statistical "tests" such as chi-square tests. The second problem is within the rubric of estimation. More will be said on these topics later. For the present, we note that a direct estimate of P(E) can be obtained by using the principle of probability theory known as the law of large numbers.

(9.8.2) Law of Large Numbers. Let N be the number of repetitions referred to in the principle of applied probability. Then, if each repetition is truly described by the basic finite probability model, as $N \to \infty$ the relative frequency of occurrence of event E approaches its expectation, namely, P(E).

The empirical meaning usually ascribed to this principle is that for large N, one takes the observed relative frequency as a direct estimate of the unknown probability.

9.9. Conditional Probability. Often we need to compute or hypothesize a probability subject to a condition. This "recomputation" is called conditional probability. It amounts to relativizing to a smaller portion of the set of logically possible outcomes, hypothetically introducing the condition "only things in this subset are possible" and recomputing. In the corresponding direct estimate based on data, we have to consider only those repetitions in which the condition is satisfied. Thus, in model and in data, a "recomputation" occurs based on the condition. Here are some data examples:

Example. (a) The fraction of high school graduates who go to college in a certain specified population is 55 percent. The fraction of black high school graduates going on to college in this population is 21 percent. Thus, the event is

E occurs   iff   high school graduate goes on to college,

and the condition is

The high school graduate is black.

Let us write

F occurs   iff   high school graduate is black,

a little artificially regarding "black" as an event. Then we have recomputed the empirical (direct) estimate of the probability of E, given that F occurs.

Example. (b) In a certain population, the proportion of those eligible who vote (in a certain type of election) is 65 percent. But, on the condition that the eligible person is in the "middle class," the proportion becomes 85 percent. Here the recomputation led to an increase, while in (a) it led to a decrease.

In the formal probability model we need a corresponding operation to effect a change in probability, owing to the conceptualized condition. Consider the Venn diagram of Figure 9.9.1,

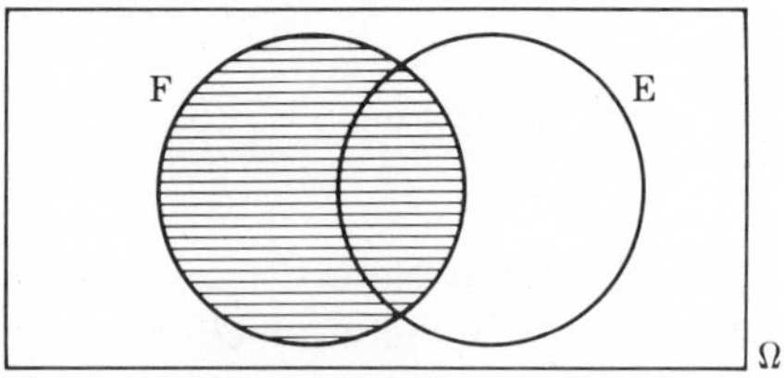

Figure 9.9.1

where F is shaded to show that we impose the condition that "suppose F has occurred." Then, clearly, the only possibilities for E are those that are also realizing F; that is, the set of possible outcomes realizing the event E is restricted to E ∩ F, the overlap. Now the area model says that

$$\frac{\alpha(E \cap F)}{\alpha(F)}$$

is the proportionate area of such possibilities for E. This suggests that we take

$$\frac{P(E \cap F)}{P(F)}$$

as a definition of the idea of "conditional probability" (with condition F). This requires, then, that $P(F) \neq 0$.

(9.9.1) Definition. By the conditional probability of event E, given condition F, we mean

$$\frac{P(E \cap F)}{P(F)}$$

where F is understood as an event with nonzero probability. Also, we adopt the notation

$$P(E|F)$$

which is read "probability of E, given F," for this conditional probability.

Example. (c) A sociometric test is given to a child, who is asked to name his favorite among three acquaintances, say A = {a, b, c}. Suppose he chooses as if at random. The model is in Figure 9.9.2.

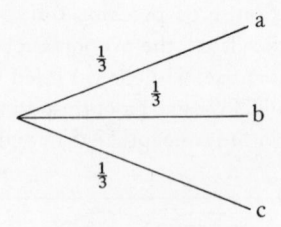

Figure 9.9.2

Consider the event E that he chooses b. Naturally, because E can only be realized one way, E = {b}. Consider the event F that he does not choose c. This is realized by either b or a. Thus, F = {b, a}. Hence, E ∩ F = {b} ∩ {b, a} = {b}. Now from the model,

$$P(E) = P(\{b\}) = \tfrac{1}{3}$$
$$P(F) = P(\{b, a\}) = P(\{b\}) + P(\{a\}) = \tfrac{2}{3}$$
$$P(E \cap F) = P(\{b\}) = \tfrac{1}{3}$$

Thus, if we ask for the probability that he chooses b, given he does not choose c, we have,

$$P(E|F) = \frac{P(E \cap F)}{P(F)} = \frac{1/3}{2/3} = \frac{1}{2}$$

which is just what we would say intuitively: one chance in two, after you discard the c possibility.

Example. (d) Consider the same situation but a different model, given by Figure 9.9.3.

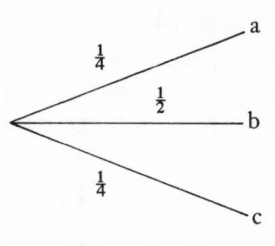

Figure 9.9.3

Then, from this model,

$$P(E) = P(\langle b \rangle) = \tfrac{1}{2}$$
$$P(F) = P(\langle b, a \rangle) = P(\langle b \rangle) + P(\langle a \rangle) = \tfrac{3}{4}$$
$$P(E \cap F) = P(\langle b \rangle) = \tfrac{1}{2}$$

therefore,

$$P(E \,|\, F) = \frac{P(E \cap F)}{P(F)} = \frac{1/2}{3/4} = \frac{2}{3}$$

which shows that the intuition "one chance in two, after you discard c" rests upon the tacit assumption of the equally-likely model.

We now state some properties of conditional probability:

(1) $P(F \,|\, F) = 1$

Proof. Replace "E" by "F" in definition (9.9.1). Then since $F \cap F = F$, we obtain (1).

(2) $P(\emptyset \,|\, F) = 0$.

Proof. Replace "E" by "$\emptyset$" in definition (9.9.1). Then since $\emptyset \cap F = \emptyset$, $P(\emptyset \cap F) = P(\emptyset) = 0$. Thus (2) follows.

(3) If $E_1 \cap F$ and $E_2 \cap F$ are disjoint,

$$P(E_1 \cup E_2 \,|\, F) = P(E_1 \,|\, F) + P(E_2 \,|\, F)$$

An appropriate diagram is given in Figure 9.9.4.

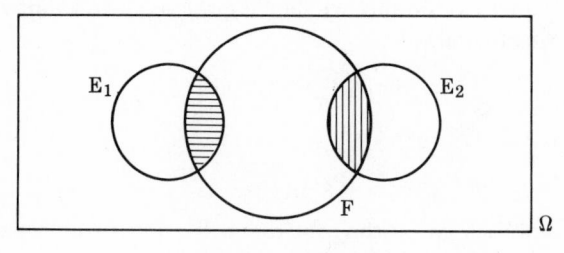

Figure 9.9.4

The horizontal shading shows $E_1 \cap F$: the set of logically possible outcomes for E_1, once we restrict ourselves to F. Similarly, $E_2 \cap F$ is shown by vertical shading. Now the area interpretation suggests a summation, as usual. This is just what (3) says. To practice set algebra and appeal to formal definitions, verify the steps in the following:

Proof.

$$P(E_1 \cup E_2 \,|F) = \frac{P[(E_1 \cup E_2) \cap F]}{P(F)} \qquad \begin{array}{l}\text{[definition (9.9.1), with} \\ E = E_1 \cup E_2]\end{array}$$

But (see Table 3.4.1),

$$(E_1 \cup E_2) \cap F = (E_1 \cap F) \cup (E_2 \cap F)$$

for any sets, so that,

$$P(E_1 \cup E_2 \,|F) = \frac{P[(E_1 \cap F) \cup (E_2 \cap F)]}{P(F)}$$

Now the hypothesis of the conditional (3) is that the two sets $E_1 \cap F$, $E_2 \cap F$ are disjoint, so we can conclude

$$P(E_1 \cup E_2 \,|F) = \frac{P(E_1 \cap F) + P(E_2 \cap F)}{P(F)}$$

$$= \frac{P(E_1 \cap F)}{P(F)} + \frac{P(E_2 \cap F)}{P(F)}$$

$$= P(E_1 \,|F) + P(E_2 \,|F)$$

These results are merely exemplifications of one basic idea:

(9.9.2) Principle of Conditional Probability. For a fixed condition, say F, the conditional probability $P(E \,|F)$ as E varies over the event representations in the basic outcome space, itself satisfies all the principles of probability.

Example. (e) In all generality find an expression for $P(\overline{E} \,|F)$. But $P(\overline{E}) = 1 - P(E)$. Thus, $P(\overline{E} \,|F) = 1 - P(E \,|F)$, remembering to preserve the condition.

Example. (f) Find an expression for the conditional probability of E_1, given E_2, all on the condition F. To do this, we should make our job easier by the purely contextual change of notation:

$$\text{let } P_F(\_) \equiv P(\_|F)$$

Then we want

$$P_F(E_1 \,|E_2)$$

with the "F moved back up" at the conclusion. We note that we must preserve the condition F in writing out the meaning of the "inner" conditioning:

$$P_F(E_1 | E_2) = \frac{P_F(E_1 \cap E_2)}{P_F(E_2)}$$

This follows from (9.9.1) and (9.9.2). Now, restoring notation,

$$P_F(E_1 \cap E_2) = P(E_1 \cap E_2 | F)$$
$$= \frac{P(E_1 \cap E_2 \cap F)}{P(F)} \qquad \text{[definition (9.9.1)]}$$

and,

$$P_F(E_2) = P(E_2 | F)$$
$$= \frac{P(E_2 \cap F)}{P(F)} \qquad \text{[definition (9.9.1)]}$$

Thus,

$$P_F(E_1 | E_2) = \frac{P(E_1 \cap E_2 \cap F)}{P(F)} \cdot \frac{P(F)}{P(E_2 \cap F)}$$

(9.9.3)
$$= \frac{P(E_1 \cap E_2 \cap F)}{P(E_2 \cap F)}$$

But, now looking at the definition (9.9.1) on the right side, we identify $E_2 \cap F$ as a condition, and so,

$$P_F(E_1 | E_2) = P(E_1 | E_2 \cap F)$$

Thus, the conditional probability of E_1, given E_2, all on the condition of F is given by the conditional probability of E_1, given $E_2 \cap F$, which can be directly estimated by using (9.9.3), with appropriate relative frequencies.

When we construct a finite probability model, we are essentially thinking of the various branch probabilities as subject to certain conditions having been realized earlier in a given path. This assignment of parameters to branches then receives a formal conditional probability interpretation.

Example. (g) Consider the learning model of section 9.4, example (c), as continued in section 9.5, where we arrived at the tree shown in Figure 9.9.5.

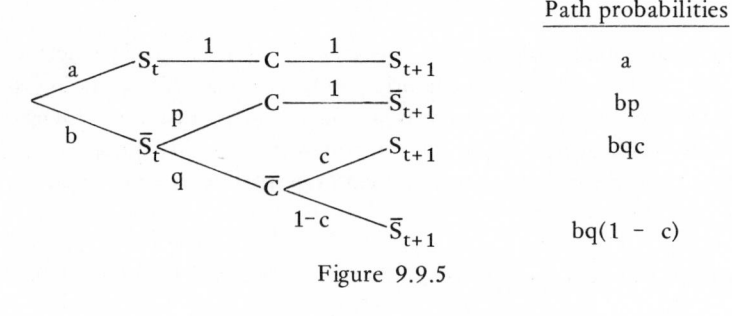

Figure 9.9.5

This is a transition from time t to time t + 1, so for the present purpose we use time subscripts on the states. Now this diagram is created in the context of model construction ("phase 1," in terms of chapter 1). In this context, our thought is "suppose this occurs, and then that, then something . . . will occur with a certain probability." For instance, suppose a person begins in state $\overline{S}$ and suppose the person then makes an error ($\overline{C}$ occurs), then the chance of learning is c. But this should be nothing else than

$$(9.9.4) \qquad\qquad c = P(S_{t+1}|\overline{S}_t\overline{C})$$

In fact, we can show that our constructive tree diagram, with its product rule for path probabilities (9.5.3), implies (9.9.4). We have

$$P(S_{t+1}|\overline{S}_t\overline{C}) = \frac{P(\overline{S}_t\overline{C}S_{t+1})}{P(\overline{S}_t\overline{C})}$$

Now the event $\overline{S}_t\overline{C}S_{t+1}$ is the event uniquely realized by the particular path with probability bqc. The event $\overline{S}_t\overline{C}$, on the other hand, occurs if either of two (mutually exclusive) paths is realized: $\overline{S}_t\overline{C}S_{t+1}$ and $\overline{S}_t\overline{C}\overline{S}_{t+1}$. These two paths have probabilities bqc and bq(1 − c), respectively. Thus, by the addition rule, the probability of $\overline{S}_t\overline{C}$ is bqc + bq(1 − c) = bq. Hence,

$$P(S_{t+1}|\overline{S}_t\overline{C}) = \frac{bqc}{bq} = c$$

This shows that c has a formal interpretation as a conditional probability.

The general point is that the probability parameters directly assigned to the branches of a tree diagram in model construction are interpretable as conditional probabilities, with the condition being the realization of the path up to the point where the parameter is assigned.

9.10. Independence. Sometimes it is useful to impose a special hypothesis in the model construction phase in order to develop analytical consequences of the properties of the model with this (simplifying) assumption; namely, the independence assumption is defined

$$P(E|F) = P(E)$$

for the given condition F and event E. This says that contrary to the whole idea of recomputation on the basis of a condition, in this case no recomputation is necessary: the probability of E given F is just the original probability. Indeed, in very simple probabilistic systems it is sometimes the case that the special hypothesis is even true of the concrete system's repeated behavior or of an aggregate of such concrete systems.

Example. (a) Consider the experiment "a card is chosen at random." What is the chance of getting an ace, given we know we have a red card? Letting C = clubs, S = spades, H = hearts, D = diamonds, we have

$$A \quad \text{occurs} \quad \text{iff} \quad \text{ace shows up}$$
$$A = \{A_C, A_S, A_H, A_D\}$$
$$R \quad \text{occurs} \quad \text{iff} \quad \text{red card is drawn}$$
$$R = \{A_H, A_D, 2_H, 2_D, \ldots\}$$

Hence,

$$P(A|R) = \frac{P(A \cap R)}{P(R)}$$
$$= \frac{P(\{A_H, A_D\})}{P(\{A_H, A_D, 2_H, 2_D, \ldots\})}$$
$$= \frac{2/52}{26/52}$$
$$= \tfrac{1}{13}$$

However,

$$P(A) = P(\{A_C, A_S, A_H, A_D\}) = \tfrac{4}{52} = \tfrac{1}{13}$$

so indeed no recomputation was required. Thus, A and R are said to be independent.

Example. (b) The chance of being correct, given "you have learned" is, in the learning model discussed above, $P(C|S_t) = 1$. But

$$P(C) = P(S_t C S_{t+1}) + P(\bar{S}_t C \bar{S}_{t+1})$$
$$= a + bp$$
$$= a + (1 - a)p$$
$$\neq 1$$

in general. Thus, C is not independent of S_t, as is clear from the terminology alone.

Consider the set of all events in a given model such that the event has nonzero probability. Then the relation of independence is symmetric on this set. To prove this, suppose E is independent of F, so that

$$P(E) = P(E|F) = \frac{P(E \cap F)}{P(F)}$$

Then,

$$P(E)P(F) = P(E \cap F)$$

Next, since $P(E) \neq 0$, we write,

$$P(F|E) = \frac{P(E \cap F)}{P(E)} = \frac{P(E)P(F)}{P(E)} = P(F)$$

so that F is independent of E.

The difficulty that $P(E|F)$ cannot even be defined if $P(F) = 0$ is resolved by exploiting the symmetry of the independence relation. Note that

$$P(E|F) \; = \; \frac{P(E \cap F)}{P(F)}$$

yields

(9.10.1) $P(E \cap F) \; = \; P(F)P(E|F)$

and that

$$P(F|E) \; = \; \frac{P(E \cap F)}{P(E)}$$

yields

(9.10.2) $P(E \cap F) \; = \; P(E)P(F|E)$

so that with independence both (9.10.1) and (9.10.2) yield

(9.10.3) $P(E \cap F) \; = \; P(E)P(F)$

Conversely, if we start with (9.10.3), then if $P(E) \neq 0$, we can divide by it:

$$\frac{P(E \cap F)}{P(E)} \; = \; P(F)$$

Then the left-side is $P(F|E)$, and so, F is independent of E. Similarly, if $P(F) \neq 0$, we show that $P(E|F) = P(E)$, and so, E is independent of F. If, on the other hand, either term vanishes, we can still appeal to (9.10.3) as follows:

(9.10.4) Definition.   Two events E and F of a probability model are independent if, and only if,

$$P(E \cap F) \; = \; P(E)P(F)$$

Independence plays a role in all three basic stages of probability model-building. In the setting-up phase it can be used as a regulative principle. This usage is characteristic of LSA—latent structure analysis (see Lazarsfeld, 1954; Lazarsfeld and Henry, 1968). The basic idea is that a person is given an analytical ground system description at two levels: the state-space level and the observable level. Let X be the state-space variable and let R_1 and R_2 be two response variables. Assume that a probabilistic model is to be imposed on this ground system. Then LSA amounts to one fundamental regulative assumption, called "local independence." Namely:

(9.10.5) $P(R_1 = r \text{ and } R_2 = s | X = x) = P(R_1 = r | X = x) \, P(R_2 = s | X = x)$

which says that, conditioning on the value of the state-space variable, the various responses are independent in the probability sense.

Example. (c) Suppose a child must answer two arithmetic questions, one on addition and the other on multiplication. Let R_1 be a variable measuring the addition response: for example, its latency (i.e., waiting time to provide the response from the onset of the stimulus question). Let R_2 be a similarly defined variable for multiplication. Let X be an unobservable state-space variable interpreted as arithmetic ability. Then with X = x, the chance that the latency is 5 seconds for addition ($R_1 = 5$) and 10 seconds for multiplication ($R_2 = 10$) is the product

$$P(R_1 = 5 | X = x) P(R_2 = 10 | X = x)$$

Later we shall indicate how one of the fundamental derived formulas of LSA follows naturally, using this example, from the Local Independence Assumption and the principles of probability.

At the stage of model-analysis, one has a constructed probability model and one may investigate analytically any possible independence of events of interest, possibly subject to some special initial condition.

Example. (d) As part of a social psychological experiment two teams compete in two successive games. The experimenter assumes that each win-lose configuration of outcomes has the same chance as any other. Does this commit him to the assumption of probabilistic independence of successive games?

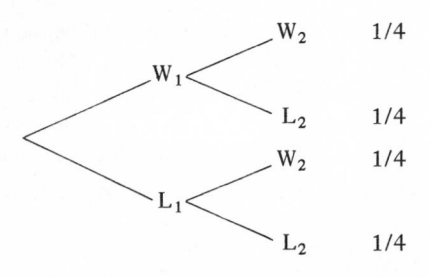

Figure 9.10.1

Figure 9.10.1 is the investigator's model. Here W_i signifies some named one of the two teams, say Team A, wins the ith play (i = 1, 2). We see that

$$P(W_1 \cap W_2) = \tfrac{1}{4} \qquad \text{(top path)}$$

while

$$P(W_1) = P(W_1 \cap W_2) + P(W_1 \cap L_2)$$
$$= \tfrac{1}{2}$$
$$P(W_2) = P(W_2 \cap W_1) + P(W_2 \cap L_1)$$
$$= \tfrac{1}{2}$$

Hence,

$$P(W_1 \cap W_2) = P(W_1)P(W_2)$$

so that W_1 and W_2 are independent. In particular, then, the probability of winning following a win is no different than the probability of winning following a loss. What is really implied by this one assumption of "global equally-likelihood" is not only competitively equivalent teams (as the experimenter had intended) but no effect of the outcomes of plays upon following plays (probably not intended by the experimenter).

Finally, independence plays its role in the application phase of model-building. Here it is an essential component of estimation and goodness-of-fit. The mathematics of this gets a bit rugged—we postpone it for a while and only say that among the "auxiliary statistical assumptions" that are made in connecting models to data there generally appears the following (tacit) assumption, really part of the meaning postulates for the notion of "repetition" of a process, namely: each repetition (each concrete system: person, person-pair, group, whatever) realizes the events of interest independently of the others. This is reasonable for most soundly conducted social psychological experiments where, for example, subjects in distinct repetitions of the experiment do not communicate. It is not an unreasonable assumption in surveys, in which each household responds independently of any other in the intuitive sense that they collectively form a mere aggregate. But as soon as we consider social relational systems with some nontrivial linkages among the social units, the assumption of independence of behavior is implausible. In fact, Coleman (1964) takes nonindependence of the repeated processes—high school students' getting involved in sports, or buying records, or adopting some fad; union shop members' voting in union elections—as the basis for a novel methodology for the analysis of social processes. This methodology will be treated in Chapter 13.

CHAPTER TEN   MARKOV CHAINS AND RELATED PROBABILITY TOPICS

10.1. Aim of the Chapter. In the last chapter, we developed the elementary ideas and techniques for a probabilistic approach to a phenomenon. In this chapter, we do an "advanced course" in such techniques, focusing on process. Thus, the abstract process language of Chapter 8 will be exemplified in the focus on Markov chains. We will see examples of chains, their connection with matrices and graphs, the classification of chains, and other topics that arise naturally in studying probability processes. To keep the treatment of processes relatively elementary, the chapter is restricted to discrete-time and discrete-state Markov chains; in fact, we usually assume a finite number of states. From a substantive point of view this chapter serves as an introduction to Chapters 12-16, which deal with Markov social processes.

10.2. Law of Total Probability and the State-Space Approach. An important aspect of the state-space approach (see section 8.5) in the context of probabilistic systems is that the ground system of states—mental states, in the case of individuals—is used to partition the outcome space of the model. To see how this works, let us first treat the partitioning of an outcome space of an arbitrary finite probability model and then apply the result in the context of probabilistic systems employing the state-space approach.

Let Ω be the outcome space—the set of possible outcomes—in some model. Let $F_1, F_2, \ldots, F_m$ be m events (and so represented as subsets of Ω) such that the family $\{F_i\}$ forms a partition of Ω. That is,

$$\Omega = \bigcup_{i=1}^{m} F_i \text{ and } F_i \cap F_j = \emptyset \qquad (\text{if } i \neq j)$$

An appropriate picture to have in mind is Figure 10.2.1. Thus, Ω is expressed as a disjoint union of the family of events $\{F_i\}$. By the assumption that Ω is the outcome space of a finite probability model, we conclude that

$$1 = P(\Omega) = P(\bigcup_{i=1}^{m} F_i) = \sum_{i=1}^{m} P(F_i)$$

Thus, the total probability of Ω, which is 1, is expressed as a sum over the probabilities of the members of the disjoint union.

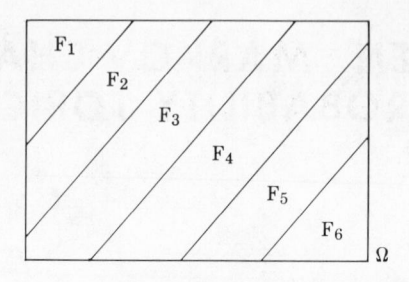

Figure 10.2.1

Example. (a) In a learning experiment comprised of 25 trials, by trial 10 the subject either has learned (state S) or not (state $\bar{S}$); thus, the representation rule for events implies that the set of all possible paths that the subject might take in the experiment is partitioned into $\{S_{10}, \bar{S}_{10}\}$, where S_{10} contains all paths in which the subject has learned by trial 10 and $\bar{S}_{10}$ contains all others.

Example. (b) A person is sampled at random from an adult population. This population forms the outcome space of the model for this sampling of one person, which assigns equal probability to each member of the population. Then any classification of this population can be taken as the partition system of events.

Now suppose that in this context of having a certain partition system of events $\{F_i\}$ we are interested in a certain event E. Then the appropriate picture is Figure 10.2.2.

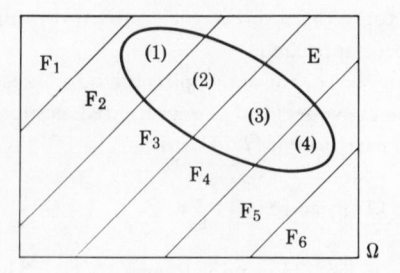

Figure 10.2.2

Thus, E is represented by some subset of Ω. From the figure, we see that E can be expressed as a disjoint union, using the partition system $\{F_i\}$, namely,

$$E = (1) \cup (2) \cup (3) \cup (4) \qquad \text{(as shown)}$$

But,

$$
\begin{aligned}
(1) &= F_2 \cap E \\
(2) &= F_3 \cap E \\
(3) &= F_4 \cap E \\
(4) &= F_5 \cap E
\end{aligned}
$$

Thus,

$$E = \bigcup_{i=2}^{5} (F_i \cap E)$$

Moreover,

$$F_1 \cap E = \emptyset, \qquad F_6 \cap E = \emptyset$$

so no harm is done if we set-theoretically add ("union") them into the overall disjoint union:

$$E = \bigcup_{i=1}^{6} (F_i \cap E)$$

Thus, applying the probability function to both sides,

$$
\begin{aligned}
P(E) &= P[\bigcup_{i=1}^{6} (F_i \cap E)] \\
&= \sum_{i=1}^{6} P(F_i \cap E)
\end{aligned}
$$

where the last step follows because we have a disjoint union. Now from the definition (9.9.1) of conditional probability, for each i,

$$P(F_i \cap E) = P(E|F_i)P(F_i)$$

and so,

$$P(E) = \sum_{i=1}^{6} P(E|F_i)P(F_i)$$

In general, if $\{F_i\}$ is a partition of Ω and if E is any event, we say that the valid identity

(10.2.1) $$P(E) = \sum_{i} P(E|F_i)P(F_i)$$

expresses "the law of total probability." This total probability law plays an enormous role in the probabilistic systems developed by the state-space technique.

A routine example of (10.2.1) is given first:

Example. (c) Sample a child at random from a certain class containing 30 boys and 10 girls. Here, the model is

$$\Omega = \{\alpha_1, \alpha_2, \ldots, \alpha_{40}\} \qquad \text{(the class as a set)}$$

$$P(E) = \frac{N_E}{40} \qquad \text{(for any } E \subseteq \Omega)$$

where N_E is the number of children in set E. Also the partitioning is,

$$F_1 = \{\alpha_i: \alpha_i \text{ is a boy}\}$$
$$F_2 = \{\alpha_i: \alpha_i \text{ is a girl}\}$$

and so,

$$P(F_1) = .75, \ P(F_2) = .25$$

Now suppose that, of the boys, 90 percent passed a certain test, while only 75 percent of the girls did so. Let us define event E such that

$$E \text{ occurs} \qquad \text{iff} \qquad \text{sampled child passed the test,}$$

and so,

$$P(E|F_1) = .90$$
$$P(E|F_2) = .75$$

Hence, by (10.2.1),

$$\begin{aligned} P(E) &= P(E|F_1)P(F_1) + P(E|F_2)P(F_2) \\ &= (.90)(.75) + (.75)(.25) \\ &= .675 + .1875 \\ &= .8625 \end{aligned}$$

Example. (d) A less routine example involves the learning situation. The grounding of the probabilistic system involves a state-space with two states on trial 10: $S_{10}, \overline{S}_{10}$. Now if we partition the outcome space—of paths of state and response through successive trials—we have the picture given in Figure 10.2.3.

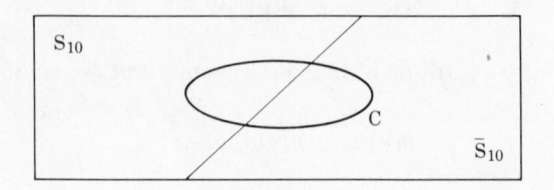

Figure 10.2.3

The event C of interest is

$$C \text{ occurs} \quad \text{iff} \quad \text{person gives correct response on trial 10.}$$

Thus,

(10.2.2) $$P(C) = P(C|S_{10})P(S_{10}) + P(C|\overline{S}_{10})P(\overline{S}_{10})$$

by taking $F_1 = S_{10}$, $F_2 = \overline{S}_{10}$ in the law of total probability. Now the basic conditions describing this process include

$$P(C|S_t) = 1$$

meaning that if you are in the learned state at any trial t, then with probability 1 you give a correct response. But

$$P(C|\overline{S}_t) = p$$

a "guessing parameter." Thus, (10.2.2) becomes

$$P(C) = P(S_{10}) + pP(\overline{S}_{10})$$

where since $\overline{S}_{10}$ is the complement of S_{10},

$$P(\overline{S}_{10}) = 1 - P(S_{10})$$

Thus,

$$P(C) = P(S_{10}) + p[1 - P(S_{10})]$$

This expresses the response probability (at a given time) in terms of the present state probability and a parameter.

Example. (e) In Bernard C. Cohen's model for the Asch conformity situation (cf. section 8.5), we find a state-space approach: at the observable response level, conformity (C) occurs or not ($\overline{C}$) on any trial n; and at the mental (state-space) level a person is in one of four possible states denoted M_1, M_2, M_3, M_4. Also, note that

$$P(C|M_i) = 1 - P(\overline{C}|M_i) \qquad (i = 1, 2, 3, 4)$$

by applying the principle of conditional probability (9.9.2). Thus, we need only specify how the probability of $\overline{C}$ depends on the state. Cohen defines the state-space by a structure whose presently relevant ideas include

$$P(\overline{C}|M_1) = 1$$
$$P(\overline{C}|M_2) = 1$$
$$P(\overline{C}|M_3) = 0$$
$$P(\overline{C}|M_4) = 0$$

Thus, the total probability of the nonconforming response (on any trial n) is given by

$$P_n(\overline{C}) = \sum_{i=1}^{4} P(\overline{C}|M_{i,n})P(M_{i,n})$$

where $M_{i,n}$ occurs if and only if the subject is in state M_i on trial n. Hence,

$$P_n(\overline{C}) = P(M_{1,n}) + P(M_{2,n})$$

gives the observable response probability in terms of the state probabilities holding at the given time.

Example. (f) Recall in our discussion of LSA [see (9.10.5)] that we said that our next look at LSA would involve exploiting the local independence axiom to generate a formula of basic importance in this model. This can now be accomplished with the following substitutions into the law of total probability (10.2.1):

$$E = E_1 \cap E_2$$

where

$$E_1 \text{ occurs } \quad \text{iff} \quad R_1 = r$$
$$E_2 \text{ occurs } \quad \text{iff} \quad R_2 = s$$

and we are interested in $P(E)$, the observable probability that a person gives both response r (in dimension R_1) and response s (in dimension R_2). Thus, take

$$F_x \text{ occurs } \quad \text{iff} \quad X = x$$

assuming x varies over a finite set of values—the latent classes counting as the composition of the state-space. Then we note that an immediate application of total probability gives

$$P(E) = P(E_1 \cap E_2) = \sum_x P(E_1 \cap E_2|F_x)P(F_x)$$

Now the axiom of local independence amounts to

$$P(E_1 \cap E_2|F_x) = P(E_1|F_x)P(E_2|F_x)$$

and so,

$$P(E_1 \cap E_2) = \sum_x P(E_1|F_x)P(E_2|F_x)P(F_x)$$

which may be considered the fundamental formula of LSA. In the original notation,

$$P(R_1 = r \text{ and } R_2 = s) = \sum_x P(R_1 = r|X = x)P(R_2 = s|X = x)P(X = x)$$

yielding a joint or multidimensional response probability in terms of the state probabilities $P(X = x)$ and certain parameters, namely,

$$P(R_1 = r|X = x), P(R_2 = s|X = x)$$

Example. (g) Models of the probabilistic type are readily constructed by using a combination of ideas: the state-space approach, the law of total probability, and the Markov concept. We will deal with the latter shortly. But here is an example of a model for an experiment in which we have a group process described, at the observable level, in Bales' 12 categories (see Bales, 1950). At the state-space level, let the description be in terms of four "phases" in which the group might find itself at a given time (cf. Parsons, Bales, and Shils 1953)

α: adaptation phase
γ: goal-attainment phase
ι: integrative phase
λ: latent tension-reduction and pattern maintenance phase.

Thus, if we take a probabilistic-system standpoint, we have four possible states in the analytical ground system, with twelve possible responses per group member. Let N be the group size. Let R_i be the response variable for member i, i = 1, ..., N. Assume the LSA axiom of local independence holds for the system of N responses during a small time interval or episodic duration:

$$P(R_1 = r_1, R_2 = r_2, \ldots, R_N = r_N | X = x) = P(R_1 = r_1 | X = x)P(R_2 = r_2 | X = x)$$
$$\cdots P(R_N = r_N | X = x) = \prod_i P(R_i = r_i | X = x)$$

where now x is a logical variable with domain $\{\alpha, \gamma, \iota, \lambda\}$ and each r_i ranges over the twelve response categories. Thus, by generalization of the LSA theorem on total response probability from 2 to N responses, we have

$$P(R_1 = r_1, R_2 = r_2, \ldots, R_N = r_N) = \sum_x [\prod_i P(R_i = r_i | X = x)] P(X = x)$$

which expresses the group's configuration of observable responses in a short duration in terms of the state of the group—which is here interpreted in terms of phases—and certain parameters of the form

$$P(R_i = r_i | X = x)$$

that is, the chance that person i makes response r_i, given the group phase is x. This is a shift in the usual intended interpretation of local independence. A common interpretation is that there are N items answered by one person in ability state x, and the basic consequence is an expression of the chance for each configuration of responses by the individual in terms of his state. In the group process interpretation, instead of N items we have N "parts" of the single entity (the group). The usual idea is that, conditional on the ability state (of a person), the observable dimensions of response (of that person) are independent. Here the idea is that the observable responses (by different individuals), conditional on the state (of the group), are independent.

Example. (h) Here is a somewhat more concrete example of the state-space approach. Each of two persons is in an expectation state: either E or $\overline{E}$, where E means "person is oriented to regarding other as better" (at a given task) and $\overline{E}$ means "this is not the case." The pair-state is an ordered pair of person states: (E, E), (E, $\overline{E}$) ($\overline{E}$, E), ($\overline{E}$, $\overline{E}$). Suppose that the relation

$$pDo \quad \text{iff} \quad \text{person defers to other}$$

is observable at a given trial or time. Assume that on each trial,

$$\text{either } pDo \quad \text{or} \quad oDp \quad \text{(but not both)}$$

and we write D, $\overline{D}$ to mean, respectively: p defers to o, or not (and then this means o defers to p). Thus, the grounding for our probabilistic system is

$$\{EE, E\overline{E}, \overline{E}E, \overline{E}\overline{E}\} \qquad \text{(state-space)}$$
$$\{D, \overline{D}\} \qquad \text{(responses)}$$

Clearly, we will introduce the idea that conditional on the pair-state, D has a certain chance of occurring. Thus,

$$P(D \text{ on trial } n) = \sum_e P(D|e)P(e \text{ on } n)$$

where e ranges over the state-space of pairs of expectation states. This equation is the expression of the response probability in terms of the state probabilities at the given time and certain parameters of the form $P(D|e)$. An expectation-states model is studied in detail in Chapter 14.

10.3. Markov Chains: Examples. In this paragraph we present a first look at Markov chains. Here we are dealing with processes in terms of probabilistic systems, but first we need a simple example of the basic idea of a Markov chain. The basic process rule of thinking in the small and deducing in the large can be exploited if one can make a definite assumption about process in the small that will activate well-defined and well-known mathematical techniques such that the trajectory of the process, through time, is deduced. One such mathematical assumption is that the transition from the present state to the next state does not depend on the path by which the system arrived at the present state. This assumption is exploited in the setting-up stage of a model by envisioning an arbitrary time at which the transition takes place, picking up the system at its present state, and disregarding system behavior prior to the contemporary state. This description makes it evident that the Markov idea is just a definite probabilistic version of the very concept of the state of the system. (Recall our discussion in section 8.10.)

Here are some examples of setups that assume a Markov chain.

Example. (a) Consider a pair of persons and a series of equally spaced times $t_0, t_1, t_2, \ldots, t_n, \ldots$, measured, say, in minutes. We want to indicate that the chance of an interact (I) at t_{n+1} depends upon whether or not ($\bar{I}$) they are interacting at t_n. Thus, we draw two trees, as in Figure 10.3.1.

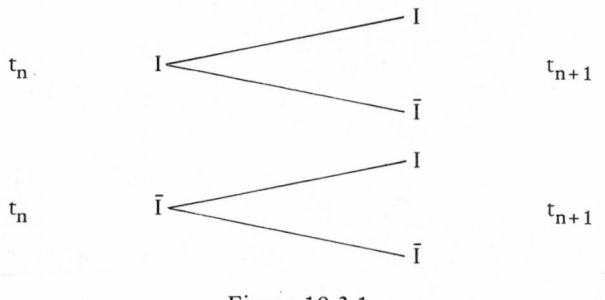

Figure 10.3.1

We are now picturing our system "in the small," and taking a probabilistic standpoint toward the ground system, $\{I, \bar{I}\}$. We argue as follows.

Axiom 1. If at t_n the pair is in state I, then at t_{n+1} the pair is still in state I with probability c.

Axiom 2. If at t_n the pair is in state $\bar{I}$, then at t_{n+1} the pair is in I with chance p. Then the trees in Figure 10.3.1 become those of Figure 10.3.2.

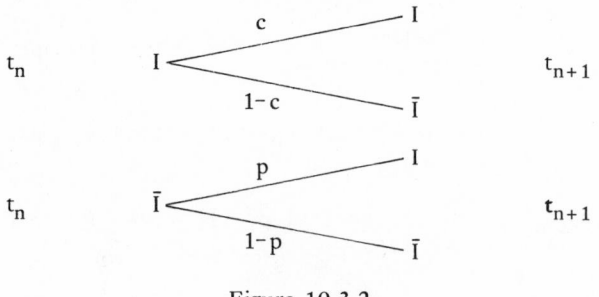

Figure 10.3.2

We can call c the continuation probability and call p the initiation probability—the chance of an interact starting (from $\bar{I}$ at t_n). Now two aspects of this setup should be noted:

1. The failure to mention anything about $t_1, t_2, \ldots, t_{n-1}$ and, so, the indifference (systematically) as to how long the pair has been in a given state, how many changes of state have occurred, and so forth.

2. The explicit assumption, for each possible state at t_n, of what the pair may or may not do in the way of transition to t_{n+1}.

The two trees of Figure 10.3.2—one per possible present state—express the

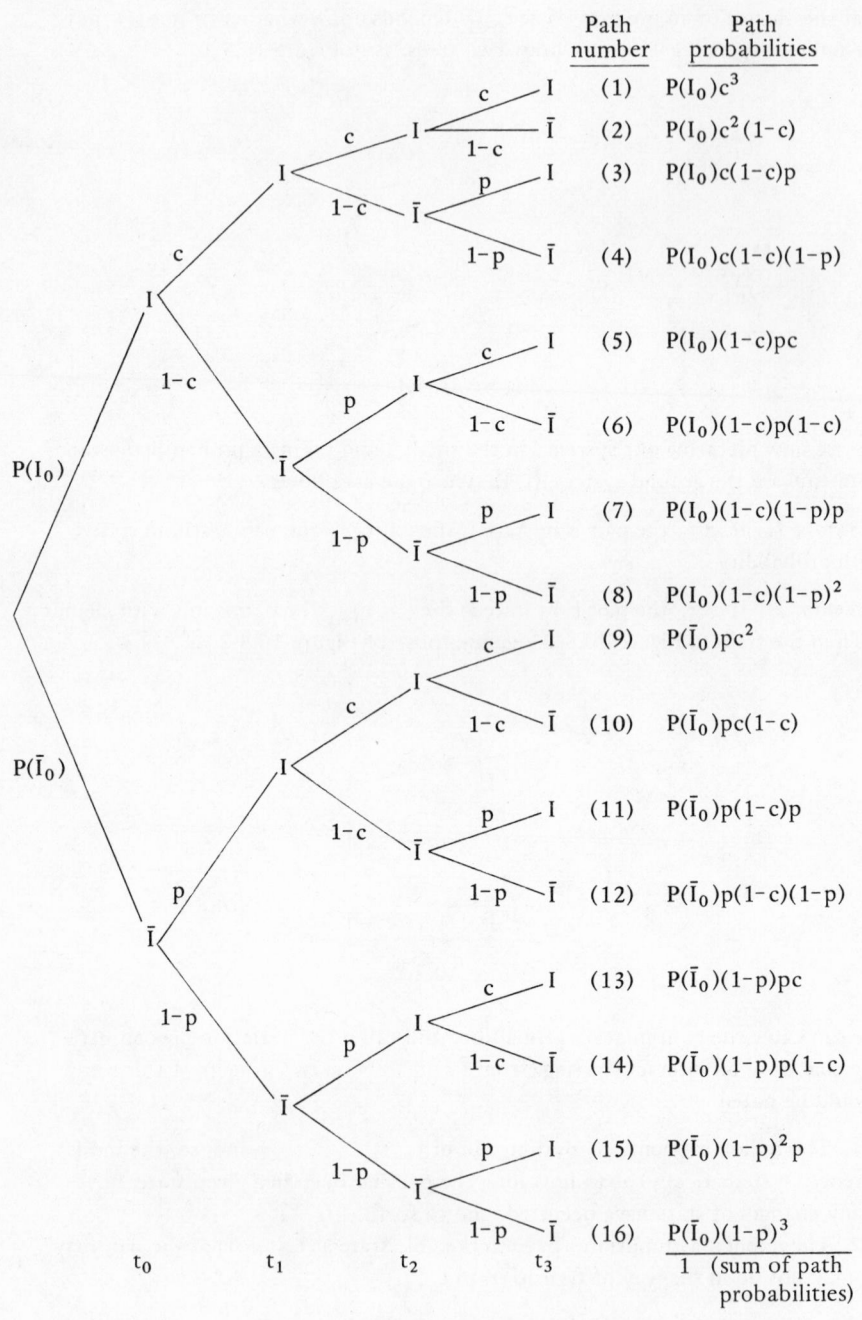

Figure 10.3.3

transition probabilities; these are the chances of state change in an interval of time $t_{n+1} - t_n$. To see the process in the large consider a tree diagram with the initial condition of chance $P(I_0)$ that the individuals are in interaction at the initial time of observation, called t_0, and let us calculate the probability of various interaction sequences (or trajectories) in the time from t_0 to t_3 as in Figure 10.3.3. The tree in Figure 10.3.3 shows by the product rule for path probabilities (9.5.3) the probabilities of the possible interaction sequences. In other words, the two items—the setup in the small, given by Figure 10.3.2 and repeatedly applied, and the initial condition, $P(I_0)$—together generate a finite probability model over interaction sequences over time; as shown in the partial tree of Figure 10.3.4.

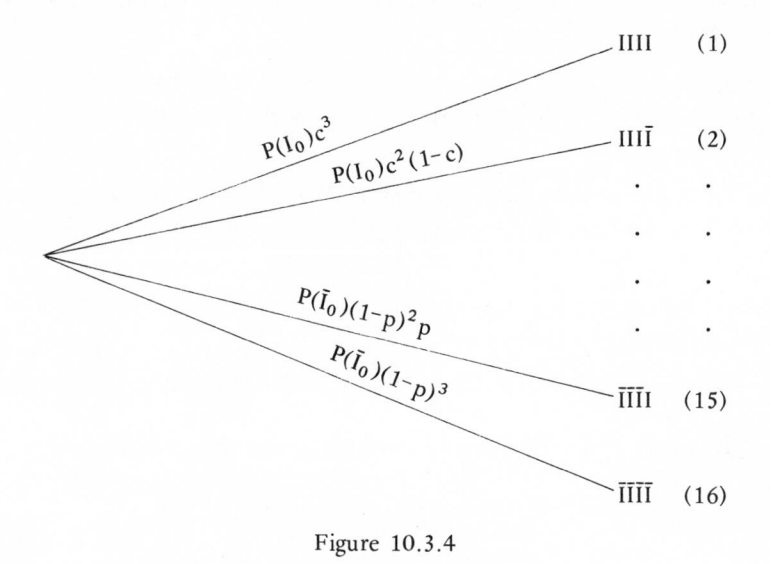

Figure 10.3.4

For any event occurring in time, its probability is computed in the usual way: (a) apply the representation principle of events to represent the event as a set of paths (possible realizations of the process in which one would say the event occurred if one of these possibilities is realized) and (b) apply the addition rule, adding the path probabilities of these possible realizations.

For instance, if

$$E \text{ occurs} \quad \text{iff} \quad \text{they interact at } t_1 \text{ and } t_2$$

then the set of paths for the representation of E is

$$E = \{(1), (2), (9), (10)\}$$
$$= \{\text{IIII}, \text{III}\bar{\text{I}}, \bar{\text{I}}\text{III}, \bar{\text{I}}\text{II}\bar{\text{I}}\}$$

Thus,

$$
\begin{aligned}
P(E) &= P(1) + P(2) + P(9) + P(10) \\
&= P(I_0)c^3 + P(I_0)c^2(1-c) + P(\bar{I}_0)pc^2 + P(\bar{I}_0)pc(1-c) \\
&= P(I_0)c^2[c + (1-c)] + P(\bar{I}_0)pc[c + (1-c)] \\
&= P(I_0)c^2 + P(\bar{I}_0)pc
\end{aligned}
$$

Example. (b) At time t_n, arbitrary individual α (in some specified set of individuals) holds a certain disposition to vote a certain way in a forthcoming election. At this time he is in definite "contact" with two other individuals also holding some such attitude. For simplicity, assume the attitude is "for" some given candidate (A) or not ($\overline{A}$). During the interval of time from t_n to t_{n+1} (namely, $t_{n+1} - t_n$) he is subjected to two influence inputs from the others, one from each. The following assumptions are made:

Axiom 1. If α receives an influence input from an individual and that individual is in the same (attitude) state, then no change of attitude is created by this influence input.

Axiom 2. If α receives an influence input from an individual and that individual is in the complementary state, then α changes to that complementary state with probability γ.

Axiom 3. Influence-induced events stemming from distinct individuals, in a given time interval, are independent.

Axiom 4. (Initial condition) α is in state A at time t_0 and the two others are in state $\overline{A}$.

The partial graph of the relational system is shown in Figure 10.3.5.

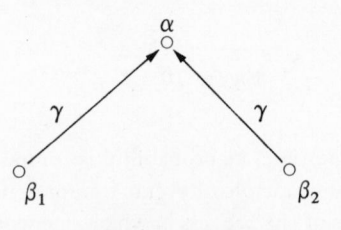

Figure 10.3.5

Here β_1 and β_2 are the others; the graph is a representation of an artificial relational system with domain $\{\alpha, \beta_1, \beta_2\}$, because there is no indication of the connectedness from α back to others or among the others. We shall introduce this aspect of the process subsequently in treating the work of James Coleman (1964). At present, we can treat β_1 and β_2 as fixed in state $\overline{A}$.

Now the tree-diagram technique for setting up the model uses the rule: each possible state at t_n has its own diagram (see Figure 10.3.6).

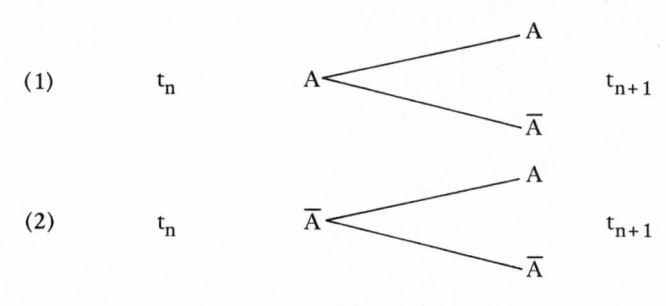

Figure 10.3.6

The problem is now to utilize the axioms to fill in the transition probabilities.

In the (small) time interval from t_n to t_{n+1} we have the following mutually exclusive outcomes with respect to the influence process (given that α in A at t_n):

(a) β_1 induces α to $\overline{A}$, β_2 does not
(b) β_2 induces α to $\overline{A}$, β_1 does not
(c) β_1 and β_2 induce α to $\overline{A}$
(d) neither β_1 nor β_2 induce α to $\overline{A}$.

Now the event that β_1 induces α to $\overline{A}$ is independent of the event that β_2 does not induce α to $\overline{A}$, by axiom 3, so that

$$P(a) = P(\beta_1 \text{ induces } \alpha \text{ to } \overline{A})P(\beta_2 \text{ does not induce } \alpha \text{ to } \overline{A})$$
$$= \gamma(1 - \gamma)$$

applying axiom 2. Similarly,

$$P(b) = \gamma(1 - \gamma)$$
$$P(c) = \gamma^2$$
$$P(d) = (1 - \gamma)^2$$

Note that

$$P(a) + P(b) + P(c) + P(d) = \gamma^2 + 2\gamma(1 - \gamma) + (1 - \gamma)^2$$
$$= [\gamma + (1 - \gamma)]^2$$
$$= 1$$

as required. Now the two possible transition events of tree (1) in Figure 10.3.6 can be treated in terms of the representation principle: which of (a), (b), (c), and (d) realize $A \rightsquigarrow \overline{A}$ (for α)? Clearly, all except (d).

Thus, we have the assignment shown in Figure 10.3.7 (1). Similarly, given α is in state $\overline{A}$ at t_n, we apply first axiom 3 and then axiom 1: the conclusion is that α stays in $\overline{A}$ as shown in tree (2) of Figure 10.3.7. Thus, once again we have set up the model in the small by looking at what can happen in each possible state, under the given inputs.

In the large, then, the induced probability of various possible realizations (paths)

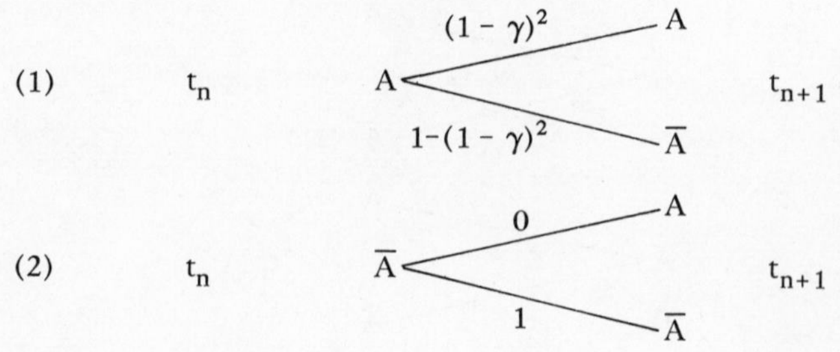

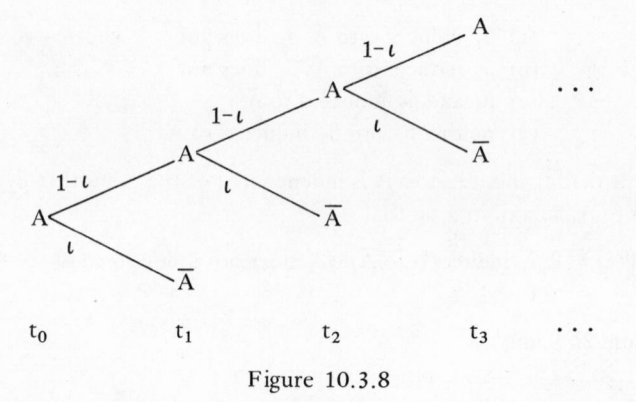

Figure 10.3.7

of the process [taking $\iota = 1 - (1 - \gamma)^2$ for the influence-induced shift probability] is shown in Figure 10.3.8, keeping in mind that by axiom 4, the starting state is A.

Figure 10.3.8

Here we have a state ($\overline{A}$) such that once it is reached it is preserved, under the given initial and boundary conditions. Thus, A is an equilibrium state. The tree structure makes clear that if X is the time of entry into $\overline{A}$, we have, by computing path probabilities,

$$P(X = 0) = 0$$
$$P(X = 1) = \iota$$
$$P(X = 2) = (1 - \iota)\iota$$
$$P(X = 3) = (1 - \iota)^2 \iota$$
$$\cdot$$
$$\cdot$$
$$\cdot$$

(10.3.1) $P(X = n) = (1 - \iota)^{n-1}\iota$ $(n = 1, 2, \ldots)$

Also, the probability that at time t_n the process is still in state A is the probability of the uppermost path in Figure 10.3.8, namely, $(1-\iota)^n$. Hence, the probability that at time t_n the process has entered the equilibrium state is $1 - (1 - \iota)^n$, which approaches 1 as n grows large. Hence, the equilibrium state is approached in this probability sense. In probability theory, such a state is called absorbing.

The outcome space of the process is given by all paths that consist of a string of A's followed by a $\overline{A}$:

$$\Omega = \left\{A\overline{A}, AA\overline{A}, AAA\overline{A}, \ldots\right\}$$

with the probability assignment easily read off from Figure 10.3.8. Thus, the axioms of the process, including the initial conditions described in axiom 4, induce a probability model. This model differs from that generically described by the axioms for a finite probability model in that Ω is not finite. This is because for every integer n (n = 1, 2, . . .) there exists an outcome in Ω, where X = n, with an assigned probability. Such an outcome space is called countably infinite. The set of all possible realizations of the process, which is Ω, forms a countably infinite domain with a probability assignment.

10.4. Concept of a Random Variable. The variable X in (10.3.1) is an example of a scientific variable with empirical domain also having a probability assignment. In such a case we say that X is a random (scientific) variable. To be clear in the conceptual connection to measurement theory, note that the manner in which we introduced X makes it clear that X is a mapping with domain

$$\Omega = \left\{A\overline{A}, AA\overline{A}, AAA\overline{A}, \ldots\right\}$$

and range the real numbers (since the integers are a proper subset of the real numbers). Thus,

$$\Omega \xrightarrow{\ \ X\ \ } \Re$$

is given by

$$X(\underbrace{AA \cdots A\overline{A}}_{n\text{ A's}}) = n$$

Thus, X is on an absolute scale; since it assigns numbers by counting, the assignment is unique. The tree in Figure 10.3.8 gives the probability assignment over the domain Ω of this scientific variable. Then we compute, for each possible value of X, the probability that it assumes that value using the probability assignment over the empirical domain. Thus, (10.3.1) is the induced probability distribution of the variable X.

In general, let Ω be the domain of possible outcomes of a probability model. Then any mapping into the real numbers is called a random variable. It immediately takes its values with probabilities, and we say that it has a probability distribution. The notation "X = n" means "an outcome obtains that is mapped into n under X."

Example. (a) The context need not be dynamic. Consider a status mapping (see section 5.7) of the form, s: $A \to \Re$. Let α be in A. Suppose α's next interact can be construed to occur as if he sampled $A - \{\alpha\}$ randomly. Then we have a probability model with $\Omega = A - \{\alpha\}$ and equally-likely assignment of probabilities. It follows that variable s, as restricted to Ω, becomes a random variable. Suppose there are three possible status values: 1, 2, and 3, representing an empirical system of invidious distinctions ordinally. Then let X be the random variable given by

$$X(\beta) = s(\beta) \qquad \text{iff} \qquad \alpha \text{ samples } \beta \in \Omega$$

The possible outcomes in Ω are each assigned a number, and so, X is a random variable.

To say that X = 1 means an event occurs: namely, $s(\beta) = 1$. But since β is chosen at random, we really want to use the representation of the event by a subset of all β such that $s(\beta) = 1$:

$$X = 1 \text{ is the event} \{\beta \in \Omega: s(\beta) = 1\}$$
$$X = 2 \text{ is the event} \{\beta \in \Omega: s(\beta) = 2\}$$
$$X = 3 \text{ is the event} \{\beta \in \Omega: s(\beta) = 3\}$$

and so if n_1, n_2, and n_3 are the respective sizes of the status levels (excluding α), we have

$$P(X = 1) = P(\{\beta \in \Omega: s(\beta) = 1\}) = \frac{n_1}{n_1+n_2+n_3}$$

$$P(X = 2) = P(\{\beta \in \Omega: s(\beta) = 2\}) = \frac{n_2}{n_1+n_2+n_3}$$

$$P(X = 3) = P(\{\beta \in \Omega: s(\beta) = 3\}) = \frac{n_3}{n_1+n_2+n_3}$$

because the assumption of random sampling assigns each β a chance $\dfrac{1}{n_1+n_2+n_3}$ of being chosen by α. Thus, we obtain the probability distribution of X from the model based on Ω.

Example. (b) Another example is given by Figure 10.3.3, letting

$$X = \text{the number of times at which I obtains.}$$

That is, X maps each path of the process into the number of I's in that path. X is scaled absolutely because it is a counting variable. Several terms of the probability distribution of X are given by

$P(X = 4) = P(\langle(1)\rangle) = P(I_0)c^3$ [where (1),(2), ... refer to path labels]

$P(X = 3) = P(\langle(2), (3), (5), (9)\rangle)$

$\qquad = P(I_0)c^2(1-c) + P(I_0)c(1-c)p$

$\qquad + P(I_0)(1-c)pc + P(\bar{I}_0)pc^2$

where $\langle(1)\rangle$ represents the event that we realize four successive interacts, and $\langle(2), (3), (5), (9)\rangle$ represents the event that we realize exactly three interacts.

Some other brief examples of random variables are easily cited: the length of the waiting line at a counter in a post office, the time to terminate a "busy" signal in a telephone call, the number of children born on a particular day in a particular town, the velocity of a molecule in a gas, the position of a molecule in a gas, and so forth, provided that the phenomenon is represented by some probabilistic model. The general rule is to consider any scientific variable, imagine that it is applied over a concrete empirical domain and that a probability model is constructed with this domain as outcome space: then the scientific variable in that context becomes a random variable. Formally, a mathematician only looks at two things: the existence of a model, (Ω, β, P), and the existence of a mapping $\Omega \overset{X}{\to} \Re$; together they prompt him to say that X is a random variable no matter what interpretation Ω or X may have. In particular, the scale-type of X is not considered in pure probability theory.

10.5. Working with Random Variables. Given a random variable X, we can briefly describe some basic notions associated with it; these notions are familiar to sociologists because of their common use in statistics.

(10.5.1) Definition. The expectation of a random variable X is given by

$$E(X) = \sum_x xP(X = x)$$

where x is a logical variable with domain the image of Ω under X: the set of possible values for X.

Example. (a) Suppose X is such that it counts the number of interacts in two time intervals in a model as shown in Figure 10.5.1.

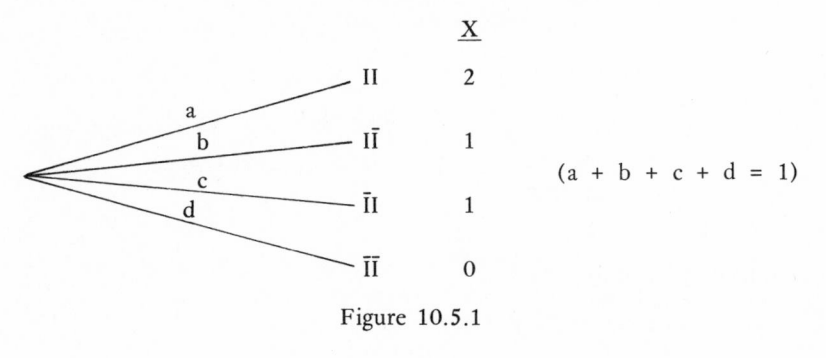

Figure 10.5.1

Then from the tree we see that

$$P(X = 2) = a$$
$$P(X = 1) = b + c$$
$$P(X = 0) = d$$

Then,

$$E(X) = \sum_x xP(X = x) \qquad (x \in \{2, 1, 0\})$$
$$= 2 \cdot P(X = 2) + 1 \cdot P(X = 1) + 0 \cdot P(X = 0)$$
$$= 2a + b + c$$

Example. (b) Let X record the status level of the sampled individual in example (a) of section 10.4. Then

$$E(X) = \sum_x xP(X = x) \qquad (x \in \{1, 2, 3\})$$
$$= 1 \cdot P(X = 1) + 2 \cdot P(X = 2) + 3 \cdot P(X = 3)$$
$$= \frac{n_1}{n_1 + n_2 + n_3} + \frac{2n_2}{n_1 + n_2 + n_3} + \frac{3n_3}{n_1 + n_2 + n_3}$$
$$= \frac{n_1 + 2n_2 + 3n_3}{n_1 + n_2 + n_3}$$

From a purely mathematical viewpoint this expectation is well defined. It is, for definite values of the parameters n_1, n_2, and n_3, a definite number. However, in this case it is empirically meaningless, because it involves (weighted) summation over the values of an ordinally scaled variable. (See section 7.6.)

Expectations are numbers existing in a model. They stand to arithmetic means as probabilities stand to relative frequencies.

Example. (c) A learning experiment is conducted with 30 subjects independently realizing the learning process and is modeled such that the probability of 0, 1, or 2 errors is calculated in the model (following estimation of certain parameters) as

$$P(X = 0) = .10$$
$$P(X = 1) = .50$$
$$P(X = 2) = .40$$

Thus,

$$E(X) = \sum_x xP(X = x) \qquad (x = 0, 1, 2)$$
$$= 0 \cdot P(X = 0) + 1 \cdot P(X = 1) + 2 \cdot P(X = 2)$$
$$= .50 + .80$$
$$E(X) = 1.3$$

On the side of observations, let

$$X_i = \text{number of errors in the protocol of subject i} \qquad (i = 1, 2, \ldots, 30)$$

Then,

$$\overline{X} = \frac{\sum\limits_i X_i}{30}$$

is the observed mean. The predicted value of $\overline{X}$ is $E(X) = 1.3$. If, say, $\overline{X} = 12$, this would be evidence that the assumptions of the model need adjustment, assuming a sound identification and estimation procedure, because $E(X)$ is based on terms of the form $P(X = x)$, which in turn are based on summations over path probabilities.

Some basic properties of expectation are now discussed.

(10.5.2) Properties of E Operator.

$$
\begin{aligned}
&\text{(a)} \ \ E(X + Y) = E(X) + E(Y) \\
&\text{(b)} \ \ E(cX) = cE(X) \qquad (c \in \Re) \\
&\text{(c)} \ \ E(c) = c
\end{aligned}
$$

The general use of these properties is in abstract calculations in the analysis phase of a model. For example, X is the number of errors on trial t_n and Y is the number of errors on trial t_{n+1}. Then $X + Y$ is the number of errors cumulated. The expected value of $X + Y$, according to (a), is just the sum of the expected number of errors on each trial. To illustrate (b), suppose we change the unit of a money variable X from dollars to hundreds of dollars, by dividing by 100. Then the expected value of the amount of money in hundreds of dollars is just the original expected value divided by 100. These are the same operations that are valid for the arithmetic mean, of course.

Another important idea associated with random variables is the idea of independence, extended now from events (such as $X = x$ and $Y = y$) to variables as such.

(10.5.3) Definition. Two random variables X, Y defined on the same outcome space are independent iff for all possible values x, y

$$P(X = x \text{ and } Y = y) = P(X = x)P(Y = y)$$

In other words, X and Y are independent variables iff the events $X = x$ and $Y = y$ are independent, as x and y are varied.

This definition generalizes: $X_1, X_2, \ldots, X_N$ are independent iff

$$P(X_1 = x_1, X_2 = x_2, \ldots, X_N = x_N) = P(X_1 = x_1)P(X_2 = x_2) \cdots P(X_N = x_N)$$

provided the X_i are defined over a common outcome space.

Now the important point about the "expectation operator" E, as indicated by the absence of any condition in (10.5.2), is that the expectations add even if the variables are not independent. This is a fact often exploited in model-building, for a generalization of property (a) of (10.5.2) is

$$E(\Sigma_i X_i) = \Sigma_i E(X_i)$$

even if the X_i are dependent.

Just as in the summary of data both the mean and the variance (where empirically meaningful) are computed, so a parallel calculation occurs in a probability model. Let X be a random variable and let E(X) be its expectation. We define the deviation from the expectation by

$$d_X = X - E(X)$$

and, since taking its expectation yields

$$
\begin{aligned}
E(d_X) &= E[X - E(X)] \\
&= E(X) - E[E(X)] \qquad \text{[property (10.5.2)(a)]} \\
&= E(X) - E(X) \qquad \text{[property (10.5.2)(c) with c = E(X)]} \\
&= 0
\end{aligned}
$$

we take the expectation of the square:

(10.5.4) Definition. $VAR(X) = E(d_X^2) = E[X - E(X)]^2$

Example. (d) From two successive occasions in each of which a specified interaction (I) may occur, we have a tree with four possible outcomes. Suppose that the model is numerical as in Figure 10.5.2.

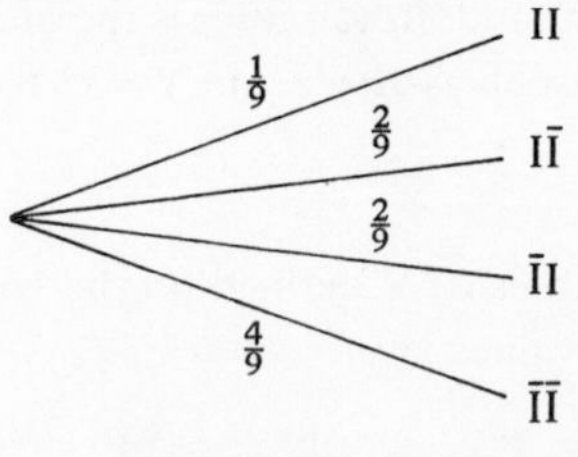

	X	d_X	d_X^2
II	2	$\frac{4}{3}$	$\frac{16}{9}$
I$\bar{\text{I}}$	1	$\frac{1}{3}$	$\frac{1}{9}$
$\bar{\text{I}}$I	1	$\frac{1}{3}$	$\frac{1}{9}$
$\bar{\text{I}}\bar{\text{I}}$	0	$-\frac{2}{3}$	$\frac{4}{9}$

Figure 10.5.2

From the tree, we see that

$$E(X) = 2 \cdot \tfrac{1}{9} + 1 \cdot \tfrac{4}{9} = \tfrac{6}{9} = \tfrac{2}{3}$$

is the expectation of X, which counts the number of interactions. This allows us to compute d_X, for each outcome, as shown. Then we square to obtain d_X^2. At this point d_X^2 is just another mapping from outcomes to numbers:

$$II \mapsto \tfrac{16}{9}$$
$$I\bar{I} \mapsto \tfrac{1}{9}$$
$$\bar{I}I \mapsto \tfrac{1}{9}$$
$$\bar{I}\bar{I} \mapsto \tfrac{4}{9}$$

which means it is a random variable. Taking its expectation,

$$
\begin{aligned}
VAR(X) &= E(d_X^2) = \sum_x xP(d_X^2 = x) \qquad (x \in \{\tfrac{16}{9}, \tfrac{4}{9}, \tfrac{1}{9}\}) \\
&= \tfrac{16}{9} \cdot P(d_X^2 = \tfrac{16}{9}) + \tfrac{1}{9} \cdot P(d_X^2 = \tfrac{1}{9}) + \tfrac{4}{9} \cdot P(d_X^2 = \tfrac{4}{9}) \\
&= \tfrac{16}{9} \cdot \tfrac{1}{9} + \tfrac{1}{9} \cdot \tfrac{4}{9} + \tfrac{4}{9} \cdot \tfrac{4}{9} \\
&= \tfrac{1}{81}(16 + 4 + 16) \\
&= \tfrac{36}{81} \\
VAR(X) &= \tfrac{4}{9}
\end{aligned}
$$

We define the standard deviation σ_X as the positive square root of $VAR(X)$. Thus, in the above example, $\sigma_X = \tfrac{2}{3}$.

Just as $E(X)$ corresponds to $\overline{X}$, so also $VAR(X)$ corresponds to the variance of data, denoted s^2:

$$
s^2 = \frac{1}{N-1} \sum_{i=1}^{N} (X_i - \overline{X})^2
$$

The model predicts that the value of $\overline{X}$ is $E(X)$, except for random variation already rationalized in the probability model. Also, it predicts that the value of s^2 is $VAR(X)$.

From the definition of $VAR(X)$ we obtain the basic properties of the variance "operator" VAR:

(10.5.5) Properties of VAR(X).

(a) $VAR(X) = E(X^2) - [E(X)]^2$

(b) $VAR(X + c) = VAR(X)$ \qquad (any constant c)

(c) $VAR(cX) = c^2 VAR(X)$

(d) If $X_1, X_2, \ldots, X_N$ are independent random variables defined on the same outcome space,

$$VAR(\Sigma_i X_i) = \Sigma_i VAR(X_i)$$

A little practice is obtained in the calculus of the E operator by deducing property (a), the "working formula" for the variance:

$$
\begin{aligned}
VAR(X) &= E(d_X^2) & \text{(definition)} \\
&= E[X - E(X)]^2 & \text{(definition)} \\
&= E[X^2 - 2X \cdot E(X) + [E(X)]^2] & \text{(algebraic expansion)} \\
&= E(X^2) - E[2X \cdot E(X)] + E([E(X)]^2) & \text{(10.5.2)(a)}
\end{aligned}
$$

$$= E(X^2) - 2E(X)E(X) + [E(X)]^2 \qquad (10.5.2)(b)(c)$$
$$= E(X^2) - 2[E(X)]^2 + [E(X)]^2$$
$$= E(X^2) - [E(X)]^2$$

The topic of random variables essentially is one of unending subtopics: any function or operation applied to one or more random variables may induce a problem for investigation, because this operation will define a new random variable. For example, any statistic is, from the standpoint of mathematical statistics, a random variable because it is an operation on a set of (observed) random variables.

Here is one topic that is often useful to know about in the analysis of a probability model: the distribution of a sum of two independent random variables.

Let X and Y both be defined on the same outcome space. Then X + Y is well defined, and we have

$$P(X + Y = z) = \sum_X P(X = x \text{ and } Y = z - x)$$

where on the right side x varies over all possible values of X, but z is held fixed because we want to find the chance that the sum is z. We do this by considering all the possible ways X and Y sum up to z. Now, if X and Y are independent,

$$P(X = x \text{ and } Y = z - x) = P(X = x)P(Y = z - x)$$

because we can treat the "and" here as specifying the intersection of two numerical events. This yields the following:

(10.5.6) Proposition on Sums of Independent Random Variables. If X and Y are two independent random variables on an outcome space, then the distribution of their sum is

$$P(X + Y = z) = \sum_X P(X = x)P(Y = z - x)$$

Example. (e) Two fathers each are succeeded by a random number of sons, denoted X and Y. Suppose that

$$P(X = 0) = \tfrac{1}{2}, P(Y = 0) = \tfrac{1}{2}$$
$$P(X = 1) = \tfrac{1}{4}, P(Y = 1) = \tfrac{1}{4}$$
$$P(X = 2) = \tfrac{1}{4}, P(Y = 2) = \tfrac{1}{4}$$

To find the total number of sons, assume the births are independent. Then (10.5.6) yields

$$P(X + Y = 0) = P(X = 0)P(Y = 0) = \tfrac{1}{4}$$
$$P(X + Y = 1) = P(X = 0)P(Y = 1) + P(X = 1)P(Y = 0)$$
$$= \tfrac{1}{2} \cdot \tfrac{1}{4} + \tfrac{1}{4} \cdot \tfrac{1}{2}$$
$$= \tfrac{1}{4}$$

$$P(X + Y = 2) = P(X = 0)P(Y = 2) + P(X = 1)P(Y = 1) + P(X = 2)P(Y = 0)$$
$$= \frac{1}{2} \cdot \frac{1}{4} + \frac{1}{4} \cdot \frac{1}{4} + \frac{1}{4} \cdot \frac{1}{2} = \frac{5}{16}$$
$$P(X + Y = 3) = P(X = 1)P(Y = 2) + P(X = 2)P(Y = 1)$$
$$= \frac{1}{4} \cdot \frac{1}{4} + \frac{1}{4} \cdot \frac{1}{4} = \frac{1}{8}$$
$$P(X + Y = 4) = P(X = 2)P(Y = 2) = \frac{1}{4} \cdot \frac{1}{4} = \frac{1}{16}$$

Checking, with $z \in \{0, 1, 2, 3, 4\}$,

$$\sum_z P(X + Y = z) = \frac{1}{4} + \frac{1}{4} + \frac{5}{16} + \frac{1}{8} + \frac{1}{16} = 1$$

as required for a probability distribution.

We sometimes deal with a phenomenon that has the structure of Figure 10.5.3,

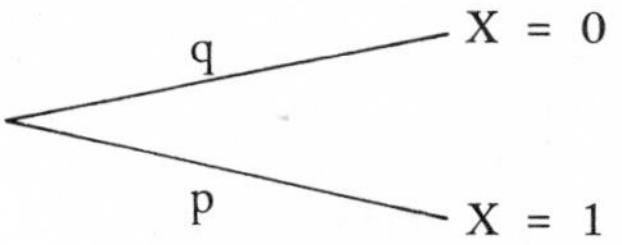

Figure 10.5.3

and that is repeated N times independently. Thus, for i = 1, 2, . . . , N, we have Figure 10.5.4.

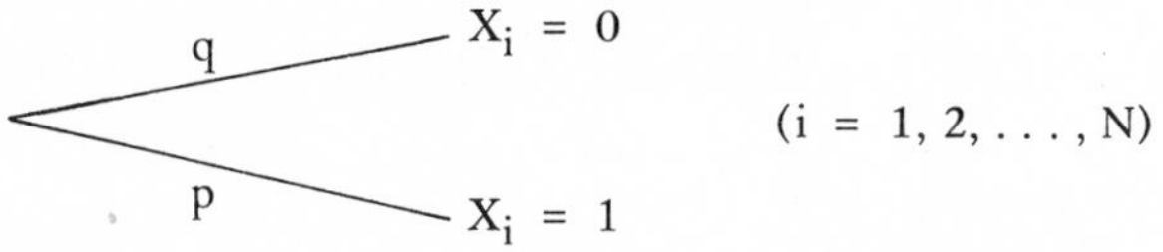

Figure 10.5.4

Then we want to compute the distribution of the frequency of occurrence of the event $X = 1$, which stands for some empirical event.

Example. (f) Consider N married couples each of whom gets a divorce in the time period of one specified year with probability $1/15$. Then, we have Figure 10.5.5 as a special case of Figure 10.5.4.

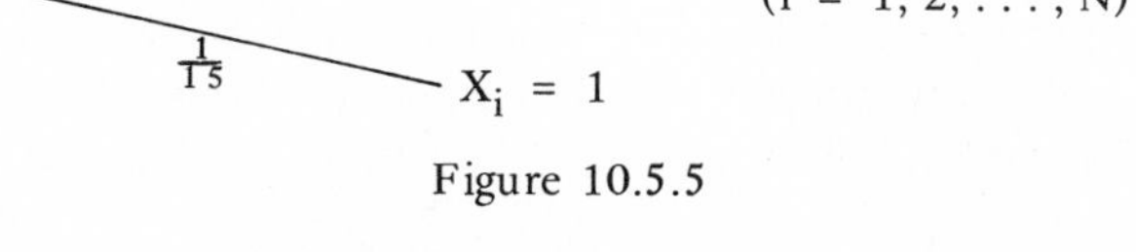

Figure 10.5.5

We want the probability distribution of the number of couples who get a divorce.

Example. (g) Each group in a specified set of N groups will experience a "change of structure" event specified by some other model. Assuming independence and a parameter p for change, the same for each of the groups, Figure 10.5.4 applies. We want to calculate the probability that k groups have a change of structure, k = 0, 1, . . . , N.

Thus, in each case we want the distribution of the sum

$$S = \sum_{i=1}^{N} X_i$$

Note that

$$E(S) = \sum_{i=1}^{N} E(X_i) = \sum_{i=1}^{N} (1 \cdot p + 0 \cdot q) = \sum_{i=1}^{N} p = Np$$

which is equivalent to

$$p = \frac{E(S)}{N} = E\left(\frac{S}{N}\right)$$

Hence, p is the expected relative frequency of the event coded X = 1. This is the formal basis for the principle of applied probability, proposition (9.8.1).

To find the distribution of the sum, imagine the variables written in the list

$$X_1 \ X_2 \ X_3 \ \cdots \ X_N$$

An outcome is a list of zeros and ones, as in

$$0 \ 1 \ 1 \ \cdots \ 0$$

To obtain a sum of k, exactly k of the N places (variables) must show a 1. Hence, we need to know the number of subsets of k places in a set of N places.

(10.5.7) Proposition. If A is any set of N elements then the number of subsets of size k contained in A is given by

$$\binom{N}{k} = \frac{N!}{k!(N-k)!}$$

This proposition applies to our problem: there are $\binom{N}{k}$ subsets of k places in the list of N variables where we could have a 1 with the remainder zero. Each such outcome has a probability, which we need to compute. Consider a concrete outcome, with associated probabilities:

$$0 \ 0 \ 1 \ 1 \ 0 \ 1 \qquad (N = 6, k = 3)$$
$$q \ q \ p \ p \ q \ p$$

By independence, the probability of this outcome is $p^3 q^3$. A rearrangement leads to a different concrete outcome but the same probability:

$$0 \; 1 \; 1 \; 1 \; 0 \; 0$$
$$q \; p \; p \; p \; q \; q$$

In general, each of the $\binom{N}{k}$ outcomes realizing the event that $S = k$ has the same probability. And for any one of them the probability is of the form: $p^k q^{N-k}$. Hence, to find the probability that $S = k$ we add up $\binom{N}{k}$ terms, each of the form $p^k q^{N-k}$. This leads to the desired formula:

$$(10.5.8) \quad P(S = k) = \binom{N}{k} p^k (1 - p)^{N-k} \qquad (k = 0, 1, 2, \ldots, N)$$

This is the binomial distribution with parameters N and p.

Example. (h) Each of N cities "has a riot" in a given summer; that is, it has at least one riot. Suppose the N cities have the same probability p of at least one riot and that riots occur independently in distinct cities. Then the number S of cities having a riot has probability distribution given by (10.5.8).

Example. (i) There are M individuals in a given town. During one day each pair of distinct persons has the same chance of interacting (call it p). If the interactions occur independently, then to use (10.5.8) we need the number of terms in the sum S. Now let $X_i = 1$ if pair i interacts and $X_i = 0$ otherwise. Since $S = \Sigma_i X_i$, we need to know how many (unordered) pairs there are. But this is the number of subsets of size two; hence, by proposition (10.5.7),

$$\binom{M}{2} = \frac{M(M-1)}{2}$$

Taking this as N, formula (10.5.8) gives the probability distribution of the number S of interactions in a given day.

10.6. Formal Definition of the Markov Chain Concept. We can apply the random variable idea to define an important mathematical concept: Markov chain in discrete time. We know that our task is to formalize the idea of state-determined system (see section 8.10) in the probabilistic context, although the definition will deal only with the special case of a discrete-time and discrete-state process from the variants discussed in section 8.10.

(10.6.1) Definition. Let $X_0, X_1, X_2, \ldots, X_t, \ldots$ be a sequence of discrete random variables, defined on the same outcome space. Then we say that this

sequence forms a (discrete-state, discrete-time) Markov chain if and only if for any t (t = 1, 2, . . .),

$$P(X_t = x_t | X_{t-1} = x_{t-1}, X_{t-2} = x_{t-2}, \ldots, X_1 = x_1, X_0 = x_0)$$
$$= P(X_t = x_t | X_{t-1} = x_{t-1})$$

for any possible values $x_t, x_{t-1}, x_{t-2}, \ldots, x_1, x_0$.

Example. Let $\{1, 2, 3\}$ be the set of possible values of a status variable X. Let X_t be the actor's status at time t, construed as arising in some probability sense (e.g., fluctuations in the underlying evaluations). Then X_0 is the initial status, and $X_0, X_1, \ldots, X_t, \ldots$ constitutes the status sequence, over time, of this actor. This sequence forms a Markov chain if the general condition of definition (10.6.1) is satisfied. Among the specific conditions comprehended by this general condition, consider time t = 2. Then the condition becomes

$$P(X_2 = x_2 | X_1 = x_1, X_0 = x_0) = P(X_2 = x_2 | X_1 = x_1)$$

for any possible values x_2, x_1, x_0. The possible values of x_2, x_1, and x_0 are three, in each case, namely, 1, 2, or 3. Hence, there are 3 x 3 x 3 = 27 special cases subsumed under this one condition. For example, the three cases,

$$P(X_2 = 3 | X_1 = 3, X_0 = 3) = P(X_2 = 3 | X_1 = 3)$$
$$P(X_2 = 3 | X_1 = 3, X_0 = 2) = P(X_2 = 3 | X_1 = 3)$$
$$P(X_2 = 3 | X_1 = 3, X_0 = 1) = P(X_2 = 3 | X_1 = 3)$$

cover the situation in which the actor is in an arbitrary state at the initial time, in state 3 at time 1, and stays in state 3 at time 2. This is illustrated by taking $P(X_2 = 3 | X_1 = 3) = .9$ in the left-hand tree diagram in Figure 10.6.1. A contrasting non-Markov case appears on the right in this figure. It is seen that on the left, the Markov type, the probability of staying in state 3 from time 1 to time 2 does not depend on which of the three paths through $X_1 = 3$ is considered. We say that this constitutes independence of path. Another situation altogether appears on the right, where the staying probability increases depending on the initial status:

$$P(X_2 = 3 | X_1 = 3, X_0 = 1) = .7$$
$$P(X_2 = 3 | X_1 = 3, X_0 = 2) = .8$$
$$P(X_2 = 3 | X_1 = 3, X_0 = 3) = .9$$

It is clear that these conditions violate the general Markov assumption given in definition (10.6.1).

The concept of Markov chain must not be confused with homogeneity in the sense of identity of dynamic law as time passes (see section 8.9). It is entirely possible that the precise values of the probabilities shift in time while independence of path is preserved. The distinction can be reflected in notation. From

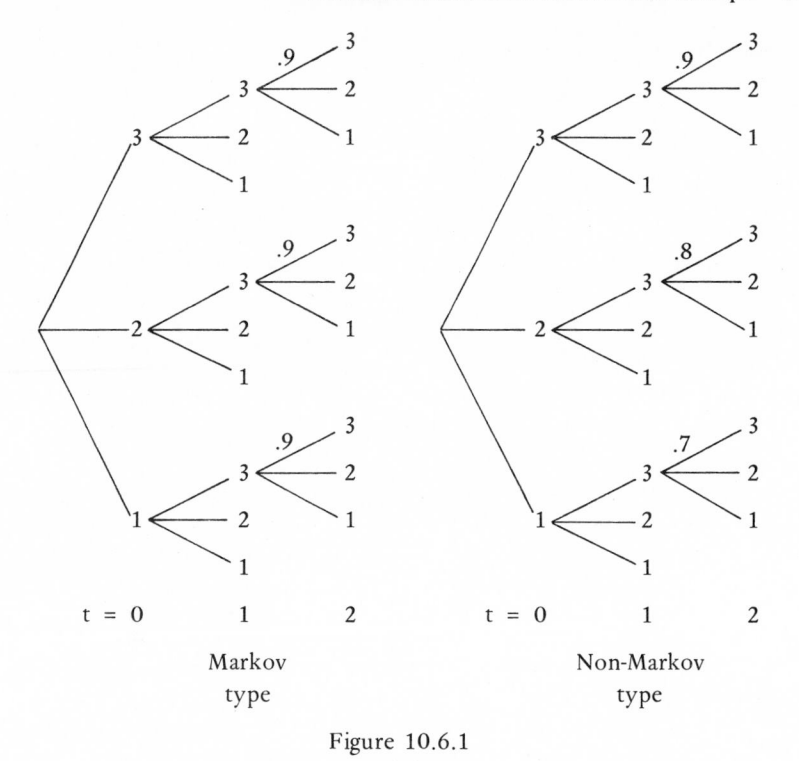

Figure 10.6.1

the definition, we see that the key quantities in a Markov chain are the transition probabilities,

$$P(X_t = x_t | X_{t-1} = x_{t-1})$$

Now x_t, x_{t-1} are mere logical variables, place holders for possible values of the random variables. Hence, since only two of them are now needed we can arbitrarily replace them with j and i, respectively. Then the transition probability is denoted

$$P(X_t = j | X_{t-1} = i)$$

It is clear that this is a conditional probability, hence, a number. But this number depends on time t as well as values i and j. When it does not vary with t, we say that the chain is time-homogeneous. Figure 10.6.2 gives an illustration, using only the relevant paths of the left tree in Figure 10.6.1. The diagram on the left shows that

$$P(X_2 = 3 | X_1 = 3) = P(X_1 = 3 | X_0 = 3)$$

so that in the general form $P(X_t = 3 | X_{t-1} = 3)$ there is invariance with respect to

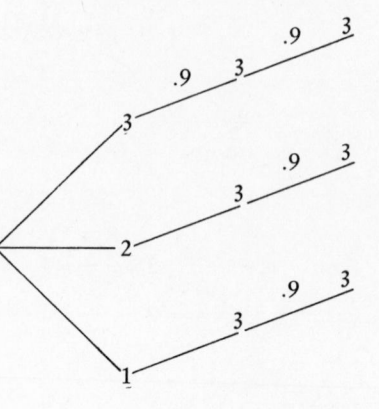
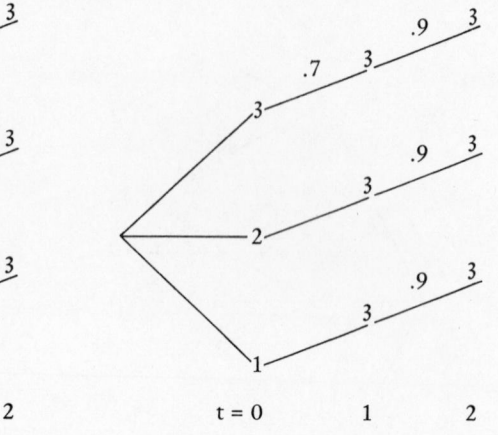
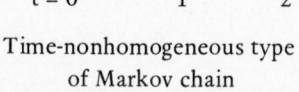

t = 0   1   2   t = 0   1   2

Time-homogeneous type   Time-nonhomogeneous type
of Markov chain   of Markov chain

Figure 10.6.2

the time. This is time-homogeneity. On the right in Figure 10.6.2, on the other hand, we have

$$P(X_1 = 3|X_0 = 3) = .7$$
$$P(X_2 = 3|X_1 = 3) = .9$$

and, so, no such time invariance. Note that both trees exemplify the Markov property because the probability at any given time t of a transition (from 3 to 3, taken as an instance) is independent of the path leading up to the occupied state at that time t.

We summarize these ideas in the following:

(10.6.2) Definition. A Markov chain $X_0, X_1, X_2, \ldots, X_t, \ldots$ is said to be time-homogeneous if, and only if, for all times s and t,

$$P(X_s = j|X_{s-1} = i) = P(X_t = j|X_{t-1} = i)$$

for all possible values i and j.

For notational simplicity, we can now eliminate the time index in the transition probabilities of time-homogeneous chains because only states i and j matter. Some convenient notational variants are p_{ij}, $P(i,j)$, $P(i \to j)$, $P(j|i)$, each meaning "probability of a transition, in one step, from state i to state j."

A final note on notation: when the variable X is merely nominal, it is frequently more convenient to work directly with the empirical domain rather than the numerical representation. In such cases, instead of p_{ij} or other such notation, one finds direct conditional probability statements. For instance,

$$P(D \text{ at } t | R \text{ at } t-1)$$

might be used to refer to the probability that a voter shifts from Republican at time t–1 to Democrat at t. These frequently more intuitively appealing notations are not to be eschewed in favor of a rigid, more formalized notation; the latter, we have seen, becomes essential in the process of rigorously defining the concepts involved, Markov chain and time-homogeneity. However, in the constructive, model-building use of these concepts, notation should be flexibly adapted to the case at hand in a manner that simplifies the setting up, analysis, and interpretation of the model. In this spirit, we use a multiplicity of notational variants for transition probabilities and other process entities in this and later chapters.

10.7. Example of a Finite Probability Model: Setup and Analysis. In this section an extended example of a simple setup and analysis of a probability model is provided in order to draw together certain of the ideas introduced above.

Consider a learning model that is set up, via the two tree diagrams, as a Markov chain (see Figure 10.7.1).

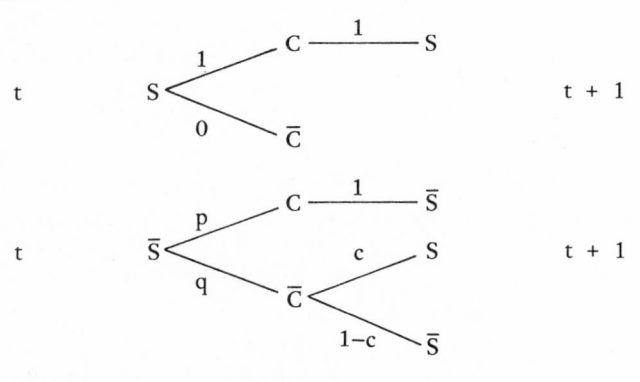

Figure 10.7.1

The axioms of the model are as follows, relative to a single person:

Axiom 1. (State space) There are two possible states, denoted S and $\overline{S}$.

Axiom 2. (Response) There are two possible responses, C and $\overline{C}$.

Axiom 3. (Trials) There is a sequence of trials t (t = 0, 1, . . .) such that at each trial, the person begins in a certain state, makes a certain response, and terminates in a new state that is the starting state for the next trial.

Axiom 4. (Response output) (a) If the person is in state S at the start of a trial, then he makes a C response with probability 1. (b) If a person is in state $\overline{S}$ at the start of a given trial, then he makes a C response with probability p and a $\overline{C}$ response with probability q = 1 − p.

Axiom 5. (Response feedback) (a) If a person makes response C on any trial, then he stays in the state in which he entered the trial. (b) If a person makes response $\overline{C}$ on any trial, then with probability c he enters state S and with probability 1 − c he stays in state $\overline{S}$.

Axiom 6. (Initial condition) The initial state, on trial 0, is $\overline{S}$.

We may term the first three axioms structure conditions. They introduce the abstract structure of states and responses (defining them as binary in this case) as well as the structure of time itself, which in this case is a sequence of trials. (Variant possibilities: continuous space, discrete response, discrete time; continuous space, discrete response, continuous time; and so forth.) Axiom 4 generates observable responses from states. Axiom 5 is the basis for the state transition role and the system dynamics because it lets the response term feedback to create a change of state under some conditions.

The response C is the correct response, while $\overline{C}$ is incorrect. State S is the solution, or learned state, while $\overline{S}$ is the nonsolution, or unlearned state. The person is envisaged as fluctuating randomly in the observable response until, after some error, he hits upon the solution and then makes the correct response from that time forward. This is the qualitative picture of the process; the next step is to illustrate a detailed analysis based on this setup (for a more thorough analysis of this model, see Atkinson, Bower, and Crothers, 1965, chapter 2).

Define

$$R_t = \begin{cases} 1 & \text{if realized path has } \overline{C} \text{ on trial t} \\ 0 & \text{otherwise} \end{cases}$$

for t = 0, 1, 2, ... Now R_t is a random variable defined on the paths of the process. We fail to include a terminal trial in order to illustrate a simplicity introduced by working on a countably infinite outcome space (of process realizations, in this case). Now our objective is to deduce the probability distribution of the R_t—for each t we have a distinct random variable and, so, a distinct probability distribution—and use this to compute the expected value of the total number of errors. We let a subscript on an event stand for the trial (time) of the event. The logic of computing $P(R_t = 1)$ for all t is now given in utter detail to illustrate related points made earlier. Note that with two possible values for R_t the calculation requires only one term per trial variable because

$$P(R_t = 0) = 1 - P(R_t = 1).$$

The calculation of the distribution of R_t is here set forth in three steps:

Step 1. Partition the outcome space by the two events

S_t iff   system in state S at t
$\bar{S}_t$ iff   not the case that system in state S at t.

Here "system" means the subject for this example. Let

$$R_t = 1$$

be represented as a set of possible paths of the process: thus, a subset of the outcome space. The Venn diagram is shown in Figure 10.7.2.

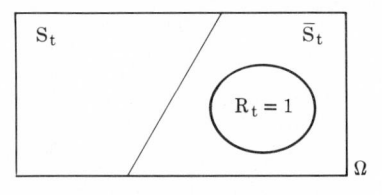

Figure 10.7.2

The picture is valid because the error event occurs only if $\bar{S}$ holds, at a given time. Thus, the general law of total probability as applied here becomes

$$P(R_t = 1) = P(R_t = 1|\bar{S}_t)P(\bar{S}_t) + P(R_t = 1|S_t)P(S_t)$$
(10.7.1) $$= P(R_t = 1|\bar{S}_t)P(\bar{S}_t)$$

Step 2. Find a general formula expressing $P(\bar{S}_t)$ in terms of the parameters of the process and the initial conditions. This is an application of the general process strategy and is now made concrete. The procedure used here typifies a large class of similar cases:

Step 2a. Write down a formula for $P(\bar{S}_{t+1})$ in terms of $P(\bar{S}_t)$. Why should one think of this? Essentially, it is part of the setup in the small, which deals precisely with transitions in the small from t to t + 1. To activate this setup (e.g., the tree diagram) think of partitioning by the events $\{S_t, \bar{S}_t\}$. The picture is shown in Figure 10.7.3.

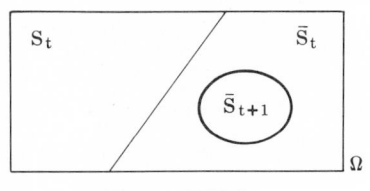

Figure 10.7.3

This picture is valid because $\overline{S}_{t+1} \subset \overline{S}_t$: if the system is in state $\overline{S}$ at $t+1$, then it is in state $\overline{S}$ at the prior time. Contrapositively, if in S at t, then not in $\overline{S}$ at $t+1$. This is clear from Figure 10.7.1, based on axioms 4 and 5. Thus, applying the law of total probability amounts to

$$P(\overline{S}_{t+1}) = P(\overline{S}_{t+1}|\overline{S}_t)P(\overline{S}_t) + P(\overline{S}_{t+1}|S_t)P(S_t)$$
$$= P(\overline{S}_{t+1}|\overline{S}_t)P(\overline{S}_t)$$

But now we use the setup in the small (the tree diagram in Figure 10.7.1 for the transition from t to t+1) to write

$$P(\overline{S}_{t+1}|\overline{S}_t) = 1 - P(S_{t+1}|\overline{S}_t) \qquad \text{(principle of conditional probability)}$$
$$= 1 - qc \qquad \text{(Figure 10.7.1)}$$

Therefore,

(10.7.2) $$P(\overline{S}_{t+1}) = (1 - qc)P(\overline{S}_t)$$

We see that $P(\overline{S}_t)$ is transformed into $P(\overline{S}_{t+1})$ under formula (10.7.2), which is our general change-of-state equation with initial condition

(10.7.3) $$P(\overline{S}_0) = 1$$

A trajectory of $P(\overline{S}_t)$, t = 0, 1, . . . , satisfying (10.7.2) with the initial condition (10.7.3) is called a solution of the equation.

Step 2b. Solve the change-of-state equation by first simplifying notation:

(10.7.4) $$q_{t+1} = \alpha q_t \qquad (t = 0, 1, \ldots)$$
$$q_0 = 1$$

so that $q_t = P(\overline{S}_t)$, all t, and $\alpha = 1 - qc$. The solution has two parts, discovery and proof:

(1) Discovery phase: The basic idea here is to substitute the first few values of t in (10.7.4) to get an idea of what the q_t look like when expressed in terms of the initial condition (and parameter).

If t = 0 in (10.7.4), then
$$q_1 = \alpha q_0 = \alpha$$
If t = 1 in (10.7.4), then
$$q_2 = \alpha q_1 = \alpha\alpha = \alpha^2$$
If t = 2 in (10.7.4), then
$$q_3 = \alpha q_2 = \alpha\alpha^2 = \alpha^3$$

By now we perceive a pattern or law of the form

(10.7.5) $$q_t = \alpha^t \qquad (t = 0, 1, \ldots)$$

(2) Proof stage: Strictly speaking the perception that (10.7.5) is the pattern of

the first few terms, or even the first n terms, for any finite n, is not a proof that for all t the law holds. This is because (10.7.5) is really a countably infinite set of propositions, one for each nonnegative integer. Thus, in claiming (10.7.5) from our discovery phase we are not saying that (10.7.5) is a logical consequence of the results of the discovery phase but a rather compelling conjecture based on the discovery phase. In most model-building work, one would terminate with (10.7.5) as if it were proved, so compelling is the pattern seen in the first few terms. But logically a proof of (10.7.5) is required, namely, a proof that for each t = 0, 1, 2, . . . the proposition $q_t = \alpha^t$ is valid.

To do this we assume the following rule of inference justified by logic:

(10.7.6)   Rule of Inference for Induction Proofs.   Let $\{P_t\}$ be a family of propositions, one for each value of t, where t is a variable with domain $\{0, 1, 2, . . .\}$. Then, from P_0 and $P_t \Rightarrow P_{t+1}$, one may conclude: $\forall t,\ P_t$. (That is, conclude that for any t in the infinite domain, P_t holds.)

To apply this rule to our conjecture, we denote the various propositions as follows:

$$P_0: q_0 = \alpha^0 = 1$$
$$P_1: q_1 = \alpha$$
$$P_2: q_2 = \alpha^2$$
$$\cdot\quad\cdot$$
$$\cdot\quad\cdot$$
$$\cdot\quad\cdot$$
$$P_t: q_t = \alpha^t$$

Thus, to utilize rule (10.7.6) we need to show the two conditions P_0 and $P_t \Rightarrow P_{t+1}$ hold. But P_0 is the assumed initial condition, and so, it holds. Thus, we have only to show that $P_t \Rightarrow P_{t+1}$. To do so, we assume P_t holds; then we try to "force out" the result P_{t+1}. This will prove that if P_t, then P_{t+1}, as desired. Assuming P_t is "making the induction hypothesis." Thus, assume

$$q_t = \alpha^t$$

Now we want to conclude

$$q_{t+1} = \alpha^{t+1} \qquad \text{(namely, } P_{t+1})$$

The only way of getting from the assumed P_t to the desired P_{t+1} is by using prior results in the model. In particular, (10.7.4) is crucial because it contains the term q_{t+1}. Thus,

$$
\begin{aligned}
q_{t+1} &= \alpha q_t && \text{(previously shown)} \\
&= \alpha \alpha^t && \text{(using the induction hypothesis)} \\
&= \alpha^{t+1} && \text{(the desired conclusion)}
\end{aligned}
$$

Now we have shown

$$q_0 = 1 = \alpha^0$$

and

$$\text{if } q_t = \alpha^t, \text{ then } q_{t+1} = \alpha^{t+1}$$

Rule (10.7.6) justifies the purely logical step

$$\text{For all } t = 0, 1, 2, \ldots$$
$$q_t = \alpha^t$$

This digression into the logic of induction proofs indicates that in the proof stage one uses rule (10.7.6) to justify impeccably the compelling conjecture arrived at in the discovery phase. (The reminder is in order that this proof stage is often eliminated with the tacit understanding that the reader sees that the validity of the conjecture could readily be shown by proof by induction.)

To return to the model, we now have a solution (10.7.5) satisfying the "dynamic" equation (10.7.4), which was shorthand for equation (10.7.2). Thus, we have the result

$$(10.7.7) \qquad P(\overline{S}_t) = (1 - qc)^t \qquad (t = 0, 1, 2, \ldots)$$

We recall that step 2 above began with the objective of finding an explicit expression for $P(\overline{S}_t)$. This we have now accomplished with (10.7.7).

Step 3. We return to the basic results of the first two steps (in this derivation of the distribution of R_t):

$$P(R_t = 1) = P(R_t = 1|\overline{S}_t)P(\overline{S}_t) \qquad \text{(all t)}$$
$$P(\overline{S}_t) = (1 - qc)^t \qquad \text{(all t)}$$

Hence,

$$P(R_t = 1) = P(R_t = 1|\overline{S}_t)(1 - qc)^t \qquad \text{(all t)}$$

Now from the setup given in the tree diagram of Figure 10.7.1 we can replace the conditional probability of $R_t = 1$, given the state $\overline{S}_t$, with q,

$$(10.7.8) \qquad P(R_t = 1) = q(1 - qc)^t \qquad \text{(all t)}$$

Our analysis of this model (recall this is phase 2 of model-building, see Chapter 1) has taken us to the desired distribution (10.7.8) in terms of the parameters of the system. This was the first of our two objectives for the analysis. The second objective is to use this first result to obtain a formula for the expected total number of errors.

Now R_t is a counting variable, coding errors as "1." Thus, summing over R_t gives the cumulative number of errors by trial N,

$$(10.7.9) \qquad T_N = \sum_{t=0}^{N} R_t$$

Now we see from (10.7.9) that we can write

$$E(T_N) = E\left(\sum_{t=0}^{N} R_t\right) \qquad \text{(definition of } T_N)$$

$$(10.7.10) \qquad\qquad = \sum_{t=0}^{N} E(R_t). \qquad \text{(property of E operator)}$$

Now, for any random variable,

$$E(X) = \sum_{x} xP(X = x)$$

so that

$$E(R_t) = 1 \cdot P(R_t = 1) + 0 \cdot P(R_t = 0)$$
$$(10.7.11) \qquad\qquad = P(R_t = 1)$$

Thus, for a zero-one variable, expectation is just probability. Hence, substituting (10.7.11) into (10.7.10),

$$E(T_N) = \sum_{t=0}^{N} P(R_t = 1)$$

and using (10.7.8),

$$(10.7.12) \qquad\qquad E(T_N) = \sum_{t=0}^{N} q(1 - qc)^t$$

This can be written,

$$(10.7.13) \qquad\qquad E(T_N) = q \sum_{t=0}^{N} \alpha^t \qquad (\alpha = 1 - qc)$$

Anticipating a large number of trials (say $N \geqslant 25$) we want to take limits because formulas ordinarily assume a simpler form after limiting operations are applied. To do this, we assume the following:

(10.7.14) Sum of Infinite Geometric Series. Let

$$\sum_{t=0}^{\infty} \alpha^t$$

mean

$$\lim_{N \to \infty} \sum_{t=0}^{N} \alpha^t$$

that is, the number that the sums approach as $N \to \infty$. Then,

$$\sum_{t=0}^{\infty} \alpha^t = \frac{1}{1-\alpha}$$

provided that $|\alpha| < 1$. For example, let $\alpha = 1/2$. Then

$$\sum_{t=0}^{\infty} (\tfrac{1}{2})^t$$

means

$$\lim_{N \to \infty} \sum_{t=0}^{N} (\tfrac{1}{2})^t$$

In particular, when $N = 0$,

$$\sum_{t=0}^{N} (\tfrac{1}{2})^t = (\tfrac{1}{2})^0 = 1$$

when $N = 1$,

$$\sum_{t=0}^{N} (\tfrac{1}{2})^t = (\tfrac{1}{2})^0 + (\tfrac{1}{2})^1 = 1 + \tfrac{1}{2} = \tfrac{3}{2}$$

when $N = 2$,

$$\sum_{t=0}^{N} (\tfrac{1}{2})^t = (\tfrac{1}{2})^0 + (\tfrac{1}{2})^1 + (\tfrac{1}{2})^2 = \tfrac{3}{2} + \tfrac{1}{4} = \tfrac{7}{4}$$

when $N = 3$,

$$\sum_{t=0}^{N} (\tfrac{1}{2})^t = (\tfrac{1}{2})^0 + (\tfrac{1}{2})^1 + (\tfrac{1}{2})^2 + (\tfrac{1}{2})^3 = \tfrac{7}{4} + \tfrac{1}{8} = \tfrac{15}{8}$$

and the sequence of sums

$$(1, \tfrac{3}{2}, \tfrac{7}{4}, \tfrac{15}{8}, \ldots)$$

appears to be converging toward the number 2. In fact, by (10.7.14), since $\alpha = 1/2$ and $|1/2| = 1/2 < 1$, we conclude immediately that

$$\sum_{t=0}^{\infty} (\tfrac{1}{2})^t = \frac{1}{1-(1/2)} = \frac{1}{1/2} = 2$$

as anticipated concretely.

Returning now to (10.7.13),

$$\lim_{N \to \infty} E(T_N) = \lim_{N \to \infty} q \sum_{t=0}^{N} \alpha^t \qquad (\alpha = 1 - qc)$$

The q is a constant that can be "factored out" of the limit operation, and so,

$$\lim_{N \to \infty} E(T_N) = q \sum_{t=0}^{\infty} \alpha^t$$

(10.7.15)
$$= \frac{q}{1-\alpha}$$

provided $|\alpha| < 1$, which implies $|1 - qc| < 1$. This requires that $qc \neq 1$ or, in other words, that either $0 \leqslant q < 1$ or $0 \leqslant c < 1$. This is only to say that we are not at the "extremes" of the model, where, for example, error is certain to occur on any trial. Then, since $1 - \alpha = qc$, formula (10.7.15) yields

(10.7.16)
$$\lim_{N \to \infty} E(T_N) = \frac{1}{c}$$

which exhibits, in its simplicity, the reason why applied mathematicians make use of limiting operations although actual experiments and processes of empirical inquiry are finite.

In this section, we have examined the two mathematically crucial phases of model-building for a particular probabilistic model: setup, exhibiting the state-space approach concretized with the Markov chain idea in a probabilistic system, and analysis, in which one exploits all the relevant mathematics available to grind out results. This example exhibited in the analysis stage, the use of random variables, the law of total probability, Venn diagrams, the discovery of pattern by concrete elaboration of a recursion equation, proof by induction, expectation-operator properties, and properties of certain infinite series. The general impression should be that the more background mathematics one has internalized, the greater the power of analysis given a setup. The setup itself relies upon certain heuristics and prior mathematical background but ordinarily involves considerably less formal skill than the analysis phase. It is a mistake to leave the analysis to mathematicians, preserving setups as the scientific task. For the mathematician, the analysis of a particular model may be a bore. One model, analyzed by routine application of known techniques, is of no interest to him. Thus, the sociologist must be prepared to do his own analyses as well as his own setups of models.

10.8. Graph and Matrix Representations for Markov Chains. A Markov chain can be represented by two useful objects: a graph and a matrix.

First, let us introduce X_t as a (random) state variable. Thus, X_t represents the state at time t. In the example of section 10.7, we could let

$$X_t = \begin{cases} 2 & \text{if } S_t \\ 1 & \text{if } \bar{S}_t \end{cases}$$

but any pair of distinct numbers would do because the state space is only a classification (of the system's dispositions). Then the transition behavior on the state space is a passage from, say $X_{t-1} = j$ to $X_t = i$, as discussed in section 10.6.

The Markov process has a state-space that in the ground sense is non-numerical but in the representational sense of X_t is some subset of the real numbers. Let S be this subset of real numbers. For instance, $S = \{1, 2\}$ above.

A graph can be used to represent the process by representing states by nodes, labeling the nodes with the names of the states, and by representing possible transitions by directed lines, weighted by the transition probability.

Example. (a) Let $S = \{1, 2, 3, 4\}$ be the state-space of a process with tree diagrams (with $p \neq 0$, $q \neq 0$, $r \neq 0$, $s \neq 0$) shown in Figure 10.8.1.

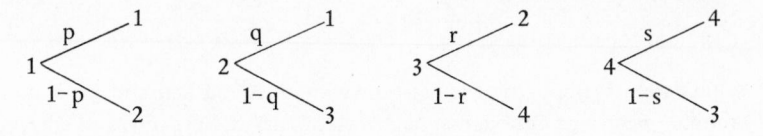

Figure 10.8.1

Then, the graph is shown in Figure 10.8.2.

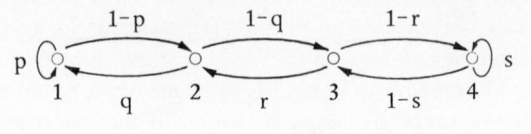

Figure 10.8.2

By the "qualitative analysis" of the chain we mean an analysis based only on "the topology" shown by the graph: the implied relational system involving the states rather than the transition probability magnitudes. (This qualitative analysis is useful in determining which theorems of a reference book on Markov chains applies as we shall see in a later section.) The topology of this system is shown in Figure 10.8.3.

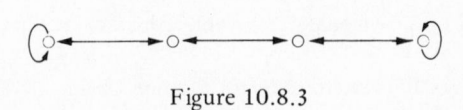

Figure 10.8.3

From this graph we conclude: starting in any state, we can reach any other state. Since the process consists in the repeated application of the basic in-the-small diagrams, or assumptions, this means that if the process is run long enough, each of the states is "visited" or "occupied," starting in any given state.

Example. (b) A process with state-space $\{1, 2, 3, 4\}$ has the graph and associated topology of Figure 10.8.4.

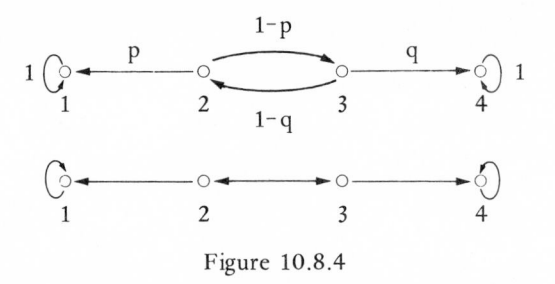

Figure 10.8.4

From the topological graph we can see that (1) if the system starts in state 1 or in state 4, it stays where it starts; (2) if the system starts in state 2 or state 3, it reaches state 1 or state 4 (eventually). Thus, (1) shows that we have two equilibrium states. As mentioned earlier, they are known as absorbing states in Markov chain theory. Property (2) shows that these states are unstable relative to each other: beginning outside either one, we need not go into it. But the subset of states $\{1, 4\}$ is stable: beginning outside this subset we eventually are led back into it.

A chain is called ergodic if each state is reachable from each other state. This means, in terms of the probabilities, that there is a nonzero probability of reaching any given state, no matter which state is the initial state. The graph, showing the qualitative structure, is the best tool for determination of whether the chain is ergodic: trace from each node, successively.

Example. (c) Is the chain of example (b) ergodic? Trace from state 1. No others are reached. The chain is not ergodic, since a state exists such that starting from it one cannot reach all other states.

Example. (d) A process is represented qualitatively by Figure 10.8.5.

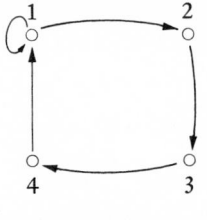

Figure 10.8.5

Is it ergodic? From 1 we have the tracing, $1 \to 2 \to 3 \to 4$, and all others are reached. Similarly, one can trace a path from any node reaching all other nodes. Hence, the chain is ergodic.

The second major representational tool in Markov chain theory is the matrix representation. Let the state-space S have N states. Then we form an N x N matrix each row of which is the array of path probabilities to next possible states, starting from the row state. In other words, a tree starting at the state corresponding to the row induces the corresponding row of probabilities.

Example. (e) Figure 10.8.6 shows

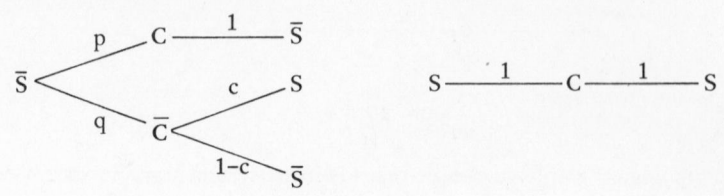

Figure 10.8.6

the two tree diagrams we use in the setup of the model of section 10.7. This provides the two rows:

		t + 1	
		$\overline{S}$	S
	$\overline{S}$	$1 - qc$	qc
t	S	0	1

The matrix

$$P = \begin{pmatrix} 1 - qc & qc \\ 0 & 1 \end{pmatrix}$$

is called the (one-step) transition probability matrix of the chain. The graph of the system is shown in Figure 10.8.7.

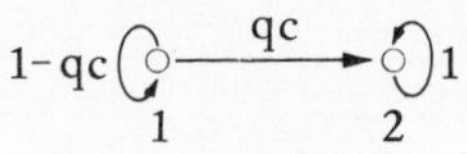

$$1-qc \;\bigcirc\!\!\!\!\!\overset{qc}{\longrightarrow}\!\!\!\!\!\bigcirc 1$$
$$\quad\;\; 1 \qquad\qquad 2$$

Figure 10.8.7

Here node 1 represents state $\overline{S}$; node 2 represents state S. Clearly this chain is not ergodic: tracing from node 2 gets us nowhere else. The state 2 (i.e., S) is absorbing.

10.9. Classification of Finite Markov Chains. The intuitions of the prior section may be coupled with the formalism of the reachability relation outlined in section 6.4 in terms of a matrix representation to generate an elegant standard, or "canonical," form of transition matrix. This form then provides the key to the proper theorems and techniques that apply in the analysis of the chain. This material is based on the book by Kemeny and Snell (1960), which introduces this canonical form and uses it for the organization of the mathematical techniques and theorems.

We know from section 6.4 that from the reachability relation R_* obtained from R we can define an equivalence relation of mutual reachability and thereby partition the domain of relation R. To apply this idea in this context, we first define

(10.9.1) $\qquad\qquad\qquad\qquad$ iRj $\quad$ iff $\quad$ $P(i \to j) > 0$

That is, i and j are in relation R if and only if there is a positive one-step transition probability that the system goes from state i to state j. This was the basis of the topological graphs of the prior section.

Correspondingly, the matrix of transition probabilities is replaced, in the qualitative analysis for classification purposes, by the adjacency matrix of R. This is matrix A_R.

Next, we define reachability relation R_* on the basis of this relation R:

(10.9.2) $\quad$ iR$_*$j $\quad$ iff $\quad$ a path in the graph of R leads from i to j.

Also, we adopt the convention (as discussed in section 6.4)

$$iR_* i \qquad \text{(all i)}$$

to make R_* reflexive. The corresponding adjacency matrix is denoted A_{R_*}. Then the communication relation C, an equivalence relation, is defined as mutual reachability,

(10.9.3) $\qquad\qquad\qquad\qquad$ iCj $\quad$ iff $\quad$ iR$_*$j and jR$_*$i

And A_C is the adjacency matrix of C, which was shown in section 6.4 to satisfy

(10.9.4) $$A_C = A_{R_*} \times A_{R_*}^T$$

where the multiplication is elementwise.

Example. Given a process with transition matrix,

$$P = \begin{pmatrix} .1 & .2 & 0 & .4 & .3 \\ 0 & 0 & .4 & .6 & 0 \\ 0 & 0 & .1 & 0 & .9 \\ 0 & .2 & .3 & .5 & 0 \\ 0 & 0 & .7 & 0 & .3 \end{pmatrix}$$

we define R as in (10.9.1), to obtain the graph of Figure 10.9.1, where state numbers correspond to row (column) numbers.

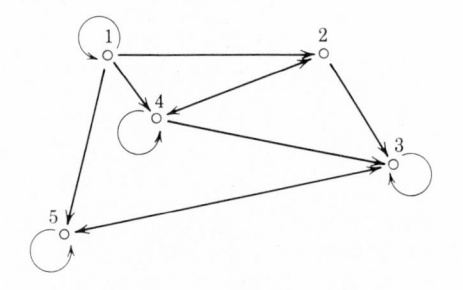

Figure 10.9.1

The adjacency matrix, readily obtained by changing the nonzero entries of **P** to 1, is

$$A_R = \begin{pmatrix} 1 & 1 & 0 & 1 & 1 \\ 0 & 0 & 1 & 1 & 0 \\ 0 & 0 & 1 & 0 & 1 \\ 0 & 1 & 1 & 1 & 0 \\ 0 & 0 & 1 & 0 & 1 \end{pmatrix}$$

Using definition (10.9.2) and Figure 10.9.1 we compute the reachability matrix by successive tracing from each node to note which nodes may be reached in some (directed) path. Then, because of the convention of self-reachability, we obtain the matrix with ones along the main diagonal:

$$A_{R*} = \begin{pmatrix} 1 & 1 & 1 & 1 & 1 \\ 0 & 1 & 1 & 1 & 1 \\ 0 & 0 & 1 & 0 & 1 \\ 0 & 1 & 1 & 1 & 1 \\ 0 & 0 & 1 & 0 & 1 \end{pmatrix}$$

Computing the adjacency matrix of the communication relation, as in formula (10.9.4),

$$A_C = A_{R*} \times A_{R*}^T = \begin{pmatrix} 1 & 1 & 1 & 1 & 1 \\ 0 & 1 & 1 & 1 & 1 \\ 0 & 0 & 1 & 0 & 1 \\ 0 & 1 & 1 & 1 & 1 \\ 0 & 0 & 1 & 0 & 1 \end{pmatrix} \times \begin{pmatrix} 1 & 0 & 0 & 0 & 0 \\ 1 & 1 & 0 & 1 & 0 \\ 1 & 1 & 1 & 1 & 1 \\ 1 & 1 & 0 & 1 & 0 \\ 1 & 1 & 1 & 1 & 1 \end{pmatrix} = \begin{pmatrix} 1 & 0 & 0 & 0 & 0 \\ 0 & 1 & 0 & 1 & 0 \\ 0 & 0 & 1 & 0 & 1 \\ 0 & 1 & 0 & 1 & 0 \\ 0 & 0 & 1 & 0 & 1 \end{pmatrix}$$

Now we plot the results of A_C into the graph shown in Figure 10.9.2, omitting the reflexive loops assumed present on all nodes. As we see from the graph, there are three communication classes.

The states are mapped into classes under equivalence relation C. Hence, $i \to \bar{i}$ for all i in state-space $\mathcal{S}$. (Recall that $\bar{i}$ is the class containing i, as in Part 1.)

We call a class $\bar{i}$ closed if and only if

$$P(i' \to j) = 0 \text{ whenever } i' \text{ in } \bar{i} \text{ and } j \text{ not in } \bar{i}.$$

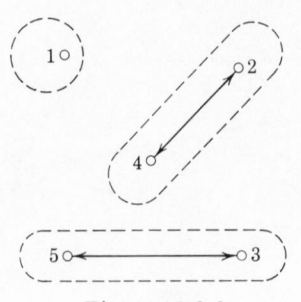

Figure 10.9.2

In other words, it is not possible to leave a closed class. A class that is not closed is called open.

To return to our example, in mapping states into classes, we have

State	Class
1	$\bar{1} = \{1\}$
2	$\bar{2} = \{2, 3\}$
3	$\bar{3} = \{3, 5\}$
4	$\bar{4} = \{2, 4\}$
5	$\bar{5} = \{3, 5\}$

Next we classify each communication class itself as open or closed, using matrix **P**.

Class	Category	Reason
$\{1\}$	open	$P(1 \to 2) > 0$ and $2 \notin \bar{1}$
$\{2, 4\}$	open	$P(2 \to 5) > 0$ and $5 \notin \bar{2}$
$\{3, 5\}$	closed	$P(3 \to j) = P(5 \to j) = 0$ for all $j \notin \bar{3}$

Having arrived at a categorization of the communication classes, we rearrange the matrix **P** so as to put all the closed classes in the upper left, as in the rearrangement of **P** in the above example,

$$
\begin{array}{c}
 \\
3 \\
5 \\
2 \\
4 \\
1
\end{array}
\begin{pmatrix}
\begin{array}{cc|cc|c}
.1 & .9 & 0 & 0 & 0 \\
.7 & .3 & 0 & 0 & 0 \\
\hline
.4 & 0 & 0 & .6 & 0 \\
.3 & 0 & .2 & .5 & 0 \\
\hline
0 & .3 & .2 & .4 & .1
\end{array}
\end{pmatrix}
\begin{array}{c}
3 \quad 5 \quad 2 \quad 4 \quad 1
\end{array}
$$

This is an example of putting **P** into canonical form—that is, a standard form in which any finite transition matrix may be put by the procedure based on reachability. The abstract form is:

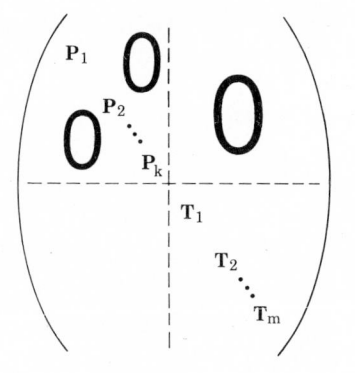

Blank areas are matrices of numbers of no particularly simple character. The matrices P_i ($i = 1, 2, \ldots, k$) are those of corresponding closed classes (for a chain may have more than one closed class). The matrices T_j ($j = 1, 2, \ldots, m$) contain internal transition probabilities within a given open class. The rows of the P_i sum to unity because each alone is a transition matrix. This is not true of the T_j.

In our example, the canonical form has the abstract structure,

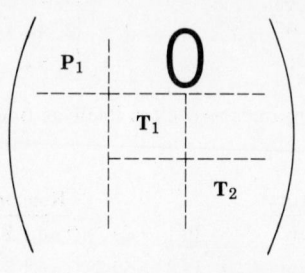

Note row sums of the T_j ($j = 1, 2$) matrices are not one.

The canonical form is used to classify finite Markov chains as shown in Figure 10.9.3.

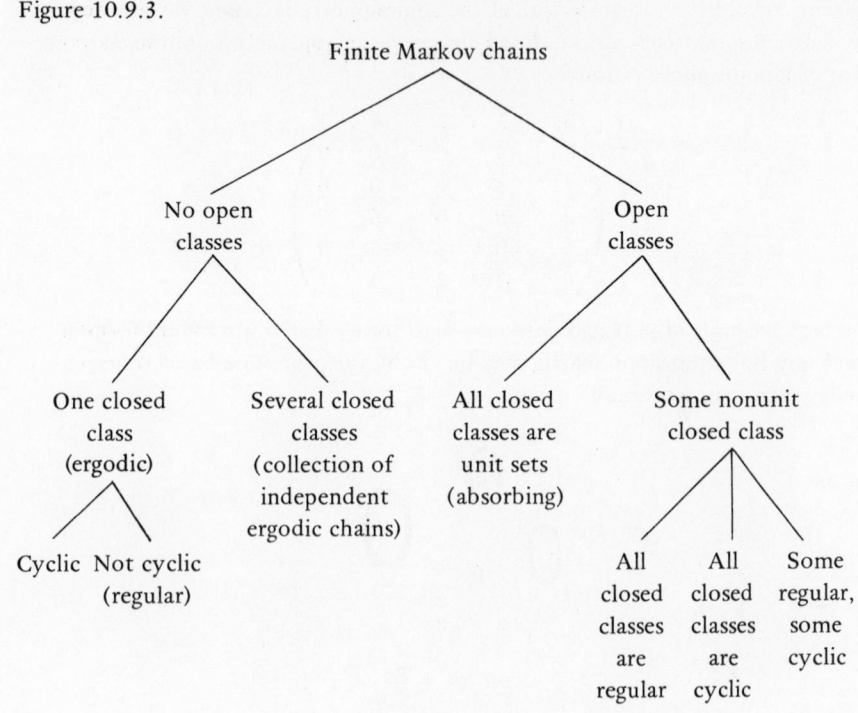

Figure 10.9.3

We begin with an arbitrary chain. Either it has an open class or not. If not, we classify it by its closed classes: either it has only one and, so, forms an ergodic chain, in which every state is reachable from every other, or it has several closed sets, in which case it is nothing but a set of unrelated ergodic processes. If it is ergodic, it is possible that it is cyclic, meaning there is a "period" for visiting states such that some states simply cannot be reached at certain times (e.g., a frog hops back and forth between lily pad A and lily pad B; starting from A, the frog cannot be in A at odd times). If it is not cyclic, then an ergodic chain is called regular. To return to the question of whether there are any open classes—if the answer is yes, then we classify the chain as to the character of the closed classes as follows: if each closed class is nothing but a single state (so that it loops into itself with probability one), we call the chain absorbing. The form of an absorbing chain is

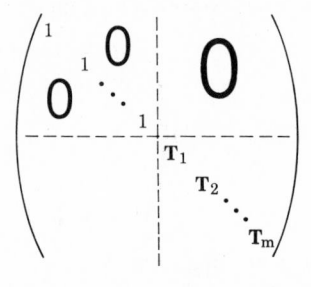

If the chain has some nonunit closed set, as in our example, then it is not absorbing and the classification is on the basis of whether these nonunit classes are all regular, all cyclic, or some mixture of the two. There are no special names for these latter classes.

The main types of chains are ergodic and absorbing, and the major subtype of the ergodic type is regular. The reader can use the chart in Figure 10.9.3 as a guide as to which chapter in Kemeny and Snell (1960) applies to his matrix, after first putting it into canonical form.

10.10. Operations with the Transition Matrix. We saw in Chapter 6 that a basic use of a matrix is to carry vectors into vectors by matrix multiplication. This operation is used very frequently in the analysis of Markov chains.

Let $\mathbf{P}$ be an N x N transition matrix. At any time or "step," now denoted n in this matrix context, the state of the system is X_n with possible values in $\mathcal{S} = \{1, 2, \ldots, N\}$. We write

$$(10.10.1) \quad p(n) = (P(X_n = 1), P(X_n = 2), \ldots, P(X_n = N)) \qquad (n = 0, 1, 2, \ldots)$$

and call (10.10.1) the probability state vector. Its components collectively form the probability distribution of X_n.

We want to show that

(10.10.2) $p(n + 1) = p(n)\mathbf{P}$ $(n = 0, 1, 2, \ldots)$

This says that $\mathbf{P}$ takes the probability state vector at n into the probability state vector at n + 1. Because so many conditional probabilities are involved in Markov theory, it is important to realize that p(n) contains "absolute" probabilities. For example, in the learning model of section 10.7, p(n) has two components because S and $\overline{S}$ are the underlying states and the first component (referent to S) gives the value $P(S_n)$, the probability of being in state S at time n.

To prove (10.10.2), we utilize the law of total probability. At n = 0, formula (10.10.2) reads

$$p(1) = p(0)\mathbf{P}$$

To prove this we write out the matrix product on the right:

$$(P(X_0 = 1), P(X_0 = 2), \ldots, P(X_0 = N)) \begin{pmatrix} P(1 \to 1) & P(1 \to 2) & \ldots & P(1 \to N) \\ P(2 \to 1) & P(2 \to 2) & \ldots & P(2 \to N) \\ \cdot & \cdot & & \cdot \\ \cdot & \cdot & & \cdot \\ \cdot & \cdot & & \cdot \\ P(N \to 1) & P(N \to 2) & \ldots & P(N \to N) \end{pmatrix}$$

The entries $P(i \to j)$ stand for, according to section 10.6,

$$P(i \to j) = P(X_{n+1} = j | X_n = i)   (n = 0, 1, 2, \ldots)$$

Hence, at n = 0,

$$P(i \to j) = P(X_1 = j | X_0 = i)$$

Therefore, the dot product of the row vector with the typical column j yields

$$\sum_{i=1}^{N} P(X_0 = i)P(X_1 = j | X_0 = i)$$

which is the same as

$$\sum_{i=1}^{N} P(X_0 = i \text{ and } X_1 = j)$$

and since the N states exhaust the possibilities (see Figure 10.10.1), the sum is just $P(X_1 = j)$. Of course, this is all an instantiation of the law of total probability (10.2.1): the states at time n = 0 partition the event that at time n = 1 the state is j.

We see, then, that

$$p(0)\mathbf{P} = (P(X_1 = 1), P(X_1 = 2), \ldots, P(X_1 = N)) = p(1)$$

which was to be shown.

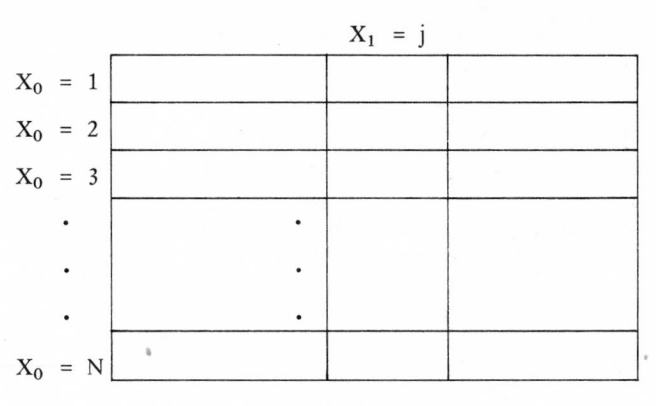

$$X_1 = j$$

Figure 10.10.1

The argument for transition from arbitrary n to n+1 has precisely the same form, substituting n for 0 and n+1 for 1.

Hence, (10.10.2) holds. It says that to transform the probability state vector to obtain its value next time, postmultiply by the transition matrix. Thus, **P** maps p(n) into p(n+1), but notationally this is not conformable to the usual: f(x) = y for x mapped into y by mapping f. On the other hand, in column form, we have,

(10.10.3) $$\mathbf{p}^T(n+1) = \mathbf{P}^T\mathbf{p}^T(n)$$

so that $\mathbf{P}^T$ maps the column vector of probabilities at time n into the column vector of probabilities at time n + 1.

In some references, the transition matrix is defined columnwise. Call it **Q**. Then one will see an expression,

$$q(n+1) = \mathbf{Q}q(n)$$

with the **q** vectors understood to be in column form. This corresponds to (10.10.3). Then transposition of this gives

$$\mathbf{q}^T(n+1) = \mathbf{q}^T(n)\mathbf{Q}^T$$

which result corresponds to (10.10.2).

At this point let us return to example (e) of section 10.8 with the chain of Figure 10.8.7. The matrix of one-step transition probabilities was shown to correspond to the tree diagrams in the setup: row i emerges from the setup phase in postulating what happens, when in state i, to take us to the next time. We found that

$$\mathbf{P} = \begin{pmatrix} 1-qc & qc \\ 0 & 1 \end{pmatrix}$$

Now tracings on the graph persuade us that the path

$$1 \to 1 \to 1$$

has probability $(1-qc)^2$; the path

$$1 \to 2 \to 1$$

has probability zero; thus in two steps, the transition from 1 to 1 has total probability $(1-qc)^2$. Thus,

$$
\begin{aligned}
P(1 \to 1 \text{ in two steps}) &= P(1 \to 1 \to 1 \text{ or } 1 \to 2 \to 1) \\
&= P(1 \to 1 \to 1) + P(1 \to 2 \to 1) \\
&= P(1 \to 1)P(1 \to 1) + P(1 \to 2)P(2 \to 1)
\end{aligned}
$$

and this shows a dot product of two vectors, namely,

$$(10.10.4) \qquad (P(1 \to 1), P(1 \to 2))\begin{pmatrix} P(1 \to 1) \\ P(2 \to 1) \end{pmatrix}$$

using row representation for the first and column representation for the second. Similarly,

$$
\begin{aligned}
P(1 \to 2 \text{ in two steps}) &= P(1 \to 1 \to 2 \text{ or } 1 \to 2 \to 2) \\
&= P(1 \to 1 \to 2) + P(1 \to 2 \to 2) \\
&= P(1 \to 1)P(1 \to 2) + P(1 \to 2)P(2 \to 2)
\end{aligned}
$$

and we can regard this as a dot product

$$(10.10.5) \qquad (P(1 \to 1), P(1 \to 2))\begin{pmatrix} P(1 \to 2) \\ P(2 \to 2) \end{pmatrix}$$

Also,

$$
\begin{aligned}
P(2 \to 1 \text{ in two steps}) &= P(2 \to 1 \to 1) + P(2 \to 2 \to 1) \\
&= P(2 \to 1)P(1 \to 1) + P(2 \to 2)P(2 \to 1)
\end{aligned}
$$

and this can be regarded as the dot product

$$(10.10.6) \qquad (P(2 \to 1), P(2 \to 2))\begin{pmatrix} P(1 \to 1) \\ P(2 \to 1) \end{pmatrix}$$

Finally,

$$
\begin{aligned}
P(2 \to 2 \text{ in two steps}) &= P(2 \to 1 \to 2) + P(2 \to 2 \to 2) \\
&= P(2 \to 1)P(1 \to 2) + P(2 \to 2)P(2 \to 2)
\end{aligned}
$$

and in dot product conception,

$$(10.10.7) \qquad (P(2 \to 1), P(2 \to 2))\begin{pmatrix} P(1 \to 2) \\ P(2 \to 2) \end{pmatrix}$$

Now combining the products (10.10.4) through (10.10.7),

$$\begin{pmatrix} P(1 \to 1) & P(1 \to 2) \\ P(2 \to 1) & P(2 \to 2) \end{pmatrix} \begin{pmatrix} P(1 \to 1) & P(1 \to 2) \\ P(2 \to 1) & P(2 \to 2) \end{pmatrix}$$

and the four dot products we want from this juxtaposition of arrays are precisely those we get by the matrix product. Conclusion: the matrix product P^2 contains in the (i, j) position the probability of going from i to j in two steps.

In general, P^n gives in the (i, j) position the chance of going from state i to state j in n steps.

In these matrix calculations we see that dot products are useful in the calculation of the chances of various intermediate paths.

To sum up, the product $P^n = PP \ldots P$ (n times), where P is the one-step transition matrix, is the n-step transition matrix. The conditional probability interpretation of its typical entry $p_{ij}(n)$ is

$$p_{ij}(n) = P(\text{in state j at time n} \mid \text{in state i at time 0})$$

10.11. Process Dynamics and Markov Chains. Let us return now to equation (10.10.2):

(10.11.1) $p(n+1) = p(n)P$ $(n = 0, 1, \ldots)$

which is the basic change-of-state equation derived from the in-the-small setup for any Markov chain in discrete time. We suppose the generic initial condition

(10.11.2) $p(0)$

and we want the solution of (10.11.1) meaning a trajectory given by the form

$$p(n) = f(p(0), \text{parameters})$$

where "parameters" means the parameters in the setup, now appearing in P.

By attempting the logic of a proof by induction, we try in the discovery stage to find a formula, using the first few values of n in (10.11.1):

$$p(1) = p(0)P$$
$$p(2) = p(1)P = p(0)PP = p(0)P^2$$
$$p(3) = p(2)P = p(0)P^2P = p(0)P^3$$

which suggests

(10.11.3) $p(n) = p(0)P^n$ $(\text{for } n = 0, 1, \ldots)$

provided we define

$$P^0 = I$$

Now to prove the conjecture (10.11.3) we need to appeal to rule (10.7.6). First, for n = 0,

$$p(0) = p(0)P^0 = p(0)I = p(0)$$

which is true. Suppose that (10.11.3) is true for integer n, this being the induction hypothesis. We have to show it is true for n+1; that is, we want to conclude:

(10.11.4) $p(n+1) = p(0)P^{n+1}$

Now, to show this, we have to rely on (10.11.1) and the induction hypothesis. Thus,

$$
\begin{aligned}
p(n) &= p(0)P^n &\text{(induction hypothesis)} \\
p(n+1) &= p(n)P &\text{(10.11.1)} \\
&= [p(0)P^n]\,P &\text{(using the hypothesis)} \\
&= p(0)P^{n+1}
\end{aligned}
$$

The basic result is that for any step n, the state of the probabilistic system—that is, $p(n)$—can be obtained by applying the matrix P^n on the right to the initial condition $p(0)$. Of course, this was anticipated in the preceding section.

Now this raises the question: how can we obtain P^n from P? If P is numerical, hand computation or computers do the job. If not, as is the case in theoretical work, two techniques are available: spectral analysis and z-transform. We shall not treat these in this book (but see, for spectral theory, Bailey, 1964; for z-transform, see Howard, 1971).

Example. (a) Given

$$P = \begin{pmatrix} .7 & .3 \\ .2 & .8 \end{pmatrix} \quad p(0) = (.4, .6)$$

then

$$p(2) = p(0)P^2 = (.4, .6)\begin{pmatrix} .7 & .3 \\ .2 & .8 \end{pmatrix}\begin{pmatrix} .7 & .3 \\ .2 & .8 \end{pmatrix}$$

so that

$$
\begin{aligned}
p(2) &= (.4, .6)\begin{pmatrix} .49 + .06 & .21 + .24 \\ .14 + .16 & .06 + .64 \end{pmatrix} \\
&= (.4, .6)\begin{pmatrix} .55 & .45 \\ .30 & .70 \end{pmatrix} \\
&= (.4(.55) + .6(.30), .4(.45) + .6(.70)) \\
&= (.220 + .180, .180 + .420) \\
&= (.40, .60)
\end{aligned}
$$

Note that in this case,

$$p(2) = p(0)$$

The corresponding tree diagram is given in Figure 10.11.1.

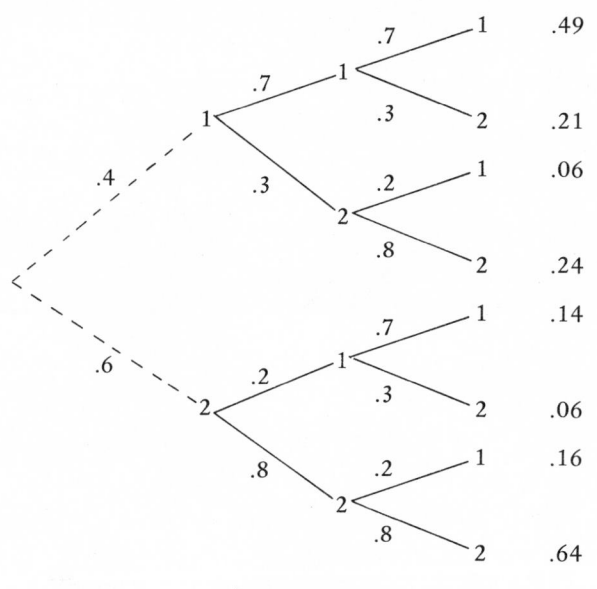

Figure 10.11.1

The initial condition $p(0)$ is shown by dotted lines. Adding all path probabilities (excluding the dotted branches) terminating in state 1 we get $.49 + .06 = .55$ if we start in state 1, and $.14 + .16 = .30$ if we start in state 2. Note that this gives the pair $(.55, .30)$ shown as the first column vector of $\mathbf{P}^2$: entries into state 1, so a column. When these entries are weighted by the initial distribution, we obtain $.4(.55) + .6(.30) = .22 + .18 = .40$, as shown by the matrix computations.

So far we have shown in matrix terms, several elements of process:

(1) the transition rule, or dynamic law, given by (10.11.1), which carries the probability state vector at time (step) n into that at the next step;

(2) the initial condition $p(0)$, stated in (10.11.2);

(3) the deduced in-the-large equation, given in (10.11.3), which gives the probability state vector at any step in terms of the initial condition and the nth power of the matrix $\mathbf{P}$, which corresponds to n iterations of the same dynamic law operating over time.

Next we consider equilibrium. According to our process language, an equilibrium state is a state of a process such that if the process is in that state, it stays there, according to the dynamic law. For a probabilistic system, this equilibrium will be a probability state vector, say p_e, such that

(10.11.5) $$p_e = p_e\mathbf{P}$$

and then,

$$p_e P^2 = (p_e P)P = p_e P = p_e$$

and so forth, so that

$$p_e = p_e P^n \qquad \text{(any n)}$$

We saw in example (a) above an instance of an equilibrium vector, since

$$(.4, .6) = (.4, .6) \begin{pmatrix} .7 & .3 \\ .2 & .8 \end{pmatrix}$$

Hence, for the process whose dynamic law is given by matrix

$$\begin{pmatrix} .7 & .3 \\ .2 & .8 \end{pmatrix}$$

the equilibrium probability state is given by

$$p_e = (.4, .6)$$

In general, to find the equilibrium probability state vector we solve the system (10.11.5), treating p_e as a set of unknowns.

If we did not know that $(.4, .6)$ is the equilibrium vector, we would set up the system

$$\begin{aligned} (x, 1-x) &= (x, 1-x) \begin{pmatrix} .7 & .3 \\ .2 & .8 \end{pmatrix} \\ &= (.7x + .2(1-x), .3x + .8(1-x)) \\ &= (.5x + .2, .8 - .5x) \end{aligned}$$

This yields

$$\begin{aligned} x &= .5x + .2 \\ .5x &= .2 \\ x &= .4 \\ 1-x &= .6 \end{aligned}$$

and the solution vector $(.4, .6)$ is the equilibrium.

Example. (b) Given

$$P = \begin{pmatrix} .5 & .5 \\ .1 & .9 \end{pmatrix}$$

we find the equilibrium vector:

$$\begin{aligned} (x, 1-x) &= (x, 1-x) \begin{pmatrix} .5 & .5 \\ .1 & .9 \end{pmatrix} \\ &= (.5x + .1(1-x), .5x + .9(1-x)) \\ &= (.4x + .1, .9 - .4x) \end{aligned}$$

Hence,

$$
\begin{aligned}
x &= .4x + .1 \\
.6x &= .1 \\
x &= \tfrac{1}{6} \\
1-x &= \tfrac{5}{6}
\end{aligned}
$$

the solution is $p_e = (1/6, 5/6)$.

The following two theorems summarize two basic aspects of equilibrium analysis of a Markov chain: whether it has a unique equilibrium and whether it is stable.

(10.11.6) *Theorem.* If P is regular, then (a) there exists a unique equilibrium probability vector denoted p_e, so that the equation

$$
p_e = p_e P
$$

has a unique solution; and (b) for any initial condition p(0),

$$
p(0)P^n \to p_e \qquad \text{as } n \to \infty
$$

that is, the equilibrium p_e is stable (in the asymptotic sense).

Part (b) of the theorem means that no matter what the initial distribution over states, the process gravitates toward the equilibrium probability distribution. The condition that P is regular was discussed earlier in section 10 (see Figure 10.9.3). Recall that it means the chain is ergodic—mutual reachability holds for all states— and there is no periodic aspect. Not all chains in sociological applications are regular, but many of those that are not regular are absorbing (we treat these in the next section).

To apply theorem (10.11.6) it is convenient to have a matrix method of recognition of the regular property.

(10.11.7) *Proposition.* A transition matrix P is regular if and only if for some n, the matrix P^n has no zero entries.

Returning to example (a) above, we ask if the equilibrium probability distribution is stable. We check P and see that all entries of P are nonzero. Hence, P is regular. Then $p_e = (.4, .6)$ is a stable equilibrium: no matter what the initial probability state p(0), the process moves towards p_e.

The matrix P need not contain all nonzero entries; it suffices that for some n, this be true, not necessarily n = 1.

To sum up our work in this section: we have instantiated the general process language and illustrated in the context of discrete-time Markov chains the concepts of (probability) state, initial condition, transition rule or dynamic law, deduced trajectory of the process, equilibrium (probability vector), and stability of equilibrium.

10.12. Absorbing Chains. If a finite Markov chain is absorbing, then we can use a procedure suggested by Kemeny and Snell (1960). The procedure requires that one put such a chain into canonical form, as follows, for example. Note that we need not go through the adjacency matrix route, as presented in section 10.9, because it is obvious from the transition matrix when a chain is absorbing.

Example. (a)

$$\mathbf{P} = \begin{pmatrix} 0 & .50 & .50 \\ .25 & .75 & 0 \\ 0 & 0 & 1 \end{pmatrix}$$

$$\text{Canonical form of } \mathbf{P} = \left(\begin{array}{c|cc} 1 & 0 & 0 \\ \hline 0 & .75 & .25 \\ .50 & .50 & 0 \end{array} \right)$$

Example. (b)

$$\mathbf{P} = \begin{pmatrix} 0 & .5 & .5 & .0 \\ 0 & 1 & 0 & 0 \\ .7 & 0 & .2 & .1 \\ 0 & 0 & 0 & 1 \end{pmatrix}$$

$$\text{Canonical form of } \mathbf{P} = \left(\begin{array}{cc|cc} 1 & 0 & 0 & 0 \\ 0 & 1 & 0 & 0 \\ \hline .5 & 0 & 0 & .5 \\ 0 & .1 & .7 & .2 \end{array} \right)$$

The general canonical form of an absorbing chain is

(10.12.1)
$$\left(\begin{array}{c|c} \mathbf{I} & \mathbf{O} \\ \hline \mathbf{R} & \mathbf{Q} \end{array} \right)$$

where $\mathbf{I}$ is a $k \times k$ identity matrix (for the k absorbing states); $\mathbf{R}$ is an $(n-k) \times k$ matrix showing transition probabilities from nonabsorbing to absorbing states; $\mathbf{O}$ is an all-zero matrix of order $k \times (n-k)$, since from an absorbing state one cannot go to any other state; and, finally, $\mathbf{Q}$ is a $(n-k) \times (n-k)$ matrix of transition probabilities among the nonabsorbing states. These states are also called transient. (Note that we are not distinguishing between various open classes with their own $\mathbf{T}_j$ matrices, as in the general canonical form.)

Because of the special form of $\mathbf{P}$, we can show that

(10.12.2)
$$\mathbf{P}^n = \left(\begin{array}{c|c} \mathbf{I} & \mathbf{O} \\ \hline \sum\limits_{k=0}^{n-1} \mathbf{Q}^k \mathbf{R} & \mathbf{Q}^n \end{array} \right) \qquad (n = 1, 2, \dots)$$

This result becomes obvious from so-called block multiplication of $\mathbf{P}$ with itself, in canonical form, in which we treat the four submatrices as elements of an algebra. Thus,

$$\mathbf{P}^2 = \begin{pmatrix} I & O \\ R & Q \end{pmatrix}\begin{pmatrix} I & O \\ R & Q \end{pmatrix} = \begin{pmatrix} I^2 + O \cdot R & I \cdot O + O \cdot Q \\ R \cdot I + Q \cdot R & R \cdot O + Q \cdot Q \end{pmatrix}$$

$$= \begin{pmatrix} I & O \\ R + QR & Q^2 \end{pmatrix}$$

This is valid not only because of the algebra of matrices (elements: matrices; operations: addition and multiplication) but because the various matrices are conformable for multiplication when multiplied and are of the same order when added. Continuing,

$$\mathbf{P}^3 = \mathbf{P}^2\mathbf{P} = \begin{pmatrix} I & O \\ R + QR & Q^2 \end{pmatrix}\begin{pmatrix} I & O \\ R & Q \end{pmatrix}$$

$$= \begin{pmatrix} I^2 + O \cdot R & I \cdot O + O \cdot Q \\ (R + QR)I + Q^2 R & (R + QR)O + Q^3 \end{pmatrix}$$

$$= \begin{pmatrix} I & O \\ R + QR + Q^2 R & Q^3 \end{pmatrix}$$

This block multiplication procedure results in the discovery of (10.12.2) as a conjecture. The proof is by induction, because the conjecture is that (10.12.2) is valid for all positive integers. For n = 1, (10.12.2) becomes (10.12.1), the canonical form. Assume (10.12.2) holds for n (induction hypothesis). Then

$$\mathbf{P}^{n+1} = \mathbf{P}^n\mathbf{P} = \begin{pmatrix} I & O \\ \sum_{k=0}^{n-1} Q^k R & Q^n \end{pmatrix}\begin{pmatrix} I & O \\ R & Q \end{pmatrix}$$

$$= \begin{pmatrix} I & O \\ \sum_{k=0}^{n-1}(Q^k R + Q^n R) & Q^{n+1} \end{pmatrix}$$

$$= \begin{pmatrix} I & O \\ \sum_{k=0}^{n} Q^k R & Q^{n+1} \end{pmatrix}$$

which is (10.12.2) for n+1. Thus, the conjecture is proved.

Let us see if simplicity is introduced in (10.12.2) if we use a limiting process. From experience, this often occurs. Besides, the sum resembles

$$\sum_{k=0}^{m} q^k r \qquad (m = n-1)$$

where q and r are just numbers. Then factor out r and take the limit as $m \to \infty$:

$$r \lim_{m \to \infty} (1 + q + q^2 + q^3 + \dots) = r \cdot \frac{1}{1-q} \qquad \text{(if } |q| < 1)$$

$$= r(1 - q)^{-1}$$

according to proposition (10.7.14).

By analogy,

$$\sum_{k=0}^{m} Q^k R \qquad (m = n-1)$$

might first be treated by factoring out **R**, to seek a result $\mathbf{R}(\mathbf{I} - \mathbf{Q})^{-1}$; however, since matrix multiplication is not commutative, we need to hold **R** on the right:

(10.12.3)
$$(\mathbf{Q}^0 \mathbf{R} + \mathbf{Q}^1 \mathbf{R} + \mathbf{Q}^2 \mathbf{R} + \dots + \mathbf{Q}^m \mathbf{R})$$
$$= (\mathbf{I} + \mathbf{Q} + \mathbf{Q}^2 + \dots + \mathbf{Q}^m)\mathbf{R}$$

Next, suppose we take limits, as $m \to \infty$. Kemeny and Snell show that since **Q** is the sort of matrix it is (of probabilities with row sums $\leqslant 1$), it satisfies a condition somewhat like $|q| < 1$. The series then converges with

$$\mathbf{I} + \mathbf{Q} + \mathbf{Q}^2 + \dots = (\mathbf{I} - \mathbf{Q})^{-1}$$

Thus,

(10.12.4)
$$\lim_{m \to \infty} \sum_{k=0}^{m} \mathbf{Q}^k \mathbf{R} = (\mathbf{I} - \mathbf{Q})^{-1} \mathbf{R}$$

and Kemeny and Snell call $(\mathbf{I} - \mathbf{Q})^{-1}$ the fundamental matrix, labeling it **N**. We note that the entry n_{ij} in **N** has i and j ranging only over the transient states of the process.

We now derive a probabilistic interpretation of **N**, followed by a similar interpretation for $\mathbf{NR} = (\mathbf{I} - \mathbf{Q})^{-1} \mathbf{R}$. Let us note that we seem to be taking into account the amount of time spent in the set of transient states ($\mathbf{Q}^k$'s) before absorption. This suggests a counting variable for states i and j, which are transient:

$$X_{ijn} = \begin{cases} 1 & \text{if in state j at time n, starting in i} \\ 0 & \text{if not} \end{cases}$$

Then

$$T_{ijm} = \sum_{n=0}^{m} X_{ijn}$$

gives the amount of time in transient state j, up to time m, starting in i. Thus,

$$E(T_{ijm}) = \sum_{n=0}^{m} E(X_{ijn})$$

follows from the fact that the expectation of a sum (even of dependent variables) is the sum of the expectations. Since

$$
\begin{aligned}
E(X_{ijn}) &= \sum_x xP(X_{ijn} = x) \\
&= 1 \cdot P(X_{ijn} = 1) + 0 \cdot P(X_{ijn} = 0) \\
&= P(X_{ijn} = 1)
\end{aligned}
$$

we obtain,

$$
E(T_{ijm}) = \sum_{n=0}^{m} P(X_{ijn} = 1)
$$

But,

$$
P(X_{ijn} = 1) = P(\text{state } j \text{ at } n \mid \text{state } i \text{ at } 0)
$$

is the (ij) entry in Q^n. Hence, forming a matrix,

$$
\begin{aligned}
(E(T_{ijm})) &= \left(\sum_{n=0}^{m} P(X_{ijn} = 1) \right) \\
&= \sum_{n=0}^{m} (P(X_{ijn} = 1)) \\
(10.12.5) \qquad &= \sum_{n=0}^{m} Q^n
\end{aligned}
$$

Now,

$$
\begin{aligned}
N = (I - Q)^{-1} &= \lim_{m \to \infty} \sum_{n=0}^{m} Q^n \qquad &\text{(definition of } N) \\
&= \lim_{m \to \infty} (E(T_{ijm})) \qquad &(10.12.5) \\
&= (\lim_{m \to \infty} E(T_{ijm})) \qquad &[\lim A = (\lim a_{ij}), \text{ by definition}]
\end{aligned}
$$

So the n_{ij} entry in the fundamental matrix gives the expected total number of times the system is in state j before absorption, given it starts in state i. Finally, by summing over j we obtain the expected total number of times the system is in a transient state, starting from i: the time to absorption.

To interpret $(I - Q)^{-1}R$ we use another method to exhibit another application of the event representation ideas of Chapter 9. Define

$$
\begin{array}{lll}
E_{ij} \text{ occurs} & \text{iff} & \text{a one-step transition } i \to j \text{ occurs} \\
A_{ik} \text{ occurs} & \text{iff} & \text{absorption into k from i occurs}
\end{array}
$$

Here i is a transient state.

The event A_{ik} can occur in two mutually exclusive ways: a direct transition

$i \rightarrow k$ occurs (E_{ik}) or some indirect absorption occurs via first a transition event E_{ij}, j transient, and then A_{jk} occurs, for some j. Hence,

$$A_{ik} = \underbrace{E_{ik}}_{\text{direct}} \cup [\underbrace{\underset{j}{\cup}(E_{ij} \cap A_{jk})}_{\substack{\text{indirect} \\ \text{via j}}}] \qquad \text{(j transient)}$$

Hence,

$$P(A_{ik}) = P(E_{ik}) + P[\underset{j}{\cup}(E_{ij} \cap A_{jk})]$$

$$= P(E_{ik}) + \underset{j}{\Sigma} P(E_{ij} \cap A_{jk})$$

$$= P(E_{ik}) + \underset{j}{\Sigma} P(E_{ij}) P(A_{jk}) \qquad \text{(j transient)}$$

Now let $b_{ik} = P(A_{ik})$, so that we have shown

$$b_{ik} = p_{ik} + \underset{j}{\Sigma} p_{ij}b_{jk} \qquad \text{(j transient)}$$

In matrix form, then, since both i and j are transient and k is absorbing,

$$\mathbf{B} = (b_{ik}) = \mathbf{R} + \mathbf{QB}$$

Hence,

$$\mathbf{B} - \mathbf{QB} = \mathbf{R}$$
$$\mathbf{IB} - \mathbf{QB} = \mathbf{R}$$
$$(\mathbf{I} - \mathbf{Q})\mathbf{B} = \mathbf{R}$$
$$\mathbf{B} = (\mathbf{I} - \mathbf{Q})^{-1}\mathbf{R}$$

This shows, then, that the matrix $(\mathbf{I} - \mathbf{Q})^{-1}\mathbf{R}$ of (10.12.4) is interpretable as the matrix of absorption probabilities: the typical entry is the probability of the event A_{ik} that the system absorbs into state k, starting from i.

Thus, this interpretation provides a real motivation for the trouble involved in the inversion procedure. By applying the inversion procedure to $(\mathbf{I} - \mathbf{Q})$ we learn a good deal about how the process behaves before it enters the absorbing state, and obtain the various ultimate absorption probabilities.

Example. The classic example of an absorbing Markov chain that is manageable mathematically is based on Bernard C. Cohen's study of the responses in the Asch situation (Cohen, 1963). (See sections 8.5 and 10.2.) We recall that the state-space approach to this behavior led Cohen to four mental states and a Markov chain that has two absorbing states, nonconforming and conforming. A graph of the system is given in Figure 10.12.1.

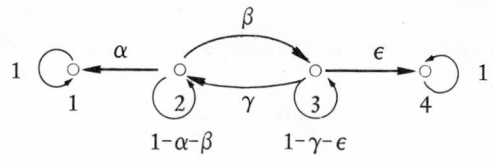

Figure 10.12.1

Thus, in canonical form,

$$
\begin{array}{c}
 \begin{array}{ccccc} 1 & 4 & \vert & 2 & 3 \end{array} \\
\mathbf{P} = \begin{array}{c} 1 \\ 4 \\ 2 \\ 3 \end{array}
\left(
\begin{array}{cc|cc}
1 & 0 & 0 & 0 \\
0 & 1 & 0 & 0 \\ \hline
\alpha & 0 & 1-\alpha-\beta & \beta \\
0 & \epsilon & \gamma & 1-\gamma-\epsilon
\end{array}
\right)
\end{array}
$$

Hence,

$$
\mathbf{I} = \begin{pmatrix} 1 & 0 \\ 0 & 1 \end{pmatrix} \qquad \mathbf{O} = \begin{pmatrix} 0 & 0 \\ 0 & 0 \end{pmatrix}
$$

$$
\mathbf{R} = \begin{array}{c} 2 \\ 3 \end{array}\begin{array}{c}\begin{array}{cc}1 & 4\end{array}\\ \begin{pmatrix} \alpha & 0 \\ 0 & \epsilon \end{pmatrix}\end{array} \qquad \mathbf{Q} = \begin{array}{c} 2 \\ 3 \end{array}\begin{array}{c}\begin{array}{cc}2 & 3\end{array}\\ \begin{pmatrix} 1-\alpha-\beta & \beta \\ \gamma & 1-\gamma-\epsilon \end{pmatrix}\end{array}
$$

Then,

$$
\mathbf{I} - \mathbf{Q} = \begin{array}{c} 2 \\ 3 \end{array}\begin{array}{c}\begin{array}{cc}2 & 3\end{array}\\ \begin{pmatrix} \alpha+\beta & -\beta \\ -\gamma & \gamma+\epsilon \end{pmatrix}\end{array}
$$

We apply inversion to $\mathbf{I} - \mathbf{Q}$ in the abstract, an important aspect of the theoretical use of the Kemeny-Snell fundamental matrix method that works when $(\mathbf{I} - \mathbf{Q})$ does not have too many nonzero entries. Since $(\mathbf{I} - \mathbf{Q})$ is 2 x 2, we use (6.3.10). Thus,

$$
\Delta = (\alpha+\beta)(\gamma+\epsilon) - \beta\gamma
$$

and

$$
\mathbf{N} = (\mathbf{I} - \mathbf{Q})^{-1} = \frac{1}{\Delta}\begin{array}{c}\begin{array}{cc}2 & 3\end{array}\\ \begin{pmatrix} \gamma+\epsilon & \beta \\ \gamma & \alpha+\beta \end{pmatrix}\end{array}\begin{array}{c} 2 \\ 3 \end{array}
$$

Since Cohen assumed his subjects began in state 2 (temporary nonconforming), we need only examine the row of $\mathbf{N}$ that is associated with this state, the first. Thus,

$$n_{22} = \frac{\gamma+\epsilon}{\Delta} = \text{expected number of times in state 2 (temporary nonconforming)}$$

$$n_{23} = \frac{\beta}{\Delta} = \text{expected number of times in state 3 (temporary conforming)}$$

$$\frac{\beta+\gamma+\epsilon}{\Delta} = \text{expected time to absorption.}$$

The response behavior is a function of the state:

$$
\begin{aligned}
\text{state 1} &\mapsto \text{nonconforming} \\
\text{state 2} &\mapsto \text{nonconforming} \\
\text{state 3} &\mapsto \text{conforming} \\
\text{state 4} &\mapsto \text{conforming}
\end{aligned}
$$

Hence,

$$\frac{\gamma+\epsilon}{\Delta} = \text{expected number of nonconforming responses before absorption}$$

$$\frac{\beta}{\Delta} = \text{expected number of conforming responses before absorption.}$$

However, in this model there is no observable response that signals absorption and, so, steady-state behavior ever after. A string of 10 nonconforming responses could be generated in state 2—but the probability grows small as we lengthen our string-length that the subject is merely in state 2 and not in state 1. Thus, string-length of 10 or 15 steady conforming or steady nonconforming responses reasonably could be taken to indicate absorption has occurred. But at which point in the steady sequence? The answer is that we do not know, but we do not have to know to make predictions: all we need is a way to estimate parameters and an indicator that the subject is not yet absorbed at a specified time. The latter is readily detected from the responses: any switch shows the subject is still vacillating and, so, in state 2 or 3. Thus, all times up to the last observed switch are times the subject was in the transient set. Thus, some correspondences can be set up: enough identifications can be made of the notions in the model to permit estimation and testing. (For an extensive discussion of estimation and testing of the Cohen model, see Bartos, 1967.)

CHAPTER ELEVEN   PROBABILISTIC CHOICE THEORY

11.1. Introduction. In sections 3.8-3.12 a set-theoretic formulation of the choice situation of higher organisms was presented, based on the work of Frank Restle (1961). In this chapter, we first review and amplify the conceptual basis of this formulation and then explore the characterization of choice behavior it provides. Secondly, having constructed a conceptual foundation for the probabilistic theory of choice behavior, we examine the consequences of a single powerful axiom proposed by Luce (1959). Finally, we apply a measurement model for occupational prestige to empirical data, where the model emerges from an application of the consequences of Luce's axiom. For the social scientist, this choice framework, particularly the theory based on Luce's axiom, provides an opportunity to see an important general problem in the study of human behavior—how people choose—imbedded in a deductively rich probabilistic framework. The construction of this framework, the analysis of its properties by the strict proof of theorems, and, finally, its application to a sociological measurement problem should be regarded as of interest in their own right as well as exemplifications of a general methodology of formal theory construction.

11.2. Mathematical Formulation of the Restle Foundations. We can summarize and extend the Restle foundations of sections 3.8-3.12 as in Table 11.2.1. In reading this table, recall that for any set X, by 2^X we mean the collection of all subsets of X.

The basic psychological postulate of the system connects P, the response probability, to m, the measure on relevant aspects, as in (3.10.2). Since the relevant aspects are not identifiable ordinarily and since m is a representation of a subjective measure on such aspects (e.g., in terms of salience or differential evaluation), the main function of the postulate is to relate the scientifically accessible behaviors in R to the unobservable phenomena that we believe are the basis for the choice. This is done in detail for the special case of two alternative

Table 11.2.1 Summary of the Concepts of the Restle Foundations

Name of Concept	Mathematical Representation	Symbol	Relationships to Other Concepts and Remarks
Set of situations	Set	$\mathcal{S}$	$s \in \mathcal{S}$ is a situation; $\mathcal{S}$ is a primitive notion.
Stimulus variable	Set	v	v is a partition of $\mathcal{S}$.
Situation complex	Ordered pair	$(\mathcal{S}, \mathcal{P})$	$\mathcal{P}$ is the set of all partitions of $\mathcal{S}$ and each partition is a stimulus variable.
Aspect	Set	a	a is a part in some partition in $\mathcal{P}$; hence, a is a "cell" of a variable v.
Set of all aspects	Set	A^*	$A^* = \{a\colon a \in v, \exists\, v\}$
Aspects of situations	Mapping	A	$\mathcal{S} \overset{A}{\to} 2^{A^*}$ and we write $A_s = A(s) = \{a\colon s \in a\}$ By theorem (3.9.1) A is 1-1.
Relevant aspects of situations	Mapping	A'	$\mathcal{S} \overset{A'}{\to} 2^{A^*}$ and we write $A'_s = A'(s)$ and $A'_s \subseteq A_s$, so that if $a \in A'_s$, then $s \in a$. A' is a primitive notion.
Ideal situations	Set	I	$I \subseteq \mathcal{S}$; each $s \in I$ is termed an ideal situation. I is a primitive notion.
Response set	Set	R	Each $R_i \in R$ is a possible response. R is a primitive notion.
Response schema	Mapping	$\mathcal{s}$	The domain is I and the range is R. $\mathcal{s}$ is a primitive notion.
Measure on aspects	Measure	m	m has domain 2^{A^*} and satisfies measure axioms stated in section 11.3. m is a primitive notion.
Response probability	Probability measure	P	P is defined over R, i.e., R is the outcome set of a finite probability model. P is a primitive notion.

responses by Restle, but a full range of complex relations between R and the A'_s are not treated. For the simplest case, then, the concepts specialize to:

Some $s \in \mathcal{S}^*$ is the focal situation

$R = \{R_1, R_2\}$	two possible responses
$I = \{s_1, s_2\}$	two assumed ideal situations
$\mathcal{s}(s_i) = R_i\ (i = 1, 2)$	two assumed parts to the schema
$A'_i \equiv A'_{s_i}\ (i = 1, 2)$	the two sets of relevant aspects for the s_i in I
A'_s	the relevant aspects of s.

Hence, in this case,

$$P(R_1) + P(R_2) = 1$$

because P is a probability measure, and so, there is just one independent response probability, say $P(R_1)$. We define

set (1) $= A'_s \cap A'_1 \cap \overline{A}'_2$ relevant aspects of s common with s_1 but not s_2
set (2) $= A'_s \cap \overline{A}'_1 \cap A'_2$ relevant aspects of s common with s_2 but not s_1

For intuitive clarity, consult Figure 3.10.1.

Since each of the above sets has its measure, m(1) and m(2) exist, and although not identified directly they can be estimated in a manner to be shown subsequently. The postulate,

$$(11.2.1) \qquad\qquad P(R_1) = \frac{m(1)}{m(1) + m(2)}$$

restated here for convenience, is based on all the primitive notions of the foundations, because R_1 connects to the current situation s only via the schema—perhaps a norm, perhaps a stored trace of prior response modes, or perhaps with some other interpretation—relating ideal situation s_1 to the response R_1. Moreover, this connection is exhibited in terms of the present relevant aspects and the subjective (say, valuation or salience) measure defined over these aspects.

Before proceeding, we pause to prove the property of mapping A stated earlier as theorem (3.9.1): namely, A is a 1-1 (or injective) mapping. The proof is a simple exercise in scrupulous attention to the mathematical details. Recall that to say that A is injective (see section 5.3) means that for any two situations s, s',

(1)   if s $\neq$ s', then $A_s \neq A_{s'}$

Since the contrapositive (see section 2.11 to recollect this point) is given by

(2)   if $A_s = A_{s'}$, then s = s'

and is tautologically equivalent to (1), we can prove either (1) or (2) to show the injective property. We prove (2) as follows: Assume that $A_s = A_{s'}$. This means

$$(11.2.2) \qquad\qquad a \in A_s \quad \text{iff} \quad a \in A_{s'} \qquad (a \in A^*)$$

Now,

$$a \in A_s \quad \text{iff} \quad s \in a$$

and so, with a = $\{s\}$, this instantiates to

$$\{s\} \in A_s \quad \text{iff} \quad s \in \{s\}$$

From this we conclude

$$(11.2.3) \qquad\qquad \{s\} \in A_s$$

because obviously $s \in \{s\}$. From (11.2.2) and (11.2.3), we conclude

$$(11.2.4) \qquad\qquad \{s\} \in A_{s'}$$

But then because

$$a \in A_{s'} \quad \text{iff} \quad s' \in a$$

from (11.2.4) we obtain

$$s' \in \{s\}$$

and consequently,

$$s = s'$$

which was to be shown. That is, A is 1-1 from situations to sets of aspects.

However, this is not generally true of A' because the primitive character of A' allows any subset of A_s to serve as A'_s. Thus, we can easily have

$$s \neq s' \text{ but } A'_s = A'_{s'}$$

with the result that two distinct situations are identical for the subject.

Recall that for any mapping, we can ask if it is surjective (onto its range). This would mean in the case of A that for any subset of aspects of A^*, there is some situation in $\mathcal{S}$ such that these form its set of aspects. Symbolically,

for any $x \in 2^{A^*}$, there exists an $s \in \mathcal{S}$ such that $A_s = x$.

This is not generally true. We assume that $\mathcal{S}$ has at least three situations in it. For a counterexample to the surjective property, we take the set

$$x = \{\{s\}\}$$

Note that since

$$\{s\} \in A^*$$

under the variable

$$v = \{\{s\}, \mathcal{S} - \{s\}\}$$

we know that $x \subseteq A^*$. Hence, $x \in 2^{A^*}$. Now we show that no s goes into x under A. If $s \mapsto A_s = x$, we would have

(11.2.5) $A_s = \{\{s\}\}$

that is, the only aspect of s would be $\{s\}$. However, taking situation s' in $\mathcal{S}$ with $s' \neq s$, we have the stimulus variable

$$\{\{s'\}, \mathcal{S} - \{s'\}\}$$

with $s \in \mathcal{S} - \{s'\}$. Hence,

$$\mathcal{S} - \{s'\} \in A_s$$

From (11.2.5), this implies

$$\mathcal{S} - \{s'\} = \{s\}$$

Then S has only two members, which contradicts our assumption. Thus $\{s\}$ cannot be the image of s under A. Hence, in general, A is not surjective. Thus, the properties of A and A' are (1) A is injective but not surjective, and (2) A' is generally neither injective nor surjective.

11.3. Measures, Metrics, and the Aspect Differential. Let us take this foundation to the point at which we make a connection with the idea that the aspects are differentially evaluated. We need an elaboration first on the idea of measure, denoted m, in Table 11.2.1.

The concept of measure is based on repeated experience in applied mathematics with structures of sets over which a numerical mapping is defined. The idea of measure is neutral as to the interpretation and, so, can be said to have emerged by abstraction. Here are three prime "concrete" measures from which the abstract idea was obtained:

(1) Number: Write N(A) for the number of elements in set A. For a domain of sets, say $\mathfrak{B}$, then

$$A \mapsto N(A) \in \mathfrak{R}$$

for $A \in \mathfrak{B}$. That is, each set is mapped into the real numbers (not surjectively, because only the nonnegative integers are in the image set under the number mapping). Clearly,

$$N(A \cup B) = N(A) + N(B) \qquad (\text{if } A \cap B = \emptyset)$$
$$N(A) \geqslant 0$$

for all A, B in the domain $\mathfrak{B}$ of sets. The first property says that the number property is additive and the second that it is nonnegative.

(2) Probability: Write P(A) for the probability of A, where A is any subset (representing an event) of an outcome space. By the earlier definition of a finite probability model, we know that

$$P(A \cup B) = P(A) + P(B) \qquad (\text{if } A \cap B = \emptyset)$$
$$P(A) \geqslant 0$$

for all A, B in the domain $\mathfrak{B}$ of sets.

(3) Area: Write $\alpha(A)$ for the area of a subset of points in the plane $\mathfrak{R}^{(2)} = \mathfrak{R} \times \mathfrak{R}$. (Strictly, area is a mapping defined on flat aspects of bodies, but the flat aspects are represented as regions in $\mathfrak{R}^{(2)}$ just as events are represented as subsets of possible outcomes). Then,

$$\alpha(A \cup B) = \alpha(A) + \alpha(B) \qquad (\text{if } A \cap B = \emptyset)$$
$$\alpha(A) \geqslant 0$$

for all regions A, B in the domain $\mathfrak{R}^{(2)}$.

The abstraction from number, probability, and area (among other concrete measures) leaves:

$$(11.3.1) \quad \begin{array}{ll} \text{(a)} \ m(A \cup B) = m(A) + m(B) & \text{(if } A \cap B = \emptyset) \\ \text{(b)} \ m(A) \geqslant 0 \end{array}$$

for all sets A, B in some domain of sets. Expressions (11.3.1) (a) and (b) define the concept of measure. Then, by this definition the concrete cases (1), (2), and (3) are exemplifications of the abstract measure concept.

Restle employs a measure m without giving it a fixed interpretation in terms of any measurement model. This is somewhat contrary to common experience: number itself has a "measurement model" based on the idea of counting; probability has several measurement models, the most important being counting relativized to a maximum count or relative frequency; and area has its measurement model based on length. In each case, the measurement model functions to deprimitivize a numerical primitive notion and, so, to base the concept upon some type of empirical system and make it interpretable for the analysis of phenomena.

In the case of Restle measure m, on the other hand, the logic of fundamental measurement does not apply, because the empirical systems in question are the "private facts" A'_s of subjects. Thus, we are taken into a mode of indirect measurement. Note that R, the set of responses, is interpretatively public; that is, it is referent to features of the behavior of the subject. The fact that R is identifiable means that $P(R_i)$ will be capable of estimation. In other words, the analytical ground system for the probabilistic process on the state-space side—A'_s facts—is not observable, but on the response side it is, because we are agreed in science to use the concept of response to refer to public facts.* Hence, the logic of measurement is as follows:

(1) postulate m over unobservable states;

(2) postulate R, as related to states, so that measure P is obtainable via a measurement model, that is, counting in the relative frequency sense;

(3) then derive the values m(1) and m(2) via the lawlike relation

$$P(R_1) = \frac{m(1)}{m(1) + m(2)}$$

This procedure may be summarized by noting that what is psychologically prior (m) is in measurement derived rather than fundamental. Braithwaite (1956) seems to have had this sort of common scientific procedure in mind in speaking of "fastening the zipper [connecting theory and actualities] from the bottom up."

The concrete interpretation of m then depends upon the context of identification of such concepts as: $\mathcal{S}$, I, R, and $\mathcal{S}$. Thus, in some contexts, as m receives its meaning from "the bottom up" it will be thought of as salience. In other contexts,

---

*In social science, such a public fact is essentially a consequence of a coding system held in common by "subject" and "observer." For example, both interpret a certain sound as a token of the letter "A."

those of interest to us at present, we will think of it as value. But no word can perform interpretative magic: the "meaning" of an object in a model is determined by the systematic general framework governing the model construction and by specific and concrete identifications referent to the actualities to which all constructions ultimately refer.

Whatever the interpretation of m, the general theory of measure on sets shows that once we have a measure we also have a metric in the following sense: a metric space is a set M together with a function (call it d) with domain the set of all pairs in M (i.e., domain M x M) and range the real line such that for all x, y $\in$ M,

(11.3.2)

(D1) $d(x, x) = 0$
(D2) $d(x, y) \geqslant 0$
(D3) $d(x, y) = d(y, x)$
(D4) $d(x, y) + d(y, z) \geqslant d(x, z)$

Axiom D1 says that a point has distance 0 from itself. If the condition

$$d(x, y) = 0 \text{ implies } x = y$$

is added, we have the full-fledged metric space concept; in its absence, so that distance can be 0 between distinct points, the metric is sometimes called a pseudometric. Axiom D2 says distance is never negative. Axiom D3 says it is symmetric. D3 is termed the triangle inequality (see Figure 11.3.1).

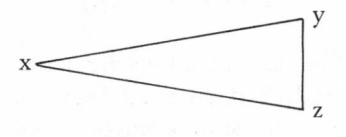

Figure 11.3.1

As in the case of measure, the concept of distance has been abstracted from its many concrete exemplifications.

For our purposes, an important metric arises as soon as one considers a family of subsets of a set, independently of any other interpretation, and a measure exists on this family. To be technical, the family must contain the union, intersection, and complement of any sets in the family. It is then called an algebra of sets. (One example is the collection of all subsets of a given set as we used it in the concept of a finite probability model in section 9.7.) The reason for requiring this kind of "closure" is that we want to define for every pair of sets the following binary operation, called the symmetric difference:

(11.3.3) $A \Delta B = (A - B) \cup (B - A)$

The relevant Venn diagram is shown in Figure 11.3.2, where the ruled portion represents $A \Delta B$.

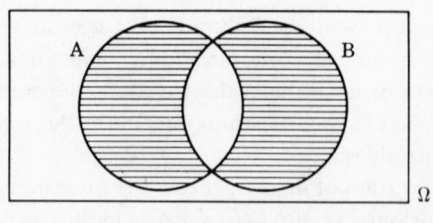

Figure 11.3.2

An alternative way of defining symmetric difference is clear from the diagram; the union minus the overlap:

$$A \triangle B = (A \cup B) - (A \cap B)$$

Note that

$$
\begin{aligned}
A \triangle B &= (A - B) \cup (B - A) \\
&= (B - A) \cup (A - B) \\
&= B \triangle A
\end{aligned}
$$

Thus, the name makes sense: the binary operation $\triangle$ is commutative on the subsets of Ω. The following development is standard mathematics (see, for instance, Royden, 1963): we define:

(11.3.4) $d(A, B) = m(A \triangle B)$

(11.3.5) Theorem. The function $d(A, B)$ is a metric—satisfying, that is, (11.3.2)—on the space over which m is defined, assuming that for any A, B the union, intersection, and complement are in the space whenever A and B are in the space.

Proof. If $A = B$, then $A \triangle B = A \triangle A = \emptyset$, by (11.3.3). Now by (11.3.1) (a),

$$m(\emptyset) = m(\emptyset \cup \emptyset) = m(\emptyset) + m(\emptyset) = 2m(\emptyset)$$

and so, $m(\emptyset) = 0$. Hence, $d(A, B) = 0$ if $A = B$, satisfying D1 of (11.3.2).

Since $d(A, B) = m(A \triangle B)$ and, by (11.3.1)(b), $m(A \triangle B) \geqslant 0$, it follows that $d(A, B) \geqslant 0$, satisfying D2 of (11.3.2).

Because $A \triangle B = B \triangle A$, $m(A \triangle B) = m(B \triangle A)$, and so, $d(A, B) = d(B, A)$. Hence, D3 of (11.3.2) is satisfied.

Finally, we come to the triangle inequality D4 of (11.3.2). Let A, B, and C be in the space. The general Venn diagram is shown in Figure 11.3.3. And we see

(1) $= \overline{A} \cap B \cap C$		(5) $= A \cap \overline{B} \cap \overline{C}$	
(2) $= A \cap \overline{B} \cap C$		(6) $= \overline{A} \cap B \cap \overline{C}$	
(3) $= A \cap B \cap C$		(7) $= \overline{A} \cap \overline{B} \cap C$	
(4) $= A \cap B \cap \overline{C}$		(8) $= \overline{A} \cap \overline{B} \cap \overline{C}$	

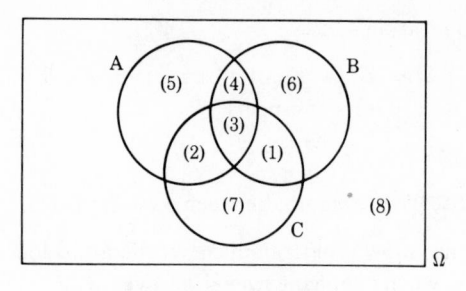

Figure 11.3.3

We must show that

$$d(A, B) + d(B, C) \geqslant d(A, C)$$

which by definition (11.3.4) means

(11.3.6) $$m(A \, \Delta \, B) + m(B \, \Delta \, C) \geqslant m(A \, \Delta \, C)$$

Using the fact that the two parts of the symmetric difference are disjoint, property (11.3.1) (a) of measure implies

$$m(A \, \Delta \, B) + m(B \, \Delta \, C) = m[(A{-}B) \cup (B{-}A)] + m[(B{-}C) \cup (C{-}B)]$$
(11.3.7) $$= m(A{-}B) + m(B{-}A) + m(B{-}C) + m(C{-}B)$$

Now we note that in the notation of Figure 11.3.3, we have the disjoint unions, with measure additive, yielding

$$A - B = (5) \cup (2) \quad \text{so} \quad m(A - B) = m(5) + m(2)$$
$$B - A = (6) \cup (1) \quad \text{so} \quad m(B - A) = m(6) + m(1)$$
$$B - C = (6) \cup (4) \quad \text{so} \quad m(B - C) = m(6) + m(4)$$
$$C - B = (7) \cup (2) \quad \text{so} \quad m(C - B) = m(7) + m(2)$$

Substitution of these expressions into the right side of (11.3.7) yields

$$m(A \, \Delta \, B) + m(B \, \Delta \, C) = m(1) + 2m(2)$$
(11.3.8) $$+ \, m(4) + m(5) + 2m(6) + m(7)$$

We must compare this with

$$m(A \, \Delta \, C) = m(A - C) + m(C - A)$$
$$= m[(5) \cup (4)] + m[(7) \cup (1)]$$
(11.3.9) $$= m(5) + m(4) + m(7) + m(1).$$

Clearly, since m is nonnegative, when we subtract (11.3.9) from (11.3.8) we obtain a nonnegative number. This proves (11.3.6). Hence, D4 is satisfied and $d(A, B)$ is a metric in the sense of (11.3.2).

Thus, under the application,

$$\Omega = A^* = \text{set of all aspects of s in } \mathcal{S}$$
$$A, B = \text{sets of aspects}$$

we conclude that

$$d(A, B) = \text{distance between sets of aspects.}$$

To sum up, Restle has shown that a familiar set-theoretic construction, the symmetric difference, when coupled with a primitive measure over sets representing situations in terms of aspects, induces a metric space over the set of aspects. The distance between two sets of aspects, representing two situations, is the measure of their "aspect-differential," the aspects that differentiate them.

11.4. Concept of Betweenness. The above construction can now be carried a step further to define a "betweenness relation" between situations as represented in terms of aspects. Consider the expression (11.3.6) above. We see that by the calculations based on the Venn diagram in Figure 11.3.3

$$m(A \, \Delta \, B) + m(B \, \Delta \, C) = m(A \, \Delta \, C)$$

if, and only if,

$$m(2) + m(6) = 0$$

Looking at our Venn diagram we realize that because of disjointness, this could be achieved if

$$m[(2) \cup (6)] = 0$$

and so if (although not only if) $(2) \cup (6) = \varnothing$

and so if $(2) = \varnothing$ and $(6) = \varnothing$. Thus, (referring to Figure 11.3.3 for visualization) if something is in $A \cap C$ it has to be in B: $A \cap C \subseteq B$. And, also, if something is in $\overline{A} \cap \overline{C}$, it has to be in $\overline{B}$: $\overline{A} \cap \overline{C} \subseteq \overline{B}$. Since the intersection of the complements is the complement of the union (see Table 3.4.1), the latter condition becomes $\overline{A \cup C} \subseteq \overline{B}$. Thus, if $x \in \overline{A \cup C}$, then $x \notin B$. Contrapositively, if $x \in B$, then $x \notin \overline{A \cup C}$, so $x \in A \cup C$. Thus, $B \subseteq A \cup C$. We conclude:

$$A \cap C \subseteq B \subseteq A \cup C$$

is a condition that ensures that

$$m(A \, \Delta \, B) + m(B \, \Delta \, C) = m(A \, \Delta \, C)$$

In other words, when B is "between" A and C, then the distance from A to C is the sum of the distances from A to B and B to C.

(11.4.1) Definition. We say that set B is between sets A and C if, and only if,

$$A \cap C \subseteq B \subseteq A \cup C$$

Example. (a) Consider Figure 11.4.1.

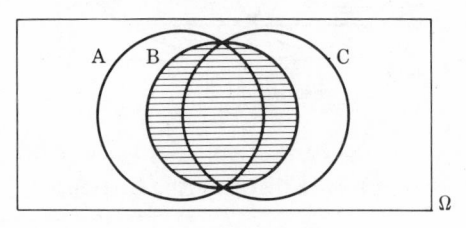

Figure 11.4.1

Let B be the shaded region. Then note that $A \cap C \subseteq B$. Thus, B "covers" $A \cap C$. Also, however, $B \subseteq A \cup C$, so that $A \cup C$ covers B. Thus, B is between A and C.

Example. (b) Let us consider the Figure 11.4.2.

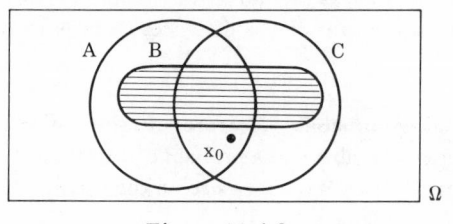

Figure 11.4.2

Here B is not between A and C, because although $B \subseteq A \cup C$, we have $x_0 \in A \cap C$, and $x_0 \notin B$. That is, $A \cap C$ is not contained in B.

Example. (c) Consider Figure 11.4.3.

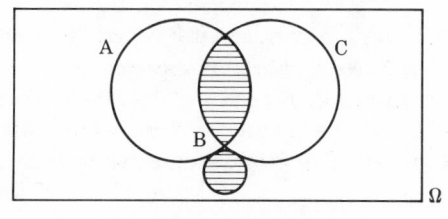

Figure 11.4.3

Let B be the shaded region. Here, clearly, B has a part unrelated to A and C. It is false that $B \subseteq A \cup C$. Hence, B is not between A and C.

Example. (d) Consider Figure 11.4.4.

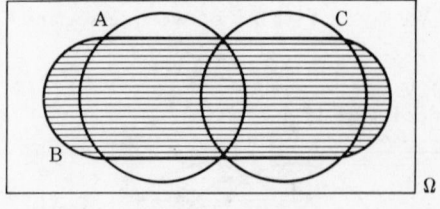

Figure 11.4.4

Let B be the shaded region. Then $A \cap C \subseteq B$, so B covers $A \cap C$. But note that $A \cup C$ does not cover B, so it is false that $B \subseteq A \cup C$. Hence B is not between A and C.

The whole motivation for considering the betweenness relation on a domain of sets was a result we noticed in considering distance between aspect-differentials. The general set-theoretic result is:

(11.4.2) Theorem. Suppose an algebra of sets is supplied with a measure m. If set B is between A and C, then in the induced distance (11.3.4) we have: the distance from A to C is the sum of the distances from A to B and from B to C.

11.5. An Example Using Probability Measure. In this paragraph we exemplify the purely set theoretic ideas above in a probability context. Let $(\Omega, \mathcal{B}, P)$ be a finite probability model. Then P is a measure on the events. Then a distance between events is defined, namely,

(11.5.1) $d(E, F) = P(E \,\Delta\, F)$

and betweenness is defined on events; namely, event F is between events E and G if and only if

$$E \cap G \subseteq F \subseteq E \cup G$$

so that if both E and G occur, F occurs, and if F occurs, then either E or G occur. In probability theory, $E \,\Delta\, F$ has a probability meaning, since $(E \cup F) - (E \cap F)$ is the event that exactly one of E, F occurs. Thus, the distance between two events is the chance that exactly one of them occurs. Also, because $0 \leqslant P(E) \leqslant 1$ for any E, the distance is bounded above by unity in a probability model.

Example. A concrete example can be given by considering a person about to choose two people, as in sociometry. Let the population from which the two are drawn be small for the sake of concrete illustration, say, $\{a, b, c, d\}$ and, so, six possible outcomes. Two models, shown as left (L) and right (R) tree diagrams on the same nodes, are given in Figure 11.5.1.

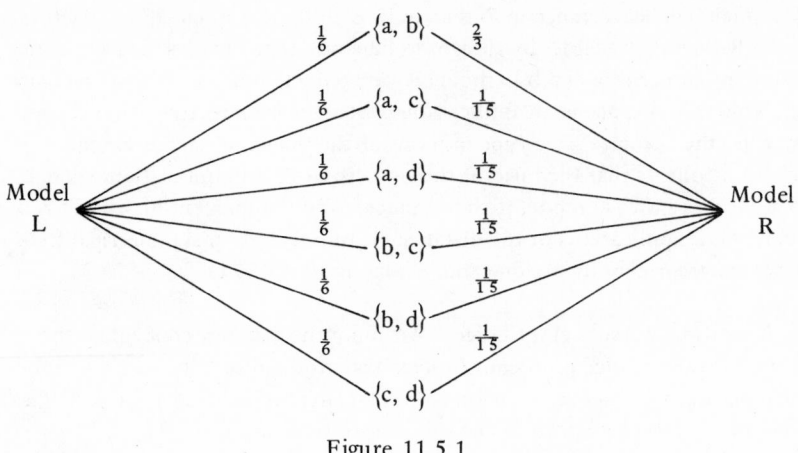

Figure 11.5.1

Consider the events

$$E \text{ occurs} \quad \text{iff} \quad a \text{ is chosen}$$
$$F \text{ occurs} \quad \text{iff} \quad b \text{ is chosen.}$$

Then

$$d(E, F) = P(E \triangle F) \qquad \qquad \text{[using (11.5.1)]}$$
$$(11.5.2) \qquad \qquad = P(E \cup F) - P(E \cap F) \qquad \text{(definition of } \triangle\text{)}$$

The principle of representation for events in a probability model yields

$$E = \big\{\{a, b\}, \{a, c\}, \{a, d\}\big\}$$
$$F = \big\{\{a, b\}, \{b, c\}, \{b, d\}\big\}$$

Thus,

$$E \cup F = \big\{\{a, b\}, \{a, c\}, \{a, d\}, \{b, c\}, \{b, d\}\big\}$$
$$E \cap F = \big\{\{a, b\}\big\}.$$

In the two models, we compute

$$P(E \cup F) = \begin{cases} \frac{5}{6} & \text{in model L} \\ \frac{14}{15} & \text{in model R} \end{cases}$$

$$P(E \cap F) = \begin{cases} \frac{1}{6} & \text{in model L} \\ \frac{2}{3} & \text{in model R} \end{cases}$$

Thus, substituting into (11.5.2) for these E, F,

$$d(E, F) = \begin{cases} \frac{5}{6} - \frac{1}{6} = \frac{2}{3} & \text{in model L} \\ \frac{14}{15} - \frac{2}{3} = \frac{4}{15} & \text{in model R} \end{cases}$$

With a much smaller distance in R than in L, E and F are much closer together in model R than in model L: by design, to illustrate that the bias in favor of the concomitant choice of a and b in model R gets reflected in the distance measurement. Now this distance is, in applications, obtained from relative frequencies. Thus, it has the absolute scale-type in terms of the theory of measurement. From this it follows that the ratio of the two distances is empirically meaningful. Thus, we can say that in model R the distance is only 40 percent of what it is in model L. (For applications of the distance measure based on symmetric difference, see the treatment by Majone and Sanday in Kay, 1971).

11.6. Valuation Measure and Choice. We apply the distance concept to the following abstract choice problem. A person is informed or believes that a mode of behavior, R_1, will lead to situation s_1 and an alternative behavior, R_2, will lead to situation s_2. His problem is to choose a response from

$$R = \left\{R_1, R_2\right\}$$

The nature of the problem dictates the schema

$$\mathcal{S}(R_1) = s_1$$
$$\mathcal{S}(R_2) = s_2$$

It is reasonable that the choice situation s be said to include these two possible situations (the "ideals" s_1, s_2 of the general apparatus) and that a reasonable idealization of the situation confines the representation, in terms of aspects, to those relevant to the judgment between s_1 and s_2. Hence, one is led to

$$s_1 \mapsto A_1'$$
$$s_2 \mapsto A_2'$$
$$s \mapsto A_1' \cup A_2' = A_s'$$

and the Venn diagram in Figure 11.6.1.

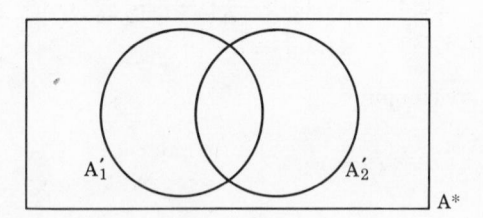

Figure 11.6.1

In this case the sets (1) and (2) entering into the postulate (11.2.1) become

$$\text{set (1)} = A_1' \cap \overline{A}_2' = A_1' - A_2'$$
$$\text{set (2)} = \overline{A}_1' \cap A_2' = A_2' - A_1'$$

For example, set (1) arises thus:

$$\begin{aligned}
\text{set (1)} &= A_s \cap A_1' \cap \overline{A}_2' \\
&= (A_1' \cup A_2') \cap A_1' \cap \overline{A}_2' \\
&= A_1' \cap \overline{A}_2'
\end{aligned}$$

The measure m is thought of as a valuation of these aspects as a whole. Then the distance $d(A_1', A_2')$ is the evaluative nearness of the two alternatives.

Clearly, then, postulate (11.2.1) becomes

$$(11.6.1) \qquad P(R_1) = \frac{m(A_1' - A_2')}{m(A_1' - A_2') + m(A_2' - A_1')} = \frac{m(A_1' - A_2')}{d(A_1', A_2')}$$

because $m(1) + m(2)$ is given by

$$\begin{aligned}
m(1) + m(2) &= m(A_1' - A_2') + m(A_2' - A_1') \\
&= m(A_1' \, \Delta \, A_2') \\
&= d(A_1', A_2')
\end{aligned}$$

In other words, (11.6.1) says that the probability of choice R_1 under the given circumstances is a ratio of the valuation $m(A_1' - A_2')$ of the distinctive aspects of situation s_1 connected with R_1 (as the outcome of R_1 via the schema) to the evaluative distance $d(A_1', A_2')$ separating the alternatives.

Example. (a) Suppose that one alternative includes all the relevant aspects of the other. For instance, $A_2' \subseteq A_1'$, and so, $A_2' - A_1' = \emptyset$. Then in (11.6.1) we find that numerator and denominator are identical. Thus, $P(R_1) = 1$, as intuitively required.

Example. (b) Here is a more concrete example. A person must choose between two career situations. The number of relevant aspects is large, and there is, we suppose, sizeable overlap. For instance, s_1 = lawyer, s_2 = doctor. Let us suppose that

$$\begin{aligned}
m(A_1' - A_2') &= 2 \\
m(A_2' - A_1') &= 1
\end{aligned}$$

representing, respectively, a two-unit valuation of the aspects of the legal career not present in a medical career and a one-unit valuation of the aspects present in a medical career not present in a legal career. Then,

$$d(s_1, s_2) = 3$$

For the schema, we require a mode of observable behavior meaningfully connected with these two possible future situations. It is reason to think of this as

$$R_1 = \text{the institutionalized mode of behavior culminating}$$
$$\text{in the designation ``lawyer''}$$

and R_2 similarly related to s_2 = doctor.

Then, because of the valuations, the probability that the person chooses to undertake R_1 rather than R_2 is

$$P(R_1) = \frac{m(A_1' - A_2')}{d(A_1', A_2')} = \tfrac{2}{3}$$

For an application we require a measurement model of the derived type, as mentioned earlier. Since one person makes this choice but once, there is no opportunity to obtain N repetitions except over distinct persons in reasonably similar circumstances. The auxiliary assumptions are as follows:

1. each such person has the same pair of parameters in terms of the valuation: $u_1 = m(A_1' - A_2')$ and $u_2 = m(A_2' - A_1')$;
2. they choose independently;
3. the number of repetitions (i.e., persons N) is large enough to assume the reasonableness of the estimate of $P(R_1)$ by the relative frequency of R_1 choices.

Letting f_1 be the observed frequency of R_1 choices,

$$\frac{f_1}{N} \cong \frac{u_1}{u_1 + u_2}$$

from (11.6.1) using u_1 and u_2 for the valuations. Dividing through by u_2, on the right,

$$\frac{f_1}{N} \cong \frac{\nu}{1 + \nu}$$

where $\nu = u_1/u_2$. Solving for all ν and calling it $\hat{\nu}$ (for estimate of ν),

$$\hat{\nu} = \frac{f_1}{N - f_1}$$

the ratio of the frequency of chosen legal training paths to chosen medical training paths. But after estimating ν, we cannot recover u_1 and u_2. Thus, the method of estimation establishes a derived scale-type for the measure m, as interpreted and applied in this manner; namely, m is a derived scale of the ratio type. This is because we cannot distinguish empirically between

$$\frac{u_1}{u_1 + u_2} \quad \text{and} \quad \frac{cu_1}{cu_1 + cu_2}$$

the latter under the transformation $u_i \mapsto cu_i$. Thus, the very method of applying the model in this way induces a measurement level on the measure m. In general, if response probability is given its usual empirical interpretation in terms of counts based on repetitions of the phenomenon, then the valuation measure must be interpreted as a ratio-level construct.

11.7. On the Need for a Choice Principle. Restle takes his analysis further by considering larger sets of alternatives, but we shall leave his analysis behind at this point. For a sociologist, the main gain from the Restle foundation is a systematic qualitative basis for the abstract representation of how people objectify their situations to themselves in terms of aspects and how a valuation defined over such aspect-representations can be a basis for the idea that choice is a probabilistic phenomenon. This is to say that inherent in the way the person represents a choice situation—mentally, if unconsciously—there arises a difference and a distance between alternatives that defines choice probabilities. From the standpoint of this foundation, only in very special cases can we interpret choice in deterministic terms, although the deterministic approach offers itself as an idealization that may provide a suitable basis in many contexts in which a probabilistic model would be too complex to set up at the outset.

Yet, the Restle approach lacks a powerful principle that will make the formalism "come to life" in the sense that one can grind out consequences that follow from real or hypothetical observed conditions. Lacking this kind of principle, the Restle foundation becomes a conceptual background for theories that do have richer consequences. The best single example of a probabilistic choice theory is Luce's theory (1959). Luce takes for granted the foundational idea that choice behavior is probabilistic and then formulates a powerful axiom. This axiom provides a strong theoretical foundation for formulas expressing choice probability as ratios of valuation measures, since it does not assume at the outset that observable response probabilities have this kind of relation to any underlying measures. On the contrary, Luce's axiom states a condition on observable probabilities and in effect permits a deduction of the sort of measure assumed by Restle. Thus, Luce's theory will permit us to illustrate an extremely important point about mathematically formulated theoretical sciences: measurement is part of theory-building. Whether fundamental or derived, measurement that relates to basic knowledge has its source in theory. No theory, no measurement.

Finally, we note that in passing to Luce's axiom we no longer strictly adhere to the language and notation of Table 11.2.1. Rather than impose Restle's concepts upon the axiom, it is preferable to let the language of Luce prevail. In this language, a chooser simply chooses one from a set of alternatives: aspects, schema, and so forth are not features explicitly introduced. But the usefulness of these foundational ideas is that they inform us of the reasons why choice phenomena are well construed in probabilistic terms. Thus, they provide scientific significance for the task of formulating a general and deductively fertile probabilistic choice principle. This is the task that Luce undertook, with the many results to be noted in our next sections.

It is at this point that a knowledge of axiomatic probability is important. What Luce does is append a new axiom to probability theory. First let us recall that by a "finite probability model" we mean a triple $(\Omega, \mathcal{B}, P)$ such that Ω is a finite set, $\mathcal{B}$ is the collection of all subsets of Ω and P is a measure on $\mathcal{B}$ such that $P(\Omega) = 1$. The last formulation on P exploits our general measure concept to say that P is just "normed" measure.

11.8. The Question of Invariance in Choice Behavior. We want to deal with the probability of a certain choice. A choice is the selection of an element from a finite set of elements, termed alternatives. On most occasions, higher organisms are forced by nature or by artificial experimental manipulation to make their selection from some proper subset of all potential alternatives. For example, a person may be asked to communicate one of several specified messages to a specified other; he may be required to select a verdict from a preselected list of possible verdicts, a word from a list of words, a category to which to assign a phenomenon, from some subset of possible categories; and so forth. Many occasions are related to each other by a structural relation between the possible alternatives; for instance, a man who, on a rainy day, will either drive his car or take a bus is in a situation that is structurally related to another situation, an ordinary day when he will drive, take a bus, or walk. Also, on some other occasion, he may admit staying home as an alternative. Thus, if choice from a set of alternatives is common, it is most commonly found under the guise of some imposed constraint on the alternative set. Luce's favorite example is the choice of a food from a menu. The menu is the set of real and pertinent alternatives; but the choice of food, on another occasion, may involve—indeed, for ordinary purposes, will involve—a different menu. Thus, a theory of choice might well begin in the attempt to solve the problem of how choice probabilities from different menus are related. More generally, how do choice probability models, with outcome spaces all drawn from some superset of alternatives, relate to each other? Is there an invariant?

Intuitively, we feel certain that there is an invariant. For example, something remains invariant in choosing from different menus: one's taste for various dishes. This menu formulation is an illustration of the intuitive conviction we have about ourselves: we have values, tastes, preferences that we carry "across" occasions. The valuation aspect, then, is what appears to be invariant amid occasion-dependent boundary conditions, which establish one or another list of alternatives as the set from which a selection is to be made.

Luce's strategy, then, is to formulate a condition on the probabilities from different models; that is, he imposes an axiomatic connection between all the various possible sets of alternatives that might appear on an occasion, where these sets are from some homogeneous universe of alternatives (which, as a totality, may never be the actual set of alternatives confronting a person because it is too big).

11.9. The Axiom: Explication and Examples. Having attempted an intuitive motivation for the strategy that Luce proposes, let us now turn to the axiom. The first thing to ask is from where the axiom is to come. It will, of course, come from experience in the systematic study of choice behavior: this experience is the empirical source of the axiom (for data and other background, see Luce, 1959).

Let U be a primitive set, interpreted as a universe of objects of a certain homogeneity; for example, U may be interpreted as a set of foods, a set of electrical appliances, a set of lists of commodities, a set of jobs, a set of actions, a set of punctuation marks, and so forth.

The collection of all finite subsets of U may be denoted $\mathfrak{F}_U$. The preliminary setup is as follows: Let T be an arbitrary set in $\mathfrak{F}_U$. The interpretation is that T is a finite set of alternatives. Let us assume prototheoretically that for every subset S of T there exists a probability measure P_S defined over subsets of S, with the interpretation: if S were the set of alternatives, then choices from S would occur in accordance with the probability measure P_S.

Example. (a) Let U be the set of all occupations listed in a certain "Dictionary of Occupational Titles" in the contemporary United States. Then $\mathfrak{F}_U$ is the collection of all finite subsets of such jobs. T is some arbitrary such finite set of jobs. For explicitness,

$$T = \{\text{lawyer, doctor, carpenter}\}$$

Since T has three alternatives, there are the following subsets (S) containing at least two alternatives:

$$
\begin{aligned}
S_1 &= \{\text{lawyer, doctor, carpenter}\} = T \\
S_2 &= \{\text{lawyer, doctor}\} \\
S_3 &= \{\text{lawyer, carpenter}\} \\
S_4 &= \{\text{doctor, carpenter}\}
\end{aligned}
$$

Then S has these four subsets as possible values. The prototheoretical assumption amounts to assuming probability functions P_1, P_2, P_3, P_4, as in the trees of Figure 11.9.1.

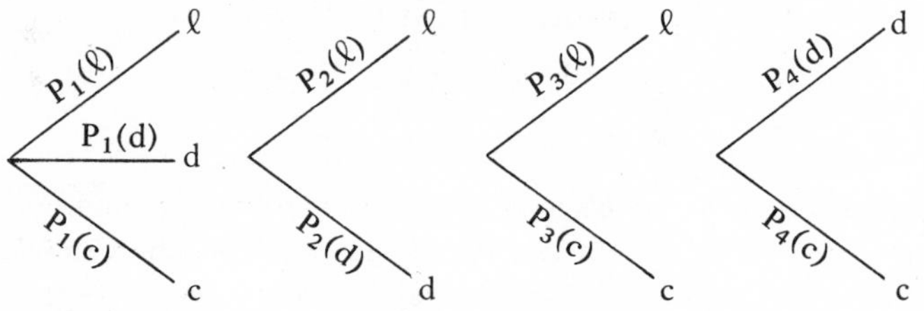

Figure 11.9.1

The interpretation, to repeat, is that if S_i were the set of alternatives, then the particular tree diagram with subsets of S_i as its endpoints applies. Thus, there are four finite probability models, given T,

$$(S_i, \mathfrak{B}_i, P_i) \qquad (i = 1, 2, 3, 4)$$

We do not attend to magnitudes here; the P_i are arbitrary probability functions except as the choice axiom then rules out a whole class of P_i as incompatible with each other.

Before stating the axiom, we need additional notation. We subscript a probability measure with the name of the set of alternatives forming the outcome space. Thus, if S is the outcome space, then

$$(S, \mathcal{B}_S, P_S)$$

would denote a finite probability model over S. In the above example we should have written P_{S_1}, P_{S_2}, P_{S_3}, and P_{S_4}, respectively.

Example. (b) If T = {doctor, lawyer, carpenter}, then for a choice from T, the abstract model has the form $(T, \mathcal{B}_T, P_T)$ where if $E \in \mathcal{B}_T$, $P_T(E)$ is the probability of choice event E in the occasion where choice is from T. But if S = {doctor, lawyer}, then $P_S(E)$, assuming $E \subseteq S$, is the probability of choice event E when the set of alternatives for the occasion is S.

One additional bit of notation: if S contains only two elements, say S = {x, y}, then we define

$$p(x, y) = P_S(\{x\})$$

so that p(x, y) is the probability that x is chosen when the only alternative to x is y. Finally, we never intend to pit x against x, but for reasons of mathematical convenience we set p(x, x) = 1/2 for any x.

With this notation, we are ready to state the axiom.

(11.9.1) Luce's Choice Axiom. Let T be a finite set of alternatives from $\mathcal{F}_U$. Let S be any subset of T.

(1) If for all x and y in T, $p(x, y) \neq 0$ or 1, then for any $E \subseteq S$,

$$P_T(E) = P_T(S)P_S(E)$$

(2) For any x in T, if p(x, y) = 0 for some y in T, then

$$P_T(S) = P_{T-\{x\}}(S - \{x\})$$

Now let us look at these two parts of the axiom in terms of the interpretation. Part (1) begins with the condition that for any pair of alternatives, should they be pitted against each other, the choice probability would not be perfect (i.e., 0 or 1). Under the hypothesis that this is the case for set T of alternatives, part (1) then says that if you consider a choice event, say E, which may occur when $S \subseteq T$ is the set of alternatives (and note that by transitivity of containment, then $E \subseteq T$), then the probability of E when T is the choice-set for the occasion [denoted $P_T(E)$] is given by the product of two probabilities, one being the probability of S as a choice event in the occasion where T applies $[P_T(S)]$, the other being the probability of E when S is the choice set for the occasion $[P_S(E)]$.

Examples for Part (1). (c) Let U be a set of actions. Let T = $\{a, b, c\}$ be the possible actions in an occasion. Suppose that were the occasion confined to $\{a, b\}$, then p(a, b) ≠ 0, 1; that were the occasion confined to $\{a, c\}$, then p(a, c) ≠ 0, 1; and that were the occasion confined to $\{b, c\}$, then p(b, c) ≠ 0, 1. (This is ∀ x, y in T, p(x, y) ≠ 0, 1.) Now consider some subset S ⊆ T. For example, let S = $\{a, b\}$. Now consider some subset E of S. For example, let E = $\{a\}$. Thus, we want to instantiate Part (1) for $P_T(\{a\})$. This is the left side. It denotes the probability that a is chosen when T is the set of possible actions. Also, now $P_T(S) = P_T(\{a, b\})$. This denotes the probability of the choice of a or b as an action when T is the set of possible actions for the occasion. Finally, we have

$$P_S(E) \;=\; P_{\{a,b\}}(\{a\})$$

which we agreed to write as p(a, b): this is the probability that action a would be chosen were b the only alternative to a in the occasion. The axiom says in (1),

$$P_{\{a,b,c\}}(\{a\}) \;=\; P_{\{a,b,c\}}(\{a, b\})p(a, b)$$

One intuitive way to see the meaning of this is to define

S occurs   iff   action type 1 is taken

and suppose a and b both realize this type, while c does not. Suppose the chooser decides what to do in two stages: first he decides on a type (say 1, 2); then, having a type, he "applies the probabilities" that are appropriate when action is confined to that type.

Example of Part (1). (d) Here is a more concrete example. Let U be a universe of occupations. Let T = $\{$doctor, lawyer, carpenter$\}$. Suppose a person is asked to choose the best occupation in T, on grounds of its "general standing in the community." Define,

S occurs   iff   he chooses a "professional" position

so that, by the usual representation principle of events,

$$S \;=\; \{\text{doctor, lawyer}\}$$

Abbreviate these jobs: d, ℓ, c. Now

$$T \;=\; \{d, \ell, c\}$$
$$S \;=\; \{d, \ell\}$$

Let

E occurs   iff   he chooses "doctor."

Then

$$E \;=\; \{d\}$$

Thus, by axiom (11.9.1) (1), assuming the nonperfect choice probabilities in pairs from T,

$$P_{\{d,\ell,c\}}(\{d\}) = P_{\{d,\ell,c\}}(\{d, \ell\})p(d, \ell)$$

now thought of as follows: the chooser regards "doctor" as best in {doctor, lawyer, carpenter} with a probability that is equal to the probability that he first chooses "professional" (thus ruling out "carpenter") and then chooses doctor over lawyer. This last probability, moreover, is not a mere conditional probability. It is the probability that would be estimated, at the given time, were the chooser constrained in the first place to choose between the two professions. The empirical side of Part (1), then, can be seen in what it would require to obtain a direct test: namely, two choice situations, one with T as the set of alternatives and one with S. Separate estimates, by the usual independent repetitions, would be required. If this were done, then apart from chance fluctuations, to be expected to some calculable extent, the estimates would satisfy Part (1).

The resemblance between Part (1) and conditional probability is worth noting and highlights the fact that we are using for empirically relevant purposes the modern concept of abstract probability model. By conditional probability reasoning, we have for any event E over outcome space T

$$P_T(E) = P_T(S)P_T(E|S) + P_T(\bar{S})P_T(E|\bar{S})$$

which is the law of total probability. Note the constant T, as assumed in all textbook discussions of probability. In fact, $T = \Omega$ of the earlier part of this book. If we now impose the hypothesis that $E \subseteq S$, then $P_T(E|\bar{S}) = 0$. Thus, the above expression becomes

(11.9.2) $$P_T(E) = P_T(S)P_T(E|S)$$

In this form, the resemblance to Part (1) of the axiom is obvious. The difference is between

$$P_T(E|S) = \text{the probability of event E, given S occurs,}$$
$$\text{when T is the outcome space}$$

and

$$P_S(E) = \text{the probability of event E, when S is the}$$
$$\text{outcome space}$$

Compare formula (11.9.2) with Part (1) of axiom (11.9.1). If $P_T(S) \neq 0$, we can, under the condition for Part (1) conclude that the axiom implies

(11.9.3) $$P_T(E|S) = P_S(E)$$

In fact, Luce shows (his lemma 2) that the conditions of Part (1) imply $P_T(S) \neq 0$. We can therefore conclude that (11.9.3) is another way of expressing the identity of Part (1). In this form it has clear empirical import. The left side is estimated in principle as follows: Consider N independent repetitions of the choice situation with T as the set of alternatives. Let N_S be the number of times S occurs. Let $N_{E \cap S}$ be the number of times both E and S occur. Then the left side is estimated by

$$\frac{N_{S \cap E}/N}{N_S/N} = \frac{N_{S \cap E}}{N_S}$$

For the right side of (11.9.3), let M be a number of independent repetitions of the choice situation, with S as the set of alternatives. Then let M_E be the number of times E occurs. Then the estimate of $P_S(E)$ is

$$\frac{M_E}{M}$$

and for N and M large, we expect the approximate identity

$$\frac{N_{S \cap E}}{N_S} \simeq \frac{M_E}{M}$$

This whole idea of doing repetitions of distinctly different experiments is abstractly anticipated in the axiomatic concept of a probability model. The first experiment has the model

$$(T, \mathcal{B}_T, P_T)$$

while the second has the model

$$(S, \mathcal{B}_S, P_S)$$

and for any $E \subseteq S$, the two models are connected by Luce's axiom: what is merely conditional probability in the T-experiment has the interpretation, because of the axiom, in terms of a different choice situation, covered by the second model.

Now let us consider Part (2) of the axiom. The condition "the pairwise choice would be imperfect, for any pair in T" is imposed on Part (1). But suppose this is not the case. Then for some pair, the corresponding pairwise probability is perfect. Then with this as the condition, Part (2) applies. It says that if x is certain to be rejected in favor of y, then the probability of any choice event S is obtainable by deleting x from both T and S.

Example of Part (2). (e) Let $T = \{a, b, c\}$ be the set of three actions from example (c). Suppose that contrary to the condition of Part (1), we have $p(a, c) = 1$. Thus, $p(c, a) = 0$. Now suppose we consider $S = \{b, c\}$. Then $T - \{c\} = \{a, b, c\} - \{c\} = \{a, b\}$ and $S - \{c\} = \{b, c\} - \{c\} = \{b\}$. Thus, according to Part (2),

$$P_{\{a,b,c\}}(\{b, c\}) = P_{\{a,b\}}(\{b\}) = p(b, a)$$

Therefore, the chance that b or c is chosen from T is reduced to the chance that b is chosen over a, because, by Part (2), we eliminate c from the scene once we learn there is something available such that it is inevitably chosen over c.

Example of Part (2).   (f) Consider the occupations of example (d) above. Here T = {doctor, lawyer, carpenter} = {d, ℓ, c}. Suppose, contrary to the condition of Part (1) that both p(d, c) = 1 and p(ℓ, c) = 1. Thus, p(c, d) = 0, p(c, ℓ) = 0. Here we find by Part (2) that for S = {d, c},

$$P_{\{d,\ell,c\}}(\{d, c\}) = P_{\{d,\ell\}}(\{d\}) = p(d, \ell)$$

and that for S = {ℓ, c},

$$P_{\{d,\ell,c\}}(\{\ell, c\}) = P_{\{d,\ell\}}(\{\ell\}) = p(\ell, d)$$

Here, "carpenter" is deleted from the expression for the two probabilities because it will be rejected in favor of a profession with certainty. Thus, choosing a best job is reduced to choosing between professions.

11.10.   Logical Consequences: Proofs with Explications.   Our next task is to show that Luce's axiom has many and fertile consequences, some of which admit of empirical test and which amplify the intuitive content of the axiom. We shall set out the basic results in the form of a series of propositions, each followed by a logical proof drawn from Luce's work and an interpretative statement.

First, in reference to (11.9.3) it was claimed that the axiom implies, under the conditions of Part (1), that $P_T(S) \neq 0$. Let us show this.

(11.10.1) Proposition.   If p(x, y) ≠ 0 or 1, for all x and y in T, then for any $S \subseteq T$, $P_T(S) > 0$.

Proof.   In fact for each $x \in T$, $P_T(x) > 0$ will imply, because $S \subseteq T$, that $P_T(S) = \sum_{x \in S} P_T(x) > 0$. Suppose, on the contrary, that for some $x_0 \in T$, we had $P_T(x_0) = 0$. Now Part (1) applies under the hypothesis of this proposition to give

$$0 = P_T(x_0) = P_T(\{x_0, y\})p(x_0, y)$$

for any $y \neq x_0$. But, by additivity of probability measure, we obtain

$$0 = [P_T(x_0) + P_T(y)]\, p(x_0, y) = P_T(y)p(x_0, y)$$

and since $p(x_0, y) \neq 0$, by hypothesis, it follows that $P_T(y) = 0$ for any y. Thus, if there is at least one element whose probability is 0, all elements have probability 0. So for any $x \in T$, $P_T(x) = 0$. But then,

$$1 = P_T(T) = \sum_{x \in T} P_T(x) = 0$$

which is a contradiction. Thus, there is no element with probability 0, and so,
$P_T(S) = \sum\limits_{x \in S} P_T(x) > 0$ for any $S \subseteq T$.

(11.10.2) Proposition. If $p(x, y) \neq 0$ or 1 for all x and y in T, then for any $S \subseteq T$ that has x and y as elements,

$$\frac{p(x, y)}{p(y, x)} = \frac{P_S(x)}{P_S(y)}$$

Proof. By (1) of the axiom,

$$P_S(x) = P_S(\{x, y\})p(x, y)$$

By additivity of P_S over the event $\{x, y\}$,

$$P_S(x) = [P_S(x) + P_S(y)]\, p(x, y)$$

This yields,

$$P_S(x) - P_S(x)p(x, y) = P_S(y)p(x, y)$$
$$P_S(x)[1 - p(x, y)] = P_S(y)p(x, y)$$
$$P_S(x)p(y, x) = P_S(y)p(x, y).$$

But $P_S(y) > 0$, by proposition (11.10.1), and $p(y, x) \neq 0$, so the conclusion follows.

The important point of this proposition is that the ratio of choice probabilities of two alternatives, x and y, is independent of the set S of wider alternatives in which they are embedded.

Example. Let T = {lawyer, doctor, carpenter, reporter} and suppose the axiom and the hypothesis of proposition (11.10.2) hold over T for a chooser. Then suppose the chooser faces at time t_1 a set of S_1 = {lawyer, carpenter, reporter} of possibilities, while at t_2 he faces S_2 = {lawyer, doctor, reporter}. Then, with ℓ, c, r, and d denoting these occupations, proposition (11.10.2) says,

$$\frac{p(\ell, r)}{p(r, \ell)} = \frac{P_{S_1}(\ell)}{P_{S_1}(r)} = \frac{P_{S_2}(\ell)}{P_{S_2}(r)}$$

even though the concrete probabilities may be distinct in each of the three situations: ℓ vs. r only; ℓ vs. r vs. c; and, ℓ vs. d vs. r. The ratio of the choice probabilities of ℓ and r is what remains invariant as the choice "context" (i.e., the space of alternatives including ℓ and r) varies.

In this example, the variation in time introduces the question of whether dynamic shifts in the various probability functions are involved here. This is most definitely not the case. For suppose $\{P_S^{(1)}, S \subseteq T\}$ is the family of all probability measures that exist at t_1: if at t_1, S were the outcome space, then $P_S^{(1)}$ would apply. Let $\langle P_S^{(2)}, S \subseteq T \rangle$ be the family of all probability measures that exist at t_2: if at t_2, S were the outcome space, then $P_S^{(2)}$ would apply. In constructing this example, it was tacitly assumed that

$$P_S^{(1)} = P_S^{(2)} \qquad \text{(for all } S \subseteq T)$$

so that probability measure was stationary. If it is allowed to change, then we are in a context of dynamic analysis, and Luce's axiom can still apply but with the possibility of distinct families of probability measures at distinct times. The point is that the axiom is a constraint on a fixed system of probability measures. If the axiom is correct, it is simply not possible to construct a real chooser who can hold a system of measures $\langle P_S, S \subseteq T \rangle$ not constrained by the binding or connection law constituting the axiom.

(11.10.3) Proposition. Let $T = \langle x, y, z \rangle$. If $p(x, y) = 1$ and $p(y, z) = 1$, then $p(x, z) = 1$.

Proof. With $p(y, x) = 0$ and $p(z, y) = 0$, by hypothesis of the proposition, Part (2) applies, giving

$$\begin{aligned} P_T(x) \equiv P_T(\langle x \rangle) &= P_{T-\langle y \rangle}(\langle x \rangle - \langle y \rangle) \\ &= P_{\langle x, z \rangle}(\langle x \rangle) \\ &= p(x, z) \end{aligned}$$

and also, z can be deleted, using Part (2), to obtain

$$\begin{aligned} P_T(x) \equiv P_T(\langle x \rangle) &= P_{T-\langle z \rangle}(\langle x \rangle - \langle z \rangle) \\ &= P_{\langle x, y \rangle}(\langle x \rangle) \\ &= p(x, y) \\ &= 1 \end{aligned}$$

by hypothesis, and so, since we have shown that $p(x, z) = p(x, y)$, we obtain the conclusion of the proposition.

This proposition is an elegant probabilistic version of transitivity. It will be recalled that a relation R is transitive if for all x, y, z,

$$\text{if } xRy \text{ and } yRz, \text{ then } xRz.$$

We can let R be given by,

$$xRy \qquad \text{iff} \qquad x \text{ chosen over } y \text{ when } \langle x, y \rangle \text{ is the choice set,}$$

then we can let $r = p(x, z)$ and represent the situation in the tree diagram of Figure 11.10.1.

This is the picture of the situation before we apply Luce's axiom. Applying it (i.e., supposing the system of probabilities conforms to it), we can immediately deduce from proposition (11.10.3) that $r = 1$, thus ruling out all other logically possible probability assignments on the tree of choice from

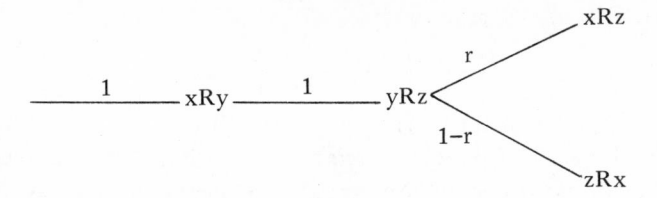

Figure 11.10.1

$\{x, z\}$. If a chooser will select x over y with certainty and y over z with certainty, then he will select x over z with certainty.

(11.10.4) Proposition. If $p(x, y) \neq 0$ or 1 for all x and y in T, then

$$P_T(x) = \frac{1}{\displaystyle\sum_{y \in T} \frac{p(y,x)}{p(x,y)}} \qquad \text{(including in the sum, } y = x\text{)}$$

Proof.

$$\sum_{y \in T} \frac{p(y,x)}{p(x,y)} = \sum_{y \in T} \frac{P_T(y)}{P_T(x)} \qquad \text{[proposition (11.10.2)]}$$

$$= \frac{1}{P_T(x)} \sum_{y \in T} P_T(y) \qquad \text{(holding x fixed)}$$

from which we obtain the conclusion of proposition (11.10.4), because

$$\sum_{y \in T} P_T(y) = P_T(T) = 1$$

The main point of (11.10.4) is that when imperfect choice probabilities exist over T in the pairwise sense, then the probability of a choice from T can be expressed purely in terms of the pairwise choice probabilities. Thus, for any choice event E when T comprises the set of alternatives, $P_T(E)$ is a function only of the pairwise choice probabilities.

(11.10.5) Proposition. Let $T = \{x, y, z\}$. If $p(x, y) \neq 0$ or 1 for all x and y in T, then $p(x, y)p(y, z)p(z, x) = p(x, z)p(z, y)p(y, x)$.

Proof. Since the measure P_T never assumes a zero value, we can write the identity,

$$\frac{P_T(x)\,P_T(y)\,P_T(z)}{P_T(y)\,P_T(z)\,P_T(x)} = 1$$

Applying (11.10.2) to each ratio—where S = T in (11.10.2)—we get

$$\frac{p(x, y)}{p(y, x)} \frac{p(y, z)}{p(z, y)} \frac{p(z, x)}{p(x, z)} = 1$$

and then the conclusion of (11.10.5) is obtained directly.

This proposition can be looked at in terms of the likelihood of loops. On the left side of the identity stated in the proposition, we have the intransitive loop of Figure 11.10.2 (1) and on the right the intransitive loop of Figure 11.10.2 (2).

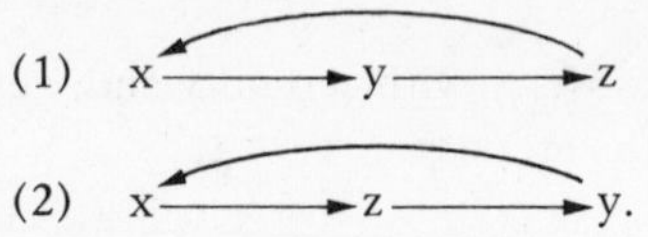

Figure 11.10.2

The proposition says the two intransitive loop configurations have the same probability, obtained by pairwise probabilities. In the choice theory literature, (11.10.5) is called the multiplication rule.

11.11. The Derived Measurement Model: Ratio Scale Valuation. In this section, we show that Luce's axiom logically implies the existence of a ratio scale over the set T. Also, under suitable auxiliary assumptions, the various scales as T varies over the wider universe U of alternatives can be "tied together" to conclude that a single valuation mapping on U exists that is unique in the ratio sense. This work exemplifies derived measurement rather than fundamental measurement. As was discussed in Chapter 7, the terms "fundamental" and "derived" are technical in measurement theory: it is not the case that a fundamental measurement in this technical sense is theoretically more important than a derived measurement. The distinction is that a derived scale, in contrast to a fundamental scale, takes its values by relation to other, given numerical variables. In a fundamental scale, the scale is obtained by direct assignment of numbers to things; this is important and indispensable but not very common. Most of scientific measurement involves building new numerical assignments from given numerical assignments.

In the case of Luce's theory we have the structure,

$$\frac{\{P_S, S \subseteq T\} \text{ with } T \subseteq U, \text{ and } T \text{ finite,}}{\text{Hence, v representing } \{P_S\} \text{ exists and is ratio-level}}$$

where the inference is justified by showing that the axiom implies the conclusion that v exists and that v is ratio-level. Thus, the axiom implies two basic theorems that we said in Chapter 7 make up a measurement theory: an existence or representation theorem and a uniqueness theorem. Thus, in this section, we want to derive a measurement theory in this sense from Luce's axiom.

(11.11.1) Existence Theorem.   If $p(x, y) \neq 0$ or 1 for all x and y in T, then there exists a positive real-valued function, call it v, with domain T such that for any $S \subseteq T$,

$$P_S(x) \;=\; \frac{v(x)}{\displaystyle\sum_{y \in S} v(y)}$$

Proof. By Part (1) of the axiom,

$$P_T(x) \;=\; P_T(S)\, P_S(x)$$

Therefore,

$$P_S(x) \;=\; \frac{P_T(x)}{P_T(S)} \;=\; \frac{P_T(x)}{\displaystyle\sum_{y \in S} P_T(y)}$$

Now we can let v be a mapping defined over T by

$$v(x) \;=\; k P_T(x)$$

for each x, where k is any number greater than zero. Then,

$$P_S(x) \;=\; 1\,\frac{P_T(x)}{\displaystyle\sum_{y \in S} P_T(y)} \;=\; \frac{k}{k}\,\frac{P_T(x)}{\displaystyle\sum_{y \in S} P_T(y)} \;=\; \frac{k P_T(x)}{\displaystyle\sum_{y \in S} k P_T(y)} \;=\; \frac{v(x)}{\displaystyle\sum_{y \in S} v(y)}$$

which was to be shown.

The next question is, How unique is v? In other words, how is any other such scale, say u, related to it?

(11.11.2) Uniqueness Theorem.   Let u be a positive function on T such that

$$P_S(x) \;=\; \frac{u(x)}{\displaystyle\sum_{y \in S} u(y)} \qquad \text{(for any } S \subseteq T)$$

Then there exists a positive constant, say c, such that $v(x) = cu(x)$, all $x \in T$. Thus, $v = cu$.

Proof. We have

$$v(x) \;=\; k P_T(x) \qquad\qquad (k > 0)$$

$$=\; k\,\frac{u(x)}{\displaystyle\sum_{y \in T} u(y)} \qquad \text{(letting } S = T \text{ in the assumption)}$$

$$=\; \frac{k}{\displaystyle\sum_{y \in T} u(y)}\, u(x)$$

Now $\sum\limits_{y \in T} u(y)$ is a sum of positive numbers and, hence, a positive number. Thus,

$$c = \frac{k}{\sum\limits_{y \in T} u(y)} > 0$$

can be defined, and

$$v(x) = cu(x) \qquad (\text{all } x \in T)$$

so $v = cu$. Thus, v is a scale on T whose scale-type is ratio.

The formula in theorem (11.11.1) for $S = \{x, y\}$ becomes

$$(11.11.3) \qquad p(x, y) = \frac{v(x)}{v(x) + v(y)}$$

which is analogous to Restle's formula (11.6.1),

$$(11.11.4) \qquad P(R_1) = \frac{m(A_1' - A_2')}{m(A_1' - A_2') + m(A_2' - A_1')}$$

The differences are both methodological and interpretative. First, the measure function is a primitive numerical notion in Restle's system. On the other hand, the function v is induced and so defined with the context given by Luce's axiom. The estimation situation is somewhat the same, because in both cases the choice probabilities are thought of as estimated directly by independent repetitions, while the m and the v functions are not observables in this sense—but their values can be estimated once we estimate choice probabilities.

Interpretively, a correspondence between Luce's formula and Restle's formula of the form

$$x \mapsto A_1'$$
$$y \mapsto A_2'$$
$$v \mapsto m$$
$$p(x, y) \mapsto P(R_1)$$

yields the following:

$$v(x) = m(A_1') = m(A_1' - A_2') + m(A_1' \cap A_2')$$
$$= m(A_1' - A_2') + f(x, y)$$

where $f(x, y)$ is a function of x and y, interpretable as the measure of their similarity $A_1' \cap A_2'$. Similarly,

$$v(y) = m(A_2') = m(A_2' - A_1') + m(A_1' \cap A_2')$$
$$= m(A_2' - A_1') + f(x, y).$$

Thus,

$$(11.11.5) \qquad \begin{aligned} m(A_1' - A_2') &= v(x) - f(x, y) \\ m(A_2' - A_1') &= v(y) - f(x, y) \end{aligned}$$

and so, substituting formulas (11.11.5) into formula (11.11.4) we obtain

$$(11.11.6) \qquad P(R_1) = \frac{v(x) - f(x, y)}{v(x) + v(y) - 2f(x, y)}$$

Thus, in general, Restle's theory would lead to a different prediction for the response probability than would the Luce formula. Note that $f(x, y) = 0$ implies $P(R_1) = p(x, y)$, so in the special case of no aspect overlap, the formulas agree.

There are three criteria by which we can evaluate these competing choice formulations:

(1) one formulation covers the other as a special case in a given type of choice situation (pairwise); Restle's formula (11.11.6) is in this sense to be preferred to Luce's formula (11.11.3), provided that,

(2) empirical tests indicate that the function $f(x, y)$ measuring stimulus similarity is really "needed"—that is, the Restle formula yields fitting predictions where the Luce model does not—although

(3) apart from this two-alternative special case, one formulation is deductively more comprehensive than the other: Luce's axiom generates a whole battery of consequences applicable to choice situations that involve more than two alternatives.

By criterion (2), recent evidence (see Rumelhart and Greeno, 1971) indicates that we cannot rule out $f(x, y)$ on grounds of simplicity; it makes an empirical difference, in terms of predictive accuracy, whether or not the stimulus similarity term $f(x, y)$ is present in the formula. Hence, combining (1) with (2) leads to a preference for the Restle formula (11.11.6) over the Luce formula. On the other hand, criterion (3) shows that we should take into account that the Luce formula is a mere single proposition in a rich system, whereas the Restle system is largely conceptual apart from this one formula. It is clear, then, that the next step in choice theory could be taken by some modification of the Luce axiom that allows formula (11.11.6) to be derived with formula (11.11.3) as a special case. Having noted this empirical criticism of Luce's choice theory, we will continue to outline its structure both for pedagogical reasons and because it is more likely to be modified than completely abandoned in future developments in probabilistic choice theory.

The next problem Luce treats (Luce, 1959, pp. 24–27) extends the scale to all of the universe U. This is like passing from a scale of mass on some subset of bodies to a proof that mass is a ratio scale defined over all bodies. We will let the reader take up this topic in Luce's monograph.

The next questions are, first, whether (and how) we can estimate v, and, second, whether we can invent a dynamic theory for changes in valuation. In regard to the first query, the basic estimation theory has been worked out. In section 11.13, we will give an application of the v-scale approach that makes use of this theory, and at that point in section 11.14 an indication of the statistical procedures in the application of Luce's theory will be made clear.

In regard to the second query, the answer has been given by Luce himself in his various learning models. Learning is seen as systematic modification of the response probabilities, and if Luce's axiom is assumed to hold at each instant of time, one can study learning dynamics by postulating a law for changing valuation in the face of experience. This theory is remarkable for its illustration of the power of a metatheory. Descriptions of the learning model may be found both in Luce (1959, Chap. 4) and in Atkinson, Bower, and Crothers (1966, pp. 290-300).

11.12. Application to the Theory of Ranking Behavior. The next query is sociologically natural: Can Luce's axiom be applied to ranking behavior? For sociologists this is a crucial field of application of the formalism, since as our treatment of status mappings (sections 5.7-5.10) made manifest, ranking is intrinsic to many of the phenomena of sociological interest. An indication that Luce's axiom could be applied to ranking behavior would justify the time and energy needed to master Luce's rather abstract approach.

Fortunately, Luce has already treated this question. He notes an important metatheoretical result: it is not the case that a unique probability function over rankings follows from probability functions in a family $\{P_S, S \subseteq T\}$ satisfying Luce's axiom. This means that an independent assumption must be introduced. The assumption that Luce formulated after exploring some alternatives will now be stated and its consequences explored.

The following notation will be used. If a ranking phenomenon is considered, all possible rankings of set S (without ties in the present theory) form an outcome space. Let R_S be the probability measure over events defined on this space. For example,

$$E \text{ occurs} \quad \text{iff} \quad x \text{ ranked over } y$$

specifies an event, then represented in terms of the logically possible rankings with x before y, and $R_S(E)$ is the probability of E in the ranking of set S.

For any x in S, the finite set to be ranked, let σ_x denote any ranking of the set $S - \{x\}$. For example,

$$S = \{A, B, C\}$$
$$S - \{B\} = \{A, C\}$$

and σ_B ranges over

$$\{(A, C), (C, A)\}$$

the two possible rankings of $S - \{B\}$. Thus, σ_x is a ranking of all the remaining alternatives, x omitted.

Rankings of x over y are written as ordered pairs, (x, y). We also allow the notation (x, σ_x) to mean the ranking of x over σ_x: this means the complete ranking simply begins with x and then has the ranking σ_x of all other alternatives.

We assume Luce's axiom holds for T, so that the chooser is posited to be holding some family $\{P_S, S \subseteq T\}$ satisfying the constraints of the axiom and, so, all the consequences noted above. For intuitive clarity, the following postulate is stated in two parts, although the first part is actually a special case of the second part.

(11.12.1)   Luce's Ranking Postulate.

(1) $R_{\{x,y\}}(x, y) = p(x, y)$ (all $x, y \in T$)

(2) $R_S(x, \sigma_x) = P_S(x)R_{S-\{x\}}(\sigma_x)$ (any $S \subseteq T$)

Part (1) says that the probability that x is ranked over y when the set $\{x, y\}$ is to be ranked is the same as the probability that x is chosen when the alternatives are limited to x and y.

Part (2) says that the probability that the chooser sets up the ranking with x on top, followed by σ_x, is given by postulating a two-step process: First, he picks x from S, and then, he ranks the remainder. In other words, the event

$$(x, \sigma_x)$$

occurs if and only if he chooses x and then ranks $S - \{x\}$ as σ_x. This amounts to saying that the chooser recursively generates any ranking.

Example. (a) Let T = {duplex, townhouse, apartment} and the chooser is going to rank the types of residential dwellings in T in accordance with their "desirability." For contextual notation let us write T = {d, t, a}.

The possible rankings of T are:

(1) (d, t, a)
(2) (d, a, t)
(3) (t, d, a)
(4) (t, a, d)
(5) (a, d, t)
(6) (a, t, d)

The ranking (1) could be referred to as (d, σ_d), where $\sigma_d = (t, a)$, while ranking (2) is (d, σ_d), where $\sigma_d = (a, t)$, thus exemplifying the notation of the postulate. The ranking that occurs is thought of as generated by successive applications of part (2) until part (1) applies. A tree diagram is shown in Figure 11.12.1. We check that we have a finite probability model with outcome space all the possible rankings by summing the column of path probabilities in Figure 11.12.1:

$$P(1) + P(2) = P_T(d)$$
$$P(3) + P(4) = P_T(t)$$
$$P(5) + P(6) = P_T(a)$$

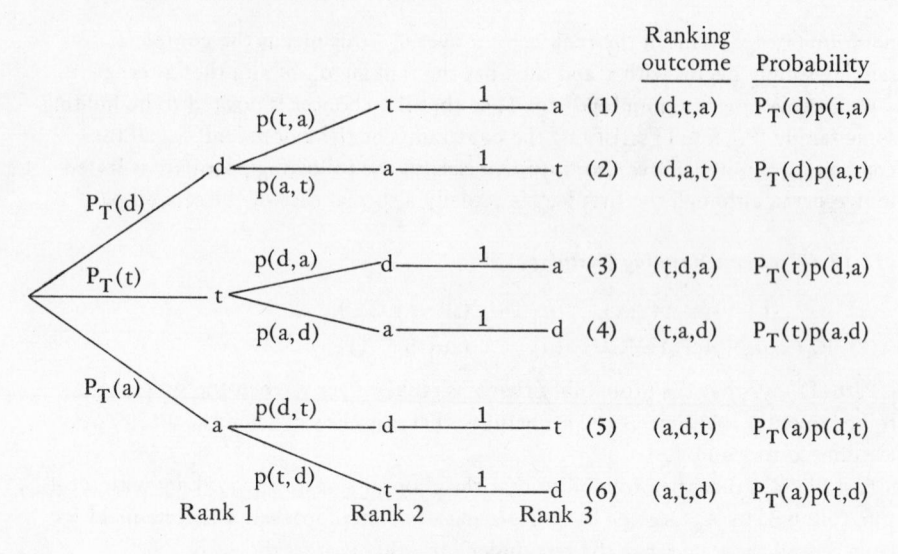

Figure 11.12.1

The sum of these partial sums is over all the elements in T: thus, it is 1.

Example. (b) Let us consider a more complex case, a set of four objects to be ranked. For example, a boy is ranking his acquaintances by extent to which he plays with them. Let the acquaintances be a, b, c, and d. Thus,

$$T = \{a, b, c, d\}$$

The ranking, according to the postulate, proceeds as follows: part (2) is employed for the first choice. For example,

$$R_T(a, \sigma_a) = P_T(a)R_{\{b,c,d\}}(\sigma_a)$$

where σ_a ranges over the six possible orderings of $T - \{a\} = \{b, c, d\}$. Similarly,

$$R_T(b, \sigma_b) = P_T(b)R_{\{a,c,d\}}(\sigma_b)$$
$$R_T(c, \sigma_c) = P_T(c)R_{\{a,b,d\}}(\sigma_c)$$
$$R_T(d, \sigma_d) = P_T(d)R_{\{a,b,c\}}(\sigma_d)$$

Now part (2) is employed again to the smaller sets of three; in other words, the boy chooses from the remaining three. Thus, if $\sigma_a = (b, c, d)$, then

$$R_{\{b,c,d\}}(b,c,d) = P_{\{b,c,d\}}(b)R_{\{c,d\}}(c,d)$$

and by part (1), $R_{\{c,d\}}(c,d) = p(c,d)$. Thus, the ordering (a,b,c,d) has probability $P_T(a)P_{\{b,c,d\}}(b)p(c,d)$. A tree diagram shows the complete work. (See Figure 11.12.2, where the last branch in each path has probability 1, as in Figure 11.12.1.)

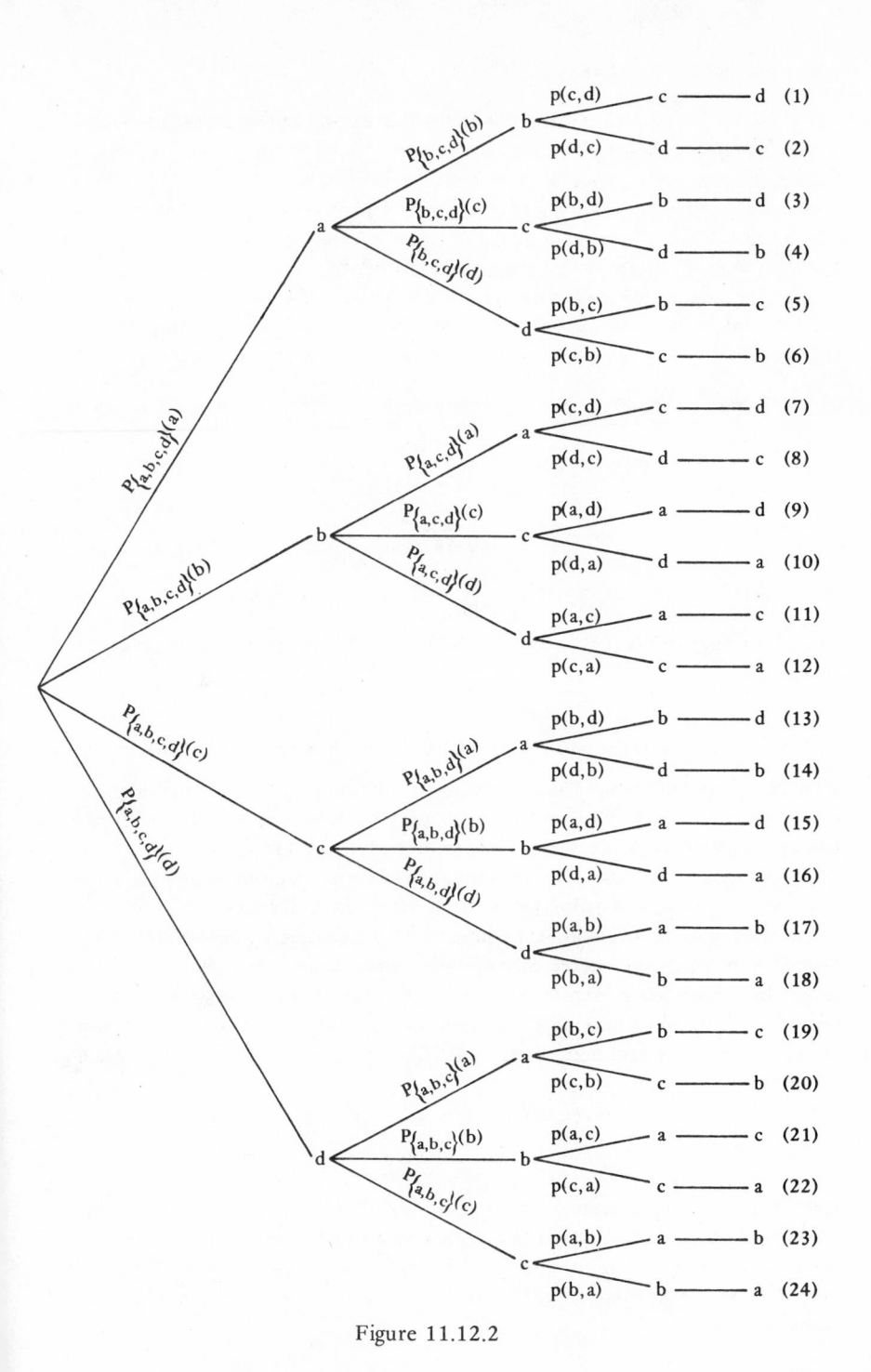

Figure 11.12.2

As Figure 11.12.2 makes clear, the chooser satisfying Luce's ranking postulate makes his rank-1 choice, say x, from $\{a, b, c, d\}$; then his rank-2 choice from $\{a, b, c, d\} - \{x\}$, say y; then his rank-3 choice from $\{a, b, c, d\} - \{x\} - \{y\}$, a choice between two objects so that part (1) applies. His last ranking object then follows with probability 1. Thus, each of the 24 possible rankings is assigned a probability, this assignment following from the ranking postulate.

Since the valuation exists, assuming Luce's axiom, these ranking probabilities can be expressed in terms of the valuations. We see that by proposition (11.11.1), for instance,

$$P_{\{a, b, c, d\}}(a) = \frac{v(a)}{v(a) + v(b) + v(c) + v(d)}$$

$$P_{\{b, c, d\}}(d) = \frac{v(d)}{v(b) + v(c) + v(d)}$$

$$P_{\{b, c\}}(b) = \frac{v(b)}{v(b) + v(c)}$$

$$P_{\{c\}}(c) = \frac{v(c)}{v(c)} = 1$$

hence, the outcome (a,d,b,c), which is path (5) in Figure 11.12.2, has ranking probability

$$R(a, d, b, c) = \frac{v(a)}{v(a) + v(b) + v(c) + v(d)} \frac{v(d)}{v(b) + v(c) + v(d)} \frac{v(b)}{v(b) + v(c)}$$

in terms of the valuation. Thus, if v were already known by prior theory and experiment, we could calculate the ranking probabilities directly. In fact, this is what we shall do in the application in sections 11.13-11.15.

An important point about Luce's ranking postulate is that it theoretically justifies a certain method of estimating pairwise preference probabilities, as follows.

The repetitions of a ranking experiment serve to provide a direct count of the frequency of times x is ranked over y, when the ranked set includes more than $\{x, y\}$. But this relative frequency has as its expectation the probability of x ranked before y when ranking T: R_T (x before y). The event "x before y" can be realized in many rankings, so one must sum up over all rankings, say σ, having x before y.

$$R_T(x \text{ before } y) = \sum_{\substack{\sigma \text{ such} \\ \text{that } x \\ \text{before } y}} R_T(\sigma)$$

and this is the expected relative frequency with which x is ahead of y. The direct count over observed rankings is an estimate of R_T (x before y). But we really want the pairwise probability $p(x, y)$. The tacit assumption is made, in countless everyday cases, that these are the same. This is justified if one accepts Luce's theory:

(11.12.2) Proposition. Let Luce's axiom and ranking postulate hold for every subset $S \subseteq T$, where T is a finite set of alternatives. Then if $p(x, y) \neq 0$ or 1 for all x and y in T,

$$p(x, y) = R_T(x \text{ before } y)$$

Proof. See Luce (1959, pp. 72-73).

Thus, an estimate of R_T (x before y) based on counting the relative frequency of the event that x is before y is also an estimate of $p(x, y)$, the probability that x is chosen over y when they are pitted against each other alone.

Example. (c) Each of N people ranks the set T = $\{$AMA, NAACP, AFL$\}$ according to the perceived prestige of the associations in that set. Assume: (1) for each person, the axiom, the ranking postulate, and imperfect probabilities, as in the hypothesis of (11.12.2), hold; (2) if α is a person holding probability $P_S^{(\alpha)}$ and β is a person with $P_S^{(\beta)}$, for any $S \subseteq T$, then $P_S^{(\alpha)} = P_S^{(\beta)}$, so that the N people are homogeneous in this probability sense. Then we have N repetitions of the basic experiment. If f is the number of people ranking the AMA over the AFL, then f/N is the estimate of p(AMA, AFL), the probability that any one of them regards the AMA as more prestigious than the AFL.

11.13. An Empirical Application: Job Prestige. In this section, the aim is to show the logic of application of the axiom and the ranking postulate. Having outlined the analysis of the Luce model for individual choice behavior, we turn now to an example of the third phase: empirical application. The various subphases of this empirical phase are

(1) identification of the set of alternatives,
(2) estimation of parameters (the valuations),
(3) assessment of the goodness of fit of the model to the data.

The data used in this section were collected in the course of a pilot study designed to evaluate the potential use of this procedure in a N.O.R.C.-type job prestige measurement on a national level. (A more detailed discussion of the empirical background is given in Skvoretz, Windell, and Fararo, 1972). For this reason, a smaller sample of jobs was selected (9 rather than 90), and no attention was given to selecting subjects from a population to which we wished to generalize. The probabilistic model applies to any individual and, hence—with a suitable homogeneity (consensus) assumption—to any set of subjects. We were not concerned to test the hypothesis that Americans, in general, accord prestige to these jobs in a particular manner. This is a matter to be settled at the N.O.R.C.-level of a national sample. Our concern is with the question, When people are asked to differentially evaluate jobs, do they choose in accordance with the axiom? When people rank jobs, do they rank as the ranking postulate proposes? If the answer is positive in terms of the empirical test, then the conjecture is that the next stage of attempting to utilize these analytic procedures at the national sampling level is justified.

The identification of U, the universe of alternatives, is in terms of a set of jobs whose titles are not unfamiliar to our subjects. (The subjects were male undergraduate engineering students taking a special social science course.) The set T drawn from U was selected from the N.O.R.C. listing to maximize the a priori likelihood that $p(x, y) \neq 0$ or 1 for all pairs of jobs x, y. The set T comprised nine job titles, shown in Table 11.13.1. This table also shows the estimated valuations. Details on the estimation procedure will appear shortly.

Table 11.13.1.   The Job Titles Used in the Empirical Study

No.	Title	v-Scale Estimate	N.O.R.C. Score in 1963
1	Author of novels	.243	78
2	Musician in symphony orchestra	.162	78
3	Newspaper columnist	.135	73
4	Public-school teacher	.118	81
5	Official of an international labor union	.112	77
6	Electrician	.069	76
7	Welfare worker for a city government	.064	74
8	Trained machinist	.055	75
9	Carpenter	.041	68

Although there were several tasks, including rankings, for the present we will deal with only one task: each of 79 subjects completed a paired comparisons task. This means a choice was made by subject k(k = 1, 2, . . . , 79) for each of $\binom{9}{2} = 36$ pairs of jobs. The instructions were as follows:

> Following is a list containing several pairs of occupations. From each pair please choose that occupation which, *in general,* you accord a higher social standing. Indicate your choice by placing an "X" beside the chosen occupation.

To explain how the v-scale scores in Table 11.13.1 were obtained, we next outline the derivation of the estimation formula. The method uses the technique known as maximum likelihood. Readers unfamiliar with this technique may omit the following section or study it after first reading Chapter 15, in which the technique of maximum likelihood is explained.

11.14. Bradley-Terry Estimation Method. The basic statistical method employed was developed by Bradley and Terry in a series of specialized statistical papers

(Bradley and Terry, 1952; Bradley, 1954a, 1954b, 1955). The basic method of derivation of an estimation formula for the v-scale values involves maximum likelihood. A number of auxiliary assumptions are made in order to justify the steps of the derivation.

Auxiliary Assumption 1. (Subject homogeneity) Each subject chooses x from pair $\{x, y\}$ of jobs with the identical probability $p(x, y)$.

Auxiliary Assumption 2. (Subject independence) Choice events for any given subject are independent of those for any other subjects.

Auxiliary Assumption 3. (Pair independence) For any given subject, the event of choosing from any given pair is independent of the event of choosing from any other pair.

The first assumption is a consensus-type of notion. Note that one could be wrong: any premise of the whole statistical procedure may be false. However, prior job-evaluation research (see, for instance, Hodge, Siegel, and Rossi, 1966) indicates that even in large heterogeneous populations surprisingly substantial consensus exists in job evaluations.

The second assumption could be wrong if subjects cooperated in making choices. Proper precautions in the administration of the instrument will make this assumption true.

The third assumption could be wrong if the subject attempted to make his choices consistent in a self-conscious way. Again, proper administration of the instrument—for example, putting a tight time limit on the task, while requiring that the pairs each be quickly evaluated—makes this assumption acceptable but worth noting as a possible source of error.

Let $x_1, x_2, \ldots, x_t$ be the t alternatives in set T. Let v_i be the v-scale value of alternative $x_i (i = 1, 2, \ldots, t)$. By the Luce model,

$$(11.14.1) \qquad p(x_i, x_j) = \frac{v_i}{v_i + v_j} \qquad (i, j = 1, 2, \ldots, t)$$

We introduce the random variables

$$r_{ijk} = \begin{cases} 1 & \text{if } x_i \text{ chosen over } x_j \text{ by subject k} \\ 2 & \text{if } x_j \text{ chosen over } x_i \text{ by subject k} \end{cases}$$

for $i, j = 1, 2, \ldots, t$ and $k = 1, 2, \ldots, N$, where there are N subjects. Note that if $r_{ijk} = 1$, then $r_{jik} = 2$, while if $r_{ijk} = 2$, then $r_{jik} = 1$. Hence,

$$(11.14.2) \qquad r_{ijk} + r_{jik} = 3 \qquad (\text{all } i, j, k)$$

For a single subject k and single pair x_i, x_j, the probability function associated with r_{ijk} may be written

$$P(r_{ijk} = 1) = \frac{v_i}{v_i + v_j}, P(r_{ijk} = 2) = \frac{v_j}{v_i + v_j}$$

If r_{ijk} is the datum for subject k, the likelihood of this datum is given by

(11.14.3) $$L_k(v_i, v_j) = \left(\frac{v_i}{v_i + v_j}\right)^{2-r_{ijk}} \left(\frac{v_j}{v_i + v_j}\right)^{2-r_{jik}}$$

Note that if $r_{ijk} = 1$ (and so $r_{jik} = 2$) then (11.14.3) on the right becomes $P(r_{ijk} = 1)$. On the other hand, if $r_{ijk} = 2$ (so that $r_{jik} = 1$), the right side of (11.14.3) becomes $P(r_{jik} = 1) = P(r_{ijk} = 2)$.

From (11.14.3) we obtain, using expression (11.14.2),

(11.14.4) $$L_k(v_i, v_j) = v_i^{2-r_{ijk}} \, v_j^{2-r_{jik}} \, (v_i + v_j)^{-1}$$

Now consider the data $\{r_{ijk}, k = 1, 2, \ldots, N\}$ for all N subjects on the fixed pair x_i, x_j. By auxiliary assumption 2, the likelihood $L(v_i, v_j)$ of these data is the product of the likelihoods (11.14.4) over all $k = 1, 2, \ldots, N$. Hence,

$$L(v_i, v_j) = \prod_{k=1}^{N} L_k(v_i, v_j)$$
$$= \prod_{k=1}^{N} v_i^{2-r_{ijk}} \, v_j^{2-r_{jik}} \, (v_i + v_j)^{-1}$$

This yields the result:

(11.14.5) $$L(v_i, v_j) = v_i^{2N-\sum_k r_{ijk}} \, v_j^{2N-\sum_k r_{jik}} \, (v_i + v_j)^{-N}$$

We obtain the likelihood of all the data—considering all $\binom{t}{2}$ pairs x_i, x_j as i and j vary—by taking the product of the likelihoods for pairs. This follows from the independence assumed in auxiliary assumption 3. Hence, since this depends on $v_1, v_2, \ldots, v_t$, we have

$$L(v_1, v_2, \ldots, v_t) = \prod_{i<j} L(v_i, v_j)$$

where the product is over all pairs of indices (i, j) where $i<j$. To see how we obtain this product, assume that $t = 3$. Then

$$L(v_1, v_2, v_3) = L(v_1, v_2) \, L(v_1, v_3) \, L(v_2, v_3)$$
$$= v_1^{2N-\sum_k r_{12k}} \, v_2^{2N-\sum_k r_{21k}} \, (v_1 + v_2)^{-N}$$
$$\times v_1^{2N-\sum_k r_{13k}} \, v_3^{2N-\sum_k r_{31k}} \, (v_1 + v_3)^{-N}$$
$$\times v_2^{2N-\sum_k r_{23k}} \, v_3^{2N-\sum_k r_{32k}} \, (v_2 + v_3)^{-N}$$
$$= v_1^{4N-\sum_k r_{12k}-\sum_k r_{13k}} \, v_2^{4N-\sum_k r_{21k}-\sum_k r_{23k}} \, v_3^{4N-\sum_k r_{31k}-\sum_k r_{32k}}$$
$$\times (v_1 + v_2)^{-N} (v_1 + v_3)^{-N} (v_2 + v_3)^{-N}$$

Note that

$$\sum_k r_{12k} + \sum_k r_{13k} = \sum_k (r_{12k} + r_{13k}) = \sum_j' \sum_k r_{1jk}$$

where $\sum_j'$ means a sum over all $j \neq 1$, and similarly,

$$\sum_k r_{21k} + \sum_k r_{23k} = \sum_j' \sum_k r_{2jk}$$

where $\sum_j'$ means a sum over all $j \neq 2$, and finally,

$$\sum_k r_{31k} + \sum_k r_{32k} = \sum_j' \sum_k r_{3jk}$$

where $\sum_j'$ means a sum over all $j \neq 3$. Hence,

$$(11.14.6) \qquad L(v_1, v_2, v_3) = \prod_{i=1}^{3} v_i^{4N - \sum_j' \sum_k r_{ijk}} \prod_{i<j} (v_i + v_j)^{-N}$$

The exponent 4N arises from two exponents of the form 2N. In the general case, for arbitrary finite t, the exponent 2N will appear t-1 times, giving rise to a v_i exponent equal to 2N(t-1). Hence, in general, we have the likelihood function

$$(11.14.7) \qquad L(v_1, v_2, \ldots, v_t) = \prod_{i=1}^{t} v_i^{2N(t-1) - \sum_j' \sum_k r_{ijk}} \prod_{i<j} (v_i + v_j)^{-N}$$

Having the likelihood function—the probability of the data $\{r_{ijk}, i, j = 1, 2, \ldots, t, k = 1, 2, \ldots, N\}$ in terms of the parameters—we next determine the values of the v_i that maximize this function. Differentiation is best done after transformation to logarithmic form, since the logarithm of a product is the sum of the logarithms of the factors in the product, yielding

$$(11.14.8) \qquad \ln L = \sum_{i=1}^{t} \ln \left(v_i^{a_i} \right) + \sum_{i<j} \ln (v_i + v_j)^{-N}$$

where $a_i = 2N(t-1) - \sum_j' \sum_k r_{ijk}$.

Further simplification of (11.14.8) arises when one recalls that $\ln (x^c) = c \ln x$, any exponent c. Hence,

$$(11.14.9) \qquad \ln L = \sum_{i=1}^{t} a_i \ln v_i - N \sum_{i<j} \ln (v_i + v_j)$$

Now differentiation of (11.14.9) with respect to argument v_m yields, recalling that $(d/dx) \ln x = 1/x$,

$$(11.14.10) \qquad \frac{\partial \ln L}{\partial v_m} = \frac{a_m}{v_m} - N \sum_{i<j}^{*} \frac{1}{v_i + v_j}$$

Here $\sum\limits_{i<j}^{*}$ means the sum over all pairs (i, j) where $i<j$ and either $i=m$ or $j=m$.

Hence, m will appear precisely $t-1$ times in this sum: once with every other integer from 1 to t. In short, we can write

$$(11.14.11) \qquad \frac{\partial \ln L}{\partial v_m} = \frac{a_m}{v_m} - N \sum_{\substack{j=1 \\ j \neq m}}^{t} \frac{1}{v_j + v_m} \qquad (m = 1, 2, \ldots, t)$$

Setting (11.14.11) to zero to locate the point $(\hat{v}_1, \hat{v}_2, \ldots, \hat{v}_t)$ at which each of the t derivatives in (11.14.11) vanishes, we have after replacing index m by i

$$(11.14.12) \qquad \frac{a_i}{\hat{v}_i} - N \sum_{\substack{j=1 \\ j \neq i}}^{t} \frac{1}{\hat{v}_j + \hat{v}_i} = 0 \qquad (i = 1, 2, \ldots, t)$$

where

$$(11.14.13) \qquad a_i = 2N(t-1) - \sum_{j}' \sum_{i} r_{ijk}$$

This result was obtained by Bradley and Terry in their 1952 paper. We note that we can impose the constraint,

$$\sum_{i=1}^{t} v_i = 1$$

because in normalizing all terms to their sum we are multiplying each of them by a fixed constant, an admissible transformation of the v-scale.

Formula (11.14.12) can be written in the form

$$\hat{v}_i = a_i \left(\sum_{\substack{j=1 \\ j \neq i}}^{t} \frac{N}{\hat{v}_j + \hat{v}_i} \right)^{-1}$$

This suggests an iterative procedure to obtain the $\hat{v}_i$, $i=1, 2, \ldots, t$. Namely, place initial values $\hat{v}_j^{(0)}$ $(j = 1, 2, \ldots, t)$ on the right and compute a new value $\hat{v}_i^{(1)}$ for all i. This gives a new set of values for the right, yielding $\hat{v}_i^{(2)}$, and so forth, with convergence expected. The iterative form may be written for iteration n

$$(11.14.14) \qquad v_i^{(n+1)} = a_i \left(\sum_{\substack{j=1 \\ j \neq i}}^{t} \frac{N}{\hat{v}_j^{(n)} + \hat{v}_i^{(n)}} \right)^{-1}$$

The iteration may be stopped when

$$\sum_{i=1}^{t} |\hat{v}_i^{(n+1)} - \hat{v}_i^{(n)}| < \epsilon$$

for some small ϵ. The final values are then the maximum likelihood estimates of the parameters $v_1, v_2, \ldots, v_t$. To use this procedure, initial guesses $v_i^{(0)}$ ($i = 1, 2, \ldots, t$) are required and should be chosen so as to assure rapid convergence. Fortunately, for small t, appropriate initial values may be found by using tables appearing in Bradley (1954a), using the procedure described in that paper.

11.15. Goodness-of-Fit. In our job prestige study, we have $t = 9$ jobs and $N = 79$ subjects. The terms a_i, $i = 1, 2, \ldots, t$ are statistics, that is, computed functions of the r_{ijk} data. The initial values $\hat{v}_1^{(0)}, \hat{v}_2^{(0)}, \hat{v}_3^{(0)}, \hat{v}_4^{(0)}, \hat{v}_5^{(0)}$ were obtained by using Bradley's tables. The iterative formula (11.14.14) was employed with a cut-off given by $\epsilon = .0002$. This yielded the results shown in Table 11.13.1. In comparing N.O.R.C. scores with our v-scale estimates one should note the following points:

(1) The N.O.R.C. "scale" is not based on a measurement model; if its "weights" assigned to rating categories are regarded as only ordinal, for instance, then the example of section 7.4 shows that the N.O.R.C. scale is not even nominal. That is, equivalent jobs in one numerical representation are not equivalent under an admissible scale transformation (see Skvoretz, 1971). On the other hand, assuming the weights represent the rating categories on an interval-scale, then the N.O.R.C. prestige scale is interval as well. But in this case no testable measurement model was involved in the N.O.R.C. procedure, so that the measurement was "by fiat," to use the term of Suppes and Zinnes (1963).

(2) Taking only the job ordering into account, there is lack of agreement between the N.O.R.C. job orderings and our orderings. This is shown rather dramatically in the case of the "public-school teacher." According to N.O.R.C. standing, this job ranks over "author of novels." Our engineering students differentially evaluated "author of novels" and "public-school teacher" in the ratio of about 2:1. Comparing "author of novels" to "carpenter," the ratio is about 6:1. These statements could not be made about the N.O.R.C. measurements even if we granted N.O.R.C. interval-level status, for ratios are not invariant under arbitrary linear transformations. Finally note how the v-scale provides a sharp break between jobs $x_1, x_2, \ldots, x_5$ and jobs x_6, x_7, x_8, x_9. Apart from the "welfare worker," this marks a break between nonmanual and manual jobs. Of course, we cannot know how much of this is special to the student engineers.

Finally, we remind the reader that taking scale values seriously—interpreting them, as above—assumes a fitting measurement model, which we have not yet demonstrated.

In fact, however, given the estimates $\hat{v}_i$ ($i = 1, 2, \ldots, 9$), the full set of $\binom{9}{2} = 36$ probabilities for the pairwise choices can be calculated using formula (11.14.1). Note that the constraint $\sum_{i=1}^{9} v_i = 1$ means that 8 (not 9) degrees of freedom are lost from the full 36 available. Hence, a goodness-of-fit test can be done, with 28 degrees of freedom. We are comparing observed and expected frequencies among independent observations, according to the auxiliary assumptions.* Hence, χ^2 is appropriate. The data (shown in Skvoretz, Windell, and Fararo, 1972) yield the result that $\chi^2 = 32.46$ and with 28 degrees of freedom, the chance of a χ^2 value at least this large is .25. We conclude that the amount of deviation between the observed values and the expected values is not unlikely if the model is correct. This gives some empirical support to the hypothesis that choice probabilities are constrained by the valuations as proposed in formula (11.14.1).

It should be pointed out, however, that as a test of Luce's axiom this is not very direct. Although Luce's axiom logically implies (11.14.1) and, hence, is confirmed (under the logic of falsifiability discussed in section 2.11), it is not unusual to see formula (11.14.1) in contexts having little to do with Luce's choice theory. In other words, other assumptions lead to this formula. In a deeper analysis of Luce's axiom we might look at the multiplication rule or some other logical consequences making predictions about the data. However, our present interest is in Luce's theory as a basis for job-prestige measurement. Hence, the next question for us is, What about rankings? A second task, completed by 155 of the engineering students, involved ranking five jobs. Using proposition (11.12.2) of the Luce theory we know that

$$p(x_i, x_j) = R_S(x_i \text{ before } x_j)$$

where x_i and x_j are jobs among our set of nine jobs and R_S (x_i before x_j) is the probability of the event "x_i before x_j" in a ranking of set S containing x_i and x_j. Hence, we obtain "pairwise frequencies" from the ranking data, use the estimation procedure described in section 11.14 to estimate the scale values of the ranked jobs and, then, finally, return to the rankings to predict the proportion of times each job is ranked first, second, . . . , last. Two sets of jobs were used. Here, the results involving one set are shown. We have

$$S = \{x_1, x_3, x_5, x_6, x_9\}$$

where for x_i the index i refers to the number shown in Table 11.13.1. We obtain the estimates,

$$\hat{v}_1 = .37, \hat{v}_5 = .26, \hat{v}_3 = .22, \hat{v}_6 = .11, \hat{v}_9 = .05,$$

and from these the probability of every possible ranking of S, which is then used to generate the probability of any ranking event. Table 11.15.1 shows the observed and predicted ranking results.

---

*Bradley (1954b) develops a likelihood ratio statistic for this test and shows that the distribution of the statistic can be approximated by the χ^2 distribution, thus providing a rigorous foundation for the goodness-of-fit test.

Table 11.15.1 Proportions of Subjects Assigning a Given Rank to a Given Job (Expected Values in Parentheses)

Rank	x_1 (Author)	x_5 (Labor Official)	x_3 (Columnist)	x_6 (Electrician)	x_9 (Carpenter)
1	.49 (.37)	.31 (.26)	.09 (.22)	.07 (.11)	.04 (.05)
2	.17 (.29)	.24 (.27)	.37 (.24)	.16 (.14)	.06 (.06)
3	.16 (.21)	.22 (.25)	.36 (.26)	.16 (.19)	.10 (.09)
4	.07 (.10)	.10 (.17)	.12 (.21)	.48 (.34)	.23 (.18)
5	.11 (.02)	.13 (.05)	.07 (.08)	.12 (.22)	.57 (.63)

A χ^2 test with 16 degrees of freedom yields $\chi^2 = 146$, and this is a very rare event if the model is correct, because $P(\chi^2 > 146) \ll .01$. Hence, the model involving Luce's axiom, subject homogeneity and other auxiliary assumptions, and the ranking postulate is falsified for job rankings.

We saw earlier that the model exclusive of the ranking postulate yielded a favorable empirical assessment based on goodness-of-fit. Augmenting the model with the ranking postulate led to (a) an inversion of x_3 and x_5 —columnist and labor official—in the scale values assigned them and (b) the poor fit to the ranking data, even using the scale values based on those data. Hence, we are led to isolate the ranking postulate as the primary source of the empirical difficulties.

The conclusion is that the subjects did not rank jobs in the manner proposed by the ranking postulate. From this conclusion, at least two distinct directions for further research are clear: (1) develop alternatives to the ranking postulate within Luce's theory and return to the data to test these alternatives; and (2) retain the postulate, but redesign the empirical situation to make the postulate true. The gain of tactic (2) is that one can then employ the entire analytic procedure to scale jobs using a well-developed theory, whereas (1) might not be easy to do at all. For instance, if each job title were printed on a small card, the subject could be given the deck of cards and asked to choose the job he accords the highest standing. After he has done this, one removes the card corresponding to that job and again asks the subject to choose the job he accords the highest standing, this time from the reduced set. This empirical procedure should mean that the ranking postulate can be regarded as true by empirical research design and, so, in a class with some of the auxiliary assumptions. Hence, it should be possible—on the basis of a good fit with the earlier data—to utilize the measurement model in the ranking situation to obtain, on the one hand, scale values for each job, and, on the other hand, an empirical test of formula (11.14.1) validating the model in the situation.

This chapter has presented a conceptual framework for choice theory, a rather rich deductive system based on Luce's axiom and at least two empirical cautionary notes on the adequacy of the theory: first, Restle's more general two-alternative choice formula seems to yield better predictions than Luce's formula (11.14.1) and second, even for subjects and choice alternatives fitting the Luce pairwise choice model, additional empirical difficulties arise in applying the ranking postulate.

In this section, we have given our first detailed example of a goodness of fit evaluation. The impression should not be left that the matter is a simple one of hypothesis-testing with a special significance level appropriate for the type of problem. Most frequently, goodness-of-fit involves questions of intuition based on visual examinations of graphs and on many sets of statistics for which a model makes predictions. To get some feel for these aspects of goodness-of-fit and model evaluation, the reader might consult Bush (1963) and Lewis (1971).

SUGGESTIONS FOR FURTHER READING (PART TWO)

Chapter 7. A good introduction to measurement theory from a mathematical standpoint is given by Pfanzagl (1968), who bases his work upon that of Suppes and Zinnes (1963). In Hempel (1952) one finds the same topics covered more discursively and in terms of their significance for the development of general scientific systems. For a recent point of view on measurement in relation to social theory, see Abell (1968, 1969).

Chapter 8. The systems viewpoint that informs this chapter has been outlined in several places in more or less technical detail. An informative set of conceptual contributions has been compiled by Buckley (1968). Bellman (1961) gives a survey without much technical detail. More mathematical treatments are usually specialized (e.g., probabilistic systems in Howard, 1971; linear systems in Schwarz and Friedland, 1965; differential equations in Sanchez, 1968; difference equations in Goldberg, 1961). Highly relevant for sociologists is an essay on the mathematics of change by Coleman (1968).

Chapter 9. A good introduction to finite probability is given by Goldberg (1960). A treatment emphasizing tree diagrams is given by Kemeny, Snell, and Thompson (1966). A readable introduction may also be found in Gelbaum and March (1969).

Chapter 10. Both Kemeny and Snell (1960) and Kemeny, Snell, and Thompson (1966) are good references for Markov chains. For random variables, an introduction is provided in Goldberg (1960), as well as in Parzen (1960). For more on learning models, see Atkinson, Bower, and Crothers (1965).

Chapter 11. A brief introduction to Luce's choice axiom is found in Chapter 4 of Atkinson, Bower, and Crothers (1965). Of course, Restle (1961) and Luce (1959) should also be consulted.

3 SOCIAL PROCESS AND SOCIAL STRUCTURE

CHAPTER TWELVE MARKOV SOCIAL PROCESSES: IMAGES OF STRATIFICATION

12.1. Introduction. The aim of this chapter is to show how a mathematical formulation of a process of image development over time leads to several consequences that agree with some well-known empirical findings—for example, as found in *Deep South* (Davis, Gardner, and Gardner, 1941). In this sense, the model explains preexisting empirical generalizations. We will speak, then, of an arbitrary actor in an arbitrary social system. An ordering of the actors exists that is (1) multidimensional as to content, (2) nevertheless a single ordering in generic form, (3) based upon characteristics relevant and evaluated among such actors so that (4) they use these characteristics as cues in interaction process leading to (5) the development of an "image," a more-or-less imperfect mapping of the actual multidimensional ordering that (6) reflects their positions in that system in ways revealed by empirical research. In regard to point (6), the typical actor makes finer discriminations in nearby positions than in positions more distant in the actual ordering.

A typical example of an image in relation to the actual stratification system and the actor's location in it, is shown in Figure 12.1.1. This diagrammatic form was used in *Deep South*, for instance. The actor whose image is depicted is termed the focal actor. He is represented in terms of a profile or position in the system of stratification: white, not old family, not wealthy. His image contains only four classes as compared with the finer distinctions that could be made among the "society" people (e.g., with and without wealth) and that could be made among "blacks." On the other hand, the nearest similar class of whites, those who are not old family but have money, are discriminated in the image.

Our aim, then, is to represent an arbitrary actual stratification system and the flux of images of this system in the interaction process such that the basic properties of the diagram are generated: (a) the image is an order-preserving partition of the

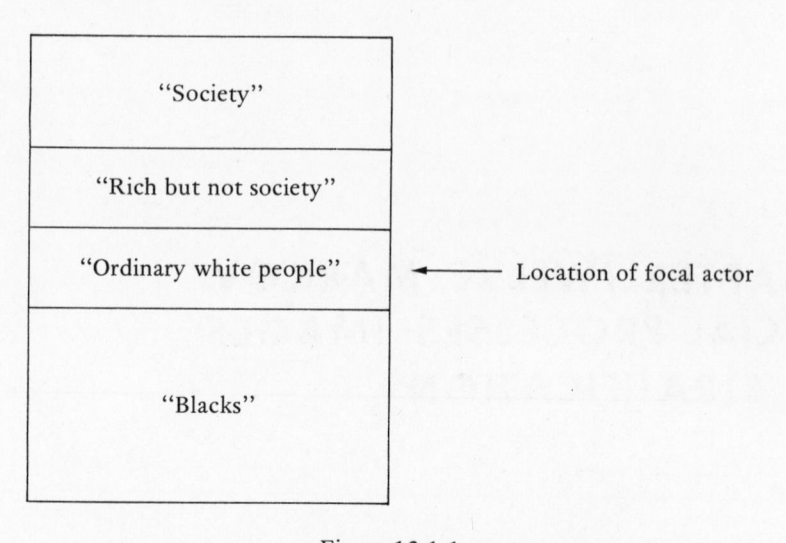

Figure 12.1.1

actual system, (b) the image depends on the location in the actual system, and (c) the fineness of discrimination depends on the nearness of the other classes: the greater the social distance, the less clearly are fine distinctions made. If we can do this, we can take the generating process described by the axioms as a more-or-less satisfactory explanation of these properties of the images of stratification.

12.2. Axioms of the Theory. The theory is given by a set of axioms that both specify its scope and state its principal assumptions. We present the theory in a semiformal mode, mixing empirical language and purely mathematical concepts for the sake of intelligibility. However, it is not difficult to be more formal once a semiformal version has been specified (see, for example, Fararo, 1970 b, c).

Axiom 1. There exists a multidimensional stratification system S over a set A of actors in a time-domain T; this means:

(1) there are n characteristics $C_1, C_2, \ldots, C_n$ and each C_i is a linearly ordered set representing a uniform differential evaluation in A (see section 5.7),

(2) there is an ordering of the n characteristics,

(3) based on (1) and (2), the set of all possible positions in S is ordered "lexicographically."

In part 3 of axiom 1 the term "position" refers to an ordered n-tuple, where the ith component is a state of characteristic C_i, $i = 1, 2, \ldots, n$. The idea of "lexicographic order" exploits the double ordering, that of each C_i and that of the C_i themselves, to create a single linear order which "composes" the various n linear orders: there is one resulting order of stratification, but it has many component linear orders. Rather than describe the lexicographic, or dictionary, order principle in formal terms (see, however, Fararo, 1970b), we let the following discussion more concretely convey its meaning.

To illustrate the meaning of this axiom, suppose we specify to the special model with n = 2 and each C_i has only two evaluatively distinct states, say H_i and L_i. Then let C_1 be ordered "before" C_2 under (2). According to (3) we have the two-dimensional ordering,

(12.2.1)
1. $H_1 H_2$
2. $H_1 L_2$
3. $L_1 H_2$
4. $L_1 L_2$

If we assume, instead, that C_2 is before C_1 in part (2), then (3) yields

(12.2.2)
1. $H_2 H_1$
2. $H_2 L_1$
3. $L_2 H_1$
4. $L_2 L_1$

In a concrete system, we may have the identifications

$$C_1 = \{\text{aristocrat, commoner}\}$$
$$C_2 = \{\text{wealthy, not wealthy}\}$$

and then C_1 "before" C_2 is interpreted as relative importance in that system and the lexicographic ordering (12.2.1) is

aristocrat, wealthy
aristocrat, not wealthy
commoner, wealthy
commoner, not wealthy.

On the other hand, the ordering (12.2.2), while identical with this order at top and bottom, reverses the positions of nonwealthy aristocrats and wealthy commoners.

In the model's later axioms, the ordering (2) is made "active"—that is, interpreted in such a way that it functions in the interaction process and has consequences for images.

Axiom 2.   S is stable over T.

This means that while the images of actors in S will be treated as dynamical (time-varying), we restrict the scope of our arguments to stratification systems that not only may be represented as in axiom 1 but are such that the identical characteristics and orderings exist over time. Note that this contributes to an implicit definition of the time domain: in systems undergoing rapid stratification change, T must be correspondingly small.

Axiom 3.   There exists a subset of actors in A such that their positions in S are time-invariant. Any such actor may be chosen for the role of focal actor in the analysis—that is, as the actor whose image is developing over time.

Axiom 4.   The focal actor has a series of interacts with members of A in time domain T. Denote points of T, in which such interacts occur for the focal actor, by $t = 0, 1, \ldots$. At $t = 0$ the actor has an *initial image* of S.

Axiom 5.   Given an interact of focal actor and some alter, the probability that alter is in a certain position in S is not time-dependent, and it depends only on the position of the focal actor and that of alter.

For example, let t be an interact time and let α be the focal actor. Assume α is located in $H_1 L_2$. Then the tree diagram of Figure 12.2.1 represents axiom 5. Here p_j^2 is the probability, given an interaction involving α in $H_1 L_2$ —class 2 in the ordering given by (12.2.1)—that alter is a member of a specified class (which is j^{th} in the ordering). Thus $\{p_j^2, j = 1, 2, 3, 4\}$ forms a probability distribution. Similarly, for each possible fixed class location of the focal actor, say i ($i = 1, 2, \ldots$), we have a distribution $\{p_j^i, j = 1, 2, \ldots\}$. The important points of axiom 5 are that the probabilities vary by class and that they are time-invariant, so that the intuitive idea of a fixed stratification system is set up not only in axiom 2 (where "stratification" refers to a morphological property) but also in reference to patterns of interaction (an additional component of the intuitive meaning of stratification).

Axiom 6.   Let the C_i be labeled so that the ordering of axiom 1 (2) is given by the integer subscript of $C_1, C_2, \ldots, C_n$. Then, when α interacts with alter β,

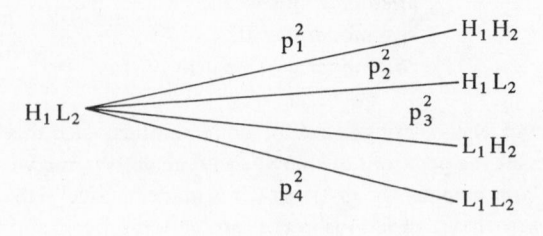

Figure 12.2.1

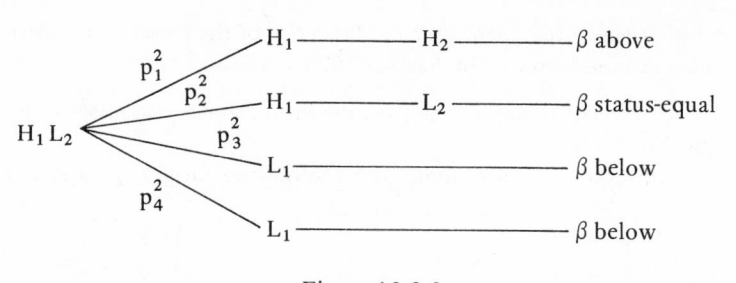

Figure 12.2.2

a process of cue sampling occurs as follows: α first samples β's state of C_1, then β's state of $C_2, \ldots,$ with sampling terminated as soon as a relative status determination occurs: "β above," "β below," or (if all states are same as α's) sampling continues to the C_n stage, where it terminates in "β a status-equal."

To illustrate, take $C_1 = \{H_1, L_1\}$, $C_2 = \{H_2, L_2\}$ and let the position of α be $H_1 L_2$. Let t be an occasion of interaction and suppose that β has position $H_1 H_2$. Then the sampling process is as follows:

		alter	
actor	C_1	C_2	
$H_1 L_2$ ———————	H_1 ———————	H_2	β above

Considering all possibilities we obtain the tree diagram of Figure 12.2.2. Note that the cue "L_1" on the first sampled dimension—the "most important" dimension of axiom 1 (2)—provides the basis for the relative status definition by focal actor α, and so, α does not sample the second component of alter's position. This means that for α, the second dimension is salient only in the nearby class locations.

Changing α to $L_1 H_2$, we have the diagram of Figure 12.2.3. Here again, the "second dimension" is the "secondary dimension" invoked in interaction to define the status situation when required but not given priority in cue sampling.

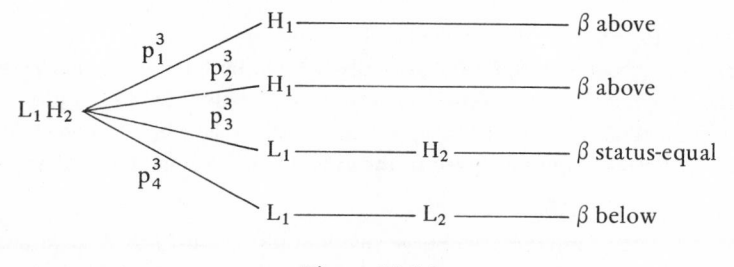

Figure 12.2.3

The possibility of perceptual bias is outside the scope of the present formulation but is included in a model set up by Fararo (1970c).

Axiom 7.   At the start of any interaction, the focal actor has an image of S, which is a partition of S.

To illustrate, an example of a system S and a lower-class image (i.e., α in a low class in S) is

| S | Image of S |

The interpretative distinction is that S is a pattern partially definitive of the social systemic character of the set of actors. In a word, it is an objective pattern in A. On the other hand, an image of S is a pattern in the mind of a focal actor, built up in interaction process with other members of A. Hence, it is a property of an actor, while S is a property of the system of actors. Because of the consequences of our axioms, in particular, axiom 9 below, an image in this sociological sense will be a homomorphic image of S: an order-preserving map that may (and usually does) lump distinctions in S to make them nondiscriminated (for that actor).

The following axioms propose that this image undergoes successive transformations, so that the actor does not possess any fixed imagery of the system until stable equilibrium is reached.

Axiom 8.   At t = 0, the image of S is a single entity,

$$C\ell(\alpha)$$

where $C\,\ell(\alpha)$ is the position in S of the focal actor α.

Thus, according to axiom 8, t = 0 is partially defined by the initial undeveloped state of the image: it is homogeneous. "Everyone is like me" is the tacit proposition represented by the initial image. Concretely, α might be a young child. This initial condition is not necessary to the model but helps make its analysis simpler.

Axiom 9.   Given an interact, the image that α holds is transformed according to the following, where $C\ell(\beta)$ is the sampled part of the position of β in S:

(1)   if $C\ell(\beta)$ is represented in the image held by α, no change occurs;

(2)   if $C\ell(\beta)$ is above the highest class in the image, then $C\ell(\beta)$ becomes the new highest class;

(3)   if $C\ell(\beta)$ is lower than the lowest class in the image then $C\ell(\beta)$ becomes the new lowest class;

(4)   if $C\ell(\beta)$ is between some pair of classes in the image, then it is "inserted" between them in the new image.

This process may be thought of as a rapid "scanning" in which the map held in memory is checked against the cues, presented by β, but only those cues sampled by α in the process of relative status determination.

To illustrate, let

$$C\ell(\alpha) \; = \; H_1 L_2 \qquad \text{(class 2)}$$

and suppose that we use a tree diagram to represent the possible transformations of an image at t given by,

$H_1 L_2$
L_1

Namely, the possible transformations are as indicated in Figure 12.2.4. Here we see that if α interacts with a β from class $H_1 H_2$, then he samples H_1 and finds equality, so that he samples H_2, finds β above but the class $H_1 H_2$ not part of his map or image, and so, by axiom 9 (2) he modifies his image as shown. On the other hand, if β is from any of the other three classes, no image reconstruction is needed. Note that the image of the lower class is homogeneous, lumping the

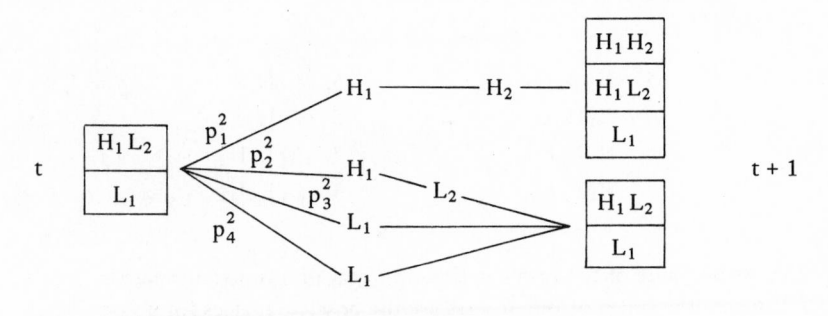

Figure 12.2.4

actual distinctions in the second dimension. Thus, depending upon the location of alter, the focal actor enters interact $t + 1$ with a "transformed image."

12.3. Analysis of a Model. On the basis of these general axioms, different properties of the process may be explored. The aim in the general exploration would be to prove theorems about the image-building process and its outcomes without loss of generality due to either (a) specification to a particular number of characteristics, particular numbers of levels in each characteristic, and so forth; or (b) identification of the abstract notions in some concrete system. For our purposes, however, we want to show that by a suitable (if somewhat arbitrary) choice of specification in the sense of (a) above, the abstract empirical generalizations about images are reproduced. Thus, a particular abstract model is needed from the class of all models defined by the above axioms. When a model is chosen, we can ask the following types of questions:

(1)   questions related to the transient aspect of the process: the dynamics of shifting imagery in terms of its form;

(2)   questions about stable states of the process; in particular, Does the model have a unique stable equilibrium?

(3)   questions about the full panoply of all stable images, all classes (of focal actors) taken into account: what does this array of stable homomorphic images of S look like? What appear to be its properties?

Let $n = 3$ and let

$$
\begin{aligned}
C_1 &= \{H_1, L_1\} \\
C_2 &= \{H_2, L_2\} \\
C_3 &= \{H_3, L_3\}
\end{aligned}
$$

with the order of axiom 1 (2) given by the subscripting of the C_i. Then, by axiom 1 (3), the stratification system S is given by:

1.	$H_1 H_2 H_3$	5.	$L_1 H_2 H_3$
2.	$H_1 H_2 L_3$	6.	$L_1 H_2 L_3$
3.	$H_1 L_2 H_3$	7.	$L_1 L_2 H_3$
4.	$H_1 L_2 L_3$	8.	$L_1 L_2 L_3$

Each possible value of the image defines one row of a transition matrix from t to $t + 1$, where the states of the process are the possible images for a given focal actor. We derive this matrix by starting from the homogeneous initial image.

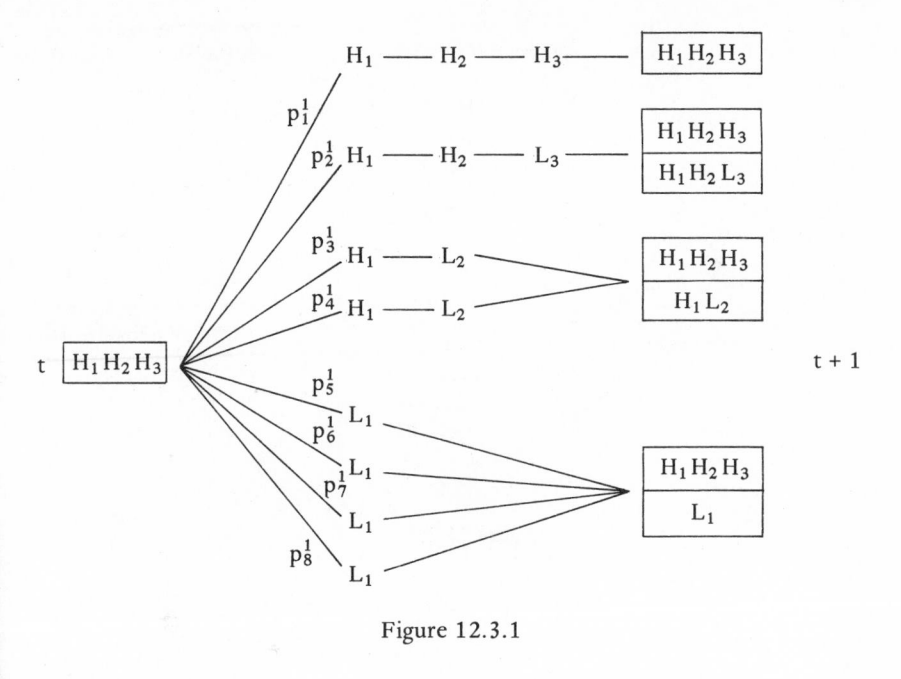

Figure 12.3.1

For example, $C\ell(\alpha) = H_1 H_2 H_3$ (class 1) yields Figure 12.3.1.

This yields the first row of Table 12.3.1. For instance, the probability of the transformation

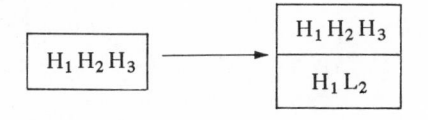

is the probability of the event that α interacts with a β in actual class $H_1 L_2 H_3$ or $H_1 L_2 L_3$, namely, $p_3^1 + p_4^1$.

Examining each of the possible new images in the same way, we have the tree diagrams of Figures 12.3.2, 12.3.3, and 12.3.4. (In these and subsequent tree diagrams we omit the obvious branch probabilities, which are the same as those in Figure 12.3.1.)

We have traced out possible transformations of the possible transforms of the initial image. These lead to the second, third, and fourth rows of Table 12.3.1, respectively.

Considering images arising in these three diagrams that were not previously considered, we obtain three more tree diagrams, shown in Figures 12.3.5, 12.3.6, and 12.3.7.

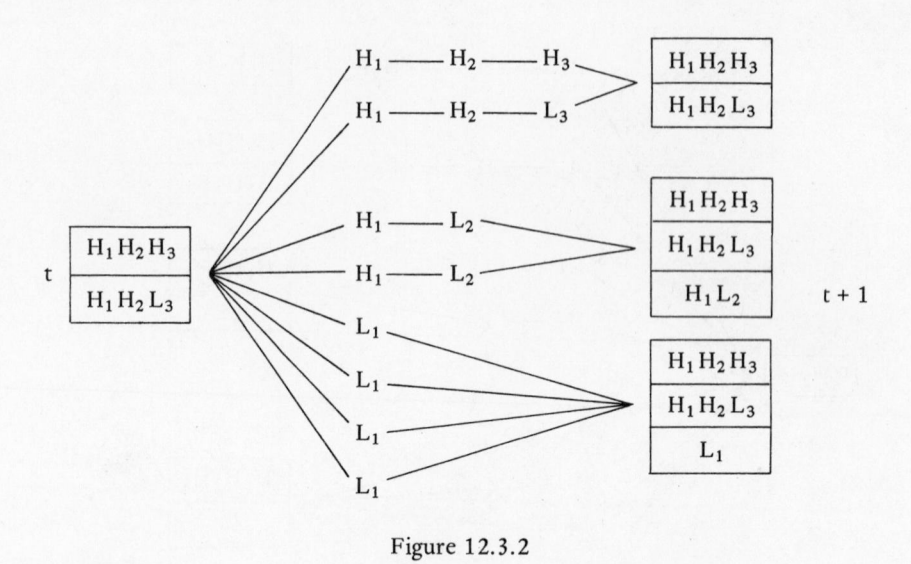

Figure 12.3.2

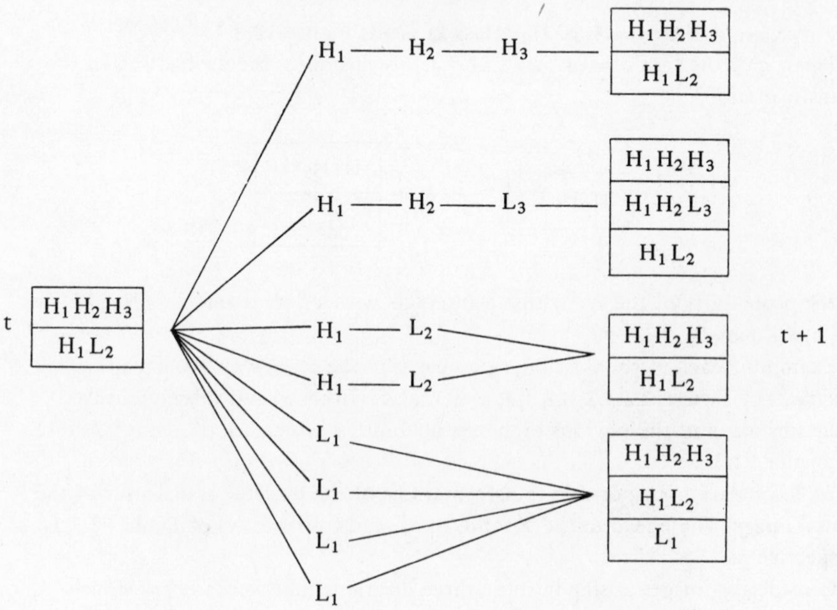

Figure 12.3.3

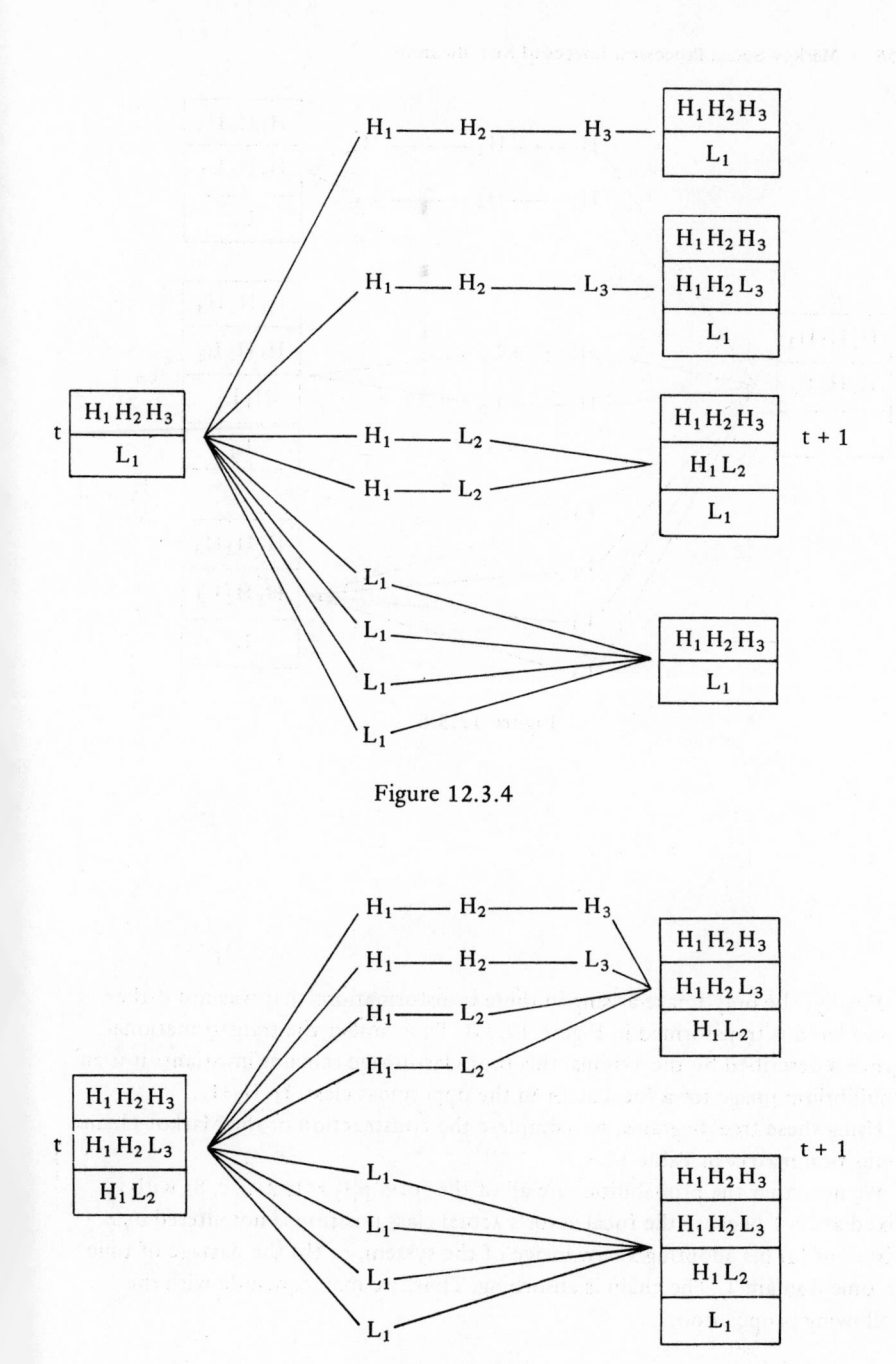

Figure 12.3.4

Figure 12.3.5

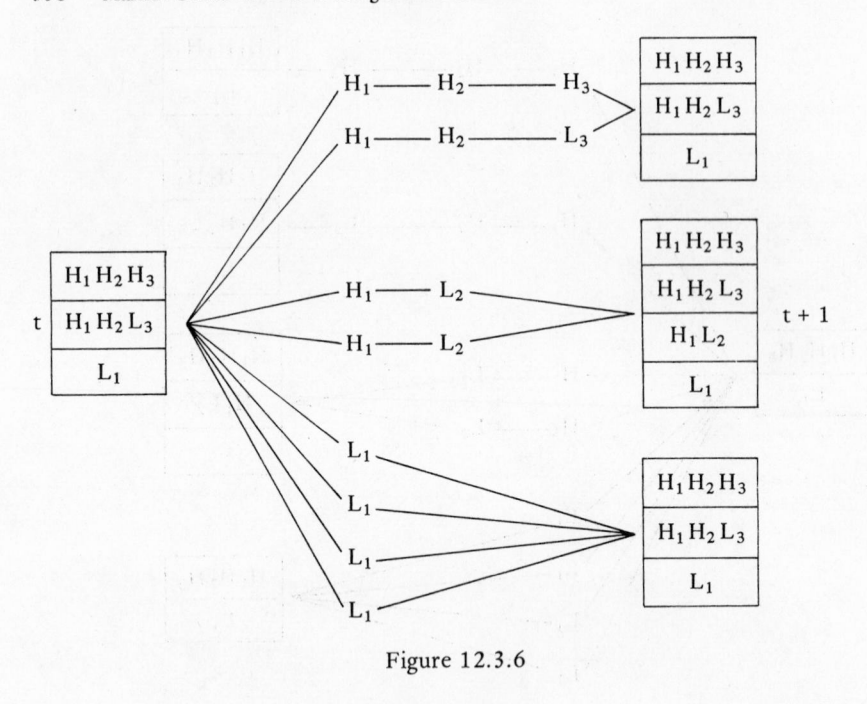

Figure 12.3.6

Finally, the only image arising in these transformations that was not earlier considered is transformed in Figure 12.3.8. Thus, under the transformational process described by the axioms, this four-class image remains invariant: it is an equilibrium image for a focal actor in the uppermost class, $H_1 H_2 H_3$.

Using these tree diagrams, we complete the construction of the Markov chain transition matrix in Table 12.3.1.

We note that the probabilities are all of the form $p_j^i (j = 1, 2, \ldots, 8)$ with i fixed at i = 1 because the focal actor's actual class position is not altered by virtue of (a) his adopting a new image of the system, or (b) the passage of time in time domain T. The chain is absorbing. Thus, we may conclude with the following proposition:

(12.3.1) Proposition. If the interaction probabilities $p_j^1 (j = 1, 2, \ldots, 8)$ are nonzero, then with probability one a focal actor in class $H_1 H_2 H_3$ develops an

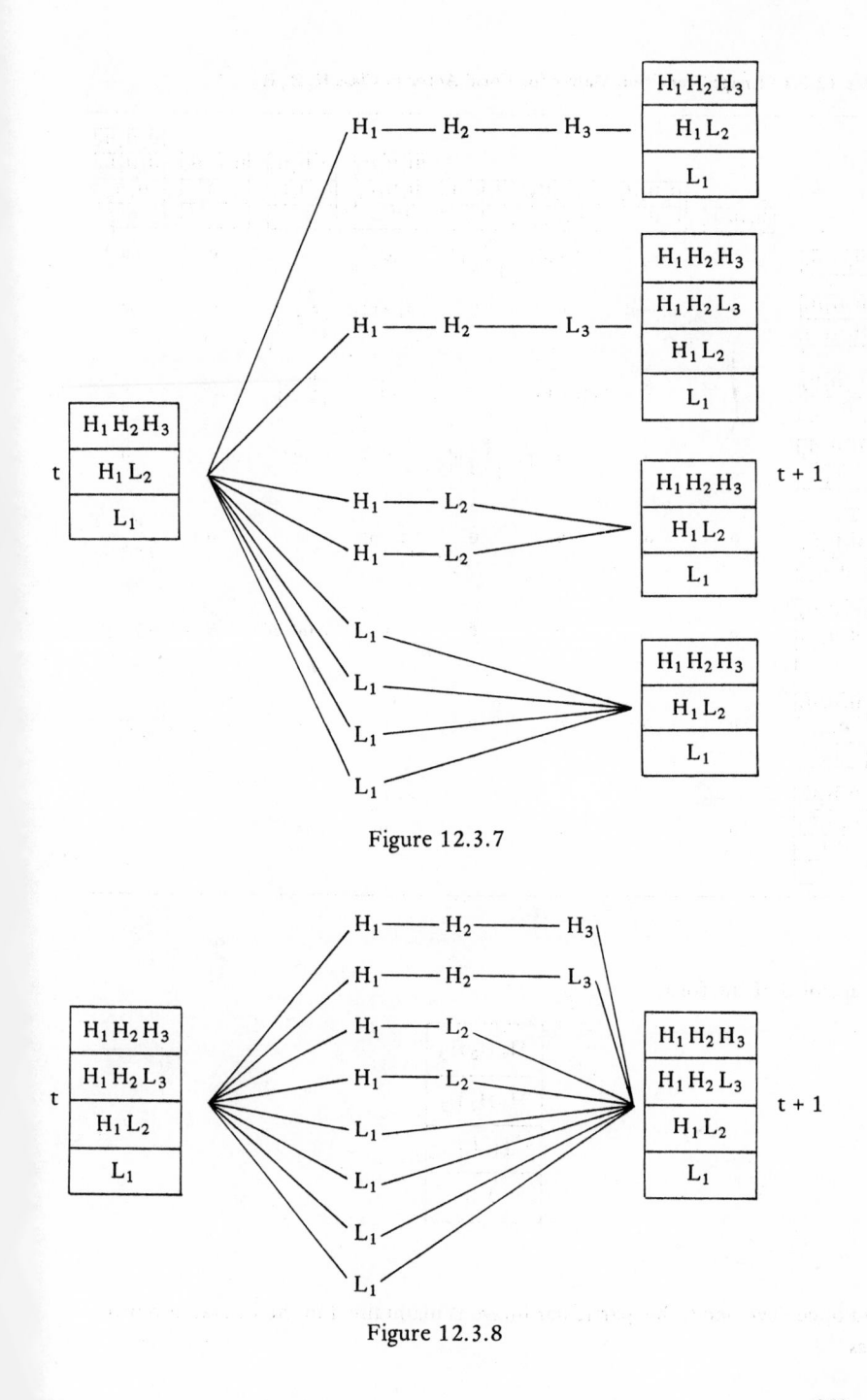

Figure 12.3.7

Figure 12.3.8

Table 12.3.1 Image Transition Matrix for Focal Actor in Class $H_1H_2H_3$

	$H_1H_2H_3$	$H_1H_2H_3$ / $H_1H_2L_3$	$H_1H_2H_3$ / H_1L_2	$H_1H_2H_3$ / L_1	$H_1H_2H_3$ / $H_1H_2L_3$ / H_1L_2	$H_1H_2H_3$ / $H_1H_2L_3$ / L_1	$H_1H_2H_3$ / H_1L_2 / L_1	$H_1H_2H_3$ / $H_1H_2L_3$ / H_1L_2 / L_1
$H_1H_2H_3$	p_1^1	p_2^1	$p_3^1+p_4^1$	$\sum_{j=5}^{8} p_j^1$	0	0	0	0
$H_1H_2H_3$ / $H_1H_2L_3$	0	$p_1^1+p_2^1$	0	0	$p_3^1+p_4^1$	$\sum_{j=5}^{8} p_j^1$	0	0
$H_1H_2H_3$ / H_1L_2	0	0	$p_1^1+p_3^1+p_4^1$	0	p_2^1	$\sum_{j=5}^{8} p_j^1$	0	0
$H_1H_2H_3$ / L_1	0	0	0	$p_1^1+\sum_{j=5}^{8} P_j^1$	0	p_2^1	$p_3^1+p_4^1$	0
$H_1H_2H_3$ / $H_1H_2L_3$ / H_1L_2	0	0	0	0	$\sum_{j=1}^{4} P_j^1$	0	0	$\sum_{j=5}^{8} P_j^1$
$H_1H_2H_3$ / $H_1H_2L_3$ / L_1	0	0	0	0	0	$1-(p_3^1+p_4^1)$	0	$p_3^1+p_4^1$
$H_1H_2H_3$ / H_1L_2 / L_1	0	0	0	0	0	0	$1-p_2^1$	p_2^1
$H_1H_2H_3$ / $H_1H_2L_3$ / H_1L_2 / L_1	0	0	0	0	0	0	0	1

image of S of the form

$H_1H_2H_3$
$H_1H_2L_3$
H_1L_2
L_1

and once developed, this particular image is maintained in the interaction process.

If a tree-diagram analysis is made for each other possible class location, a series of matrices is obtained, one for each class, and in each case the Markov chain follows the rule: if there are nonzero interaction probabilities, then an absorbing state of the image construction process exists. In this way, our second proposition is obtained.

(12.3.2) Proposition. If the probabilities p_j^i are nonzero, then focal actors develop stable images that depend on their class locations as shown in Figure 12.3.9.

Thus, we have four distinctive homomorphic images of the actual stratification ordering shown on the left marginal (not eight, because actors who are distinguishable only on characteristic C_3 do not develop distinct stable images apart from self-location).

12.4. Concrete Examples and Evaluation. We will provide two examples to give two concrete identifications of the abstract structure of proposition (12.3.2) and use these examples to point out additional features of the structure of images.

Example. (a) C_1 = {white, nonwhite} = {w, $\bar{w}$}
   C_2 = {old family, not old family} = {f, $\bar{f}$}
   C_3 = {rich, not rich} = {r, $\bar{r}$}

Then we have the system of images shown in Figure 12.4.1.

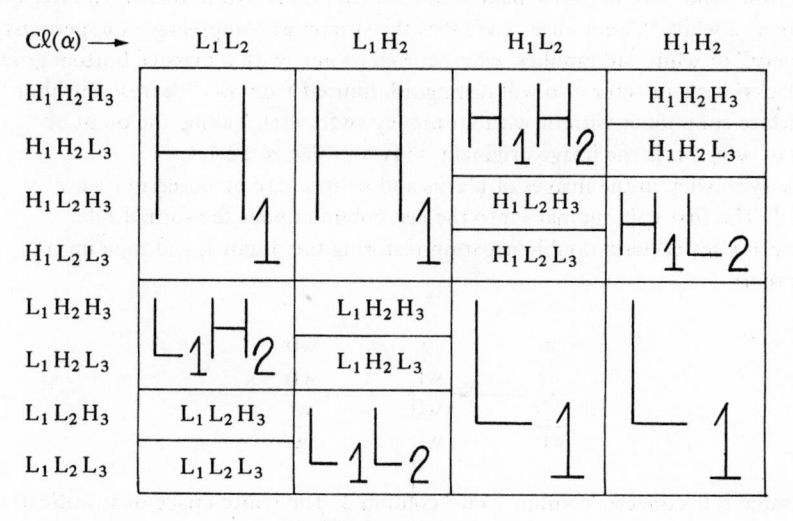

Figure 12.3.9

	$\overline{w}\overline{f}$	$\overline{w}\overline{f}$	wf	wf
wfr			wf	wfr
$wf\overline{r}$	w	w	"society"	$wf\overline{r}$
$w\overline{f}r$	"whites"	"whites"	$w\overline{f}r$	$w\overline{f}$
$w\overline{f}\,\overline{r}$			$w\overline{f}\,\overline{r}$	"common folk"
$\overline{w}fr$	$\overline{w}f$ "black aristocrats"	$\overline{w}fr$		
$\overline{w}f\overline{r}$		$\overline{w}f\overline{r}$	$\overline{w}$	$\overline{w}$
$\overline{w}\,\overline{f}r$	$\overline{w}\,\overline{f}r$	$\overline{w}\,\overline{f}$	"blacks"	"blacks"
$\overline{w}\,\overline{f}\,\overline{r}$	$\overline{w}\,\overline{f}\,\overline{r}$			

Figure 12.4.1

The logic of this identification is as follows. If the three status characteristics based on race, family background, and wealth are ordered as shown; if people located in the various positions construct images of this system as our theory assumes; and if interaction probabilities are nonzero, then the stable structure of class images should be as shown in the diagram. Notably, blacks and whites lump each other and only begin to make refined distinctions within their own race. For instance, a white "commoner" ($w\overline{f}$) sees the system as comprising an uppermost "society" of white old families, whether rich or not, with a massive bottom group of blacks; such an actor also will distinguish himself from his "nearest neighbor," the white commoner with or without money ($w\overline{f}r$, $w\overline{f}\,\overline{r}$). Taking the point of view of $w\overline{f}r$, this is the image originally shown in Figure 12.1.1.

The symmetry in the images of blacks and whites may be noted in Figure 12.4.1. The first column maps into the last column under the formal rule: negate any letter, with double negation restoring the original, and then invert the order:

$$
\begin{array}{ccc}
w & \overline{w} & wfr\\
\overline{w}f & \overline{w}\overline{f} & wf\overline{r}\\
\overline{wfr} & \overline{wfr} & \overline{wf}\\
\overline{wfr} & wfr & \overline{w}
\end{array}
$$

The same rule converts column 2 into column 3. The white image of stratification is in this sense an isomorphic transformation of "the dual" of the black image, and both are homomorphic images of the actual stratification system. By "dual" we mean the replacement of high by low and low by high everywhere in the ordering. (For more on dual isomorphism see section 18.7).

Example. (b) Let us confine our attention to the white caste based on the characteristics of example (a), but we add a fourth characteristic

$$C_4 = \left\{ p, \bar{p} \right\}$$

where p = "prestigious job." Then, with C_1 fixed at w, we have the white stratification system and its image structure given by Figure 12.4.2.

In this example, we see that w cancels out as a discriminating factor, but that p or $\bar{p}$ permits "refinement" of the white images of example (a). For instance, consider a focal actor α with $C\ell(\alpha)$ = wfrp. Then by the prior example and the present example, the full four-dimensional image is

$$wfrp$$
$$wfr\bar{p}$$
$$wf\bar{r}$$
$$w\bar{f}$$
$$\bar{w}$$

Here we see that the prestige characteristic allows the actor to discriminate himself from his "nearest neighbor" with wfrp.

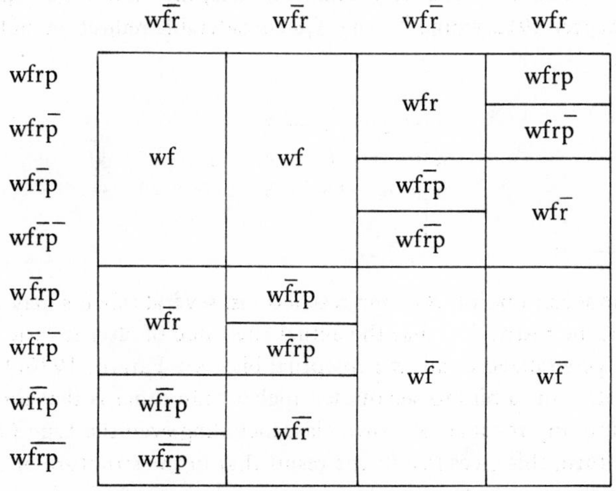

	w$\bar{f}$r	w$\bar{f}$r	w$\bar{f}$r	wfr
wfrp	wf	wf	wfr	wfrp
wfr$\bar{p}$	wf	wf	wfr	wfr$\bar{p}$
wf$\bar{r}$p	wf	wf	wf$\bar{r}$p	wf$\bar{r}$
wf$\bar{r}\bar{p}$	wf	wf	wf$\bar{r}\bar{p}$	wf$\bar{r}$
w$\bar{f}$rp	w$\bar{f}$r	w$\bar{f}$rp	w$\bar{f}$	w$\bar{f}$
w$\bar{f}$r$\bar{p}$	w$\bar{f}$r	w$\bar{f}$r$\bar{p}$	w$\bar{f}$	w$\bar{f}$
w$\bar{f}\bar{r}$p	w$\bar{f}\bar{r}$p	w$\bar{f}\bar{r}$	w$\bar{f}$	w$\bar{f}$
w$\bar{f}\bar{r}\bar{p}$	w$\bar{f}\bar{r}\bar{p}$	w$\bar{f}\bar{r}$	w$\bar{f}$	w$\bar{f}$

Figure 12.4.2

A great deal of additional research needs to be done, both theoretical and empirical, in the area of image construction. Among the topics for research at the theoretical level, we may mention proving general theorems based on general questions about the process and relaxing various axiomatic restrictions.

In regard to the first point for instance, we may ask how the number of classes in a stable image depends upon the number of actual classes. Consider n characteristics; then the number of actual classes is the number

$$N(C_1 \times C_2 \times \ldots \times C_n) = N(C_1) N(C_2) \ldots N(C_n),$$

so in the special case where $N(C_i) = 2$, all i, we have 2^n actual classes. Now consider an arbitrary position of the form

$$x_1 x_2 \ldots x_n$$

where $x_i = H$ or L. Write $\bar{x}_i$ for the category that is not x_i. Then, in the described interactions leading to relative status "decisions," we have a series of n binary choices terminating at any one of these choices or continuing depending on the position of alter, as shown in Figure 12.4.3.

The parenthetical terms show the sampled part of the position of alters when focal actor has position $x_1 x_2 \ldots x_n$. Hence, there are n + 1 distinctive sampled alters. In stable equilibrium it is not possible to interact and not locate alter; hence, in stable equilibrium, the image must contain all n + 1 image classes. This shows that in the special case of binary characteristics, the "homomorphic reduction" (see Chapter 19) is from 2^n to n + 1, a considerable reduction, indeed, if n is fairly large:

$$
\begin{aligned}
n &= 2: & 4 &\rightarrow 3 \\
n &= 3: & 8 &\rightarrow 4 \\
n &= 4: & 16 &\rightarrow 5
\end{aligned}
$$

and so forth.

In regard to the second point, relaxing restrictions, several aspects may be mentioned. First, the restriction that the actual cue value of alter is veridically perceived can be generalized to allow perceptual bias (see Fararo, 1970c). One consequence of allowing a bias to see oneself higher than other is that there are image classes below any focal actor's own class including even the type $C\ell(\alpha) = L_1 L_2 \ldots L_n$. In turn, this gives rise to the result that in the structure of images, greater refinements are made looking down in the system than looking up.

Second, the restriction that the ordering of characteristics of axiom 1 (2) be the ordering of sampling in axiom 6 can be generalized to allow distinctive

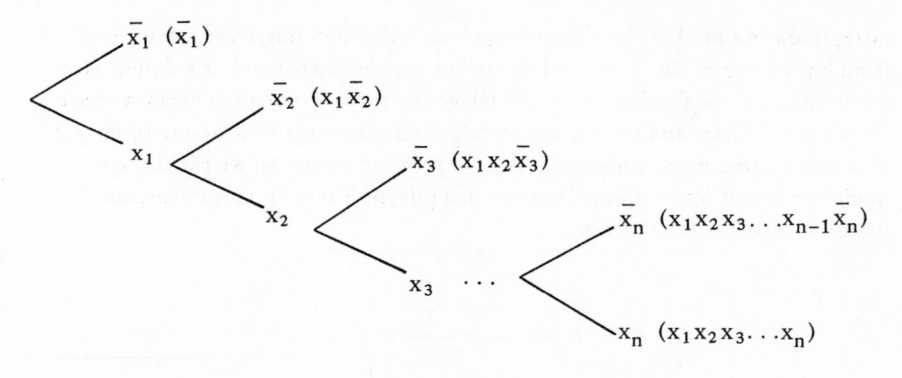

Figure 12.4.3

sampling orders, depending on class location. This would better represent the varying salience of characteristics of varying groups in the system.

Clearly, an empirical research methodology is needed to test these models, building upon the empirical work of such investigators as Davis, Gardner, and Gardner (1941) and Davies (1967).

Looking at this family of models more critically, we may raise several questions. First, axiom 1 assumes a finite set of states on each characteristic C_i. Thus, if C_i is a continuum, it is already partitioned by the sociologist at this point. But what cutting points should be used and with what justification? This is an old problem with a consequence for the interpretation of the generated images. For instance, if wealth is partitioned as rich/not-rich this is quite different from not-poor/poor in terms of the interpretation of Figure 12.4.1. Axiom 1 (1) demands knowledge as to how actors in A use C_i categories.

Secondly, the use of a binary system of characteristics is not essential, at least as far as the theory presented in Section 12.2 is concerned. But, if a set of characteristics with larger numbers of states is considered, the image becomes very large: too large for a fit to the phenomenon, it would seem. For instance, as Teuter (1969) indicates, the number of classes in a stable image in a system with n characteristics where C_i has r_i states, $i = 1, 2, \ldots, n$, is

$$1 + \sum_{i=1}^{n} (r_i - 1)$$

For example, with $n = 3$ and $r_i = 3$ ($i = 1, 2, 3$), we obtain 7 image classes, which is not unreasonable, but if $r_i = 6$ ($i = 1, 2, 3$), we obtain 16 image classes.

All this suggests a need for a method of deriving or constructing the objective system from the image systems given as data. The rule should be, Construct S so that any given image is a homomorphic image of S. This means that S must be interpreted as a kind of "universal image" such that any image can be deduced from knowledge of the "universal image" of the sociologist and of a datum as to the location of the focal actor in that image. To be objective then means to have the universal image and the associated rule for generating the structure of images of actors located in the universal image. In terms of mapping, we can let our model, denote it M, be interpreted as a mapping of S into an image structure I, itself associated with actors in A:

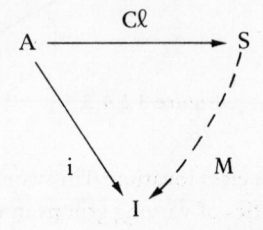

Here each actor α in A is characterized by the sociologist as $C\ell(\alpha)$, a location in S, the universal image called the objective stratification system. Also, each actor α in A is associated with some image $i(\alpha)$ in I, the structure of images. Then the model M, interpreted as a mapping, shows that

$$i(\alpha) = M[C\ell(\alpha)] \qquad \text{(for all } \alpha)$$

More compactly,

$$i = M(C\ell)$$

which says that any image is a function of a universal image class using the associated universal mapping M, which is the model constructed by the sociologist. (In terms of abstract algebra, M makes the image mapping diagram commutative.) The reader may wish to study category theory in Chapter 19, and then return to this discussion. For a mathematical treatment of "universal objects" in algebra, see MacLane and Birkhoff (1967).

Another set of criticisms of this model has to do with its over determinacy in regard to images. No account of variability is given. Variability may be introduced into the model in many ways: replace any definite choice by a probabilistic choice. For instance, the order of sampling in axiom 6 may be randomized by interpreting the ordering of this axiom stochastically, say each C_i has weight v_i and the probability that C_i is chosen first is $v_i/\sum_i v_i$. In other words, we use Luce's ranking postulate (see section 11.12) to obtain a C_i-ranking in a given interaction. This means the v_i need to be estimated from the data in the application phase. Yet this will drastically alter the images produced, yielding from a formal point of view a very unwieldy model. In these matters it is wise to recall that the introduction of too much realism into models in the setup phase will yield its costs at the analysis stage.

Finally, we may look briefly at the consequences of dropping the limitation imposed in axiom 8 that the initial condition be a homogeneous or one-class image. The general notion of image is ordered partition of S. If S has N classes (positions, ordered n-tuples), then the number of such partitions is 2^{N-1}. To see this, construct the partition along the following lines: (a) let class 1 be in partition cell P_1; (b) let class 2 be in P_2, which is P_1 or a new cell; (c) let class 3 be in P_3, again $P_3 = P_2$ or it is new; then continue until all classes are exhausted. For each class except the first a binary decision is involved. Hence, there are $N-1$ binary decisions, which yield 2^{N-1} possible partitions. For the case of n binary characteristics, we know that $N = 2^n$. For instance, there were $N = 2^3 = 8$ actual classes in the model of section 4.3. The number of possible images if we drop axiom 8 is then $2^{N-1} = 2^7 = 128$ rather than the 8 shown in the Markov transition matrix of Table 12.3.1, constructed using axiom 8. Any other initial condition will presumably generate its own limited set of reachable states, thus reducing the full 128 in some manner. But then other stable images may be possible—those which contain more information than needed according to axiom 6. In short, once the initial homogeneity is abandoned, further problems of analysis and interpretation arise that should be explored.

Some of these problems may be solvable in the context of a reinterpretation of the general theory; namely, by giving distinct interpretations to abstract characteristic C_j ($j = 2, 3, \ldots$) in distinct subpopulations determined by the earlier portions of the profile. For instance, let C_1 be interpreted as race. Then there are two subpopulations, corresponding to H_1 and L_1. In the H_1 subpopulation we let C_2 be the primary basis for more refined stratification, say family background. In the L_1 subpopulation, we let C_2 again be the primary basis for stratification within this subset of persons, say wealth. Then the following are given:

$$H_1 H_2: \quad \text{white, old family}$$
$$H_1 L_2: \quad \text{white, not old family}$$
$$L_1 H_2: \quad \text{black, wealthy}$$
$$L_1 L_2: \quad \text{black, not wealthy}$$

Under this more general interpretation, the equilibrium structure of images still makes sense. Moreover, the cue sampling continues to be interpretable: among blacks, for example, cue samplings determining relative status pay no attention to family background. In a third characteristic, the lower-class blacks would sample for different information than would the upper-class blacks.

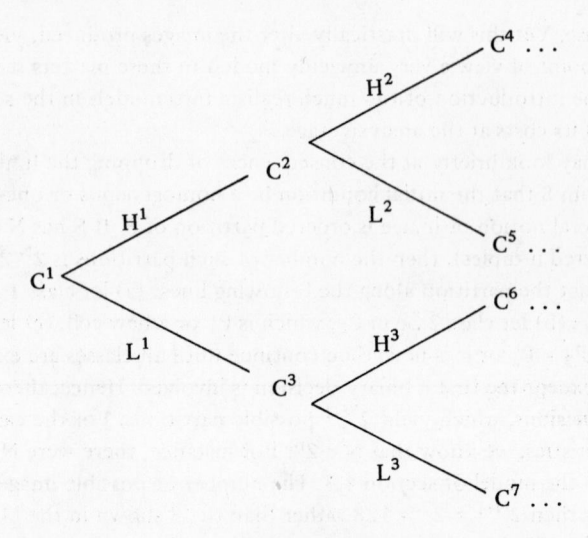

Figure 12.4.4

Rather than leaving this development to the purely interpretative aspect of the theoretical models, we can incorporate it into the basic axioms. Omitting details, the basic idea is to label the abstract characteristics in some new way; for instance,

$$C^1, C^2, C^3, \ldots$$

Then the structural relation among the C^j can be shown in the tree diagram of Figure 12.4.4. In this notation, C^1 stratifies the whole of the population; call it primary in A. Then C^2 is primary in H^1, the high subpopulation of A, while C^3 is primary within L^1, and so forth. (Note that as a special case, we can have $C^2 = C^3$. The earlier treatment is embedded within this revision.) A system of successive binary distinctions, each based on a feature which is primary within a given sector of the stratification system, defines a single complex ordering of the entirety of A. This sort of idea seems to correspond to one interpretation of caste systems in India provided by Dumont (1970). According to Dumont, the system is generated by one basic principle of hierarchy: religious purity. However, it receives distinct application in distinct subpopulations, so that the salient feature varies structurally. Nevertheless, by successive binary distinctions with varying features, nested as in Figure 12.4.4, one can generate a hierarchy.

CHAPTER THIRTEEN MARKOV SOCIAL PROCESSES: COLEMAN METHODOLOGY

13.1. A Continuous-Time Formulation. In his seminal work on mathematical sociology, James Coleman (1964) introduced into sociological analysis the theory of continuous-time finite state Markov chains. But he did more than simply interpret a preexisting mathematical system for sociological problems; he produced an entire abstract methodology for the analysis of the phenomenon of social influence. In this first section the aim is to introduce the reader to a simple continuous-time model; then in the second section we treat the Coleman methodology for the analysis of networks of social influence based on a continuous-time stochastic framework. The aim is not to be comprehensive but to elucidate the basic logic of the approach.

The first model we treat has the following character: at each moment of time t, the system has a state X_t that is a binary random variable taking values 0 or 1. If the process were in discrete time, it would have the transition matrix

$$
\begin{array}{c}
 & \multicolumn{2}{c}{t+1} \\
 & 0 \qquad\qquad 1 \\
\begin{array}{c} 0 \\ t \\ 1 \end{array}
\begin{pmatrix}
1 - q_{01} & q_{01} \\
q_{10} & 1 - q_{10}
\end{pmatrix}
\end{array}
$$

In continuous time the matrix takes the form of a transition from the state at time t to that at $t + \Delta t$ with transition probabilities proportional to Δt:

$$t + \Delta t$$

$$
\begin{array}{c}
 \\
t
\end{array}
\begin{array}{cc}
 & \begin{array}{cc} 0 & \qquad\qquad 1 \end{array} \\
\begin{array}{c} 0 \\[1.5em] 1 \end{array} &
\left(
\begin{array}{cc}
1 - q_{01}\Delta t & q_{01}\Delta t \\[1em]
q_{10}\Delta t & 1 - q_{10}\Delta t
\end{array}
\right)
\end{array}
$$

Applying the law of total probability as in the discrete chain case, we obtain

$$
\begin{aligned}
P(X_{t+\Delta t} = 1) &= P(X_t = 0)P(X_{t+\Delta t} = 1 \mid X_t = 0) + P(X_t = 1)P(X_{t+\Delta t} = 1 \mid X_t = 1) \\
&= P(X_t = 0)q_{01}\Delta t + P(X_t = 1)(1 - q_{10}\Delta t)
\end{aligned}
$$

Next we introduce the notation,

$$
\begin{aligned}
p_0(t) &= P(X_t = 0) \\
p_1(t) &= P(X_t = 1)
\end{aligned}
$$

so that we have,

$$p_1(t + \Delta t) = p_0(t)q_{01}\Delta t + p_1(t)(1 - q_{10}\Delta t)$$

Rearranging terms,

$$p_1(t + \Delta t) - p_1(t) = p_0(t)q_{01}\Delta t - p_1(t)q_{10}\Delta t$$

Since in a continuous-time model the term Δt refers to an arbitrary interval of time, we obtain the basic equation of the setup phase of the dynamic model by first writing the last equation as a rate and then letting Δt approach zero:

$$\lim_{\Delta t \to 0} \frac{p_1(t + \Delta t) - p_1(t)}{\Delta t} = p_0(t)q_{01} - p_1(t)q_{10}$$

And letting the left side be denoted $\dot{p}_1(t)$, a time rate of change, we obtain the equation

$$\dot{p}_1(t) = p_0(t)q_{01} - p_1(t)q_{10}$$

Since $p_0(t) = 1 - p_1(t)$, we can rewrite this equation as

$$(13.1.1) \qquad\qquad \dot{p}_1(t) + (q_{01} + q_{10})p_1(t) = q_{01}$$

and this is a linear first-order differential equation that has the following solution (see any text on differential equations, e.g., Brauer and Nohel, 1967):

$$(13.1.2) \quad p_1(t) = \frac{q_{01}}{q_{01} + q_{10}} [1 - e^{-(q_{01} + q_{10})t}] + p_1(0)e^{-(q_{01} + q_{10})t}$$

Equation 13.1.2 gives the behavior of the probabilistic system over time as a function of the transition parameters, q_{10} and q_{01}, and the initial condition, $p_1(0) = P(X_0 = 1)$.

For equilibrium analysis, we set the rate $\dot{p}_1(t)$ equal to zero in (13.1.1), obtaining the equilibrium state of the probabilistic system:

$$(13.1.3) \quad\quad\quad\quad\quad p_{1e} = \frac{q_{01}}{q_{01} + q_{10}}$$

The meaning of this quantity is that if the initial probability is p_{1e} —that is, if $p_1(0) = P(X_0 = 1) = p_{1e}$ —then the system stays with this distribution. This can be verified by (13.1.2) since if we set $p_1(0) = p_{1e}$, we obtain p_{1e} on the right.

For analysis of the stability of equilibrium, we ask if the probabilistic system approaches the equilibrium p_{1e} if it begins "outside" this distribution. Thus, we study the limit of (13.1.2) as $t \to \infty$,

$$\lim_{t \to \infty} p_1(t) = \frac{q_{01}}{q_{01} + q_{10}} \lim_{t \to \infty} [1 - e^{-(q_{01} + q_{10})t}] + p_1(0) \lim_{t \to \infty} e^{-(q_{01} + q_{10})t}$$

$$= \frac{q_{01}}{q_{01} + q_{10}}$$

since the exponential function is a decay function that approaches zero in the limiting process. From this we conclude (p_{0e}, p_{1e}) is a stable equilibrium probability distribution of the two-state continuous-time Markov chain.

Note that if we represent the "instantaneous" transition rates between the two states as shown in Figure 13.1.1, then the intuitive condition for equilibrium would be that the flow into each state is the same, $p_{0e}q_{01} = p_{1e}q_{10}$,

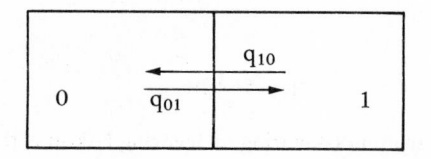

Figure 13.1.1

where $p_{0e} = 1 - p_{1e}$, so that if we solve for p_{1e}, we obtain (13.1.3) directly. In other words, the intuitive picture of equal flows across the boundary gives the identical result as that based on formally setting the rate of change of the probability equal to zero.

The application phase of this sort of model typically involves the following assumption: Each of n specified persons (more generally, concrete systems) independently realizes the continuous-time Markov process given above, with identical parameters q_{01}, q_{10}.

Then, taking expected values in formulas (13.1.2) and (13.1.3), we obtain, on the one hand, the trajectory of the expected number of concrete systems in state 1 as time flows and, on the other hand, the number of such systems in state 1 when the probabilistic system is in equilibrium.

Example. (a) At a given time, we observe that in an aggregate of 100 persons, exactly 75 favor the policies of the current president of the United States, say Nixon. Taking,

$$X_t = \begin{cases} 1 & \text{if person favors Nixon policies} \\ 0 & \text{if not} \end{cases}$$

We assume that (1) each person shifts attitude over time in accordance with the Markov chain outlined above; (2) each person has the same rate parameters, q_{01}, q_{10}; and (3) the probabilistic system is in equilibrium. This last assumption means that although attitudes may still be changing, the probabilities have stabilized. Thus, a fixed expected proportion of persons favor Nixon policies over a time domain in which we assume equilibrium exists. Hence, estimation of p_{1e} by the corresponding fraction in the observations yields

$$\hat{p}_{1e} = .75$$

so that (13.1.3) gives

$$\frac{3}{4} = \frac{q_{01}}{q_{01} + q_{10}}$$

or

$$q_{01} = 3q_{10}$$

Thus, the rate of shift from not favoring to favoring Nixon is three times the reverse rate from favoring to not favoring. But we cannot conclude from this that eventually everyone favors Nixon's policies, because the conclusion is reached on the assumption of equilibrium, which means that these relative rates produce a constant proportion (.75) in favor of Nixon policies. Note that in the absence of further data, say at other times, we are not able to test any of the three assumptions of the application. That is, (1) the actual attitude change process may not be Markov with fixed rates (it may have time-varying rates or it may not be Markov), (2) people in the sample may vary in their rate parameters, and (3) the process may well be Markov with constant rates and with identical rates for all individuals but it may not be in equilibrium. Because of these cautions, we can only say that such an application is an example of the model but not a test.

To return to the model, we may briefly indicate Coleman's treatment of the problem of variability in rates between individuals. Essentially, the problem is one of applied comparative dynamics, which one may picture as shown in Figure 13.1.2. In other words, the same process works itself out under distinct parametric conditions in distinct populations.

One aspect of Coleman's method is to consider the two populations that arise when one divides a main population into two classes: those that have some attribute and those that do not. Then "having the attribute" is postulated to produce an increment in the rate from state 1 to state 0 (or vice versa). Thus,

Population 1: nonpossession of the attribute
Rates: q_{01}
q_{10}

Population 2: possession of the attribute
Rates: $q'_{01} = q_{01}$
$q'_{10} = q_{10} + \alpha$

Here α is called the effect parameter. In applications, one labels states suitably to reflect the idea that possession of a certain attribute means a higher transition rate than would otherwise exist.

Example.   (b) It is suspected that Republicans will have a stronger commitment to Nixon policies than Democrats. Thus, the transition rate from state 1

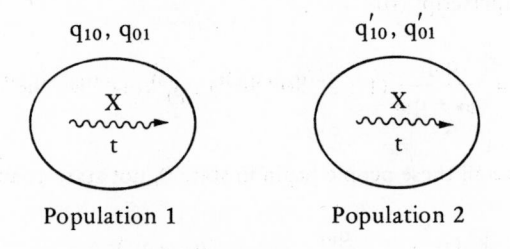

Population 1   Population 2

Figure 13.1.2

(favoring) to state 0 (not favoring) is likely to reflect a "party effect." In any small time interval Δt, the probability $q_{10}'\Delta t$ that a particular Democrat shifts from a positive to a negative attitude will be greater than the probability $q_{10}\Delta t$ that a particular Republican shifts. Hence, the attribute is "Democrat." We have

$$\begin{aligned}
&\text{Population 1:} \quad \text{Republicans} \\
&\qquad\qquad \text{Rates:} \quad q_{01} \\
&\qquad\qquad\qquad\qquad q_{10} \\
&\text{Population 2:} \quad \text{Democrats} \\
&\qquad\qquad \text{Rates:} \quad q_{01}' = q_{01} \\
&\qquad\qquad\qquad\qquad q_{10}' = q_{10} + \alpha
\end{aligned}$$

We indicate now a method of parameter estimation for the probabilistic system whose temporal behavior is given by (13.1.2). Assume data of the form

$$\begin{array}{cc}
 & \text{time } t \\
 & \begin{array}{cc} 0 & \quad 1 \end{array} \\
\text{time 0} \quad \begin{array}{c} 0 \\ 1 \end{array} & \begin{pmatrix} n_{00}(t) & n_{01}(t) \\ n_{10}(t) & n_{11}(t) \end{pmatrix}
\end{array}$$

so that there is a "turnover table" with the first time point labeled 0 and

$n_{ij}(t) =$ number of persons initially in state i who are in state j at time t

Also, let

$$\begin{aligned}
N_0 &= n_{00}(t) + n_{01}(t) \\
N_1 &= n_{10}(t) + n_{11}(t)
\end{aligned}$$

be the initial number in states 0 and 1, respectively.

Now, in (13.1.2) we first relativize the process to those initially in state 0 and indicate this by a superscript (0):

$$p_1^{(0)}(t) = \frac{q_{01}}{q_{01} + q_{10}} (1 - e^{-(q_{01} + q_{10})t}) + p_1^{(0)}(0)e^{-(q_{01} + q_{10})t}$$

But $p_1^{(0)}(0) = 0$, since all these people begin in state 0, not state 1, and so

$$(13.1.4) \qquad\qquad p_1^{(0)}(t) = \frac{q_{01}}{q_{01} + q_{10}} (1 - e^{-(q_{01} + q_{10})t})$$

Similarly, write superscript (1) for the process among those who start in state 1 to obtain, as before,

$$p_1^{(1)}(t) = \frac{q_{01}}{q_{01} + q_{10}} [1 - e^{-(q_{01}+q_{10})t}] + p_1^{(1)}(0)e^{-(q_{01}+q_{10})t}$$

and, with $p_1^{(1)}(0) = 1$,

(13.1.5) $$p_1^{(1)}(t) = \frac{q_{01}}{q_{01} + q_{10}} [1 - e^{-(q_{01}+q_{10})t}] + e^{-(q_{01}+q_{10})t}$$

If we subtract (13.1.4) from (13.1.5) we obtain,

(13.1.6) $$p_1^{(1)}(t) - p_1^{(0)}(t) = e^{-(q_{01}+q_{10})t}$$

and so, taking natural logarithms,

$$\ln[p_1^{(1)}(t) - p_1^{(0)}(t)] = -(q_{01} + q_{10})t$$

Thus,

(13.1.7) $$q_{01} + q_{10} = -\frac{1}{t} \ln[p_1^{(1)}(t) - p_1^{(0)}(t)]$$

Since (13.1.6) allows us to replace the exponential term in (13.1.4) by an expression in the probabilities and (13.1.7) allows us to replace the sum $q_{01} + q_{10}$ by such an expression, when we substitute in (13.1.4) in this way, we have

$$p_1^{(0)}(t) = q_{01} \left(\frac{-t}{\ln[p_1^{(1)}(t) - p_1^{(0)}(t)]} \right) \left(1 - [p_1^{(1)}(t) - p_1^{(0)}(t)] \right)$$

If we let $p_0^{(1)}(t) = 1 - p_1^{(1)}(t)$, we can solve for q_{01} to obtain

(13.1.8) $$q_{01} = p_1^{(0)}(t) \frac{-\ln[p_1^{(1)}(t) - p_1^{(0)}(t)]}{t[p_0^{(1)}(t) + p_1^{(0)}(t)]}$$

The estimate of q_{01} can be obtained by replacing the probabilities with corresponding proportions from the turnover data:

$$\hat{p}_1^{(0)}(t) = \frac{n_{01}(t)}{N_0} = \frac{n_{01}(t)}{n_{00}(t) + n_{01}(t)}$$

$$\hat{p}_1^{(1)}(t) = \frac{n_{11}(t)}{N_1} = \frac{n_{11}(t)}{n_{10}(t) + n_{11}(t)}$$

$$\hat{p}_0^{(1)}(t) = \frac{n_{10}(t)}{N_1} = 1 - \hat{p}_1^{(1)}(t)$$

To obtain an expression for q_{10}, we return to (13.1.7), writing it in the form

$$(13.1.9) \qquad q_{01} = -q_{10} - \frac{1}{t} \ln\,[p_1^{(1)}(t) - p_1^{(0)}(t)]$$

Replacing the exponential in (13.1.4) by (13.1.6) and q_{01} by (13.1.9), we obtain, after solving for q_{10},

$$(13.1.10) \qquad q_{10} = p_0^{(1)}(t) \frac{-\ln[p_1^{(1)}(t) - p_1^{(0)}(t)]}{t[p_1^{(0)}(t) + p_0^{(1)}(t)]}$$

Alternatively, we could have interchanged 0 with 1 throughout (13.1.8). In any case, (13.1.10) provides the basis for the second parameter estimate.

Example. (c) A sample of 200 students are administered a questionnaire on attitudes toward political affairs, first as entering freshmen and then six months later. Attitudes are labeled as "favorable to current administration" or not. The data are as follows:

		6 months later	
		Favorable	Not
	Favorable	75	25
Entering			
	Not	10	90

Let "favorable" be coded 0, and "not favorable," 1. Here we take $t = 6$ and note that

$$\hat{p}_1^{(0)}(6) = \frac{n_{01}(6)}{n_{00}(6) + n_{01}(6)} = \frac{25}{100} = .25$$

$$\hat{p}_1^{(1)}(6) = \frac{n_{11}(6)}{n_{11}(6) + n_{10}(6)} = \frac{90}{100} = .90$$

$$\hat{p}_0^{(1)}(6) = 1 - \hat{p}_1^{(1)}(6) = .10$$

Hence, using (13.1.8) and (13.1.10), respectively,

$$\hat{q}_{01} = \hat{p}_1^{(0)}(6) \frac{-\ln[\hat{p}_1^{(1)}(6) - \hat{p}_1^{(0)}(6)]}{6[\hat{p}_0^{(1)}(6) + \hat{p}_1^{(0)}(6)]} = .25 \frac{-\ln(.90 - .25)}{6(.10 + .25)} = .05$$

$$\hat{q}_{10} = \hat{p}_0^{(1)}(6) \frac{-\ln[\hat{p}_1^{(1)}(6) - \hat{p}_1^{(0)}(6)]}{6[\hat{p}_0^{(1)}(6) + \hat{p}_1^{(0)}(6)]} = .10 \frac{-\ln(.90 - .25)}{6(.10 + .25)} = .02$$

If this process continues under the same conditions, eventually we have the stable equilibrium proportion "not favorable" given by (13.1.3):

$$\hat{P}_{le} = \frac{\hat{q}_{10}}{\hat{q}_{10} + \hat{q}_{01}} = \frac{.02}{.07} = .29$$

To illustrate estimation of an effect parameter, if the students are divided into two subpopulations, male and female, it might be hypothesized for some reason that the males experience an extra effect in the direction of "not favorable." Assume the data take the form

	Males 6 months later			Females 6 months later	
Entering	$\begin{pmatrix} 40 \\ \\ 2 \end{pmatrix}$		$\begin{pmatrix} 20 \\ \\ 48 \end{pmatrix}$	Entering $\begin{pmatrix} 35 \\ \\ 8 \end{pmatrix}$	$\begin{pmatrix} 5 \\ \\ 42 \end{pmatrix}$

The parameters for females are denoted q_{01} and q_{10}, for males q_{01}', q_{10}' with the hypothesis,

$$q_{10}' = q_{10}, \quad q_{01}' = q_{01} + \alpha$$

Estimation for females gives,

$$\hat{p}_1^{(0)}(6) = \frac{1}{8}, \quad \hat{p}_1^{(1)}(6) = \frac{21}{25}, \quad \hat{p}_0^{(1)}(6) = \frac{4}{25}$$

Hence, for females, using (13.1.8) and these estimates,

$$\hat{q}_{01} = \frac{1}{8} \frac{-\ln(21/25 - 1/8)}{6(4/25 + 1/8)} = \frac{1}{8} \cdot \frac{.335}{6(.285)} = .02$$

Using (13.1.10),

$$\hat{q}_{10} = \frac{4}{25} \frac{-\ln(21/25 - 1/8)}{6(4/25 + 1/8)} = \frac{4}{25} \cdot \frac{.335}{6(.285)} = .03$$

For males the data yield

$$\hat{p}_1^{(0)}(6) = \frac{1}{3}, \ \hat{p}_1^{(1)}(6) = \frac{24}{25}, \ \hat{p}_0^{(1)} = \frac{1}{25}$$

Thus,

$$\hat{q}_{01}' = \frac{1}{3} \frac{-\ln(24/25 - 1/3)}{6(1/25 + 1/3)} = .07$$

$$\hat{q}_{10}' = \frac{1}{25} \frac{-\ln(24/25 - 1/3)}{6(1/25 + 1/3)} = .01$$

Hence, bypassing any formal statistical test, we note that

$$\hat{q}_{10}' \approx \hat{q}_{10}$$

as hypothesized and the estimate of α is given by

$$\hat{q}_{01}' = \hat{q}_{01} + \hat{\alpha}$$
$$\hat{\alpha} = .07 - .02 = .05$$

The "effect" of the male attribute is thereby measured as .05 per month; that is, over and above the transition rate that holds for all students as such, there is an extra transition in the "unfavorable" direction for males and it has the magnitude of a 5 percent loss of "favorables" per month. In this as in the preceding example, one must recall the cautionary note as to estimation and fit. One may be able to estimate parameters, but this is not the same as testing the model. As Coleman shows, if sufficient temporal data exist, one may not only estimate parameters but also evaluate the fit of the model. Strictly speaking, only if a good fit exists is it wise to think of the parameter estimates as measuring something of substantive interest.

The last few remarks show that we can interpret "the effect parameter method" as a method of measurement, and it is in this sense, it seems, that much of Coleman's work attains heightened significance: intrinsically "continuous notions" (such as strength of effect) are measured by parameter estimation in a model whose intrinsic analytical grounding is only nominal or ordinal in the state variable. Thus, beginning with classification and counting, coupled to a stochastic picture of transition behavior, the investigation culminates in a numerical continuous measure whose interpretability is given by its "location" in the systematic pattern that is described by the model. Note how this method differs from a fundamental measurement approach toward complex concepts in that it produces a derived measurement validated by the fit of the stochastic process model. For additional remarks on this point, the reader may consult Coleman (1964, Chap. 2). Another example of this methodology is given in section 15.4.

13.2. Networks of Social Influence. The representation of individual dynamics by a two-state continuous-time stochastic process is now extended to a network of individuals. Whereas in the preceding section a number n of individuals merely form an aggregate in which each person independently realizes a process, in the present case the n individuals form a relational system and the process for any one of them depends upon the state of the whole network of relationships. The aim now is to illustrate this basic principle with a minimum of mathematical complexity.

Thus, consider the relational system represented by the graph in Figure 13.2.1. Each individual may be in state A or in state $\overline{A}$ of a two-state continuous-time Markov process. This process is constructed by appeal to the rule that influence is potentially present between each pair of related persons and that a parameter γ measures this potentiality as is indicated by weighting the lines (see Figure 13.2.2). We will give γ a transition rate interpretation shortly. The second rule is that transition from state $\overline{A}$ to state A by any one individual in the network occurs either (a) because of the common conditions in which the network of individuals finds itself, with rate α, or (b) because of social influence from within the net, with rate a function of γ and the state of the network in regard to $\overline{A}$ and

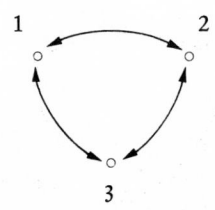

Figure 13.2.1

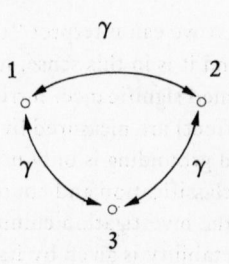

Figure 13.2.2

A (that is, the distribution of individuals in the two states, A and $\overline{A}$). Analogously, transition from A to $\overline{A}$ occurs because of (a) the common conditions, with rate β, or (b) the influence within the net, with rate depending on γ and the state of the network. Also, we assume that in a sufficiently small time interval of length Δt, at most one event will occur.

For instance, suppose that two individuals (1, 2) are in state A and one (3) in $\overline{A}$, which we represent by Figure 13.2.3.

Since at most one event occurs in small duration Δt, let us analyze the new network states arising under each possible transition event. The reader should focus on the graphs of Figures 13.2.2 and 13.2.3 in reading the description of various events in the network. The following are the possible mutually exclusive events in small Δt:

$E_{1,1}$:   node 1 receives an influence input from node 3, with probability $\gamma \Delta t$, and this shifts node 1 to state $\overline{A}$, the state of the "influencer,"

$E_{1,2}$:   node 1 shifts without the influence input, with chance $\beta \Delta t$,

$E_{2,1}$:   node 2 receives an influence input from node 3 and shifts into $\overline{A}$, with chance $\gamma \Delta t$,

$E_{2,2}$:   node 2 shifts without the influence input, with chance $\beta \Delta t$,

$E_{3,1}$:   node 3 receives an influence input from node 1 and shifts to state A, with chance $\gamma \Delta t$,

$E_{3,2}$:   node 3 receives an influence input from node 2 and shifts to state A, with chance $\gamma \Delta t$,

$E_{3,3}$:   node 3 shifts without the influence input, with chance $\alpha \Delta t$,

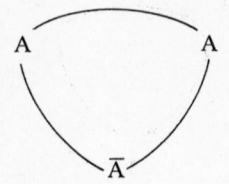

Figure 13.2.3

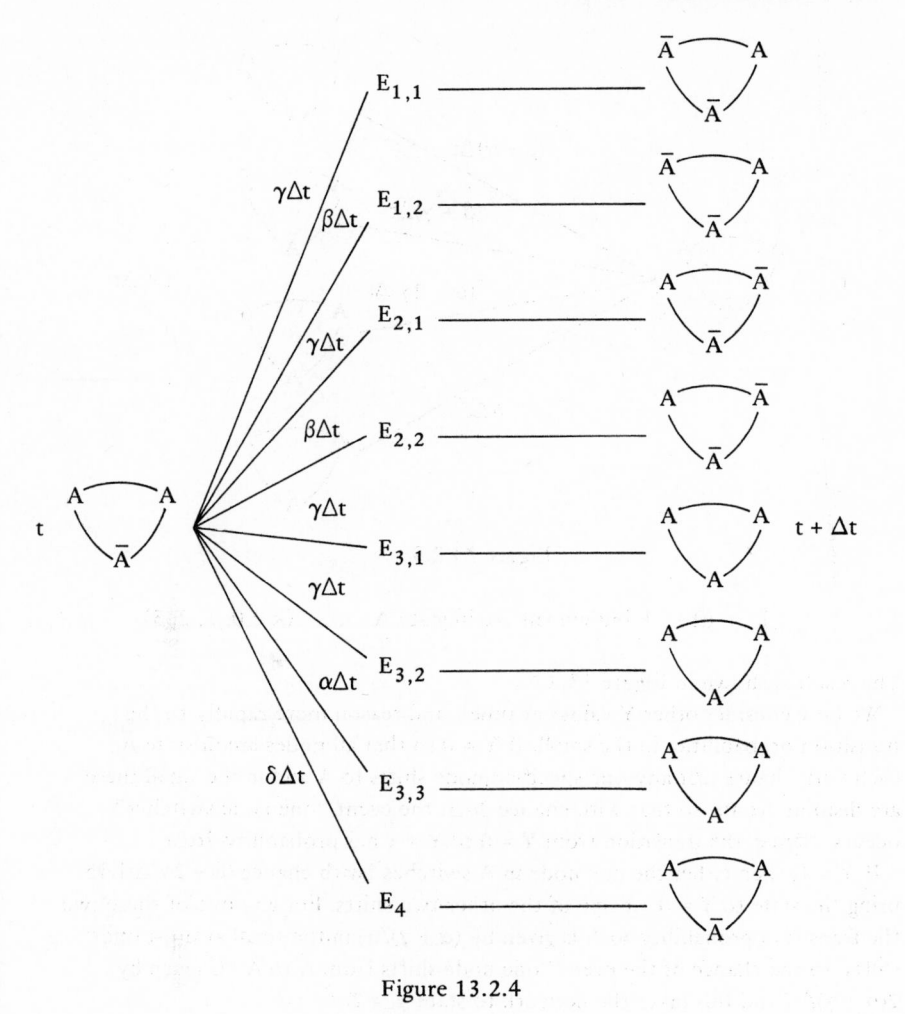

Figure 13.2.4

E_4:   none of the above events occur, with chance $\delta \Delta t$.

Accordingly, the tree diagram for transition of the network from t to t + Δt is given in Figure 13.2.4.

Hence, by summing over distinct paths leading to the same outcome we arrive at the transition probabilities shown in Figure 13.2.5.

Finally, we abstract from the individuals to obtain the state of the network: the sheer distribution of persons in states. This is done by defining

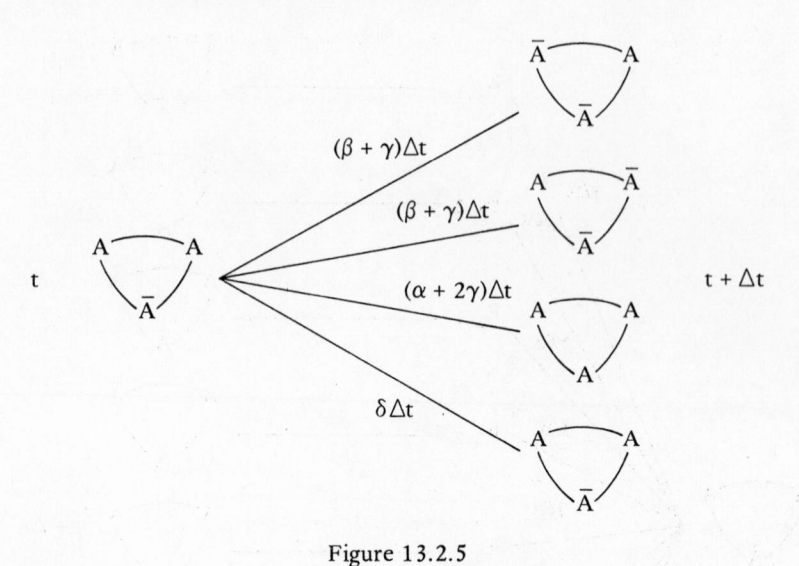

Figure 13.2.5

$$Y = k \quad \text{iff} \quad k \text{ individuals are in state A} \quad (k = 0, 1, 2, 3)$$

The result is shown in Figure 13.2.6.

We now consider other Y values at time t and reason more rapidly to the transition probabilities in the small. If $Y = 0$ so that all nodes are in state $\bar{A}$, then with chance $\alpha\Delta t$ any one specified node shifts to A and in the small these are disjoint events, so that with chance $3\alpha\Delta t$ the event "one node switches" occurs. Hence, the transition from $Y = 0$ to $Y = 1$ has probability $3\alpha\Delta t$.

If $Y = 1$, then either the one node in A switches [with chance $(\beta + 2\gamma)\Delta t$] to bring the state to $Y = 0$, or one of the other two shifts. For any one of these two the transition probability to A is given by $(\alpha + \gamma)\Delta t$; in the small at most one shifts, so the chance of the event "one node shifts from $\bar{A}$ to A" is given by $2(\alpha + \gamma)\Delta t$, and this takes the network to state $Y = 2$.

If $Y = 3$, then all nodes are in A, each shifts to $\bar{A}$ with chance $\beta\Delta t$; hence, the event that "one node shifts" has chance $3\beta\Delta t$, and this is the transition probability from $Y = 3$ to $Y = 2$.

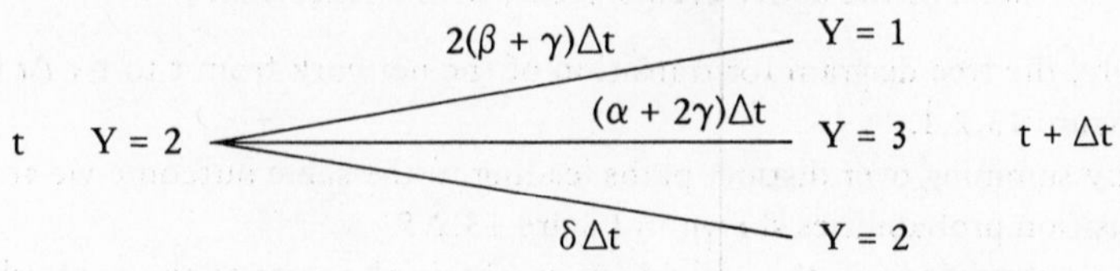

$$t \quad Y = 2 \quad \begin{array}{l} \overset{2(\beta + \gamma)\Delta t}{\diagup} \quad Y = 1 \\ \overset{(\alpha + 2\gamma)\Delta t}{\longleftarrow} \quad Y = 3 \quad t + \Delta t \\ \underset{\delta\Delta t}{\diagdown} \quad Y = 2 \end{array}$$

Figure 13.2.6

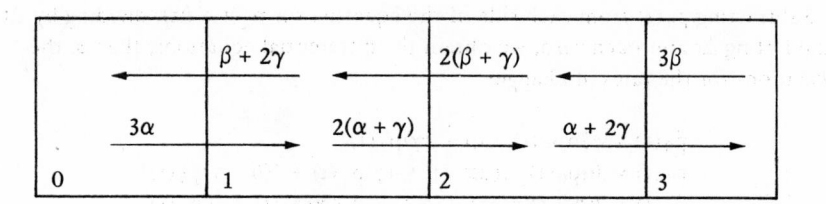

Figure 13.2.7

The above calculations yield the diagram for the network process shown in Figure 13.2.7.

When this diagram is obtained, the setup phase of the network model is completed. The remaining questions fall in two classes:

(1) What are the properties of the network process?
   (a) deduce the trajectory of the probabilistic system (i.e., the flow of probabilities of the state of the network over time),
   (b) determine the equilibrium probability distribution,
   (c) determine the stability of equilibrium,
   (d) analyze how the stable equilibrium depends on the parameters.
(2) To what actual social processes does this scheme apply?
   (a) identify the abstractions in a particular phenomenon,
   (b) estimate parameters,
   (c) measure the goodness-of-fit of the model.

In the following discussion we treat some of these remaining aspects of the model-construction paradigm, as evidenced in Coleman's model.

The transition diagram may be brought into the form of a matrix,

$$\mathbf{P} = \begin{pmatrix} * & 3\alpha\Delta t & 0 & 0 \\ (\beta + 2\gamma)\Delta t & * & 2(\alpha + \gamma)\Delta t & 0 \\ 0 & 2(\beta + \gamma)\Delta t & * & (\alpha + 2\gamma)\Delta t \\ 0 & 0 & 3\beta\Delta t & * \end{pmatrix}$$

where Δt is assumed very small and the diagonal terms are quantities that make rows sum to 1. Let $p_k(t) = P(Y = k \text{ at } t)$, $k = 0, 1, 2, 3$. Then by the law of total probability,

$$p_0(t + \Delta t) = p_0(t)(1 - 3\alpha\Delta t) + p_1(t)(\beta + 2\gamma)\Delta t$$
$$p_1(t + \Delta t) = p_0(t)3\alpha\Delta t + p_1(t)[1 - (\beta + 2\gamma + 2\alpha + 2\gamma)\Delta t] + p_2(t)2(\beta + \gamma)\Delta t$$
$$p_2(t + \Delta t) = p_1(t)2(\alpha + \gamma)\Delta t + p_2(t)[1 - (2\beta + 2\gamma + \alpha + 2\gamma)\Delta t] + p_3(t)3\beta\Delta t$$
$$p_3(t + \Delta t) = p_2(t)(\alpha + 2\gamma)\Delta t + p_3(t)(1 - 3\beta\Delta t)$$

Subtracting $p_k(t)$ from each side of the equation for $p_k(t + \Delta t)$, dividing by Δt and letting Δt approach zero, we obtain the differential equations, that is, the equations for the rates of change:

$$(13.2.1) \quad \begin{aligned} \dot{p}_0(t) &= -3\alpha p_0(t) + (\beta + 2\gamma)p_1(t) \\ \dot{p}_1(t) &= 3\alpha p_0(t) - (2\alpha + \beta + 4\gamma)p_1(t) + 2(\beta + \gamma)p_2(t) \\ \dot{p}_2(t) &= 2(\alpha + \gamma)p_1(t) - (\alpha + 2\beta + 4\gamma)p_2(t) + 3\beta p_3(t) \\ \dot{p}_3(t) &= (\alpha + 2\gamma)p_2(t) - 3\beta p_3(t) \end{aligned}$$

This is a linear system of the form,

$$(13.2.2) \quad \begin{aligned} \dot{p}_0(t) &= Ap_0(t) + Bp_1(t) \\ \dot{p}_1(t) &= Cp_0(t) + Dp_1(t) + Ep_2(t) \\ \dot{p}_2(t) &= Fp_1(t) + Gp_2(t) + Hp_3(t) \\ \dot{p}_3(t) &= Jp_2(t) + Kp_3(t) \end{aligned}$$

where A, B, . . . , K are abbreviated notations for the constants in (13.2.1).

In turn, this system may be put in matrix form,

$$(13.2.3) \quad \dot{p}(t) = Rp(t)$$

where $\dot{p}(t) = [\dot{p}_0(t), \dot{p}_1(t), \dot{p}_2(t), \dot{p}_3(t)]^T$, $p(t) = [p_0(t), p_1(t), p_2(t), p_3(t)]^T$ and

$$R = \begin{pmatrix} A & B & 0 & 0 \\ C & D & E & 0 \\ 0 & F & G & H \\ 0 & 0 & J & K \end{pmatrix}$$

Note that, returning to the original constants,

$$(13.2.4) \quad R^T = \begin{pmatrix} -3\alpha & 3\alpha & 0 & 0 \\ \beta + 2\gamma & -(2\alpha + \beta + 4\gamma) & 2(\alpha + \gamma) & 0 \\ 0 & 2(\beta + \gamma) & -(\alpha + 2\beta + 4\gamma) & \alpha + 2\gamma \\ 0 & 0 & 3\beta & -3\beta \end{pmatrix}$$

and compare R^T with matrix P above. Apart from the * quantities, and the Δt terms, the matrices are identical. The matrix R^T is called the instantaneous transition matrix. Note that it contains the rates of transition in the off-diagonal entries and the diagonals are terms making the rows sum to 0, not 1. This peculiarity arises, as we have seen, from the passage to rates of change based on the matrix of transition in the small, which is P.

Having obtained the instantaneous matrix from more intuitive considerations, we will not actually use it but sketch the path to a solution to this first problem of analysis—finding the trajectory of the probabilistic system. The interested reader should consult Bartholomew (1967) for extensive applications of the method.

Given

$$\dot{p}(t) = Rp(t)$$

the equation is a matrix generalization of the one variable equation,

$$\dot{x}(t) = rx(t)$$

which has the solution,

$$x(t) = x_0 e^{rt}$$

Now,

$$e^{rt} = 1 + rt + \frac{(rt)^2}{2!} + \cdots + \frac{(rt)^n}{n!} + \cdots$$

This suggests that for the matrix equation the solution might take the form

$$p(t) = [I + Rt + \frac{(Rt)^2}{2!} + \cdots + \frac{(Rt)^n}{n!} + \cdots]\, p(0)$$

or, abbreviating the series,

$$p(t) = e^{Rt}p(0)$$

Then in suitable cases, R can be put in the form,

$$R = \sum_{i=1}^{4} \lambda_i A_i$$

where the A_i are matrices and the λ_i are the eigenvalues of R (see section 8.7). Then, according to Bartholomew (1967, p. 79), the solution can be put in the form

(13.2.5)
$$p(t) = \sum_{i=1}^{4} e^{\lambda_i t} A_i p(0)$$

and the limiting behavior depends upon the λ_i.

Unfortunately, in most cases the labor involved in actually obtaining the λ_i and interpreting the solution—which describes the trajectory—is formidable. For this reason, we have only sketched the logic to show the reader the form of the approach and to hint at its complexity. Following Coleman, we now turn to the equilibrium problem.

By defintion of the concept of equilibrium, we require probability vector p_e satisfying the condition that the left side of (13.2.3) is 0. Hence, vector p_e satisfying

(13.2.6) $$\mathbf{R}p_e = 0$$

is an equilibrium distribution. This is the same as setting each equation of (13.2.1) equal to 0. We obtain:

(1) $-3\alpha p_{0e} + (\beta + 2\gamma)p_{1e} = 0$

(2) $3\alpha p_{0e} - (2\alpha + \beta + 4\gamma)p_{1e} + 2(\beta + \gamma)p_{2e} = 0$

(3) $2(\alpha + \gamma)p_{1e} - (\alpha + 2\beta + 4\gamma)p_{2e} + 3\beta p_{3e} = 0$

(4) $(\alpha + 2\gamma)p_{2e} - 3\beta p_{3e} = 0$

Solving (1) for p_{1e} in terms of p_{0e},

(1)′ $$p_{1e} = \frac{3\alpha}{\beta + 2\gamma} p_{0e}$$

Substituting (1)′ into (2) and solving for p_{2e} in terms of p_{0e},

(2)′ $$p_{2e} = \frac{3\alpha(\alpha + \gamma)}{(\beta + 2\gamma)(\beta + \gamma)} p_{0e}$$

Substituting (2)′ into (4) and solving for p_{3e} in terms of p_{0e},

(3)′ $$p_{3e} = \frac{\alpha(\alpha + \gamma)(\alpha + 2\gamma)}{\beta(\beta + \gamma)(\beta + 2\gamma)} p_{0e}$$

Adding p_{0e} and (1)′, (2)′, and (3)′, which sum to 1, we obtain

$$\left[1 + \frac{3\alpha}{\beta + 2\gamma} + \frac{3\alpha(\alpha + \gamma)}{(\beta + 2\gamma)(\beta + \gamma)} + \frac{\alpha(\alpha + \gamma)(\alpha + 2\gamma)}{\beta(\beta + \gamma)(\beta + 2\gamma)} \right] p_{0e} = 1$$

Solving for p_{0e}, we obtain,

(13.2.7) $$p_{0e} = \frac{\beta(\beta + \gamma)(\beta + 2\gamma)}{\beta(\beta + \gamma)(\beta + 2\gamma) + 3\alpha\beta(\alpha + \beta + 2\gamma) + \alpha(\alpha + \gamma)(\alpha + 2\gamma)}$$

In anticipation of using the mean and variance of an observed distribution to estimate parameters, we realize we will be able to obtain estimates of only two parameters. Hence, we define

$$a = \frac{\alpha}{\alpha + \beta}, \quad c = \frac{\gamma}{\alpha + \beta}$$

so that $\beta/\alpha + \beta = 1 - a$. Then p_{0e} is expressed in terms of a and c by dividing numerator and denominator by $(\alpha + \beta)^3$ in (13.2.7):

$$p_{0e} = \frac{(1 - a)(1 - a + c)(1 - a + 2c)}{(1 - a)(1 - a + c)(1 - a + 2c) + 3a(1 - a)(1 + 2c) + a(a + c)(a + 2c)}$$

(13.2.8)

Then, after dividing numerator and denominator by $\alpha + \beta$, $(\alpha + \beta)^2$, and $(\alpha + \beta)^3$, respectively, (1)$'$, (2)$'$, and (3)$'$ yield

$$p_{1e} = \frac{3a}{1 - a + 2c} p_{0e}$$

(13.2.9)

$$p_{2e} = \frac{3a(a + c)}{(1 - a + c)(1 - a + 2c)} p_{0e}$$

$$p_{3e} = \frac{a(a + c)(a + 2c)}{(1 - a)(1 - a + c)(1 - a + 2c)} p_{0e}$$

To compute the mean and variance of the equilibrium distribution we proceed as follows:

$$E(Y) = p_{1e} + 2p_{2e} + 3p_{3e}$$

Substituting from (13.2.9), we obtain

$$E(Y) = \left[\frac{3a}{1 - a + 2c} + \frac{6a(a + c)}{(1 - a + c)(1 - a + 2c)} + \frac{3a(a + c)(a + 2c)}{(1 - a)(1 - a + c)(1 - a + 2c)} \right] p_{0e}$$

Algebraic manipulation then yields the simple result, using (13.2.8),

(13.2.10) $$E(Y) = 3a$$

For the variance, we require

$$VAR(Y) = E(Y^2) - [E(Y)]^2$$

where

$$E(Y^2) = p_{1e} + 4p_{2e} + 9p_{3e}$$

$$= \left[\frac{3a}{1 - a + 2c} + \frac{12a(a + c)}{(1 - a + c)(1 - a + 2c)} + \frac{9a(a + c)(a + 2c)}{(1 - a)(1 - a + c)(1 - a + 2c)} \right] p_{0e}$$

Then, using the expression for p_{0e} in (13.2.8), we obtain after algebraic manipulation,

(13.2.11) $$VAR(Y) = \frac{3a(1 - a)(1 + 3c)}{1 + c}$$

Solving (13.2.10) and (13.2.11) for a and c in terms of the theoretical mean and variance and writing $\sigma^2 = VAR(Y)$, we obtain

$$a = \frac{E(Y)}{3}$$

$$c = \frac{\sigma^2 - 3a(1 - a)}{9a(1 - a) - \sigma^2}$$

which provide estimation formulas for a and c by the method of moments (see Chapter 15), by which we estimate $E(Y)$ by the observed mean $\overline{Y}$ and estimate σ^2 by the observed variance s^2:

$$\hat{a} = \frac{\overline{Y}}{3}$$

(13.2.12)

$$\hat{c} = \frac{s^2 - 3a(1 - a)}{9a(1 - a) - s^2}$$

If $\hat{a}$ and $\hat{c}$ are used in (13.2.8) and (13.2.9), we obtain the numerical values of the equilibrium distribution. The distribution has the one constraint of

$$\sum_{i = 0}^{3} p_{ie} = 1$$

and we have estimated two parameters from the observed distribution. Hence, three degrees of freedom are lost; one degree of freedom remains.

Let us recapitulate the logic of the above analysis before moving on to the application phase. Each individual in a network of size 3 is postulated to shift dynamically between two possible mental states in accordance with a continuous-time Markov chain with time-invariant rates. However at any time the rate that applies to a person depends upon the state of the whole network. Passing to a characterization of the dynamics of the state of the network, we are able to describe it as a stochastic process whose state is the number of persons in a given mental state. The trajectory of this process is described or traced out by the time-dependent vector p(t) as time flows continuously.

The equilibrium of the network, as a probabilistic system, is given by a distribution that remains time-invariant if attained. In other words, the process drives the distribution into itself continuously, even though in accordance with the stochastic description of the individuals, the actual individuals in one or another state are fluctuating. The expected value and variance of this equilibrium distribution were derived. Then, in anticipation of application, we derived two estimation formulas. The original model is actually a family of models of the abstract character,

$$\mathfrak{M}\,[p(0),\,\alpha,\,\beta,\,\gamma]$$

but with $a = \alpha/\alpha + \beta$, $c = \gamma/\alpha + \beta$, we have $b = \beta/\alpha + \beta = 1 - a$, so that there are only two parameters:

$$\mathfrak{M}\,[p(0),\,a,\,c]$$

Assuming that the equilibrium is stable, when we return to the data interpreted as equilibrium data, we have a model with two unknown parameters, since there is independence from the initial condition,

$$\mathfrak{M}(a,\,c)$$

Coleman applied this model—constructed in a general form for n-node rather than three-node networks—to shops with n members in the New York Typographical Union elections (Coleman, 1964:347). The concept of a binary state $(A, \overline{A})$ was identified with the voting intention of a union member for or against a given candidate, and the actual vote was assumed to reflect the intention directly. The data are of the form, based on 129 shops,

Number of individuals in network who voted for candidate A	Proportion of networks with given number voting for candidate A
0	.25
1	.12
2	.23
3	.40

Let us note the auxiliary assumptions being made here:

(1) Each three-man shop independently realizes the network process.
(2) The parameters are the same in each such shop.
(3) Each shop is in an equilibrium with respect to the network process.

Hence, the logical consequence is that each shop is characterized by the distribution p_e, which describes the equilibrium distribution of Y. Hence, each shop is an independent replication from the point of view of the equilibrium probability model. It follows that the data observed have expected values given by the p_{ie} (i = 1, 2, 3, 4).

Using the mean and variance of the observed distribution and (13.2.12), Coleman computed the estimates

$$\hat{a} = .60, \qquad \hat{c} = 1.09$$

which means that, approximately,

$$\frac{\alpha}{\alpha + \beta} = .60, \qquad \frac{\gamma}{\alpha + \beta} = 1.09$$

so that γ is about 1.8 times as great as α: the internal flow of influence in the network is this much greater than the magnitude of forces creating changes of mind independently of internal group influence. Having estimated parameters, Coleman replaced the unknowns in formulas (13.2.8) and (13.2.9) with $\hat{a}$ and $\hat{c}$ to compute the predicted proportion of networks with a given number of men in state A (i.e., favoring candidate A) in equilibrium. Actually, he did this for other than 3-person groups, and Figure 13.2.8 constitutes the results for networks of size 3, 4, and 5. It may be seen that the fit, by visual inspection, is very good. Coleman includes a statistical discussion of the fit for the overall body of data, concluding that for larger groups the model fails to provide as good a fit. This makes sense in terms of the graph representation of the underlying relational system: the γ influence rates exist uniformly between all pairs

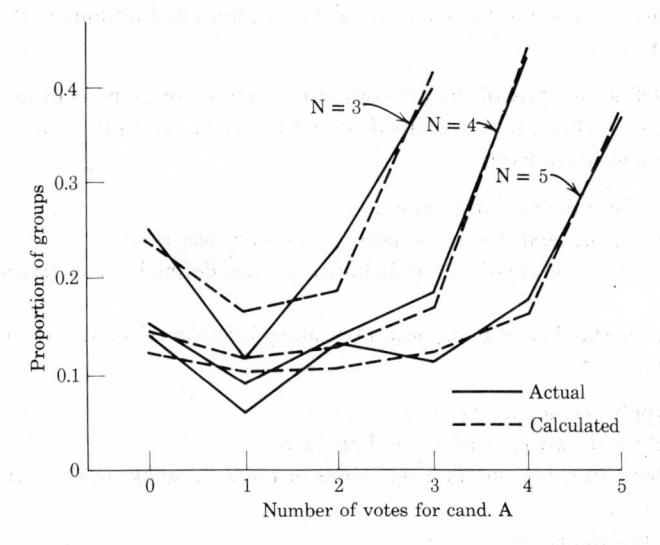

Figure 13.2.8 From Coleman: 1964, Figure 11.5(a), Page 346.

of nodes. It is reasonable to suppose that with larger networks (a) heterogeneity of influence rates is prominent and (b) the relational system is no longer complete in the sense that for some pairs of persons, no direct relationship exists to mediate the influence process. For additional discussion of the problems arising in the application phase and the problems for research they suggest, see Coleman (1964, pp. 347–353). For a review of the entire array of stochastic processes of the continuous-time type developed and applied by Coleman, see Jaeckel (1971).

13.3. Summary. We may summarize the basic logic of Coleman's process methodology as applied to networks of social influence by the list of steps below. Note that the sequence of steps begins with setting-up steps, passes to analysis, and finally to application. Also note, however, that the analysis phase of this sequence is ordinarily so complex that attention may be concentrated on the equilibrium solution. Also, general theorems on Markov processes may be useful in various steps (see, for instance, Karlin, 1966).

Step 1. Postulate some individual process involving two* possible states.

Step 2. Embed n such individuals in a complete* network with uniform* weights or "values" of the links, γ.

*Generalization of the method would be natural by modification of this restriction, yet obtaining the present results as a special case.

Step 3. Postulate that the individual processes are now modified to include network-induced transition behavior, using the weights γ as additions to the transition rates.

Step 4. Define the state of the network as the number of individuals in one of the two states. Then use the results of step 3 to derive a transition rate description at the network level.

Step 5. Analyze the network model:
(a)   determine the behavior of the network process over time,
(b)   find the equilibrium of the probabilistic system defined at the network level,
(c)   check the stability of equilibrium by taking limits in (a) to approach the terms in (b).

Step 6. Apply the network model:
(a)   identify K networks with n members each,
(b)   use observations to obtain proportions of the K networks in the various network states,
(c)   estimate parameters,
(d)   check goodness-of-fit of the predicted distributions with the data in (b).

CHAPTER FOURTEEN MARKOV SOCIAL PROCESSES: AN EXPECTATION-STATES MODEL

14.1. Introduction. In this part of our introduction to several dynamic probabilistic models, we set up and analyze a model proposed by a group of sociologists at Stanford (see, particularly, Berger et al., 1966, Chaps. 2 and 3). A typical problem that they have treated is as follows. Consider small-group situations in which persons bring external statuses to the situation. These external status characteristics frequently differentiate the individuals by what might be a quite irrelevant criterion so far as the task of the group is concerned. Nevertheless, observable power and prestige in the group may reflect these ostensibly irrelevant, externally based statuses. A body of literature exists that has shown some of the effects of prior statuses in a variety of concrete settings (for instance, see the jury studies by Strodtbeck, 1956, 1957). The "Stanford methodology" involves as an essential step the framing of an abstract empirical generalization based on these findings, followed by (a) a theoretical picture or model of the dynamics of the group leading to the observed results and (b) experimental studies to test the time path as well as the equilibrium predictions of the model.

The abstract empirical generalization in this case is framed as follows:

> When task groups are differentiated with respect to some status characteristic external to the task situation, this differentiation determines the observable power and prestige order within the group, whether or not the external characteristics are related to the group task (Berger et al., 1966, p. 31).

Note that implicitly there is "universal quantification" in the logical sense of section 2.12: for any task group, for any external status characteristic, the generalization holds. This statement goes well beyond what has been observed, of course, but so does every scientific generalization. The next step in this methodology involves construction of a theoretical model to deal with the dynamics of status and power, to explain this generalization.

Actually the logic is as follows: bodies of empirical data prompt descriptive empirical generalizations, the generalizations prompt the statement of the one abstract empirical generalization, and it in turn prompts a theoretical model to explain it; the model is detailed enough to return to the data with the aim of reproducing the data that prompt the generalization. In practice, new studies are conducted with new data, but of the same generic content, so that we have the following sequence of stages:

Stage 1

Substantive hypothesis$_1$ $\longrightarrow$ Empirical study$_1$ $\longrightarrow$ Data$_1$ $\longrightarrow$ Generalization$_1$

Substantive hypothesis$_2$ $\longrightarrow$ Empirical study$_2$ $\longrightarrow$ Data$_2$ $\longrightarrow$ Generalization$_2$

$\cdot \qquad \cdot \qquad \cdot \qquad \cdot$

Stage 2

$\left.\begin{array}{l} \text{Generalization}_1 \\ \text{Generalization}_2 \\ \quad \cdot \\ \quad \cdot \\ \quad \cdot \end{array}\right\} \longrightarrow$ Abstract empirical generalization

Stage 3

Abstract empirical generalization $\longrightarrow$ Theoretical model

Stage 4

Theoretical model

$\qquad\qquad\qquad\qquad$ Predictions

Empirical study$_T$ $\longrightarrow$ Data$_T$

For instance, "generalization$_1$" of stage 1 might refer to the influence of sex and job on jury decisions, while, as we have seen, the abstract generalization of stage 2 eliminates this reference to instances of the universality claimed for the phenomenon. In stage 3 this generalization leads to a theoretical model and in stage 4 to a fresh body of experimental data especially collected to conform to the type of data needed for a sharp test of the model. In this fourth stage, "empirical study$_T$" differs from the typical earlier "empirical study$_i$" (i = 1, 2, . . .) that led to one of the lower-level descriptive generalizations. Primarily the function of "empirical study$_T$" is to serve as a device for focusing on the question of the truth-value to be assigned to various theoretical premises in an argument leading to a prediction. Hence, it is central that theoretically non-crucial premises not be in doubt—for instance, that certain definite initial conditions be known to exist. We shall say more about this later.

In stage 4, then, if the predictions are accurate, we may say that the basic kind of observations that led to the abstract empirical generalization have been explained. This means that the generalization itself is explained. (Actually, the scope of explanation in the present case is restricted by conditions limiting the generality of the theoretical model; see section 14.3.) One important thing to see is that one does not envision a mere verbal exercise in logical deduction of the generalization from other similar generalizations: one generates the form of the processes whose observation led to the generalization.

Implicit is the notion that were the data for the original studies in proper form—in terms, say, of time series; then, with suitable tailoring of the basic theoretical model, these data could also be derived. Note that with the passage to universality of formulation, although stage 4 involves data it need not involve the same concrete content as in stage 1 for any concrete content that instantiates the abstract concepts has the potential to show the model is wrong. A main reason for experimental analysis at stage 4 is to allow the collection of extensive data over time, to subject the dynamics proposed by the model to a severe test.

In outlining one example of this methodology, we focus on the stages involving the theoretical model. Essentially, this is an application of ideas from "expectation-states theory," developed by Joseph Berger and his colleagues at Stanford and elsewhere, using one basic principle: balance. We introduce balance theory first and then return to the expectations-states formulation involving its sociological application.

14.2. Balance Theory. The basic ideas are Heider's (1958). Several primitive notions are involved:

   (1) "Entities," which are of two types,
     (a) persons,
     (b) nonpersons;
   (2) "unit-formation," which is an operation performed by a person upon two entities so as to construct a third entity, said to form a "unit" out of the two

original entities (like addition, once we say it applies to two entities, unit-formation applies to three or more entities by successive unit-formations);

Example. (a) ● ●   ● ●   ● ●   (6 entities, 3 units formed)

(3) "sentiment," which relates a person to an entity and has a sign, positive or negative.

Example. (b) Let

$$p = \text{a person}$$
$$o = \text{another person}$$

Then

$$p \xrightarrow{+} o$$

represents a positive sentiment from p to o.

Heider's theory is "p-centric," meaning that one person is chosen, called p, and all units and sentiments are outcomes of operations and feelings of p. In particular, the diagram in Figure 14.2.1 means: p likes o, p likes x, and p perceives that o likes x. Also, we write p′ to mean p as an object to himself. Thus,

$$p \xrightarrow{+} p'$$

means that p has a positive self-evaluation.

From a mathematical point of view, we have

E	an arbitrary (abstract) set (the entities)
P	a specified subset of E (the persons)
p	a specified member of P (the central person for p-centric analysis)
∪	a binary operation with domain pairs of members of E (the unit formations by p)
S^+	a subset of P × E (positive sentiment relation)
S^-	a subset of P × E (negative sentiment relation)

satisfying the meaning postulates:

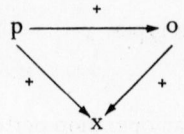

Figure 14.2.1

Axiom 1. P is nonempty.

Axiom 2. ∪ is commutative: x ∪ y = y ∪ x

Axiom 3. $S^+ \cap S^- = \phi$

We define

$$S = S^+ \cup S^-$$

so that

xSy   iff   x has a sentiment toward y   (in p-centric analysis)

where, in this relation, x is a person and y is any entity.

A Heider triad may be defined as a triple (p, x, y) where x and y are entities in E and where both pSx and pSy hold. We explicate Heider's principles in the context in which, in addition, x and y form a unit for p. Formally, x and y are in the domain of the unit-forming operation; hence, x ∪ y is an entity in E since ∪ combines two entities into another entity. There are four such triads, as shown in Figure 14.2.1. Since x ∪ y is an entity, the question arises as to whether (p, x ∪ y) is in S: Is there a unique sentiment toward the unit? And if so, what is its sign? In the case where x and y have positive signs for p, we clearly can postulate a positive sign for the unit; where x and y both have negative signs, we can postulate a negative sign for the unit. But the mixed cases yield the intuition that p will feel more or less discomfort in being unable to induce a unique sentiment relationship to the unit. This leads to Heider's definition:

(14.2.1) Definition. A Heider triad (p, x, y) in which x and y form a unit is balanced if and only if both x and y have the same sign.

Then triads (1) and (4) are balanced, (2) and (3) are not.

The basic principles of Heider's theory are as follows:

Principle 1. A balanced state is an equilibrium state.

Principle 2. If a triad is not balanced, then pressures toward a balanced state arise.

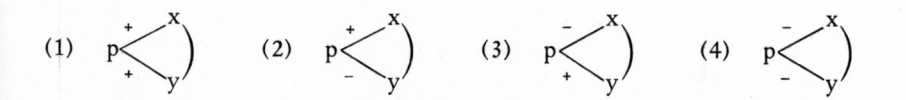

Figure 14.2.1

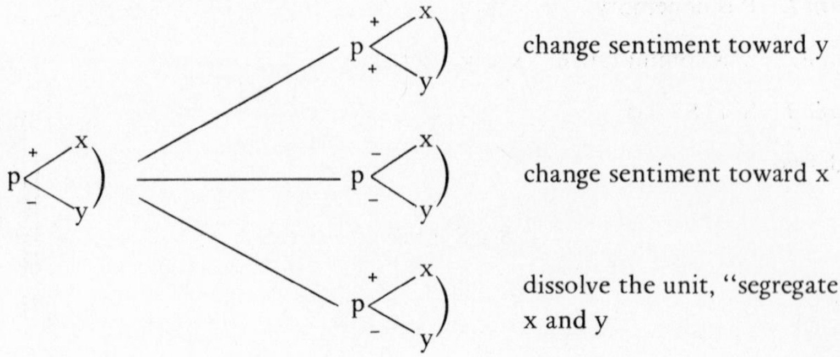

change sentiment toward y

change sentiment toward x

dissolve the unit, "segregate" x and y

Figure 14.2.2

Principle 3. If a triad is not balanced and if no change is possible under the given conditions, then tension arises.

For instance, from the initial condition that triad (2) of Figure 14.2.1 obtains, we can say that, by the definition and principle 2, assuming change is possible, that the possible changes are shown in Figure 14.2.2, assuming the sentiments might change. (This also assumes the in-the small analysis in which only one relationship may change.) To allow the third of these branches under principle 2, we would want to extend the definition of balance to allow this situation, in which the sign incompatibilities no longer present a problem, to be balanced.

This can be done if we first agree to put a "+" sign on the graph to represent unit formation, as a formal device for generalizing the definition. Then the triads are shown in Figure 14.2.3. Define a cycle in a graph to be a path of lines that returns to its starting node with no line or node repeated. Thus, each of (1) through (4) contains one* cycle, but the graph in the lowest branch of Figure 14.2.2, obtained by a dissolution of a unit, has no cycle.

Note that in Figure 14.2.3 a triad is balanced if and only if the product of the signs of the lines in its cycle is positive. Taking this as the clue to generalization, we introduce the following

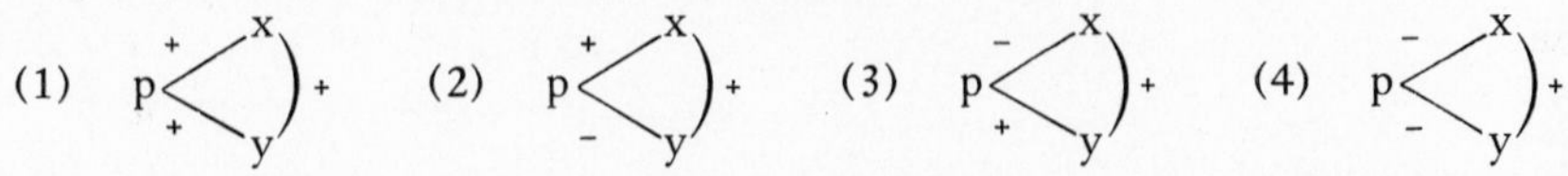

Figure 14.2.3

*The cycle p-x-y-p may be regarded as the same cycle as x-y-p-x, and so forth.

(14.2.2) Definition. We say that the cycle condition is satisfied if and only if each cycle of a graph has a positive product.

Hence, a Heider triad is balanced if and only if the cycle condition is satisfied: this merely describes Figure 14.2.3. The basic idea of using graphs and the cycle condition to formalize balance theory comes from Cartwright and Harary (1956). Now consider the case where x and y do not form a unit for p. There is no cycle in the graph representation. But then the cycle condition is still satisfied because it means "for any path of lines, if the path is a cycle, then it has a positive product." But the hypothesis "the path is a cycle" is false in this case, no matter which path we consider, and so, the conditional statement is true. Hence, formally this triad may be considered balanced. We call this vacuous balance.

A time-varying triad (p, x, y) has five possible states: vacuous balance and the four configurations of Figure 14.2.3. In the matrix below, these are labeled $S_1, S_2, \ldots, S_5$ as indicated. We give it a Markov representation in order to show how the principles of balance are reflected in a dynamic process model. According to the first principle, a balanced state is an equilibrium state: if the system is in such a state, it stays there. This means the chain is absorbing, with the possible balanced triads as the absorbing states:

	$p\!<^{x}_{y}$	$p\!<^{+\,x}_{+\,y}$ +	$p\!<^{-\,x}_{-\,y}$ +	$p\!<^{+\,x}_{-\,y}$ +	$p\!<^{-\,x}_{+\,y}$ +
$S_1: \; p\!<^{x}_{y}$	1	0	0	0	0
$S_2: \; p\!<^{+\,x}_{+\,y}$ +	0	1	0	0	0
$S_3: \; p\!<^{-\,x}_{-\,y}$ +	0	0	1	0	0
$S_4: \; p\!<^{+\,x}_{-\,y}$ +	p_{41}	p_{42}	p_{43}	p_{44}	0
$S_5: \; p\!<^{-\,x}_{+\,y}$ +	p_{51}	p_{52}	p_{53}	0	p_{55}

Here the notation

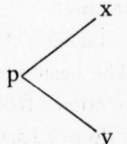

means the class of all vacuously balanced triads,

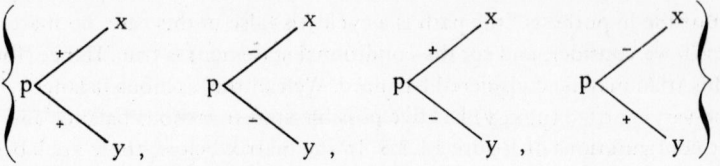

Assume that for any p_{ij}, we have: $0 < p_{ij} < 1$.

A few remarks about the equilibrium character of a balanced state are necessary. From the standpoint of general process language, the S_i states exhibit change over time because of the intrinsic character of the process and because of inputs. In writing down a Markov matrix in this manner, we are implicitly treating a theoretically isolated system. That is, we are asking for the trajectory, the equilibrium states, and the like, which are intrinsic to the process of balance. If we consider inputs, these can be treated as disturbances or "shocks" that shift the system state. Such a perturbation may be due to the simple fact that the given triad is part of a larger system. For instance, suppose the system state at a given time is S_2. This is an equilibrium state. But suppose that entity x forms a unit with another entity, say z, and that, for whatever reason, p develops a negative sentiment toward z. We have at this time:

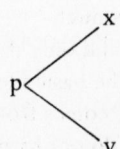

The imbalance in the triad (p, z, x) may be resolved by changing the sentiment to x from positive to negative, thereby imbalancing the triad (p, x, y). This puts the system under analysis in state S_5, and the balancing process now yields a transformation into some state S_i in accordance with the probabilities in the final row of the matrix. The general point being made is that the concept of balance should not be thought of in purely static structural terms. In systems within larger networks of elements, the incessant shifting of signs induces continuous state-to-state transformations quite compatible with the idea that for a theoretically isolated triad there exist equilibrium states and these are the balanced states.

The second principle of balance receives its interpretation in terms of the probabilities of shifting to a balanced state from the imbalanced states. In particular, in this representation an analysis of the chain by the fundamental matrix method of section 10.12 yields the result that matrix $\mathbf{B}$ of absorption probabilities gives, for each possible balanced state, the chance that the system "resolves" an initially imbalanced condition that particular way. Also, this yields the result that no balanced state in itself is a stable equilibrium: for example, if the system is in balanced state S_2, then a perturbation to S_4 may resolve itself in states S_1 or S_3, rather than in S_2. On the other hand, the entire set of balanced states forms a stable set because the matrix $\mathbf{B}$ has rows that sum to unity: with probability 1, the system eventually becomes a balanced triad.

The third principle of balance in regard to tension may be interpreted as a cumulation function that increases with the number of times the system stays in an imbalanced state. We may interpret p_{44} and p_{55} as measures of the difficulty of change. (For a somewhat different treatment of a probabilistic dynamic system dealing with balance, see Flament, 1963.)

Given the Heider psychological background, one can see the origins of the mathematical concept of balance defined (by Cartwright and Harary, 1956) as follows: Let G be a graph with or without directed lines and with each line having a sign (+ or −). If G has undirected lines, define a cycle to be any path of lines returning to its starting point without using the same line or point more than once. If G has directed lines, neglect the direction in determining possible paths satisfying the definition of cycle. Then define G to be balanced if and only if the cycle condition of definition (14.2.2) is satisfied. Note that this now includes longer cycles.

Examples.

(c)

One cycle: product positive

(d)

Three cycles: (1) $a \overset{-}{=} b \overset{+}{-} c \overset{-}{=} d \overset{+}{-} a$: positive
(2) $a \overset{-}{=} b \overset{+}{-} d \overset{+}{-} a$:   negative
(3) $b \overset{+}{-} c \overset{-}{=} d \overset{+}{-} b$:   negative

Here (c) is balanced but (d) is not, since it contains two negative cycles—hence, at least one such cycle, which negates the cycle condition.

The following theorem proved by Cartwright and Harary is a statement of the graph theory of balance that aids in determining balance in complex graphs:

(14.2.3) Structure Theorem. A graph is balanced if and only if the set of all its points can be partitioned into two disjoint sets (one of which may be empty)

so that every positive line connects two points in the same set and every negative line connects two points from different sets.

In example (c) above, the structure theorem is exemplified by the following:

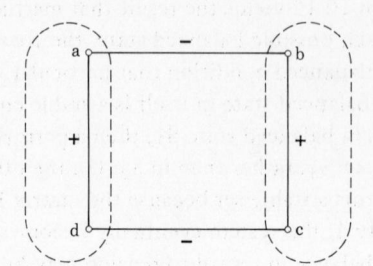

In example (e) the two sets are less easily seen.

Example. (e)

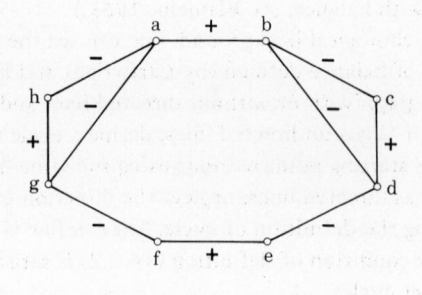

The two sets are

$$A = \{a, b, f, e\}, \quad \overline{A} = \{c, d, h, g\}$$

as a rearrangement makes clear:

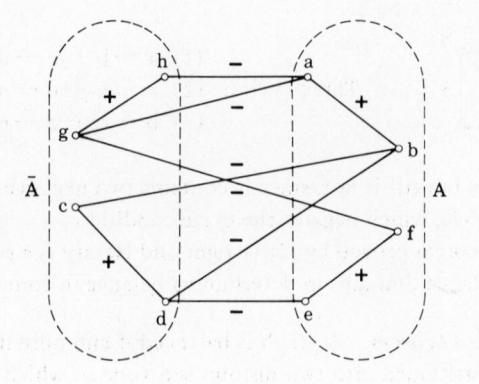

Note that A is not a "clique" in the sense that there are points not positively connected in A; in fact, there are two disconnected subsets of points in A: {a, b} and {e, f}. The same remarks hold for $\bar{A}$. Thus, the structure theorem should not be "overinterpreted" as saying that a system is balanced only when two opposing cliques, or "alliances," exist. (If it is assumed, however, that a line connects every pair of points, then the clique interpretation makes sense.)

14.3. Balance and Expectations: Framework and Setup of the Theoretical Model. We return now to the treatment of the problem of explanation of section 14.1. At this point, we can set down the basic outline of both the experimental design and the theoretical model but, for our purposes, will put central emphasis on the model. Keeping in mind the methodological remarks made earlier, the reader may see that we enter stages 3 and 4. In detail, in these phases one envisions a pair of sequences

$$s_0, s_1, s_2, \quad \ldots$$
$$\mathfrak{M}_0, \mathfrak{M}_1, \mathfrak{M}_2, \quad \ldots$$

where s_0 is the initial (simplest) experimental situation in which the first (simplest) model $\mathfrak{M}_0$ will be tested. Assuming a favorable evaluation of $\mathfrak{M}_0$ in s_0, then $\mathfrak{M}_1$ generalizes $\mathfrak{M}_0$, and s_1 is a situation constructed to test $\mathfrak{M}_1$. Assuming a favorable evaluation of $\mathfrak{M}_1, \ldots,$ and so forth. If $\mathfrak{M}_i$ is not favorably evaluated, then it is revised and a new version of s_i is used to test it. This process of model extension might, for example, occur by first taking two-person interactions, then three; first restricting the model to one external characteristic, then two, and so forth.

To get some idea of how $\mathfrak{M}_0$ compares with the range of possibilities requiring theoretical analysis requires some preliminary framework. This we may briefly give by listing a collection of some of the primitive ideas drawn from Berger et al. (1966, Chap. 2). These components of the framework are given in Table 14.3.1. The vocabulary is essentially that of Berger and colleagues, but the particular symbols are altered somewhat, as in Fararo (1972b). We shall not use much of this complex framework apparatus here; its main function is that of allowing us to reveal quite starkly the limited nature of the model that is later to be constructed. However, it should be emphasized that the framework has the capability for deployment in other explanatory contexts (for instance, see Zelditch et al., 1966).

The intuitive meaning of these components must be noted here. The actors in A are the human individuals or groups who possess attributes such as the status characteristics D_i. Such a characteristic (e.g., sex, race, or job) has several levels or states; for instance, "male" is a state of the sex characteristic. What makes these attributes status characteristics is the existence of differential evaluations of the various states, such evaluations held by the particular actors. (See sections 5.7-5.10.) Each characteristic D_i has its own ordering by each actor (so

Table 14.3.1   Components of the Expectation-States Framework

Name of Concept	Symbol	Mathematical Object	Remarks
(1) Actors	A	set	α in A is an actor.
(2) Diffuse status characteristics	D_i	n sets $(i = 1, 2, \ldots, n)$	x in D_i is a state.
(3) Differential evaluations	$>_\alpha^i$	order relation on D_i indexed in A $(i = 1, 2, \ldots, n)$	For every α, $x >_\alpha^i y$ iff α regards x as above y $(x, y \in D_i)$
(4) General expectation states	G_i	n sets	G_i is in 1–1 correspondence with D_i $(i = 1, \ldots, n)$.
(5) Characterizations	d_α^i	n mappings: $A \xrightarrow{d_\alpha^i} D_i$ indexed in A $(i = 1, 2, \ldots, n)$	d_α^i is one way α sees members of A; $d_\alpha^i(\beta)$ is the state (of D_i) assigned to β by α.
(6) Task outcomes	O	set	O_k in O is an outcome.
(7) Outcome preference orderings	$\blacktriangleright_\alpha$	order relations on O indexed in A	For every α in A, $O_1 \blacktriangleright_\alpha O_2$ iff α prefers outcome O_1 to outcome O_2.
(8) Specific performance characteristics	C_j	m sets $(j = 1, 2, \ldots, m)$	z in C_j is a performance capacity.
(9) Expectation mappings	e_{ij}	nm mappings $D_i \xrightarrow{e_{ij}} C_j$ $(i = 1, 2, \ldots, n;$ $j = 1, 2, \ldots, m)$	$e_{ij}(x)$ is the performance level expected of an actor assigned x.
(10) Specific performance expectations	c_α^j	m mappings $A \xrightarrow{c_\alpha^j} C_j$ indexed in A $(j = 1, 2, \ldots m)$	$c_\alpha^j(\beta)$ is the expectation of a specific performance capacity for β by α on "dimension" C_j.
(11) Instrumentality	o_j	m mappings $C_j \xrightarrow{o_j} O$ $(j = 1, 2, \ldots, m)$	$o_j(z)$ is the task outcome most likely when the capacity level is z in C_j.
(12) Time	T	ordered set	$t \in T$ is a "trial."

$>_\alpha^i$ depends on α as well as i). A general expectation is a "halo effect" associated with a status level. We may think of its arising when people hold implicit rules in their minds of the form, for instance, "the better the occupation, the better . . . ," where ". . ." represents the generality of the rule. It is "applied" in specific situations when an actor characterizes the actors in the situation in terms of one or more status characteristics. The characterizations are operations performed by actors, so that in general they need not be in agreement; hence,

d_α^i depends on α as well as i. In task-oriented situations, there are possible out-comes for which actors hold preference orderings, not necessarily in agreement (so that $\succ_\alpha$ depends on α). In such situations, specific performance character-istics (skills, abilities) may be instrumental to the realization of certain outcomes. Such skills and abilities might (or might not) be such that the people in the situ-ation expect them of people they have characterized in status terms. For instance, a person characterized as "female" may be expected to perform at a certain level in sewing, cooking, and other household tasks not as a matter of direct know-ledge of that person's skills but simply as part of an expectation mapping linking the status to the performance capacity. Finally, all these phenomena occur over time. The theory deals with time as a series of events that occur in order; it is not concerned with the "waiting time" between events. Experimentally, the assumption is made that nothing of importance for the phenomena under study occurs between those events, called trials, that were created to study the pheno-mena in detail.

The "expectation-states framework" is given by these components together with one fundamental principle that we state in inexact working form as follows:

(14.3.1) *Basic Principle of the Expectation-States Framework.* Expectations satisfy balance principles.

A comparable statement in physical science is, Light travels in straight lines. It is a statement that helps to define the way in which a class of phenomena are to be investigated. Our next step is to set up one particular model based on this framework, in which we follow the logic of treating a simple class of situations in the earliest stage of working within this framework.*

The class of situations within the possible scope of the theoretical model to be constructed—and from which experimental situation s_0 is to be chosen—satisfies the following conditions as to these components:

  (1) A has two members, denoted p and o.

  (2) There is only one diffuse status characteristic, say D, and it has two states.

  (3) The actors p and o have uniform differential evaluations over D: $>_p = >_o$.

  (4) The two general expectation states are denoted $+$ and $-$.

  (5) The characterizations are uniform: $d_p = d_o$ (and we write d) and provide a basis for discrimination of actors: $d(p) \neq d(o)$.

  (6)-(7) Outcome preferences are uniform: $\succ_p = \succ_o$.

  (8) There is only one specific performance characteristic, say C, and it has two states.

---

*From other perspectives, these situations may be far from simple. The term "simple" refers to the background spectrum of possibilities given by Table 14.3.1 and the way in which these are restricted.

(9) The expectation mapping is not defined: there are no prior expectations linking states of D to this particular performance capacity.

(10) The levels of C are differentially instrumental to outcomes.

(11) There is a sequence of trials such that all the above boundary conditions are satisfied on each trial.

(12) Initially, at t = 0, specific performance expectations are not defined: actors do not initially know their relative capacities in terms of C.

The model is largely concerned with how the specific performance expectations are constructed and stabilized. If actor p attributes high ability to actor o, then if this is a stable attribution it is not surprising that o dominates the task-oriented interaction and, so, has higher power and prestige.

The actual experimental scene has the following additional properties. The experimenter presents a stimulus situation on each trial and:

(13) On each trial t, there are two stages:
      Stage 1: each actor independently (and without communication) chooses one of two responses,
      Stage 2: each actor is informed of the other's response and may now change his response.

(14) On each trial t, each actor is told in stage 2 that the other disagreed (i.e., gave the alternative response) at stage 1.

(15) Each actor is informed that exactly one of the two possible responses is correct (i.e., leads to the preferred outcome) but is not told which of the two is correct.

(16) Each actor believes that only the second response counts toward the attainment of the preferred outcome.

(17) Each actor is indifferent as to who originates the response leading to the preferred outcome: the situation is "task-oriented."

For example, each actor may be presented with a stimulus pattern to be classified as I or II. Each responds with a first coding of the pattern. The experimenter relays to each actor not the actual response of the other but the opposite of what was used as the coding response. Hence, if the subject says I, then the experimenter informs him other said II. Then the subject may switch to II or remain with I. It is clear from this description that with experimental manipulation to achieve disagreement (condition 14 above), it is not really necessary to have a pair of subjects. From the standpoint of one subject, p, there is another person, o, whose responses he learns indirectly by a communication from the experimenter, but o need not really exist.

Given these conditions, a second response is called a p-response if the subject p does not switch; it is called an o-response if the subject does switch. Clearly Berger and his colleagues intend to interpret the (stabilized) relative frequency of o-responses as the measure of influence on p. Contrasting two aggregates of subjects, one in which p is higher in status than o and the other in which p is lower in status than o, the obvious hypothesis is that the latter group will show the higher measure of influence on p. But the theoretical model aims to do more than recover this hypothesis as a deduction: it aims to reproduce the actual relative frequencies of o-responses both in the transient part of the process (before the frequency stabilizes) and in the stable state of the process. It will do this by constructing a process which generates the observable process of successive p- or o-responses. The generating process deals with states of expectation.

We note that in the framework of Table 14.3.1 three types of expectations exist: (1) general expectations, $D_i \rightarrow G_i$, which are associated with the status levels of various characteristics; (2) expectation mappings, $D_i \rightarrow C_j$, which link the status levels of various status characteristics to specific performance levels; and (3) specific performance expectations, $A \rightarrow C_j$, which predicate or attribute performance levels to specific actors. Under condition 9 above, the second type of expectation does not exist in our present context. Hence expectations are composites of the other two types. Our notation will be as follows:

General expectations: $\{+, -\}$
  + means that p is assigned the higher level of D
  − means that p is assigned the lower level of D.
Specific performance expectations: $\{+, -, 0\}$
  + means that p attributes the higher level of C to himself
  − means that p attributes the lower level of C to himself
  0 means that p has not formed one of the two definite expectations.

Forming the Cartesian product of these two sets of possible expectations, we obtain the state space of the process that will generate the observable process:

$$\{(+, +), (+, -), (+, 0), (-, +), (-, -), (-, 0)\}$$

According to the initial condition 12, the subject begins in state (+, 0) or (−, 0), depending on his D-status. For instance, a freshman informed that the other involved, whose responses he will learn on each trial, is a senior begins in state (−, 0). This follows from the various conditions, and if one or more of them is violated, this initial condition may not obtain (for example, if the freshman fails to differentially evaluate "freshman" and "senior" or if he believes that "actually" the other is also a freshman).

Applying the basic principle (14.3.1),

Axiom 1. (Balance) States (+, −), and (−, +) cannot occur.

A simple graph to show the imbalance is shown in Figure 14.3.1. This is a Heider triad involving two nonperson entities (L_C, the state of C regarded as low and negatively evaluated: H_D, the state of D regarded as high and positively evaluated). The unit-formation of L_C and H_D occurs because p characterizes himself in terms of these two attributes: they are "together." The triad is not balanced. As we saw in section 14.2, it is possible to represent imbalanced states as transient states of a process. Instead, Berger and his colleagues choose to rule them out altogether. The important methodological point, then, is that a basic framework-level principle may be represented in diverse ways and some of these

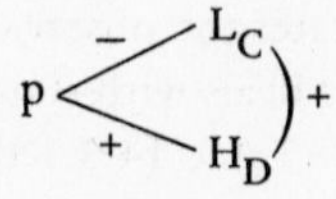

$$p \left\langle \begin{matrix} - \diagup L_C \\ + \diagdown H_D \end{matrix} \right) +$$

Figure 14.3.1

representations may lead to different predictions about observables. Since we are outlining here the basic structure of the model created by Berger and his colleagues, we will work with axiom 1.

Given axiom 1, our state space now reduces to four states, which may be labeled as follows:

$$S_1: \; (+, 0)$$
$$S_2: \; (+, +)$$
$$S_3: \; (-, 0)$$
$$S_4: \; (-, -)$$

The next two axioms deal with the response rule, which relates these unobservable states to observable responses, and with the feedback from response to state, creating a possible change in state.

Axiom 2. (Response rule)   On any trial t, if p is in state S_i, then p makes a p-response with probability α_i ($i = 1, 2, 3, 4$), and

$$\text{(a)} \quad \alpha_1 < \alpha_2$$
$$\text{(b)} \quad \alpha_4 < \alpha_3$$

The two inequalities arise from considering the likely behavioral meaning of a passage from $(+, 0)$ to $(+, +)$ and from $(-, 0)$ to $(-, -)$. For instance, in the latter case, if p is low status and has no specific expectations, then acquires an expectation that other is better at the skill required, then this means a greater susceptability to defer to the other and, so, a lower probability of a p-response (i.e., $\alpha_4 < \alpha_3$).

Axiom 3. (Feedback)   On any trial t, independently of events prior to t,

(a) if p is in state S_1 and makes an o-response, then p stays in S_1,
(b) if p is in state S_1 and makes a p-response, then p moves to S_2 with probability r and stays in S_1 with probability $1 - r$,
(c) if p is in state S_3 and makes a p-response, then p stays in S_3,

(d) if p is in state S_3 and makes an o-response, then p moves to S_4 with probability d and stays in S_3 with probability $1 - d$,

(e) if p is in state S_2, p remains in S_2,

(f) if p is in state S_4, p remains in S_4.

The phrase "independently of events prior to t" is used in this axiom to indicate that a Markov assumption is being made.

The theoretical justification for the various parts of this axiom stems from repeated application of the basic framework principle (14.3.1). We draw signed graphs to represent the situation, then "induce" a result conformable with a balanced graph.

Specifically, there are various entities related to each other and to person p via unit-formation and evaluations. As in the earlier discussion, let p' be person as an object to himself, let o be the other person, let R be a response, and let H_C be the high state of C. Assume that p is characterized by H_D, the high state of the diffuse status characteristic.

To illustrate the use of the balance concept, suppose that at the beginning of a specified trial, p is in state S_1 —namely, $(+, 0)$—where he has not yet formed a specific performance expectation. From this condition we know that p is embedded in a system of the form in Figure 14.3.2. Note that p characterizes himself (p') as high on D but does not characterize himself as high on C (no line from p' to H_C). Following the convention of the earlier treatment, we represent both unit-formations (characterizations) and sentiments (evaluations) by signed lines of the same type.

A series of steps may now be envisioned that induce the self-characterization $p' \xrightarrow{+} H_C$, as follows:

Step 1. p makes a first response, say R, and learns that o does not make this response: Figure 14.3.3.

Step 2. With probability α_1, p makes a p-response (i.e., retains R as his second response): Figure 14.3.4.

Step 3. p completes the triads (p, p', R) and (p, o, R) in a balanced manner: Figure 14.3.5.

Step 4. With probability r, p completes the triad (p, p', H_C) in a balanced manner, so that $p' \xrightarrow{+} H_C$; and completes the triad (p, o, H_C) in a balanced manner, so that $o \xrightarrow{-} H_C$ (other is not characterized by H_C). Only the relevant parts of the graph are shown: Figure 14.3.6.

Hence, at the start of trial $t + 1$, the expectation state of p is $(+, +)$, that is, S_2. The two events required to do this have compound probability $\alpha_1 r$.

To justify part (a) of axiom 3, note that if at step 2 p makes an o-response, with probability $1 - \alpha_1$, then we have the sequence in Figure 14.3.7. If p were now to

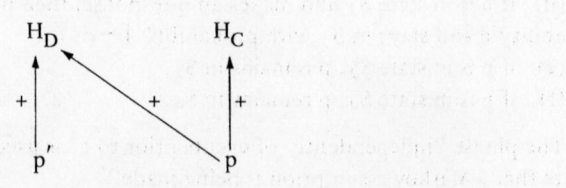

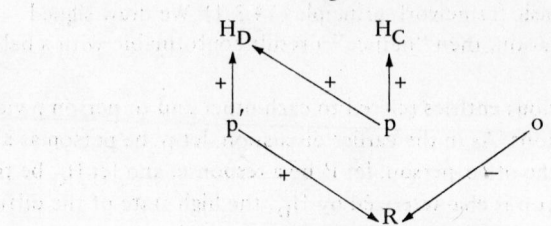

Figure 14.3.2

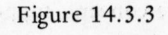

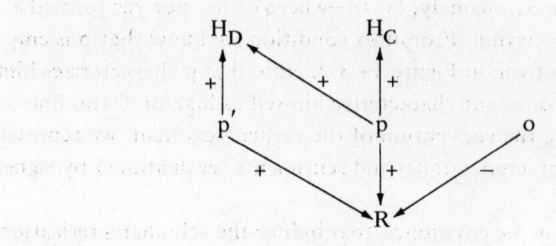

Figure 14.3.3

Figure 14.3.4

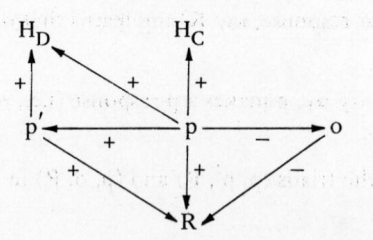

Figure 14.3.5

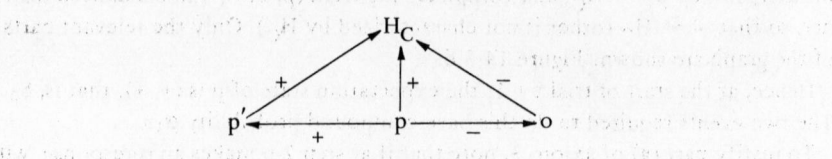

Figure 14.3.6

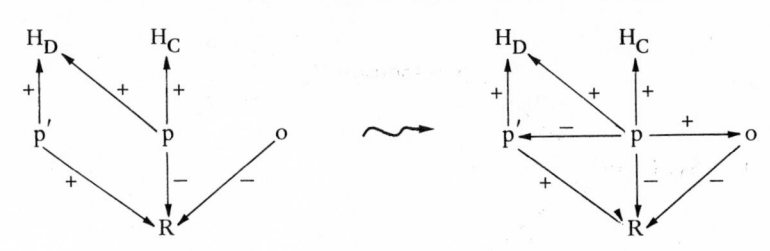

Figure 14.3.7

complete the triad (p, p$'$, H$_C$) to make it balanced, the line would be p$' \overset{-}{\to}$ H$_C$, and so, the new state would be (+, −) in which p cognizes himself (p$'$) as generally high (H$_D$) but specifically low (not H$_C$). This violates axiom 1. Hence, if p makes an o-response, it is assumed that no specific expectation is formed: this in the interests of preserving the "governing" property of the general expectation in the situation and of consistency with axiom 1. It might be noticed that in step 3 the triad completion is deterministic rather than, as in step 4, probabilistic. This assumption is specifically stated by the authors: the self-other evaluations are surely formed, but the performance characterizations only "tend" to be formed (in a balanced manner).

The reader should consult Berger et al. (1966, Chap. 3) for a more detailed justification of this axiom. Clearly, (e) and (f) apply the basic principle (14.3.1) and translate "balanced state" into "absorbing state," as in our treatment in section 14.2.

We note one variant of the process described in steps 1-4: in step 1 it is already possible that the subject p completes triad (p, p$'$, H$_D$) in a balanced manner. This induces p $\overset{+}{\to}$ p$'$. Then the triad (p, p$'$, H$_C$) is ready for completion, which induces p$' \overset{+}{\to}$ H$_C$. Hence, in this variant of the completion process, the actor is self-characterized as high on the specific performance characteristic without any response evaluations at all. This is noted in Berger et al. (1966, p. 59, Assumption 3.6*), in which they mention that a tendency in this direction (if it occurs) can be detected by comparing estimates of α_1 and α_3. Under this process $\alpha_1 > \alpha_3$, whereas in the original formulations $\alpha_1 = \alpha_3$ seems likely.

From the axioms, we may construct tree diagrams. For example, let the subject p be in state (+, 0) on trial t. Then, we have the diagram of Figure 14.3.8. Hence,

$$P(S_1 \text{ at } t + 1 | S_1 \text{ at } t) = 1 - \alpha_1 r$$

In this way, one obtains the one-step transition matrix,

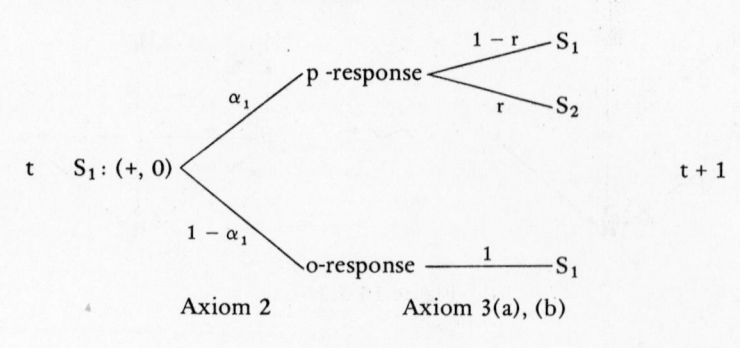

$$
\begin{array}{ccc}
 & & 1-r \quad S_1 \\
 & p\text{-response} & \\
\alpha_1 & & r \quad S_2 \\
t \quad S_1: (+, 0) & & \\
1-\alpha_1 & & \\
 & o\text{-response} & \underline{\quad 1 \quad} S_1 \\
\text{Axiom 2} & & \text{Axiom 3(a), (b)}
\end{array}
$$

t + 1

Figure 14.3.8

$$
t+1
$$

$$
\begin{array}{c}
 & S_1 & S_2 & S_3 & S_4 \\
\begin{array}{c} S_1 \\ S_2 \\ t \quad S_3 \\ S_4 \end{array} &
\left(\begin{array}{cccc}
1-\alpha_1 r & \alpha_1 r & 0 & 0 \\
0 & 1 & 0 & 0 \\
0 & 0 & 1-\bar{\alpha}_3 d & \bar{\alpha}_3 d \\
0 & 0 & 0 & 1
\end{array}\right)
\end{array}
$$

where $\bar{\alpha}_3 = 1 - \alpha_3$. Clearly, there are two distinct processes here, not one: the $S_1 - S_2$ process occurs if the subject p is high-status, while the $S_3 - S_4$ process occurs if the subject is low-status.

Relativizing to the high-status subject, then, the generating process is given by a two-state absorbing Markov chain with unobservable states, and there is an associated rule linking these states to observables:

$$
t+1
$$

(14.3.2)

$$
\begin{array}{c}
 & S_1 & S_2 & \quad P(p\text{-response} \,|\, S_i) \\
\begin{array}{c} t \quad S_1 \\ S_2 \end{array} &
\left(\begin{array}{cc}
1-\alpha_1 r & \alpha_1 r \\
0 & 1
\end{array}\right) &
\begin{array}{c} \alpha_1 \\ \alpha_2 \end{array}
\end{array}
$$

The initial condition is $P(S_1 \text{ at } 0) = 1$.

This concludes the setup of the model. The next phase involves analysis, including concern with estimation: if the S_i are not observable, will it be possible to estimate the parameters and thereby test the model? As we have seen in section 10.7, the presence of an unobservable state-space does not necessarily preclude estimation, but it often does make the analysis leading to estimation formulas more complex than it might otherwise be.

14.4. Analysis and Estimation. The first task of the analysis follows the logic of deriving the trajectory, the behavior of the system over time, from the setup in the small, which in this instance is given by the transition matrix and response rules of (14.3.2).

By raising the matrix of (14.3.2), call it **P**, to higher powers one obtains the conjecture,

$$(14.4.1) \qquad \mathbf{P}^t = \begin{pmatrix} (1 - \alpha_1 r)^t & 1 - (1 - \alpha_1 r)^t \\ 0 & 1 \end{pmatrix} \qquad (t = 1, 2, \ldots)$$

The verification is by induction. For $t = 1$, matrix (14.4.1) is the given one-step matrix (14.3.2). Letting (14.4.1) be the induction hypothesis for t,

$$\mathbf{P}^{t+1} = \mathbf{P}^t \mathbf{P} = \begin{pmatrix} (1 - \alpha_1 r)^t & 1 - (1 - \alpha_1 r)^t \\ 0 & 1 \end{pmatrix} \begin{pmatrix} 1 - \alpha_1 r & \alpha_1 r \\ 0 & 1 \end{pmatrix}$$

$$= \begin{pmatrix} (1 - \alpha_1 r)^{t+1} & 1 - (1 - \alpha_1 r)^{t+1} \\ 0 & 1 \end{pmatrix}$$

Since the initial condition is that the process starts in state S_1 with probability 1, this means that (14.4.1) yields

$$(14.4.2) \qquad P(S_1 \text{ at } t) = (1 - \alpha_1 r)^t \qquad (t = 0, 1, 2, \ldots)$$

To obtain the probability of the observable responses at time t, we first introduce the random variable

$$X_t = \begin{cases} 1 \text{ if p-response at } t \\ 0 \text{ if o-response at } t \end{cases}$$

Hence, $P(X_t = 1 \mid S_i) = \alpha_i$ $(i = 1, 2)$.

To determine $P(X_t = 1)$ we conditionalize by the state of the system at t:

$$P(X_t = 1) = P(S_1 \text{ at } t)P(X_t = 1 \mid S_1 \text{ at } t) + P(S_2 \text{ at } t)P(X_t = 1 \mid S_2 \text{ at } t)$$

$$= (1 - \alpha_1 r)^t \alpha_1 + [1 - (1 - \alpha_1 r)^t] \alpha_2$$

From this expression, we obtain the desired result:

(14.4.3) $P(X_t = 1) = \alpha_2 + (\alpha_1 - \alpha_2)(1 - \alpha_1 r)^t$ $(t = 0, 1, \ldots)$

We see from (14.4.3) that three properties hold. First,

$$P(X_0 = 1) = \alpha_1 \qquad \text{(consequence of the initial condition)}$$

Second,

$$P(X_{t+1} = 1) > P(X_t = 1)$$

We show this by first writing,

$$P(X_{t+1} = 1) = \alpha_2 + (\alpha_1 - \alpha_2)(1 - \alpha_1 r)^{t+1}$$
$$\underline{P(X_t = 1)\quad = \alpha_2 + (\alpha_1 - \alpha_2)(1 - \alpha_1 r)^t}$$
(14.4.4) $P(X_{t+1} = 1) - P(X_t = 1) = (\alpha_1 - \alpha_2)[(1 - \alpha_1 r)^{t+1} - (1 - \alpha_1 r)^t]$

But by axiom 2(a),

$$\alpha_1 - \alpha_2 < 0$$

and since $0 < 1 - \alpha_1 r < 1$, assuming that $\alpha_1 \neq 0$, $r \neq 0$,

$$(1 - \alpha_1 r)^{t+1} - (1 - \alpha_1 r)^t < 0$$

whereby the product of the two factors in (14.4.4) is positive; that is,

$$P(X_{t+1} = 1) - P(X_t = 1) > 0$$

Finally, the third property is:

$$\lim_{t \to \infty} P(X_t = 1) = \alpha_2$$

Hence, the function starts at α_1 and increases monotonically to α_2 (see Figure 14.4.1).

In the absorbing state, of course, the p-response probability is α_2. The distribution of the time to absorption can be obtained by thinking in terms of the graph of Figure 14.4.2. We start in S_1. The time to absorption is 1 if we take the path to S_2 immediately; it is 2 if we loop once around S_1 and then go to S_2; it is 3 if we loop twice around S_1 and then go to S_2; and so on. In general, the

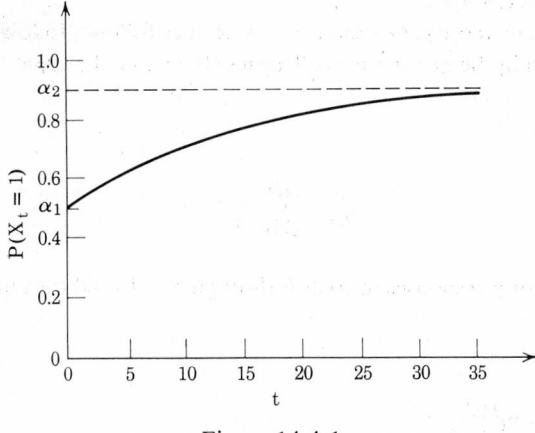

Figure 14.4.1

Figure 14.4.2

time to absorption is k if we loop $k - 1$ times around S_1 and then go to S_2.
Hence, the probability that the time to absorption is k is $(1 - \alpha_1 r)^{k-1} \alpha_1 r$. Thus,
letting T be the time-to-absorption random variable,

$$P(T = k) = (1 - \alpha_1 r)^{k-1} \alpha_1 r \qquad (k = 1, 2, \ldots)$$

This is a geometric distribution whose expected value is given by using the
technique of summing infinite series based on (10.7.14),

(14.4.5)
$$E(T) = \frac{1}{\alpha_1 r}$$

We may deduce this formula even more rapidly if we recall from section 10.12,
that the fundamental matrix $(I - Q)^{-1}$ also provides the expected time to absorption. Here, using (14.3.2),

$$\begin{pmatrix} I & O \\ R & Q \end{pmatrix} = \begin{pmatrix} 1 & 0 \\ \hline \alpha_1 r & 1 - \alpha_1 r \end{pmatrix}$$

Hence, $\mathbf{I} - \mathbf{Q}$ becomes $1 - (1 - \alpha_1 r) = \alpha_1 r$. Then $(\mathbf{I} - \mathbf{Q})^{-1} = (\alpha_1 r)^{-1}$, in agreement with formula (14.4.5).

To estimate parameters α_1, α_2, and r we proceed as follows, following the argument outlined by Berger and his colleagues (Berger et al., 1966, Chap. 3, Appendix).

We first define

$$S_N = \sum_{t=0}^{N-1} X_t$$

the total number of p-responses in trials 0 through $N - 1$ (and thus in the first N trials). Then,

$$E(S_N) = E\left(\sum_{t=0}^{N-1} X_t \right)$$

$$= \sum_{t=0}^{N-1} E(X_t)$$

$$= \sum_{t=0}^{N-1} P(X_t = 1)$$

$$= \sum_{t=0}^{N-1} [\alpha_2 + (\alpha_1 - \alpha_2)(1 - \alpha_1 r)^t] \qquad \text{(using 14.4.3)}$$

$$= \sum_{t=0}^{N-1} \alpha_2 + (\alpha_1 - \alpha_2) \sum_{t=0}^{N-1} (1 - \alpha_1 r)^t$$

$$= N\alpha_2 + (\alpha_1 - \alpha_2) \sum_{t=0}^{N-1} (1 - \alpha_1 r)^t$$

To obtain the sum over the terms $(1 - \alpha_1 r)^t$, we note that

$$\sum_{t=0}^{N-1} (1 - \alpha_1 r)^t = \sum_{t=0}^{\infty} (1 - \alpha_1 r)^t - \sum_{t=N}^{\infty} (1 - \alpha_1 r)^t$$

$$= \frac{1}{\alpha_1 r} - (1 - \alpha_1 r)^N \sum_{t=N}^{\infty} (1 - \alpha_1 r)^{t-N} \qquad \text{[using (10.7.14)]}$$

$$= \frac{1}{\alpha_1 r} - (1 - \alpha_1 r)^N \sum_{t-N=0}^{\infty} (1 - \alpha_1 r)^{t-N}$$

$$= \frac{1}{\alpha_1 r} - \frac{(1 - \alpha_1 r)^N}{\alpha_1 r}$$

Hence,

(14.4.6)
$$\sum_{t=0}^{N-1} (1 - \alpha_1 r)^t = \frac{1 - (1 - \alpha_1 r)^N}{\alpha_1 r}$$

Hence,

(14.4.7)
$$E(S_N) = N\alpha_2 + (\alpha_1 - \alpha_2) \frac{1 - (1 - \alpha_1 r)^N}{\alpha_1 r}$$

The left-hand side is estimated by the method of moments (see Chapter 15): replace $E(S_N)$ by $\bar{S}_N$, where $\bar{S}_N$ is the calculated mean number of p-responses in the first N trials taken over a number of subjects. (Here N is the number of trials; the number of subjects enters the analysis only by way of an assumed application in which there are, say, K independent repetitions of the experimental situation and, so, computed means over subjects for any observable quantity in the model). Hence, (14.4.7) is the first of three equations needed if we are to estimate the three parameters.

Next we note that an initial run of o-responses of a certain length n has a probability as follows:

Run	n	Probability
p $\ldots$	0	α_1
op $\ldots$	1	$(1 - \alpha_1)\alpha_1$
oop $\ldots$	2	$(1 - \alpha_1)^2 \alpha_1$
$\cdot$	$\cdot$	$\cdot$
$\cdot$	$\cdot$	$\cdot$
$\cdot$	$\cdot$	$\cdot$

This is based upon the fact, evident from Figure 14.3.8, that following an o-response, the system stays in state S_1, and so, the probability of a p-response following an o-response is α_1.

If we let

$$L = \text{length of initial run of o-responses}$$

then

$$P(L = n) = (1 - \alpha_1)^n \alpha_1 \qquad (n = 0, 1, \ldots)$$

so that again we have a geometric distribution with expectation given by

(14.4.8)
$$E(L) = \frac{1 - \alpha_1}{\alpha_1}$$

Replacing $E(L)$ by $\bar{L}$, the mean length of the initial o-response run, we have our second equation for estimation purposes. The derivation of (14.14.8) proceeds as follows:

$$E(L) = \sum_{n=0}^{\infty} nP(L = n)$$

$$= \sum_{n=1}^{\infty} n(1 - \alpha_1)^n \alpha_1$$

$$= (1 - x) \sum_{n=1}^{\infty} nx^n \qquad \text{(letting } x = 1 - \alpha_1)$$

$$= (1 - x)x \sum_{n=1}^{\infty} nx^{n-1}$$

$$= (1 - x)x \sum_{n=1}^{\infty} \frac{d}{dx} x^n$$

$$= (1 - x)x \frac{d}{dx} \sum_{n=1}^{\infty} x^n$$

$$= (1 - x)x \frac{d}{dx} \left(\frac{1}{1 - x} - 1 \right)$$

$$= (1 - x)x \frac{1}{(1 - x)^2}$$

$$= \frac{x}{1 - x}$$

$$= \frac{1 - \alpha_1}{\alpha_1}$$

For the third estimation equation, consider counting the number of consecutive pairs of p-responses over an observed sequence. Define

$$u_t = \begin{cases} 1 & \text{if } X_t = 1 \text{ and } X_{t+1} = 1 \\ 0 & \text{otherwise} \end{cases}$$

for $t = 0, 1, \ldots, N - 2$, where there are N trials labeled $0, 1, 2, \ldots, N - 1$. For instance, if the observed sequence consists of

$$0 \ 1 \ 0 \ 1 \ 1 \ 1$$

then we have

$$u_0 = 0, \ u_1 = 0, \ u_2 = 0, \ u_3 = 1, \ u_4 = 1$$

Define

$$u = \sum_{t=0}^{N-2} u_t$$

so that in the example, we find $u = 2$ consecutive pairs of p-responses. Note that the two pairs overlap, but this will not present any problem.

The model allows us to find $E(u)$ in several steps as follows:

$$(1) \quad E(u) = \sum_{t=0}^{N-2} E(u_t)$$

$$(2) \quad \quad = \sum_{t=0}^{N-2} P(u_t = 1)$$

$$(3) \quad \quad = \sum_{t=0}^{N-2} P(X_t = 1 \text{ and } X_{t+1} = 1)$$

$$(4) \quad \quad = \sum_{t=0}^{N-2} \left[\sum_{i=1}^{2} P(X_t = 1 \text{ and } X_{t+1} = 1 \mid S_i \text{ on trial } t) P(S_i \text{ at } t) \right]$$

Here, to obtain (1), linearity of expectation is applied to the definition of u; to obtain (2), the fact that the expected value of a $0 - 1$ random variable is the probability it takes the value 1 is applied to $E(u_t)$; to obtain (3), the definition of u_t is used; and to obtain (4), we use the technique of conditionalizing by the state of the system, using the law of total probability. To compute the bracketed terms of (4), we note from the tree of Figure 14.4.3 that

$$P(X_{t+1} = 1, X_t = 1 \mid S_1 \text{ at } t) = \alpha_1^2(1 - r) + \alpha_1 \alpha_2 r$$

Similarly,

$$P(X_{t+1} = 1, X_t = 1 \mid S_2 \text{ at } t) = \alpha_2^2$$

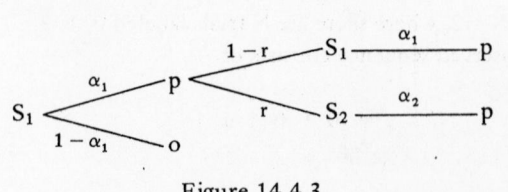

Figure 14.4.3

Hence, using formula (14.4.2) for the probability of being in state 1 at t,

$$
\begin{aligned}
P(X_{t+1} = 1, X_t = 1) &= P(S_1 \text{ at } t)P(X_{t+1} = 1, X_t = 1 | S_1 \text{ at } t) \\
&\quad + P(S_2 \text{ at } t)P(X_{t+1} = 1, X_t = 1 | S_2 \text{ at } t) \\
&= (1 - \alpha_1 r)^t [\alpha_1^2(1 - r) + \alpha_1 \alpha_2 r] + [1 - (1 - \alpha_1 r)^t] \alpha_2^2 \\
&= \alpha_2^2 + [\alpha_1^2(1 - r) + \alpha_1 \alpha_2 r - \alpha_2^2] (1 - \alpha_1 r)^t
\end{aligned}
$$

Returning to step (3), with the above result we write

$$
\begin{aligned}
E(u) &= \sum_{t=0}^{N-2} (\alpha_2^2 + [\alpha_1^2(1 - r) + \alpha_1 \alpha_2 r - \alpha_2^2] (1 - \alpha_1 r)^t) \\
&= (N - 1)\alpha_2^2 + [\alpha_1^2(1 - r) + \alpha_1 \alpha_2 r - \alpha_2^2] \sum_{t=0}^{N-2} (1 - \alpha_1 r)^t
\end{aligned}
$$

If we use (14.4.6) with N replaced by N − 1 we can rewrite the sum on the right in this last expression to conclude that,

$$
(14.4.9) \quad E(u) = (N - 1)\alpha_2^2 + [\alpha_1^2(1 - r) + \alpha_1 \alpha_2 r - \alpha_2^2] \frac{1 - (1 - \alpha_1 r)^{N-1}}{\alpha_1 r}
$$

Formulas (14.4.7), (14.4.8), and (14.4.9) give expected values for three observable quantities. Replacing the left hand sides with means taken from the data on K subjects, we have a system of three nonlinear equations in three unknowns: α_1, α_2, and r, since N is the observed number of trials. In fact, we can use (14.4.8) to replace α_1 by its estimate $\hat{\alpha}_1$,

$$
\hat{\alpha}_1 = \frac{1}{1 + \bar{L}}
$$

where $\bar{L}$ is the observed mean length of the initial run of o-response. Then the problem reduces to solving a pair of simultaneous nonlinear equations in two unknowns, α_2 and r. This can be done by computer methods to obtain $\hat{\alpha}_2$ and $\hat{r}$.

With these estimates, the model yields specific numerical predictions for any function of the data (i.e., for any statistic). Where analytical methods break down, the process can be simulated to obtain the predicted quantities. In this way, using selected statistics, one can test the goodness-of-fit of the model. For example, formula (14.4.3) now becomes

$$P(X_t = 1) = \hat{\alpha}_2 + (\hat{\alpha}_1 - \hat{\alpha}_2)(1 - \hat{\alpha}_1 \hat{r})^t \qquad (t = 0, 1, \ldots)$$

which is a definite curve to compare with the observed proportion of subjects giving a p-response on trial t. Note that not only does the model make the stable equilibrium, or asymptotic, prediction that $P(X_t = 1)$ approaches $\hat{\alpha}_2$ but it also makes a prediction of the specific transient behavior of the probabilistic system. The latter is significant because competing theoretical models might yield the same equilibrium predictions but differ in their predictions about the nonequilibrium behavior of the system.

For an example of the phase of model application emphasizing questions of goodness-of-fit and the comparison of alternative expectation-states models, see Berger, Conner, and McKeown (1969). The monograph edited by Berger, Connor, and Fisek (forthcoming) should also be relevant.

14.5. Concluding Discussion. The reader may note that the effort to offer a theoretical explanation of the abstract empirical generalization about status and influence has been followed through to the point of the stage in which the theoretical model makes predictions about data from an empirical study, called stage 4 in section 14.1. A decision of goodness-of-fit, then, has theoretical import in regard to the explanatory task of the theory. If the predictions are inadequate, any one of a considerable number of premises may be wrong. To list the most important premises:

Premise 1. The three axioms are satisfied by the person in the experimental situation, including the Markov assumption.

Premise 2. The initial condition is that p is characterized by the high state of D.

Premise 3. All of the boundary conditions listed in section 14.3 hold in the experimental situation. For instance, there are no prior expectations linking D to C.

To see how difficult the inferential task is at this point, note that with disagreement on every trial as one of the experimental conditions, one may be led to wonder how other conditions might be affected by the presence of, say, an implausible string of 20 consecutive disagreements. (That such worries are not idle may be seen in the experiment conducted by Berger, Conner, and McKeown, 1969: no less than 32 of 95 subjects had to be eliminated at the data analysis phase because they "became suspicious of one or more deceptions.")

Suppose that we are able to narrow down the difficulties to premise 1. What about the Markov assumption? It is clear from experience that scientists working in the mode of a probabilistic state-space approach are loathe to modify the Markov assumption. Our general discussion on dynamics (see Chapter 8) should make this reaction understandable: the search for a state-description is intimately related to the idea of a Markov process. Hence, if a clear decision can be made that premise 1 is wrong, the investigators are likely to look at the way in which the basic framework assumption (14.3.1) was applied in the construction of the model. Alternatives exist, of course. The relation between the general principle and the model is not deductive: it is "cybernetic" in the sense that the model is constructed to conform to the principle, to exemplify it, but the principle no more determines the specific axioms than the principle that light travels in straight lines determines the specific laws of optics (for a discussion of this point, see Toulmin, 1953). But the main point is this: the concept of "expectation states" evolves under the meaning postulate that it be a state-space concept. Since the approach is probabilistic, in the detailed models, this amounts to making the Markov property part of the meaning of the theoretical concepts. Hence, when a Markov model fails to fit, effort is made to revise the "substantive" assumptions—that is, those assumptions specifically implementing the basic balance principle within the context of the probabilistic state-space approach.

In this context, the same question may be raised about the Coleman methodology of Chapter 13. When a Coleman model fails to fit, where do we seek the problem? Here a basic distinction is clear: the models of Chapter 13 combine "state" and "response." That is, they identify the state of the system with the observable behavior. For example, the state of a voting intention was assumed in one-to-one correspondence with the actual vote. Coleman drops this assumption in various other contexts (see Coleman, 1964, Chap. 13) in favor of what he calls "response uncertainty." This is abstractly the same as the state-space approach with a distinction between state and observable behavior. The result is that Coleman's methodology extends so as to exemplify the general dynamic system notions centering around our two methodological rules of Chapter 8: "setup in the small" and "state-space approach." In this extended Coleman methodology, the Markov assumption again becomes part of the meaning of the unobservable state concept.

CHAPTER FIFTEEN ESTIMATION
METHODS IN MODEL-BUILDING

15.1 Concept of Estimation. We have seen in the preceding chapters that an important aspect of the application of a model concerns parameter estimation. In this chapter, various techniques for parameter estimation are gathered together and explained in an elementary manner (for mathematical details, see Mood and Graybill, 1963).

By numerical data we mean a collection of values of observed random variables. Intrinsic to this conceptualization is the idea that the values are occurring with probabilities in some specified or unspecified probability model. Here, we assume a definite model, and the problem is to use the data to estimate parameters of the model. Our generic notation is that the model is given as an object depending on several parameters

$$\mathfrak{M}(\alpha, \beta, \ldots)$$

and we desire to replace each parameter with a number that estimates it.

Recall that by a statistic we mean any function of data. (See section 7.6.) For instance, if X_1 and X_2 are data, then $X_1 + X_2$ is a statistic.

By a decision function one means a mapping with domain the possible data, based on a certain number of observable random variables, and range some set of actions. We write

$$a = d(X_1, X_2, \ldots)$$

where each of the X_i is a random variable and a is an action in an action space (i.e., a set of possible actions).

For an estimation problem the action consists in behaving as if a certain statistic were a parameter. Let θ be an arbitrary parameter. Call the statistic $\hat{\theta}$. Then we can let

$$\hat{\theta} = d(X_1, X_2, \ldots)$$

Table 15.1.1   Probability Distribution of Data
Under Model $\mathfrak{M}(p)$

Possible Data		Probability of the Data
X_1	X_2	
1	1	p^2
1	0	$p(1-p)$
0	1	$p(1-p)$
0	0	$(1-p)^2$

be the function mapping the data into the statistic. Any such function is called an estimator of parameter θ.

Example.   A coin is to be tossed repeatedly, say K times. Let X_t be the random variable whose value is 1 if the outcome is heads on toss t and 0 otherwise. Assume that there is a fixed probability p of heads and that the tosses are independent. This defines "a model of one parameter" (i.e., a model depending on one parameter),

$$\mathfrak{M}(p)$$

in which we envision the actual behavior of the coin as generated by a probabilistic mechanism of the type described. Hence, the mechanism, including its characteristic p, determines the distribution of the possible data. For K = 2, we have the distribution shown in Table 15.1.1.

For several possible decision functions that can serve as estimators of parameter p, we let $\hat{p}_i$ be the estimate under decision function d_i:

$$\hat{p}_1 = d_1(X_1, X_2) = \frac{X_1 + X_2}{2}$$

$$\hat{p}_2 = d_2(X_1, X_2) = X_1$$

$$\hat{p}_3 = d_3(X_1, X_2) = \frac{X_1 X_2}{2}$$

We see that we have the following estimates as functions of the data, using these possible estimators:

Possible data		$\hat{p}_1$	$\hat{p}_2$	$\hat{p}_3$
1	1	1	1	.5
1	0	.5	1	0
0	1	.5	0	0
0	0	0	0	0

Note that since each $\hat{p}_i$ is a function of the data, it takes definite values with definite data. Thus, each $\hat{p}_i$ is a statistic. Hence, each $\hat{p}_i$ has a probability distribution, a mean and a variance. For $\hat{p}_1$ we have,

$$P(\hat{p}_1 = 1) = P(X_1 = 1, X_2 = 1) = p^2$$
$$P(\hat{p}_1 = .5) = P(X_1 = 1, X_2 = 0) + P(X_1 = 0, X_2 = 1) = 2p(1 - p)$$
$$P(\hat{p}_1 = 0) = P(X_1 = 0, X_2 = 0) = (1 - p)^2$$

The expectation of $\hat{p}_1$ is,

$$E(\hat{p}_1) = \sum_x x P(\hat{p}_1 = x)$$
$$= 1\, P(\hat{p}_1 = 1) + .5\, P(\hat{p}_1 = .5)$$
$$= p^2 + .5[2p(1 - p)]$$
$$= p$$

For $\hat{p}_2$ we obtain

$$P(\hat{p}_2 = 1) = p^2 + p(1 - p) = p$$
$$P(\hat{p}_2 = 0) = p(1 - p) + (1 - p)^2 = 1 - p$$

Hence,

$$E(\hat{p}_2) = p$$

For $\hat{p}_3$ we obtain

$$P(\hat{p}_3 = .5) = p^2$$
$$P(\hat{p}_3 = 0) = 1 - p^2$$

Hence,

$$E(\hat{p}_3) = .5p^2$$

It is clear that a desirable property of an estimator is that its expected value agree with the actual value of the parameter generating the data. This is true of $\hat{p}_1$ and $\hat{p}_2$ but not of $\hat{p}_3$. We say that $\hat{p}_3$ is a biased estimator, while $\hat{p}_1$ and $\hat{p}_2$ are unbiased.

To choose between $\hat{p}_1$ and $\hat{p}_2$, we consider the variances. We have

$$E(\hat{p}_1^2) = p^2 + .25[2p(1 - p)] = \frac{p(p + 1)}{2}$$

Hence,

$$VAR(\hat{p}_1) = E(\hat{p}_1^2) - [E(\hat{p}_1)]^2$$
$$= \frac{p(p + 1)}{2} - p^2$$
$$= \frac{p(1 - p)}{2}$$

For $\hat{p}_2$,

$$E(\hat{p}_2^2) = p$$

Hence,

$$\begin{aligned} VAR(\hat{p}_2^2) &= E(\hat{p}_2^2) - [E(\hat{p}_2)]^2 \\ &= p - p^2 \\ &= p(1 - p) \end{aligned}$$

Hence,

$$VAR(\hat{p}_1) = \frac{1}{2} VAR(\hat{p}_2)$$

If we use low variance as a desirable property of an estimator, we prefer $\hat{p}_1$ to $\hat{p}_2$, or, alternatively decision rule d_1 to decision rule d_2.

Hence, d_1 is the best estimator among the three competing decision rules that could serve as an estimator of p. In fact, there are many other rules that could be formulated, but d_1 is still best among them in the sense described above. That is, the estimator

$$\hat{p}_1 = d_1(X_1, X_2) = \frac{X_1 + X_2}{2}$$

is a minimum-variance unbiased estimator: among all possible rules for computing $\hat{p}$ that yield an unbiased estimator, this is the best in the sense that it yields minimum variance. Accordingly, in this problem we behave as if the average of the X_t (t = 1, 2) is the parameter p, calling it $\hat{p}$. If it is objected that "we know that p is actually about .5," the whole point has been lost: our model had an unspecified parameter p. We began with $\mathfrak{M}(p)$, not $\mathfrak{M}(.5)$. Beginning with $\mathfrak{M}(.5)$, where p = .5, there is no estimation problem. But there may be a hypothesis-testing problem: Is $\mathfrak{M}(.5)$ the model generating the data? We assume the reader is acquainted with such questions and concentrate on the relatively unfamiliar notion of estimation and on estimation techniques.

Only one further general point needs to be made. When we have a probability model $\mathfrak{M}(\theta)$ for a phenomenon, the method of application consists in (a) identification of K independent repetitions of the phenomena, and (b) estimation and goodness-of-fit. For instance, in part (a) we might have:

(1) K groups and Coleman's model for influence in social networks: each group is thought of as an independent realization of the process described by the model;

(2) K subjects and Berger's expectation-states model: each subject is thought of as an independent realization of the expectation-states process;

(3) K people and the images of stratification model: each person is regarded as an independent realization of the process.

Thus, the following auxiliary assumptions typically* make their appearance at the application phase:

(a) Each of K concrete systems is an independent realization of the process.
(b) Each such system has the same parameter values.

From this we conclude that we have the equivalent of the assumption of random sampling in hypothesis-testing situations familiar to the reader. In the latter case, the formal assumption of a random sample is justified (more or less) by the empirical method of sampling. Here, also, the auxiliary assumptions are satisfied (more or less) by selection and control procedures. For instance, forbidding communications between subjects who serve as repetitions (replications from a design viewpoint) serves to satisfy the assumption of independence. In the Coleman methodology, choosing groups that have little contact serves to validate the independence assumption. If the intergroup network had nonzero influence parameters, for example, independence would fail. Hence, for purposes of testing the model, one wants "nonsociological" social forms at some point, where interaction is minimized.

In any case, the general point is that the auxiliary assumptions enter into the body of premises generating model predictions. To the extent that they are unreasonable, the model's failure to fit is difficult to place. Is it the auxiliary assumptions that are at fault or the basic picture of the process merely repeated by the various concrete systems?

In what follows, we assume that the data are generated by empirical procedures of inquiry that make the auxiliary assumptions plausible. Then the formal estimation technique uses the mathematical equivalent of these assumptions to derive the estimator. Once again, the main point: estimation involves making one or more auxiliary assumptions referent to the repetitions of the process of interest. These assumptions are the model-building equivalent of random-sampling assumptions in hypothesis-testing procedures familiar to sociologists.

Some rather well-known methods exist for finding estimators that have desirable properties. In the following sections, these are reviewed from the standpoint of the formal technique involved rather than of the deeper questions of mathematical statistics concerning these estimators.

15.2 Maximum Likelihood.   Introduce the notation

$$L = P(\text{data})$$

---

*The second assumption is frequently a cause of difficulties, so that efforts are made to introduce nonhomogeneity in parameter values. See Chapter 16 for examples.

so that L is the probability of the data. In general, in model $\mathfrak{M}(\theta)$, L depends on θ, and we use the rule that we choose as $\hat{\theta}$ the value that maximizes L compared with all competing values. That is,

Rule. Where θ' is any possible estimate of θ, find $\hat{\theta}$ such that

$$L(\hat{\theta}) \geqslant L(\theta')$$

In other words, maximizing the probability of the data is the rule.

Example. (a) Using the previous example, let us use Table 15.1.1 to find the maximum likelihood estimator for p. Here,

$$L(p) = p^{\Sigma X_t}(1-p)^{2-\Sigma X_t}$$

is the formula that summarizes Table 15.1.1. For instance, in row 1,

$$\Sigma X_t = X_1 + X_2 = 2$$

so that

$$L(p) = p^2(1-p)^{2-2} = p^2$$

while in row 2,

$$\Sigma X_t = X_1 + X_2 = 1 + 0 = 1$$

so that

$$L(p) = p^1(1-p)^{2-1} = p(1-p)$$

Here the application of the maximum likelihood rule instructs us to find that value, say $\hat{p}$, that maximizes $L(p)$. To do this, we first transform $L(p)$ into $\ln L(p)$ because the logarithmic transformation preserves order (and so maxima and minima), while at the same time it makes the subsequent analysis simpler:

$$L^*(p) = \ln L(p) = (\Sigma X_t)\ln p + (2 - \Sigma X_t)\ln(1-p)$$

To maximize a function of one variable, we realize that at the maximum the graph of the function has a horizontal tangent line—that is, the slope is zero.* Hence, we first find the slope, which is given by the derivative

$$\frac{d}{dp}L^*(p) = \frac{\Sigma X_t}{p} - \frac{(2 - \Sigma X_t)}{1-p}$$

using elementary properties of differentiation. Next, we set this equal to 0 and solve for p, calling the solution value $\hat{p}$:

---

*This is also true at a minimum, so we ordinarily check the second derivative. If the second derivative is negative, then indeed we have a maximum: the slope is positive to the left of the extreme point and negative to the right.

$$0 = \frac{\Sigma X_t}{p} - \frac{2 - \Sigma X_t}{1 - p}$$

$$\frac{\Sigma X_t}{p} = \frac{2 - \Sigma X_t}{1 - p}$$

$$(\Sigma X_t)(1 - p) = (2 - \Sigma X_t)p$$

$$\Sigma X_t = (2 - \Sigma X_t)p + (\Sigma X_t)p = 2p$$

$$\hat{p} = \frac{\Sigma X_t}{2} = \frac{X_1 + X_2}{2}$$

We see that the maximum likelihood method yields the minimum-variance unbiased estimator of p. In general, with K arbitrary rather than K = 2, in K independent trials of the 0–1 type with constant parameter p, the relative frequency $\Sigma X_t / K$ is the maximum likelihood estimator of the parameter p.

Example. (b) In the expectation-states model, of Chapter 14, suppose it is known that each of 10 subjects is in state S_1 from trials 0 to 2. The data are the 10 subject protocols of observed responses:

Subject	Data	Probability of the Data
1	0 0 1	$\alpha_1(1 - \alpha_1)^2$
2	1 1 0	$\alpha_1^2(1 - \alpha_1)$
3	0 0 1	$\alpha_1(1 - \alpha_1)^2$
4	0 0 0	$(1 - \alpha_1)^3$
5	0 1 0	$\alpha_1(1 - \alpha_1)^2$
6	0 1 1	$\alpha_1^2(1 - \alpha_1)$
7	0 0 0	$(1 - \alpha_1)^3$
8	0 1 0	$\alpha_1(1 - \alpha_1)^2$
9	1 0 0	$\alpha_1(1 - \alpha_1)^2$
10	1 0 0	$\alpha_1(1 - \alpha_1)^2$

The problem is to use these data to obtain a maximum likelihood estimate of the parameter α_1. The likelihood of each row of data has been calculated in terms of the parameter α_1, as shown above, since once we know the subject is in S_1 on trials 0, 1, and 2, we know that α_1 is the p-response (coded 1) probability. Note how this calculation uses auxiliary assumption (b) of section 15.1: the same α_1 applies in each row. Next, we need to compute the likelihood of the whole body of data, which under the assumption of independent repetitions—auxiliary assumption (a)—is given by the product of the 10 row probabilities. Note that for subject i,

$$P(X_{i1} X_{i2} X_{i3}) = \alpha_1^{\Sigma_j X_{ij}} (1 - \alpha_1)^{3 - \Sigma_j X_{ij}} \qquad (i = 1, 2, \ldots, 10)$$

Hence,

$$L = P(\text{data}) = \prod_{i=1}^{10} \alpha_1^{\Sigma_j X_{ij}} (1 - \alpha_1)^{3 - \Sigma_j X_{ij}}$$

$$= \alpha_1^{\Sigma_i \Sigma_j X_{ij}} (1 - \alpha_1)^{30 - \Sigma_i \Sigma_j X_{ij}}$$

(15.2.1)
$$= \alpha_1^{10}(1 - \alpha_1)^{20}$$

Hence,

$$\hat{\alpha}_1 = \frac{10}{30} = \frac{1}{3}$$

which the reader can verify by taking logarithms in (15.2.1), differentiating and setting the resulting expression to 0 to solve for $\hat{\alpha}_1$. Note that the L function is of the abstract form: 10 successes in 30 independent trials. For another, rather complex, example of maximum likelihood estimation, see section 11.14.

Example. (c) Events occur at random on the time-axis according to a Poisson process (see section 16.4 for an example). This means, among other things, that the time interval between a given event and the next event is a random variable T whose distribution is given by

$$f_T(t) = \lambda e^{-\lambda t} \qquad (t > 0)$$

where $f_T(t)\Delta t$ is the chance of the next event's occurring in a small (Δt) time interval t time units from the given event. Let us suppose the problem is to estimate the parameter λ on the basis of data on six successive events, E_i:

E_1	E_2		E_3	E_4		E_5		E_6

t_1	t_2	t_3	t_4	t_5

The t_i are the time intervals between event E_i and the next event E_{i+1}, $i = 1, 2, \ldots, 5$.

The model has the property that the times between events are independent random variables, all with the common distribution of T. Hence, the likelihood of the data is given by

$$L = \lambda e^{-\lambda t_1} \lambda e^{-\lambda t_2} \lambda e^{-\lambda t_3} \lambda e^{-\lambda t_4} \lambda e^{-\lambda t_5} = \lambda^5 e^{-\lambda(\Sigma_i t_i)}$$

Using logarithms,

$$L^* = \ln L = 5 \ln \lambda - \lambda(\Sigma_i t_i)$$

and differentiating,

$$\frac{d}{d\lambda} L^* = \frac{5}{\lambda} - \Sigma_i t_i$$

Setting the derivative to 0 and solving,

$$\hat{\lambda} = \frac{5}{\Sigma_i t_i}$$

Hence, we note that $1/\hat{\lambda}$ is the average time between events. In fact, one may show (by integration) that $E(T) = 1/\lambda$, providing a direct interpretation of λ as the reciprocal of the expected time to the next event.

15.3 Method of Moments. By the nth moment of a random variable X one means the quantity $E(X^n)$, where $n = 1, 2, \ldots$. Hence, the expected value of X is the first moment. The variance is the second moment of $X - E(X)$. Using the data, one may compute corresponding properties of the observed distribution of X based on K repetitions. The decision to use the corresponding moment from the data in place of a moment of X is called the method of moments.

The simplest statement applies only to the mean:

Rule. Replace $E(X)$ by $\bar{X}$, where $\bar{X} = (\Sigma_i X_i)/K$ is the observed mean of X based on K repetitions.

Example. (a) In a learning model (see section 10.7), the expected total number of errors is calculated in the model to be

$$E(T) = \frac{1}{c}$$

where c is a learning parameter. If 50 subjects are run through the basic learning process under analysis and if $\bar{T} = 100$, then

$$\hat{c} = 1/\bar{T} = .01,$$

where $\hat{c}$ is the estimate of c based on the method of moments.

Example. (b) In studying Coleman's model (see Chapter 13) for social influence in three-node networks, we found that the two parameters a and c could be estimated because we could find the theoretical mean and variance of the equilibrium distribution. That is, we expressed $E(Y)$ and $VAR(Y)$ in terms of a and c (and the known N, number of groups) and then inverted to solve for a and c. Replacing $E(Y)$ and $VAR(Y)$ by the mean and variance from the data yielded $\hat{a}$ and $\hat{c}$, an application of the method of moments.

Example. (c) Even in the case of the expectation-states model (Chapter 14), where the states were not observable, we found that thinking in terms of the method of moments was fruitful. To estimate three parameters (in this case, α_1, α_2, and r), we tried to derive three moments of observable random variables.

Having done so, the logic of the method of moments is to replace the theoretical $E(X)$ by $\overline{X}$ in each case and to solve the simultaneous system for the three unknown parameters: these are the estimates.

Example. (d) Kemeny and Snell (1962) used the method of moments to "rationalize" the estimation procedure associated with Cohen's model (see section 10.12). In his analysis, Cohen had used a computerized procedure with a minimum χ^2 criterion (for a discussion of this method of estimation, see Cohen, 1963, and Atkinson, Bowers, and Crothers, 1965). Because the procedure did not have an explicit mathematical justification, it was not clear that the resulting estimates were good estimates. Kemeny and Snell tried to find four equations of the type expressing $E(X)$ in terms of the four parameters, where they studied four distinct observable random variables X. Having deduced four such expressions, they inverted them to express the parameters in terms of the expected values. Then they replaced these expected values with observed means to obtain the four estimates (for a further discussion, see Bartos, 1967).

15.4 Least-Squares Estimation in Coleman Methodology. This method is used in "linear models." In a typical case, n values of a variable x are chosen, say x_1, $x_2, \ldots, x_n$. For each x_i, an observable random variable Y is postulated to satisfy

$$E(Y_i) = \alpha + \beta x_i \qquad (i = 1, 2, \ldots, n)$$

where the Y_i are uncorrelated and have identical variance σ^2. The ideal textbook case is one in which an experimenter selects n values of x and then measures Y at each level of x. Although there is observable scatter in Y at each level, the model claims that the expected value increases linearly with x. The problem is to find the "least-squares line," the parameters of which are α and β. We form the function

$$f(\alpha, \beta) = \sum_{i=1}^{n} (Y_i - \alpha - \beta x_i)^2$$

and the problem is to find values, say $\hat{\alpha}$, $\hat{\beta}$, such that

$$f(\hat{\alpha}, \hat{\beta}) \leqslant f(\alpha', \beta')$$

where α', β' are any possible parameter values. Thus, $\hat{\alpha}$ and $\hat{\beta}$ minimize the sum of the squared deviations of the Y_i from the expected value line postulated to exist. This expected value line is the so-called regression line. One obtains the well-known results

$$\hat{\alpha} = \overline{Y} - \hat{\beta}\overline{x}$$

$$\hat{\beta} = \frac{\Sigma(Y_i - \overline{Y})(x_i - \overline{x})}{\Sigma(x_i - \overline{x})^2}$$

An application of the least-squares method to a nonroutine problem may be illustrated by reference to another phase of Coleman's methodology (Coleman, 1964; Chap. 6). The model is a generalization of the idea, covered in section 13.1, that a process of continuous-time attitudinal transition occurs at different rates in different subpopulations. If three dichotomous attributes are considered eight such populations are specified. Coleman presents a scheme relating the various rates as follows: Let X_t be the dynamical (random) variable of interest, where

$$X_t = \begin{cases} 1 & \text{if person has + attitude at t} \\ 0 & \text{if not} \end{cases}$$

Let the transition rates be q_{01} and q_{10} for any given individual. Let x, y, and z be the three independent variables, each dichotomous, each 0-1. Then in a table we have rates q_{01} and q_{10} as follows, using Coleman's notation:

			x			
			1		0	
			y		y	
			1	0	1	0
z	1	q_{01}	$\alpha + \beta + \gamma + \epsilon_1$	$\alpha + \gamma + \epsilon_1$	$\beta + \gamma + \epsilon_1$	$\gamma + \epsilon_1$
		q_{10}	ϵ_2	$\beta + \epsilon_2$	$\alpha + \epsilon_2$	$\alpha + \beta + \epsilon_2$
	0	q_{01}	$\alpha + \beta + \epsilon_1$	$\alpha + \epsilon_1$	$\beta + \epsilon_1$	ϵ_1
		q_{10}	$\gamma + \epsilon_2$	$\beta + \gamma + \epsilon_2$	$\alpha + \gamma + \epsilon_2$	$\alpha + \beta + \gamma + \epsilon_2$

Hence, the forms of the two rates are seen to be

$$\text{(15.4.1)} \quad \begin{aligned} q_{01} &= x\alpha + y\beta + z\gamma + \epsilon_1 \\ q_{10} &= (1-x)\alpha + (1-y)\beta + (1-z)\gamma + \epsilon_2 \end{aligned} \quad (x, y, z = 0, 1)$$

The sum of the rates is invariant (i.e., the same in each subpopulation),

$$q_{01} + q_{10} = \alpha + \beta + \gamma + \epsilon_1 + \epsilon_2$$

a fact we shall use in the estimation procedure.

In Coleman's application x, y, and z have the following meanings:

$$x = \begin{cases} 1 & \text{if person has at least high school education} \\ 0 & \text{if not} \end{cases}$$

$$y = \begin{cases} 1 & \text{if person is over 25,} \\ 0 & \text{if not} \end{cases}$$

$$z = \begin{cases} 1 & \text{if person is a noncom} \\ 0 & \text{if not (= private)} \end{cases}$$

and the positive attitude ($X_t = 1$) is "wanting to use one's civilian skills" with $X_t = 0$ meaning "not wanting to use them." (The data he uses are from Stouffer et al., *The American Soldier,* 1949, vol I.)

Apart from the relationships among rates, we have eight distinct parameter conditions in which the fundamental two-state continuous process works itself out. Thus, any analytical result previously deduced holds in each cell. In particular, in equilibrium we know (see section 13.1) that

$$p_{1e} = \frac{q_{01}}{q_{01} + q_{10}}$$

holds in each cell, where p_{1e} is the equilibrium probability for the process involving X_t. Hence, using the eight rates in (15.4.1), we formulate eight equations. For example,

$$\text{cell 000:} \qquad p_{1e,000} = \frac{\epsilon_1}{\alpha + \beta + \gamma + \epsilon_1 + \epsilon_2}$$

As was noted above, the term $q_{01} + q_{10}$ is a constant, the same in each cell. Hence, we divide the rates by this invariant sum to reduce the number of parameters to four. We write e_1 to indicate ϵ_1 divided by the sum, a for α divided by the sum, b for β divided by the sum, and c for γ divided by the sum. Also, we let

$$p_{xyz} = p_{1e,xyz}$$

so that p_{xyz} is the probability of state 1 in the equilibrium of the process taking place under conditions xyz. Thus, the eight equilibrium equations are

$$
\begin{aligned}
p_{000} &= e_1 \\
p_{100} &= a + e_1 \\
p_{010} &= b + e_1 \\
p_{001} &= c + e_1 \\
p_{110} &= a + b + e_1 \\
p_{101} &= a + c + e_1 \\
p_{011} &= b + c + e_1 \\
p_{111} &= a + b + c + e_1
\end{aligned}
$$

(15.4.2)

Note that the generic form is

(15.4.3) $p_{xyz} = xa + yb + zc + e_1 \qquad (x, y, z = 0, 1)$

Consider now a set of individuals assumed to be repetitions of the basic process in each category xyz: in the application considered by Coleman, a certain number of soldiers of each of the eight categories is considered. The number may vary with the category. Then a certain proportion $\hat{p}_{xyz}$ in each category are in state 1 at the time of measurement, assumed to be a point when all eight processes are in equilibrium.

Coleman forms the sum

$$\sum_{xyz} (\hat{p}_{xyz} - p_{xyz})^2$$

and initiates an analytical procedure to find values of a, b, c, and e_1 that minimize this sum of squares. The values that in this way minimize the expression will be the estimates $\hat{a}$, $\hat{b}$, $\hat{c}$, and $\hat{e}_1$.

Calling the sum $f(a, b, c, e_1)$ to make clear that it is a function of the parameters, we use (15.4.3) to obtain

(15.4.4) $$f(a, b, c, e_1) = \sum_{xyz} (\hat{p}_{xyz} - xa - yb - zc - e_1)^2$$

We have to minimize this function of four quantities. The necessary condition is that the gradient be zero, where the gradient is the vector

$$\left(\frac{\partial f}{\partial a}, \frac{\partial f}{\partial b}, \frac{\partial f}{\partial c}, \frac{\partial f}{\partial e_1} \right)$$

Differentiating with respect to a we have

$$\frac{\partial f}{\partial a} = -2 \sum_{xyz} (\hat{p}_{xyz} - xa - yb - zc - e_1)x$$

$$= -2 \sum_{yz} (\hat{p}_{1yz} - a - yb - zc - e_1)$$

since x = 0, 1. Setting $\partial f / \partial a = 0$ and simplifying yields

$$2a + b + c + 2e_1 = \frac{1}{2} \sum \hat{p}_{1yz}$$

where it is understood that the sum extends over y = 0, 1 and z = 0, 1.

Proceeding in a similar manner to differentiate (15.4.4) with respect to b, c, and e_1 and in each instance setting the resulting derivative to 0, we obtain the system

$$2a + b + c + 2e_1 = \frac{1}{2} \sum \hat{p}_{1yz}$$

$$a + 2b + c + 2e_1 = \frac{1}{2} \sum \hat{p}_{x1z}$$

$$a + b + 2c + 2e_1 = \frac{1}{2} \sum \hat{p}_{xy1}$$

$$a + b + c + 2e_1 = \frac{1}{4} \sum \hat{p}_{xyz}$$

The solution of this system is a vector $(\hat{a}, \hat{b}, \hat{c}, \hat{e}_1)$, which constitutes the least-squares estimate of the parameters.

To solve this system we first write it in matrix form:

$$(15.4.5) \quad \begin{pmatrix} 2 & 1 & 1 & 2 \\ 1 & 2 & 1 & 2 \\ 1 & 1 & 2 & 2 \\ 1 & 1 & 1 & 2 \end{pmatrix} \begin{pmatrix} a \\ b \\ c \\ e_1 \end{pmatrix} = \begin{pmatrix} \frac{1}{2}\Sigma\,\hat{p}_{1yz} \\ \frac{1}{2}\Sigma\,\hat{p}_{x1z} \\ \frac{1}{2}\Sigma\,\hat{p}_{xy1} \\ \frac{1}{4}\Sigma\,\hat{p}_{xyz} \end{pmatrix}$$

Next, we use the "tableau method" for solution by elimination, outlined by Kemeny, Snell, and Thompson (1966, Chap. 5). This means using the coefficient matrix and constant vector in the form

$$\left(\begin{array}{cccc|c} 2 & 1 & 1 & 2 & A \\ 1 & 2 & 1 & 2 & B \\ 1 & 1 & 2 & 2 & C \\ 1 & 1 & 1 & 2 & D \end{array}\right)$$

where A, B, C, and D are the constants of the vector on the right in (15.4.5). We now aim to reduce the left side of this tableau to an identity matrix by the same type of operations we perform on equations to eliminate variables; in fact, this is the same procedure, but with the variables not explicitly shown to make the work easier. When the left side is an identity, the right side contains the solution vector. Proceeding with the eliminations with the explanation at the right,

$$\left(\begin{array}{cccc|c} 2 & 1 & 1 & 2 & A \\ 1 & 2 & 1 & 2 & B \\ 1 & 1 & 2 & 2 & C \\ 1 & 1 & 1 & 2 & D \end{array}\right) \xrightarrow{(1)} \left(\begin{array}{cccc|c} 1 & \frac{1}{2} & \frac{1}{2} & 1 & \frac{A}{2} \\ 1 & 2 & 1 & 2 & B \\ 1 & 1 & 2 & 2 & C \\ 1 & 1 & 1 & 2 & D \end{array}\right)$$

$$\xrightarrow{(2)} \left(\begin{array}{cccc|c} 1 & \frac{1}{2} & \frac{1}{2} & 1 & \frac{A}{2} \\ 0 & \frac{3}{2} & \frac{1}{2} & 1 & B - \frac{A}{2} \\ 0 & \frac{1}{2} & \frac{3}{2} & 1 & C - \frac{A}{2} \\ 0 & \frac{1}{2} & \frac{1}{2} & 1 & D - \frac{A}{2} \end{array}\right) \xrightarrow{(3)} \left(\begin{array}{cccc|c} 1 & 0 & \frac{1}{3} & \frac{2}{3} & \frac{2A-B}{2} \\ 0 & 1 & \frac{1}{3} & \frac{2}{3} & \frac{2B-A}{3} \\ 0 & 0 & \frac{4}{3} & \frac{2}{3} & \frac{3C-A-B}{3} \\ 0 & 0 & \frac{1}{3} & \frac{2}{3} & \frac{3D-A-B}{3} \end{array}\right)$$

(1) Divide first row by 2.

(2) Subtract first row from each other row.

(3) Divide second row by 3/2 and subtract second row from each other row.

$$
\overset{(4)}{\searrow}
\begin{pmatrix}
1 & 0 & 0 & \dfrac{1}{2} & \dfrac{3A-B-C}{4} \\[2mm]
0 & 1 & 0 & \dfrac{1}{2} & \dfrac{3B-A-C}{4} \\[2mm]
0 & 0 & 1 & \dfrac{1}{2} & \dfrac{3C-A-B}{4} \\[2mm]
0 & 0 & 0 & \dfrac{1}{2} & \dfrac{4D-A-B-C}{4}
\end{pmatrix}
\overset{(5)}{\longrightarrow}
\begin{pmatrix}
1 & 0 & 0 & 0 & A-D \\[2mm]
0 & 1 & 0 & 0 & B-D \\[2mm]
0 & 0 & 1 & 0 & C-D \\[2mm]
0 & 0 & 0 & 1 & \dfrac{4D-A-B-C}{2}
\end{pmatrix}
$$

(4) Divide third row by 4/3 and subtract third row from each other row.

(5) Divide fourth row by 1/2 and subtract fourth row from each other row.

Replacing the letters in the solution column of the tableau with the expressions for which they stand, we obtain

$$\hat{a} = A - D = \frac{1}{2}\Sigma\,\hat{p}_{1yz} - \frac{1}{4}\Sigma\,\hat{p}_{xyz}$$

$$\hat{b} = B - D = \frac{1}{2}\Sigma\,\hat{p}_{x1z} - \frac{1}{4}\Sigma\,\hat{p}_{xyz}$$

$$\hat{c} = C - D = \frac{1}{2}\Sigma\,\hat{p}_{xy1} - \frac{1}{4}\Sigma\,\hat{p}_{xyz}$$

$$\hat{e}_1 = \frac{4D-A-B-C}{2} = \frac{1}{2}\ \Sigma\hat{p}_{xyz} - \frac{1}{2}\Sigma\hat{p}_{1yz} - \frac{1}{2}\Sigma\hat{p}_{x1z} - \frac{1}{2}\Sigma\hat{p}_{xy1}$$

The right-hand sides can be simplified. For instance,

$$
\hat{a} = \frac{1}{2}\,(\hat{p}_{100} + \hat{p}_{110} + \hat{p}_{101} + \hat{p}_{111} - \frac{1}{2}\hat{p}_{100} - \frac{1}{2}\hat{p}_{110} - \frac{1}{2}\hat{p}_{101} - \frac{1}{2}\hat{p}_{111} - \frac{1}{2}\hat{p}_{000}
$$
$$
- \frac{1}{2}\hat{p}_{010} - \frac{1}{2}\hat{p}_{001} - \frac{1}{2}\hat{p}_{011})
$$
$$
= \frac{1}{2}\,(\frac{1}{2}\hat{p}_{100} + \frac{1}{2}\hat{p}_{110} + \frac{1}{2}\hat{p}_{101} + \frac{1}{2}\hat{p}_{111} - \frac{1}{2}\hat{p}_{000} - \frac{1}{2}\hat{p}_{010} - \frac{1}{2}\hat{p}_{001} - \frac{1}{2}\hat{p}_{011})
$$
$$
= \frac{1}{4}\,(\Sigma\hat{p}_{1yz} - \Sigma\hat{p}_{0yz})
$$
$$
= \frac{1}{4}\,\underset{yz}{\Sigma}\,(\hat{p}_{1yz} - \hat{p}_{0yz})
$$

Proceeding in a similar manner with the other expressions, we obtain the solution in simplest terms:

$$\hat{a} = \frac{1}{4}\Sigma(\hat{p}_{1yz} - \hat{p}_{0yz})$$

(15.4.6)

$$\hat{b} = \frac{1}{4}\Sigma(\hat{p}_{x1z} - \hat{p}_{x0z})$$

$$\hat{c} = \frac{1}{4}\Sigma(\hat{p}_{xy1} - \hat{p}_{xy0})$$

$$\hat{e}_1 = \frac{1}{4}(2\hat{p}_{000} + \hat{p}_{001} + \hat{p}_{100} + \hat{p}_{010} - \hat{p}_{111})$$

This least-squares solution, as Coleman notes, is intuitively appealing in its form, for the "effect" of each of the three independent variables is expressed in terms of an average of four differences in equilibrium probabilities: "with state 1 of the variable versus without that state," under four different conditions. For instance, â estimates the effect of high school education on attitude: obtained by averaging over four different subpopulations, comparing in each case the probability of a positive attitude among those with a high school education $(\hat{p}_{1yz})$ with the probability among those who do not have such an education $(\hat{p}_{0yz})$.

The data used by Coleman (1964, Table 6.1) may be put in the following form:

		Education (x)			
		High School (1)		Grade School (0)	
		Age (y)		Age (y)	
		25+ (1)	25− (0)	25+ (1)	25− (0)
Rank (z)	Noncom (1)	.58 $(\hat{p}_{111})$	.46 $(\hat{p}_{101})$	.40 $(\hat{p}_{011})$	.37 $(\hat{p}_{001})$
	Private (0)	.56 $(\hat{p}_{110})$	.44 $(\hat{p}_{100})$	.45 $(\hat{p}_{010})$	.32 $(\hat{p}_{000})$

Here we have indicated the proportions of various men who want to use their civilian skills in the army, interpreted as estimates of the equilibrium probabilities. Using the first equation of (15.4.6) and these data, we have

$$\hat{a} = \frac{1}{4}[(.58 - .40) + (.46 - .37) + (.56 - .45) + (.44 - .32)] = \frac{.50}{4} = .125$$

In the same way, one obtains the other estimates:

$$\hat{b} = .100, \quad \hat{c} = .010, \quad \hat{e}_1 = .330$$

Finally, the parameter corresponding to ϵ_2, call it e_2, must be such as to yield a sum of 1:

$$\hat{e}_2 = 1 - (\hat{a} + \hat{b} + \hat{c} + \hat{e}_1) = .435$$

The interested reader will find in Coleman's subsequent discussion, given these estimates, questions concerning goodness-of-fit, statistical significance, generalizations to a larger number of attributes, and other matters relevant to the methodology. Our aim here has been to utilize this example only to illustrate the least-squares method of estimation in a model-building context.

For other model-building-oriented discussions of estimation methods, see Bush (1963), Bush and Mosteller (1955), Cox and Lewis (1966), and Suppes and Atkinson (1960). For an introduction to the mathematical statistics viewpoint toward estimation, written especially for social scientists, see Gelbaum and March (1969).

CHAPTER SIXTEEN   SOCIAL MOBILITY

16.1. Introduction. In this chapter, the aim is to acquaint the reader with a sample of the wide variety of existing mathematical conceptualizations of social mobility. The phenomenon of mobility has been treated by model-builders perhaps more than any other single sociological topic. The very extensive literature forces us to be very selective in exhibiting various models and techniques. Not every model can be treated, nor each in equal depth. For reviews of mobility models, see Pullum (1970), McFarland (1970), and Mayer (1972). For another introductory exposition, see Bartos (1967).

We will begin in section 16.2 with one of the earliest models, which at the same time presents us with a relatively straightforward model of intergenerational mobility. The model is a discrete-time finite-state Markov chain of the regular type. In the following two sections, the classic work of Blumen, Kogan, and McCarthy (1955) is outlined; the location of the problem of underestimation of the "inertial," or staying, element in the process is a central result, followed by efforts to solve the problem by more-elaborate models going beyond the simple discrete-time regular chain. This "semi-Markov" elaboration is treated in section 16.4.

In the last three sections, we survey three distinct approaches that emerge out of a sense of dissatisfaction with the early discrete-time Markov models. Thomas Mayer's efforts to extend the continuous-time stochastic process conceptualization, first introduced in the field in any extensive way in the work of James Coleman (see Chapter 13), are treated in section 16.5. To avoid treatment of the analytical complexities that arise in the estimation phase, the effort of this review of Mayer's models is to acquaint the reader with the setup of the models; in particular, one aim is to see how an in-the-small transition picture leads naturally to the "instantaneous generator" matrices that Mayer uses to represent the basic mechanisms of the models.

In section 16.6, the path model developed by Blau and Duncan in the context of their study of the American occupational structure (Blau and Duncan, 1967) is introduced. This provides an opportunity to introduce the reader to the logic of path models, via a definite set of sequential steps corresponding to the generic methodology of setup, analysis, and application. The use of random variables and expectations leads to a relatively compact introduction to the mathematics of path analysis, leading into the Blau-Duncan model.

Finally, in section 16.7 the conceptualization of the mobility phenomenon shifts to the analysis of movement of vacancies rather than people. The idea was introduced by Harrison White and is treated in great detail in his monograph *Chains of Opportunity* (1970). Our aim is to exhibit the logic of the approach, its assumptions and mathematical techniques, rather than to review the associated empirical methodology of tracing chains of vacancies.

Finally, in section 16.8 we mention some of the other model-building efforts in this area that were not covered in our survey, doing so in the context of a discussion of the state-space problem in mobility theory.

16.2. A Discrete-Time Model.

In the study of intergenerational social mobility, an often cited set of data is reported by Glass and Hall (1954) in their analysis of social mobility in Great Britain. They divided occupations into seven classes:

C_1: professional and high administrative
C_2: managerial and executive
C_3: supervisory and nonmanual (higher grade)
C_4: supervisory and nonmanual (lower grade)
C_5: skilled manual and routine nonmanual
C_6: semiskilled manual
C_7: unskilled manual

They obtained a sample of nearly 3,500 pairs of fathers and sons and, so, data on the frequency of movement from class C_i to class C_j intergenerationally. A mathematical analysis starting from the intergenerational mobility matrix (16.2.1) was conducted by Prais (1955).

Son

		C_1	C_2	C_3	C_4	C_5	C_6	C_7
	C_1	.388	.147	.202	.062	.140	.047	.016
	C_2	.107	.267	.227	.120	.207	.053	.020
(16.2.1)	C_3	.035	.101	.188	.191	.357	.067	.061
Father	C_4	.021	.039	.112	.212	.431	.124	.062
	C_5	.009	.024	.075	.123	.473	.171	.125
	C_6	.000	.013	.041	.088	.391	.312	.155
	C_7	.000	.008	.036	.083	.364	.235	.274

The interpretation of the matrix entries is straightforward. For instance, the entry .008 in cell (C_7, C_2) means that less than 1 percent of the sons of men in unskilled manual occupations entered managerial and executive positions. Note that each row of matrix (16.2.1) sums to unity.

To treat these data in terms of a mathematical model, we make the following assumptions:

Assumption 1. The population forms a set of family lines, where only one son of a given father is considered in tracing the line.

Assumption 2. The probability $P(C_i \rightarrow C_j$ from n to n + 1) of a transition from C_i at time n to C_j at time n + 1 for a particular family is independent of the path of classes previously occupied by that family.

Assumption 3. The transition probability of assumption 2 itself is time-invariant, so that in any generation there is a probability $P(C_i \rightarrow C_j)$ of moving from C_i to C_j that is not changing over generations.

Assumption 4. The probability $P(C_i \rightarrow C_j)$ is the same for any family line.

Assumption 5. The families trace out their occupational paths independently.

Of these five assumptions, the first is a representation axiom that maps the population (spread over a more-or-less large time domain, such as a century) into family lines. Since distinct sons of a man may be in different occupational classes, the single-son treatment is an idealization. If we try to include all sons, then assumption 5 would undoubtedly fail. The second assumption is the Markov assumption of path independence and assumption 3 makes it time-homogeneous. Finally, assumptions 4 and 5 together state that the 3,500 families under investigation "form a random sample" of the theoretical process: each family process has the same parameters, and they are mutually independent. We can say that each family's occupational history is generated by the identical process operating independently over distinct families.

Under assumptions 2 and 3, there exists a one-step Markov transition matrix, say **P**, of order 7×7. Then entry $p_{ij} = P(C_i \rightarrow C_j)$ may be estimated by the method of maximum likelihood treated in section 15.2. Namely,

$$(16.2.2) \qquad L = P(C_i, C_j \text{ data}) = p_{ij}^{n_{ij}} (1 - p_{ij})^{n_i. - n_{ij}}$$

where n_{ij} is the number of fathers in C_i who have sons in C_j and $n_{i.}$ is the number of fathers in C_i. Expression (16.2.2) arises because of assumptions 4 and 5. Without assumption 4 we would have

$$L = p_{ij}^{(1)} p_{ij}^{(2)} \cdots p_{ij}^{(n_{ij})} (1 - p_{ij})^{(n_{ij}+1)} (1 - p_{ij})^{(n_{ij}+2)} \cdots (1 - p_{ij})^{n_i.}$$

where $p_{ij}^{(k)}$ is $P(C_i \rightarrow C_j)$ for family k. We see that L has the form of n_{ij} successes in $n_{i.}$ independent trials, so that by the technique of section 15.2 it is easy to show that

(16.2.3)
$$\hat{p}_{ij} = \frac{n_{ij}}{n_{i.}}$$

These are the entries in the data matrix (16.2.1), now thought of as estimates of the transition parameters of the Markov model, $\hat{P} = (\hat{p}_{ij})$.

The general classification of Markov chains given in section 10.9 shows that P should be considered regular on the basis of these estimates. Even though $\hat{P}$ is not all positive, $\hat{P}^2$ will be; for instance,

$$P(C_7 \rightarrow C_3 \rightarrow C_1) = P(C_7 \rightarrow C_3) P(C_3 \rightarrow C_1) = (.036)(.035) > 0$$
$$P(C_6 \rightarrow C_3 \rightarrow C_1) = P(C_6 \rightarrow C_3) P(C_3 \rightarrow C_1) = (.041)(.035) > 0$$

so that the two-step transition probabilities from C_7 to C_1 and from C_6 to C_1 are nonzero (of course, C_3 is only one intermediate class but sufficient to show nonzero probabilities).

Hence, we assume as part of the model that P is regular. The general results of section 10.11 apply here:

(a) $p(n) = p(0)P^n$ (10.11.3)

(b) $p_e = p_e P$ (10.11.5)

(c) $\lim_{n \to \infty} p(n) = p_e$ (10.11.8)

Expression (a) says that the distribution of families in occupational classes at any time n measured in generations, with the initial set of fathers called generation 0, is given by postmultiplying the initial distribution by the n-generation transition matrix, whose entries are

$$P(C_i \rightarrow C_j \text{ in n generations})$$

Expression (b) is the definition of the equilibrium state of the probabilistic system. Since the chain is regular, theorem (10.11.8) implies that this equilibrium exists, is unique, and satisfies condition (c), meaning that it is a stable equilibrium. In particular, after a sufficient number of generations have passed, the system should be quite near the equilibrium distribution p_e.

Prais did the necessary calculations to yield the solution of the equilibrium equation (b), obtaining (based on $\hat{P}$),

(16.2.4) $\hat{p}_e = (.023, .041, .088, .127, .410, .182, .129)$

If the process is assumed to be in equilibrium, then the distribution of fathers and of sons should be nearly alike and with common expected values in (16.2.4). In fact:

(16.2.5)
$$p(\text{fathers}) = (.037, .043, .098, .148, .432, .131, .111)$$
$$p(\text{sons}) \;\;\; = (.029, .046, .094, .131, .409, .170, .121)$$

It is apparent that the vectors in (16.2.5) are quite similar to the theoretical expectation (16.2.4). Bartholomew (1967) notes that there may be a trend here in classes C_6 and C_7. The probabilities do move up from fathers to sons and also "from sons to equilibrium," suggesting that the process may be not yet in equilibrium but still approaching it.

To sum up this first model, it is defined by six conditions—assumptions 1–5 listed above and the regular chain assumption. These assumptions yield an estimate $\hat{\mathbf{P}}$ of the transition matrix $\mathbf{P}$. Then general theory for regular chains implies that a particular equilibrium vector $\mathbf{p}_e$ must exist and that solving with estimate $\hat{\mathbf{P}}$ yields the estimated equilibrium vector $\hat{\mathbf{p}}_e$, which agrees well with the data. What is the logic of the application? It seems we are saying that if conditions 1–5 obtain and the chain is regular, then it is nearly in equilibrium at the time of observation. The estimate $\hat{\mathbf{P}}$ suggested the regular assumption. For the moment we take assumption 1 as an idealized representation in terms of family lines. We take assumption 5 as true because these 3500 families were probably only rarely connected in some way in a relational system that might form a Coleman-type network of influence rather than a set of independent loci of events. Assumptions 2–4 are open: nothing in the data or in theory compels us to accept them. On the contrary the main import of the application phase should be a close test of these assumptions, based on both extensive over-time data and more detailed analysis of subgroups.

Since assumptions 2–4 could be wrong but the equilibrium calculations yield a good fit, we can say that this first application gives good reason for pursuing the more refined data analysis problems with a view to more direct tests of the problematic assumptions. Of course, a universal assumption of time-stationarity or of family-homogeneity is not intended: there is no denying that in some social conditions, one or more of these uniformity assumptions may fail. The function of the tests is to discover the scope of application of the simplest models of this kind and, more important, to subject selected features of these models to tests that have the outcome of producing more valid representations.

Let us return to assumption 1 referent to family lines. Of course, a man has a random number of sons, 0, 1, 2, . . ., so that enormous idealization is involved. But even if we consider only family lines in which the assumption is not too far from wrong, there is a further difficulty in terms of the way generations of men have been thought of. If each family line were literally traced and, when the data were aggregated, step 1 in the process referred to all sons of the men of step 0, step 2 referred to all sons of the sons at step 1, and so forth, no additional problem would arise. But the typical way in which generations are thought of is different: the men born in, say, 1930 form one generation. Their fathers form another, their sons a third, This leads to thinking of the steps in the Markov chain as 30-year intervals, each of which has an associated "generation."

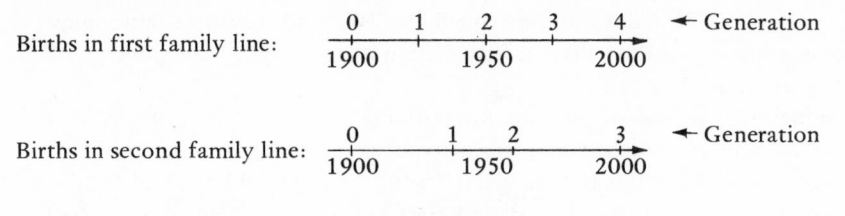

Births in first family line:

Births in second family line:

Figure 16.2.1

However, Duncan (1966) has shown that this idea is fallacious. For instance, in data collected in the context of the analysis of the American occupational structure (Blau and Duncan, 1967), Duncan shows that the fathers of men born in the years 1925–1934 scatter in their birthdays over 11 decades, with only the modal decade being 30 years back. Conversely, fathers born in 1900 have sons with birthdays scattered over many decades. The problem is shown in terms of family lines in Figure 16.2.1, where from an arbitrary 1900 starting birth date of two men we record the birthdays of their sons, their sons' sons, and so forth. In the year 2000 one family line is in its fifth generation, while the other is in its fourth. Not only are the fathers of the year-2000 sons different in their years of birth but the chains at year 2000 do not coordinate to a single step n in the Markov sense: one process is at step 4, the other step 3. Except in equilibrium conditions, this results in lack of coherence when the data are pooled to form sample estimates of the distribution of the "generation" born in year 2000. The same argument applies to the ostensible equivalence in 1900, but there we might agree counting begins with a pool of "starters" in tracing family lines.

The conclusion is that "intergenerational mobility" may be a Markov chain but that in the application of the model the counting process to arrive at parameter estimates must be based on careful pooling of comparable data and not on largely fictitious "generations" in the sense in which they are identified as if they were cohorts.

Given Duncan's critique of the Markov chain approach to intergenerational mobility, it is natural to turn to intragenerational mobility as a source of new data and insights into representational problems in the study of social mobility.

16.3.   The Mover-Stayer Model. With the publication of their monograph in 1955, Blumen, Kogan, and McCarthy (BKM) initiated a whole line of research ideas based upon a dichotomous classification of jobholders as "movers" or "stayers." The initiation of this distinction came from a very close study of over-time transition data based on social security records giving occupational identifications at quarterly intervals. In all, 12 consecutive quarters were the time points at which data existed. Moreover, BKM divided their sample of over 16,000 individuals into age and sex groups, computing matrices separately in each category. The industries were coded into 10 classes; a class of "unemployed" was also included, so that the distribution vectors have 11 components and the matrices are 11 × 11.

In their first analysis of the data, BKM attempted to fit a Markov model with the usual time and individual homogeneity assumptions. In fact, assumptions 2–5 of section 4.2 define their model if "family" is replaced by "individual." Hence, there is one theoretical matrix $\mathbf{P}$ for quarterly transitions. The quarterly data (transition) matrices may be pooled to yield a maximum likelihood estimate $\hat{\mathbf{P}}$ of $\mathbf{P}$. According to Markov theory, the matrices for longer transitions are simply powers of $\mathbf{P}$. For example, consider

$$P(C_i \to C_j \text{ from } n = 0 \text{ to } n = 8)$$

The probability that an individual moves from job class C_i to class C_j in eight quarters from the initial time point is given by the (i, j) entry of $\mathbf{P}^8$. When BKM computed the eight-quarter-data transition matrix and compared it with $\hat{\mathbf{P}}^8$ they found a pattern of discrepancy that has been replicated since that time in other data: the main diagonal entries in $\hat{\mathbf{P}}^8$ tend to underestimate the observed proportions. In other words, the Markov model they employed predicted more movement than actually occurred or, equivalently, predicted less "inertia" or "staying" than observed in each category.

One example of such data and predictions for the main diagonal only (i.e., for the probability of staying) is shown in Table 16.3.1. This table makes clear the regularity of the result and its substantial magnitude in most instances. This defines an important phenomenon in the very process of showing the failure of the baseline Markov model, and this regularity is the starting point for much additional work that aims to produce a better model.

Table 16.3.1 Observed and Predicted Proportion Staying in Given Job Class

(8 quarters; males; 20–24)

Class	Observed	Prediction
C_1	.000	.002
C_2	.449	.144
C_3	.461	.176
C_4	.459	.218
C_5	.489	.276
C_6	.440	.166
C_7	.491	.261
C_8	.439	.158
C_9	.339	.105
C_{10}	.048	.006
C_{11}	.482	.346

To make this finding logically clear, let us rather formally list the premises and the conclusion:

Premise 1.   The labor mobility process is Markov (i.e., path independent).

Premise 2.   The one-step transition matrix is time invariant.

Premise 3.   Each individual in a given age-sex group has the same parameter values.

Premise 4.   Each individual independently traces out the process.

Conclusion.   The predicted "staying" is too low.

Hence, by contrapositive reasoning (see section 2.11) we conclude that one or more of the four premises is at fault. Suppose we want to preserve the Markov approach, and that independence (premise 4) makes sense in terms of the way the sample of individuals is selected (so that they are an aggregate whose actions are statistically independent). Also, it appears that BKM, with transition matrices available at many time points, satisfied themselves that such matrices are all estimates of one matrix P (see Anderson and Goodman, 1955, for a treatment of tests of such statistical hypotheses). Hence, the "culprit" is the assumed homogeneity of individuals if we maintain the process is Markov.

BKM initiated from this background, then, the idea that the workers are not homogeneous but of two types: movers and stayers. The latter have a transition matrix of the form

$$
\begin{array}{c}
\\
C_1\\
C_2\\
C_3\\
\vdots\\
C_N
\end{array}
\begin{array}{ccccc}
C_1 & C_2 & C_3\ldots & & C_N\\
\left(\begin{array}{c}1\\0\\0\\ \\ \\ \\0\end{array}\right. & \begin{array}{c}0\\1\\0\\ \\ \\ \\0\end{array} & \begin{array}{c}0\ldots\\0\ldots\\1\ldots\\ \\ \\ \\0\ldots\end{array} & & \left.\begin{array}{c}0\\0\\0\\ \\ \\ \\1\end{array}\right)
\end{array}
$$

while the movers have a standard unconstrained transition matrix, say $M = (m_{ij})$

At any time, each class C_i includes a fraction s_i of stayers, who therefore stay in that class in the next transition. It also includes a fraction $1 - s_i$ of movers who, in accordance with matrix M, may move from C_i or happen to stay in C_i in the next transition. BKM define a matrix

$$(16.3.1) \qquad \mathbf{S} = (s_i \delta_{ij}) = \begin{pmatrix} s_1 & & & & \\ & s_2 & & & \mathbf{O} \\ & & \cdot & & \\ & \mathbf{O} & & \cdot & \\ & & & & s_N \end{pmatrix}$$

The fractions of movers are then given by

$$(16.3.2) \qquad \mathbf{I} - \mathbf{S} = ((1 - s_i)\, \delta_{ij}) = \begin{pmatrix} 1 - s_1 & & & & \\ & 1 - s_2 & & & \mathbf{O} \\ & & \cdot & & \\ & \mathbf{O} & & \cdot & \\ & & & & 1 - s_N \end{pmatrix}$$

Let us consider now the transition from time n to n + 1. Consider an arbitrary worker in class C_i at n. Then, with probability s_i, the worker is a stayer and, so, stays in C_i at n + 1; with probability $1 - s_i$, the worker is a mover and goes to C_j with probability m_{ij}. Hence, for all i, j, the transition probability p_{ij} is given by

$$(16.3.3) \qquad p_{ij} = \begin{cases} s_i + (1 - s_i)m_{ii} & \text{(if } i = j) \\ (1 - s_i)m_{ij} & \text{(if } i \neq j) \end{cases}$$

In generic form,

$$p_{ij} = s_i \delta_{ij} + (1 - s_i)m_{ij} \qquad \text{(all i, j)}$$

In matrix terms, this yields

$$(p_{ij}) = (s_i \delta_{ij}) + ((1 - s_i)m_{ij})$$

and using (16.3.1) and (16.3.2), with $\mathbf{P} = (p_{ij})$,

$$(16.3.4) \qquad \mathbf{P} = \mathbf{S} + (\mathbf{I} - \mathbf{S})\mathbf{M}$$

From the standpoint of the BKM empirical study, this is a representation of the observable one-quarter matrix $\mathbf{P}$ in terms of the unobservable components $\mathbf{S}$ and $\mathbf{M}$.

For two-step transitions, we have the situation depicted in the tree diagram of Figure 16.3.1, using N = 2 to illustrate the ideas and taking a worker in C_1 at time n. We see that by adding the path probabilities of all paths terminating in C_1 at n + 2

$$p_{11}^{(2)} = P(C_1 \longrightarrow C_1 \text{ in 2 steps}) = s_1 + (1 - s_1)(m_{11}^2 + m_{12}m_{21})$$

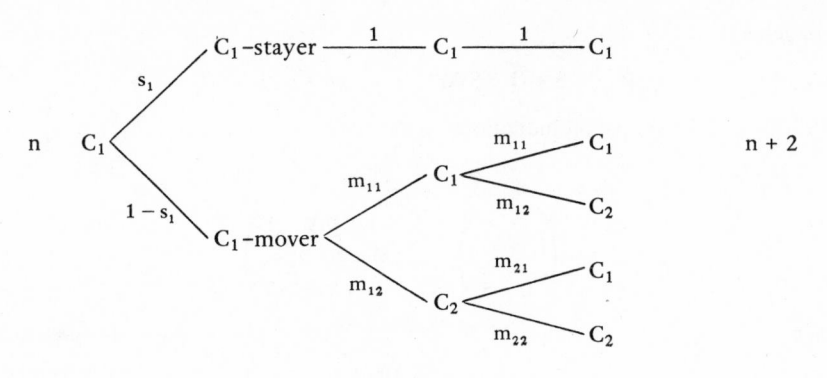

Figure 16.3.1

Also,

$$p_{12}^{(2)} = P(C_1 \longrightarrow C_2 \text{ in 2 steps}) = (1 - s_1)(m_{11}m_{12} + m_{12}m_{22})$$

Similarly,

$$p_{21}^{(2)} = P(C_2 \longrightarrow C_1 \text{ in 2 steps}) = (1 - s_2)(m_{22}m_{21} + m_{21}m_{11})$$

$$p_{22}^{(2)} = P(C_2 \longrightarrow C_2 \text{ in 2 steps}) = s_2 + (1 - s_2)(m_{22}^2 + m_{21}m_{12})$$

In matrix form these four equations can be arranged as follows:

$$
\begin{pmatrix} p_{11}^{(2)} & p_{12}^{(2)} \\ p_{21}^{(2)} & p_{22}^{(2)} \end{pmatrix} = \begin{pmatrix} s_1 + (1 - s_1)(m_{11}^2 + m_{12}m_{21}) & (1 - s_1)(m_{11}m_{12} + m_{12}m_{22}) \\ (1 - s_2)(m_{22}m_{21} + m_{21}m_{11}) & s_2 + (1 - s_2)(m_{22}^2 + m_{21}m_{12}) \end{pmatrix}
$$

$$
= \begin{pmatrix} s_1 & 0 \\ 0 & s_2 \end{pmatrix} + \begin{pmatrix} (1 - s_1)(m_{11}^2 + m_{12}m_{21}) & (1 - s_1)(m_{11}m_{12} + m_{12}m_{22}) \\ (1 - s_2)(m_{22}m_{21} + m_{21}m_{11}) & (1 - s_2)(m_{22}^2 + m_{21}m_{12}) \end{pmatrix}
$$

$$
= \begin{pmatrix} s_1 & 0 \\ 0 & s_2 \end{pmatrix} + \begin{pmatrix} 1 - s_1 & 0 \\ 0 & 1 - s_2 \end{pmatrix} \begin{pmatrix} m_{11}^2 + m_{12}m_{21} & m_{11}m_{12} + m_{12}m_{22} \\ m_{22}m_{21} + m_{21}m_{11} & m_{22}^2 + m_{21}m_{22} \end{pmatrix}
$$

Hence,

$$\mathbf{P}^{(2)} = \mathbf{S} + (\mathbf{I} - \mathbf{S})\mathbf{M}^2$$

where $\mathbf{P}^{(2)} = (p_{ij}^{(2)})$.

In general,

(16.3.5) $$\mathbf{P}^{(n)} = \mathbf{S} + (\mathbf{I} - \mathbf{S})\mathbf{M}^n \qquad (n = 0, 1, \ldots)$$

And this may be shown by induction.

Example. Let

$$\mathbf{S} = \begin{pmatrix} .4 & 0 \\ 0 & .5 \end{pmatrix} \qquad \mathbf{M} = \begin{pmatrix} .9 & .1 \\ .3 & .7 \end{pmatrix}$$

Then

$$\mathbf{P} = \mathbf{S} + (\mathbf{I} - \mathbf{S})\mathbf{M}$$

$$= \begin{pmatrix} .4 & 0 \\ 0 & .5 \end{pmatrix} + \begin{pmatrix} .6 & 0 \\ 0 & .5 \end{pmatrix}\begin{pmatrix} .9 & .1 \\ .3 & .7 \end{pmatrix}$$

$$= \begin{pmatrix} .4 & 0 \\ 0 & .5 \end{pmatrix} + \begin{pmatrix} .54 & .06 \\ .15 & .35 \end{pmatrix}$$

$$= \begin{pmatrix} .94 & .06 \\ .15 & .85 \end{pmatrix}$$

Hence,

$$\mathbf{P}^2 = \begin{pmatrix} .94 & .06 \\ .15 & .85 \end{pmatrix}\begin{pmatrix} .94 & .06 \\ .15 & .85 \end{pmatrix} = \begin{pmatrix} .89 & .11 \\ .27 & .73 \end{pmatrix}$$

Also,

$$\mathbf{M}^2 = \begin{pmatrix} .9 & .1 \\ .3 & .7 \end{pmatrix}\begin{pmatrix} .9 & .1 \\ .3 & .7 \end{pmatrix} = \begin{pmatrix} .84 & .16 \\ .48 & .52 \end{pmatrix}$$

Hence,

$$\mathbf{P}^{(2)} = \mathbf{S} + (\mathbf{I} - \mathbf{S})\mathbf{M}^2$$

$$= \begin{pmatrix} .4 & 0 \\ 0 & .5 \end{pmatrix} + \begin{pmatrix} .6 & 0 \\ 0 & .5 \end{pmatrix}\begin{pmatrix} .84 & .16 \\ .48 & .52 \end{pmatrix}$$

$$= \begin{pmatrix} .4 & 0 \\ 0 & .5 \end{pmatrix} + \begin{pmatrix} .50 & .10 \\ .24 & .26 \end{pmatrix}$$

$$= \begin{pmatrix} .90 & .10 \\ .24 & .76 \end{pmatrix}$$

$$\neq \mathbf{P}^2$$

As we see in this example, the intended effect of introducing the mover-stayer representation is achieved: the main diagonal terms of $\mathbf{P}^{(2)}$ exceed those of $\mathbf{P}^2$.

An assumption is now made that is based on the form of the data: $\mathbf{M}$ is regular. Hence, $\mathbf{M}$ has an equilibrium vector and the powers $\mathbf{M}^n$ approach a matrix all of whose rows contain the equilibrium vector, because of the asymptotic stability (and so independence of the row starting state).

Hence, since

(16.3.6)
$$\lim_{n \to \infty} \mathbf{P}^{(n)} = \mathbf{S} + (\mathbf{I} - \mathbf{S}) \lim_{n \to \infty} \mathbf{M}^n$$

it follows that the rows of $\lim_{n \to \infty} \mathbf{P}^{(n)}$ are not generally identical: the equilibrium distribution depends on the class C_i via the stayer-mover composition.

Given $\mathbf{M}$ of the previous example, let us find $\lim_{n \to \infty} \mathbf{P}^{(n)}$. We first find the equilibrium vector of $\mathbf{M}$:

$$(x_1, x_2) = (x_1, x_2) \begin{pmatrix} .9 & .1 \\ .3 & .7 \end{pmatrix}$$

$$= (.9x_1 + .3x_2, .1x_1 + .7x_2)$$

Since $x_2 = 1 - x_1$,

$$x_1 = .9x_1 + .3(1 - x_1) = .3 + .6x_1$$

which yields $x_1 = .75$. Thus,

$$\lim_{n \to \infty} \mathbf{M}^n = \begin{pmatrix} .75 & .25 \\ .75 & .25 \end{pmatrix}$$

but, using (16.3.6),

$$\lim_{n \to \infty} \mathbf{P}^{(n)} = \begin{pmatrix} .4 & 0 \\ 0 & .5 \end{pmatrix} + \begin{pmatrix} .6 & 0 \\ 0 & .5 \end{pmatrix} \begin{pmatrix} .75 & .25 \\ .75 & .25 \end{pmatrix}$$

$$= \begin{pmatrix} .4 & 0 \\ 0 & .5 \end{pmatrix} + \begin{pmatrix} .450 & .150 \\ .375 & .125 \end{pmatrix}$$

$$= \begin{pmatrix} .850 & .150 \\ .375 & .625 \end{pmatrix}$$

Hence, in equilibrium those who started in C_1 are still 85 percent concentrated in C_1, while those who started in C_2 are only 62.5 percent concentrated in C_2. The difference results from the difference in m_{11} and m_{22} in M as well as in the difference in stayer fractions.

Even though S and M are not observable, BKM were able to obtain numerical predictions to assess the extent to which they had solved the main diagonal problem. Table 16.3.2 reports some of the results. On the whole, the predictions are excellent, as compared with Table 16.3.1. However, in other sets of data, the predictions were not quite so good. This led BKM to a more elaborate model that we introduce in the next section. For a treatment of the estimation problem in the mover-stayer model, see Goodman (1961).

16.4. The Semi-Markov Mover-Stayer Model.   In the above work, it was assumed that an individual is always a mover or always a stayer. A relaxation of this assumption lets this condition of an individual be time-varying. To visualize this, let a person be in one of two states in each of m subintervals of a given quarter:

$$X_i = \begin{cases} 1 & \text{if in state "mover" in subinterval i} \\ 0 & \text{if in state "stayer" in subinterval i} \end{cases} \qquad (i = 1, 2, \ldots, m)$$

Let

$$X = \sum_{i=1}^{m} X_i$$

be the number of times the person is in the mover state. Let

$$r_x = P(X = x) \qquad (x = 0, 1, \ldots, m)$$

be the probability distribution of X.

Now if a person enters the mover state X times in a given quarter, then M is multiplied X times to yield the X-step transition probabilities:

$$M^X = \text{matrix of X-step transition probabilities.}$$

Table 16.3.2 Observed and Predicted Proportion Staying in Given Job Classes

(8 quarters; males; 20–24)

Class	Observed	Mover-Stayer Model Prediction
C_1	.000	.003
C_2	.449	.442
C_3	.461	.464
C_4	.459	.474
C_5	.489	.512
C_6	.440	.444
C_7	.491	.489
C_8	.439	.446
C_9	.339	.338
C_{10}	.048	.049
C_{11}	.482	.536

Hence, taking an expected value,

$$E(M^X) = \sum_x M^x \, P(X = x) = \sum_x r_x M^x$$

Example. Suppose that m = 2 and we assume that with chance p = .5 the person enters state "mover" independently in each subinterval, then

$$r_0 = P(X = 0) = \frac{1}{4}$$

$$r_1 = P(X = 1) = \frac{1}{2}$$

$$r_2 = P(X = 2) = \frac{1}{4}$$

Hence,

$$E(M^X) = r_0 M^0 + r_1 M + r_2 M^2 = .25I + .50M + .25M^2$$

If, as in earlier examples,

$$M = \begin{pmatrix} .9 & .1 \\ .3 & .7 \end{pmatrix}$$

then,

$$E(M^X) = \begin{pmatrix} .25 & 0 \\ 0 & .25 \end{pmatrix} + \begin{pmatrix} .45 & .05 \\ .15 & .35 \end{pmatrix} + \begin{pmatrix} .21 & .04 \\ .12 & .13 \end{pmatrix} = \begin{pmatrix} .91 & .09 \\ .27 & .73 \end{pmatrix}$$

We see that at the end of the first quarter, the transition matrix $\mathbf{P}$ is given by $E(\mathbf{M}^X)$. In the special case that all the probability on X is concentrated at X = 1 we retrieve the simple Markov model, $\mathbf{P} = E(\mathbf{M}) = \mathbf{M}$. Note that the matrix $\mathbf{S}$ no longer plays a role, because there is no fixed body of individuals called "stayers." Rather "stayer" is a state that may characterize an individual in opposition to the alternative "mover."

If we consider n quarters, then there are nm subintervals, and so, X has possible values x = 0, 1, . . ., nm,

$$(16.4.1) \qquad \mathbf{P}^{(n)} = E(\mathbf{M}^X) = \sum_{x=0}^{nm} r_x \mathbf{M}^x$$

Assume in our example that in each subinterval the event that the person is in state "mover" occurs with chance p independently of earlier events in this series, so that the sum X is binomially distributed with parameters p and nm. Thus,

$$(16.4.2) \qquad \mathbf{P}^{(n)} = \sum_{x=0}^{nm} \binom{nm}{x} p^x (1-p)^{nm-x} \mathbf{M}^x$$

Taking p = 1/2,

$$\mathbf{P}^{(n)} = \sum_{x=0}^{nm} \binom{nm}{x} \left(\frac{1}{2}\right)^{nm} \mathbf{M}^x$$

For instance, if n = 2, m = 2, then

$$\mathbf{P}^{(2)} = \frac{1}{16}(\mathbf{I} + 4\mathbf{M} + 6\mathbf{M}^2 + 4\mathbf{M}^3 + \mathbf{M}^4)$$

The idea of a subinterval of definite length during which a mover-event occurs suggests a more natural representation along the following lines. Let exposure events occur at random in time and each time such an event occurs, the person moves with probabilities taken from the mover matrix $\mathbf{M}$. Instead of a series of m independent trials in a given quarter, we substitute the idea of an arbitrary number of exposure events as determined by a random process.

The process is postulated by BKM to be Poisson. We explicate the meaning of this postulate as follows. Let E be the recurrent exposure event that a worker might experience 0, 1, 2, . . . times in a given time period. Define

$$X(t) = \text{number of times E occurs by time t} \qquad (t \geqslant 0)$$
$$r_x(t) = P[X(t) = x] \qquad (x \geqslant 0, t \geqslant 0)$$

To say that the process is Poisson means it is like a series of independent trials with fixed probability of an event except that the time interval of occurrence of the event is not a preconceived "trial." We say that in the small, for small Δt,

(16.4.3)

    (1) there is a probability $a\Delta t$ of the occurrence of E, which probability remains the same for all intervals $(t, t + \Delta t)$ no matter what the actual time t, and

    (2) events in disjoint intervals of time are independent.

Here a is a fixed exposure rate.

We aim to derive the probability distribution $r_x(t)$, for any t.

With reference to the two mutually exclusive ways, shown in Figure 16.4.1, in which there can be x occurrences of E by time $t + \Delta t$,

$$(16.4.4) \qquad r_x(t + \Delta t) = r_x(t)(1 - a\Delta t) + r_{x-1}(t)a\Delta t \qquad (x \geqslant 1)$$

We try to set up a differential equation for $r_x(t)$. We subtract $r_x(t)$ from both sides of (16.4.4):

$$r_x(t + \Delta t) - r_x(t) = -ar_x(t)\Delta t + ar_{x-1}(t)\Delta t$$

Now dividing by Δt and taking limits,

$$(16.4.5) \quad r_x'(t) = \lim_{\Delta t \to 0} \frac{r_x(t + \Delta t) - r_x(t)}{\Delta t} = -ar_x(t) + ar_{x-1}(t) \qquad (x \geqslant 1)$$

Similarly, for x = 0, we obtain for the probability of no exposure by time $t + \Delta t$,

$$r_0(t + \Delta t) = r_0(t)(1 - a\Delta t)$$

This yields the differential equation,

$$(16.4.6) \qquad\qquad r_0'(t) = -ar_0(t)$$

The system of equations (16.4.6) and (16.4.5) is solved as follows:

$$\frac{r_0'(t)}{r_0(t)} = -a$$

$$(16.4.7) \qquad \int_0^T \frac{r_0'(t)}{r_0(t)}\,dt = \int_0^T -a\,dt \qquad \text{(integrate from time 0 to time T)}$$

We recognize that

$$\frac{r_0'(t)}{r_0(t)} = \frac{d}{dt}\ln r_0(t)$$

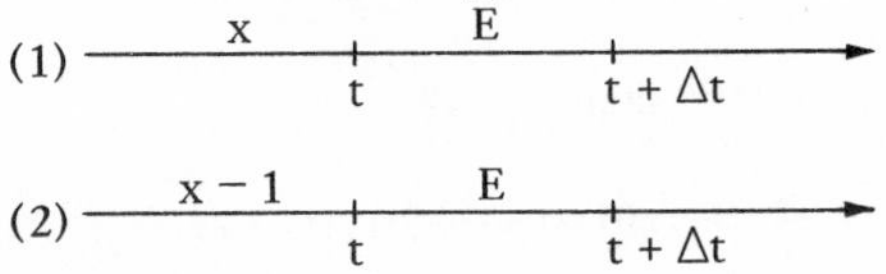

Figure 16.4.1

and so the fundamental theorem of calculus implies

$$\int_0^T \frac{r_0'(t)}{r_0(t)} \, dt = \ln r_0(t) \Big|_0^T = \ln r_0(T) - \ln r_0(0)$$

Hence, (16.4.7) becomes

$$\ln r_0(T) - \ln r_0(0) = -aT$$

$$\ln \frac{r_0(T)}{r_0(0)} = -aT$$

$$\frac{r_0(T)}{r_0(0)} = e^{-aT}$$

$$r_0(T) = r_0(0)e^{-aT}$$

However,

$$r_0(0) = 1 = \text{probability of no event by time 0, when we start counting.}$$

Hence, replacing T by t,

(16.4.8) $$r_0(t) = e^{-at} \qquad (t \geqslant 0)$$

Next, we solve (16.4.5) with $x = 1$,

$$r_1'(t) = -ar_1(t) + ar_0(t) = -ar_1(t) + ae^{-at}$$

Rearranging,

(16.4.9) $$r_1'(t) + ar_1(t) = ae^{-at}$$

We now seek to express the left side of (16.4.9) in the form

$$\frac{d}{dt}[r_1(t)f(t)]$$

so that we can integrate directly on both sides to obtain $r_1(t)$. The reason we think this will be possible is the product rule of differentiation:

$$\frac{d}{dt}[r_1(t)f(t)] = r_1'(t)f(t) + r_1(t)f'(t)$$

and we see that if $f'(t) = af(t)$, then

$$\frac{d}{dt}[r_1(t)f(t)] = f(t)[r_1'(t) + ar_1(t)]$$

This suggests a function $f(t)$ should be used to multiply both sides of (16.4.9) and that f(t) should satisfy

$$f'(t) = af(t)$$

But this is (16.4.6) all over again, so that $f(t) = e^{at}$ does the job. We term e^{at} an integrating factor. Returning to (16.4.9), having tried to provide the intuition behind the next step,

$$r_1'(t)e^{at} + r_1(t)ae^{at} = ae^{-at}e^{at} \qquad \text{[using integrating factor } f(t) = e^{at}]$$

$$\frac{d}{dt}[r_1(t)\,e^{at}] = a$$

Hence, integrating both sides from 0 to T,

$$\int_0^T \frac{d}{dt}[r_1(t)e^{at}]\ dt = \int_0^T a\,dt$$

$$r_1(t)e^{at}\,\bigg|_0^T = aT$$

$$r_1(T)e^{at} - r_1(0) = aT$$

However, $r_1(0) = 0$. Hence, letting t replace T,

(16.4.10) $$r_1(t) = ate^{-at} \qquad (t \geqslant 0)$$

Returning to (16.4.5) and letting x = 2,

$$r_2'(t) = -ar_2(t) + ar_1(t) = -ar_2(t) + a^2te^{-at}$$

$$r_2'(t) + ar_2(t) = a^2te^{-at}$$

$$\frac{d}{dt}[r_2(t)e^{at}] = a^2t \qquad \text{[using integrated factor } f(t) = e^{at}]$$

$$\int_0^T \frac{d}{dt}[r_2(t)e^{at}]\,dt = \int_0^T a^2t\,dt$$

$$r_2(t)e^{at}\,\bigg|_0^T = a^2\frac{t^2}{2}\,\bigg|_0^T$$

$$r_2(T)e^{aT} - r_2(0)e^{aT} = \frac{(aT)^2}{2}$$

But $r_2(0) = 0$, so that after replacing T by t,

(16.4.11) $$r_2(t) = \frac{(at)^2}{2}e^{-at} \qquad (t \geqslant 0)$$

In general, and this can be shown by induction,

(16.4.12) $r_x(t) = \dfrac{(at)^x}{x!} e^{-at}$ $(x = 0, 1, \ldots; t \geqslant 0)$

This is the Poisson distribution with parameter at time t. It gives the probability distribution of the number of exposures by time t, for any t.

The expected value of the number X(t) of exposure events by time t is given by

$$E[X(t)] = \sum_{x=0}^{\infty} xP[X(t) = x]$$

$$= \sum_{x=1}^{\infty} xr_x(t)$$

$$= \sum_{x=1}^{\infty} x \frac{(at)^x}{x!} e^{-at}$$

$$= e^{-at} at \sum_{x=1}^{\infty} \frac{(at)^{x-1}}{(x-1)!}$$

$$= e^{-at} at\, e^{at}$$

$$= at$$

since for any z,

$$e^z = \sum_{n=0}^{\infty} \frac{z^n}{n!} \qquad \text{(take } n = x - 1 \text{ in the above)}$$

Hence, the expected number of exposures is a linear function of time.

The matrix for one-quarter transitions—namely, P—is given as the expected value of the powers of the mover matrix, which is applied X(1) times if X(1) exposures occur by t = 1. Hence,

(16.4.13) $P = E(M^{X(1)}) = \sum_{x} r_x(1)M^x$

The two-quarter transition matrix requires that we consider the average based on the number of exposures by t = 2. Hence,

$$P^{(2)} = \sum_{x} r_x(2)M^x$$

In general, the n-quarter transition matrix is given by,

(16.4.14) $P^{(n)} = \sum_{x} r_x(n)M^x$ $(n = 1, 2, \ldots)$

where, using (16.4.12),

$$(16.4.15) \qquad r_x(n) = \frac{(an)^x}{x!} e^{-an} \qquad (x = 0, 1, \ldots)$$

A process of this kind has two central features: (a) transitions occur in accordance with a Markov transition matrix M, and (b) the times between transitions are random, or, put another way, the time spent in any state is random. It need not be the case that in (b), a Poisson process is assumed. This was a special case. The name for a process with these two features is semi-Markov. They are treated in greater detail in Bartholomew (1967). The abstract structure of the type conceived by BKM has an additional feature: the process is observed at periodic intervals separated by time τ (see Figure 16.4.2). This observation schedule introduces an artificial element that the fundamental picture of the process should not include: that states are only observed at selected time points. The ongoing process is continuous. In this respect, the semi-Markov model continues the line of thought of the Coleman methodology in which continuous time stochastic processes are seen as the basic natural rendition of ongoing social processes. The artificial element of observation at $t = 0, \tau, 2\tau, \ldots$ was seen in the above work in the derivation of the exposure process determining the distribution of X at any time t and the resulting well-defined character of $E(M^{X(t)})$ for any time t, although we then "looked at it" at times 1, 2, . . . (taking $\tau = 1$ quarter).

Both BKM and Bartholomew show that in the semi-Markov model yielding (16.4.14) and (16.4.15),

$$(16.4.16) \qquad P^{(n)} = P^n$$

a property we had "lost" in first passing to a mover-stayer representation. Bartholomew gives the more general form in terms of unspecified τ, whereas we have followed BKM and taken $\tau = 1$.

However, the result (16.4.16) is at first somewhat embarrassing, for it yields the prediction that the eight-quarter matrix should be the eighth power of the one-quarter matrix, a result already disconfirmed as shown in Table 16.3.1. In fact, Bartholomew argues that there are only two processes of exposure events that yield (16.4.16): the regular, nonrandom schedule with which we began and the totally random process assumed above. Just as the main diagonal regularity disconfirms the simple chain, it disconfirms the first semi-Markov process we have introduced.

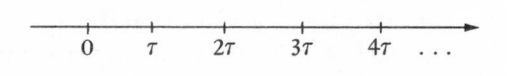

Figure 16.4.2

Both BKM and later, Bartholomew, solve this problem by introducing hetero-geneity. This is easily done in the semi-Markov model by letting the rate param-eter a have a distribution (i.e., treating it as a random variable). Let f(a)da be the probability that an individual's parameter lies between a and a + da. Then, averaging over individuals,

$$r_x(t) = E[r_x(t; a)]$$

where $r_x(t; a)$ is the Poisson form in (16.4.12) but where a is a random variable, so we have to take expectations. This yields,

$$(16.4.17) \qquad r_x(t) = \int_0^\infty r_x(t; a) \, f(a)da$$

According to Bartholomew (1967: p. 32) we can let $r_x(t; a)$ be Poisson, as as-sumed here, without incurring the undesirable consequence (16.4.16).

Finally, we note that Bartholomew demonstrates for a more general semi-Markov process the following result on limiting behavior of the process:

(16.4.18) Proposition. If matrix **M** is regular, then $\lim_{n \to \infty} \mathbf{P}^n = \lim_{n \to \infty} \mathbf{M}^n$ where **P** is given by

$$E(M^{X(1)}) = \sum_x r_x(1)M^x$$

In other words: the limit matrix based on the observation points is the same as the limiting matrix $\lim_{n \to \infty} \mathbf{M}^n$, where **M** applies whenever an exposure occurs. **M** is regular, so in $\lim \mathbf{M}^n$ the rows are identical and contain the components of the equilibrium vector. Hence, by proposition (16.4.18) we can compute the equilibrium distribution directly from **P**, the one-quarter observable transition matrix. Bartholomew reports data from BKM that show a good agreement be-tween predicted and observed equilibrium distributions, assuming the process studied by BKM reached equilibrium in the observation period.

As Bartholomew shows, it is easy to generalize and complicate the model still further; the evidence is that further work with this kind of model will be quite worthwhile. For a more recent discussion of semi-Markov models, see Ginsberg (1971), whose work we will touch upon again in section 16.8. For other appli-cations of the Poisson model, see Coleman (1964), Land (1971), and Spilerman (1970).

16.5. Continuous-Time Models. Working with cohort data from the Blau and Duncan (1967) study, T. Mayer (1972) has been formulating a series of models all based on the conceptualization of intragenerational career mobility as a stochastic process in continuous time, with a finite number of states. The states are termed status classes or strata. At each moment of time an individual is in a unique status class (determined, in the applications, by the prestige of his job) and changes of state occur at arbitrary times.

Within this rather broad framework, many different kinds of models may be studied by adding additional conditions. A convenient assumption to begin with is that of time-invariance as to the in-the-small probability transition rule. More exactly there is a single instantaneous matrix—recall section 13.2, and in particular, matrix (13.2.4.). As indicated in the special case of section 13.2, the system of equations obtained from the in-the-small transition analysis has the form, with $p(t)$ a column vector,

$$(16.5.1) \qquad\qquad \dot{p}(t) = Rp(t)$$

where R is the transpose of the instantaneous matrix of the continuous-time process. The example below illustrates a derivation of (16.5.1) in a particular mobility case.

Example. Assume four ordered status classes C_1, C_2, C_3, and C_4 and in-the-small transition chances confined to the nearest class above or below a given class. Denote such a chance by $r_{ij}\Delta t$. Then the matrix form is as follows:

$$t + \Delta t$$

$$
\begin{array}{c c}
 & \begin{array}{cccc} C_1 & C_2 & C_3 & C_4 \end{array} \\
t \begin{array}{c} C_1 \\ C_2 \\ C_3 \\ C_4 \end{array} &
\begin{pmatrix}
* & r_{12}\Delta t & 0 & 0 \\
r_{21}\Delta t & * & r_{23}\Delta t & 0 \\
0 & r_{32}\Delta t & * & r_{34}\Delta t \\
0 & 0 & r_{43}\Delta t & *
\end{pmatrix}
\end{array}
$$

The * diagonal entries are such as to make the rows sum to unity. Next, we use the law of total probability relative to the event "in C_i at $t + \Delta t$" partitioned by the classes at t. The notation is

$$p_i(t) = \text{probability in } C_i \text{ at time t} \qquad (t \geqslant 0)$$
$$(i = 1, 2, 3, 4)$$

Hence,

(1) $p_1(t + \Delta t) = p_1(t)(1 - r_{12}\Delta t) + p_2(t)r_{21}\Delta t$

(2) $p_2(t + \Delta t) = p_1(t)r_{12}\Delta t + p_2(t)(1 - r_{21}\Delta t - r_{23}\Delta t) + p_3(t)r_{32}\Delta t$

(3) $p_3(t + \Delta t) = p_2(t)r_{23}\Delta t + p_3(t)(1 - r_{32}\Delta t - r_{34}\Delta t) + p_4(t)r_{43}\Delta t$

(4) $p_4(t + \Delta t) = p_3(t)r_{34}\Delta t + p_4(t)(1 - r_{43}\Delta t)$

Subtracting $p_i(t)$ from equation (i), where i = 1, 2, 3, 4; dividing by Δt; and letting Δt go to 0 in the limit yields in column vector form:

$$
\begin{pmatrix} \dot{p}_1(t) \\ \dot{p}_2(t) \\ \dot{p}_3(t) \\ \dot{p}_4(t) \end{pmatrix} = \begin{pmatrix} -r_{12}p_1(t) + r_{21}p_2(t) \\ r_{12}p_1(t) - (r_{21} + r_{23})p_2(t) + r_{32}p_3(t) \\ r_{23}p_2(t) - (r_{32} + r_{34})p_3(t) + r_{43}p_4(t) \\ r_{34}p_3(t) - r_{43}p_4(t) \end{pmatrix}
$$

Hence, the form (16.5.1) is

$$
\begin{pmatrix} \dot{p}_1(t) \\ \dot{p}_2(t) \\ \dot{p}_3(t) \\ \dot{p}_4(t) \end{pmatrix} = \begin{pmatrix} -r_{12} & r_{21} & 0 & 0 \\ r_{12} & -(r_{21} + r_{23}) & r_{32} & 0 \\ 0 & r_{23} & -(r_{32} + r_{34}) & r_{43} \\ 0 & 0 & r_{34} & -r_{43} \end{pmatrix} \begin{pmatrix} p_1(t) \\ p_2(t) \\ p_3(t) \\ p_4(t) \end{pmatrix}
$$

The instantaneous matrix is

$$
\mathbf{R}^T = \begin{pmatrix} -r_{12} & r_{12} & 0 & 0 \\ r_{21} & -(r_{21} + r_{23}) & r_{23} & 0 \\ 0 & r_{32} & -(r_{32} + r_{34}) & \overset{*}{r}_{34} \\ 0 & 0 & r_{43} & -r_{43} \end{pmatrix}
$$

Here we see once again that in the instantaneous matrix the rows sum to 0, not 1. Clearly we could take a shortcut to the instantaneous matrix $\mathbf{R}^T$ by first setting up the in-the-small transition matrix, omitting the Δt multipliers of the rates r_{ij}, and replacing $*$ by whatever is required to make the row sum to 0.

The example given above instantiates a birth-and-death-process model of mobility. The main conceptual idea is the in-the-small assumption of transition limited to adjacent status classes. This is a kind of continuity argument: it is not possible to go from status level C_1 to C_3 without "passing through" C_2; or, in a ladder image, the ladder must be climbed one rung at a time, where "at a time" refers to status jumps occurring at arbitrary times.

The main problem associated with such a model—and indeed with all of Mayer's continuous-time models—is the formidable one of estimation. Once the instantaneous matrix and an initial condition $\mathbf{p}_0$ are given, equation (16.5.1) shows that the process is determined: the trajectory of $\mathbf{p}(t)$ is the solution of the system. But observation of mobility yields estimates of higher-order transition matrices, analogous to the quarterly transition matrices in the semi-Markov process of section 16.3, rather than direct estimates of the instantaneous "generator" $\mathbf{R}^T$. In this instance, Mayer shows how to obtain an estimate $\hat{\mathbf{R}}^T$ (see Mayer, 1972, Appendix C) and then goes on to examine predictions of the values in the observed transition matrix from "first job" to "present job." Interestingly enough, the model yields the basic result of the simple discrete-time

Markov model applied by BKM: an underestimation of the amount of staying in any given class, in this case between first and present job (where, depending on the age group, this represents a time interval of from 12 to 42 years).

Mayer treats the birth-and-death model as the baseline of the continuous-time approach in the sense that it paves the way for more complex and more appropriate models. Two of these are treated in his paper, and we shall sketch their basic aspects. One model tries to represent the effect of age-associated decline in mobility by a decaying rate

$$r_{ij}(t) = r_{ij}e^{-ct} \qquad \text{(all i, j, } t \geqslant 0)$$

so that when $c = 0$, we recover the baseline model described above. With this exponential decay model the process studied is no longer time-invariant in its basic in-the-small transition rule. Mayer proposes a solution to the estimation problem and again derives the predicted first job to present job transition matrix. An improvement in fit is to be expected and is achieved.

In the second model, Mayer uses an idea first initiated by Cohen in his model for conformity (see sections 8.5 and 10.12). As shown in the example of three classes in Figure 16.5.1, Mayer treats each observed class C_i as composed of two states: transient (T_i) and absorbing (A_i). A person is in C_i if and only if in T_i or in A_i, but we cannot tell from observable C_i whether the state is T_i or A_i. (In Figure 16.5.1, reflexive loops of staying in the given state have not been shown.) The in-the-small transition matrix is as follows:

$$t + \Delta t$$

$$
t \quad
\begin{array}{c}
A_1 \\ A_2 \\ A_3 \\ \\ T_1 \\ T_2 \\ T_3
\end{array}
\left(
\begin{array}{ccc:ccc}
A_1 & A_2 & A_3 & T_1 & T_2 & T_3 \\
* & 0 & 0 & & & \\
0 & * & 0 & & \mathbf{O} & \\
0 & 0 & * & & & \\
\hdashline
a_{11}\Delta t & 0 & 0 & * & b_{12}\Delta t & b_{13}\Delta t \\
0 & a_{22}\Delta t & 0 & b_{21}\Delta t & * & b_{23}\Delta t \\
0 & 0 & a_{33}\Delta t & b_{31}\Delta t & b_{32}\Delta t & *
\end{array}
\right)
$$

Once again, * means a quantity that yields a row sum of unity. As we noted in the analysis of the birth-and-death-process example, to convert the in-the-small transition matrix into the instantaneous matrix replace each * by a quantity that makes the row sum to 0, after eliminating Δt in each occurrence. Hence,

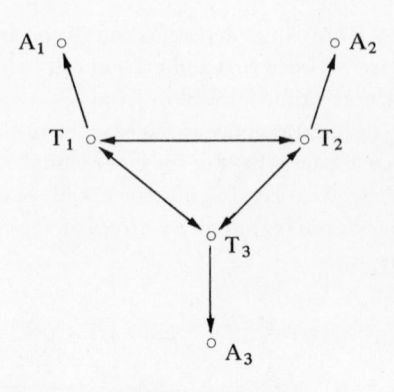

Figure 16.5.1

$$
\mathbf{R}^T =
\begin{array}{c}
\\
A_1 \\
A_2 \\
A_3 \\
\\
T_1 \\
T_2 \\
T_3
\end{array}
\begin{pmatrix}
A_1 & A_2 & A_3 & T_1 & T_2 & T_3 \\
 & & & & & \\
 & \mathbf{O} & & & \mathbf{O} & \\
 & & & & & \\
\hdashline
a_{11} & 0 & 0 & b_{11} & b_{12} & b_{13} \\
0 & a_{22} & 0 & b_{21} & b_{22} & b_{23} \\
0 & 0 & a_{33} & b_{31} & b_{32} & b_{33}
\end{pmatrix}
$$

where $b_{11} = 1 - (a_{11} + b_{12} + b_{13})$, $b_{22} = 1 - (a_{22} + b_{21} + b_{23})$, and $b_{33} = 1 - (a_{33} + b_{31} + b_{32})$. This shows a simple structure Mayer is able to exploit in his analysis. However, the unobservable character of the state-space carries its penalty: the estimation of parameters problem, already rather intricate for these continuous-time models, becomes formidable indeed. Mayer makes some headway in solving the problem, however, and is able to provide a tentative favorable evaluation of the model based on predictions of the quantities in the "first job" to "present job" transition matrix.

The model has a family resemblance to the "commitment" model proposed by Herbst (1963) in his analysis of data on workers' length of stay in employing organizations. Figure 16.5.2 shows the structure of transition in the small in Herbst's model.

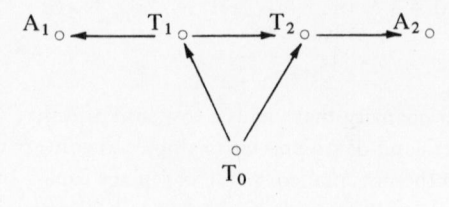

Figure 16.5.2

A new employee enters the organization in a state of indecision (T_0) about his commitment to it. He can move to a temporary or tentative commitment state (T_1) or to a decision to leave (T_2). From T_1 he can change his mind and decide to leave (entering T_2) or he can develop a full or permanent commitment (A_1). On the other hand, once he decides to leave, then he passes into the "left" state (A_2) after a time determined by the continuous-time Markov process. The instantaneous matrix, therefore, has the form

$$
\mathbf{R}^T =
\begin{array}{c}
\\ A_1 \\ A_2 \\ T_1 \\ T_2 \\ T_0
\end{array}
\begin{pmatrix}
A_1 & A_2 & \vdots & T_1 & T_2 & T_0 \\
 & & \vdots & & & \\
\mathbf{O} & & \vdots & & \mathbf{O} & \\
\hdashline
a_{11} & 0 & \vdots & b_{11} & b_{12} & 0 \\
0 & a_{22} & \vdots & 0 & b_{22} & 0 \\
0 & 0 & \vdots & b_{01} & b_{02} & b_{00}
\end{pmatrix}
$$

We see that once the worker enters one of the transient states (T_1, T_2) the process becomes a Mayer absorbing chain with two classes. Analysis of this model is given in detail by Bartholomew (1967). It may be mentioned that the model provided excellent fits to two sets of data on length of stay with employing firms (data from Hedberg, 1961).

16.6. The Blau-Duncan Path Model.

We can put the above type of work into the context of dynamic systems (see Chapter 8) by introducing a numerical variable.

$$X(\alpha) = i \qquad \text{iff} \qquad \text{unit } \alpha \text{ in class } C_i \qquad (i = 1, 2, \ldots, N)$$

Thus, X merely labels the status class, the labeling ordinal if the classes are ordered, as in Mayer's treatment, or simply nominal if they are not (as in the BKM study).

For a given unit, for every moment of time t, where $t \geqslant 0$, we have a value X_t. Hence, X_t forms a one-dimensional dynamic system. Of course, in the above work, what is treated in dynamic terms is

$$\mathbf{p}(t) = [P(X_t = 1), P(X_t = 2), \ldots, P(X_t = N)]$$

the probability state vector.

If, however, we return to the analytical ground (see Section 8.5), we see that confinement of attention to one dimension and to discrete values is a limitation not inherent in the ongoing process of social mobility. Instead, we realize that a person may be characterized in several ways that are status relevant—for instance, his job, education, and residence. Deciding upon a multidimensional representation of status leads to a status vector $\mathbf{X} = (X_1, X_2, \ldots, X_k)$. Moreover, many of these variables are naturally represented—even if by only crude measurement—as higher than ordinal in scale-type. For example, this is true of occupational status scores such as those used by Blau and Duncan (1967) in their analysis of the American occupational structure.

For definiteness, let us define the variables of the Blau-Duncan model. They are

$$X_1 = \text{educational level}$$
$$X_2 = \text{occupational status}$$

(Details of the scales are found in Blau and Duncan, 1967, Chap. 4 and 5.)

Next, we treat these in terms of a dynamic system: at every moment of time t, where $t \geqslant 0$ is measured from some observational initial age, we have the vector

$$X_t = (X_{1t}, X_{2t}) \qquad (t \geqslant 0)$$

This vector gives the person's position in the two-dimensional sense.

There will be a particular starting vector X_0 containing the two components: education level at $t = 0$ and job status at $t = 0$. Thus, it is convenient to let this be the time of first full-time entry into the job market. Note that, in general, the education variable may still be "dynamical" at this time since further educational attainment may accompany the job history.

Without specifying the dynamics of the process, let us assume that some equilibrium position

$$X_e = (X_{1e}, X_{2e})$$

exists for a given individual (for treatments of the dynamics, see Fararo, 1972a and Doreian and Hummon, 1972).

Now consider two individuals, father and son. Each has an equilibrium position,

$$X_e^{(f)} = (X_{1e}^{(f)}, X_{2e}^{(f)})$$
$$X_e^{(s)} = (X_{1e}^{(s)}, X_{2e}^{(s)})$$

The problem now to be treated is how the equilibrium position of the father influences the attained equilibrium of the son. This is the question of the amount of status inheritance in the society, treated by Blau and Duncan in terms of a model of the "path analysis" variety.

Denote the son's initial status vector by

$$X_0^{(s)} = (X_{10}^{(s)}, X_{20}^{(s)})$$

In the Blau-Duncan study, it seems to be assumed that educational level reached its maximal level as early as the first full-time entry into the job market, to a first approximation. This means we idealize to treat the situation in which the initial educational attainment $X_{10}^{(s)}$ is the equilibrium level $X_{1e}^{(s)}$. In this case we have a total of three two-vectors but only five distinct quantities:

$$X_{1e}^{(f)} = \text{father's education attainment (equilibrium)}$$
$$X_{2e}^{(f)} = \text{father's occupational status attainment (equilibrium)}$$
(16.6.1) $$X_{10}^{(s)} = \text{son's initial and attained education (equilibrium)}$$
$$X_{20}^{(s)} = \text{son's initial occupational status (initial condition)}$$
$$X_{2e}^{(s)} = \text{son's occupational status attainment (equilibrium)}$$

To sum up the conceptual background, position is a two-dimensional vector, X_t, that is time-varying through a career. But there is an initial position and an equilibrium of a process here unspecified. This is true of father and son, so there are two such processes. We link the two by the problematic aspect, How does the equilibrium of the son depend on the equilibrium of the father (as well as on the son's own initial position)?

Before stating the Blau-Duncan path model linking these variables, we summarize some elementary aspects of path analysis using expectations of random variables (see Chapter 10 for a fuller treatment of random variables, and see Boudon, 1968, for a detailed presentation of an expectation-notation version of path analytic problems). The material is presented very compactly, since, apart from the extensive use of expectation-calculus, it can be assumed the reader has some acquaintance with standardized variables and regression analysis (for another summary in computational notation, see Land, 1969). The basic mathematical concepts needed to explicate path analysis are compactly listed first, followed by their use in path analytic procedure. It will be useful to repeat some of the ideas of Chapter 10, often abbreviating $E(X)$ by EX here.

(16.6.2) Definition. $EX = \sum_X x\, P(X = x)$

(16.6.3) Proposition.   (a) $E(X + Y) = EX + EY$
          (b) $E(cX) = cEX$
          (c) $Ec = c$ (c any number)

(16.6.4) Definition. $VAR\ X = E(X - EX)^2$ (also denoted σ_x^2)

(16.6.5) Proposition.   (a) $VAR(cX) = c^2\,VAR\ X$
          (b) $VAR(c + X) = VAR\ X$
          (c) X, Y independent $\Rightarrow VAR(X + Y) = VAR\ X + VAR\ Y$

(16.6.6) Definition. $COV(X, Y) = E[(X - EX)(Y - EY)]$

(16.6.7) Proposition. X, Y independent $\Rightarrow COV(X, Y) = 0$

(16.6.8)   *Definition.* $r_{xy} = \dfrac{COV(X, Y)}{\sigma_x \sigma_y}$

(16.6.9)   *Proposition.*   X, Y independent $\Rightarrow r_{xy} = 0$

(16.6.10)   *Definition.*   X, Y uncorrelated   iff $r_{xy} = 0$

(16.6.11)   *Definition.* $Z = \dfrac{X - EX}{\sigma_x}$

(16.6.12)   *Proposition.*   (a) $EZ = 0$
(b) $EZ^2 = 1$
(c) $VAR\ Z = EZ^2 = 1$
(d) $r_{12} = r_{Z_1 Z_2} = E(Z_1 Z_2)$
(e) Z_1, Z_2 uncorrelated   iff $E(Z_1 Z_2) = 0$

Path analysis may be introduced in terms of the following typical steps in setting up and analyzing a path model:

Step 1.   Define variables (in standardized form).

Step 2.   State assumptions:
(a) causation among variables in the system
(b) unmeasured exogeneous causes
(c) linear relations

Step 3.   Draw diagram, label arrows with "path coefficients," p_{ij}.

Step 4.   Write down linear equations.

Step 5.   Express correlations in terms of coefficients.

Step 6.   Estimate coefficients p_{ij} linking variables in the system. (The path coefficients of the exogeneous inputs are then obtained from these, but we will not show how this is done until we reach the Blau-Duncan model.)

Step 7.   Check predictions, if any.

Example.   (a). Two-variable system:

Step 1.   Let Z_1 = occupational status
Z_2 = income

Step 2.   (a) Z_1 causes Z_2 not conversely.
(b) Other "outside" or "exogeneous" variables, collectively denoted Z_a, cause Z_1; other variables, collectively denoted Z_b, cause Z_2; Z_b uncorrelated with Z_1.
(c) All relations are linear and additive.

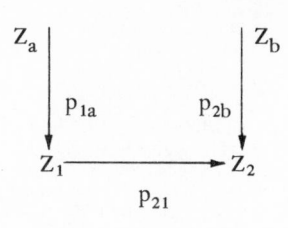

Figure 16.6.1

Step 3. See Figure 16.6.1.

Step 4. (a) $Z_1 = p_{1a} Z_a$
(b) $Z_2 = p_{21} Z_1 + p_{2b} Z_b$

Step 5.

$$Z_1 Z_2 = p_{21} Z_1^2 + p_{2b} Z_b Z_1 \qquad \text{(multiply 4(b) by } Z_1 \text{)}$$

$$r_{12} = E(Z_1 Z_2) = E(p_{21} Z_1^2 + p_{2b} Z_b Z_1) \qquad \text{(E on both sides of line above)}$$

$$= p_{21} E(Z_1^2) + p_{2b} E(Z_b Z_1) \qquad (16.6.3)$$

$$= p_{21} + p_{2b} E(Z_b Z_1) \qquad (16.6.12)(b)$$

$$= p_{21} \qquad \text{[use step 2 (b); (16.6.12)(e)]}$$

Step 6. Compute $\hat{r}_{12}$ from data. By Step 5, this is $\hat{p}_{21}$.

Step 7. No predictions: model not falsifiable.

Example. (b) Three-variable system (abstract):

Step 1. Z_1, Z_2, Z_3

Step 2. (a) Z_1 causes Z_2 causes Z_3
(b) Exogeneous Z_a causes Z_1.
Exogeneous Z_b causes Z_2 and uncorrelated with Z_1.
Exogeneous Z_c causes Z_3 and uncorrelated with Z_2, Z_1.
(c) All relations are linear and additive.

Step 3. See Figure 16.6.2.

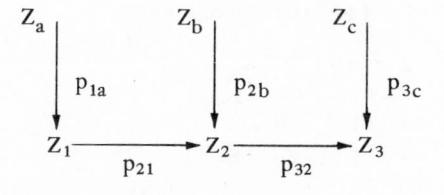

Figure 16.6.2.

Step 4.

$$Z_1 = p_{1a}Z_a$$
$$Z_2 = p_{21}Z_1 + p_{2b}Z_b$$
$$Z_3 = p_{32}Z_2 + p_{3c}Z_c$$

Step 5.

$$
\begin{aligned}
r_{21} &= E(Z_2 Z_1) \\
&= E(p_{21}Z_1^2 + p_{2b}Z_b Z_1) \\
&= p_{21} E(Z_1^2) + p_{2b} E(Z_b Z_1) \\
&= p_{21} \\
r_{32} &= E(Z_3 Z_2) \\
&= E(p_{32}Z_2^2 + p_{3c}Z_c Z_2) \\
&= p_{32} E(Z_2^2) + p_{3c} E(Z_c Z_2) \\
&= p_{32} \\
r_{31} &= E(Z_3 Z_1) \\
&= E(p_{32}Z_2 Z_1 + p_{3c}Z_c Z_1) \\
&= p_{32} E(Z_2 Z_1) \\
&= p_{32} r_{21} = r_{32} r_{21}
\end{aligned}
$$

Step 6. $\hat{p}_{21} = \hat{r}_{21}, \hat{p}_{32} = \hat{r}_{32}$

Step 7. Prediction: $r_{31} = r_{32} r_{21}$. Since the partial correlation $r_{31.2}$ between Z_3 and Z_1 with Z_2 held constant is given by

$$r_{31.2} = \frac{r_{31} - r_{32} r_{21}}{\sqrt{(1 - r_{32}^2)(1 - r_{21}^2)}}$$

the prediction is $r_{31.2} = 0$. Hence, $\hat{r}_{31.2}$ is checked against this "null hypothesis."

Having given some examples, we now apply these steps to define the path model of Blau and Duncan:

Step 1. Z_1 = standardized $X_{1e}^{(f)}$ (father's education)

Z_2 = standardized $X_{2e}^{(f)}$ (father's occupation)

Z_3 = standardized $X_{10}^{(s)}$ (son's education)

Z_4 = standardized $X_{20}^{(s)}$ (son's first job)

Z_5 = standardized $X_{2e}^{(s)}$ (son's job in 1962)

Step 2.   (a) Z_1 and Z_2 are correlated.

Z_3 is caused by Z_1 and Z_2.

Z_4 is caused by Z_2 and Z_3.

Z_5 is caused by Z_2, Z_3, and Z_4.

(b) Exogeneous Z_a causes Z_3 and is uncorrelated with Z_1, Z_2.

Exogeneous Z_b causes Z_4 and is uncorrelated with Z_1, Z_2, Z_3.

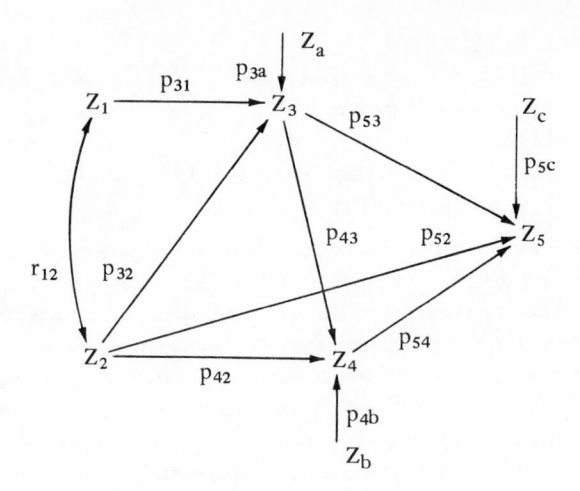

Figure 16.6.3

Exogeneous Z_c causes Z_5 and is uncorrelated with Z_1, Z_2, Z_3, Z_4.
(c) All relations linear and additive.

Step 3. See Figure 16.6.3.

Step 4.

$$Z_3 = p_{31} Z_1 + p_{32} Z_2 + p_{3a} Z_a$$
$$Z_4 = p_{42} Z_2 + p_{43} Z_3 + p_{4b} Z_b$$
$$Z_5 = p_{52} Z_2 + p_{53} Z_3 + p_{54} Z_4 + p_{5c} Z_c$$

Step 5. (1) $r_{13} = E(Z_3 Z_1)$
$\qquad\qquad = p_{31} E(Z_1^2) + p_{32} E(Z_2 Z_1) + p_{3a} E(Z_a Z_1)$
$\qquad\qquad = p_{31} + p_{32} r_{21}$

$\qquad$ (2) $r_{23} = E(Z_3 Z_2)$
$\qquad\qquad = p_{31} E(Z_1 Z_2) + p_{32} E(Z_2^2) + p_{3a} E(Z_a Z_2)$
$\qquad\qquad = p_{31} r_{12} + p_{32}$

$\qquad$ (3) $r_{24} = E(Z_4 Z_2)$
$\qquad\qquad = p_{42} E(Z_2^2) + p_{43} E(Z_3 Z_2) + p_{4b} E(Z_b Z_2)$
$\qquad\qquad = p_{42} + p_{43} r_{32}$

$\qquad$ (4) $r_{25} = E(Z_5 Z_2)$
$\qquad\qquad = p_{52} E(Z_2^2) + p_{53} E(Z_3 Z_2) + p_{54} E(Z_4 Z_2) + p_{5c} E(Z_c Z_2)$
$\qquad\qquad = p_{52} + p_{53} r_{32} + p_{54} r_{24}$

$\qquad$ (5) $r_{34} = E(Z_4 Z_3)$
$\qquad\qquad = p_{42} E(Z_2 Z_3) + p_{43} E(Z_3^2) + p_{4b} E(Z_b Z_3)$
$\qquad\qquad = p_{42} r_{23} + p_{43}$

$$(6)\ r_{35} = E(Z_5 Z_3)$$
$$= p_{52} E(Z_2 Z_3) + p_{53} E(Z_3^2) + p_{54} E(Z_4 Z_3) + p_{5c} E(Z_c Z_3)$$
$$= p_{52} r_{23} + p_{53} + p_{54} r_{34}$$
$$(7)\ r_{45} = E(Z_5 Z_4)$$
$$= p_{52} E(Z_2 Z_4) + p_{53} E(Z_3 Z_4) + p_{54} E(Z_4^2) + p_{5c} E(Z_c Z_4)$$
$$= p_{52} r_{24} + p_{53} r_{34} + p_{54}$$

Step 6. We have a linear system of seven equations with knowns being the r's; the seven unknowns are the path coefficients linking the variables in the system. This method nicely exhibits the logic of the approach. Blau and Duncan use a more direct procedure of regression successively on the equations of step 4. To sustain our interpretative background in which we are treating son's position as in equilibrium, we report the Blau-Duncan path model application to the non-farm men, age 55 to 64 in 1962, who quite reasonably may be expected to have reached their equilibrium positions. Figure 16.6.4 displays the path coefficients for this age group (from Blau and Duncan, 1967, Table 5.4). We have not shown the magnitudes of the coefficients reflecting causal influences external to the variables of the process itself. These are readily obtained, however, since each is the square root of the unexplained variance. For instance, Blau and Duncan report that the explained variance is $R^2 = .21$ in the regression equation for respondent's education, Z_3. Hence,

$$p_{3a} = \sqrt{1 - R^2} = \sqrt{.79} = .889$$

Similarly, using their data,

$$p_{4b} = \sqrt{.66} = .812$$
$$p_{5c} = \sqrt{.71} = .843$$

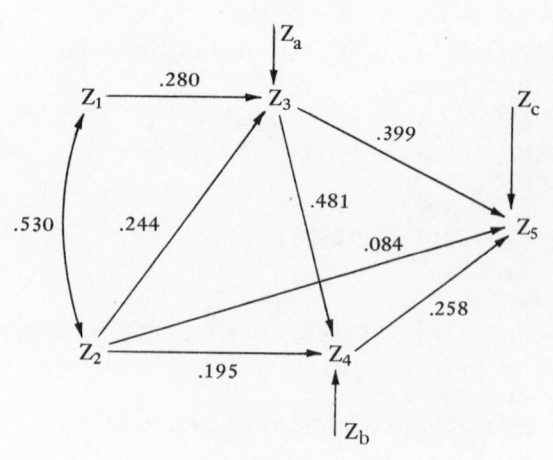

Figure 16.6.4

The magnitude of unexplained variance is not a problem in terms of the path conceptualization as interpreted in terms of process equilibria: such high residual variance merely reflects that the processes of interest are embedded in a rich causal nexus in which most of the explanation of the level of equilibrium status of the son is due to forces extrinsic to the father's level of achievement. We see that, nevertheless, the father's attained education does have a causal effect on the son's level and that the father's attained occupational level has some effect on both the educational and occupational levels of the son in both the initial condition and in the equilibrium of the son's process. The relatively small size ($p_{52} = .084$) of the causal effect of father's occupational attainment indicates that the basic role of the father's achievement is in regard to the initial education-job situation of the son.

What seems unlikely, however, is that these residual variables are uncorrelated with the independent variables in a given equation, as assumed in step 2(b) and used in the derivation of the estimation equations. Thus, we are led to at least check the ability of the model to reproduce the remaining correlations.

Step 7. We have used seven correlations to estimate the path coefficients. The remaining correlations are r_{12}, r_{14}, and r_{15}. Here $r_{12} = .530$ is the correlation of variables never taking a dependent variable role in the equations of step 4. To obtain r_{14} we use the second equation of step 4:

$$
\begin{aligned}
r_{14} &= E(Z_4 Z_1) \\
&= p_{42} E(Z_2 Z_1) + p_{43} E(Z_3 Z_1) + p_{4b} E(Z_b Z_1) \\
&= p_{42} r_{21} + p_{43} r_{31} \\
&= p_{42} r_{12} + p_{43}(p_{31} + p_{32} r_{12}) \\
&= (.195)(.530) + .481(.280 + (.244)(.530)) \\
&= .30 \\
r_{15} &= E(Z_5 Z_1) \\
&= p_{52} E(Z_2 Z_1) + p_{53} E(Z_3 Z_1) + p_{54} E(Z_4 Z_1) + p_{5c} E(Z_c Z_1) \\
&= p_{52} r_{12} + p_{53} r_{13} + p_{54} r_{14} \\
&= p_{52} r_{12} + p_{53}(p_{31} + p_{32} r_{12}) + p_{54} r_{14} \\
&= (.084)(.530) + .399(.409) + (.258)(.30) \\
&= .285
\end{aligned}
$$

The predicted value of $r_{15} = .285$ compares with the observed value of .311, but unfortunately Blau and Duncan did not have the requisite tabulation to compute r_{14} from the data (see Blau and Duncan, 1967, Table 5.3).

The interpretation of the path model is aided by the definition of magnitude of indirect effect: the difference

$$
r_{ij} - p_{ij}
$$

between the correlation and the path coefficient. For example, the direct effect of father's occupation on son's occupation is given by p_{52}, while the indirect effect is $r_{52} - p_{52}$. In the particular model of Figure 16.6.4, we find $p_{52} = .084$. Also, computation from the data yields $r_{52} = .34$ (Blau and Duncan, 1967, Table 5.3). Hence, the magnitude of indirect effect of father's attained occupational level on the son's attained level is given by

$$r_{52} - p_{52} = .34 - .084 = .256$$

In this context, the correlation is called the gross effect, while the path coefficient of course measures the direct effect.

Duncan uses the diagram to show that of the gross effect of education (Z_3) on occupational attainment (Z_5) by the son, only a very minor part of this effect can be attributed to the transmission of prior influences from father's position (Z_1, Z_2). To show this he first notes the gross effect, which in this case is $r_{35} = .576$ (where we retain our analysis in terms of data on men of nonfarm background in the age group 55–64 in 1962). Then the "transmission" is computed by finding the products of the path coefficients of all indirect paths linking Z_3 to Z_5 via Z_1 and Z_2:

Path		Product
$Z_2 \Big\langle \begin{array}{c} Z_3 \\ Z_5 \end{array}$	$p_{32}\, p_{52} = (.244)(.084)$	$= .020$
$Z_1 \longrightarrow Z_3$ $Z_2 \longrightarrow Z_5$	$p_{52}\, r_{21}\, p_{31} = (.084)(.53)(.28)$	$= .012$
$\begin{array}{cc} Z_3 & Z_5 \end{array}$ $Z_2 \longrightarrow Z_4$	$p_{54}\, p_{42}\, p_{32} = (.258)(.195)(.244)$	$= .012$
$Z_1 \longrightarrow Z_3$ $Z_2 \longrightarrow Z_4 \nearrow Z_5$	$p_{54}\, p_{42}\, r_{21}\, p_{31} = (.258)(.195)(.53)(.28)$	$= .007$
		Total $= .051$

By contrast with this transmission from father's position, we know that the direct effect of his own education on the son's eventual occupational attainment is $p_{53} = .399$. Duncan concludes, "Far from serving in the main as a factor perpetuating [father's] status, education operates *primarily* to induce variation in occupational status that is independent of [father's] status" (Duncan and Blau, 1967, p. 201).

In closing this section, we note that not "number of years of education" but its differential evaluation is the educational status component of the two-dimensional position vector used above. By empirical-meaningfulness requirements, this evaluation must be scaled on an interval level at least, to permit the additive operations of the path model. Fararo (1970c) suggests using an invariant choice theory based on Luce's axiom as treated in Chapter 11, to define the ratio-level status variable based on any characteristic, not only occupation. This is programmatic, of course, although we gave an example of one empirical study in section 11.13. Based on this idea, Fararo passes to the definition of a status vector whose components depend on the characteristics and differential evaluations generated within the social system itself, for any social system. Then, abstract theoretical models of status dynamics are constructed with a view to proving general sociological theorems in the area of status movements, defined as temporal changes in position (i.e., in vector of statuses).

A first differential equation approach was taken in Fararo (1970a) using two basic axioms drawn from Galtung (1966): (1) each person attempts to move upward on each component, and (2) the components are coupled by a balance consideration, in the sense of Chapter 14. That is, an expected state of X_{1t} is set up based on the attained X_{2t}, and deviations from this expectation act as motivational forces to "set things in line." Similarly, X_{2t} is "equilibrated" by X_{1t}. The differential equation system, with an exogeneous constraint input in each equation, yields an equilibrium that is not asymptotically stable, no matter what the parameters. Intuitively, from 100 men at a given initial position we scatter in all directions. While this would explain the fact that from a similar initial position individuals do not necessarily attain similar levels of status, it does not allow any stable attainment. This counterintuitive result is overcome in a generalization reported in Fararo (1972a), in which ascribed statuses are made to play a role in setting up expectations and, hence, balancing processes. The result is that under certain very interpretable parametric conditions, a stable equilibrium is attainable.

A fundamental conceptual difficulty of this approach is the tacit assumption that movement in social space is essentially a matter of individual choice apart from arbitrary constraint terms. The structure and process of available positions to which an actor might move is neglected. This difficulty is brought into sharp focus if a total system of positions and actors is considered. It is clear that the occupancy distribution at any time acts as either a source of, or a block to, movement by any one actor. Hence, the tracing out of a sequence of positions over time would be senseless without a simultaneous consideration of the time-varying occupancy distribution. Yet, if this is done, the interaction effects are so profound that the assumption of independence between distinct individuals— a bulwark of the entire collection of models so far discussed—would need to be abandoned. This would point to an approach analogous to Coleman methodology:

replace the concept of a set of independent actors with a network. In the special case that well-defined positions exist as individually identifiable as people, the network approach yields a model with analytical tractability. We turn to this final model in the next section.

16.7. Vacancy Chains.   Unlike the models of previous sections, the present model, drawn from White (1970), is a system model. That is, the individuals are seen as linked together in a structure of positions. Mobility means flows in networks: but there are two fundamental kinds of entities, positions and people. Fundamental to the approach of this section is the scope condition that there is a set of well-defined positions that actors occupy and that are as readily identifiable as the actors themselves. The prototypical case is that of well-defined jobs in organizations. If now we assumed independent movement by actors, the procedure would be clear: (1) construct a model for individual mobility in which individuals are constrained to occupy one of the well-defined positions, (2) obtain data on sequences of occupancies by N actors, (3) assume the N actors are independent and predict from (1) the expected occupancy distribution at any time and the equilibrium distribution. Unfortunately, this procedure only makes sense if the number of individuals and positions is but a minor part of some unspecified system. To take a concrete example, the careers of a California trucker and of a New York accountant are not in any strong sense mutually constraining; the system of jobs and people is so enormous that we can treat these as two independent realizations of an individual mobility process. Even if we select 3,000 individuals—provided we do so without an effort to make them cohere, ideally by randomly sampling—these form a trivially small fraction of the whole system.

To see the concrete meaning of the system, as opposed to the aggregate individual approach, let us consider a simple example. (The distinction between these two approaches was made explicit in the previously cited paper by T. Mayer, published in 1972; the same distinction, of course, pervades the White monograph.) Consider four jobs $\{a, b, c, d\}$ and four men $\{A, B, C, D\}$. Initially let A be in a, B in b, C in c, and D in d. Now suppose that this is a system. This means that any event relating to occupancy in one part of the structure has relatively rapid repercussions for other positions: they are coupled. Suppose that A dies, so a is vacant. Assuming that a is a job needed in the system, this vacancy creates an immediate problem. From a dynamic standpoint, it is a perturbation of the system equilibrium. Hence, activity will take place that after some time will result in the filling of position a, either from inside or outside. Suppose the first thing that happens is that B is "moved up" to a, vacating b. Then the vacancy has moved from a to b. The problem still exists. Let C move into b, vacating c. Then suppose that the D is moved to c, vacating d, and the organization abolishes job d. The chain of vacancies initiated by A's death has

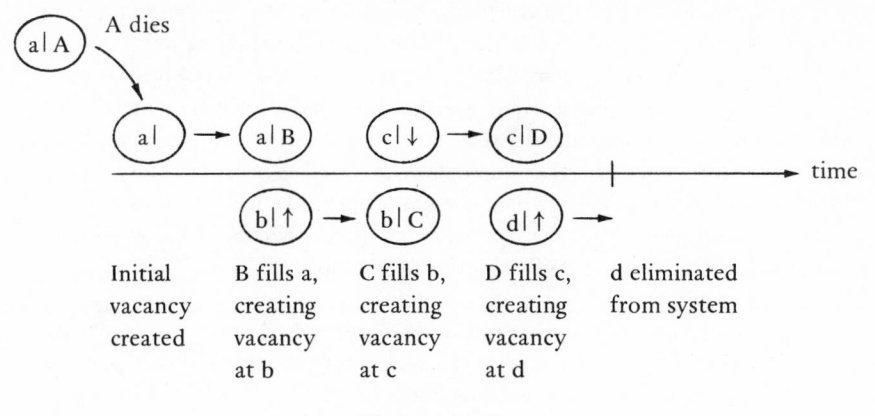

Figure 16.7.1

terminated. We show this "vacancy chain" in Figure 16.7.1, representing a job-man pair as an oval with two sides, so the initial structure has the form (x| X),
where x = a, b, c, and d. The length of a vacancy chain is given by the number of temporal arrows: durations in which a position is unfilled. Here, the length is 4.
We see in Figure 16.7.1 that the vacancy is initiated by the event of a death and terminates in a kind of "dual" event of a job's "death" (i.e., elimination). Of course, a vacancy might also be filled by recruitment from outside the system:
this also terminates the chain created by the repercussions of the initial "arrival" of a vacancy at position a, however it may have arrived (invention of a position, retirement, etc.).

Whether we allow our system to contain 40 or 40,000 or 4 million, the fundamental logic is unaltered: in every case, mobility of individuals is intertwined with the filling of vacancies in chains. Ordinarily, we are dealing with a large-scale system, and the number of chains is large. The question is, Are vacancy chains independent? White argues that to a good first approximation distinct vacancy chains are traced out independently of each other, even though each single tracing exhibits the tight linkage of positions to people and the interdependencies in the moves made by individuals.

Thus, two fundamental ideas pervade the model; one is conceptual and the other is technical. First, in place of the idea that in mobility studies the focus should be on the individual or family unit as that which traces out a status trajectory, we have the idea that in mobility analysis the units are the vacancies and individual careers are byproducts of the fundamental in-the-small dynamics of vacancy chains. Second, in place of the idea of independent realizations of the mobility process by distinct actors—an idea that defines an aggregate approach and that is valid within certain limits—we have the idea of independent vacancy chains.

Of course, at present at least, not all systems of actors and positions can be treated in this way, so that this approach has its own built-in restricted scope.

By considering many variants of the logical possibilities, White properly locates the contexts within which vacancy chain analysis should apply: namely, to "tight" systems in which the temporal delays are on the side of the positions and not the individuals. That is, when a vacancy occurs, it takes time before the position is filled or eliminated, but when actors move out of one position, they enter another in a comparatively rapid manner, treated as instantaneous. A man who leaves his job and is unemployed is not a problem: he has "died" so far as the system is concerned. But, if a man were "in limbo" between two organizational positions for any length of time, this would violate the tight system condition.

The basic mathematical structure will now be defined, where we will confine ourselves to what White calls 1–1 systems: each actor occupies a unique position and each position is occupied by at most one actor.

We treat external positions, unemployment, and the like as a single element: "the outside," a source for recruitment as well as for losses. The system is "open" because there will be an incessant interchange of actors and even positions "across the boundaries."

First, we treat the simple case where all positions, although functionally distinct, are put into one class from the point of view of strata or of other classification systems. This will define the binomial model, as White calls it.

(16.7.1) Definition.   A system of vacancy chains is said to satisfy the binomial model if and only if:

Axiom 1.   Each vacancy chain is a process in which there is a constant probability p of termination and where the termination event is independent of the length of the chain.

Axiom 2.   All chains have the same probability parameter p and are mutually independent.

Given a vacancy at some initial point we have the following:

$$\text{Initial vacancy} \xrightarrow{\ q\ } \text{New vacancy} \xrightarrow{\ q\ } \text{New vacancy} \xrightarrow{\ q\ } \cdots \xrightarrow{\ p\ } \text{Termination}$$

where p is the termination probability (i.e., the vacancy is filled from outside or the position is eliminated) and $q = 1 - p$. Hence, the random variable,

$$J = \text{length of vacancy chain}$$

has probability distribution

$$(16.7.2) \qquad P_j = P(J = j) = q^{j-1} p \qquad (j = 1, 2, \ldots)$$

To see this, examine Figure 16.7.1 again. We assume a vacancy at a. The move by B is treated as a realization of the event "new vacancy in system," the complement of termination. Similarly, the moves by C and D create new vacancies, while with probability p, the next event terminates the vacancy chain. Taking the product, using axiom 1 referent to independence, we calculate $q^3 p$ for the chain length, which is $J = 4$. This is in agreement with formula (16.7.2). We note that

$$\sum_{j=1}^{\infty} q^{j-1} p = p \sum_{j=1}^{\infty} q^{j-1} = p \sum_{j-1=0}^{\infty} q^{j-1} = \frac{p}{1-q} = 1$$

and so, with probability 1, a chain has finite length. The average length is given by

$$E(J) = \sum_{j=1}^{\infty} j P_j$$

$$= p \sum_{j=1}^{\infty} j q^{j-1} = p \sum_{j=1}^{\infty} \frac{d}{dq} q^j = p \frac{d}{dq} \sum_{j=1}^{\infty} q^j$$

$$= p \frac{d}{dq} \left(\frac{1}{1-q} - 1 \right) = p \frac{1}{(1-q)^2} = \frac{p}{p^2} = \frac{1}{p}$$

By definition,

(16.7.3) $$\lambda = \frac{1}{p}$$

is the expected chain length.

It is convenient and useful in the subsequent generalization to the differentiated system of positions to write down the matrix form of the binomial model. Let S mean "the system" (of positions and actors) and let O mean "the outside." Then axiom 1 of the binomial model amounts to saying that we have a two-state absorbing chain:

(16.7.4)

		$n+1$	
		O	S
n	O	1	0
	S	p	q

From our study of the canonical form of absorbing chain matrices in section 10.12, we see that matrix $\mathbf{Q}$ is just q. Hence, the fundamental matrix $(\mathbf{I} - \mathbf{Q})^{-1}$ reduces to $(1 - q)^{-1} = 1/p$. The row sums of the matrix $(\mathbf{I} - \mathbf{Q})^{-1}$ give the expected number of steps to absorption; hence, $1/p$ is the expected number of steps before a vacancy chain terminates. It is useful to say that each step is a move by a vacancy initiated in S at some time. Then we say that $1/p$ is the expected number of moves "by a vacancy" before it leaves the system.

Also, it is an aid to the imagination to use a birth-and-death-process language in some contexts. Entities in the system are of two types: actors and positions. An entity entering the system for the first time exemplifies a birth (for example, recruitment of a man from outside or an invention of a new job). An entity leaving the system exemplifies a death (for example, retirement by a man or abolition of a job). The duality of actors and positions is nicely expressed in this language: a vacancy chain is initiated by either a birth of a position or a death of an actor; and a vacancy chain is terminated by either a birth of an actor or a death of a position.

Now consider the collection of all vacancies arriving in the system in some time domain (e.g., one year). Then by axiom 2 of the binomial model each of these is an independent realization of the process described by matrix (16.7.4). Let J_i (i = 1, 2, . . ., F) be the number of moves in chain i, where F is the total number of vacancy chains initiated in the given time domain. Then define

$$(16.7.5) \qquad\qquad T = \sum_{i=1}^{F} J_i$$

which is the total number of moves made by vacancies in the given time domain. Hence, if we let M be the expected total number of moves,

$$(16.7.6) \qquad M = E(T) = \sum_{i=1}^{F} E(J_i) = \sum_{i=1}^{F} \lambda = F\lambda = \frac{F}{p}$$

using the linearity property of expectation and (16.7.3), since each J_i has the same expectation. It should be noted that (16.7.6) does not require the independence assumption, since the expectation of a sum of random variables is the sum of the expectations even when the variables are dependent (see sections 10.4–10.5).

White refers to the F vacancy chains initiated in the given time domain as a cohort. He treats a succession of such cohorts, as we will indicate subsequently. Formula (16.7.6) says that the total number of moves made by a cohort is just the number of chains in the cohort times the expected length of any chain in the cohort.

The generalization of the binomial model is made easy by appeal to matrix (16.7.4): we simply allow a differentiation of the system of positions into s classes, denoted $C_1, C_2, \ldots, C_s$. Each class contains a set of positions, but the classes need not be stratified. If they are, it is convenient to call them strata, as White does. The generalized process has matrix

$$
\begin{array}{c}
n+1
\end{array}
$$

		O	C_1	C_2	$\cdots$	C_s
	O	1	0	0	$\cdots$	0
(16.7.7)	C_1	q_{10}	q_{11}	q_{12}	$\cdots$	q_{1s}
	C_2	q_{20}	q_{21}	q_{22}	$\cdots$	q_{2s}
n	$\cdot$	$\cdot$	$\cdot$	$\cdot$	$\cdot$	$\cdot$
	$\cdot$	$\cdot$	$\cdot$	$\cdot$	$\cdot$	$\cdot$
	$\cdot$	$\cdot$	$\cdot$	$\cdot$	$\cdot$	$\cdot$
	C_s	q_{s0}	q_{s1}	q_{s2}	$\cdots$	q_{ss}

(16.7.8) Definition. The vacancy chain model is defined by the following:

Axiom 1. There exists a partition of the positions in the system into s classes, $C_1, C_2, \ldots, C_s$.

Axiom 2. Given a chain is initiated, it starts in class C_k with probability $f_k (k = 1, 2, \ldots, s)$.

Axiom 3. Given a vacancy is in class C_i, it moves to class C_k with probability q_{ik} $(i, k = 1, 2, \ldots, s)$ and with probability q_{i0} it terminates.

Axiom 4. Each chain in the cohort of all vacancy chains in a specified time domain has the same parameters, and the chains are mutually independent.

The generalization aspect of the model is evident by taking $s = 1$, letting $C_1 = S$, $q_{11} = q$, and $q_{10} = p$.

We see that we have one additional item, aside from the Markov matrix—namely, the probability distribution of initiated vacancies over classes, which is denoted

$$
(16.7.9) \qquad\qquad \mathbf{f} = (f_1, f_2, \ldots, f_s)
$$

In canonical form, we see that (16.7.7) yields

$$
(16.7.10) \qquad \mathbf{Q} =
\begin{pmatrix}
q_{11} & q_{12} & \cdots & q_{1s} \\
q_{21} & q_{22} & \cdots & q_{2s} \\
\cdot & \cdot & \cdot & \cdot \\
\cdot & \cdot & \cdot & \cdot \\
\cdot & \cdot & \cdot & \cdot \\
q_{s1} & q_{s2} & \cdots & q_{ss}
\end{pmatrix}
$$

The fundamental matrix $(\mathbf{I} - \mathbf{Q})^{-1}$ will contain the expected number of vacancies in class j, given a chain of vacancies starts in class i $(i, j = 1, 2, \ldots, s)$. The row sums will be the expected length of chains starting in the various classes. If we let $\mathbf{1} = (1, 1, \ldots, 1)^T$, then $(\mathbf{I} - \mathbf{Q})^{-1}\mathbf{1}$ is a column vector of row sums of $(\mathbf{I} - \mathbf{Q})^{-1}$.

Hence, if we let

(16.7.11)
$$\lambda = (\lambda_1, \lambda_2, \ldots, \lambda_s)^T$$

be the column vector of expected chain lengths by class, we have

(16.7.12)
$$\lambda = (I - Q)^{-1} 1$$

Let F be the total number of initiated chains. Then the number initiated in class C_k is a random variable F_k, which is binomially distributed with parameters f_k and F, using axiom 4, which makes each chain independent of every other and with identical parameter f_k.

The total number of moves by chains initiated in class C_k, then, may be written

(16.7.13)
$$T_k = \sum_{i=1}^{F_k} J_{k_i} \qquad (i = 1, 2, \ldots, F_k;\ k = 1, 2, \ldots, s)$$

where J_{k_i} is the length of the ith chain starting in class C_k. The sum is over a random number of terms. Let

$$E(T_k \mid F_k = n) = E\left(\sum_{i=1}^{n} J_{k_i} \right) = \sum_{i=1}^{n} E(J_{k_i}) = n\lambda_k$$

Then,

$$M_k = E(T_k) = \sum_n E(T_k \mid F_k = n)P(F_k = n)$$

$$= \sum_n n\lambda_k P(F_k = n)$$

$$= \lambda_k \sum_n nP(F_k = n)$$

Hence,

(16.7.14)
$$M_k = \lambda_k E(F_k)$$

To make our notation conform to White's, we define a vector **M**:

(16.7.15)
$$M = (M_1, M_2, \ldots, M_s) = (F_1\lambda_1, F_2\lambda_2, \ldots, F_k\lambda_k, \ldots, F_s\lambda_s)$$

where on the right it is understood that the expected value of F_k is involved, as in formula (16.7.14).

The grand total number of moves, T, is given by

$$T = \sum_{k=1}^{s} T_k$$

and so, with scalar **M** defined as in (16.7.6)

(16.7.16)
$$M = E(T) = \sum_{k=1}^{s} E(T_k) = \sum_{k=1}^{s} M_k = \sum_{k=1}^{s} F_k\lambda_k$$

If we introduce the row vector,

$$F = (F_1, F_2, \ldots, F_s)$$

Then (16.7.16) may be written as a dot product,

(16.7.17) $$M = F\lambda$$

using (16.7.11), and then, by appeal to (16.7.12),

(16.7.18) $$M = F(I - Q)^{-1} 1$$

This generalizes formula (16.7.6), in which

$$(I - Q)^{-1} = (1 - q)^{-1} = \frac{1}{p}$$

We have obtained the expected lengths λ_k by class but not the full distribution of chain length by class. We want to do this by way of generalizing (16.7.2), using the absorbing chain language.

We introduce the event notation,

E_{ik}^{j-1} occurs iff the vacancy chain starts in transient state C_i and is in transient state C_k after $j - 1$ moves.

E_{kO} occurs iff the vacancy in C_k moves out of the system to O, the outside.

Then, the event that the chain enters O from state k on the jth move, so that the length is j, is

$$E_{ik}^{j-1} \cap E_{kO}$$

and, considering all k,

$$\bigcup_k (E_{ik}^{j-1} \cap E_{kO})$$

is the event of interest. Since we have a disjoint union,

$$P[\bigcup_k (E_{ik}^{j-1} \cap E_{kO})] = \sum_{k=1}^{s} P(E_{ik}^{j-1} \cap E_{kO})$$

$$= \sum_{k=1}^{s} P(E_{ik}^{j-1}) P(E_{kO})$$

(16.7.19) $$= \sum_{k=1}^{s} q_{ik}^{(j-1)} q_{kO}$$

But,

$$Q^{j-1} = \left(q_{ik}^{(j-1)} \right)$$

because the powers of Q give the desired probabilities of transition to state C_k

from C_i in a given number of moves (steps). If we let

$$p = (q_{10}, q_{20}, \ldots, q_{sO})^T$$

then the right side of (16.7.19) may be seen to be a dot product of row i of Q^{j-1} and column vector p. Hence, as i varies, we obtain the product of the matrix and p. This is an s-component vector whose component i is the probability that a chain starting in C_i has length j. Denoting this column vector P_j we have

(16.7.20) $P_j = Q^{j-1}p$ $(j = 1, 2, \ldots)$

This generalizes formula (16.7.2). Note that

$$\lambda = \sum_{j=1}^{\infty} jP_j$$

is the vector of expected values already found via the fundamental matrix, as indicated in (16.7.12).

To find the overall distribution of chain length, we note scalar P_j is given by

$$P_j = P(J = j) = \sum_{i=1}^{s} P(\text{start in } C_i)P(J = j \mid \text{start in } C_i) \qquad (j = 1, 2, \ldots)$$

$$= \sum_{i=1}^{s} f_i P(J = j \mid \text{start in } C_i)$$

and we recognize the second term as given by (16.7.19). The sum over all i is the dot product of row vector f and column vector P_j. Hence,

(16.7.21) $P_j = fP_j$

The overall expected length is

(16.7.22) $\lambda = \sum_{j=1}^{\infty} jP_j = \sum_{j=1}^{\infty} jfP_j = f \sum_{j=1}^{\infty} jP_j = f\lambda$

To sum up the vacancy chain model: the setup is given by matrix (16.7.7), along with vector f, as stated in axioms (16.7.8). The analysis yields the following key elements:

	(1)	length distribution vector P_j	(16.7.20)
	(2)	expected length vector λ	(16.7.12)
	(3)	overall length distribution P_j	(16.7.21)
(16.7.23)	(4)	overall expected length λ	(16.7.22)
	(5)	expected number of moves vector M	(16.7.15)
	(6)	expected total number of moves M	(16.7.18)

Application of the vacancy chain model requires identification of positions, actors, and a cohort of vacancy chains. White (1970) gives enormous effort to this task to demonstrate the practical feasibility of approaching mobility by the tracing of vacancy chains. The application is, however, too intricate as to sampling and counting to report here in any detail. Hence, only a rather simplified version of this application phase (see section 1.2) of the model-building enterprise is given.

White identifies positions and actors with the various local church clergy of a national church, (e.g., the Episcopal Church) and chooses as his time domain one year. He traces vacancy chains by analysis of registers kept by the national church. Since each clergyman occupies a unique position of this kind and at most one such clergyman is in a given position (e.g., "Dean, Cathedral, Wilmington, Delaware"), the basic 1–1 requirement is satisfied. Ordinarily, vacancies have a longer duration than limbos of clergy: practically speaking, no time is spent at all in limbo between positions. Hence, the type of 1–1 system is "tight," in the sense discussed earlier. It follows that it is within the scope of a vacancy chain analysis.

Next, the abstractions of the axioms (16.7.8) are identified as follows:

(1) There are three classes, determined by size of congregation.

(2) The terms f_k are estimated by counting (from the registers) the number a_{0k} of vacancy creations in class C_k and dividing by the total number of vacancies initiated, $a_{0.}$. Hence,

$$\hat{f}_k = \frac{a_{0k}}{a_{0.}} \qquad (k = 1, 2, 3)$$

Recall that a vacancy is created or initiated by a "death" of an actor (retirement, actual death) or a "birth" of a position (new position invented).

(3) The transition probabilities q_{ik} between classes are obtained by counting (from the registers) the number a_{ik} of times a vacancy moves from class C_i to C_k and dividing by the total number of moves $a_{i.}$ from class C_i, including those terminating a chain. Hence,

$$\hat{q}_{ik} = \frac{a_{ik}}{a_{i.}} \qquad (i, k = 1, 2, 3)$$

Then $\hat{q}_{i0} = 1 - \sum_{k=1}^{s} \hat{q}_{ik}$ estimates the probability of termination from a vacancy in C_i ($i = 1, 2, 3$).

(4) The cohort identification is complex: not every chain is sampled, sampling weights are needed, etc. Hence, the evaluation is complicated by this sampling aspect (which we do not discuss here).

A typical estimated transition matrix follows (see White, 1970, Table 4.12, 1954–1955; figures rounded off here):

$$(16.7.24) \quad \begin{array}{c} \\ \text{Outside} \\ \text{Big} \\ \text{Medium} \\ \text{Small} \end{array} \begin{pmatrix} \begin{array}{c} \text{Outside} \\ \hline 1 \\ .177 \\ .455 \\ .606 \end{array} & \begin{array}{ccc} \text{Big} & \text{Medium} & \text{Small} \\ \hline 0 & 0 & 0 \\ .389 & .338 & .095 \\ .054 & .292 & .199 \\ .067 & .179 & .148 \end{array} \end{pmatrix}$$

Here C_1 = big, C_2 = medium, and C_3 = small are the three classes of positions by size of congregation.

Note that a regularity appears in the $\hat{Q}$ submatrix of (16.7.24):

$$\text{Big} \xrightarrow{.338} \text{Medium} \quad \text{vs.} \quad \text{Medium} \xrightarrow{.054} \text{Big}$$
$$\text{Big} \xrightarrow{.095} \text{Small} \quad \text{vs.} \quad \text{Small} \xrightarrow{.067} \text{Big}$$
$$\text{Medium} \xrightarrow{.199} \text{Small} \quad \text{vs.} \quad \text{Small} \xrightarrow{.179} \text{Medium}$$

In each case the transition of a vacancy is more likely in the down direction than in the up direction, where the implied ordering of the classes is by size (which no doubt has status significance in the sense of differential evaluation of clergy by clergy according to congregation size). For example, a clergyman who is pastor of a "big congregation" retires. This is an initiation of a vacancy chain. A "moderate congregation" clergyman is moved up to replace the retired man: then the vacancy moves down to "moderate" from "big." Hence, upward mobility for actors corresponds to downward mobility for vacancies.

Note now in (16.7.24) the column vector $\hat{p}$ of transition probabilities to the outside. A vacancy leaves the system if and only if an actor is born (i.e., recruited from outside) or a position dies (i.e., is eliminated). If the classes are stratified, new clergy enter "at the bottom" from the schools. Hence, the higher birth rate of actors is at the bottom, inducing a higher rate of vacancy elimination to the outside. For example, a clergyman with a small congregation "moves up" to a larger congregation. This initiates a vacancy in the "small" class, to be eliminated by moving in a recent recruit to the ministry. On the other hand, when the dean of a large congregation dies, his replacement is drawn from within the system of experienced clergy rather than from the outside; hence, the vacancy stays in the system, creating a new vacancy elsewhere in the system. This interprets the monotonic increase of the probabilities in the column vector $\hat{p}$ as congregation size decreases.

White checks goodness-of-fit by using some of the derived formulas in list (16.7.23). In particular, in Table 16.7.1 we show the predicted and observed length distribution overall, whose formula is (16.7.21) (see White, 1970, Table 4.13, 1954–1955).

To compute Table 16.7.1 one requires (16.7.21) and (16.7.20)

$$P_j = fQ^{j-1}p \qquad (j = 1, 2, \ldots)$$

Table 16.7.1   Predicted and Observed Percentage Distribution of
Lengths of Vacancy Chains

Length j	(Overall) Observed	Predicted
1	40.9	45.9
2	32.4	24.0
3	12.6	13.1
4	5.6	7.3
5	2.0	4.1
6	3.3	2.4
7	1.2	1.3
8	0.4	0.8
9	0.7	0.4
10 or more	0.9	0.57

and so the f estimate from the data (see White, 1970, Table 4.12, 1954–1955):

$$\hat{f} = (.161, .516, .323)$$

For instance, for j = 1,

$$\hat{P}_1 = \hat{f}\hat{p} = (.161 \quad .516 \quad .323)\begin{pmatrix} .177 \\ .455 \\ .606 \end{pmatrix} = .459$$

The mean chain length is estimated directly as

$$\hat{\lambda} = 2.23$$

The fit in Table 16.7.1 is good and reflects the general quality of the fit of length
distributions to other bodies of data, as reported by White. The mean length is a
little longer than the typical mean for earlier time periods but in general $\hat{\lambda}$ ranges
around 2.

White applies the model to three national churches, over five decades, comparing both over-time developments in mobility via the vacancy chain statistics and
the three churches. Indeed, he analyzes the data in great detail from many points
of view. Our aim has been the more limited one of introducing the reader to the
vacancy chain conceptualization and to illustrate the techniques and results.

16.8.   The State-Space Problem in Mobility Theory.   The idea of representing
mobility by a discrete-time Markov chain emerges out of the form of the table of
data recording shifts in occupational status from father to son, as we have seen
in section 16.2. However, as our process idea termed the state-space approach
(see Chapter 8 and especially section 8.10) emphasizes, not any observable variable can form a proper state variable in the sense of path independence. In most
cases, directly observed phenomena exhibit strong historical effects; these

variables have to be reconceptualized in some way so as to make the state-determined idea, which is central to Markov models, empirically valid. Hence, one of the most significant problems in the mobility area is the representation of the "space" within which positions exist and within which social units exhibit movement from position to position. In other words, "position" should be defined so as to make the "trajectory" of a unit analyzable by the state-space approach.

For model-builders seeking a finite-state model in discrete or continuous-time the problem specializes to the definition of strata, or classes. In this context, Mayer (1972), for example, calls this the identification problem in the theory of social mobility. That the problem persists even when vacancy chains, as in section 16.7, become the focus is evident in White's own discussion of the many ways in which one might pass from the undifferentiated binomial model to the s strata of the vacancy chain model. The problem intersects not only the conceptual notion of a state-determined system but measurement models in the stratification area. To say that there is a space of positions but then to insist on just so many strata is a strategem that avoids the problem of assigning to each position a measure of differential evaluation less coarse than the purely ordinal grouping of the stratum concept. This point is at least part of the conceptual background of the Duncan model, with its continuous occupational-prestige variable, as well as the Fararo (1972a) treatment of a vector of such differentially evaluated characteristics defining position in a continuous space, emergent from one measurement model per basic characteristic but all conceptually unified by their foundation in probablistic choice theory as outlined in Chapter 11. The fundamental idea is to represent the stratification system in terms of quantities directly reflecting the psychological bases of occupational and other social choices (see Fararo, 1970c, part II). Whereas Fararo follows the strategy of measurement followed by mapping mobility as change in measured position, Levine (1972) proposes a fruitful alternative: obtain the status measurements directly from the observed gross movements, which are in a mobility table, interpreted in terms of the continuous status construct, which is then estimated from mobility data. Unfortunately, this measurement is not linked to a process point of view. In any case, this brief discussion indicates that the identification problem of mobility theory is a crucial focus of a variety of conceptual and methodological problems in the entire field of social stratification.

One attempt to come to grips with the state-determined system idea in the discrete-state approach has been made by McGinnis (1968). He sees the underestimation of the main diagonal terms as evidence that the following "axiom of cumulative inertia" holds: The longer a person is in a certain social category, the lower the probability of his moving. But McGinnis does more than propose this axiom. He redefines the concept of "position" so as to make it a state-space concept. He does this in conformity with the axiom by treating position or state as composed of both social location in the observable sense (e.g., occupational stratum) and duration of time in that location. Essentially, this says that two positions, the same in overt social categorization, are dynamically distinctive unless they are associated with the same duration of stay. This is like the passage in physics from a focus on observable position in space to a focus on position and velocity (see the example in section 8.8) for the sake of arriving at an empirically valid formulation in which future (overt spatial) position is predictable from present state and exogeneous inputs. Until recently, the McGinnis model has been so complex that only simulations provided clues as to its properties. More recently, however, some mathematical analysis has been accomplished (see Henry, McGinnis, and Tegtmeyer, 1971). The model has been applied to human migration (see McGinnis, Myers, and Pilger, 1963, and Morrison, 1967). Even more recently, Ginsberg (1971) has subsumed aspects of the McGinnis approach under the general rubric of semi-Markov mobility processes. Following Pyke (1961), he presents an outline of semi-Markov theory in such a way as to include continuous-time Markov chains as a special case. In addition, he relates this formulation to the semi-Markov mover-stayer model treated in section 16.4.

It should not be thought that the path analysis and vacancy chain models are free of the state-space problem. In the former, a vector of evaluated social characteristics is selected: a rationale for the choice of the components lies in the direction of analysis of flux of movement in the system. The path model only connects two or more such status vectors by a linear model, but the significance of the path model is heightened if the vectors are interpretable as states of a dynamic process. In turn, as soon as the dynamic context is made explicit, there can no longer be indifference as to the representation of a person's position by some particular array of attributes as opposed to alternative arrays. Each such array defines a vector, exhibiting change in time, and so raises the question of whether it allows a state-space approach, either of the deterministic or Markov type. Similarly, in treating each move in a vacancy chain as dependent only on the current vacant position and not upon the earlier history of the vacancy chain, a state-space problem again arises. Of course, the fit of the models to the data suggests a verification of the particular state identification, but it is not a compelling verification without a direct test of the path independence assumption. White suggests that one might use "higher-order" chains in the future, should the path independence assumption not be satisfied. This is a redefinition of the state-space: it amounts to defining "state of chain" as "last two values of observable process," or some similar use of the past history of the observable process. Precisely the same idea is present in the McGinnis model with its duration-of-stay notion. Moreover, the identification of clear-cut positions for the flow of vacancies is by no means generally easy. Formal organizational networks provide the natural scope for such identifications, making the model quite restricted in scope compared with the more usual treatment of flows of people through locations. But even this restriction may be an artifact of the unsettled and chaotic procedures used to identify social locations for actors in arbitrary social systems. A coherent conceptual and technical account of how positions are to be constructed from observable relational systems linking individual actors might yield the right inputs to a vacancy chain framework, itself expanded to include more inchoate systems of categorization and expectations. White himself sees the vacancy chain methodology as just one aspect of the wider commitment to the study of social structure and process as organic systems, or networks, in which incessant flows both perturb and sustain its "skeletal" structure. In the next few chapters, we will provide the background for, and the major ideas in, this algebraic analysis of the foundations of social structure.

We close this chapter by reminding the reader that we have not been able to cover all of the representational and theoretical problems in the field of social mobility. For a specification of a set of such problems for mobility theory, see Mayer (1972), and for one example of a concept we have not treated ("equal exchange"), see Berger and Snell (1957) or Bartos (1967).

CHAPTER SEVENTEEN
INTRODUCTION TO GROUP THEORY

17.1. Aim of the Chapter. The present chapter is an interlude in our study of mathematical models of social phenomena, in which we present a purely mathematical apparatus: the theory of abstract groups. In terms of the two chapters that follow, the aim is, first, to acquaint the reader with group theory because groups do arise in the algebra of kinship; but the second aim is wider in that in the effort to extend algebraic methodology to cover a wider class of social relational systems, we find that group theory provides a basis in "abstract algebraic intuition," so that not so much the content but the spirit of group theory is brought into play. In the history of abstract algebra, group theory is old, while semigroup theory and "category theory" are relatively new: hence, they borrow conceptual devices from the group theoretic tradition and construct concepts and methods by analogy with those developed in group theory. For this reason, acquaintance with group theory is significant beyond the possibility of direct application of its detailed content. Moreover, we are able to link groups and graphs in a manner that makes the abstractions of group theory receive a visualization in concrete and familiar terms. Not surprisingly such graphs arise in the algebra of kinship, so that the group-graph correspondence developed in this chapter will play a role later in our work.

17.2. Examples of Groups. In this introductory section to the chapter we begin by restating the definition of a group and then give a physical and a geometric example. The reader should review sections 5.11–5.18 before continuing.

Recall from section 5.13 that a group is a groupoid—a set with a binary operation—satisfying certain conditions. We state these conditions in the definition below.

(17.2.1) Definition. A group is a groupoid (G, o) satisfying the following:

Axiom 1. The operation is associative:

$$(x \text{ o } y) \text{ o } z = x \text{ o } (y \text{ o } z) \qquad \text{(all } x, y, z \text{ in } G)$$

Axiom 2. There exists an identity element, say e, so that

$$e \text{ o } x = x \text{ o } e = x \qquad \text{(all } x \text{ in } G)$$

Axiom 3. Each element x has an inverse, say x^{-1}:

$$x \text{ o } x^{-1} = x^{-1} \text{ o } x = e$$

We use the same symbol, G, to designate the group and to denote the domain of elements alone. Note that axiom 3 implies directly that the inverse of x^{-1} is x.

Two variant notations are used in group theory for the binary operation. In the "additive notation" we use "+" for the binary operation. Also, the identity is denoted by "0". In the "multiplicative notation," the symbols for the elements are merely juxtaposed as in the algebra of numbers. There is variation in the multiplicative notation for the identity. Symbol "I" is often used, but some mathematicians use "e", others "1." We follow the convention of using "I." Then in terms of these notations the axioms are as follows.

	Additive	Multiplicative
Associative law:	$x + (y + z) = (x + y) + z$	$x(yz) = (xy)z$
Identity:	$0 + x = x + 0 = x$	$Ix = xI = x$
Inverse:	$x + (-x) = -x + x = 0$	$xx^{-1} = x^{-1}x = I$

In these terms, repeated operation with an element x yields,

$$nx = \underbrace{x + x + \ldots + x}_{n \text{ terms}} \qquad x^n = \underbrace{xx \ldots x}_{n \text{ terms}}$$

$$-nx = \underbrace{(-x) + (-x) + \ldots + (-x)}_{n \text{ terms}} \qquad x^{-n} = \underbrace{x^{-1}x^{-1} \ldots x^{-1}}_{n \text{ terms}}$$

Repeated operation with a series of elements $x_1, x_2, \ldots, x_n$ yields

$$\sum_{i=1}^{n} x_i = x_1 + x_2 + \ldots + x_n$$

$$\prod_{i=1}^{n} x_i = x_1 x_2 \ldots x_n$$

The group may or may not be commutative. A group which is commutative is called *abelian*. The number of elements in G is called the order of the group. If the order is a positive integer, the group is called finite.

Before proceeding with further formal developments, let us look at two intuitive examples.

Example. (a) A common household wall switch for an overhead lamp may be in one of two states: on or off. A physical transformation or change of state is definable, namely, switching:

$$on \longmapsto off$$
$$off \longmapsto on$$

This is a mapping on the set of states of the switch. Call it S. In addition, at any moment, the switch may be in state x (x = on, off) and merely left in that state:

$$on \longmapsto on$$
$$off \longmapsto off$$

This is also a mapping on the set of states of the switch. Call it I. Hence, we have two mappings. Let

$$G = \{S, I\}$$

Define a binary operation on G by the idea that we can follow one map by the other: in other words, one physical transformation may succeed another physical transformation. Use multiplicative notation, so that S^2 means "switch and then switch again." Of course, $S^2 = I$ because even though S^2 and I differ in the amount of energy expended, they have the same effect on the state of the system, and our interest is in this latter aspect of the situation. Naturally, SI = S, IS = S, and II = I. The table obtained is :

	I	S
I	I	S
S	S	I

To see that this is a group in the formal sense of the axioms, we note that "followed by" is a binary operation on G = $\{S, I\}$. It is associative, by appeal to the physical situation. There is an identity, which we denoted I. The inverse of I is I, and the inverse of S is S itself. Note, then, that $S^2 = I$ implies $S^{-1} = S$.

Example. (b) Consider an equilateral triangle ABC in the plane, shown in Figure 17.2.1. Various rotations are possible which carry the figure into itself: 120°, 240°, 360°, 480°, and so on. Indeed, any integer multiple of 120°,

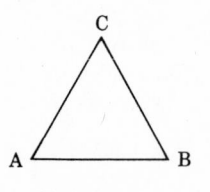

Figure 17.2.1

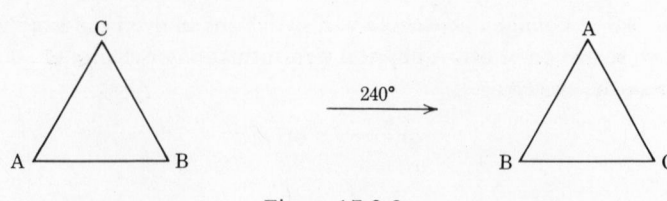

Figure 17.2.2

whether positive (counterclockwise) or negative (clockwise), yields a transformation of the figure into itself. Moreover, apart from the resulting differential locations of the labels for the vertices, we could not tell that the motion had been performed. Such a motion is called a symmetry of the figure. But only three basic rotations of this kind exist (120°, 240°, 360°) because 480°, for instance, is the same as 120° in its resulting configuration of labels. Also, 360° is really the same as 0° rotation, which in this model is like the "leave-it-alone transformation" on the states of the switch. Call the 0° rotation I. The 120° rotation may be termed r. Then the 240° rotation is shown in Figure 17.2.2. But also, using r, we have the two successive rotations in Figure 17.2.3. Hence, with the binary operation of "followed by" and multiplicative notation, we see that 240° is r^2. Similarly, $r^3 = I$. Thus, we obtain the table:

	I	r	r^2
I	I	r	r^2
r	r	r^2	I
r^2	r^2	I	r

Note that $r^2 r = I$ and that $rr^2 = I$ means that $r^2 = r^{-1}$ and $(r^2)^{-1} = r$. (To undo the effect of r^2, the 240° rotation, merely follow it by 120°: thus, r is the inverse of r^2.)

17.3. Elementary Group Algebra.   One main aim of the present section is to give the reader some insight into how the axiomatic properties of a group, listed in definition (17.2.1), are used to "calculate" consequences of given group conditions. A second aim is to review and deepen our understanding of homomorphisms (see Section 5.14) and to define the concept of a subgroup. We do all this

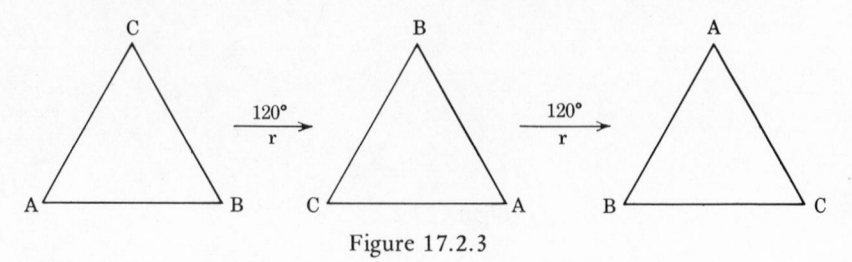

Figure 17.2.3

under the guise of proving some propositions of elementary group theory. In the previous section we aimed to show by examples physical and geometric realizations of the group concept, since all our examples in Chapter 5 were more abstract. In the present section, we want to revert to "pure" group theory to illustrate the posing of problems and the derivation of results in abstraction from any concrete realizations. This abstractness is surely one of the hallmarks of group theory.

We begin with the simple idea that we can write down various equations in an unknown x where x is in G. Hence, it is natural to ask if the ordinary algebraic solution idea is valid in group algebra. We put the matter in this way: the equation

$$(17.3.1) \qquad\qquad ax = b$$

makes sense for a group G, granted multiplicative notation. The question is, Does the equation have a solution in G?

(17.3.2) Proposition. In any group G, the equation (17.3.1) has a unique solution given by

$$x = a^{-1}b$$

To prove this proposition, apply the multiplicative form of the group axioms to ax = b on both sides:

$$
\begin{array}{ll}
ax = b & \text{(given)} \\
a^{-1}(ax) = a^{-1}b & \text{(multiply by } a^{-1}) \\
(a^{-1}a)x = a^{-1}b & \text{(associative law)} \\
Ix = a^{-1}b & \text{(inverse law)} \\
x = a^{-1}b & \text{(identity law)}
\end{array}
$$

Similarly, the equation

$$(17.3.3) \qquad\qquad xa = b$$

has a solution as follows:

$$
\begin{array}{ll}
xa = b & \text{(given)} \\
(xa)a^{-1} = ba^{-1} & \text{(multiply by } a^{-1}) \\
x(aa^{-1}) = ba^{-1} & \text{(associative law)} \\
xI = ba^{-1} & \text{(inverse law)} \\
x = ba^{-1} & \text{(identity law)}
\end{array}
$$

(17.3.4) Proposition. In any group G, equation (17.3.3) has a unique solution given by

$$x = ba^{-1}$$

If G is commutative, then the two equations present the same problem and $x = a^{-1}b = ba^{-1}$. But in the general case, the problems are distinct and the solutions exist but are not equal. Note that in multiplying, operations "on the right" or "on the left" are distinctive unless the group is commutative. Right (or "post-") multiplication of x by a^{-1} means, abstractly,

$$(x, a^{-1}) \longmapsto x \circ a^{-1} = xa^{-1}$$

while left (or "pre-") multiplication means,

$$(a^{-1}, x) \longmapsto a^{-1} \circ x = a^{-1}x$$

Apart from the commutative law we have no reason to think that the image of pair (x, a^{-1}) is equal to the image of pair (a^{-1}, x). Another application of right and left multiplication yields the following:

(17.3.5) *Proposition.* Cancellation laws are valid. That is,

(1) $ba = ca$

(2) $ab = ac$

each implies that b = c.

To see this in equation (1) use postmultiplication by a^{-1}, while in equation (2) use premultiplication by a^{-1}.

Another important algebraic property of groups is the following rule for inverses:

(17.3.6) *Proposition.* In any group G, the inverse of a product is the product of the inverses in reverse order. That is,

$$(x_1 x_2 \ldots x_n)^{-1} = x_n^{-1} x_{n-1}^{-1} \ldots x_2^{-1} x_1^{-1}$$

For n = 2, we have by definition of the inverse of $x_1 x_2$,

$$(x_1 x_2)(x_1 x_2)^{-1} = I$$
$$x_2 (x_1 x_2)^{-1} = x_1^{-1} \qquad \text{(premultiplying on both sides by } x_1^{-1})$$
$$(x_1 x_2)^{-1} = x_2^{-1} x_1^{-1} \qquad \text{(premultiplying on both sides by } x_2^{-1})$$

In a similar way the proposition is true for any n. Of course, if the group is commutative, we can neglect the reversal.

In definition (17.2.1) we stated that there exists an identity element. Yet, we seem to be employing I as if there were just one identity element. In fact, this is so, as stated in the next proposition.

(17.3.7) *Proposition.* In any group G, there is a unique identity element.

To prove this we can assume existence, because of axiom 2 of definition (17.2.1). To show uniqueness (recall section 2.14) we assume that two elements I and I$'$ are identities. Then in axiom 2, taking I as identity,

$$(1) \quad Ix = x$$

while taking I' as identity,

$$(2) \quad xI' = x$$

Now in (1) put $x = I'$ to obtain

$$(3) \quad II' = I'$$

and in (2) put $x = I$ to obtain

$$(4) \quad II' = I$$

Comparing (3) and (4), we conclude that $I' = I$. Hence, there is only one identity element.

In a similar way, in axiom 3 we assume only that an inverse exists. We state now that it is unique.

(17.3.8) Proposition. In any group G, any element has a unique inverse.

The proof takes as given the existence of at least one inverse of x, where x is in G. Let x_1^{-1} and x_2^{-1} both satisfy axiom 3. Then,

$$
\begin{aligned}
xx_1^{-1} &= I & \text{(given)} \\
x_2^{-1}(xx_1^{-1}) &= x_2^{-1}I & \text{(premultiplication by } x_2^{-1}) \\
x_2^{-1}(xx_1^{-1}) &= x_2^{-1} & \text{(axiom 2)} \\
(x_2^{-1}x)x_1^{-1} &= x_2^{-1} & \text{(axiom 1)} \\
Ix_1^{-1} &= x_2^{-1} & \text{(axiom 3, given } x_2^{-1} \text{ an inverse)} \\
x_1^{-1} &= x_2^{-1} & \text{(axiom 2)}
\end{aligned}
$$

Note that the proof uses all three axioms to show uniqueness.

We recall (from section 5.14) that a homomorphism of a groupoid (G, o) into a groupoid (G', o') was defined as a mapping,

$$f\colon\ G \longrightarrow G'$$

satisfying the rule that the image of a composition (in terms of o) is the composition of the images (in terms of o'):

$$f(xoy) = f(x)\ o'\ f(y)$$

In group theory it is customary to use uniform notation in different abstract groups so that the expressions do not become cluttered. Then, if two groups, $(G, \cdot)$, $(G', \cdot)$, are given, in multiplicative notation, a homomorphism is a map $f\colon G \to G'$ such that

$$f(xy) = f(x)f(y) \qquad \text{(all x, y in G)}$$

On the left the multiplicative notation refers to an operation in G, and on the right it refers to a possibly different operation in G′. For the moment let us denote the identity of G′ by I′.

(17.3.9) Proposition. If f: G → G′ is a homomorphism of groups, then the image of the identity is the identity

$$f(I) = I'$$

To show this, we note that

$$f(I) = f(II) = f(I)f(I)$$

because I = II and f is a homomorphism. Hence, multiplying on both sides by $[f(I)]^{-1}$, we obtain, freely using associativity,

$$I' = [f(I)]^{-1}f(I) = [f(I)]^{-1}f(I)f(I) = I'f(I) = f(I)$$

recalling that f(I), the image of I, is in G′ and, so, when multiplied by its inverse, yields I′.

(17.3.10) Proposition. If f: G → G′ is a homomorphism of groups then the image of the inverse is the inverse of the image:

$$f(x^{-1}) = [f(x)]^{-1} \qquad \text{(any x in G)}$$

To prove this, we use the prior proposition, as follows:

$$I' = f(I) = f(xx^{-1}) = f(x)f(x^{-1})$$

Hence, premultiplying by $[f(x)]^{-1}$ on both sides of $I' = f(x)f(x^{-1})$ yields the proposition.

17.4. Subgroups. We know from section 5.13 that $(\Re, +)$, the set of all real numbers with addition forms a group, but so does addition on the set of all integers, denoted $Z \subseteq \Re$. Hence, $(Z, +)$ is a group contained in another group $(\Re, +)$. Looking at an arbitrary set of numbers contained in $\Re$, under what conditions will this set form a group in its own right? Surely, the sum of any two numbers in the set must be in the set in order that we have a well-defined binary operation. This is termed closure with respect to the operation. Next, we want the identity, in this case zero, to be in the set. And finally we need the inverse of every number in the set, which in this case means we need the negative of any number in the set. If these three conditions are satisfied, then in definition (17.2.1) we have (a) axiom 1 satisfied because the elements are a subset of group elements, (b) axiom 2 satisfied because the identity is in the subset, and (c) axiom 3 satisfied because the inverse of any element in the subset is also in the subset. Thus, the three conditions suffice to formalize the intuitive idea of a subgroup.

(17.4.1) *Definition.* A subset H $\subseteq$ G forms a subgroup of group G with respect to the operation of G if and only if

(1) x, y in H $\Rightarrow$ xy in H   (closure)
(2) I is in H
(3) x in H $\Rightarrow$ x^{-1} in H

Just as we consider all subsets of a given set and the relation of containment as forming a partial order of these subsets (see section 4.13), so also the collection of all subgroups of a given group is partially ordered by the relation "is a subgroup of." To show this, we must prove that the relation is reflexive, antisymmetric, and transitive, according to definition (4.13.1).

Any group is a subgroup of itself, by taking H = G in definition (17.4.1). Hence, the relation is reflexive.

Let H and K be subgroups of G. Then, if H is a subgroup of K and K is a subgroup of H, we have both H $\subseteq$ K and K $\subseteq$ H. Hence, H = K. Thus, antisymmetry holds.

To show transitivity, let H, K, and L be subgroups of G. If H is a subgroup of K and K is a subgroup of L, then H $\subseteq$ K and K $\subseteq$ L, so that transitivity of set containment yields the result that H $\subseteq$ L. But, also, H satisfies closure (since it is a subgroup of G); it contains the same identity as L (since H and L are subgroups of G); and the inverse of any element in H is in H (since H is a subgroup of G). We conclude that H is a subgroup of L.

We sum up this reasoning in the following:

(17.4.2) *Proposition.* Given any group G, the relation "is a subgroup of" is a partial order of the collection of all subgroups of G.

We may think of G itself as the largest subgroup of G. Similarly, the group H = $\{I\}$, where I is the identity of G, is the smallest subgroup of G in that it is contained in any subgroup of G. This subgroup is called trivial. Any subgroup of G that is neither of these two extremes is called a proper subgroup.

Example. We know that Z is a subgroup of $\Re$, with respect to addition. Define the set of even integers,

$$E = \{2n, n \in Z\} = \{\ldots, -2, 0, 2, 4, \ldots\}$$

Clearly E is also a subgroup of $\Re$, since E $\subseteq$ $\Re$ and

(1) the sum of any two even integers is even,
(2) 0 is even,
(3) if 2n is even, then so is −2n.

Now we note that, in addition, E is a subgroup of Z. Note that E $\subseteq$ Z $\subseteq$ $\Re$. The trivial subgroup is $\{0\}$. Both Z and E are proper subgroups of $\Re$. The set of all subgroups of $\Re$—that is, subgroups of ($\Re$, +)—is a partially ordered set whose largest member is $\Re$ and smallest member is $\{0\}$.

17.5.   Cyclic Groups.   The additive group of integers $(Z, +)$ is such that one element plays a special role, namely, 1. For we have

$$1 = 1 \qquad\qquad -1 = -1$$
$$2 = 1 + 1 \qquad\qquad -2 = (-1) + (-1)$$
$$3 = 1 + 1 + 1 \qquad\qquad -3 = (-1) + (-1) + (-1)$$

$$\cdot \qquad\qquad\qquad \cdot$$
$$\cdot \qquad\qquad\qquad \cdot$$
$$\cdot \qquad\qquad\qquad \cdot$$

$$0 = 1 + (-1)$$

In short, any element in Z can be represented in terms of 1 and its inverse. We say that 1 generates Z. This is an example of a cyclic group.

In the switch example of section 17.2, the physical transformation S generated the group. In this case the notation is multiplicative, and we have

$$S = S$$
$$I = S^2$$

where $\{S, I\}$ comprises the group.

In the rotation example, $G = \{I, r, r^2\}$ so quite obviously r (the 120° rotation) generates the group. If a square is considered, the generator is a 90° rotation and it takes four such rotations to return to the starting position. Calling r the 90° rotation, the group is given by $\{I, r, r^2, r^3\}$ since $r^4 = I$.

All these examples have something in common and yet something that differentiates them. The common aspect is the fact that every element can be represented in terms of positive or negative "powers" of one fixed element and its inverse. The differentiating aspect is that the power of the element that equals the identity varies. The common aspect is called the cyclic property. The differentiating aspect is called the order of the cyclic group.

(17.5.1) Definition.   A group G is called cyclic if and only if there exists an element, say g, such that every element in the group can be represented as g^n for some integer n. The order of the cyclic group is the smallest positive integer N such that $g^N = I$. If no such N exists, the cyclic group is said to be of infinite order . The element g is called the generator of G.

Hence, the examples were of the form

(1) integers: infinite order
(2) switching: order two, $g^2 = I$
(3) rotations of triangle: order three, $g^3 = I$
(4) rotations of square: order four, $g^4 = I$.

It is convenient to note the rules for exponents in multiplicative notation. They are the usual rules:

$$x^n x^m = x^{n+m} = x^{m+n} = x^m x^n$$

This holds because of the meaning of x^n, namely,

$$x^n x^m = \underbrace{\underbrace{xx \ldots x}_{n \text{ times}} \; \underbrace{xx \ldots x}_{m \text{ times}}}_{n + m \text{ times}}$$

Note that the rule applies to negative exponents as well. Also, by a convention we let $x^0 = I$. For instance,

$$
\begin{aligned}
x = x^{-n} x^{n+1} &= x^{-n+(n+1)} \\
&= x^{(-n+n)+1} \\
&= x^0 x^1 \\
&= x \qquad \text{(if we set } x^0 = I)
\end{aligned}
$$

(17.5.2) Proposition. Any two cyclic groups of the same order are isomorphic. Hence, there is only one "abstract" cyclic group of a given order.

For if G and G' are of finite order N, then there exist generators g and g' such that

$$G = \left\{ I, g, g^2, \ldots, g^{N-1} \right\}$$
$$G' = \left\{ I', g', g'^2, \ldots, g'^{N-1} \right\}$$

The mapping

$$f(g^n) = g'^n \qquad (n = 0, 1, \ldots, N - 1)$$

is bijective. It is a homomorphism because,

$$
\begin{aligned}
f(g^m g^n) = f(g^{m+n}) &\qquad \text{(exponent rule)} \\
= g'^{m+n} &\qquad \text{(definition of f)} \\
= g'^m g'^n &\qquad \text{(exponent rule)} \\
= f(g^m) f(g^n) &\qquad \text{(definition of f)}
\end{aligned}
$$

Hence, f is an isomorphism. Since G and G' are arbitrary cyclic groups of order N, any two such groups are isomorphic. Hence, abstractly there is only one such group —that is, there is a unique structure—at each order N. In the infinite case the correspondence is established in a similar manner to show that abstractly (Z, +) represents any and all cyclic groups of infinite order.

Notation Conventions. Hereafter, we use the symbol $\cong$ for isomorphism. If G and G' are cyclic groups of order N, then proposition (17.5.2) asserts that $G \cong G'$. If G is any infinite cyclic group, then $G \cong Z$.

The standard notation for the abstract cyclic group of order N is C_N. The abstract infinite cyclic group is denoted C_∞. The abstract group C_∞ is easily "represented" by Z since any concrete group with C_∞ structure is isomorphic to Z.

We define an associated relational system for any cyclic group. Let the domain be G and

$$(x, y) \in R \qquad \text{iff} \qquad xg = y \qquad (x, y \text{ in } G)$$

In other words, xRy means that x is carried into y by the generator g. For instance, 2R3 holds in the group Z of integers.

Next we represent the relational system (G, R) by a graph. The nodes correspond to group elements. A directed line from a node x to node y means x is taken into y by postmultiplication by the generator. For instance, for the rotation group of an equilateral triangle,

$$G = \{I, r, r^2\}$$

the graph is shown in Figure 17.5.1. We see that the group of rotations (G) that leave the triangle in position has for its graph the triangle itself with orientations as shown. The directed line from r^2 to I means

$$r^2 r = I \qquad (r \text{ is the generator})$$

Of course, since $r^3 = I$, this is valid.

Note that

$$xg = y \Leftrightarrow x = yg^{-1}$$

Hence, $(x, y) \in R$ also means that $x = yg^{-1}$. Thus, by working backwards on the graph, we are postmultiplying by g^{-1}.

For example, in Figure 17.5.1, if we start at I, and move back one step we arrive at r^2. Hence, $Ir^{-1} = r^{-1} = r^2$.

For the rotations of the square the group is the cyclic group of order four. The graph is itself a "square" as indicated in Figure 17.5.2. We note that working back from r^3 takes us to r^2 as expected, since $r^3 r^{-1} = r^2$. Forward from r^3 means $r^3 r = r^4$. Hence, again as expected, $r^4 = I$.

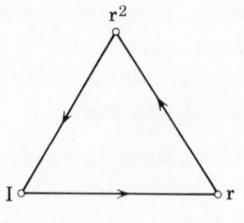

Figure 17.5.1

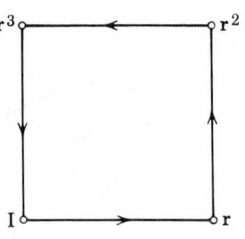

Figure 17.5.2

The cyclic group of infinite order has the graph in Figure 17.5.3. Since this group is in additive notation, we add 1 (the generator) to go forward, and going backward by one step means adding −1 (i.e., subtracting 1).

The group C_N is uniquely specified if we merely add one condition to the three axioms of group theory:

Defining Relation. There exists an element g in G such that g generates G and $g^N = I$. This is compactly written

(17.5.3) $\qquad\qquad (g: \ g^N = I)$

The form (17.5.3), meaning the generating condition added to the three axioms of group theory, is said to constitute a "defining relation." This is so because it yields C_N as a definite group in the abstract sense (because of isomorphism of all concrete N-order cyclic groups). To repeat, knowing only

(1) G is a group
(2) $(g: \ g^N = I)$

we conclude immediately as a logical consequence that $G \cong C_N$.

The first few nontrivial cyclic groups can be represented as shown in Figure 17.5.4, using the multiplication table and the graph.

Before passing on to more complex groups, we note the following propositions.

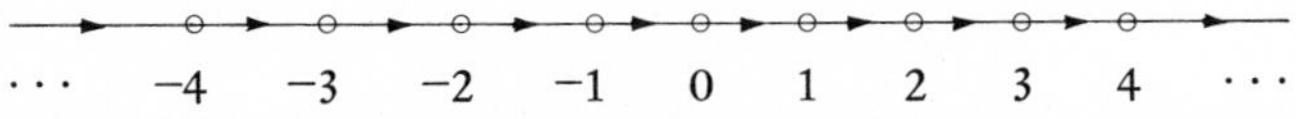

$$\cdots \quad -4 \quad -3 \quad -2 \quad -1 \quad 0 \quad 1 \quad 2 \quad 3 \quad 4 \quad \cdots$$

Figure 17.5.3

(17.5.4) Proposition. Any cyclic group is abelian (i.e., commutative).

For, given x and y in a cyclic group,

$$x = g^n \qquad \text{(some n)}$$
$$y = g^m \qquad \text{(some m)}$$

C_2: $(g:\ g^2 = I)$

	I	g
I	I	g
g	g	I

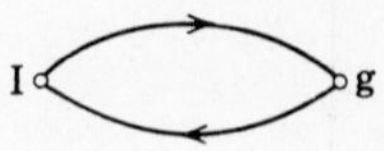

C_3: $(g:\ g^3 = I)$

	I	g	g^2
I	I	g	g^2
g	g	g^2	I
g^2	g^2	I	g

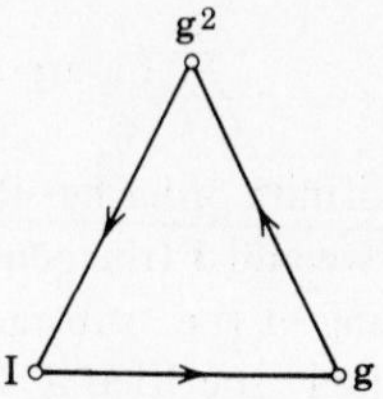

C_4: $(g:\ g^4 = I)$

	I	g	g^2	g^3
I	I	g	g^2	g^3
g	g	g^2	g^3	I
g^2	g^2	g^3	I	g
g^3	g^3	I	g	g^2

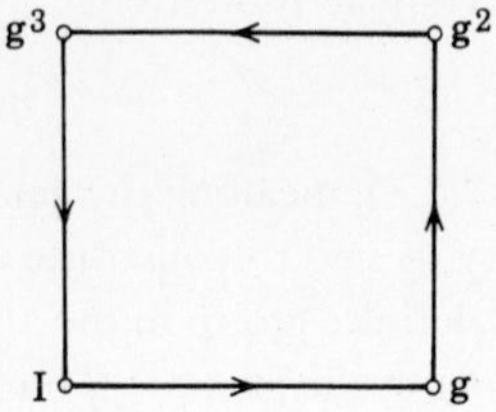

Figure 17.5.4

Hence,

$$xy = g^n g^m = g^{n+m} = g^{m+n} = g^m g^n = yx$$

For instance, in the rotation example, if we rotate the triangle first $240°$ counterclockwise (r^2) and then $120°$ clockwise (r^{-1}) the result $(r^2 r^{-1} = r = 120°$ counterclockwise) is the same as first rotating $120°$ clockwise (r^{-1}) and then $240°$ counterclockwise (r^2).

One additional fact we shall not prove is summed up in the next proposition.

(17.5.5) Proposition. Any subgroup of a cyclic group is cyclic.

For example, Z is cyclic, and the subgroup of even integers E, defined in the example following proposition (17.4.2), is also cyclic: it is generated by the number 2.

Cyclic groups play a role in group theory somewhat analogous to the role of prime numbers in ordinary arithmetic. We know that a given whole number can be "broken down" into a product of prime numbers. For instance,

$$72 = 8 \times 9 = 2^3 3^2$$
$$40 = 8 \times 5 = 2^3 5$$

In an analogous manner, complex groups may frequently be represented as "products" of cyclic groups. One must keep in mind here the passage to a new level of analysis in which groups as such become elements in operations with other such elements. (The sociological equivalent is the passage to relationships and processes between human collectivities, having based the collectivity notion on relationships between individuals.)

To explore this point of view, we first need a definition of a product of cyclic groups. In the next section, we provide this definition and represent certain complex groups by products of cyclic groups.

17.6. Product Groups and Dihedral Groups. Consider two cyclic groups C_M and C_N. Since they have domains that are sets we can form a Cartesian product of their domains. Then the elements of $C_M \times C_N$ will be pairs of the group elements. For instance, the pair (g_1, g_2) is the pair of generators: $g_1^M = I$, $g_2^N = I$. To create a binary operation on the product set we need a rule to map a pair of pairs into a new pair in the product set. An obvious definition is

$$(x_1, x_2)(y_1, y_2) = (x_1 y_1, x_2 y_2)$$

where x_1 and y_1 are in C_M and where x_2 and y_2 are in C_N. This is a binary operation on $C_M \times C_N$. The associative property holds; using the associative law in each group,

$$
\begin{aligned}
[(x_1, x_2)(y_1, y_2)] (z_1, z_2) &= (x_1 y_1, x_2 y_2)(z_1, z_2) \\
&= (x_1 y_1 z_1, x_2 y_2 z_2) \\
&= (x_1, x_2)(y_1 z_1, y_2 z_2) \\
&= (x_1, x_2)[(y_1, y_2)(z_1, z_2)]
\end{aligned}
$$

The identity can be taken as (I, I') where I is the identity of C_M and I' is the identity of C_N, since

$$(I, I')(y_1, y_2) = (I y_1, I' y_2) = (y_1, y_2)$$

The inverse of (x_1, x_2) is then (x_1^{-1}, x_2^{-1}) because

$$(x_1, x_2)(x_1^{-1}, x_2^{-1}) = (x_1 x_1^{-1}, x_2 x_2^{-1}) = (I, I')$$

(17.6.1) Theorem. If and only if the greatest common divisor of M and N is 1 (and we say M and N are coprime), the product group $C_M \times C_N$ is isomorphic with C_{MN}.

This statement will not be proved, but we give concrete cases.

Example.   (a) If M = 1, then for any N the pair is coprime and so

$$C_1 \times C_N \cong C_N$$

Example.   (b) Similarly, if N = 1, then

$$C_M \times C_1 \cong C_M$$

Example.   (c) For M = 2, N = 2, there is a number—namely, 2—dividing both M and N. Hence, $C_2 \times C_2$ is a new abstract group: a group of size four not isomorphic with C_4. To see the structure of $C_2 \times C_2$ we first let I'' denote its identity—the pair of identities (I, I'). Then let

$$a = (g, I')$$
$$b = (I, g)$$
$$c = (g, g)$$

We obtain a multiplication table by using the binary operation on pairs. For instance,

$$a^2 = (g, I')(g, I') = (g^2, I') = (I, I') = I''$$

The table is as follows:

	I''	a	b	c
I''	I''	a	b	c
a	a	I''	c	b
b	b	c	I''	a
c	c	b	a	I''

We note that since c = ab = ba, we can replace c throughout by ab. Finally, we drop the primes from I'' to obtain the isomorphic table:

		I	a	b	ab
	I	I	a	b	ab
(17.6.2)	a	a	I	ab	b
	b	b	ab	I	a
	ab	ab	b	a	I

We note the relations, each of which defines one component C_2 group in the product $C_2 \times C_2$,

$$a^2 = I, \quad b^2 = I$$

and also,

$$ab = ba$$

because both these products were c. In fact, these three relations are sufficient to define $C_2 \times C_2$ without first constructing pairs and defining a pair product. Note that the group is abelian, as are all the cyclic groups (17.5.4). It can be shown that C_4 and $C_2 \times C_2$ exhaust the abstract groups of order four. If we add to the group axioms the condition, "There are four group elements in G," then the resulting class of groups modulo isomorphism (see section 5.16) contains only two structures. One structure is C_4; the other is $C_2 \times C_2$.

Example. (d) At M = 3, the group $C_3 \times C_2$ has M = 3, N = 2 coprime. Hence, theorem (17.6.1) says that $C_3 \times C_2 \cong C_6$. Hence, from an abstract structural standpoint, no new group arises.

Example. (e) Another new group arises at M = 4, where $C_4 \times C_2$ has M = 4, N = 2 not coprime and, so, $C_4 \times C_2$ is not isomorphic to C_8. In fact, since C_4 and $C_2 \times C_2$ are distinct, it is not surprising that a third distinct product is found: $(C_2 \times C_2) \times C_2$, which is isomorphic neither with C_8 nor with $C_4 \times C_2$.

The reader can see that theorem (17.6.1) is a powerful result, allowing us to use mere arithmetic to find new abstract groups. Another important theorem is the following, which yields the structure of a class of groups in terms of cyclic groups.

(17.6.3) Theorem. If G is any finite abelian group, then G is isomorphic with some product of cyclic groups; in fact it has the form

$$G \cong C_{m_1} \times C_{m_2} \times \cdots \times C_{m_r}$$

where m_i divides m_{i+1} (i = 1, 2, . . . , r − 1) and $\prod_{i=1}^{r} m_i$ = the order of G.

For instance, suppose we are told that G is an abelian group of order 360. Then the various ways of constructing $m_1, m_2, \ldots, m_r$ (with $m_1 m_2 \ldots m_r = 360$) provide the abstractly distinct structures that G might exemplify. For instance, three possibilities are

$C_6 \times C_{60}$	(m_1 = 6, m_2 = 60)
$C_3 \times C_{120}$	(m_1 = 3, m_2 = 120)
$C_2 \times C_6 \times C_{30}$	(m_1 = 2, m_2 = 6, m_3 = 30)

In example (c) above, culminating in the multiplication table (17.6.2), we mentioned that we could generate $C_2 \times C_2$ by the relations

$$a^2 = I, \quad b^2 = I, \quad ab = ba$$

This is just a special case of the following elegant and important result of group theory.

(17.6.4) Theorem. A product group based on C_M and C_N can be specified by the defining relation method and the defining relations are

$$(g_1, g_2: \ g_1^{M} = I, \ g_2^{N} = I, \ g_1 g_2 = g_2 g_1)$$

Recall the generic meaning of this idea: if we add to the group axioms of definition (17.2.1), the defining relations that assert that there are two generators satisfying given conditions, then the four axioms (the three that define any group, plus one) define the product group $C_M \times C_N$.

Hence, theorem (17.6.4) extends the idea of generating a group from one generator to two.

Concretely, to say that g_1 and g_2 generate a group means any element in the group can be put into the form of a product of powers of g_1 and of powers of g_2. In the special case of example (c), where $g_1 = a$, $g_2 = b$, we had the remaining elements

$$I = a^2$$
$$c = ab$$

Another method of combination of groups gives rise to new abstract groups. We start from a physical or geometrical situation and imagine two distinct kinds of physical (or geometric) transformations of the state of a system and a mode of combining them: namely, "succession." Thus, we can "flip" a triangle around some fixed median and then follow it by rotation as in Figure 17.6.1, using the median from vertex A to the opposite side. The final state is now different from that which we could get by any rotation in C_3. Also, successive flips could never carry us into this state. Hence, a new transformation has been "induced" on the triangle that is nevertheless a symmetry. Applying each rotation of C_3 after the flip, we obtain three new transformations, the one in Figure 17.6.1 plus the two in Figure 17.6.2. Note that f is premultiplying r, r^2, r^3, or since $r^3 = I$, it premultiplies I, r, r^2. We can write this as fC_3. Hence, our multiplication table will be expanded to have the "block" form,

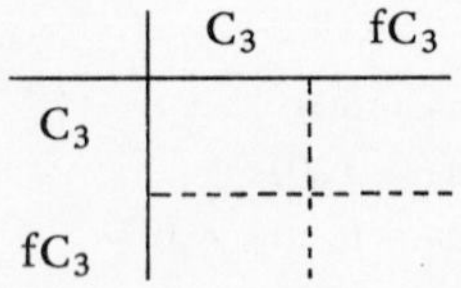

	C_3	fC_3
C_3		
fC_3		

$$C_3 = \{I, r, r^2\}$$
$$fC_3 = \{f, fr, fr^2\}$$

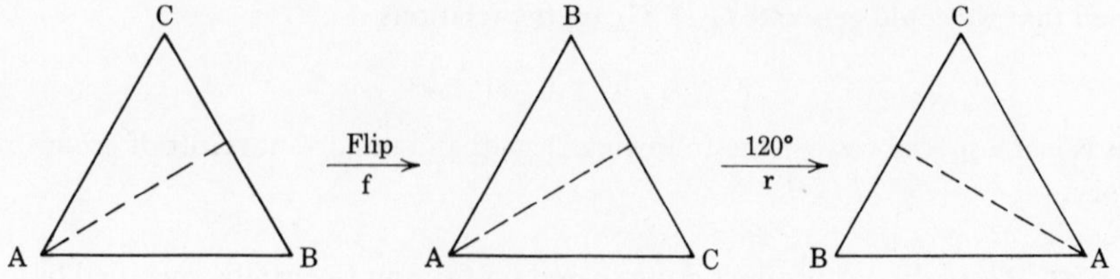

Figure 17.6.1

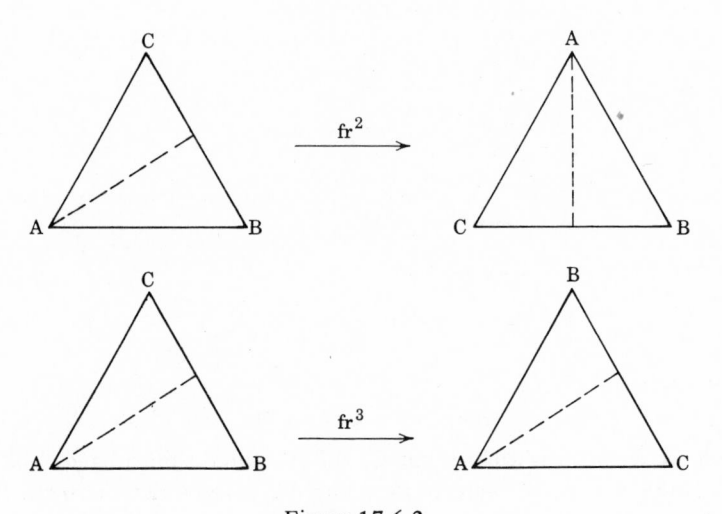

Figure 17.6.2

When this is computed the result is as follows:

	I	r	r^2	f	fr	fr^2
I	I	r	r^2	f	fr	fr^2
r	r	r^2	I	fr^2	f	fr
r^2	r^2	I	r	fr	fr^2	f
f	f	fr	fr^2	I	r	r^2
fr	fr	fr^2	f	r^2	I	r
fr^2	fr^2	f	fr	r	r^2	I

To represent this group by a graph we first define two relational systems with domain the group elements:

$$(x, y) \in R_1 \quad \text{iff} \quad xr = y$$
$$(x, y) \in R_2 \quad \text{iff} \quad xf = y$$

Then we represent R_1 by solid directed lines, R_2 by dashed directed lines. For example, since $(fr)f = r^2$, we have $(fr, r^2) \in R_2$, and so,

$$fr \; \circ\; \text{-----}\text{>}\text{-----}\; \circ\; r^2$$

and since $(fr^2)f = r$,

$$fr^2 \; \circ\; \text{-----}\text{>}\text{-----}\; \circ\; r$$

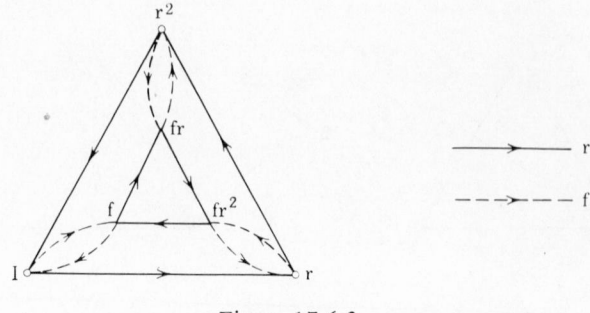

Figure 17.6.3

Similarly, since $rf = fr^2$,

$$r \circ \text{-----}> \text{-----} \circ fr^2$$

In general, we see that dashed lines link C_3 and fC_3. On the other hand, solid lines occur within the C_3 and fC_3 parts of the group. This suggests a C_3 triangle, an fC_3 triangle, both solid, and dashed connectors as shown in Figure 17.6.3.

This group is called the dihedral group with index 3, denoted D_3. The index suggests the generalization to D_N. For $N = 4$, the graph consists of two squares with "mutuality" of dashed lines connecting them, and it arises by considering C_4 and C_2 as based on two distinct geometric operations (rotate in multiples of $90°$ and flip) that can be performed in succession. In general, D_N is based on C_N and C_2.

In terms of defining D_N by the method of defining relations, the process is similar to that used for products. In the dihedral situation, we start with $g_1^N = I$ and $g_2^2 = I$ and use the graph to conjecture the required link between generators. A part of the graph of D_3 is given in Figure 17.6.4. This shows that $rfrf = I$. In fact, it is a theorem of group theory that this indeed is a correct new defining relation. As we have not defined "dihedral group" independently of the process described above, we can use the defining relations to define D_N.

(17.6.5) Definition. By a "dihedral group D_N" we mean a group presented by the defining relations

$$(g_1, g_2: \ g_1^N = I, g_2^2 = I, (g_1 g_2)^2 = I)$$

(and D_N has $2N$ elements).

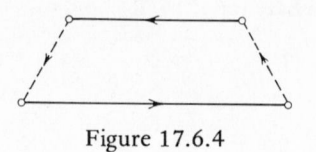

Figure 17.6.4

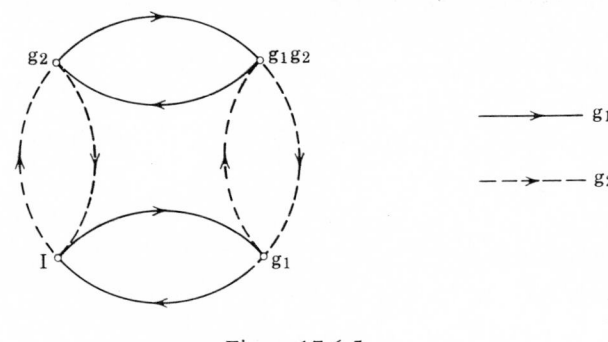

Figure 17.6.5

For D_2 we obtain by this method the graph shown in Figure 17.6.5, based on $g_1^2 = I$, $g_2^2 = I$ and $(g_1 g)^2 = I$.

We see from multiplication table (17.6.2) for $C_2 \times C_2$ that the correspondence

$$g_1 \longrightarrow a$$
$$g_2 \longrightarrow b$$

is an isomorphism of D_2 with $C_2 \times C_2$ because $a^2 = I$, $b^2 = I$, and $(ab)^2 = I$. In general, however, the products do not coincide with the dihedral groups. Representing $C_2 \times C_4$ and D_4 by graphs, for instance, yields a distinction based on the defining relations differentiating them. In the first case the generators commute [see theorem (17.6.4)], so we must have a partial graph as in Figure 17.6.6. For D_4, on the other hand, the analogous partial graph has a different pattern, shown in Figure 17.6.7. In the second case, we do not have the commutative property.

The complete graphs are juxtaposed in Figure 17.6.8. Here we show a single line for the g_2 loops connecting the inner and outer squares, a convention justified by the fact that $g_2^2 = I$ implies $g_2^{-1} = g_2$.

17.7. Abstract Groups and Their Graphs. Based on the convention that whenever $g_i^2 = I$ holds we use an undirected line to represent the generator g_i, as above, we can summarize in compact geometric form the abstract groups of small order.

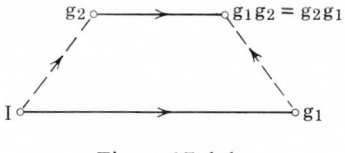

Figure 17.6.6

Table 17.7.1 Graphs of All Groups of Order N ≤ 8

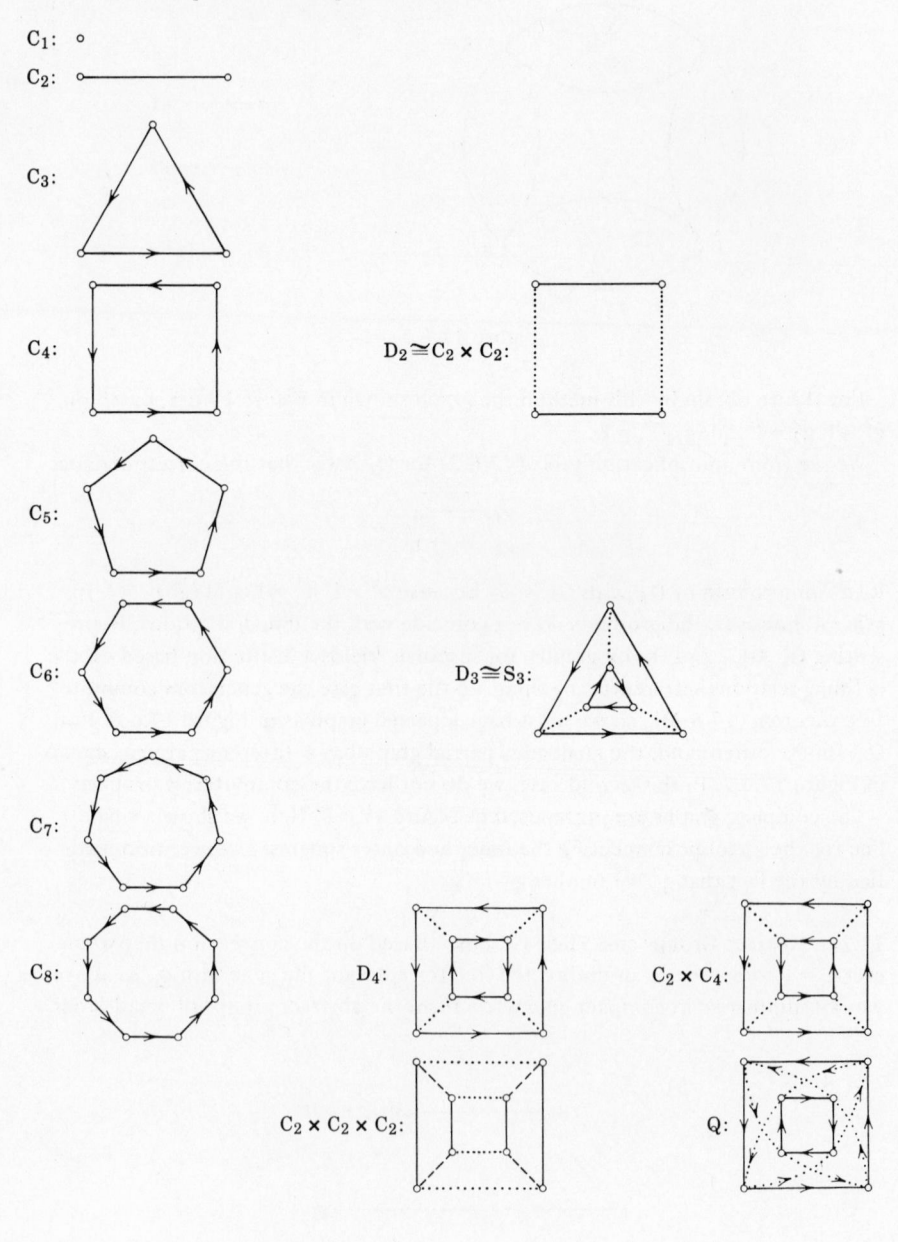

C_1:

C_2:

C_3:

C_4:

$D_2 \cong C_2 \times C_2$:

C_5:

C_6:

$D_3 \cong S_3$:

C_7:

C_8:

D_4:

$C_2 \times C_4$:

$C_2 \times C_2 \times C_2$:

Q:

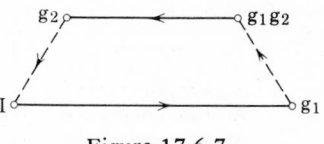

Figure 17.6.7

This is shown in Table (17.7.1). In Table (17.7.2) the same groups are described in terms of their standard notation and generating relations. (A more extensive table may be found in Coxeter and Moser, 1965; a table may also be found in Higman, 1964).

In Table 17.7.2, one notes that some groups are unique at the given N. In fact, a theorem of group theory is that if N is prime, then C_N is the only abstract group of order N. The group S_3, which is isomorphic with D_3, is the group of all permutations of a three-element set, and D_3 provides an interpretation in terms of the symmetries of an equilateral triangle. The group Q is a special structure that emerged out of a generalization of complex numbers (for a discussion of Q, including the graph representation of Q, see Grossman and Magnus, 1964).

We may summarize and extend the graph-group correspondence. A node corresponds to a group element. A directed line of a given type represents a relation between the group elements (x, y) given by $xg_i = y$, where g_i is a generator determining the type of the line.

By a "word" in the generators $g_1, g_2, \ldots, g_k$ is meant a sequence of powers of the g_i, in any order, for i ranging from 1 to k. For instance, the following are words:

$$W_1 = g_1 g_2$$
$$W_2 = g_1^2 g_2^{-1} g_3$$
$$W_3 = g_3^{-4} g_2 g_1 g_3 g_1^3 g_2$$

Once the generators of a group are given, all words are determined. A word corresponds to a path on the graph: begin at any node and "apply" the word from left to right by tracing the corresponding directed lines. For example, in D_3,

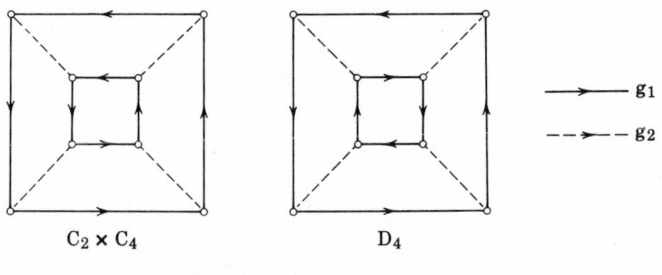

$C_2 \times C_4$ D_4

Figure 17.6.8

Table 17.7.2 All Finite Groups of Order N $\leqslant$ 8

Order	Symbol	Name	Remarks	Defining Relations
1	1	trivial	Abelian; also C_1	(a: a = I)
2	C_2	cyclic	Abelian; prime order	(a: a^2 = I)
3	C_3	cyclic	Abelian; prime order	(a: a^3 = I)
4	C_4	cyclic	Abelian	(a: a^4 = I)
	$C_2 \times C_2$		Abelian	(a,b: $a^2 = b^2$ = I, ab = ba)
5	C_5	cyclic	Abelian; prime order	(a: a^5 = I)
6	C_6	cyclic	Abelian; also, $C_3 \times C_2$	(a: a^6 = I)
	S_3	symmetric	non-Abelian; also, D_3	(a,b: $a^3 = b^2 = (ab)^2$ = I)
7	C_7	cyclic	Abelian; prime order	(a: a^7 = I)
8	C_8	cyclic	Abelian	(a: a^8 = I)
	$C_2 \times C_2 \times C_2$		Abelian	(a,b,c: $a^2 = b^2 = c^2$ = I, ab = ba, ac = ca, bc = cb)
	$C_4 \times C_2$		Abelian	(a,b: $a^4 = b^2$ = I, ab = ba)
	D_4	dihedral	non-Abelian	(a,b: $a^4 = b^2$ = I, $(ab)^2$ = I)
	Q	quaternion	non-Abelian	(a,b: $a^2 = b^2 = (ab)^2$)

Figure 17.6.3, the word $W = r^2 fr$ in the generators r and f may be traced as the path from I as shown in Figure 17.7.1. Hence, $W = fr^2$. But tracing the same word $W = r^2 fr$ from r^2 yields the path shown in Figure 17.7.2. Hence, $r^2 W = f$, and since $W = fr^2$, then $r^2 fr^2 = f$. To check this formally:

$$I = rfrf \quad \text{(a defining relation for } D_3)$$
$$f^{-1} = rfr \quad \text{(postmultiplying by } f^{-1})$$
$$f = rfr \quad (f = f^{-1} \text{ since } f^2 = I \text{ is a defining relation)}$$
$$fr^2 = rfr^3 \quad \text{(postmultiplying by } r^2)$$
$$fr^2 = rf \quad (r^3 = I \text{ is a defining relation)}$$
$$r^2 fr^2 = r^3 f \quad \text{(premultiplying by } r^2)$$
$$r^2 fr^2 = f \quad (r^3 = I)$$

Note that the formal proof uses all three defining relations, as well as general group properties.

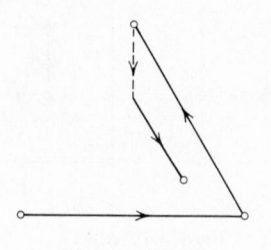

Figure 17.7.1

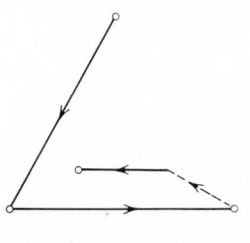

Figure 17.7.2

Since a word corresponds to a path, it may loop back and return to the origin to form a closed path: this is a word that is then of the form W = I. For instance, r^3 = I. However, we have words of the form W = I in which the closed path is achieved by retracing the path back to the origin. For instance, W = $frr^{-1}f^{-1}$ = I yields a closed path only in the sense of retracing. This corresponds to the fact that this word is the identity without appeal to the defining relations: it is the identity in every group. Hence, words of the form W = I fall into two classes:

(1) "empty words": W = I deducible from group axioms alone,
(2) W = I consequence of defining relations.

The former correspond to retracing paths; the latter correspond to closed paths that are not merely retracings. The fact that when a word W = I is applied to any element, say x, we get x again means that on the graph a certain type of path that at any one node returns to that node must do so at all nodes. This is called homogeneity. (As we shall see in the next chapter, this is a property of major importance in the algebra of kinship.)

For example, in D_3 the word W = $f^2(rf)^2$ is a certain type of path shown in Figure 17.7.3. Since f^2 = I and $(rf)^2$ = I, it follows that W = I under the defining relations. Hence, xW = x, all x in the group. Hence, no matter where the above path is started, it must form a closed path back to its origin.

$$\circ\;\text{-}\text{-}\text{-}\text{>}\text{-}\text{-}\;\circ\;\text{-}\text{-}\text{-}\text{>}\text{-}\text{-}\;\circ\;\longrightarrow\;\circ\;\text{-}\text{-}\text{-}\text{>}\text{-}\text{-}\;\circ\;\longrightarrow\;\circ\;\text{-}\text{-}\text{-}\text{>}\text{-}\text{-}\;\circ$$
$$\quad\;\;\text{f}\qquad\qquad\text{f}\qquad\quad\text{r}\qquad\;\;\text{f}\qquad\quad\text{r}\qquad\;\;\text{f}$$

Figure 17.7.3

An example of a graph that lacks homogeneity is given in Figure 17.7.4. The path $\circ\;\cdot\cdot\;\text{>}\;\cdot\cdot\;\circ\;\longrightarrow\;-\;\circ$ at a leads to b, but the same path at c leads back to c.

We saw in proposition (17.3.2) that the equation

$$ax = b$$

has the unique solution

$$x = a^{-1}b$$

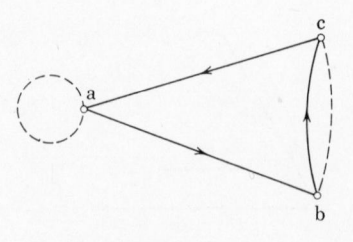

Figure 17.7.4

for any a, b in the group. In the graph representation, this means that given any a and b, there exists a path from a to b—namely, that which corresponds to applying the word $W = x = a^{-1}b$ at the node a. Hence, by proposition (17.3.2), the graph is connected in that from any node (a) there is a path ($W = a^{-1}b$) to another specified node (b), for any a and b. For example, the solution of the equation

$$r^2 x = f$$

in D_3 is given by $x = r^{-2}f = rf$. Hence, the path corresponding to $W = rf$ connects r^2 to f, as may be verified in the graph in Figure 17.6.3.

Summarizing the group-graph correspondence we have Table 17.7.3.

17.8. Cycles in Permutation Groups.

We saw in section 5.13 that the collection of all mappings on a set forms a monoid. Hence, restricting the collection to all bijective mappings yields an inverse map for every map and, so, a group. We now introduce a special notation and nomenclature for the case in which the domain set of such maps is finite. In this case, the maps are generally called permutations and such a group is called a permutation group. The nomenclature arises because if we

Table 17.7.3   The Group-Graph Correspondence

Group	Graph
Element	Node
Generator	Type of directed line
Word in generators	Path
Multiplication of elements	Tracing of path
Word $W = I$	Closed path
Solution of $ax = b$ exists	Path exists: graph is connected

are given any such map, say

$$a \longmapsto b$$
$$b \longmapsto c$$
$$c \longmapsto d$$
$$d \longmapsto a$$

we can think of it as a "cycle,"

$$a \longmapsto b \longmapsto c \longmapsto d \longmapsto a$$

that is, "a replaced by b; b replaced by c; c replaced by d; and d replaced by a." Alternatively we can always nominally label the objects under discussion by the first N integers to obtain a "rearrangement" of these integers by a permutation. In this case, a standard notation is a pair of rows:

$$\begin{pmatrix} 1 & 2 & 3 & 4 \\ 2 & 3 & 4 & 1 \end{pmatrix}$$

Here the second row is the rearrangement of the first row under the permutation. The cycle form in integers is, in this case,

$$1 \longmapsto 2 \longmapsto 3 \longmapsto 4 \longmapsto 1$$

which is usually abbreviated by

$$(1 \quad 2 \quad 3 \quad 4)$$

where it is understood that 4 is mapped into 1.

Similarly, $(1 \ 2)(3 \ 4)$ means

$$1 \longmapsto 2 \longmapsto 1 \text{ and } 3 \longmapsto 4 \longmapsto 3$$

And the form $(1 \ 3 \ 4)(2)$ means

$$1 \longmapsto 3 \longmapsto 4 \longmapsto 1 \text{ and } 2 \longmapsto 2$$

Note the geometric correspondences when we represent the set being permuted as a collection of nodes and the mapping $x \longmapsto y$ by a directed line segment from x to y, as shown in Figure 17.8.1

(1) Permutation: $\begin{pmatrix} 1 & 2 & 3 & 4 \\ 2 & 3 & 4 & 1 \end{pmatrix}$

Cycle notation: (1 2 3 4)

Graph:

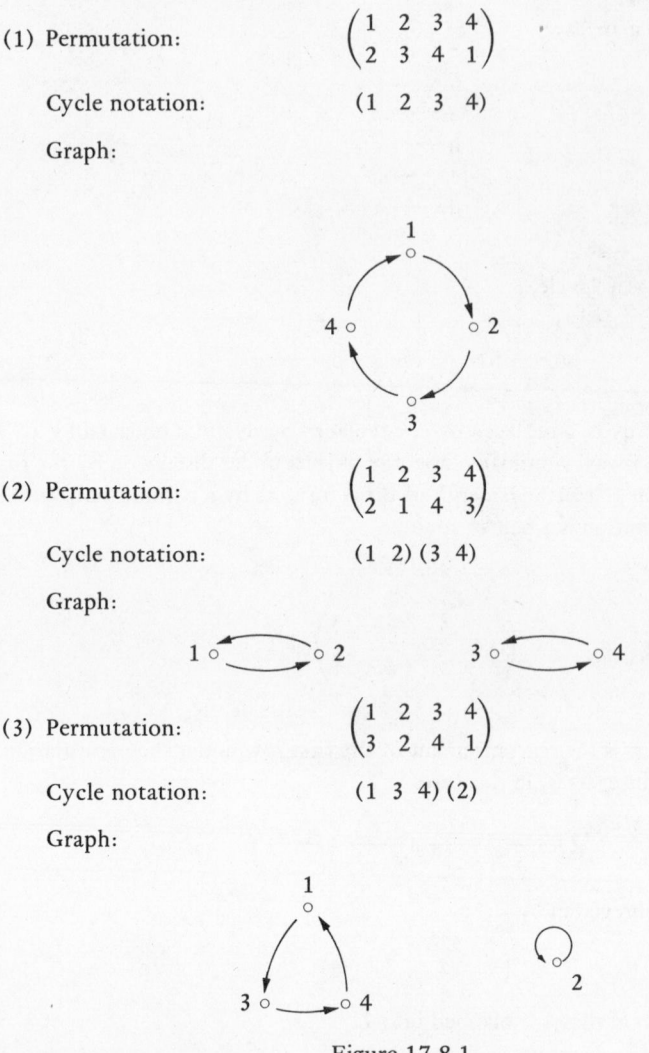

(2) Permutation: $\begin{pmatrix} 1 & 2 & 3 & 4 \\ 2 & 1 & 4 & 3 \end{pmatrix}$

Cycle notation: (1 2)(3 4)

Graph:

(3) Permutation: $\begin{pmatrix} 1 & 2 & 3 & 4 \\ 3 & 2 & 4 & 1 \end{pmatrix}$

Cycle notation: (1 3 4)(2)

Graph:

Figure 17.8.1

In general, as the graphs make evident, we can represent any permutation of N objects as a "product" of cycles in cycle notation. It is customary to eliminate the one-cycles in the notation: (1 3 4) = (1 3 4)(2). The identity permutation I = (1)(2)(3)(4) is denoted by the symbol I itself in the following discussion.

17.9. Regular Representation of a Finite Group. We note from the multiplication tables for C_2, C_3, and C_4 given in Figure 17.5.4 that each row is a permutation of the column heading. For instance, in C_3 we have the permutation associated with the second row:

$$\begin{pmatrix} I & g & g^2 \\ g & g^2 & I \end{pmatrix}$$

It is convenient to use integers and to write

$$\begin{pmatrix} 1 & 2 & 3 \\ 2 & 3 & 1 \end{pmatrix}$$

or $1 \longmapsto 2 \longmapsto 3 \longmapsto 1$ or (1 2 3), as in the preceding paragraph. Since each row permutes the set of group elements, we have a correspondence between the group elements and a set of permutations:

Element	Permutation	Cycle Notation
I	$\begin{pmatrix} 1 & 2 & 3 \\ 1 & 2 & 3 \end{pmatrix}$	I
g	$\begin{pmatrix} 1 & 2 & 3 \\ 2 & 3 & 1 \end{pmatrix}$	(1 2 3)
g^2	$\begin{pmatrix} 1 & 2 & 3 \\ 3 & 1 & 2 \end{pmatrix}$	(1 3 2)

Analyzing C_4 we obtain the following:

Element	Permutation	Cycle Notation
I	$\begin{pmatrix} 1 & 2 & 3 & 4 \\ 1 & 2 & 3 & 4 \end{pmatrix}$	I
g	$\begin{pmatrix} 1 & 2 & 3 & 4 \\ 2 & 3 & 4 & 1 \end{pmatrix}$	(1 2 3 4)
g^2	$\begin{pmatrix} 1 & 2 & 3 & 4 \\ 3 & 4 & 1 & 2 \end{pmatrix}$	(1 3)(2 4)
g^3	$\begin{pmatrix} 1 & 2 & 3 & 4 \\ 4 & 1 & 2 & 3 \end{pmatrix}$	(1 4 3 2)

Note that apart from the identity, all cycles shown in the cycle notation are of length exceeding 1. This reflects the fact that in each of the nonidentity permutations associated with a row of the multiplication table of a finite group no element is mapped into itself, a fact important in our subsequent application.

What is perhaps not apparent is that the set of permutations obtained from the rows of a multiplication table forms more than a mere list: it is a group, a subgroup of the permutation group over the domain of the original group. Note that three groups are now under discussion:

(1) a finite group, G,

(2) the permutation group over the domain of G,

(3) the set of permutations corresponding to the rows of the multiplication table of G (this is a subgroup of group 2).

This idea is part of the following major theorem of group theory.

(17.9.1) Theorem. (Representation theorem)   The set of permutations forming the rows of any multiplication table of a finite group forms a group. This group is isomorphic with the group from whose multiplication table it arises.

Rather than immediately prove this important representation theorem—it says that every finite group is isomorphic to some group of permutations—first we give an illustration for C_4. The set of permutations is evidently

$$\{I, (1\ 2\ 3\ 4), (1\ 3)(2\ 4), (1\ 4\ 3\ 2)\}$$

The binary operation is functional composition. This is a subset of the group of all permutations over the set of four integers (containing 4! = 24 permutations). Hence, to claim it is a group means that the conditions of definition (17.4.1) are satisfied: (1) the composition of any pair of these permutations is a permutation in the set of four listed, (2) the identity is in the set, and (3) the inverse of any permutation in the set is in the set. Clearly, (2) is satisfied. The remainder of the work is computational. For example, $(1\ 4\ 3\ 2)$ o $[(1\ 3)(2\ 4)]$ means the composition

$$1 \longmapsto 3 \longmapsto 2 \qquad (\text{so } 1 \longmapsto 2)$$
$$2 \longmapsto 4 \longmapsto 3 \qquad (\text{so } 2 \longmapsto 3)$$
$$3 \longmapsto 1 \longmapsto 4 \qquad (\text{so } 3 \longmapsto 4)$$
$$4 \longmapsto 2 \longmapsto 1 \qquad (\text{so } 4 \longmapsto 1)$$

Hence,

$$(1\ 4\ 3\ 2)\ \text{o}\ [(1\ 3)(2\ 4)] = (1\ 2\ 3\ 4)$$

so that the composition is in the set. Similarly, we compute the other compositions. The multiplication table turns out to be as follows:

o	I	(1 2 3 4)	(1 3)(2 4)	(1 4 3 2)
I	I	(1 2 3 4)	(1 3)(2 4)	(1 4 3 2)
(1 2 3 4)	(1 2 3 4)	(1 3)(2 4)	(1 4 3 2)	I
(1 3)(2 4)	(1 3)(2 4)	(1 4 3 2)	I	(1 2 3 4)
(1 4 3 2)	(1 4 3 2)	I	(1 2 3 4)	(1 3)(2 4)

The structure of the table should be compared with that of C_4 (see Figure 17.5.4) It is clear that they are isomorphic. The point of theorem (17.9.1) is that this always holds: the permutation group obtained from a given group multiplication table has the same structure as that table. Hence, it may be used to represent the group. It is called the regular representation of the group. It is shown in group theory that all groups, not only finite groups, have a regular representation (see, for instance, Jacobson, 1951).

The proof in the finite case is direct. Let $x_1, x_2, \ldots, x_N$ be the N elements of the group. The multiplication table at row i yields the row of products $x_i x_j$ (j = 1, 2, ..., N). Hence, the permutation corresponding to x_i is given by

$$x_j \longmapsto x_i x_j \qquad (j = 1, 2, \ldots, N)$$

If x_i and x_k are any group elements, then, the product $x_i x_k$ is associated with the permutation

$$x_j \longmapsto (x_i x_k) x_j \qquad (j = 1, 2, \ldots, N)$$

Now we must show that the permutation corresponding to the product is the product of the corresponding permutations. Applying first the x_k-permutation,

$$x_j \longmapsto x_k x_j \qquad (j = 1, 2, \ldots, N)$$

followed by the x_i-permutation,

$$x_k x_j \longmapsto x_i(x_k x_j)$$

we see that the composition is

$$x_j \longmapsto x_i(x_k x_j) = (x_i x_k) x_j$$

and indeed this is the permutation corresponding to the product $x_i x_k$. This shows that the association of the x_i with the row permutations is a homomorphism. To show that no two distinct elements permute the group identically, suppose that x_i and x_k were associated with the same permutation. Then for every j,

$$x_i x_j = x_k x_j$$
$$x_i = x_k \qquad \text{(postmultiplying by } x_j^{-1})$$

This shows that if the elements differ, so do the permutations.

Having shown that the representation is a matter of a 1–1 homomorphic correspondence between the group elements and their row permutations, we conclude that the subgroup of permutations obtained from the rows of the table is an isomorphic image of the group given by the table.

17.10. Conjugation. An operation on group elements we need in dealing with empirical systems is introduced here: conjugation.

(17.10.1) Definition. Let G be any group and let a be any element of G. Then the mapping

$$x \longmapsto axa^{-1} \qquad (x \in G)$$

is termed conjugation by a.

Here the group is mapped into itself by using a fixed element a, applied as shown. For example, in D_3 we have conjugation by generator r computed by reference to the graph in Figure 17.6.3:

$$I \longmapsto rIr^{-1} = I$$
$$r \longmapsto rrr^{-1} = r$$
$$r^2 \longmapsto rr^2 r^{-1} = r^{-1} = r^2$$
$$f \longmapsto rfr^{-1} = fr$$
$$fr \longmapsto rfrr^{-1} = rf = fr^2$$
$$fr^2 \longmapsto rfr^2 r^{-1} = f$$

The image of element x under conjugation by a is called the conjugate of x by a. For instance, in D_3 the conjugate of fr^2 by r is f.

(17.10.2) Proposition. Conjugation is an isomorphism.

Define $f(x) = axa^{-1}$, all x. If $f(x) = f(y)$, then

$$axa^{-1} = aya^{-1}$$

$ax = ay$	(postmultiplication by a)
$x = y$	(premultiplication by a^{-1})

Hence, f is an injective (1–1) mapping. It is onto the group; if $y \in G$, we need to solve for x such that

$$y = axa^{-1}$$

But then postmultiplication by a followed by premultiplication by a^{-1} yields the solution $x = a^{-1}ya$. Thus, f is onto G. To show that f is a homomorphism, we consider the image of a product:

$f(xy) = axya^{-1}$	
$= ax(a^{-1}a)ya^{-1}$	(inserting the identity $I = a^{-1}a$)
$= (axa^{-1})(aya^{-1})$	(associativity)
$= f(x)f(y)$	(definition of f)

Hence, f is an isomorphism.

We define a relation on G by the rule,

$$(x, y) \text{ are conjugates} \quad \text{iff} \quad y = axa^{-1} \text{ for some a in G.}$$

(17.10.3) Proposition. The relation "are conjugates" is an equivalence relation on G.

The relation is reflexive, because $x = IxI^{-1}$.

The relation is symmetric, since if $y = axa^{-1}$, then $x = a^{-1}ya = a^{-1}y(a^{-1})^{-1}$.

The relation is transitive. Suppose (x, y) and (y, z) are in the relation. Then there exists an element a with $y = axa^{-1}$, and there exists an element b with $z = byb^{-1}$.

Then $z = b(axa^{-1})b^{-1} = bax(ba)^{-1}$, so that (y, z) is in the relation; here we have used proposition (17.3.6). We say such a pair of elements are conjugate elements.

Accordingly, we can partition G modulo conjugate elements into conjugate classes, (recall section 4.12). Two elements are in the same conjugate class if and only if there exists a conjugation mapping one into the other. For instance, in D_3 we saw that conjugation by r carries f into fr. Hence, f and fr are in the same conjugate class. Some useful properties of conjugation are now stated.

(17.10.4) Proposition.   (a) The inverse of a conjugate is the conjugate of the inverse: $(axa^{-1})^{-1} = ax^{-1}a^{-1}$. (b) The conjugate of a product is the product of the conjugates: $axya^{-1} = (axa^{-1})(aya^{-1})$.

(17.10.5) Proposition.   In the group of all permutations of a finite set, two permutations are conjugate elements if and only if they have the same cycle type.

As a consequence of proposition (17.10.4)(b), the conjugate of a power is the power of the conjugate:

$$ax^n a^{-1} = (axa^{-1})^n$$

(The proof is simple, using induction.)

Proposition (17.10.5) will be used in a section of the next chapter in an apparently unrelated context dealing with the criterion for two "prescriptive marriage systems" to be isomorphic. The reader should be able to prove (17.10.4), but (17.10.5) is less simple. (For a more detailed treatment of conjugation, see Yale, 1968.)

When conjugation appears in the context of permutations it becomes very intuitively meaningful. Let $\{1, 2, \ldots, n\}$ be the set of first n positive integers and suppose we are given a permutation C of this set. Hence,

$$\{1, 2, \ldots n\} \xrightarrow{\ C\ } \{1, 2, \ldots, n\}$$

or

(17.10.6) $C: \begin{pmatrix} 1 & 2 & \ldots & n \\ C(1) & C(2) & \ldots & C(n) \end{pmatrix}$

Now suppose we "rename" all the objects under discussion, so that what was previously 1 is f(1), what was 2 is f(2), and so forth. This, too, is a permutation of $\{1, 2, \ldots, n\}$, applied both in the domain and range of C. Hence,

$$
\begin{array}{ccc}
\{1, 2, \ldots, n\} & \xrightarrow{\ C\ } & \{1, 2, \ldots, n\} \\
\ \downarrow f & & \ \downarrow f \\
\{1, 2, \ldots, n\} & \dashrightarrow & \{1, 2, \ldots, n\}
\end{array}
$$

The dotted line is the map C after renaming of objects, where f is the renaming. To determine the map in terms of C and f, note that we can go around the mapping diagram beginning with an arbitrary element x in the lower-left domain set: first find the old name of x, which is $f^{-1}(x)$; then apply C to obtain $C[f^{-1}(x)]$, then rename to obtain $f(C[f^{-1}(x)])$. Hence, $x \longmapsto f(C[f^{-1}(x)])$. The dotted line corresponds to fCf^{-1}, the conjugation of C by f. Hence, conjugation yields a new representation of C in the new nominal system over some set of objects. Note also that by applying f to domain and range of C in its form (17.10.6) we obtain:

$$(17.10.7) \qquad \begin{pmatrix} f(1) & f(2) & \cdots & f(n) \\ f[C(1)] & f[C(2)] & \cdots & f[C(n)] \end{pmatrix}$$

Example. Given

$$C: \begin{pmatrix} 1 & 2 & 3 & 4 \\ 4 & 3 & 2 & 1 \end{pmatrix}$$

Rename so that $1 \longmapsto 3$, $3 \longmapsto 2$, $2 \longmapsto 4$, and $4 \longmapsto 1$. This is f. Then (17.10.7) becomes

$$fCf^{-1}: \begin{pmatrix} 3 & 4 & 2 & 1 \\ 1 & 2 & 4 & 3 \end{pmatrix}$$

Then this permutation is put in standard form, yielding

$$fCf^{-1} = \begin{pmatrix} 1 & 2 & 3 & 4 \\ 3 & 4 & 1 & 2 \end{pmatrix}$$

The underlying group is the group of permutations on $\{1, 2, 3, 4\}$.

CHAPTER EIGHTEEN
ALGEBRA OF KINSHIP

18.1. Introduction. In the previous chapter, we introduced the reader to the basic elements of group theory, with special attention to topics needed for applications: the group-graph correspondence, the representation theorem, the conjugation operation. In the course of doing so, the manipulative skills based on elementary group algebra were built up to some degree. We are ready to leave pure group theory to turn to the algebraic analysis of social structures, starting with an area that has received considerable attention: kinship. In this chapter, then, we introduce the formal theory of "prescribed marriage systems." In the subsequent chapter, we will treat the notion of homomorphism (only touched upon in the preceding, introductory chapter) in considerable depth, as a prelude to the idea of mapping any social relational system into some homomorphic image revealing its basic structural features. To repeat, the next step is kinship and in particular one particular topic in kinship analysis; then we take our tools beyond kinship to an arbitrary social relational system.

18.2. Prescribed Marriage Systems. Consider a society with membership A and a partition of A into a finite set S of classes such that

(1) if man α is in class $\bar{\alpha}$ in S, then there is a unique class, say $W(\bar{\alpha})$ in which his wife must be found, according to the cultural norms (note that this rule depends on $\bar{\alpha}$, on the class and not the individual α);

(2) if man α is in class $\bar{\alpha}$ in S, then any children he may have belong to a unique class, say $C(\bar{\alpha})$ (note that the rule depends on $\bar{\alpha}$ and not on the particular individual α in $\bar{\alpha}$).

A triple (S, C, W) may be used·to represent the situation: S is the finite set of classes, C is a mapping on S into S, and W is a mapping on S into S. A further restriction that defines the scope of the efforts to follow is that C and W are permutations of S. Hence, if α and β are in distinct classes, their wives belong to distinct classes and their children belong to distinct classes. Note that C^{-1} maps the class of the child into that of the father and W^{-1} specifies the class in which women in a given class must find their husbands.

(18.2.1) Definition.   By a "prescribed marriage system" (pms) on a set A, we mean a triple (S, C, W) satisfying

  (1) S is a finite partition of A,
  (2) C is a permutation of S,
  (3) W is a permutation of S.

Note that (S, C, W) is a relational system, in which relations C and W are functions. The work to follow is based on the studies of pms's by Kemeny, Snell, and Thompson (1966) and White (1963). Later, we shall utilize some of the ideas in an important paper by Boyd (1969).

Example.   (a) Consider the two-class system given by

$\overline{\alpha}$	$C(\overline{\alpha})$	$W(\overline{\alpha})$
nobles	nobles	nobles
commoners	commoners	commoners

This is a pms in which C = W and S has just two classes. Interpretatively, it means "Children of nobles are nobles, nobles must marry nobles, commoners' children are commoners, and commoners must marry commoners."

  Let (S, C, W) and (S$'$, C$'$, W$'$) be two pms's, with or without a common set A. Then we define a concept of structural equivalence of pms's by using the notion of isomorphism of relational systems (see section 5.15).

(18.2.2) Definition.   Two pms's are isomorphic if and only if there exists a 1–1 mapping f: S $\rightarrow$ S$'$ that preserves C and W; that is,

  (1) $C(\overline{\alpha}) = \overline{\beta}$ iff $C'[f(\overline{\alpha})] = f(\overline{\beta})$
  (2) $W(\overline{\alpha}) = \overline{\beta}$ iff $W'[f(\overline{\alpha})] = f(\overline{\beta})$

In words, (1) says the children of $\overline{\alpha}$ men, in S, are in $\overline{\beta}$ if and only if the children of the corresponding class f($\overline{\alpha}$) in S$'$ are in the corresponding class f($\overline{\beta}$). Similarly, (2) says that the wives of men in $\overline{\alpha}$ are in $\overline{\beta}$ if and only if the wives of the corresponding class f($\overline{\alpha}$) are in the corresponding class f($\overline{\beta}$).

Example.   (b) The following pms is isomorphic with that of example (a)

$\overline{\alpha}$	$C'(\overline{\alpha})$	$W'(\overline{\alpha})$
warriors	warriors	warriors
others	others	others

The map f is given by

$$\text{nobles} \longmapsto \text{warriors}$$
$$\text{commoners} \longmapsto \text{others}$$

Statement (1) becomes for class $\bar{\alpha}$ = nobles: Children of nobles are nobles iff children of warriors are warriors. Here "warriors" means a clan of some sort.

Example.   (c) The following abstract pms is not isomorphic with those of examples (a) and (b):

$\bar{\alpha}$	$C(\bar{\alpha})$	$W(\bar{\alpha})$
A	B	A
B	A	B

Note that this is a pms, but since $A \longmapsto B$ under C, this makes the system structurally different from nobles $\longmapsto$ nobles and warriors $\longmapsto$ warriors.

(18.2.3)   Theorem.   Two pms's are isomorphic if and only if there exists a 1–1 mapping f: $S \to S'$ such that

$$W' = fWf^{-1} \text{ and } C' = fCf^{-1}$$

where the juxtaposition indicates functional composition.

This can be shown as follows: The conditions of the isomorphism definition may be written

$$C^{-1}(\bar{\beta}) = \bar{\alpha} \quad \text{iff} \quad C'[f(\bar{\alpha})] = f(\bar{\beta})$$
$$W^{-1}(\bar{\beta}) = \bar{\alpha} \quad \text{iff} \quad W'[f(\bar{\alpha})] = f(\bar{\beta})$$

Hence, if isomorphism holds, we have the chain of deductions

$$C'[f(\bar{\alpha})] = f(\bar{\beta})$$
$$C'(f[C^{-1}(\bar{\beta})]) = f(\bar{\beta}) \qquad \text{[substituting } C^{-1}(\bar{\beta}) = \bar{\alpha}]$$
$$C'fC^{-1} = f \qquad \text{(identity of mappings)}$$
$$C' = fCf^{-1} \qquad \text{(postmultiplying by } Cf^{-1})$$

Similarly,

$$W'[f(\bar{\alpha})] = f(\bar{\beta})$$
$$W'(f[W^{-1}(\bar{\beta})]) = f(\bar{\beta})$$
$$W'fW^{-1} = f$$
$$W' = fWf^{-1}$$

Conversely, if the conditions in theorem (18.2.3) hold, then the argument is reversed to conclude that isomorphism holds.

A good way to remember the conditions of isomorphism is the use of mapping diagrams. For W, for instance,

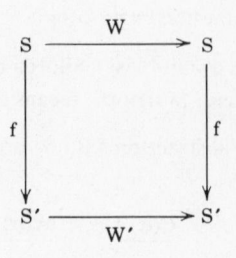

in which the two paths from S to S′ give the same results:

$$W'f = fW \qquad (\Rightarrow W' = fWf^{-1})$$

A similar diagram can be drawn for C. We say that the diagram, because of the path condition, is commutative. Hence, the isomorphisms make the diagrams commutative. Abstractly, f is a mere relabeling of a finite set. Hence, the diagram has the form of a renaming situation discussed in section 17.10:

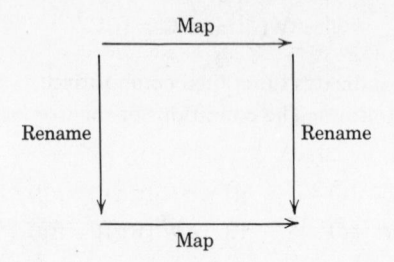

The commutative property is stated thus: the procedure of first renaming and then mapping is the same in its effects as first mapping and then renaming. For instance, the rename-then-map route is illustrated by

$$\text{nobles} \longmapsto \text{warriors} \longmapsto \text{warriors}$$

The route map-then-rename corresponding to this is

$$\text{nobles} \longmapsto \text{nobles} \longmapsto \text{warriors}$$

and the composite result is the same.

18.3. Kinship Interpretation. A special category of prescribed marriage system arises if we consider tribes in which clans are defined such that within-clan marriage is forbidden. In this case, W must not map any $\bar{\alpha}$ into itself. White (1963, pp. 34–35) has defined the basic category succinctly in a set of eight axioms, paraphrased as follows:

(18.3.1)   White's Axioms.

(1) The population of the society is divided into mutually exclusive groups, called clans.

(2) There is a rule fixing the unique clan among whose women the men of a given clan must find their wives.

(3) Men from different clans cannot marry women of the same clan.

(4) All children of a given couple are assigned to a single clan, uniquely determined by the clans of the mother and father.

(5) Children whose fathers are in distinct clans must themselves be in distinct clans.

(6) A man cannot marry a woman of his own clan.

(7) Every person has some relative by marriage and descent in each other clan, so that the society is not split into unrelated groups.

(8) Whether two people related by marriage and descent links are in the same clan depends only on the kind of relationship and not on the clans they belong to.

Axiom 1 states that there is a partition S of the actors in A.
Axiom 2 states that a mapping on S exists, call it W.
Axiom 3 states that W is a 1–1 mapping.
Axiom 4 says that a second mapping on S exists, say C.
Axiom 5 makes C a 1–1 mapping.
Axiom 6 says that $W(\bar{\alpha}) \neq \bar{\alpha}$, all $\bar{\alpha}$ in S.
Axiom 7 says that some composition of W and C, say M, relates any $\bar{\alpha}$ and $\bar{\beta}$ in S.
Axiom 8 is a homogeneity assumption we shall clarify later.

We see that White's axioms define a pms (S, C, W) with restrictions on the two mappings C and W. After developing some intuitions about systems satisfying these axioms, we shall return to formalize the special category of pms's specified by these axioms.

It will be useful to adopt the notation of kinship analysis used in anthropology:

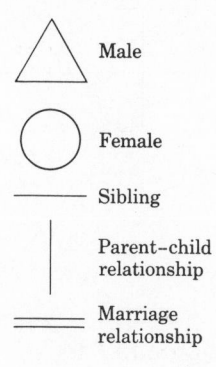

Male

Female

Sibling

Parent–child relationship

Marriage relationship

Table 18.3.1   Definition of Cousin Relations

Parental Relationship	Type of Cousin Relationship
Fathers are brothers	Patrilateral parallel
Mothers are sisters	Matrilateral parallel
Girl's mother is sister of boy's father	Patrilateral cross
Girl's father is brother of boy's mother	Matrilateral cross

For a more elaborate list of nomenclature in kinship analysis see Zelditch (1964).

To illustrate the concrete basis for the conditions described in White's axioms and the type of problem involved, we will analyze four cousin relationships with one question in mind: Under what conditions may they marry, if at all?

The four cousin relationships are defined in terms of the parental relationships in Table 18.3.1.

Clan brother and sister relations are defined in terms of sex and clan membership:

$$(18.3.2) \quad \begin{array}{lll} \alpha \text{ is a brother of } \beta & \text{iff} & \alpha \text{ is male and } \overline{\alpha} = \overline{\beta} \\ \alpha \text{ is a sister of } \beta & \text{iff} & \alpha \text{ is female and } \overline{\alpha} = \overline{\beta} \end{array}$$

for all α, β in the set A, which is partitioned into clans.

In the first type of relationship, patrilateral parallel, the kinship diagram is given in Figure 18.3.1.

Note that we label the male ("ego" in anthropological usage) by "α" and use α as the reference point. We ask whether α may marry β. If so, under what conditions on the two fundamental relationships among the clans denoted by C and W? We trace up from α, across to the girl's parents, then down to her to find the clan in which β is located, which is $\overline{\beta}$. If $\overline{\alpha} = \overline{\beta}$, then by White's axiom 6, α may not marry

Figure 18.3.1

β. If $\bar{\alpha} \neq \bar{\beta}$, then the condition of marriage is that W assigns $\bar{\beta}$ to $\bar{\alpha}$.

To proceed,* locate α in clan $\bar{\alpha}$ and trace to his parents: $C^{-1}(\bar{\alpha})$. From the clan of the father, which is $C^{-1}(\bar{\alpha})$, we know the clan of the father's brother, the girl's father, is also $C^{-1}(\bar{\alpha})$. Then the clan of this man's child is $C[C^{-1}(\bar{\alpha})]$. This is $\bar{\beta}$. But $C[C^{-1}(\bar{\alpha})] = \bar{\alpha}$. Hence, $\bar{\alpha} = \bar{\beta}$ and α may not marry β.

To sum up this first type of cousin relationship: if the axioms hold, then patrilateral parallel cousins are not allowed to marry "because they are brother and sister," in the clan sense of these two kinship concepts. Axiom 6 is the key to the conclusion. Axiom 8 tells us this conclusion does not depend on the particular clan $\bar{\alpha}$: it holds for all people in the specified patrilateral parallel cousin relationship.

Next consider the matrilateral parallel cousins diagramed in Figure 18.3.2. May α marry β? From α we determine $\bar{\alpha}$, the clan of α. From $\bar{\alpha}$ we find the father's clan $C^{-1}(\bar{\alpha})$. We see that we need the wife's clan, which is $W[C^{-1}(\bar{\alpha})]$. Horizontally tracing to the sister of the mother of α, we require the clan of her children, which depends upon the clan of her husband, namely $W^{-1}(W[C^{-1}(\bar{\alpha})])$. The children are then in clan $C[W^{-1}(W[C^{-1}(\bar{\alpha})])] = \bar{\alpha}$. Hence, β is in $\bar{\alpha}$, and with the strict rule (on W) that a man may not marry his clan sister, marriage is forbidden.

An example of a possible marriage is given in the diagram for the patrilateral cross-cousins (Figure 18.3.3). Under what conditions may α and β marry? We pass to $\bar{\alpha}$, the clan of α, then to father's clan, $C^{-1}(\bar{\alpha})$. Father's clan sister marries into clan $W^{-1}[C^{-1}(\bar{\alpha})]$ and a child of this marriage is in clan $C(W^{-1}[C^{-1}(\bar{\alpha})])$. This is $\bar{\beta}$. Hence, if α is to be allowed to marry β, a condition on W exists:

$$W = CW^{-1}C^{-1}$$

*Instead of $C^{-1}(\bar{\alpha})$ one could write $\bar{\alpha}C^{-1}$, and then $\bar{\alpha}C^{-1}C$ would be written instead of $C[C^{-1}(\bar{\alpha})]$. Our notation simply conforms to our mapping notation throughout this book, but the reader should note the reverse notation in White (1963) and Boyd (1969). White uses permutation matrices. These are just the adjacency matrices of the graphs corresponding to the C and W mappings.

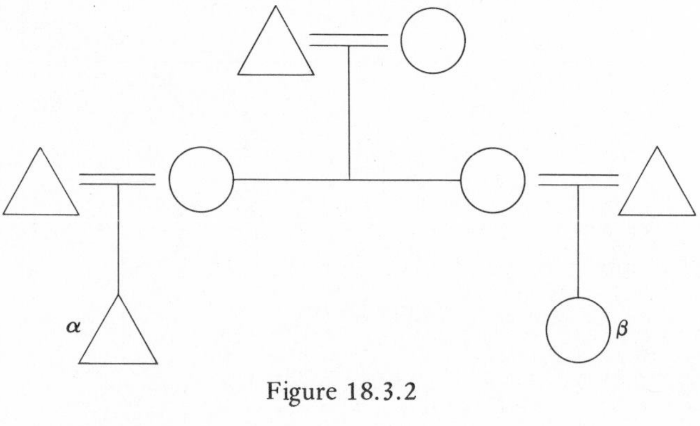

Figure 18.3.2

Figure 18.3.3

or, equivalently,

$$WC = CW^{-1}$$

Finally, for matrilateral cross-cousins, the relationship is diagramed in Figure 18.3.4. From α we find $\bar{\alpha}$, then $C^{-1}(\bar{\alpha})$, then $W[C^{-1}(\bar{\alpha})]$, and then $C(W[C^{-1}(\bar{\alpha})])$. Hence, $\bar{\beta} = C(W[C^{-1}(\bar{\alpha})])$. Thus, α may marry β if and only if the relation W satisfies

$$W = CWC^{-1}$$

or, equivalently,

$$WC = CW$$

We may summarize these four relationships in Table 18.3.2.

Figure 18.3.4

Table 18.3.2 Marriage Conditions for Cousins

Relationship	Marriage Conditions
Patrilateral parallel cousins	Not permitted
Matrilateral parallel cousins	Not permitted
Patrilateral cross-cousins	$WC = CW^{-1}$
Matrilateral cross-cousins	$WC = CW$

18.4. Axiomatic Concept of a Strict System. White's axioms add three conditions (axioms 6, 7, and 8) to the general prescribed marriage system defined earlier: exogamy, connectedness, and homogeneity. Therefore, we can define an exact concept termed "strict pms" which corresponds to White's conditions.

(18.4.1) Definition. A pms (S, C, W) is a strict pms if and only if:

Axiom S1. (Exogamy) For all $\bar{\alpha}$ in S,

$$W(\bar{\alpha}) \neq \bar{\alpha}$$

Axiom S2. (Connectedness) For any $\bar{\alpha}, \bar{\beta}$ in S, there exists some mapping M generated by C and W such that $M(\bar{\alpha}) = \bar{\beta}$.

Axiom S3. (Homogeneity) If M is any mapping generated by C and W and if there exists some $\bar{\alpha}_0$ in S with $M(\bar{\alpha}_0) = \bar{\alpha}_0$, then for all $\bar{\alpha}$ in S, we have $M(\bar{\alpha}) = \bar{\alpha}$, so that M = I.

Axiom S1 is a formal statement corresponding to White's axiom 6. Axiom S2 corresponds to White's axiom 7. The least obvious of White's eight axioms is his axiom 8, and this we state formally as axiom S3. We have illustrated the intuitive meaning of this axiom in tracing relationships in kinship diagrams. More generally, think of the graph representation of (S, C, W) with its two types of lines. A relationship M is a type of path and axiom S3 says the graph must be homogeneous as defined on page 515: if a type of path M forms a closed path back to its origin at any one node, then it does so at every node.

18.5. The Enumeration and Classification Problem. The problem that now arises is whether we can develop a method for determining the distinct structures in the category of algebraic systems defined by the axioms. The role of group theory is to provide a methodical abstract procedure for determining and classifying these structures for various sizes of the system, where size refers to the number of classes in set S. To see the contribution made by group theory, let us imagine trying to solve this problem by "straightforward" listing of the tables of (S, C, W) directly from the idea that C and W permute S.

Start with N = 2. Two permutations of S exist: the identity and the switching. By axiom S1 rule out I for W. This leaves two for C and, so, the two tables:

$\bar{\alpha}$	$C(\bar{\alpha})$	$W(\bar{\alpha})$	$\bar{\alpha}$	$C'(\bar{\alpha})$	$W'(\bar{\alpha})$
A	A	B	A	B	B
B	B	A	B	A	A

Consider axiom S2 and the first pms. Let $\bar{\alpha}$ = A, $\bar{\beta}$ = B. Then W maps A into B. Let $\bar{\alpha}$ = B, $\bar{\beta}$ = A. Then, again, W maps B to A. Hence, the first table satisfies axiom S2. Since W' = W, the second table also satisfies S2. Consider S3 and the first table. We need all the mappings generated by C and W. But C = I, W^2 = I, so W^{-1} = W and there are no other mappings generated by C and W. Testing C for the requirement of S3, we see that C(A) = A and C($\bar{\alpha}$) = $\bar{\alpha}$, all $\bar{\alpha}$, so C "passes the test." Testing W for the requirement of S3, we find that W does not map any $\bar{\alpha}$ into itself; hence, the test is not applicable. Therefore, the first table is a strict pms. In the second table, we see that W' generates W'^2 = I, which passes the test. W' is such that the test is not applicable. Since C' = W', we are finished. Hence, the second pms is strict. Also, the tables are not isomorphic. We conclude that there are exactly two strict structures for N = 2.

For N = 3, we have 3! = 6 permutations of a set of 3 objects. Hence, there are 6×6 = 36 pms's for N = 3. To proceed as with N = 2 we would have to reduce these 36 tables: first, to those that are strict and, second, to the distinct structures. In fact, there are only 3 distinct structures. For N = 4, there are 24×24 = 576 pms's, but only 6 strict structures. Clearly, even for small N, the method of writing down tables and analyzing them straightforwardly breaks down.

The contribution of the Kemeny-Snell-Thompson and the White procedures enters at this point: they apply the apparatus of group theory to study the problem of enumerating the structurally distinct strict prescribed marriage systems. Whatever the system type, "enumerating the structurally distinct. . . systems" yields a generic process. Here is the methodology in capsule form: (a) first an exact concept of some class of social or cultural systems is framed (e.g., strict pms), and then, (b) all possible structurally distinct systems of this type are enumerated.* Neither of these tasks constitutes the totality of the enterprise of constructing algebraic models, but they are a part of the larger domain of activity of studying social structure using formal methods. The enumeration problem alone is not the most fascinating or substantively most interesting. In a certain sense, it is a job for a computer programmed to generate and test pairs of permutations, but even a computer has computational limits. White (1963) managed to solve the enumeration problem for N up to 32, where the number of pms's is $(32!)^2$ but there are only 70 strict structures.

Moreover, White did more than enumerate strict structures. For each N up to 32 he found the number of such structures exemplifying one or another marriage

*Another example occurs in the context of dominance relations (see section 4.13). A counting method is given, using group theory, by Davis, 1954; the ideas concerning dominance relations and processes were introduced by Landau, 1951. See, also, Bartos, 1967.

rule for cousins. We saw earlier (Table 18.3.2) that parallel cousins cannot marry in a strict prescribed marriage system. The two cross-cousin marriage conditions generate two types of strict structures: (1) patrilateral cross-cousins marry, or (2) matrilateral cross-cousins marry. At this point, White introduces the idea of bilateral cross-cousins: the girl's mother is a sister of the boy's father (patrilateral cross; see Table 18.3.1), and also, the girl's father is a brother of the boy's mother (matrilateral cross). This means that a bilateral relationship encompasses both cross-cousin diagrams. Hence, by Table 18.3.2, the marriage conditions for bilateral cousins include both $WC = CW^{-1}$ and $WC = CW$. Hence, $CW = CW^{-1}$, and so, $W = W^{-1}$. Therefore, $W^2 = I$.

This was deduced under the hypothesis that bilateral cross-cousins exist. Hence, a necessary condition for such existence is that $W^2 = I$. Recalling the notions of "switching" and "flipping" in other contexts this identity can be seen to be interpretable as follows: clan $\bar{\alpha}$ gets its wives in $\bar{\beta}$ and clan $\bar{\beta}$ gets its wives in $\bar{\alpha}$, so that an exchange is institutionalized.

Note that if $W^2 = I$, then $CW^{-1} = CW$; and so, in Table 18.3.2 we conclude that the marriage conditions on bilateral cross-cousins become $WC = CW$.

The basic typology is generated as shown in Figure 18.5.1. The verbal descriptions by White's type number for the nonresidual categories are as follows:

1. Bilateral cross-cousins exist and marry.
2. Bilateral cross-cousins do not exist, and matrilateral cross-cousins marry.
3. Bilateral cross-cousins do not exist, and patrilateral cross-cousins marry.
4. Bilateral cross-cousins exist and do not marry.

The name "paired clans" is chosen for type 4 because when $W^2 = I$, the exchange of wives between pairs of clans occurs.

Our next step is to describe the group theoretic analysis that leads to an enumeration of the strict structures, including the enumeration of the number within each of the White types.

	Name	White Type No.
$W^2 = I$ — WC = CW	Bilateral Marriage	1
$W^2 = I$ — WC ≠ CW	Paired Clans	4
$W^2 ≠ I$ — WC = CW	Matrilateral Marriage	2
$W^2 ≠ I$ — WC ≠ CW — $WC = CW^{-1}$	Patrilateral Marriage	3
$W^2 ≠ I$ — WC ≠ CW — $WC ≠ CW^{-1}$	Residual	5

Figure 18.5.1

18.6.   Application of Group Theory to Solve the Problem.   The basic justification of the group theoretic method in this context is the following proposition.

(18.6.1)   Theorem. (Kemeny-Snell-Thompson)   A pms (S, C, W) is a strict pms if and only if the group of permutations generated by C and W is regular and W $\neq$ I.

To say that a permutation group is regular means that, apart from I, each permutation maps each object into a different object and that, for any pair of elements in the set S being permuted, there is a permutation mapping one into the other.

Formally, if G(C, W) is the group generated by C and W, then G(C, W) is regular if and only if the following hold:

(1) if f $\in$ G and f $\neq$ I, then f(x) $\neq$ x, all x $\in$ S
(2) if x, y $\in$ S, then there is an f $\in$ G(C, W), such that f(x) = y.

The contrapositive of (1) is

$$\text{if } f(x) = x, \text{ for some } x \in S, \text{ then } f = I$$

This shows that axiom S3 corresponds to (1). Axiom S2 corresponds to (2). Hence, axioms S2 and S3 together make the generated group regular. Also, axiom S1 implies that W $\neq$ I. Hence, if (S, C, W) is a strict system, then the group G(C, W) is regular and W $\neq$ I. Conversely, given a pms (S, C, W) in which G(C, W) is regular and W $\neq$ I, we conclude that axioms S2 and S3 are satisfied. We want to show that W $\neq$ I implies axiom S1. If for some $\bar{\alpha}_o$, W($\bar{\alpha}_o$) = $\bar{\alpha}_o$ then by axiom S3, W = I, a contradiction. Hence, for all $\bar{\alpha}$, W($\bar{\alpha}$) $\neq$ $\bar{\alpha}$, which is axiom S1. The theorem, in somewhat different form, was first proved by Kemeny, Snell, and Thompson (1966) in their finite mathematics book. Hence, theorem (18.6.1) will be referred to as "the KST theorem."

The word "regular" occurred in the previous chapter. Theorem (17.9.1), the representation theorem for groups, asserted that given a group multiplication table— no matter which group—the rows yield an isomorphic group of permutations. The word "regular representation" is used for this isomorphic group because it is indeed a regular permutation group. Hence, an instant source of abstract regular permutation groups exists: take any abstract, finite group (from Tables 17.7.1 and 17.7.2, for instance), write down its multiplication table, and then from the rows of the table write down the regular representation. Suppose that the group whose multiplication table is given is generated by two nonidentity elements say, g_1 and g_2. Then, by isomorphism, the same is true for its regular representation. Hence, the regular group obtained from the multiplication table will satisfy the conditions of the KST theorem. It follows that by calling the set of permutations S, the first generator C, and the second W that we have an abstract strict pms (S, C, W). If this procedure is followed for all possible pairs of generators of the regular group, a family of abstract strict pms's is obtained. To determine the distinct structures, we divide this class modulo isomorphism in the sense of the definition of isomorphic (S, C, W) given in section 18.2.

In general, an isomorphism is a mapping $f: S \rightarrow S'$ that is 1–1 and preserves the two permutations of S. But here $S' = S$. Hence, f is itself a permutation of S. If two generator pairs are (F, G) and (F', G'), then by theorem (18.2.3) they determine isomorphic pms's if and only if there exists a permutation on S such that

$$F' = fFf^{-1} \text{ and } G' = fGf^{-1}$$

since F is to be replaced by C and G by W. This means in terms of section 17.10 that F' is the image of F under the same conjugation that carries G into G'. Hence, under one renaming—namely, f—F' represents F and G' represents G.

(18.6.2) Definition. Two pairs of permutations, denoted (F, G) and (F', G'), in the group of all permutations of a set are pair conjugates if and only if there is a permutation f such that

$$F' = fFf^{-1}$$
$$G' = fGf^{-1}$$

Evidently we have the following propositions, using the ideas of section 17.10.

(18.6.3) Proposition. A necessary condition that (F, G) and (F', G') are pair conjugates is that F and F' be conjugate elements and G and G' be conjugate elements in the group of all permutations of the set being permuted.

(18.6.4) Corollary. If either F and F' or G and G' are not of the same cycle type, then the pairs (F, G) and (F', G') are not pair conjugates.

The proposition follows from the definitions. The corollary gives a tool for later analysis; it follows by applying proposition (17.10.5) to proposition (18.6.3). A caution must be added: the relations of "conjugate elements" and "pair conjugates" are defined in reference to the full permutation group and not merely the regular representation, a subgroup of the full group. This means that the necessary map f may be drawn from outside the regular group.

(18.6.5) Proposition. The relation between pairs of permutations given by "are pair conjugates" is an equivalence relation.

We must show that reflexivity, symmetry, and transitivity hold. Denote the relation by p.

(1) Reflexivity. (F, G)p(F, G) because letting f = I,

$$F = IFI^{-1}$$
$$G = IGI^{-1}$$

(2) Symmetry. If (F, G)p(F', G'), then for some f,

$$F' = fFf^{-1} \Rightarrow F = f^{-1}F'f = f^{-1}F'(f^{-1})^{-1}$$
$$G' = fGf^{-1} \Rightarrow G = f^{-1}G'f = f^{-1}G'(f^{-1})^{-1}$$

and so, f^{-1} yields $(F', G')p(F, G)$.

(3) Transitivity. If $(F, G)p(F', G')$ and $(F', G')p(F'', G'')$, then there exist f_1 and f_2 such that

$$F' = f_1 F f_1^{-1} \text{ and } F'' = f_2 F' f_2^{-1} \Rightarrow F'' = f_2(f_1 F f_1^{-1})f_2^{-1} = f_2 f_1 F(f_2 f_1)^{-1}$$
$$G' = f_1 G f_1^{-1} \text{ and } G'' = f_2 G' f_2^{-1} \Rightarrow G'' = f_2(f_1 G f_1^{-1})f_2^{-1} = f_2 f_1 G(f_2 f_1)^{-1}$$

hence,

$$(F, G)p(F'', G'')$$

It follows that the relation of pair conjugation partitions any subset of pairs of permutations of the full group of permutations. In particular, all the pairs of generators of the regular group obtained as noted above are partitioned modulo pair conjugation. Once the partition is obtained, we know that only one representative of each class is to be chosen to attain a complete set of nonisomorphic strict structures.

This entire procedure is now illustrated for the cases $N = 2$ and $N = 3$. Inasmuch as we have already analyzed the $N = 2$ case, the results of the procedure as applied in this instance are verified by an independent method.

Example. (a) $N = 2$

The only abstract group of order two is C_2, by Table 17.7.2. The multiplication table is

	I	g
I	I	g
g	g	I

Letting I' and g' be the corresponding permutations obtained from the rows we have

$$I \longmapsto I' = \begin{pmatrix} 1 & 2 \\ 1 & 2 \end{pmatrix}$$

$$g \longmapsto g' = \begin{pmatrix} 1 & 2 \\ 2 & 1 \end{pmatrix} \equiv (1\ 2)$$

in cycle notation. The pairs of permutations that generate the regular group correspond to the pairs of elements that generate C_2:

$$(I', g'), (g', g'), (g', I')$$

Our rule is to identify C with the first component and W with the second component.

Hence, since we cannot have W be the identity (by axiom S1 for strict pms's), we are left with two generator pairs:

$$(I', g'), (g', g')$$

Since I' and g' are of different cycle type, corollary (18.6.4) implies that these pairs are not conjugate. Hence, the strict pms's obtained from them are not isomorphic. The first yields $C = I$, $W^2 = I$; the other yields $C^2 = W^2 = I$.

Formally, then, we have the two strict structures:

(I)				(II)		
$\bar{\alpha}$	$C(\bar{\alpha})$	$W(\bar{\alpha})$		$\bar{\alpha}$	$C(\bar{\alpha})$	$W(\bar{\alpha})$
1	1	2		1	2	2
2	2	1		2	1	1

where

$$\text{(I)} \quad C = I, W^2 = I$$
$$\text{(II)} \quad C^2 = I, W^2 = I$$

and these relations follow from the generating relation satisfied by g of C_2 and, so, g' of its regular representation.

Finally, we note that since C_2 is commutative, so is its regular representation and so both (I) and (II) satisfy $WC = CW$. Hence, the White typology in Figure 18.5.1 yields: both (I) and (II) are bilateral marriage (type 1) systems.

Example. (b) $N = 3$

There is only one abstract group of order three, C_3:

	I	g	g^2
I	I	g	g^2
g	g	g^2	I
g^2	g^2	I	g

Hence, the regular representation is

$$I \longmapsto I' = \begin{pmatrix} 1 & 2 & 3 \\ 1 & 2 & 3 \end{pmatrix}$$

$$g \longmapsto g' = \begin{pmatrix} 1 & 2 & 3 \\ 2 & 3 & 1 \end{pmatrix} \equiv (1\ 2\ 3)$$

$$g^2 \longmapsto g'^2 = \begin{pmatrix} 1 & 2 & 3 \\ 3 & 1 & 2 \end{pmatrix} \equiv (1\ 3\ 2)$$

There are nine pairs of permutations in the regular group, but we rule out the case of I' in the second component. The six remaining pairs each generate the group because g generates C_3; hence, $g^{-1} = g^2$ generates C_3; hence either of g' and g'^2 is sufficient to generate the regular group:

$$\left\{ (I', g'), (I', g'^2), (g', g'), (g', g'^2), (g'^2, g'), (g'^2, g'^2) \right\}$$

We have to divide this set modulo pair conjugation. The first two pairs are pair conjugate because I' is mapped into I' under any conjugation and g'^2 is a conjugate of g' by noting that g' and g'^2 have the same cycle type [using proposition (17.10.5)]. However, I' cannot be a conjugate element with any nonidentity element by the fact of difference in cycle type. Hence, by corollary (18.6.4), no pair containing I' is pair conjugate to a pair without I'. Hence, one class in the division of the class of generator pairs is

$$\left\{ (I', g'), (I', g'^2) \right\}$$

Next we note that (g', g') is pair conjugate to (g'^2, g'^2) since g' and g'^2 are conjugate elements. Also, since there is an f such that $fg'f^{-1} = g'^2$, it follows that

$$fg'^2 f^{-1} = (fg'f^{-1})^2 = (g'^2)^2 = g'$$

Hence, with this f we have

$$g'^2 = fg'f^{-1}$$
$$g' = fg'^2 f^{-1}$$

and so (g', g'^2) and (g'^2, g') are pair conjugate. Tentatively, we have two more classes:

$$\left\{ (g', g'), (g'^2, g'^2) \right\}, \ \left\{ (g', g'^2), (g'^2, g') \right\}$$

We have to check if pairs in different sets conjugate. If (g', g') were pair conjugate to (g', g'^2) then for some f

$$g' = fg'f^{-1}$$
$$g'^2 = fg'f^{-1}$$

an obvious contradiction since $g' \neq g'^2$. Hence, these sets are distinct equivalence classes.

Having divided the pairs modulo pair conjugation, we choose one pair from each: (I', g'), (g', g'), and (g', g'^2). Assigning C to the first component and W to the second, we complete the analysis by writing down the defining relations and the tables for the three structures.

(III)			(IV)			(V)		
$\bar{\alpha}$	$C(\bar{\alpha})$	$W(\bar{\alpha})$	$\bar{\alpha}$	$C(\bar{\alpha})$	$W(\bar{\alpha})$	$\bar{\alpha}$	$C(\bar{\alpha})$	$W(\bar{\alpha})$
1	1	2	1	2	2	1	2	3
2	2	3	2	3	3	2	3	1
3	3	1	3	1	1	3	1	2

$$C = I$$
$$W^3 = I$$

$$C^3 = I$$
$$W^3 = I$$

$$C^3 = I$$
$$W = C^{-1}$$

Since C_3 is commutative, so is the regular representation. Hence, the permutations I', g', and g'^2 commute. It follows that when C and W are substituted into the pairs of generators, the relationship CW = WC obtains. This is true in each structure. Also, in each structure $W^2 \neq I$. Hence, we can classify these three structures as to White type: namely, type 2, matrilateral marriage.

Once we have abstract strict pms's, we can aid our intuition about them by using a graph with two types of lines. The correspondence is as follows: the nodes represent the clans in abstract set S; the two types of directed lines represent C and W, respectively. This is shown in Figure 18.6.1 for the structures found for systems with two and three clans. It is not surprising that the graphs should resemble those for groups. For instance, we obtained the (S, C, W) for N = 3 by reference to a group, the regular group of C_3. Since the regular group is isomorphic to C_3, the transformation into the three (S, C, W) systems bears traces of the C_3 origin. But

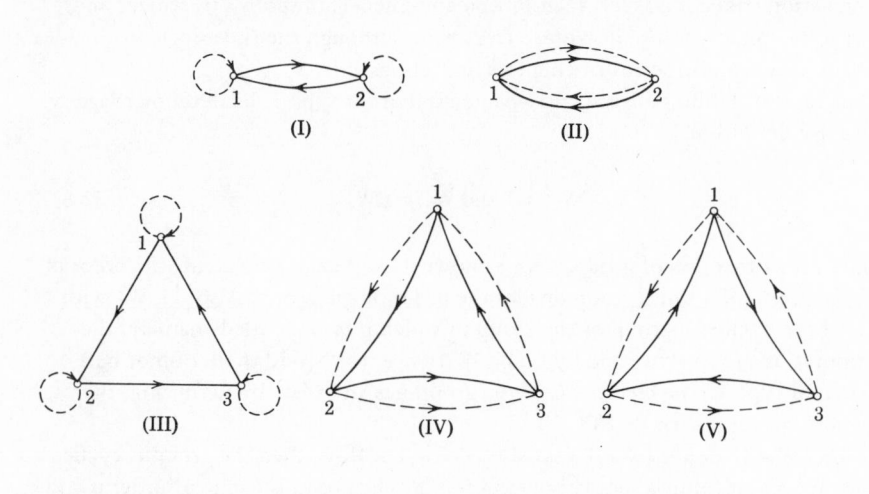

Figure 18.6.1

the one C_3 graph has become three (S, C, W) graphs. They are not isomorphic, of course, because they represent three nonisomorphic strict systems. But note that each graph is homogeneous in the sense of section 17.7 and that this reflects axiom S3.

The pair conjugation technique has an apparent limitation: we still need to compare all the generator pairs to determine the equivalence classes. For N = 4, for instance, there are two abstract groups, $C_2 \times C_2$ and C_4. In the former group we have six generator pairs. Even knowing that the pair conjugate relation is an equivalence does not save the labor of comparing a good many pairs before we construct the equivalence classes. Then we have to consider C_4 and the pairs and their classes, and finally, we need to consider possible pair conjugates drawn from distinct constructions, because they are all defined as permutations over $\{1, 2, 3, 4\}$. Hence, the reliance on a conjugation approach to the enumeration problem seems to reach a limit of easy applicability rather quickly. This is true, however, only in the "concrete" sense of the way we have applied it. In White's monograph, the conceptual essentials of the logic we have outlined play a role, but the work involved in enumeration is abbreviated through various calculational strategems. For instance, rather than fixing N and then enumerating and classifying, White fixes the type within which he then enumerates. It is through these strategems that White went well beyond the Kemeny-Snell-Thompson treatment in their finite mathematics book, where the method we outlined above was developed. Moreover, for beginners the concrete construction of the strict structures is a useful exercise. (The pair conjugation concept was left tacit in Kemeny-Snell-Thompson's treatment and does not explicitly figure in White's treatment, although the sheer logic of the idea of pms isomorphism seems to entail its definition.)

The logic of White's analysis may be illustrated for type 1, bilateral marriage systems. By definition,

$$W^2 = I \text{ and } WC = CW$$

White uses a theorem of group theory known as Lagrange's theorem: the order of any subgroup of a finite group divides the order of the group. Also, $\{I, W\}$ with $W^2 = I$ is a cyclic subgroup of the group in which it is embedded—namely, the permutation group generated by C and W. Hence, for N odd there cannot be a system of type 1. The element C of the group has an order: by definition, the smallest integer p such that $C^p = I$.

Either $W = C^j$ or not, where j satisfies $1 \leqslant j < p$. If so, then the group is cyclic. Since $W^2 = I$, it follows that $C^{2j} = I$ and p = 2j. The group is cyclic of order p = 2j. For instance, if j = 3, then the group is C_6 and $W = C^3$. From an enumeration viewpoint we start with number N, but now this must be p. Hence, j = N/2. Thus, we identify C with the generator and W with $C^{N/2}$. This is a unique system. Thus, for every even integer, we obtain a unique system of the form, $C^N = I$, $W = C^{N/2}$.

If $W \neq C^j$ $(1 \leqslant j < p)$, then W and C are generators with

$$W^2 = I, \ C^p = I, \ WC = CW$$

and so the group is a product group of order $2p = N$: namely, $C_2 \times C_p$.

Hence, there are precisely two type-I systems for every even integer and no type-I systems for odd integers. For $N = 8$, the two systems are shown in Figure 18.6.2.

The analysis for other types of systems (II, III, and IV) is carried out in White's (1963) monograph. The later part of the monograph compares the selected structures from the catalogue of structurally distinct systems with kinship data from several societies. (The basic results of the enumeration are presented in White's Table 2.3, covering $N = 2$ to $N = 32$.)

18.7. Duality.   The strict systems were interpreted in the previous sections from a patrilineal point of view. For instance, $C(\overline{\alpha})$ was interpreted as the clan of the children of a male α in clan $\overline{\alpha}$. We can obtain an associated female point of view by noting that if α is a female in clan $\overline{\alpha}$, then $W^{-1}(\overline{\alpha})$ is the clan of her husband and so $C[W^{-1}(\overline{\alpha})]$ is the clan of the children of a female α in $\overline{\alpha}$. Hence, the system (S, CW^{-1}, W^{-1}) is the "dual" of (S, C, W): it is the associated female point of view. The ideas developed here are based on White's (1963) discussion of the notion of dual structures.

(18.7.1) Definition.   The transformation

$$(S, C, W) \longmapsto (S, CW^{-1}, W^{-1})$$

is termed dualization and (S, CW^{-1}, W^{-1}) is the dual of (S, C, W).

(18.7.2) Proposition.   The dual of the dual of (S, C, W) is (S, C, W).

We first dualize (S, C, W) to obtain (S, CW^{-1}, W^{-1}). Then the dual of (S, CW^{-1},

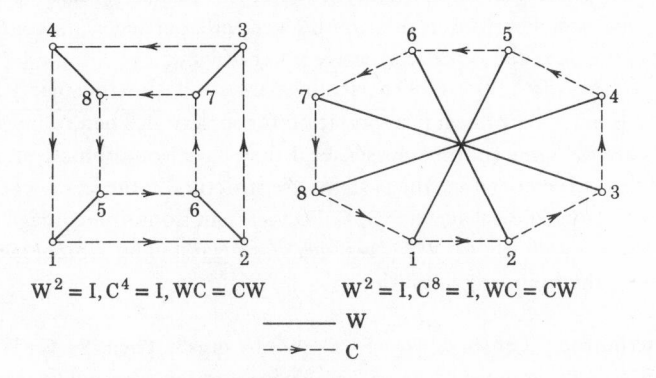

$$W^2 = I, C^4 = I, WC = CW \qquad W^2 = I, C^8 = I, WC = CW$$

——————— W

--->--- C

Figure 18.6.2

W^{-1}) is $[S, (CW^{-1})(W^{-1})^{-1}, (W^{-1})^{-1}]$ and since $(W^{-1})^{-1} = W$, this is

$$(S, CW^{-1}W, W) = (S, C, W)$$

It is this proposition that justifies the term "dual". A concept that relates isomorphism to duality is the following.

(18.7.3) Definition. A system (S, C, W) is self-dual if and only if it is isomorphic to its dual.

That is, (S, C, W) is self-dual if and only if there is a permutation f on S such that

$$CW^{-1} = fCf^{-1}$$
$$W^{-1} = fWf^{-1}$$

A simple proposition about self-duality is the following:

(18.7.4) Proposition. Let (S, C, W) be strict. Then a necessary condition for self-duality is that $C \neq I$.

If (S, C, W) is self-dual, then

$$CW^{-1} = fCf^{-1} \qquad \text{(some f)}$$

and if C = I,

$$W^{-1} = I$$

yielding

$$W = I$$

in violation of axiom S1 for strict systems.

If a female point of view interprets (S, C, W), then (S, CW^{-1}, W^{-1}) yields a male point of view: for if α is a male in clan $\bar{\alpha}$, then $W^{-1}(\bar{\alpha})$ is the clan of his wife and then $C[W^{-1}(\bar{\alpha})]$ is the clan of his children. Hence, the formalism itself is neutral in point of view: whichever interpretation is given for a particular model for a given society, the dual yields the alternative point of view.

Now suppose that (S, C, W) is given a patrilineal interpretation for society A and (S', C', W') is given a matrilineal interpretation for society B. Then rather than say the societies are the same (in the pms sense) if they have isomorphic systems, we might want to use the criterion: the systems are structurally the same when the dualization of the patrilineal system (S, C, W) yields an isomorphic image of the matrilineal society with (S', C', W'). This idea was suggested by White (1963) and is formalized in the following definition.

(18.7.5) Definition. Let (S, C, W), (S', C', W') be pms's. Then (S', C', W') is similar to (S, C, W) if and only if (S', C', W') is isomorphic with the dual of (S, C, W).

The criterion of similarity is seen to be: there exists a bijective mapping $f: S \xrightarrow{\sim} S'$ such that

$$C' = fCW^{-1}f^{-1}$$
$$W' = fW^{-1}f^{-1}$$

by definitions of isomorphism and dual. Then,

$$C' = fCf^{-1}fW^{-1}f^{-1}$$
$$= fCf^{-1}W'$$

so that

$$C'W'^{-1} = fCf^{-1}$$

Since also

$$W'^{-1} = (fW^{-1}f^{-1})^{-1} = fWf^{-1}$$

we see that similarity is a symmetric relation.

However, it fails to be reflexive and transitive. To be reflexive would mean that (S, C, W) be isomorphic to the dual of (S, C, W)—that is, be self-dual. As proposition (18.7.4) shows, however, not all systems are self-dual. Although the relation is not transitive, we have the following:

(18.7.6) Proposition.   If (S'', C'', W'') is similar to (S', C', W') and (S', C', W') is similar to (S, C, W), then (S'', C'', W'') and (S, C, W) are isomorphic.

To see this, note that (S', C', W') is isomorphic with the dual of (S, C, W), so symmetry implies that (S, C, W) is isomorphic with the dual of (S', C', W'). But, also, (S'', C'', W'') is isomorphic with the dual of (S', C', W'). Hence, (S'', C'', W'') and (S, C, W) are isomorphic with each other.

It should be possible to define and relate other concepts of structural similarity based on the notion of a prescribed marriage system. Formally, the concepts are all free from the specific kinship interpretation, and it may be that interpretations in terms of other phenomena will suggest fruitful lines of development.

In the next chapter we will want to consider dynamics of these types of systems. This is approached by way of transition rules that map systems homomorphically. Hence, the chapter will open with a deeper mathematical treatment of homomorphisms before returning to the kinship application of algebraic methods. The concept of homomorphism is the foundation of an algebraic analysis of an arbitrary social relational system, also to be considered in the next chapter.

CHAPTER NINETEEN
HOMOMORPHISMS AND SOCIAL
STRUCTURE

19.1. Introduction. Several related topics are taken up in this chapter on algebraic methods for the analysis of social structures. A key idea is homomorphism, thought of as a way to approximate one structure by another. Hence, in section 19.2, we give a more detailed study of the idea of homomorphism in the now familiar context of group theory. Then, in section 19.3, we discuss a principle of structural change over time that is based on the homomorphism concept. In section 19.4 we raise the question as to what conditions on a social relational system are sufficient to generate the kind of classificatory system we studied in the last chapter; this allows a development toward a concept of structural equivalence among individual actors in a social relational system. Finally, in sections 19.5 and 19.6, we outline and apply the mathematical theory of categories to the generic problem: given an aribitrary social relational system, to find the structural skeleton of the system by formal analysis. We close with some bibliographic references on other algebraic model-building efforts and with some thoughts on the logic of algebraic analysis in sociology.

19.2. Homomorphisms. Let N be a subgroup of a group G such that under any conjugation the elements of N remain in N. That is, if x in N and a in G, then

$$axa^{-1} \in N \qquad (\text{all } a \in G)$$

Such a subgroup N is called an invariant or normal subgroup of G.

For example, consider the group D_3, whose defining relations are

$$r^3 = I, \ f^2 = I, \ (rf)^2 = I$$

The graph is shown in Figure 19.2.1. Consider the set $\{I, r, r^2\}$. Clearly this is C_3 embedded in D_3. Hence, this is a subgroup of D_3. We ask, Is C_3 normal in D_3? We take element r and consider conjugations, using the graph correspondence in the

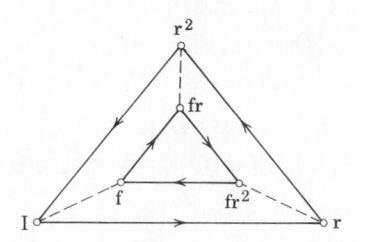

Figure 19.2.1

computations:

$$IrI^{-1} = r$$
$$rrr^{-1} = r$$
$$r^2rr^{-2} = r$$
$$frf^{-1} = r^2$$
$$(fr)r(fr)^{-1} = frrr^{-1}f^{-1} = r^2$$
$$(fr^2)r(fr^2)^{-1} = fr^3r^{-2}f^{-1} = r^2$$

Hence, the element r has its conjugate element in C_3 under any conjugation. Similarly, one shows that conjugation of r^2 leaves the conjugate element in C_3. Of course, the conjugation of I is I. Hence, C_3 is a normal subgroup of D_3.

Let us adopt the notation, for any set H of elements in group G:

$$aH = \{ax, x \in H\} \qquad (a \in G)$$

For example, if $H = \{r, r^2\}$, then

$$fH = \{fr, fr^2\}$$
$$fr^2H = \{fr^3, fr^4\} = \{f, fr\} = f\{I, r\}$$

Also, let us use the notation

$$Ha = \{xa, x \in H\} \qquad (a \in G)$$

For example, if $H = \{r, r^2\}$, then

$$Hf = \{rf, r^2f\} = \{fr^2, fr\}$$
$$Hfr^2 = \{rfr^2, r^2fr^2\} = \{fr, f\}$$

If a group is commutative, then

$$Ha = \{xa, x \in H\} = \{ax, x \in H\} = aH$$

Consider now a normal subgroup N of group G. Then

$$N = aNa^{-1} \qquad (a \in G)$$

Hence, the equivalent form is

(19.2.1) $aN = Na$ $(a \in G)$

For instance,

$$fC_3 = f\{I, r, r^2\} = \{f, fr, fr^2\}$$
$$C_3 f = \{I, r, r^2\} f = \{f, rf, r^2 f\} = \{f, fr^2, fr\}$$

so $fC_3 = C_3 f$.

Let $a \in G$ and $a \notin N$. Then

$$aN \cap N = \emptyset$$

For, if not, then some $y_0 \in aN \cap N$, and so, with $x_0 \in N$,

$$y_0 = ax_0 \quad \text{and} \quad y_0 \in N$$

Since N is a subgroup, $y_0 x_0^{-1} \in N$. But $a = y_0 x_0^{-1}$ and $a \in N$—a contradiction.

Let $b \in G$ and $b \notin N$, $b \notin aN$. Then

$$bN \cap aN = \emptyset$$

For, if not, then some $y_0 \in bN \cap aN$, and so,

$$y_0 = bx_0, \, y_0 = ax_1 \qquad (x_0, x_1 \in N)$$

Then $x_1 x_0^{-1} \in N$, and so $ax_1 x_0^{-1} \in aN$. However, we then obtain

$$ax_1 = bx_0$$
$$b = ax_1 x_0^{-1} \in aN$$

a contradiction.

In general, the algorithm of forming N, aN with $a \notin N$, bN with $b \notin N$, $b \notin aN$, ..., until all elements are exhausted forms a partition of the group G. This partition is denoted G/N, "G modulo N".

For example, to form D_3/C_3 we proceed as follows: The first element of D_3/C_3 is C_3 itself. Here $C_3 = \{I, r, r^2\}$. We arbitrarily select an element from $D_3 - C_3$, say f. Then we form

$$fC_3 = \{f, fr, fr^2\}$$

But $C_3 \cup fC_3 = D_3$, and so, we terminate the algorithm. We see that

$$D_3/C_3 = \{C_3, fC_3\}$$

Next we consider the multiplication of one class by another. Let H_1 and H_2 be subsets of G. Then

$$H_1 H_2 = \{xy, x \in H_1, y \in H_2\}$$

For instance, if

$$H_1 = \{r, r^2\}$$
$$H_2 = \{fr^2, I\}$$

then

$$H_1 H_2 = \{rfr^2, r^2 fr^2, rI, r^2 I\}$$
$$= \{fr, f, r, r^2\} = \{r, r^2, f, fr\}$$
$$H_2 H_1 = \{fr^2 r, fr^2 r^2, Ir, Ir^2\}$$
$$= \{f, fr, r, r^2\} = \{r, r^2, f, fr\}$$

Note that $H_1 H_2 = H_2 H_1$ in this instance. However, this is not a general rule for class product.

We note that class product is associative, because the product of elements is associative. Also

$$(aH_1)H_2 = \{(ax)y, x \in H_1, y \in H_2\} = a\{xy\} = a(H_1 H_2)$$

Note that

$$H_1 H_2 = \bigcup_{x \in H_1} xH_2 = \bigcup_{y \in H_2} H_1 y$$

For instance, if H_1 and H_2 are as in the preceding illustration,

$$H_1 H_2 = rH_2 \cup r^2 H_2$$
$$= r\{fr^2, I\} \cup r^2\{fr^2, I\}$$
$$= \{rfr^2, r\} \cup \{r^2 fr^2, r^2\}$$
$$= \{fr, r\} \cup \{f, r^2\}$$
$$= \{r, r^2, f, fr\}$$

Next we compute the four products of D_3/C_3 in order to fill in the table

	C_3	fC_3
C_3		
fC_3		

(1) $C_3 C_3 = \bigcup_{x \in C_3} xC_3 = IC_3 \cup rC_3 \cup r^2 C_3$

$$= C_3 \cup \{r, r^2, I\} \cup \{r^2, I, r\}$$
$$= C_3 \cup C_3 \cup C_3$$
$$= C_3$$

(2) $C_3fC_3 = \underset{x \in C_3}{\cup} xfC_3$

$\qquad = IfC_3 \cup rfC_3 \cup r^2fC_3$

$\qquad = fC_3 \cup fr^2C_3 \cup frC_3$ (using the graph)

$\qquad = fC_3 \cup fC_3 \cup fC_3$ (since $xC_3 = C_3$, if $x \in C_3$)

$\qquad = fC_3$

(3) $(fC_3)C_3 = f(C_3C_3) = fC_3$

(4) $(fC_3)(fC_3) = f(C_3fC_3) = f(fC_3) = f^2C_3 = IC_3 = C_3$

Hence,

	C_3	fC_3
C_3	C_3	fC_3
fC_3	fC_3	C_3

This is the form of C_2 where $(fC_3)^2 = I' = C_3$. Hence,

$$D_3/C_3 \cong C_2$$

That is, D_3 divided modulo C_3 is just C_2 in structure. Note that the normal subgroup C_3 is the identity in the group D_3/C_3.

The general rule is that given a normal subgroup N of a group G, we may form N, aN, bN, . . . and define

(19.2.2) $(aN)(bN) = abN$

In the instance of dividing D_3 by C_3, we note that (19.2.2) yields

$$C_3C_3 = (IC_3)(IC_3) = IC_3 = C_3$$
$$C_3fC_3 = (IC_3)(fC_3) = IfC_3 = fC_3$$
$$fC_3C_3 = (fC_3)(IC_3) = fIC_3 = fC_3$$
$$fC_3fC_3 = (fC_3)(fC_3) = f^2C_3 = C_3$$

The classes N, aN, bN, . . . are called cosets of G under division by N. And rule (19.2.2) says that the product of the coset aN and the coset bN is the coset abN.

The resulting system of cosets (and table for multiplication) forms a group called the factor group of G modulo N. This is in reference to the fact that the new group is just G/N. For instance, C_2 is the factor group of D_3 modulo C_3. We write as noted above

$$D_3/C_3 \cong C_2$$

The identity of D_3/C_3 is C_3. The identity of G/N is N, since

$$(aN)N = (aN)(IN) = (aI)N = aN$$

A group may be divided by any of its normal subgroups. Since $\{I\}$ is a trivial subgroup of any group and $\{I\}$ is normal—that is, satisfies (19.2.1)—

$$a\{I\} = \{aI\} = \{Ia\} = \{I\}a \qquad (a \in G)$$

we can divide by $\{I\}$. Then the result is that we get back the original group. Calling $\{I\}$ "1" as in Table 17.7.2,

$$G/1 \cong G$$

Also, G is a subgroup of G. It is normal because

$$aG = G = Ga \qquad (a \in G)$$

Division of G by G yields the one-element group,

$$G/G \cong 1$$

Thus, G and 1 are always factors of a group G. A group with no other factors is called simple: this means that there are no proper normal subgroups in G—normal subgroups N where set N is properly contained in G and $N \neq \{I\}$.

(19.2.3) Proposition. The mapping $\sigma\colon G \to G/N$ given by

$$\sigma(a) = aN$$

is a homomorphism from G onto G/N.
   This is so because

$$\begin{aligned} \sigma(ab) = abN &= aNbN \qquad \text{[by (19.2.2)]} \\ &= \sigma(a)\sigma(b) \end{aligned}$$

For instance, $C_2 \cong D_3/C_3$ is a homomorphic image of D_3. This is clear if we juxtapose the graphs and imagine a shrinking of each triangle to a point, first thinking of the graph as in three-space (see Figure 19.2.2). (Recall the graphs given in Table 17.7.1). Also, juxtaposing the tables:

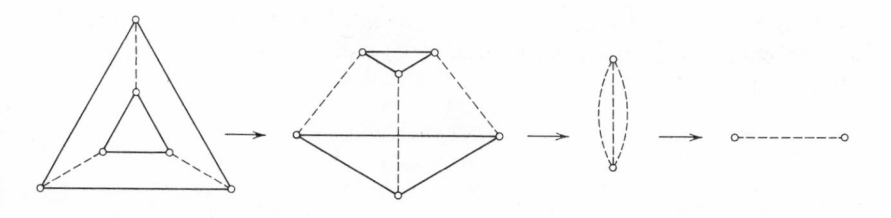

Figure 19.2.2

	I	r	r²	f	fr	fr²
I	I	r	r²	f	fr	fr²
r	r	r²	I	fr²	f	fr
r²	r²	I	r	fr	fr²	f
f	f	fr	fr²	I	r	r²
fr	fr	fr²	f	r²	I	r
fr²	fr²	f	fr	r	r²	I

$$\xrightarrow{\sigma}$$

	C_3	fC_3
C_3	C_3	fC_3
fC_3	fC_3	C_3

The mapping σ is called the canonical homomorphism or natural homomorphism. There is one such canonical map for each division of G modulo a normal subgroup. Note that σ reduces G. The image G/N is "smaller" yet structurally similar because of the homomorphism.

Let f: $G \to G'$ be a homomorphism of G into G'. Then by the "kernel" of the homomorphism f we mean the set of all points in G mapped into the identity of G':

$$K = \{ x : f(x) = I' \}$$

Thus, the kernel of $\sigma: G \to G/N$ is

$$K = \{ x : \sigma(x) = N \} = \{ x : x \in N \} = N$$

That is, the kernel of a natural homomorphism is the normal subgroup in the division. More generally, we have the following:

(19.2.4) Proposition.   The kernel of any homomorphism f: $G \to G'$ is a normal subgroup of G.

That is, in general, K is a normal subgroup of G if f: $G \to G'$ is a homomorphism with K as its kernel.

To see this, note that if a and b in K, then $f(a) = f(b) = I' (\in G')$. Hence, since f is a homomorphism,

$$f(ab) = f(a)f(b) = I'I' = I'$$

and so $ab \in K$. Also, since I is mapped into the identity I' of G', we see that $I \in K$. Since $f(b^{-1}) = [f(b)]^{-1} = (I')^{-1} = I'$, we see that $b^{-1} \in K$. Hence, K is a subgroup of G. To check the normal condition, we check to see if $xax^{-1} \in K$ whenever $a \in K$. In fact,

$$\begin{aligned} f(xax^{-1}) &= f(x)f(a)[f(x)]^{-1} \\ &= f(x)I'[f(x)]^{-1} \\ &= I' \end{aligned}$$

Hence, the subgroup K is normal.

It follows from proposition (19.2.4) that a factor group G/K is defined, and then proposition (19.2.3) implies it is a homomorphic image of G with the natural mapping σ: G → G/K.

Hence we now have two homomorphic images of G if f: G → G' is onto G': namely, G' itself and the factor group G/K, where K is the kernel of homomorphism f. The next proposition says these two homomorphic images are themselves structurally identical: they are isomorphic.

(19.2.5) Proposition. If f: G → G' is a homomorphism onto G' and K is its kernel, then G/K ≅ G'.

Thus, the factor group is isomorphic with the range group G'. In fact, the mapping diagram given in the situation is as follows:

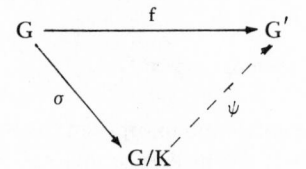

The dotted line is the required isomorphism ψ: G/K → G' to prove proposition (19.2.5). We define ψ so that it "makes the diagram commutative":

$$\psi \circ \sigma = f$$

or

$$\psi [\sigma(x)] = f(x).$$

This means, elementwise,

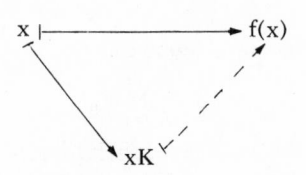

Thus, ψ maps xK (coset associated with x) into f(x). The mapping is well defined, for if y ↦ yK = xK, then y ∈ xK, so y = xa where f(a) = I'. Thus, f(y) = f(xa) = f(x)f(a) = f(x). Hence, class xK is uniquely associated with element f(x) in G'. The map is a homomorphism, since

$$\psi [(xK)(yK)] = \psi(xyK)$$
$$= f(xy)$$
$$= f(x)f(y)$$
$$= \psi(xK)\psi(yK)$$

The map ψ is onto G', because if $y \in G'$, then there is an x such that $f(x) = y$, and $\psi(xK) = y$. To show the map ψ is injective (1–1), let $\psi(xK) = \psi(yK)$. Then $f(x) = f(y)$, and

$$I' = [f(x)]^{-1} f(y) = f(x^{-1}) f(y) = f(x^{-1}y)$$

so that $x^{-1}y \in K$. Hence, for some $a \in K$, $x^{-1}y = a$, and so, $y = xa$. This shows that $y \in xK$, and so, $yK = xK$, which was to be shown.

A generalization of proposition (19.2.5) can be obtained. Let $f: G \to G'$ be any homomorphism of G into G', so that f need not be onto G'. Let $f(G)$ be the image of G under f:

$$f(G) = \{f(x), x \in G\}$$

Clearly, $f(G) \subseteq G'$.

(19.2.6) Proposition. $f(G)$ is a subgroup of G'.

(19.2.7) Theorem.   (Fundamental homomorphism theorem) If $f: G \to G'$ is any homomorphism of G into G' and K is its kernel, then $G/K \cong f(G)$.

To see that $f(G)$ is a subgroup of G', note that if $f(x)$ and $f(y)$ are in $f(G)$, then $f(xy) = f(x)f(y)$ shows that the product is in $f(G)$. Since $f(I) = I'$, the identity is in $f(G)$. Since $[f(x)]^{-1} = f(x^{-1})$, the inverse of $f(x)$ is in $f(G)$. Hence, $f(G)$ is a subgroup of G'. The isomorphism is again obtained by associating xK with $f(x)$:

Thus, $\psi(xK) = f(x)$ and the diagram is commutative:

$$\psi \circ \sigma = f$$

We leave to the reader the proof that ψ is an isomorphism.

Note that proposition (19.2.7) is called the fundamental homomorphism theorem of group theory. It says that any homomorphic image of G is some factor group of G so far as structure is concerned. The equation $\psi \circ \sigma = f$ may be written

$$\psi(\sigma) = f$$

to suggest the idea that any homomorphism out of G is a function of a canonical homomorphism. If f is onto, then $f(G)$ and G' are the same, so proposition (19.2.5) is a special case of the fundamental homomorphism theorem.

Following Boyd (1969), let us call any homomorphic image of G an approximation to G. Then the fundamental homomorphism theorem says that all possible transformations of G that yield an approximation to G are reflected in the structure of G in terms of its factor groups. Any factor group of G is a structural approximation to G. In the next section, we provide a brief indication of how this idea may be useful in the description of structural dynamics. Then, later in this chapter, we will see how these ideas of group theory are generalized and applied to arbitrary social networks.

19.3. Structural Change. We have encountered earlier in this book the notion of a temporal relational system: a domain of entities and a family of relations that are time-varying on this domain (see section 4.14). If the domain consists of human beings and if the process of interest extends over typical human lifetimes, then even the domain is taken as time-varying. Applying this concept to pms's, we arrive at the notion of a process,

$$(S_t, C_t, W_t) \qquad (t \in T)$$

of time-varying pms's, defined on a set A_t of humans, where $t \in T$ is time t in time domain T. Here S_t is the set of classes over A_t, valid at time t. It is clear that, as time varies, this temporal system may or may not change in a structural sense. The criterion for nonstructural change is this: at the later time, the system is an isomorphic image of its state at an earlier time.

Example. (a) We want to exhibit this point very concretely. Consider the process in an interval from t to $t + \Delta t$. For convenience of notation let us take $\Delta t = 1$. Assume that at $t = 0$ we have a population A_0 given by

$$A_0 = \left\{ \alpha_1, \alpha_2, \ldots, \alpha_N \right\}$$

Let the classes in S_0 be

$$\begin{aligned} S_{01} &= \left\{ \alpha_1, \alpha_2, \ldots, \alpha_j \right\} \qquad \text{(for some j, } 1 < j < N) \\ S_{02} &= \left\{ \alpha_{j+1}, \alpha_{j+2}, \ldots, \alpha_N \right\} \end{aligned}$$

Then let a patrilineal (S_0, C_0, W_0) be defined by the table:

S_{0i}	$C_0(S_{0i})$	$W_0(S_{0i})$
S_{01}	S_{01}	S_{02}
S_{02}	S_{02}	S_{01}

Hence, at this time, if α_1 is a male in this population, then his wife must come from class S_{02}.

Let the following two events occur between $t = 0$ and $t = 1$:

(1) α_1 marries α_{j+1} and they have a child, α_{N+1}
(2) α_N dies.

Then at t = 1, we have

$$A_1 = \left\{ \alpha_1, \alpha_2, \ldots, \alpha_{N-1}, \alpha_{N+1} \right\}$$

and since α_{N+1} is a child of a man in S_{01}, α_{N+1} must enter S_{01}. Hence, at t = 1 the classes are:

$$S_{11} = \left\{ \alpha_1, \alpha_2, \ldots, \alpha_j, \alpha_{N+1} \right\}$$
$$S_{12} = \left\{ \alpha_{j+1}, \alpha_{j+2}, \ldots, \alpha_{N-1} \right\}$$

Note that $S_{01} \neq S_{11}$, $S_{02} \neq S_{12}$. Suppose the system (S_1, C_1, W_1) is assumed to satisfy the table:

S_{1i}	$C_1(S_{1i})$	$W_1(S_{1i})$
S_{11}	S_{11}	S_{12}
S_{12}	S_{12}	S_{11}

The classes at t = 0 are extensionally different from those at t = 1. Of course, this means the mappings C_1 and W_1 differ from C_0 and W_0, respectively [check the discussion of definition (5.2.1)]. But there is invariance, which is exhibited explicitly when one notes the correspondence,

$$S_{01} \longmapsto S_{11}$$
$$S_{02} \longmapsto S_{12}$$

which is an isomorphism of pms's. Hence, under demographic processes that alter the concrete relations and, so, the concrete group generated by these relations, the structure of the pms's remains the same. Formally, in the notation of section 5.16,

$$St(S_0, C_0, W_0) = St(S_1, C_1, W_1)$$

because (S_0, C_0, W_0) and (S_1, C_1, W_1) are isomorphic.

Let $\left\{ (S_t, C_t, W_t), t \in T \right\}$ be a "prescribed marriage system process" (pmst) on time domain T: for each $t \in T$, there is a pms at t. We term the structure of the pms at time t—namely, $St(S_t, C_t, W_t)$—the state of the pmst. Let us denote it by X_t. Then, in the notation of Chapter 8,

$$\dot{X}_t = f(X_t, u_t)$$

expresses the abstract form of a dynamic law of the pmst: the rate of change of X_t is some function of the state X_t and the input u_t. However, the rate of change of X_t is not well defined without a measure of the difference between $X_{t+\Delta t}$ and X_t, two strict structures.

This "difference" between two strict structures is not uniquely given by our earlier concepts. We have compared two systems but only in terms of a dichotomy of isomorphism-nonisomorphism. One possibility for the measurement of difference in two strict structures is the utilization of the concept of homomorphism. For, in general, if algebraic system α_1 is a homomorphic image of a system α_2, this means similarity of structure without imposing the necessity of the strict correspondence demanded in the isomorphism concept. This, then, is the essence of the important suggestion made by Boyd (1969) in his rigorous analysis of what White (1963) has called "the anatomy of kinship": to study evolutionary change as a dynamic process in which the transformations in time—the mechanisms—are built around the homomorphism notion.* In this section, we will not be able to develop all of the mathematics needed to show the full importance of Boyd's idea. Enough will be presented in terms compatible with our earlier analyses to give the reader some notion of how Boyd's structural change idea looks in algebraic detail.

To generalize the concept of an isomorphism of two pms's, we simply drop the requirement that the mapping be one-to-one. We allow, for instance, many clans at t = 1 to correspond to one clan at t = 0 (or vice versa). For two abstract systems, consider, for instance:

$\bar{\alpha}$	$C(\bar{\alpha})$	$W(\bar{\alpha})$		$\bar{\alpha}$	$C'(\bar{\alpha})$	$W'(\bar{\alpha})$
1	1	4		A	A	B
2	2	3		B	B	A
3	3	2				
4	4	1				

We see that under the correspondence, call it f,

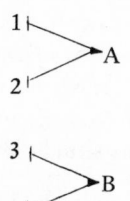

we have structure preservation. For example,

$$C(1) = 1 \quad \text{and} \quad C'[f(1)] = C'(A) = A = f(1)$$
$$W(3) = 2 \quad \text{and} \quad W'[f(3)] = W'(B) = A = f(2)$$

If the numerical system is the structure of (S_t, C_t, W_t) at t = 0 and the alphabetical system is the structure of (S_t, C_t, W_t) at t = 1, then there has been a reduction in size of the system with a maintenance of a similar abstract form. If, on the other hand, we want to make increasing complexity correspond with the forward

---

*From an algebraic standpoint, the mechanism of a linear system (see section 8.6) is also a homomorphism: input time functions are homomorphically mapped into output time functions. See Schwartz and Friedland, 1965.

direction of time, then we take the larger system as (S_1, C_1, W_1) and the smaller as (S_0, C_0, W_0). Then the larger system was already approximated at $t = 0$, where

(S_0, C_0, W_0) approximates (S_1, C_1, W_1) iff there is a homomorphism from (S_1, C_1, W_1) to (S_0, C_0, W_0).

A measure of the difference between the two structures is suggested by taking the ratio of the number of classes.* For instance, in the above illustration, the ratio is 2. A "better" approximation would have had ratio 4/3, with three classes in the homomorphic image. The "best" approximation would be 4/4 = 1, and then we have isomorphism.

A principle of structural change suggested by Boyd (1969) is that (a) change occurs via passage out of approximations—that is, the past and the present are homomorphic—and (b) this occurs in small steps, so that the approximation is always "good" in the small. Note that the methodological rule of setup in the small (see section 8.9) is being used here, where "small" is relative to the time unit for evolutionary analysis.

Example. (b) Given a system described by

$$W^2 = I, \quad C^4 = I, \quad WC = CW$$

we see that this defines a product group $C_2 \times C_4$ (Table 17.7.2). The graph is shown in Figure 19.3.1. This is an eight-clan system. In thinking about how it could have evolved, we use the principle of structural change: we try to locate structures that are homomorphic images of $C_2 \times C_4$. Graphically, we think about merging points to preserve structure. In this case, suppose we merge points connected by the W relation, the solid lines. This means the two squares become one, as in Figure 19.3.2.

Here A corresponds to $\{1, 5\}$, B to $\{2, 6\}$, C to $\{3, 7\}$, and D to $\{4, 8\}$. This correspondence is a homomorphism from $C_2 \times C_4$ onto the new structure. For instance, there is a dotted line from 1 to 2 in the structure $C_2 \times C_4$ and a corresponding dotted

*Boyd's (1969) measure is more sophisticated but would require a very lengthy explanation of the underlying algebraic concepts. The present measure is in the spirit of Boyd's suggestion, not the letter of it.

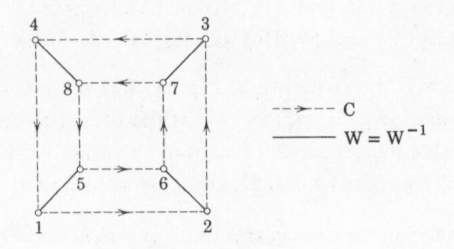

Figure 19.3.1

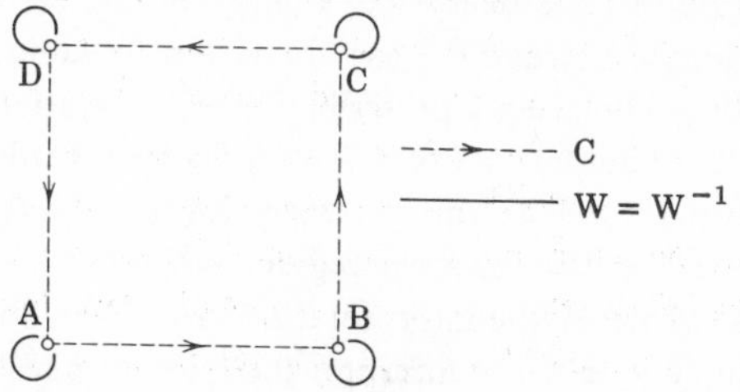

Figure 19.3.2

line from A to B in the "older" structure. Here the evolutionary step involved a binary division of each of four clans, while preserving the "principle of organization" given by $C^4 = I$. To get from the new to the old structure, one changed $W^2 = I$ to $W = I$. Conversely, the evolutionary step involved changing $W = I$ to $W^2 = I$: wives are exchanged rather than obtained from within. But $W = I$ is impossible by axiom S1 for strict pms's. Hence, this evolutionary step must be ruled out. A second possible merging of points is an identification of 5 with 8 and 6 with 7 and then folding over to identify 1 with 4 and 2 with 3. In Figure 19.3.3, A corresponds to $\{5, 8\}$, B to $\{6, 7\}$, C to $\{2, 3\}$, and D to $\{1, 4\}$. This is a homomorphism. For instance, a dotted line goes froms 3 to 4, and a corresponding dotted line goes from C to D. A solid line joins 2 and 6, and a solid line joins C and B. No line joins 4 and 7, and no line joins D and B. The reduced structure is seen to satisfy the defining relations

$$W^2 = I, \quad C^2 = I, \quad WC = CW,$$

which is $C_2 \times C_2$. In this example, we see that this second of the two four-clan systems yielded by "reduction" is a two-approximation (eight clans to four), while the first is not an allowable approximation at all, because while it does satisfy the homomorphism condition, it is ruled out on grounds of violating an axiom. We have not attempted to list every possible approximation here. A more extensive listing and corresponding measurement discussion appears in Boyd's (1969) paper.

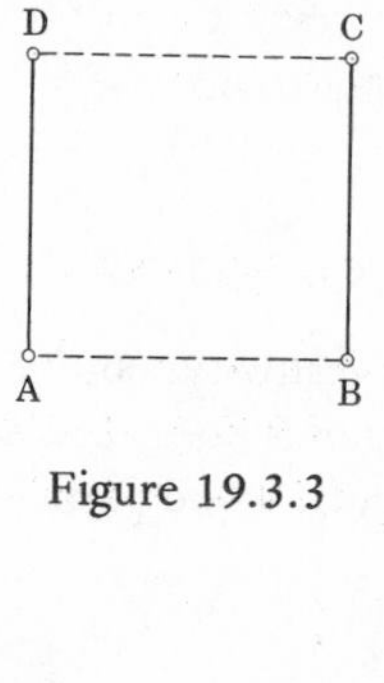

Figure 19.3.3

19.4. Construction of Classificatory Systems. Before leaving the kinship context behind us to explore less-structured social relational systems, it is worth investigating how we might construct a prescribed marriage system from a social relational system over the set of actors. In other words, the basis for a prescribed marriage system may be analyzed. If we do this in a formal way, even the kinship context is not strictly necessary in the sense that the argument shows how conditions on an abstract relational system, whatever the interpretation, lead to an abstract prescribed marriage system. For example, one might interpret the relational system in terms of formal organizations: then the work leads to conditions under which such organizations have an aspect to which the theory of strict prescribed marriage systems (S, C, W) applies. In this section, we draw upon Chapter 1 of the White (1963) monograph, in which he outlines the logic of kin-role trees. In our treatment, the emphasis is on the formal generation of the classes of actors (with appropriate relations on classes).

Recall that S is a partition of a set A of actors. Hence, concrete relations on the set A must form the real foundation for the mappings relating the classes in S. Put another way, S must be a quotient system based on some equivalence relation on the set A. Mathematically, we can think of the problem in terms of starting from a relational system with domain A and constructing (S, C, W). Then, if the constructed pms is strict, the previous theory applies; but its strictness should be a consequence of certain conditions on the relational system. The relation of equivalence on A might be thought of as an "induced alter relation" on the set of actors. To make this concrete, let $(A, K_1, K_2, \ldots, K_n)$ be the relational system, where K_i is thought of as a primitive "kinship" relation on A. Formally, K_i is a binary relation on A, for every i.

It will be convenient at this point to denote actors in A by a, b, c, and so on.

Define a specific alter relation on A as follows:

$$a\alpha_x^i b \quad \text{iff} \quad xK_i a \text{ and } xK_i b$$

Here x is thought of as ego, and a and b are two "nodes" in his "kinship role tree" (see White, 1963, Chap. 1). For instance, taking $K_1 \equiv$ "father of" on A and suposing x is father of both a and b, we have $xK_1 a$ and $xK_1 b$, which implies $a\alpha_x^1 b$: the siblings are alters. As another example, let $K_2 \equiv$ "husband of" and suppose that x has two wives, a and b. Then a and b are in relation $a\alpha_x^2 b$.

It is easy to show that for any x and i, the relation α_x^i is symmetric and transitive on A. For symmetry note that

$$xK_i a \quad \text{and} \quad xK_i b$$

is tautologically equivalent to

$$xK_i b \quad \text{and} \quad xK_i a$$

Hence, if $a\alpha_x^i b$, then $b\alpha_x^i a$. For transitivity, note that if $a\alpha_x^i b$ and $b\alpha_x^i c$, then $xK_i a$ and $xK_i c$, and so, $a\alpha_x^i c$. The relation is not necessarily reflexive, however, since we need not have $xK_i a$ for the particular x and particular K_i.

Now define the relation on A,

$$a\alpha b \qquad \text{iff} \qquad a\alpha_x^i b \qquad \text{(for some } x \in A, \text{ some } i = 1, 2, \ldots, m)$$

This is an alter relation, no longer specific to a given ego x or relation K_i. It is symmetric because α_x^i is symmetric, but it is not necessarily reflexive and transitive.

For reflexivity of α, we require that $a\alpha_x^i a$, for some x and i, or equivalently, $xK_i a$. Hence, the axiom that everyone is kin to someone assures α is reflexive (call this axiom 1).

Transitivity can be obtained if we assume the following axiom (call it axiom 2): $a\alpha_x^i b$ implies,

$$\text{if } yK_j a, \text{ then } yK_j b, \text{ and if } aK_j y, \text{ then } bK_j y \qquad \text{(any } y \in A, \text{ any } j)$$

The axiom says that if a and b are specific alters with respect to x and K_i, and so in the alter relation α, then anyone (y) related to a in some kinship relation is related to b in the same way.

To prove transitivity of α, suppose that $a\alpha b$ and $b\alpha c$. Then, for some x and y in A,

$$a\alpha_x^i b \quad \text{and} \quad b\alpha_y^j c$$

This implies $yK_j b$. But axiom 2 implies that if $a\alpha_x^i b$ and $yK_j b$, then $a\alpha_y^j b$. Since α_y^j is transitive,

$$a\alpha_y^j b \quad \text{and} \quad b\alpha_y^j c \quad \text{imply} \quad a\alpha_y^j c$$

and so, $a\alpha c$, which was to be shown. Hence, α is an equivalence relation on A. We define

$$\bar{a} = \{b \in A : b\alpha a\}$$

to be the typical equivalence class of alters in A. It is an element of A/α.

To endow A/α with algebraic structure, we need to define the kinship relations between these classes of actors. A natural definition is

$$\bar{a}\bar{K_i}\bar{b} \qquad \text{iff} \qquad aK_i b$$

This needs justification, for if we replace $a \in \bar{a}$ by $a' \in \bar{a}$ and $b \in \bar{b}$ by $b' \in \bar{b}$, we obtain

$$\bar{a}\bar{K_i}\bar{b} \qquad \text{iff} \qquad a'K_i b'$$

Hence, the definition makes sense if and only if

$$aK_i b \qquad \text{iff} \qquad a'K_i b'$$

where $a\alpha a'$ and $b\alpha b'$. But axiom 2 assumes this is so. It follows that the quotient relational system

$$(A/\alpha, \bar{K}_1, \bar{K}_2, \ldots, \bar{K}_m)$$

is "classificatory." Alters, individuals in the same class, are structurally interchangeable throughout the network.

If m = 2, we see that $(A/\alpha, \bar{K}_1, \bar{K}_2)$ is a potential prescribed marriage system, in the sense of the formal definition. The system satisfies the definition if $\bar{K}_1$ and $\bar{K}_2$ are not only relations but injective mappings with domain A/α.

But each $\bar{K}_i$ is a mapping, for suppose that $a\bar{K}_i x$ and $a\bar{K}_i y$ hold. Then $aK_i x$ and $aK_i y$ obtain, so that, by definition, x and y are specific alters, $x\alpha_a^i y$. Hence, $x\alpha y$, and so, $\bar{x} = \bar{y}$. Thus,

$$a\bar{K}_i x \text{ and } a\bar{K}_i y \Rightarrow \bar{x} = \bar{y}$$

yields the result that $\bar{K}_i$ is a mapping on A/α. We write $\bar{K}_i(\bar{a})$ for $a\bar{K}_i y$, since $\bar{y}$ is unique.

To be a permutation of A/α, mapping $\bar{K}_i$ needs to satisfy

$$\bar{a} \neq \bar{b} \Rightarrow \bar{K}_i(\bar{a}) \neq \bar{K}_i(\bar{b})$$

or contrapositively,

$$\bar{K}_i(\bar{a}) = \bar{K}_i(\bar{b}) \Rightarrow \bar{a} = \bar{b}$$

Now, if $\bar{K}_i(\bar{a}) = \bar{K}_i(\bar{b})$, then there is a class, say $\bar{y}$, such that

$$a\bar{K}_i \bar{y} \text{ and } b\bar{K}_i \bar{y}$$

and so,

$$aK_i y \text{ and } bK_i y$$

Hence, this condition, for all a, b and y in A, must imply that $a\alpha b$ (and so $\bar{a} = \bar{b}$) for the mapping $\bar{K}_i$ to be a permutation. But this is the same as $yK_i^{-1}a$ and $yK_i^{-1}b$, and this yields $a\alpha b$ provided that we extend the definition of the alter relation so that its basis is the K_i and their inverses K_i^{-1}. As we shall see in section 19.6, we have moved toward a generic concept of "structural equivalence," formally defined in (19.6.1).

To sum up, given $(A, K_1, K_2, \ldots, K_m)$, an arbitrary relational system, we induce $(A/\alpha, \bar{K}_1, \bar{K}_2, \ldots, \bar{K}_m)$ in which each $\bar{K}_i$ is a permutation of A/α, provided that axioms 1 and 2 are satisfied and the alter relation α is based on the K_i and the K_i^{-1} relations.

In the special case m = 2, the system $(A/\alpha, \bar{K}_1, \bar{K}_2)$ is a prescribed marriage system in the formal sense of definition (18.2.1), with

$$S = A/\alpha$$
$$C = \bar{K}_1$$
$$W = \bar{K}_2$$

A different treatment of this problem is given by Boyd (1969), who relates it to balance theory. We recall that the structure theorem (14.2.3) says that a balanced graph is one in which the points may be partitioned into two disjoint classes such that the negative relation is between the classes and the positive relation is within classes.

Now it is clear that the relational systems involved in balance theory are as follows. We start from

$$(A, P, N)$$

where A is finite, P and N are binary relations on A, and $P \cap N = \emptyset$. The induced system, provided that (A, P, N) is balanced, is given by $(\bar{A}, \bar{P}, \bar{N})$ where $\bar{A}$ contains two classes, $\bar{P}$ is the identity relation on the classes, and $\bar{N}$ relates the classes to each other.

Boyd's contribution is this: starting from a very general algebraic rendition of the idea of a system of relations between actors and the induced algebraic system of these relations, he proves a general theorem called the group partition theorem. Essentially, it states equivalent conditions on the relational system that guarantee a classificatory system can be formed in which a group theoretic representation is valid. Then the structure theorem is proved as a special case of the group partition theorem, just as $(A/\alpha, \bar{K}_1, \bar{K}_2, \ldots, \bar{K}_m)$ encompasses the (S, C, W) as a special case. The reader may wonder where the group representation is in balance theory. The answer is simple, for the table of sign multiplication

	+	−
+	+	−
−	−	+

is nothing more than a group—namely, a particular interpretation of C_2, the cyclic group of order two. Note that + is the identity of the group. Representing the relational system $(\bar{A}, \bar{P}, \bar{N})$ as a graph, we have Figure 19.4.1. This is the (S, C, W) with $S = \bar{A}$, $C = \bar{P}$, and $W = \bar{N}$, a prescribed marriage system in the formal sense of definition (18.2.1), satisfying $C = I$, $W^2 = I$. Of course, $W^2 = I$ corresponds to two negatives making a positive. Since C_2 in this form represents a bilateral system in White's typology, we have a formal correspondence between such a system of wife exchange between two clans and a balanced structure in any graph of the type for which the structure theorem holds. Note, of course, that this does not mean that wife exchange is a "negative" relationship in the sense of sentiment.

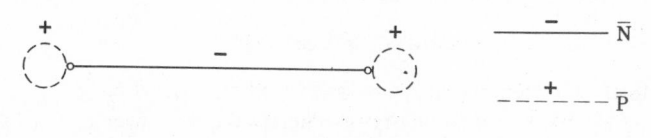

Figure 19.4.1

19.5. Category Theory. A more recent development in the analysis of relational systems employs "category theory." This approach is outlined in a paper by Lorrain and White (1971). The best way to understand the genesis of the concept of category is via the analogy:

$$
\begin{array}{ccc}
\text{Category} & \text{is to} & \text{Abstract} \\
 & & \text{Algebraic System}
\end{array}
$$

as

$$
\begin{array}{ccc}
\text{Abstract} & \text{is to} & \text{Concrete} \\
\text{Algebraic System} & & \text{Phenomena}
\end{array}
$$

In other words, an algebraic system involves some abstraction from phenomena to represent some pattern ingredient in recurrences of the phenomena. The same algebraic system may then have an enormous number of concrete sources for its abstract postulates. Similarly, a category is a mathematical system that is instantiated in any abstract algebraic system. In fact, the term "algebraic" is too narrow. It is better to say that any mathematical system may be understood as defining a category.

For example, in treating algebraic groups as a category, we first form the space of all groups and then note a relation exists on this space—namely, homomorphism. Note that this "category of groups" has groups as "units," related to each other via homomorphisms. In category theory, the latter form an example of what are called morphisms.

The abstract concept of a category $\mathcal{C}$ is given by three primitive notions (see Lang, 1965, and Cohn, 1965):

(1) A space of objects, denoted $Ob(\mathcal{C})$.

(2) A family of sets of morphisms: for each pair of objects in $Ob(\mathcal{C})$, say A and B, a set Mor (A, B) called the morphisms from A to B. For instance, if $\mathcal{C}$ is the class of groups, then

$$Ob(\mathcal{C}) = \text{class of all groups}$$

and if G_1 and G_2 are groups, then they are objects in $Ob(\mathcal{C})$ and

$$Mor(G_1, G_2) = \text{set of all homomorphisms of } G_1 \text{ into } G_2$$

(3) For any three objects A, B, and C in $Ob(\mathcal{C})$, a mapping with domain

$$Mor(B, C) \times Mor(A, B)$$

and range Mor(A, C). Hence, to each morphism pair (g, f), where $g \in Mor(B, C)$, $f \in Mor(A, B)$, we have a unique morphism denoted $g \circ f \in Mor(A, C)$. This mapping will be called "the law of composition" of the category.

For instance, the map from $\text{Mor}(G_2, G_3) \times \text{Mor}(G_1, G_2)$ to $\text{Mor}(G_1, G_3)$ in the category of groups can be taken to be composition of homomorphisms. Thus, if $g \in \text{Mor}(G_2, G_3)$, $f \in \text{Mor}(G_1, G_2)$, the situation is

$$G_1 \xrightarrow{\ f\ } G_2 \xrightarrow{\ g\ } G_3$$

and so $g \circ f$ is a homomorphism of G_1 into G_3 —that is, $g \circ f \in \text{Mor}(G_1, G_3)$.

These three primitive notions are controlled by three axioms:

Axiom C1.   Each morphism is in at least one* set $\text{Mor}(A, B)$, where $A, B \in \text{Ob}(\mathcal{C})$.

Axiom C2.   For each object A of $\mathcal{C}$ there is a morphism, denoted I_A, in $\text{Mor}(A, A)$ that acts as left and right identity for the morphisms of $\text{Mor}(A, B)$ and $\text{Mor}(B, A)$, respectively, for all objects B in $\text{Ob}(\mathcal{C})$.

For example, take a group G and a mapping of the form $x \longmapsto x$, all $x \in G$. Then I_G is this map for G, since

$$G \xrightarrow{\ I_G\ } G \xrightarrow{\ f\ } G'$$

yields the homomorphism, $f \circ I_G = f \colon G \longrightarrow G'$. So I_G is a left identity for the set $\text{Mor}(G, G')$. Also,

$$G' \xrightarrow{\ f\ } G \xrightarrow{\ I_G\ } G$$

yields the homomorphism $I_G \circ f = f \colon G' \longrightarrow G$. Hence, I_G is a right identity for the set $\text{Mor}(G', G)$, for any G' in the category (i.e., any group).

Axiom C3.   The law of composition is associative when defined—that is, given $f \in \text{Mor}(A, B)$, $g \in \text{Mor}(B, C)$, and $h \in \text{Mor}(C, D)$. Then

$$(h \circ g) \circ f = h \circ (g \circ f)$$

for all objects of $\mathcal{C}$.

A morphism of a category is sometimes termed an arrow of $\mathcal{C}$. For instance, any homomorphism of groups in an arrow of the category of groups. We write

$$\text{Ar}(\mathcal{C}) = \text{class of all arrows of } \mathcal{C}$$

Thus, a category has three fundamental entities: objects, arrows, and a law (of composition).

---

*This modifies the standard axiom given in Lang (1965) and Cohn (1965) to the effect that each morphism is in exactly one set of morphisms; a discussion of the modification is in the Lorrain-White paper. Later, still another modification, introduced by Lorrain and White, will be imposed.

Example.   (a) A relatively trivial example of a category is given by

$$Ob(\mathcal{C}) = \text{class of all sets}$$
$$Ar(\mathcal{C}) = \text{mappings}$$
$$\text{Law} = \text{composition of mappings.}$$

Let $\text{Mor}(A, B) = \{f: A \xrightarrow{f} B\}$, $\text{Mor}(B, C) = \{g: B \xrightarrow{g} C\}$, and $\text{Mor}(C, D) = \{h: C \xrightarrow{h} D\}$. Hence, if f, g, and h are in the respective morphism sets, we have the arrow diagram of mappings:

$$A \xrightarrow{f} B \xrightarrow{g} C \xrightarrow{h} D$$

and we know (see section 5.13) that the law of composition of mappings is associative. Hence, axiom C3 is satisfied. The mapping $x \longmapsto x$ ($x \in A$) on set A is in $\text{Mor}(A, A)$ and constitutes an identity in the sense of axiom C2. Finally, there is no mapping without domain and range and so every arrow f is in some set $\text{Mor}(A, B)$. Hence, axiom C1 is satisfied. This is termed the category of sets.

Just as homomorphisms relate algebraic systems, so there are mappings between categories, called functors. An analogy is as follows:

Functors   are to   Categories

as

Group
Homomorphisms   are to   Groups

Hence, let $\mathcal{C}_1$ and $\mathcal{C}_2$ be two categories containing objects and arrows—that is, $Ob(\mathcal{C}_i)$ and $Ar(C_i)$, $i = 1, 2$.

By a functor F of $\mathcal{C}_1$ into $\mathcal{C}_2$ we mean a mapping that to each object A in $\mathcal{C}_1$ associates an object $F(A)$ in $\mathcal{C}_2$ and to each morphism $f \in \text{Mor}(A, B)$ associates a morphism $F(f) \in \text{Mor}[F(A), F(B)]$, such that:

Axiom F1.   For all A in $\mathcal{C}_1$, $F(I_A) = I_{F(A)}$

Axiom F2.   If $f \in \text{Mor}(A, B)$, $g \in \text{Mor}(B, C)$, where A, B, C are any objects in $\mathcal{C}_1$, then

$$F(g \circ f) = F(g) \circ F(f)$$

Essentially, the axioms say that there is a correspondence of objects and arrows that preserves the law of the category.

Example.   (b) Let $\mathcal{C}_1$ be the category of groups and $\mathcal{C}_2$ be the category of sets:

	$\mathcal{C}_1$	$\mathcal{C}_2$
Objects:	groups	sets
Arrows:	homomorphisms	mappings
Law:	composition	composition

A functor F is defined from $\mathcal{C}_1$ to $\mathcal{C}_2$, as follows:

$$
\begin{array}{ccc}
(G, o) & \xrightarrow{\ f\ } & (G', o') \\
\downarrow & \downarrow & \downarrow \\
F(G, o) & \xrightarrow{\ \ \ \ } & F(G', o') \\
& F(f) & \\
\backslash\backslash & \parallel & // \\
G & \xrightarrow{\ f\ } & G'
\end{array}
$$

$$\left(\begin{array}{l} (G, o)\ \text{and}\ (G', o') \in \mathrm{Ob}(\mathcal{C}_1) \\ f \in \mathrm{Ar}(\mathcal{C}_1) \end{array} \right)$$

$$\left(G, G' \in \mathrm{Ob}(\mathcal{C}_2) \qquad f \in \mathrm{Ar}(\mathcal{C}_2) \right)$$

In other words, this functor F simply "strips" the groups of their structure, keeping the mere set of elements. Then is simply keeps the mapping aspect of the homomorphisms, and there is no longer a structure to be preserved by these maps. To check the two axioms, we note that the identity homomorphism $I: (G, o) \to (G, o)$ (where $x \mapsto x$, all $x \in G$) becomes the identity function on G. Hence, axiom F1 is satisfied. Axiom F2 is trivially satisfied in this instance because $F(f) = f$, any morphism in $\mathcal{C}_1$, so that

$$F(g \circ f) = g \circ f = F(g) \circ F(f)$$

All of the above interpretations of the concept of category rely upon the genesis of the concept in purely mathematical contexts. The next example is from Fararo (1968) and is more sociological in character.

Example. (c) In (5.7.3) we defined the concept of a status chain. The typical such chain is a sequence of mappings

$$A \xrightarrow{\ c\ } C \xrightarrow{\ \sigma\ } C/I \xrightarrow{\ v\ } \Re$$

where σ is the natural mapping of state x in C into equivalence class $\bar{x}$ in C/I and v is an order-preserving mapping of C/I into the real line. Recall here that C/I is really $(C/I, P^*)$, where P^* linearly orders C/I.

Since the composition $v \circ \sigma \circ c$ defines a status mapping, according to section 5.7, we denote the typical chain by s.

We define the objects of a category of status chains on A, denoted $\mathcal{C}_A$, as follows: Let $\mathrm{Ob}(\mathcal{C}_A)$ be the space of all status chains over set A (of actors). This space is determined by the axioms for the concept.

To specify our morphisms, we proceed as follows: Let s_1 and s_2 be two status chains over A:

$$s_1: \quad A \xrightarrow{\ c_1\ } C_1 \xrightarrow{\ \sigma_1\ } C_1/I_1 \xrightarrow{\ v_1\ } \Re$$

$$s_2: \quad A \xrightarrow{\ c_2\ } C_2 \xrightarrow{\ \sigma_2\ } C_2/I_2 \xrightarrow{\ v_2\ } \Re$$

We define an expectation map from s_1 to s_2 as a relational homomorphism from

C_1/I_1 to C_2/I_2; most typically the map is an order-preserving association of states in C_2/I_2 with states in C_1/I_1 ("the higher you are here in C_1, the higher you are there in C_2"). We write e: $s_1 \longrightarrow s_2$ and take

$$\mathrm{Ar}(e_A) = \text{the expectation maps}$$

Of course, our law is composition.

We verify that to each pair of objects (s_1, s_2) in $\mathrm{Ob}(e_A)$ there is a set $\mathrm{Mor}(s_1, s_2)$ containing expectation maps and that the law of composition

$$\mathrm{Mor}(s_2, s_3) \times \mathrm{Mor}(s_1, s_2) \longrightarrow \mathrm{Mor}(s_1, s_3)$$

is well defined, for if e: $s_1 \longrightarrow s_2$ and f: $s_2 \longrightarrow s_3$, then

$$C_1/I_1 \xrightarrow{\ e\ } C_2/I_2 \xrightarrow{\ f\ } C_3/I_3$$

yields a composition,

$$s_1 \xrightarrow{\ e\ } s_2 \xrightarrow{\ f\ } s_3$$

which is f o e: $s_1 \longrightarrow s_3$.

We now verify that the three axioms are satisfied. Axiom C1 is satisfied because morphism e is a mapping with domain and range two chains of the category; that is, e is in $\mathrm{Mor}(s_1, s_2)$ for some s_1, s_2.

Axiom C2 is satisfied by taking I_s to be the identity mapping on C/I in chain

$$s: \quad A \xrightarrow{\ c\ } C \xrightarrow{\ \sigma\ } C/I \xrightarrow{\ v\ } \mathfrak{R}$$

Then, under compositions of I_s with other expectations, it serves as a left and right identity; this follows from the logic of mappings.

Finally, since mapping composition is associative, the expectation-map composition is associative:

$$g \, o \, (f \, o \, e) = (g \, o \, f) \, o \, e$$

where e: $s_1 \longrightarrow s_2$, f: $s_2 \longrightarrow s_3$, and g: $s_3 \longrightarrow s_4$.

We conclude that the space of status chains over A forms a category e_A whose morphisms are expectation maps.

We see that the mapping diagram

$$
\begin{array}{ccc}
C_1/I_1 & \xdashrightarrow{\ v_1\ } & \mathfrak{R} \\
\Big\downarrow{\scriptstyle e} & & \Big\downarrow \\
C_2/I_2 & \xdashrightarrow{\ v_2\ } & \mathfrak{R}
\end{array}
$$

yields a real-valued function that represents e numerically; namely, choose the function by inverting v_1 over its range in $\mathfrak{R}$, apply e, then v_2. That is, the required function is $v_2 \circ e \circ v_1^{-1}$. Call this function E. The compositions along each chain yield, using the same notation for the composite function as for the chain,

$$s_1 = v_1 \circ \sigma_1 \circ c_1$$
$$s_2 = v_2 \circ \sigma_2 \circ c_2$$

We want a condition on the morphism $e: s_1 \longrightarrow s_2$ that yields

$$s_2 = E(s_1)$$

for the functions s_1 and s_2 obtained by composition. In fact, suppose e "fits" the pair of chains, in the sense that the diagram

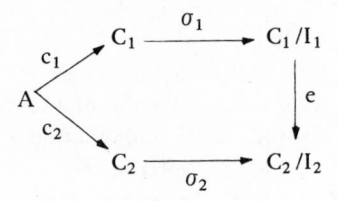

is commutative:

(19.5.1) $$e \circ \sigma_1 \circ c_1 = \sigma_2 \circ c_2$$

This means an actor in A applying e will find that his "model" (based on C_1) fits the situation (as to C_2). Then,

$$s_2 = v_2 \circ \sigma_2 \circ c_2$$
(19.5.2) $$= v_2 \circ e \circ \sigma_1 \circ c_1 \qquad \text{(using 19.5.1)}$$

Now also

$$E = v_2 \circ e \circ v_1^{-1} \qquad \text{(definition)}$$

so that

(19.5.3) $$E \circ v_1 = v_2 \circ e$$

Hence, substituting (19.5.3) into (19.5.2),

$$s_2 = E \circ v_1 \circ \sigma_1 \circ c_1 = E \circ s_1 = E(s_1)$$

We see that the numerical monotone functional relation $s_2 = E(s_1)$ represents the commutative condition on the expectations that link two chains. Other topics in status chain algebra are studied in Fararo (1968, 1972b).

19.6. The Skeletal Structure of Social Relational Systems. We now indicate the method of analysis of a general social relational system $(A, R_1, R_2, \ldots, R_m)$ proposed by Lorrain and White (1971). First they construct the semigroup of relations generated by the set $\{R_1, R_2, \ldots, R_m\}$. (The reader will find a review of section

4.16 useful at this point.) This means that the space of relations also has R_1^{-1}, R_2^{-1}, ..., R_m^{-1} and any compositions of these relations. Since relation composition is associative, by (4.16.2), and all relations may be composed because they have a common domain A, the space of relations so generated forms a semigroup [recall definition (5.13.5)].

Example.   (a)   Let relations R_1 and R_2 be defined on A by

$$xR_1y \quad \text{iff} \quad x \text{ is a friend of } y$$
$$xR_2y \quad \text{iff} \quad x \text{ is a superior of } y$$

Then the semigroup of relations generated by $\{R_1, R_2\}$ is induced by relational compositions:

$$R_1R_2 = R_1 \text{ o } R_2 \quad = \text{ "friend of a superior of"}$$
$$R_2R_1 = R_2 \text{ o } R_1 \quad = \text{ "superior of a friend of"}$$
$$R_1^2 = R_1 \text{ o } R_1 \quad = \text{ "friend of a friend of"}$$
$$R_2^{-1} = \text{ "inferior of"}$$
$$R_1R_2^{-1} = R_1 \text{ o } R_2^{-1} = \text{ "friend of inferior of"}$$
$$\begin{matrix} \cdot & & \cdot \\ \cdot & & \cdot \\ \cdot & & \cdot \end{matrix}$$

Here one may recall from section 4.16 that for any relation R, the relation R^{-1} is defined by

$$xR^{-1}y \quad \text{iff} \quad yRx$$

One should recall that despite the notation, it is not generally true that $R^{-1}R = I$ or $RR^{-1} = I$. Also, if R is symmetric, then $R^{-1} = R$.

The semigroup, call it S_A, is generated in this manner by $\{R_1, R_2, \ldots, R_m\}$. It is finite because there are only a finite number of binary relations on a set. (If A contains n actors, then since a relation is a subset of A $\times$ A, there are as many relations as there are subsets of a set of size n^2. This is 2^{n^2}.) In any case, "words" formed in S_A are thereby naturally equivalent in the case that they name the same set of ordered pairs. We note, then, that each element of S_A corresponds to a graph on nodes representing the actors in A. A typical generated relation may be denoted W, and then (A, W) has a graph representation. Of course, it also has an adjacency matrix A_W. (Recall section 6.4.)

Given S_A, we try to define an associated category,* denoted e_A, by identifying the three primitive notions of category theory as follows:

---

*A problem that arises in this attempt leads to a modified category concept, defined at the end of example (b) below.

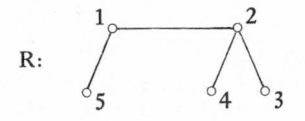

Figure 19.6.1

(1) $Ob(e_A)$ = actors in A = nodes of graphs on A
(2) $Mor(x, y) = \{ R \in S_A : xRy \}$ = graphs with a link from x to y
(3) Law of composition = relation composition

Example. (b) Consider the graph of a symmetric relation R on a set A, shown in Figure 19.6.1. Let us construct the semigroup of relations on A generated by R alone. Certainly, R is in S_A. We compute $R^2 = R \circ R$, using matrices.

Recall from section 6.4 that $A_{R \circ R}$ is obtained by matrix multiplication, row by column of A_R with itself, but the summations are Boolean. Hence,

$$A_{R \circ R} = \begin{pmatrix} 0 & 1 & 0 & 0 & 1 \\ 1 & 0 & 1 & 1 & 0 \\ 0 & 1 & 0 & 0 & 0 \\ 0 & 1 & 0 & 0 & 0 \\ 1 & 0 & 0 & 0 & 0 \end{pmatrix} \begin{pmatrix} 0 & 1 & 0 & 0 & 1 \\ 1 & 0 & 1 & 1 & 0 \\ 0 & 1 & 0 & 0 & 0 \\ 0 & 1 & 0 & 0 & 0 \\ 1 & 0 & 0 & 0 & 0 \end{pmatrix} = \begin{pmatrix} 1 & 0 & 1 & 1 & 0 \\ 0 & 1 & 0 & 0 & 1 \\ 1 & 0 & 1 & 1 & 0 \\ 1 & 0 & 1 & 1 & 0 \\ 0 & 1 & 0 & 0 & 1 \end{pmatrix}$$

Using the adjacency matrix we plot the graph of R^2 in Figure 19.6.2. Next consider $R^3 = R^2 R$, so that it contains those (x, y) such that xR^2z and zRy, for some z. We use matrices (details are not shown) to obtain the graph of R^3 given in Figure 19.6.3. Then $R^4 = R^2$, so $R^5 = R^3$, $R^6 = R^2$, ... In short, the relations generated by R are: R, R^2, R^3, and the table of the semigroup S_A is

	R	R^2	R^3
R	R^2	R^3	R^2
R^2	R^3	R^2	R^3
R^3	R^2	R^3	R^2

S_A:

Note that there is no identity element in S_A: no row of the table is the array of column headings. It is a groupoid with an associative operation of relational

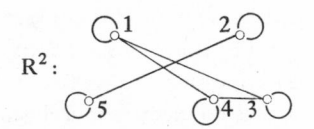

Figure 19.6.2

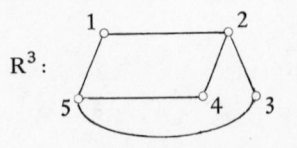

Figure 19.6.3

composition. The category e_A may be explicitly constructed. To obtain Mor(x, y), we look at each graph and ask if there is a link from x to y.

$$Ob(e_A) = \{1, 2, 3, 4, 5\}$$
$$Mor(1, 1) = \{R^2\}$$
$$Mor(1, 2) = \{R, R^3\}$$
$$Mor(1, 3) = \{R^2\}$$
$$Mor(1, 4) = \{R^2\}$$
$$Mor(1, 5) = \{R, R^3\}$$
$$Mor(2, 2) = \{R^2\}$$
$$Mor(2, 3) = \{R, R^3\}$$
$$Mor(2, 4) = \{R, R^3\}$$
$$Mor(2, 5) = \{R^2\}$$
$$Mor(3, 3) = \{R^2\}$$
$$Mor(3, 4) = \{R^2\}$$
$$Mor(3, 5) = \{R^3\}$$
$$Mor(4, 4) = \{R^2\}$$
$$Mor(4, 5) = \{R^3\}$$
$$Mor(5, 5) = \{R^2\}$$

And Mor(y, x) = Mor(x, y) because R is symmetric. There are only three distinct sets of morphisms. Hence, the category is given by

$$Ob(e_A) = \{1, 2, 3, 4, 5\}$$
$$Mor(1, 2) = Mor(1, 5) = Mor(2, 3) = Mor(2, 4) = \{R, R^3\}$$
$$Mor(3, 5) = Mor(4, 5) = \{R^3\}$$
$$Mor(x, y) = \{R^2\} \qquad \text{(all other x, y)}$$

The above example makes it clear that axiom C1 for the category concept is satisfied, since each morphism (relation in the semigroup that is not the empty relation) is in at least one of the morphism sets. Also, axiom C3 is satisfied because relation composition is associative. However, axiom C2 is not satisfied. For example, R^2 is the only morphism in Mor(1, 1). For R^2 to be an identity morphism, we require, in the particular case of Mor(1, 2), for instance, that $RR^2 = R$, since R is in Mor(1, 2). But $RR^2 = R^3 \neq R$. Hence, there is no identity morphism in Mor(1, 1).

In general, in applying the category concept to the relations generated by some primitive set of relations, it will be found that this axiom is not satisfied. One could

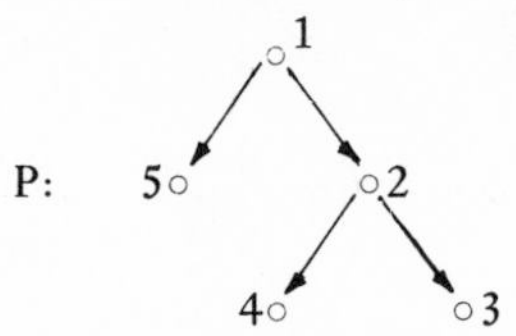

P:

Figure 19.6.4

add a "personal identity" relation in passing from the semigroup to the category: this would make axiom C2 true by the convention that every node is just itself. Formally, this puts I into the set of morphisms of the category even if I is not generated in the semigroup. A second strategy is that used by Lorrain and White: they simply drop axiom C2. To continue our exposition in connection with their work, we now stipulate that henceforward "category" means the system of entities satisfying only axioms C1 and C3. The corresponding change in the concept of functor is that of dropping axiom F1.

The next example is much more complex and a complete listing will not be given (for details on this example, see Lorrain and White, 1971).

Example. (c) Consider the graph in Figure 19.6.4. The inverse relation P^{-1} just inverts the lines. The matrix $A_{P^{-1}}$ is given by A_P^T, the transpose of A_P. Hence, to compute $P^{-1}P$, we have

$$A_{P^{-1}P} = \begin{pmatrix} 0 & 0 & 0 & 0 & 0 \\ 1 & 0 & 0 & 0 & 0 \\ 0 & 1 & 0 & 0 & 0 \\ 0 & 1 & 0 & 0 & 0 \\ 1 & 0 & 0 & 0 & 0 \end{pmatrix} \begin{pmatrix} 0 & 1 & 0 & 0 & 1 \\ 0 & 0 & 1 & 1 & 0 \\ 0 & 0 & 0 & 0 & 0 \\ 0 & 0 & 0 & 0 & 0 \\ 0 & 0 & 0 & 0 & 0 \end{pmatrix} = \begin{pmatrix} 0 & 0 & 0 & 0 & 0 \\ 0 & 1 & 0 & 0 & 1 \\ 0 & 0 & 1 & 1 & 0 \\ 0 & 0 & 1 & 1 & 0 \\ 0 & 1 & 0 & 0 & 1 \end{pmatrix}$$

Plotting the graph of $P^{-1}P$ from the matrix, we have Figure 19.6.5. Similarly, the two graphs in Figure 19.6.6 can be verified. Each such graph represents an element of the semigroup S_A. Now consider a specified pair, say (2, 5), and the content of Mor(2, 5). Into Mor(2, 5) we put all relations containing (2, 5). For example, the

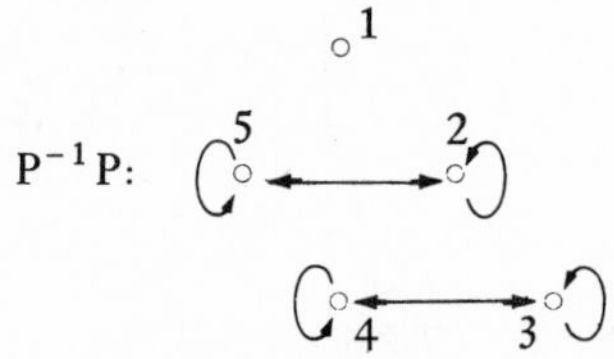

$P^{-1}P$:

Figure 19.6.5

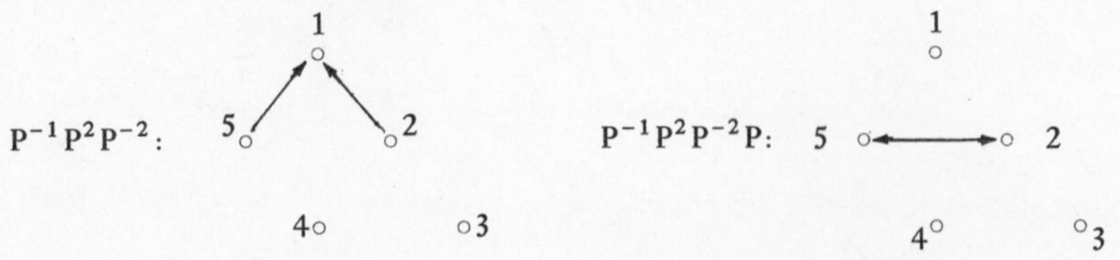

Figure 19.6.6

graphs indicate that $(2, 5) \in P^{-1}P$, but $(2, 5) \notin P$. Hence, $P^{-1}P \in \text{Mor}(2, 5)$, but $P \notin \text{Mor}(2, 5)$.

To return to our general discussion of the identification process: the objects are the nodes (in A), the morphisms are the relations (in S_A) and the law of composition is relation composition. (But $\emptyset$ is not a morphism even if it appears in S_A.)

Next, to analyze the network $(A, R_1, R_2, \ldots, R_m)$, the category is "functorially reduced." This is an idea that is based on the same group-theoretic intuition as developed by Boyd (1969) in his analysis of structures: an approximation of a system is given by a homomorphic image of that system. In category theory, "functor" corresponds to "homomorphism." Hence, we think of mapping $\mathcal{C}_A$ into a "smaller" category that approximates it in the sense of preserving structure. (It is this strong analogy that led us to include section 19.2 on homomorphisms.)

In particular, we seek a quotient system based on an equivalence relation on A determined by the morphisms (i.e., the relational patterns linking actors). We define, for a, b $\in$ A,

$$(19.6.1) \qquad a \equiv_s b \quad \text{iff} \quad \left\{ \begin{array}{l} aRx \;\Leftrightarrow\; bRx \\ xRa \;\Leftrightarrow\; xRb \end{array} \right\} \quad \text{for any R in } S_A, \text{ any x in A.}$$

That is, a and b are structurally equivalent if and only if for any relation generated by the primitive set of generating relations, $R_1, R_2, \ldots, R_m$, any actor R-related to one of a and b is R-related to the other. It is easily shown that in fact this is an equivalence relation. In terms of morphisms, the definition implies:

(19.6.2) Proposition. $\qquad a \equiv_s b \quad \text{iff} \quad \left\{ \begin{array}{l} \text{Mor}(a, x) = \text{Mor}(b, x) \\ \text{Mor}(x, a) = \text{Mor}(x, b) \end{array} \right\} \quad \text{for all x in A.}$

The system given by

Objects: classes in: $\text{Ob}(\mathcal{C}_A)/\equiv_s$
Arrows: $\text{Mor}(\bar{a}, \bar{b})$; $\bar{a}, \bar{b}$ classes in $\text{Ob}(\mathcal{C}_A)/\equiv_s$
Law: relation composition

is termed the skeleton of e_A and denoted SK_A. Hence,

$$SK_A = e_A / \equiv_s$$

In SK_A points have merged into classes based on structural equivalence. The skeleton of the social relational category on A is the system modulo structural equivalence.

Example. (d) Consider again example (b), the category with objects $\{1, 2, 3, 4, 5\}$ based on the semigroup of three relations R, R^2, R^3. The explicit list of morphisms allows us to check for structural equivalence using the morphism criterion. We find that Mor(3, x) = Mor(4, x) for x = 1, 2, . . . , 5. Hence, $3 \equiv_s 4$ in this network. But Mor(1, 3) = $\{R^2\} \neq \{R, R^3\}$ = Mor(2, 3) shows that nodes 1 and 2 are not structurally equivalent. Similarly, nodes 1 and 5 are not structurally equivalent, because Mor(1, 2) $\neq$ Mor(5, 2) = Mor(2, 5), using the symmetry of the category. Finally, comparing nodes 2 and 5 we find the Mor(3, 2) $\neq$ Mor(3, 5), and so, they are not structurally equivalent. We define,

(19.6.3) $\overline{x}\overline{R}^i\overline{y}$ iff xR^iy (for any x, y; i = 1, 2, 3)

Then the new category is given by

Objects: $\{\overline{1}, \overline{2}, \overline{3}, \overline{5}\}$ $(\overline{3} = \{3, 4\}; \overline{x} = \{x\}, x = 1, 2, 5)$
Mor($\overline{1}, \overline{2}$) = Mor($\overline{1}, \overline{5}$) = Mor($\overline{2}, \overline{3}$) = $\{\overline{R}, \overline{R}^3\}$
Mor($\overline{3}, \overline{5}$) = $\{\overline{R}^3\}$
Mor($\overline{x}, \overline{y}$) = $\{\overline{R}^2\}$ (all other $\overline{x}, \overline{y}$)

The new graphs are given in Figure 19.6.7, using (19.6.3) and Figures 19.6.1, 19.6.2, and 19.6.3. Note that a merging of points 3 and 4 into $\overline{3}$ has taken place. The skeleton SK_A therefore consists in the four nodes, the three sets of morphisms, and the law of relation composition.

However, as Lorrain and White (1971) point out, except in idealized classificatory systems, structural equivalence is ordinarily not found on the basis of the relations generated by the given social relational system. Instead, the first category e_A needs to be first reduced by some other functorial reduction, followed by the passage to the skeleton. Lorrain and White suggest two criteria for reduction: cultural and sociometric.

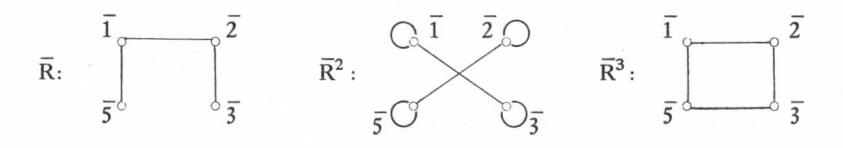

Figure 19.6.7

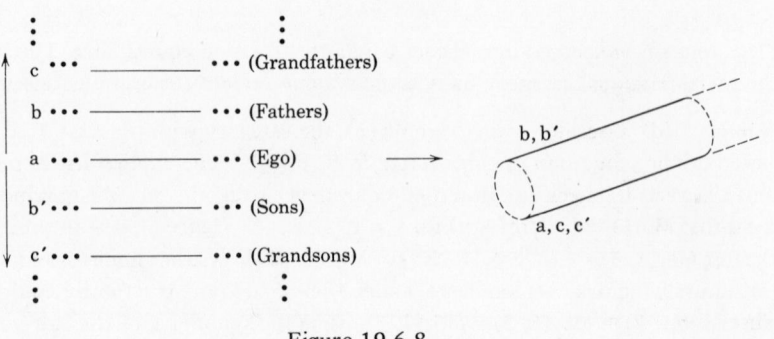

Figure 19.6.8

As an example of a cultural reduction, one might cite kinship systems in which we begin with set A and, say, a relation "is child of," C. The generational unfolding of C on A, taken to include the entire society, means we have C, C^2, C^3, . . . and C^{-1}, C^{-2}, C^{-3}, ... The sense of identification between different generations tied into the marriage system yields an equivalence on A. For instance, consider an identification involving ego and ego's grandchildren and ego's grandfather. The generational unfolding is "folded" (as shown for other cases in White, 1963, Chap. 1); see Figure 19.6.8. The horizontal lines represent people in the same generation, the vertical flow represents the unfolding of C from a in two directions: downward to a's sons, grandsons . . .; and upward to a's father, grandfather . . . . By imagining rolling up the system by identifying c with a, a with c', and so on, we have the cylinder on the right. Here C^2 is identified with I.

This illustration serves to make concrete the corresponding functorial reduction, in which starting from set A and relation C (among others), we write

$$C^n \equiv I \quad \text{or} \quad F(C^n) = I$$

to show that relation C^n generated by the relation C is to be identified with the identity of the reduced category; that is, C^n is mapped into I by the functor.

To illustrate sociometric reduction, we can use the suggestion of Lorrain and White that observed relations nearest to the identity [say, in having a small symmetric difference; see definition (11.3.3)] be identified with I. We illustrate from part of one example they use—namely, the hierarchy of example (c)—but by their first reduction already a reduced category, in which the relations are as shown in Figure 19.6.9. The relation PP^{-1} is closest to the identity. A functor F is constructed mapping objects of this category (call it $\mathcal{C}_1$) into themselves identically so that $Ob(\mathcal{C}_2) = Ob(\mathcal{C}_1) = \{1, a, b\}$. Next, we identify PP^{-1} with the identity:

$$F(PP^{-1}) = I$$

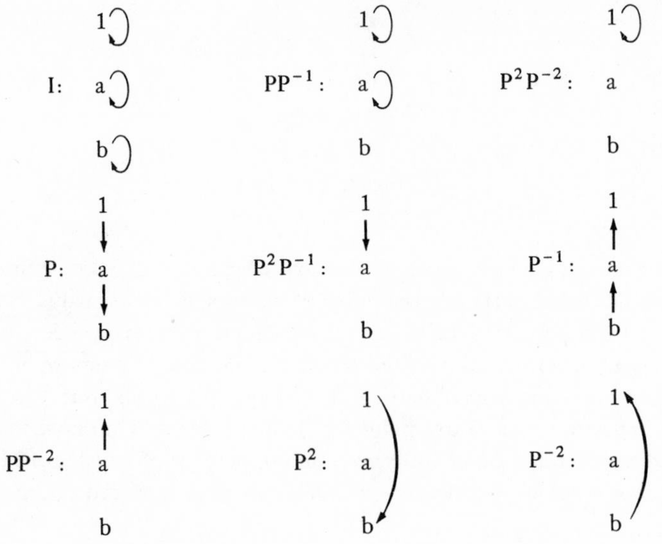

Figure 19.6.9

Hence,

$$F(P^2P^{-2}) = F(PPP^{-1}P^{-1}) = F(P)F(PP^{-1})F(P^{-1}) = F(P)F(P^{-1})$$
$$F(P^2P^{-1}) = F(PPP^{-1}) = F(P)F(PP^{-1}) = F(P)$$
$$F(PP^{-2}) = F(PP^{-1}P^{-1}) = F(PP^{-1})F(P^{-1}) = F(P^{-1})$$

Taking $F(P) = P$, $F(P^{-1}) = P^{-1}$, then

$$F(P^2P^{-2}) = PP^{-1} = I$$
$$F(P^2P^{-1}) = P$$
$$F(PP^{-2}) = P^{-1}$$

and, also,

$$F(P^2) = F(PP) = F(P)F(P) = P^2$$
$$F(P^{-2}) = F(P^{-1}P^{-1}) = F(P^{-1})F(P^{-1}) = P^{-2}$$

Hence, in the new category e_2 we have the relations of Figure 19.6.10.

Apart from the identity (of the classes), the system has been reduced to direct (P) and indirect (P^2) superordination and direct (P^{-1}) and indirect (P^{-2}) subordination. In a sense, this is indeed the "skeleton" or "anatomy" of the original five-node hierarchy.

The ease with which these minor examples were computed is deceptive, however. Ordinarily, the original network will require successive reductions, where at each reduction there are many possible functors that might be defined. Apart from sub-

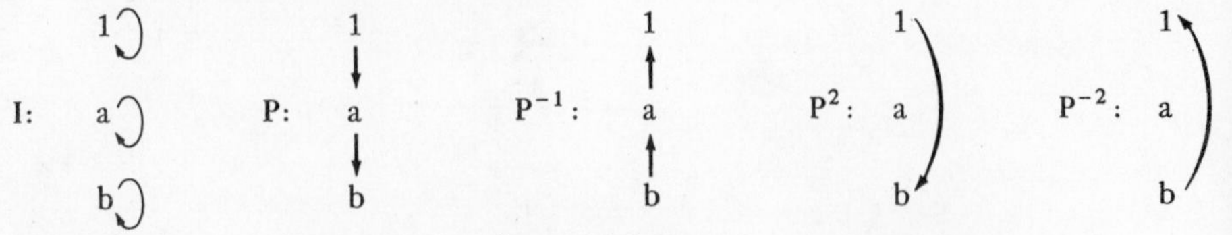

$$I: \quad \begin{matrix} 1 \circlearrowright \\ a \circlearrowright \\ b \circlearrowright \end{matrix} \qquad P: \begin{matrix} 1 \\ \downarrow \\ a \\ \downarrow \\ b \end{matrix} \qquad P^{-1}: \begin{matrix} 1 \\ \uparrow \\ a \\ \uparrow \\ b \end{matrix} \qquad P^{2}: \left. \begin{matrix} 1 \\ a \\ b \end{matrix} \right) \qquad P^{-2}: \left. \begin{matrix} 1 \\ a \\ b \end{matrix} \right)$$

Figure 19.6.10

stantive theoretical guidance, such work could degenerate into data crunching with little output knowledge. (One is reminded of misuses of factor analysis.) Moreover, we can be sure that the initial decisions as to reductions strongly affect the final outcome. There is no "canonical" reduction. All reductions are more or less intuitively justified in the interests of getting at a "simple structure." But it would be foolish to expect a new idea to emerge free from problems. (But note the continuity with the algebraic tradition of studying homomorphisms of a system.) These problems are the sources of new ideas, new techniques, and improved algorithms.

19.7. Algebraic Model-Building. It seems clear that the algebraic methods growing out of the earlier group theoretic analyses of prescribed marriage systems are destined to play a significant role in the future of the field. Moreover, this work is only part of a growing literature in which ideas like mappings, homomorphisms, categories, lattices, and Boolean algebras, play a strong role in the conceptualization and analysis of sociological problems. To give a few references:

(1) the use of semilattices to represent organizational structures (Friedell, 1967; see, also, a description and critique in Doreian, 1970);

(2) the conceptualization of levels of consensus in social systems in terms of Boolean algebra (Friedell, 1969);

(3) the treatment of problems in diffuse status theory in terms of commutative diagrams (Fararo, 1970c, part 1);

(4) the representation of role-theoretic ideas in terms of families of graphs by Oeser and Harary (1962, 1964) with the possibility of linking this work to the category theoretic method;

(5) the study of valued graphs (see, for example, Doreian, 1969), which may be brought into contact with the reduction procedures as well as with the study of the stochastic processes of social influence (see Chapter 13);

(6) symbolic psychologic (Abelson and Rosenberg, 1959), graph theoretical balance theory (Cartwright and Harary, 1956), clustering (Davis, 1967), and related algebraic analysis in connection with sociometry (Boyle, 1969, Holland and Leinhardt, 1971), as well as the generalization of balance theory (Cartwright and Harary, 1970);

(7) finally, studies of flows in networks in sociologically relevant senses (for example, vacancy chains; see section 16.7) and studies of the probabilistic description

of complex networks in random and biased net theory (see Rapoport and Horvath, 1961; Foster et al., 1963; Fararo and Sunshine, 1964).

In algebraic modeling, it is fruitful to think of two levels of model-building. At the first level, the setup phase consists in passing from intuitive considerations about some class of phenomena to a framework or space or category. The axioms on a set of primitives define the space or category of systems to be studied and applied. For example, in example (c) of section 19.5, a space of "status chains" is defined, and homomorphism relations, interpreted as expectations, are conceptualized in the space. The space of all strict pms's—that is, algebraic systems (S, C, W) satisfying a set of formal axioms—form a second example. Note that the group representation appears when we examine isomorphism relations among the systems in this space. The analysis phase of this first level has as its object the analytical study of the space of such systems, culminating in proven propositions about the entire space [for instance, the KST theorem (18.6.1)] . Also, one may classify the systems and study problems associated with a given class. Giving special attention to the strict type of pms is an example in the space of all pms's.

The application phase of the first level is the setup phase of the second level: some system or class of systems in the category is applied to some particular class of sociological phenomena. Ordinarily, additional hypotheses are needed at this point. For example, the space of status chains was applied to the problem of explaining the relationships of status to influence in small groups in Fararo (1972b). Another example is White's application of the catalogue of strict pms's to Australian kinship systems (White, 1963). This framework application, then, is the start of a model-building job of three phases, where the first phase is the tailoring of the framework for particular empirical problems. The second phase in the second level, then, consists in derivations of the properties of the model in relation to the specific issues raised in the application of the framework. For instance, if the problem is one of finding a skeleton of an observed network, specific hunches and calculations are involved. In the final phase of applying the model (as opposed to the generic framework) the problem of a criterion of evaluation arises. What outside standard exists to allow us to evaluate a given proposed skeleton of a network as more or less the "right" representation? Clearly, this can only be answered in the context of some problem of explanation.

To sum up, in algebraic modeling we see a two-level enterprise. Schematically,

Framework $\begin{cases} \text{(1) Construction of a space of systems} \\ \text{(2) Analysis of the relations among systems in the space} \\ \text{(3) Application of the framework:} \end{cases}$

                    (a) Setup of a model, using some system in the space

                    (b) Analysis of the model

                    (c) Application of the model

SUGGESTIONS FOR FURTHER READING (PART THREE)

Chapter 12. The material presented here was first published in Fararo (1970c).

Chapter 13. A guide to further study of the continuous-time models developed by Coleman (1964) has been presented by Jaeckel (1971). A short and readable Coleman paper on this style of model-building is contained in Lazarsfeld and Henry (1968): "Reward Structures and the Allocation of Effort," originally appearing in Criswell, Solomon, and Suppes (1962).

Chapter 14. For balance theory, see Cartwright and Harary (1956) or the first part of Berger et al. (1962). A different dynamic treatment appears in Flament (1963). For expectation-states theory, see the cited papers in *Sociological Theories in Progress*, vol. I (Berger et al., 1966). For a treatment that extends the theory to cover other types of situations, see Fararo (1972b).

Chapter 15. References for estimation are cited at the end of the chapter.

Chapter 16. The literature on mobility is voluminous. For the mathematics of the simple model, see Kemeny and Snell (1960). The remaining references are given in the text.

Chapter 17. On group theory, presented in an elementary manner with the graph correspondence, see Grossman and Magnus (1964).

Chapter 18. The relevant references for formal kinship analysis have been given: Kemeny, Snell, and Thompson (1966), White (1963), and Boyd (1969). Some additional papers of relevance appear in Kay (1971).

Chapter 19. For the mathematics of homomorphisms in group theory, see Grossman and Magnus (1964) or Fraleigh (1967). For category theory, in addition to the Lorrain and White (1971) paper, one should consult Lang (1965), MacLane and Birkhoff (1967), or Cohn (1965).

4 THEORY OF GAMES

CHAPTER TWENTY THE VON NEUMANN AND MORGENSTERN THEORY OF UTILITY

20.1. **The Extension Problem in Preference Systems.** We begin our treatment of the foundations of the theory of games with a summary and interpretation of the theory of utility devised by John von Neumann and Oskar Morgenstern (1947), as outlined by Luce and Raiffa (1957). Hereafter, the theory will be referred to as "NM theory."

The basic objective guiding the construction of this theory was the extension of the preference relation from a domain of strategies—from which the player in game theory makes his choice—to a domain of uncertain combinations of strategies, the so-called mixed strategies. Thus, one wanted to take the given natural meaning of "Player A prefers this strategic option to that strategic option" and conclude, on the basis of a plausible theory, that this meaning can be extended to "Player A prefers this chance situation involving strategy choices with specified probabilities to that chance situation involving strategy choices with (other) specified probabilities." The extension logic is such that wherever one begins, that beginning becomes a special part of the complete constructed object. Thus, the NM theory aimed to start with a binary relation on pure strategies and conclude with a binary relation on all (thus, even mixed) strategies, such that the original properties of preference were sustained in the extended domain.

Extension is required in many contexts. In physical theory, a meaning has to be supplied for calculations involving the masses of complex bodies having many parts. The theory of mass—in analogy with the theory of utility—has to show how one can start from "heavier than" as a binary relation on simple whole objects and extend it to a logically unlimited domain such that for any pair of objects in that domain "heavier than or equivalent to" is empirically meaningful, even if never actually utilized in any experience of an investigator. The mass notion, as a numerical concept, begins ordinally as a mere homomorphism for the ordered system consisting of objects as related by "heavier than or equivalent to"—this is the empirical beginning. But combining objects results in an extension of this relation to the combinations; at this point, mass becomes an order homomorphism for the larger ordered system, and for consistency, one must take $m(a \oplus b) = m(a) + m(b)$ so that it is additive. The induced scale-type of the mass function—with domain the extended version of the simple objects—is ratio. In NM theory, preference-indifference is to heavier-than-equivalent as utility is to mass, with similar extension logic except that the utility scale-type is interval.

20.2. The Space of Alternatives. To set up the NM theory in abstraction from the intended connection with game theory (discussed in Chapter 21) we use a general choice language. This language consists in the terms "outcome," "pure alternative," simple uncertain combination," and "compound uncertain combination." Of course, the starting point is some set of outcomes of choice in situations. This is a primitive set, say $\mathcal{O}$, whose identification varies with each application. Concerning $\mathcal{O}$, we require that it be finite (letting r be the number of outcomes in $\mathcal{O}$) and that for the chooser, there exists a weak preference relation Q on $\mathcal{O}$. Recall from section 4.17 that this means

(1) xQy and yQz implies xQz   (transitive)
(2) xQy or yQx   (strong completeness)

for all x, y, and z in $\mathcal{O}$. The interpretation is that xQy means that the chooser prefers x to y or is indifferent. Reviewing section 4.17, we see that then if any relation on a set satisfies (1) and (2), then two other relations can be defined: strict preference and indifference [see (4.17.3) and (4.17.4) and theorems (4.17.5) and (4.17.6)]. Following the tradition of utility theory, especially as elegantly expressed in the opening chapters of Luce and Raiffa (1957), we shall write $\succsim$ for Q in what follows, so that $\succsim$ has the same properties as $\geqslant$ on numbers. Also, we let

$$\mathcal{O} = \left\{A_1, A_2, \ldots, A_r\right\}$$

be the set of r outcomes, labeled such that $A_1 \succsim A_2 \succsim A_3 \succsim \ldots \succsim A_r$.

(20.2.1) Definition. The space of alternatives $\mathcal{a}$ is given by

(1) $\mathcal{O} \subset \mathcal{a}$
(2) if $\mathcal{B} \subset \mathcal{a}$ and if $\mathcal{B}$ has a finite number of elements B_i (i = 1, 2, . . . , n), then $\mathcal{B}$ with any probability distribution over $\mathcal{B}$ is an element of $\mathcal{a}$ denoted

$$(p_1 B_1, p_2 B_2, \ldots, p_n B_n)$$

Interpretatively, condition (1) says that any outcome A_i is an alternative, and we call A_i a pure alternative. Condition (2) says that any uncertain combination

of previously given alternatives forms a new alternative. From (1) and (2), we
see that

$$(p_1 A_1, p_2 A_2, \ldots, p_r A_r)$$

is an alternative, provided the p_i form a probability distribution

$$p_i \geq 0, \quad \sum_{i=1}^{r} p_i = 1$$

Any such distribution over o is termed a simple uncertain combination of alternatives. A compound uncertain combination arises from uncertain combinations of
nonpure alternatives.

Example. Let $o = \{$job, college$\}$ be a set of two outcomes for a high school
senior about to apply for college, with the outcome contingent upon whether or
not he applies to, and is accepted at, some college. By (20.2.1) (1), each of these
is an alternative. By (2), the simple uncertain combinations

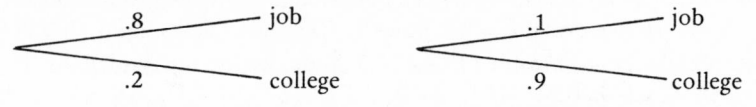

are each considered alternatives. The left simple uncertain alternative is denoted
(.8 job, .2 college), with an analogous notation for the alternative shown on the
right. While the senior may prefer (in the strict sense) college to a job, his real
options, depending on various actions he takes, might involve these two uncertain combinations of the pure alternatives of getting a job or going to college. It
is plausible to assume that if he prefers college, then the action leading to the
uncertain combination shown on the right will be his choice. Suppose that even
these actions are contingent in such a way that the first (.8 job, .2 college) alternative is seen to be about as likely as the second (.1 job, .9 college) alternative.
Then we have the compound uncertain combination

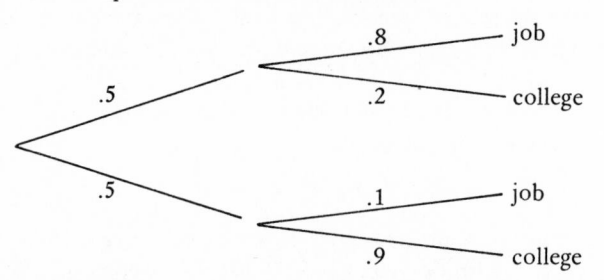

which may even be in competition with a drastically different alternative, namely,
the pure alternative "job."

The example shows that it is reasonable to assume that a chooser has the logical possibility of comparing any alternative to any other: pure vs. pure, pure vs. simple uncertain, compound uncertain vs. pure, compound uncertain vs. compound uncertain, and so forth. Also, it shows that we are employing a tacit rule: the chooser prefers the uncertain combination with the greatest chance of eventuating in what he prefers in terms of actual outcomes. It is considerations of this sort that enter into the axioms of the NM theory, stated in suitable generality.

20.3. Axioms and Elementary Consequences. In this section, the axioms of the NM theory and the major consequences will be set up. Recall we begin with the primitives $\mathcal{O}$ and $\succeq$, where the latter is a weak preference relation on $\mathcal{O}$. From $\mathcal{O}$ we obtain $\mathcal{A}$, the set of all possible alternatives including all compounds. The fact that $\succeq$ is a weak preference relation on $\mathcal{O}$ leads us to try for its extension as such on $\mathcal{A}$. Thus, we anticipate a system

$$(\mathcal{A}, \succeq)$$

where $\succeq$ is defined on $\mathcal{A}$, and turns out to be strongly complete, so that all alternatives are comparable. At the outset however, we have only $\succeq$ on $\mathcal{O}$. Recall that strict preference is definable [see (4.17.3)], and we use the symbol $\succ$ in analogy to $>$ for numbers:

(20.3.1) $x \succ y$ iff $x \succeq y$ and not-$(y \succeq x)$

Also indifference is denoted $\sim$ in analogy with $=$ for numbers:

(20.3.2) $x \sim y$ iff $x \succeq y$ and $y \succeq x$

The axioms of NM theory may be thought of as conditions that a binary relational system, denoted $(\mathcal{A}, \succeq)$, must satisfy to be termed an NM system of utility. Then identifying a particular set $\mathcal{O}$ and providing a particular interpretation for $\succeq$ will amount to asking whether the concrete system so specified forms an NM system of utility.

The first axiom of the system formally states the order character of the relation $\succeq$ on the outcome set $\mathcal{O}$, recapitulating the discussion of section 20.2.

(20.3.3) *Axiom.* (Order property of preference-indifference) The relation $\succeq$ on $\mathcal{O}$ is a weak preference relation, so that

$$A_1 \succeq A_2 \succeq \ldots \succeq A_r$$

We note that this axiom can be trivially satisfied by requiring the chooser to rank the outcomes in $\mathcal{O}$, with ties permitted. Under this operational identification, weak preference properties hold in a nonfalsifiable sense. Thus, under this identification the test of the theory must be focused on the other axioms.

(20.3.4) Axiom. (Reduction of compound uncertain combinations) Let

$$C^{(1)} = (p_1^{(1)}A_1, p_2^{(1)}A_2, \ldots, p_r^{(1)}A_r)$$
$$C^{(2)} = (p_1^{(2)}A_1, p_2^{(2)}A_2, \ldots, p_r^{(2)}A_r)$$

$$C^{(s)} = (p_1^{(s)}A_1, p_2^{(s)}A_2, \ldots, p_r^{(s)}A_r)$$

Then

$$(q_1 C^{(1)}, q_2 C^{(2)}, \ldots, q_s C^{(s)}) \sim (p_1 A_1, p_2 A_2, \ldots, p_r A_r)$$

where for $i = 1, 2, \ldots, r$

$$p_i = \sum_{j=1}^{s} q_j p_i^{(j)}$$

The interpretation of this axiom is that the chooser is indifferent as between a certain compound, as shown in Figure 20.3.1, and a particular simple uncertain combination obtained by applying the product rule for path probabilities (see section 9.5): add all the path probabilities for paths terminating in A_i: this is p_i of the reduction axiom. Thus, any compound can be reduced to a simple combination.

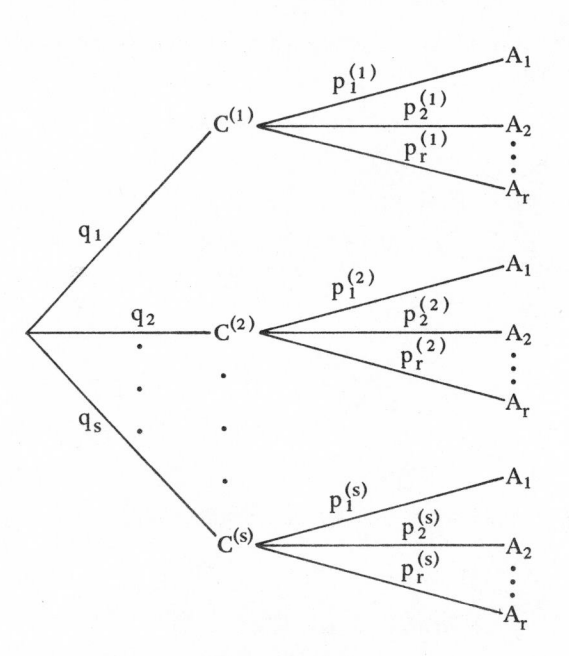

Figure 20.3.1

Example. (a) Consider the senior with the job-or-college situation. Suppose that his evaluative orientations satisfy the present axioms. Then by reduction, the following indifference relation will exist:

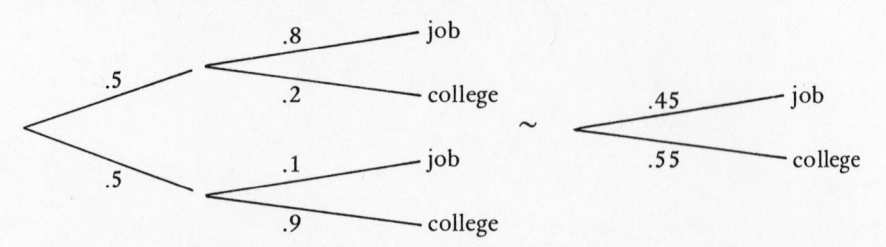

because, treating the left side by the usual path principles, we obtain a total probability of .45 for the outcome "job."

The reason the axiom is needed is that two uncertain combinations with distinct aspects from the chooser's standpoint—say, one more complex than the other—may be indistinguishable in the computed probabilities but possibly not seen as a matter of indifference by the chooser. Put another way, we do not even believe that the chooser necessarily computes anything, so the indifference relates to a felt "sense" of sameness. (This discussion is taken up again in section 20.6, dealing with methodological problems.)

(20.3.5) Axiom. (Continuity) For each $A_i \in \mathcal{O}$, there exists a probability, denoted u_i, such that

$$A_i \sim (u_i A_1, (1-u_i)A_r)$$

and we denote the right side by $\widetilde{A}_i$.

For A_1 we can think of $\widetilde{A}_1$ as

1 —— A_1
0 —— A_r

$[\text{i.e., } A_1 \sim (1A_1, 0A_r)]$

and for $\widetilde{A}_r$,

0 —— A_1
1 —— A_r

$[\text{i.e., } A_r \sim (0A_1, 1A_r)]$

Further, possible choices of p_1 for A_1 yield a whole continuum of alternatives:

p_1 —— A_1
$1-p_1$ —— A_r

$(0 \leqslant p_1 \leqslant 1)$

the endpoints being $\widetilde{A}_1$ and $\widetilde{A}_r$. The continuity axiom says that on this continuum there exists a point u_i such that the chooser is indifferent between A_i and the A_1 versus A_r choice with $p_1 = u_i$.

Example. (b) Let our senior have three possibilities after high school: college, job (civilian), or military service. Ask him to rank these in order of preference. Then label the first A_1, and so on. Thus, suppose A_1 = college, A_2 = job, and A_3 = military service. Here $r = 3$. The continuity axiom says that there exists a probability, call it u_2, such that the following indifference exists:

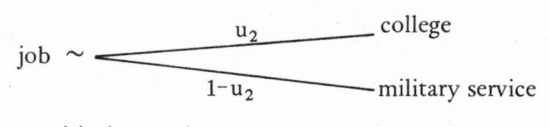

Then, by using empirical procedures, we aim to discover u_2. The point of theory, in this instance, is to say that there is something there to look for, to estimate, by some procedure. For instance, we can begin at $u_2 = 1$ and, by steps of .05, adjust u_2 downward, each time asking the person to compare two situations: the certainty of a civilian job against a $1 - u_2$ risk of military service and u_2 chance of going to college. For high values of u_2, we expect him to choose the uncertain combination because of the small risk $1 - u_2$ of military service. But "small" depends on his own evaluations; if he is terrified at the thought of the service outcome, then we will obtain immediately with the first shift from $u_2 = 1$:

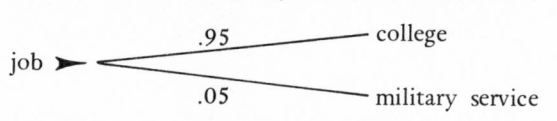

Then we have to shift u_2 upward somewhat. There must exist some $u_2 \neq 1$ for which the indifference obtains, for otherwise we would not have college $\succ$ job— at least, this is our intuitive conviction, formalized in the continuity axiom. It is precisely in making these adjustments to "home in" on the indifference point that we are learning much more about the person's evaluative orientations than the sheer order given by the ranking task.

(20.3.6) Axiom. (Substitutability) If

$$C = (p_1 A_1, p_2 A_2, \ldots, p_r A_r)$$

then for any $i = 1, 2, \ldots, r$

$$C \sim (p_1 A_1, p_2 A_2, \ldots, p_i \tilde{A}_i, \ldots, p_r A_r)$$

To interpret this axiom think of tree diagrams, for definiteness taking $r = 3$, $i = 2$, as follows, using the previous axiom:

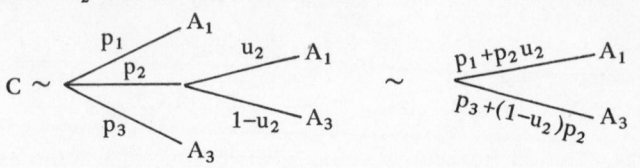

These give C and the endpoint combination to which A_2 is equivalent. Then replace $\tilde{A}_2$ where A_2 is in C:

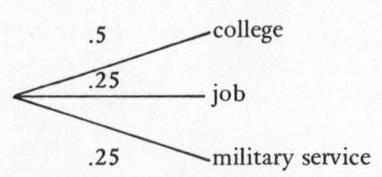

And the last indifference on the right follows from applying the reduction axiom, thought of as if it were the law of total probability as applied to any tree.

(20.3.7) Axiom. (Transitivity) The relations $\succ$ (of strict preference on α) and $\sim$ (of indifference on α) are transitive.

   This axiom postulates part of the structure we want the extended system to have, somewhat undesirably, one might say. Its empirical validity is dubious. We have to count it as a bold idealization. (The alternative is to regard it as a "normative" principle entering into the implicit definition of a rational player in game theory.)

   Example. (c) To illustrate axioms (20.3.6) and (20.3.7), consider the senior once again with A_1 = college, A_2 = civilian job, A_3 = military service. Suppose we find that

$$\text{job} \sim \begin{array}{l} \overset{.90}{\diagup} \text{college} \\ \underset{.10}{\diagdown} \text{military service} \end{array}$$

so that $u_2 = .90$. Now suppose he must deal with the following alternative:

$$\begin{array}{l} \overset{.5}{\diagup} \text{college} \\ \overset{.25}{—} \text{job} \\ \underset{.25}{\diagdown} \text{military service} \end{array}$$

According to substitutability, he is indifferent between this alternative and

$$\begin{array}{l} \overset{.725}{\diagup} \text{college} \\ \underset{.275}{\diagdown} \text{military service} \end{array}$$

because $.725 = .5 + .25 \,(.90) =$ "total probability of the college outcome," replacing the "job" outcome with its college-military service equivalent. According to the transitivy axiom (20.3.7), we can do such reasoning as :

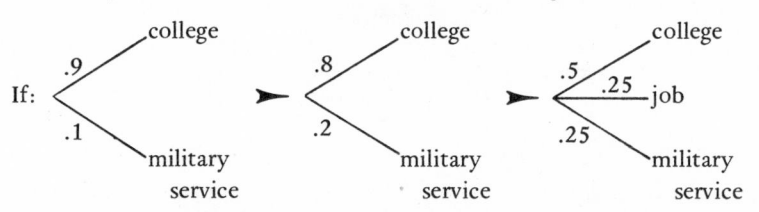

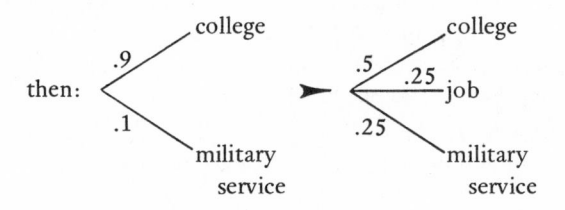

(20.3.8) Axiom. (Monotonicity) For any p and p',

$$(pA_1, (1-p)A_r) \succsim (p'A_1, (1-p')A_r) \qquad \text{iff} \qquad p \geqslant p'$$

This axiom was suggested by the type of example shown following definition (20.2.1), involving the job-or-college choice situation.

(20.3.9) Consequence.

(1) For any p and p',

$$(pA_1, (1-p)A_r) \sim (p'A_1, (1-p')A_r) \qquad \text{iff} \qquad p = p'$$

(2) For any p and p',

$$(pA_1, (1-p)A_r) \succ (p'A_1, (1-p')A_r) \qquad \text{iff} \qquad p > p'$$

Proof. By (20.3.2),

$$x \sim y \qquad \text{iff} \qquad x \succsim y \text{ and } y \succsim x$$

Taking x and y as the uncertain combinations of (20.3.8),

$$x \succsim y \qquad \text{iff} \qquad p \geqslant p'$$
$$y \succsim x \qquad \text{iff} \qquad p' \geqslant p$$

which implies that

$$x \sim y \qquad \text{iff} \qquad p \geqslant p' \text{ and } p' \geqslant p$$

and, so, consequence (1). By (20.3.1),

$$x \succ y \quad \text{iff} \quad x \succsim y \text{ and not-}(y \succsim x)$$

Again taking x and y as the uncertain combinations,

$$x \succ y \quad \text{iff} \quad p \geqslant p' \text{ and not-}(p' \geqslant p)$$
$$\text{iff} \quad p \geqslant p' \text{ and } p > p'$$

and so, consequence (2) holds.

Example. (d) Interpret $\mathcal{O}$ as a collection of occupations and let $\succsim$ on $\mathcal{O}$ be a ranking on the basis of the subject's estimate of the "relative standing" of the jobs in the United States. Interpret a tree diagram, or in other words, an uncertain combination of jobs as follows in the simple case: $(p_1 A_1, p_2 A_2, \ldots, p_r A_r)$ means the subject is presented with a description of a "situation of a person" comprising certain chances (the specified p_i) for occupying each of the various positions A_i. Suppose that in his ranking of the jobs, the subject placed "architect" at the top and "book salesman" at the bottom. Then we can let

$$A_1 = \text{architect}$$
$$A_r = \text{book salesman}$$

assuming r jobs in $\mathcal{O}$. Now, by consequence (2) of (20.3.9), since $p = .9 > .4 = p'$, we conclude immediately that he will regard the situation

$$= (.9A_1, .1A_r)$$

as better than the situation

$$= (.4A_1, .6A_r)$$

Next, we show that there is a canonical form for alternatives, so that axiom (20.3.8) can be applied to any two alternatives to determine the extended preference-indifference relation.

(20.3.10) Consequence. Let C be any alternative in $\mathcal{C}$. Then

$$C \sim (pA_1, (1-p)A_r)$$

for some p.

Proof. As before, we can let

$$A_1 \sim (1A_1, 0A_r) = \tilde{A}_1$$
$$A_r \sim (0A_1, 1A_r) = \tilde{A}_r$$

Then by the continuity axiom (20.3.5) for the other pure alternatives A_i we have, in general, for $i = 1, 2, \ldots, r$,

(20.3.11) $$A_i \sim (u_i A_1, (1-u_i)A_r) = \tilde{A}_i$$

and so, (20.3.10) is certainly valid for all the pure alternatives. Now consider an arbitrary uncertain combination C. According to the axiom on the reduction of compound uncertain combinations (20.3.4), we can get C into the form of a simple combination:

$$C \sim (p_1 A_1, p_2 A_2, \ldots, p_r A_r)$$

Applying the substitutability axiom (20.3.6) successively,

$$
\begin{aligned}
C &\sim (p_1 A_1, p_2 A_2, p_3 A_3, \ldots, p_r A_r) \\
&\sim (p_1 \tilde{A}_1, p_2 A_2, p_3 A_3, \ldots, p_r A_r) \\
&\sim (p_1 \tilde{A}_1, p_2 \tilde{A}_2, p_3 A_3, \ldots, p_r A_r) \\
&\quad \vdots \\
&\sim (p_1 \tilde{A}_1, p_2 \tilde{A}_2, p_3 \tilde{A}_3, \ldots, p_r \tilde{A}_r)
\end{aligned}
$$

and line-by-line transitivity (20.3.7) yields

$$C \sim (p_1 \tilde{A}_1, p_2 \tilde{A}_2, p_3 \tilde{A}_3, \ldots, p_r \tilde{A}_r)$$

where $\tilde{A}_i$ is given by (20.3.11). Now the right side is a compound, and so, we can apply reduction again:

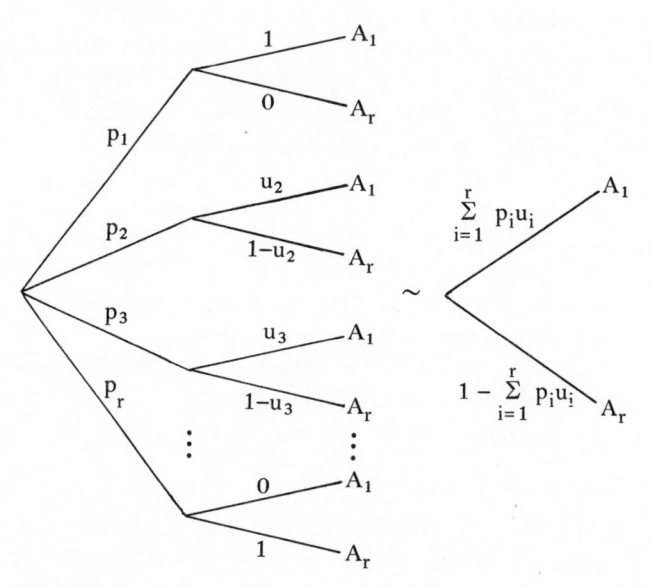

Thus, by reduction,

$$C \sim (pA_1, (1-p)A_r)$$

where

(20.3.12)
$$p = \sum_{i=1}^{r} p_i u_i$$

given that

$$C \sim (p_1 A_1, p_2 A_2, \ldots, p_r A_r)$$

20.4. Summary. The preceding section culminates in the result that the relation of weak preference $\succsim$ given on θ is defined and strongly complete on α; that is, any pair of alternatives are comparable. This follows from consequences (20.3.9) and (20.3.10): if C and C$'$ in α, put each in canonical form and then compare the corresponding probabilities of A_1, say p and p$'$, as in (20.3.9). Clearly, either we conclude $C \succ C'$, or $C' \succ C$, or $C \sim C'$. Thus, with

$$C \succsim C' \quad \text{iff} \quad C \succ C' \text{ or } C \sim C'$$

we conclude that

$$C \succsim C' \text{ or } C' \succsim C$$

for any C and C$'$ in α. Also, the relation is transitive because, as axiom (20.3.7) postulated, both $\succ$ and $\sim$ are transitive. Now an evaluative relation over a set is a weak preference relation if and only if it is transitive and strongly complete. Thus, $\succsim$ is a weak preference relation on α; $\succ$ is the corresponding strict preference on α; and $\sim$ is the corresponding indifference relation on α. The relations each specialize, when restricted to θ, to what we began with: a weak preference over θ. Thus, an extension of preference relations from θ to α is assured by the axioms.

20.5. Measurement Viewpoint: Interval Scale Utility. We now look at the axioms from the standpoint of measurement theory. We want an existence (representation) theorem and a uniqueness theorem. Since $(\alpha, \succsim)$ is an order system, we expect the representation theorem to claim that there exists a numerical (order) system to which this NM system is homomorphic; that is, we anticipate the mapping will be an order homomorphism.

Thus, we expect a mapping, say u, of the form

$$u: \alpha \to \Re$$

such that u preserves $\succsim$: $u(C) \geqslant u(C')$ iff $C \succsim C'$, all C, C$' \in \alpha$. Next we know from general measurement theory that this mapping is not the only possible way of obtaining a numerical system to represent $(\alpha, \succeq)$. Thus, we think of the mapping diagram

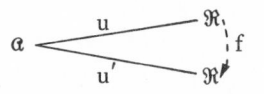

and ask what are the properties of the functions f that link any two such representations, $u' = f(u)$. This is the uniqueness theorem that will tell us the scale-type of the scientific variable u defined in the existence theorem.

One other point governing the construction and analysis of the NM system $(\alpha, \succeq)$ needs to be mentioned. Recall that this system is constructed as an auxiliary theoretical device relative to the construction of the theory of games. Now games have outcomes. Thus, in connection with game theory, the set θ can be interpreted as the set of possible outcomes of the game. Next, players have preferences vis-à-vis outcomes that determine how they behave. For example, most players prefer "win" to "lose," although any father will recall that in playing with his child his preference occasionally switches to "lose" over "win." Thus, in any concrete case, whatever the outcome set associated with the game, a relation $\succeq$ is assumed to be defined over it. (As a special case, θ may contain monetary "rewards.") We shall see that in certain games a player, satisfying certain player assumptions, will not directly employ a strategy but use a chance device to select from among strategies. The result is that the space of outcomes of the game becomes the outcome space in the sense of probability theory. Thus, players may then be faced with choices between strategic options involving different probability assignments: the NM system $(\alpha, \succeq)$ assures that a "standard" player will be able to decide, in the weak preference sense, between any two such strategic options.

In this description, the word "payoff" has not been used. For a technical usage of the term, we will later say that a payoff function associated with a game is any function over the possible outcomes of the game such that this function represents player preferences. Thus, the payoff function is the order homomorphism we anticipate will be obtained from the NM theory concerning the system $(\alpha, \succeq)$. Now when players use "randomized" strategies, so that the outcomes of the game form an outcome space in the sense of probability theory, the representation of preferences in numerical form—the payoff function—becomes a random variable.

Thus, it has an expectation, and von Neumann and Morgenstern wanted to treat this expectation, induced by the probability assignment, as a valuation of the strategic option with that probability assignment. In other words, they wanted the expectation of the payoffs to be the value of the order homomorphism preserving the relation $\succeq$ on α. This rather subtle argument can be summed up by saying that they wanted the utility of an uncertain combination (of outcomes) to be the expectation of the utilities of the outcomes under the given randomization of strategies. Thus, this last requirement goes beyond the desire to set up axioms that extend $\succcurlyeq$ to $\mathfrak{a}$ from $\mathfrak{o}$. It requires that one be able to cast the system $(\mathfrak{a}, \succcurlyeq)$ into a numerical form making valid the statement "The utility of an uncertain combination of outcomes is the expected utility of the outcomes." This last statement means that "abstract" calculations about preferences (using the empirical relations $\succcurlyeq$, $\succ$, $\sim$) can be completely replaced by concrete calculations involving expectations. Nothing new is thereby added, but presumably we all find the familiar numerical language less cumbersome than the language referent directly to the actualities of interest. Because of the proven scale-type—it is interval—we can be assured that the additive calculations of "expected utility" are empirically meaningful.

With this background, we can now formally state the two theorems that follow for a system $(\mathfrak{a}, \succcurlyeq)$ satisfying the NM axioms.

(20.5.1) Theorem. (Representation) Let $(\mathfrak{a}, \succcurlyeq)$ be a NM system—that is, a relational system satisfying axioms (20.3.3)-(20.3.8). Then there exists a positive real-valued function, call it u, with domain $\mathfrak{a}$, such that for any C, C' in $\mathfrak{a}$,

(1) $u(C) \geqslant u(C')$ iff $C \succcurlyeq C'$ (order-homomorphism)
(2) $u(qC, (1-q)C') = qu(C) + (1-q)u(C')$ (expected utility condition)

To see that part (1) of this theorem is valid, recall consequence (20.3.10), which says that for any C in $\mathfrak{a}$, there is a number p, $(0 \leqslant p \leqslant 1)$, such that

$$C \sim (pA_1, (1-p)A_r)$$

This induces a mapping from $\mathfrak{a}$ to the real numbers,

$$C \mapsto p$$

because (1) such a p exists for each C in $\mathfrak{a}$, so the domain of the mapping is $\mathfrak{a}$ and (2) such a p is clearly unique by consequence (20.3.9), for if also

$$C \mapsto p' \neq p$$

then

$$C \sim (pA_1, (1-p)A_r)$$
$$C \sim (p'A_1, (1-p')A_r)$$

and by the symmetry of $\sim$ seen from its definition

$$(pA_1, (1-p)A_r) \sim C \sim (p'A_1, (1-p')A_r)$$

implies by transitivity (20.3.7),

$$(pA_1, (1-p)A_r) \sim (p'A_1, (1-p')A_r)$$

and so, by consequence (20.3.9), p = p', a contradiction. Thus, C gets associated with a unique number. As claimed

$$C \mapsto p \qquad (C \in \mathfrak{a})$$

is a mapping; call it u. We have u (C) = p. By consequence (20.3.9) it is an order-homomorphism. For

$C \succeq C'$	iff	$C \succ C'$ or $C \sim C'$	(definition of $\succeq$)
	iff	$p > p'$ or $p = p'$	(canonical form)
	iff	$p \geqslant p'$	(definition of $\geqslant$)
	iff	$u(C) \geqslant u(C')$	(definition of u)

To show that the expected utility condition holds for any C and C' in $\mathfrak{a}$, recall that in proving the existence of a canonical form for any C, with associated p, we obtained formula (20.3.12)

$$p = \sum_{i=1}^{r} p_i u_i \qquad [= u(C)]$$

where

$$C \sim (p_1 A_1, p_2 A_2, \ldots, p_r A_r)$$
$$A_i \sim (u_i A_1, (1-u_i)A_r) \qquad (i = 1, 2, \ldots, r)$$

with the agreement that $u_1 = 1$, $u_r = 0$. In thinking of arbitrary C and C', we first apply the reduction axiom. Then with C represented as above, for C', we write

$$C' = (p_1' A_1, p_2' A_2, \ldots, p_r' A_r)$$

and so,

$$p' = \sum_{i=1}^{r} p_i' u_i, \qquad [= u(C')]$$

Let

$$\alpha = (qC, (1-q)C')$$

Using substitutability, we now have for α:

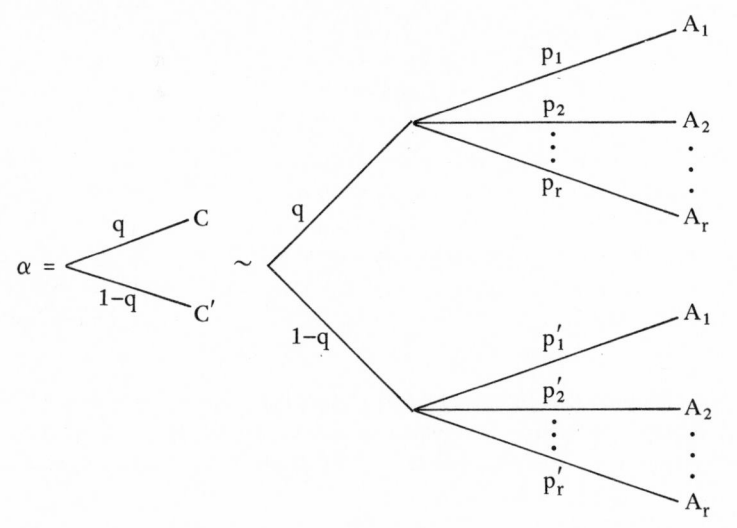

and by reduction (thought of as if the law of total probabilities applies), this yields the alternative

$$\alpha \sim ([qp_1 + (1-q)p_1'] A_1, [qp_2 + (1-q)p_2'] A_2, \ldots, [qp_r + (1-q)p_r'] A_r)$$

which we put into canonical form,

$$\alpha \mapsto p = u(\alpha)$$

by

$$
\begin{aligned}
u(\alpha) = p &= \sum_{i=1}^{r} [qp_i + (1-q)p_i'] u_i \\
&= \sum_{i=1}^{r} qp_i u_i + \sum_{i=1}^{r} (1-q)p_i' u_i \\
&= q\sum_{i=1}^{r} p_i u_i + (1-q) \sum_{i=1}^{r} p_i' u_i \\
&= qu(C) + (1-q)u(C')
\end{aligned}
$$

Since $\alpha = [qC, (1-q)C']$ this shows that the expected utility condition holds in all generality over α. Thus, the representation theorem is proved.

(20.5.2) *Theorem.* (Uniqueness) Let u and u$'$ be any two real-valued mappings with domain α that satisfy the conditions of the representation theorem. Then there exist constants a $>$ 0, and b, such that

$$u'(C) = au(C) + b$$

for any C in α.

Thus, by the uniqueness theorem, we see that the admissible transformations for the (scientific) variable u defined in the existence theorem are linear functions with positive slopes. Thus, in the diagram,

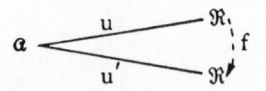

we have $\{f\}$ the class of positive linear transformations. For some such f, we have u$'$ = f(u). Also, given u, applying such an f leads to a u$'$ representing $(\alpha, \succeq)$— although perhaps not as computationally nice as u. This theorem will not be proved here. The reader is referred to NM directly (von Neumann and Morgenstern, 1947) and to the related axiomatic work reported in Luce and Suppes (1965).

Example. Let a person satisfy the NM axioms over an outcome set θ. Thus, we have $(\alpha, \succeq)$ representing his evaluative orientations: the possibilities for choice realization on occasions involving various alternatives in α. Suppose that

$$0 = \{A_1, A_2, A_3, A_4\}$$

where $A_1 \succ A_2 \succ A_3 \succ A_4$. Then here is a "problem for calculation."
Given the data

$$A_2 \sim (.6A_1, .4A_4)$$
$$A_3 \sim (.2A_1, .8A_4)$$

find for the pair of alternatives

$$C = (.25A_1, .25A_2, .25A_3, .25A_4), C' = (.15A_1, .50A_2, .15A_3, .20A_4)$$

which of the following holds

$$C \succ C' \quad \text{or} \quad C' \succ C \quad \text{or} \quad C \sim C'$$

(The axioms guarantee that exactly one of these relationships obtains.)

Procedure. We find u(C), u(C') and check numerical relationships:

$$u(C) > u(C') \Rightarrow C \succ C'$$
$$u(C') > u(C) \Rightarrow C' \succ C$$
$$u(C) = u(C') \Rightarrow C' \sim C$$

To find u(C), u(C'), we use expected utility,

$$u(C) = \sum_{i=1}^{4} p_i u_i$$

where $p_i = 1/4$ (all i). By the data, and the two stipulations of 0 and 1 at the endpoints,

$$u_1 = u(A_1) = 1$$
$$u_2 = u(A_2) = .6$$
$$u_3 = u(A_3) = .2$$
$$u_4 = u(A_4) = 0$$

Hence,

$$u(C) = \tfrac{1}{4} (1 + .6 + .2) = \tfrac{1.8}{4} = .45$$

Similarly,

$$u(C') = \sum_{i=1}^{4} p_i' u_i$$
$$= .15(1) + .50(.6) + .15(.2) + .20(0)$$
$$= .48$$

Thus, because

$$u(C') = .48 > .45 = u(C)$$

we conclude by the fact that u is an order-homomorphism that

$$C' \succ C$$

Thus, if we present this person with a choice between alternatives C and C', we predict he will choose C'. This prediction has a series of premises: (1) the NM theory, (2) the identification of o, (3) the identification of the generic idea of uncertain combination, (4) the premises of procedure, in the determination of u_2, u_3 (e.g., the identification of the relation $\sim$ yielding the data). A failure of prediction—or, better, a failure of a series of predictions—can be used to infer by the falsifiability logic outlined in section 2.11 that at least one of these premises is at fault. Note that it need not be the general theory.

20.6. Methodological Problems. In an enlightening discussion on this theory that bears upon measurement theories in general, Luce and Raiffa (1957) point out some important difficulties and reminders about difficulties in general in basic theoretical sc.ence.

First, they mention the criticism sometimes made that there are an infinity of paired comparisons involved in a NM system $(\alpha, \succ)$: thus, no finite set of data is ever really adequate to test the theory. Luce and Raiffa point out, however, that the very extension process governing the construction of the theory yields a method of testing it: obtain data of the form $\{u_i\}$, as indicated in the last example, then compute predictions. True, the theory is not verifiable for any given person because there are an infinite number of predictions in principle; yet, it is clearly falsifiable because a finite set of paired comparisons can yield a number that do not agree with the predictions. In this trait, the NM theory is similar to any theory that represents empirical systems not as bodies of realized data but as bodies of potentially realizable data. If $(\alpha, \succ)$ is the empirical system associated with a chooser, this does not mean he observably displays the relation $\succ$ on an actual set α; it means that he has a certain highly complex "capacity" or "potentiality" for displaying the relational system, in suitably re-stricted subsets, when the occasion calls for it. Thus, the extension of $\succ$ to α from o is a mathematical construction parallel to a philosophical extension from finite realization to infinite possibilities for finite realization. The point is that the theory remains falsifiable.

A second difficulty sometimes considered extremely damaging to the NM theory is the existence of empirical examples in which the relations $\succ$ and $\sim$ are not transitive. Two remarks are relevant here. First, one cannot expect a theory to literally mirror a certain aspect of a concrete system. Think, for example, of the delay in the progress of physics if Kepler had really taken the data too seriously. Surely, "the orbit" is a succession of orbital paths, and as observed, these paths show perturbations from the clean, smooth curve called an ellipse. Thus, that the ellipse "represents" the orbital path is itself an idealization; in modern probabilistic theories, one may deal with a "stochastic ellipse" and so achieve a representation in greater fidelity to the data. But for many purposes, and especially for broad understanding, the ellipse is fine. Similarly, "intransitive triples" can be regarded, for many purposes, as perturbations of a "smooth" structure: a relational system having definite order properties. Especially in theoretical work, the argument may be made that too early attention to detailed deviations from global characterizations that are productive of ideas is a mistake. Indeed, the unwillingness to idealize is the basic property of an atheoretical point of view. Second, game theory utilizes this idealization.

Finally, we return to the serious conceptual difficulty in the NM theory: namely, the notion of probability is uncritically taken as if it were to be identified by the estimation procedure based on independent repetitions. In this case, on the one hand, no scale problem arises because the scale-type is absolute. But what does this have to do with this person here and now choosing between C and C'? There are no repetitions in this choice; it is what it is. Thus, the "probabilities" are referent, not to expected relative frequencies in repeated trials of the phenomenon, but to some subjective meanings involving chance, likelihood, and the like. Thus, if the probabilities are given, as when one is asked to choose between

then the numerical things ".9," ".7," and so on figure as entities transformed by some subjective transformation rules. In other words, they are data for the subject who applies some model to them. For instance, reading "p_1 greater than p_2" the subject may discard all information except "A_1 more likely than A_2."

It is more interesting and of wider import to deal with the phenomenon of "likelihood relations" on a set than to build a theory of how people transform stimulus numerals interpreted as "probabilities" or "chances." These relations stand to events as preferences stand to alternatives.

A theory of individual choice, in other words, needs to investigate the formal structure of the naive likelihood relation a person holds over events, giving rise to a subjective probabilistic relational system. This is an empirical system. Thus, it can be mapped into a numerical system by a scientist, in order to substitute numerical calculations (e.g., as on utilities) for more abstract manipulations (e.g., as on ➤ and ~). Thus, we anticipate that if the form

$$(p_1 A_1, p_2 A_2, \ldots, p_r A_r)$$

is treated in these terms, the numbers p_i will be images under some homomorphism from an empirical system representing a likelihood relation over the

A_i, for a specified person. But, then, we anticipate that some other mapping could also represent the empirical system, and so, we are led to doubt that the p_i will be on an absolute scale. But, then, will the very definition of the utility function

$$u(C) = \Sigma\, p_i u_i$$

make sense? Thus, in questioning the conceptualization of the probabilities in the NM theory one is led to question the basic consequence of the theory: the expected utility condition and the interval scale character of the utility function. There is no need to regard this criticism as a total condemnation of this theory. On the contrary, the location of such a problem has had desirable results: investigators have been led to new theories that illuminate the inner, and previously neglected, structure of the phenomenon. A review of many such subsequent treatments of the theory will be found in elegant form in Luce and Suppes (1965). They show that in its revised form the theory of utility becomes the theory of preference and subjective probability.

20.7. Concrete Example of Application Techniques.   This section presents an illustration of how the previously stated ideas can be empirically applied. This is a mere illustration, not a full-fledged empirical study. In particular, this application of the NM system attempts to exploit the idea that by going a little beyond rankings to obtain the $\{u_i\}$ data, one could learn a good deal more about the differential evaluation of jobs.

The subject was a woman of age 36 and with a son of age 8. This latter fact was exploited to try to increase the meaningfulness of the task. The inquiry began with five occupational titles, and the subject was asked to arrange them in the order corresponding to the position she would like to have her son occupy "when he grew up." The term "position," meant a job, such as "dentist." Subsequently, its meaning was extended to include a "field" of possible jobs, an abstract spot or situation in which a person had certain chances for certain futures. The idea was to measure the u_i for five jobs in part one, and then predict preferences for positions in the generalized sense. The set $\Theta = \{A_1, A_2, A_3, A_4, A_5\}$ was identified as follows: A_1 = concert pianist, A_2 = social worker, A_3 = electrician, A_4 = dentist, A_5 = salesman, with $A_1 \succ A_2 \succ A_3 \succ A_4 \succ A_5$, by her ranking. Stipulatively,

$$u(A_1) = u(\text{pianist}) = 1$$
$$u(A_5) = u(\text{salesman}) = 0$$

Thus, the problem was to estimate $u_2 = u(\text{social worker})$, $u_3 = u(\text{electrician})$, and $u_4 = u(\text{dentist})$. For this purpose, a sure position had to be compared against a generalized position. For instance, to determine u_2, one had to manipulate p in the comparison:

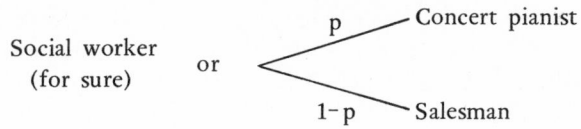

Social worker
(for sure)   or

For instance, "Would you rather see him be a social worker for sure or be in a position where he had a 100 p%-chance to go on to become a concert pianist, but a 100 (1–p)%-chance to end up a salesman?" One began with p = .95 and worked down. When a switch of preference direction occurred, the p value was raised somewhat to try to "home in" on the u_i. By this method, for "social worker," the sequence of p's was: .95, .90, .85, .80 (switch), .82; and at p = .82, indifference was expressed. Hence,

$$u_2 = u(\text{social worker}) = .82$$

Similarly,

$$u_3 = u(\text{electrician}) = .80$$
$$u_4 = u(\text{dentist}) = .38$$

With the $\{u_i\}$ data available, predictions could be made in paired comparisons of different generalized positions. In tree form, the positions described to the subject (in pairs, for comparisons) were:

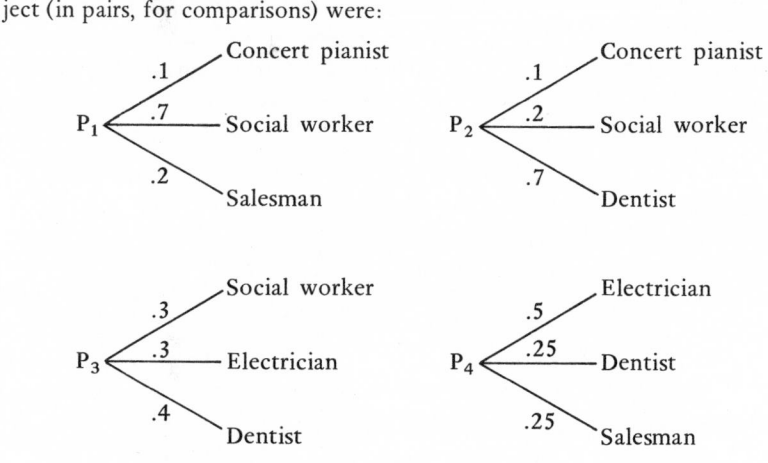

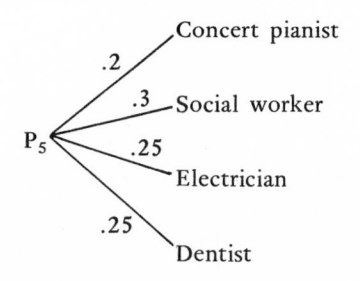

The position P_i was interpreted for the subject in the form "In a position such that the chances of becoming . . . are . . ." and so forth. The paired comparison data were compared with the predictions based on the NM model, calculated by the expected utility technique. These calculations yield:

$$u(P_1) = .674$$
$$u(P_2) = .530$$
$$u(P_3) = .638$$
$$u(P_4) = .495$$
$$u(P_5) = .741$$

and, so, the predictions:

$$P_5 \succ P_1 \succ P_3 \succ P_2 \succ P_4$$

This was also the form of the data except that for the subject we find $P_2 \succ P_3$, contradicting the prediction. The fact that P_3 gives zero chance for the favorite job seems significant as compared with P_2 giving a nonzero chance of this occupation. Yet P_2 vs. P_4 is the same in this regard, but the prediction is correct. Subtleties of relative chances are at work here that are simply not functioning in the way in which NM theory tacitly assumes. Despite this fact, however, the major trend of the data is in agreement with the theoretical model.

Thus, in this example, admittedly not a rigorous experimental study, an attempt has been made to acquaint the reader with the way in which this measurement model can be related to data. With the exception of the probability aspect, the NM model is a sound approach to tapping deeper sources of evaluative orientation than that given merely by rankings.

CHAPTER TWENTY-ONE INTERACTIVE CHOICE STRUCTURE AND THE CONCEPT OF A GAME

21.1. Interactive Choice Situations and Their Representation. An actor chooses alternatives in accordance with his preferences. This is an inexact way of saying that he acts selectively in order to realize the alternatives he prefers. This is clear in the NM language: an alternative is either an outcome or an uncertain combination based upon outcomes. Clearly actors can only "choose outcomes" in an indirect manner: they choose actions leading to the outcomes. When we consider the individual choice problem, we do not consider the fundamental difficulty that exists in real-life choices: the actor selects an action, on the basis of some outcome evaluation, but the outcome may depend upon more than this single actor's action. If the factors contributing to the outcome may be regarded as chance events, or natural nonpurposive forces, no real conceptual difficulty arises for the theory of individual choice. But if the outcome depends upon the actions taken by other actors, there are two distinct possibilities of orientation for the individual actor:

(1) the actor does not attend explicitly to the dependency of the outcome upon joint action rather than individual action;

(2) the actor does attend explicitly to the dependency of the outcome upon joint action rather than individual action.

If orientation (1) holds for the actor, then individual choice theory applies directly. If orientation (2) holds for the actor, choice theory does not apply directly, because the actor has cognitive understanding that although he prefers alternatives, one to another, his realized alternative depends not only upon his choice but upon the choices made by others. In this case, the actor must still choose (as a condition of the definition of the situation) but his choice has a more complex aspect to it. Moreover, from the point of view of the onlooking theoretician, the plurality of actors faces the identical individual "dilemma": more-or-less autonomous actors will determine a joint selection of actions— perhaps acting in concert, perhaps not—and only then will each of them "have" an outcome. Thus, the focus of analysis passes from the individual's choice situation to the interactive choice situation.

A certain symmetry obtains in the interactive choice situation: the actors each may have the explicit orientation (2), in which the dependency is a cognized element of the situation, so that each realizes that the situation is structured and that every other actor knows it is structured. Finally, the outcomes of concern to the actors may be realized not immediately but only after a succession of interactions. Thus, the situation evolves toward an outcome. In the nature of the situation and of the actors' apprehensions of it, each actor is oriented to the choice situation of others.

Example. (In conjunction with Figure 21.1.1) In a certain Department of Sociology, in order for a graduate student (actor 1) to qualify for continued work toward the Ph.D., he must at some point take a comprehensive examination. An examination committee (actor 2) is formed, conducts an examination and arrives at a decision, "pass" or "fail." If the result is "fail," the student can appeal the decision. Then an appeals committee (actor 3) is formed to decide whether to sustain the original decision or not. Also, this committee may be formed if any faculty member (actor 4) appeals the "pass" decision of the examination committee. If the appeals committee votes not to sustain the original decision, a second examination committee (actor 5) is formed, conducts an examination, and arrives at a decision that is not subject to appeal. This whole structure is shown in Figure 21.1.1. A tree diagram shows the possible paths that joint action may take. Each node is a decision-point, labeled with the name of the actor who is required to choose one of the actions (lines out of the point toward the right). Finally, at the possible terminal points, where the structure is completed, we label such points $A_1, A_2, \ldots, A_8$, the possible outcomes of the entire interactive choice situation.

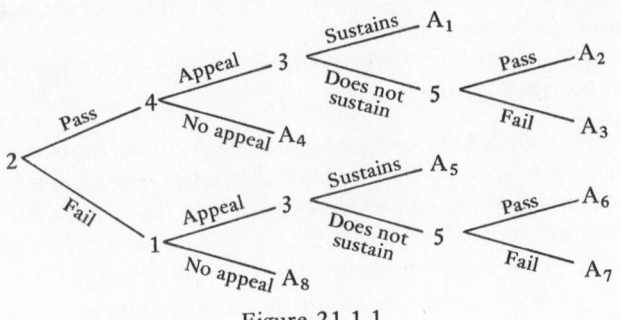

Figure 21.1.1

Here, in this last example, we see the various objects that constitute an interactive choice situation as well as an element we neglected in our earlier discussion: namely, the actions available to an actor are defined for him by the rules governing that "type" of situation. Briefly, this situation is not only structured in the sense of dependency of outcomes for each actor upon joint actions, but in fact it is even, one might say, "metastructured" by the preexisting rules establishing possible paths of joint action. Thus, although not every social situation has this character, we recognize in its rule-governed character the true element of the sociological character of the interactive choice situation. By "metastructured," we mean that, examining the basic objects of an interactive choice situation, we see that the idea of "rule-governed" does not require any new object: it defines the objects. A rule-governed interactive choice situation, then, is an interactive choice situation in which human individuals or groups find themselves assigned to various roles, as values are assigned to logical variables.

The mathematical theory of rule-governed interactive choice situations is game theory. Games, in the ordinary sense, are merely special exemplifications of situations that concern the game theoretician. To make the game exemplification more explicit consider ticktacktoe as generated sequentially, with a partial tree diagram of Figure 21.1.2. The start of the game involves a chance move (not depicted in the tree) to determine which person is first (X) or second (O). The first move by person X is to place an X in any of nine positions, labeled by matrix notation (p_{ij}) ($i, j = 1, 2, 3$). Then person O moves and must place O in any remaining position. In Figure 21.1.2, the action possibilities at move 1, assigned to X, are shown. He is assumed to place an X at position p_{22}. Then O moves, with eight possible actions, choosing p_{33} to place an O. Thus, lines represent alternative actions at a move, as in the social structural example above, and "moves," or sets of alternatives at a given decision-point, are assigned to players (actors). The eventual outcome (X wins, O wins, or draw) quite clearly

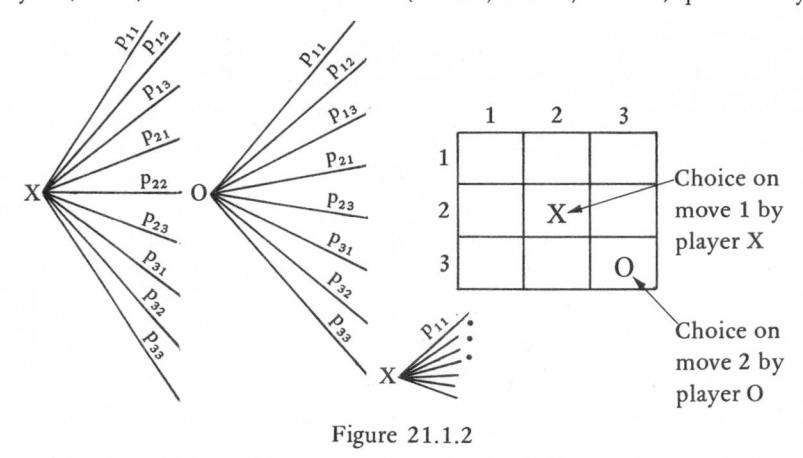

Figure 21.1.2

depends on the joint action of both persons. Thus, the game possesses the basic elements of an interactive choice situation, and in addition, these elements arise by rules, as in the social structural case. The point of this example is that while game theory applies to games, in the conventional sense, games are just special rule-governed interactive choice situations.

Moreover, it must not be forgotten that the theory of games did not arise because of any intense desire to discover the anatomy of ordinary games. Rather, it was seen from the outset by NM (1947) that the phenomena to be represented were the rule-governed interactive choice situations. Thus, in a certain sense, the nomenclature "game" is misleading: it suggests only a special subclass of the phenomena. In another sense, however, the special subclass plays a deep, even profound, role. This we realize if we recall the notion of a representation theorem in mathematics, outlined earlier in connection with measurement theory. An appropriate picture here is shown in Figure 21.1.3,

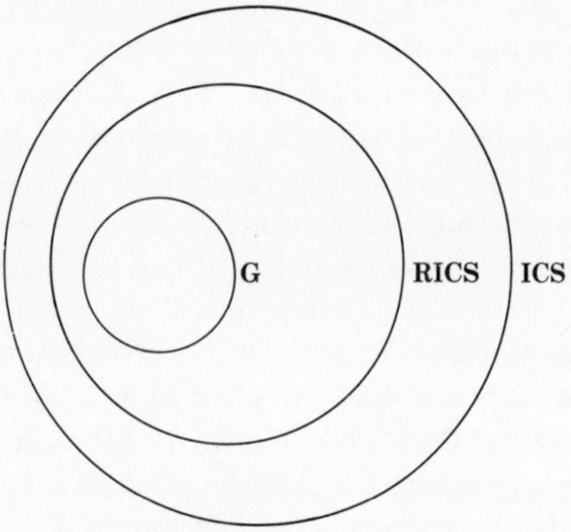

Figure 21.1.3

where "ICS" means the class of all interactive choice situations, "RICS" means the subclass in which the situation is defined by rules, and "G" means a subclass of these—the "games" in the ordinary sense, particularly "games of strategy" rather than "games of chance." NM had in mind the hypothesis, as it were, that any situation in **RICS** could be represented by a suitable game (of strategy). This is the logic of a representation theorem, except for one significant difference: there is no proof for this hypothesis because one concept, game, is to be constructed exactly, while the other concept, rule-governed interactive choice situation, is allowed to remain intuitive or empirical. Thus, the logic is the logic of model-building rather than the logic of mathematical proof. To show that any x in **RICS** is representable by some game in **G**, one constructs and verifies a model of x, using some game in **G**.

Thus, several steps are involved in this theoretical program. First, the class **G** must be exactly specified once the idea is understood that the game concept is to provide the class of models for representational purposes. This involves several substeps: defining the notion of game by a set of axioms; providing examples of the exact concept that convince us intuitively that indeed the intuitive notion has been formalized; defining the additional theoretical problems for the pure theory of games, showing their relation to the class **RICS** (or even **ICS**); proving important theorems that answer or in some sense provide solutions for the theoretical problems; and relating these solutions to the class **RICS**. This gives a very brief sketch of the ideas that lie behind the founding of the theory of games as a mathematical discipline. Our next step is to provide the mathematical foundations for the definition that NM sought to frame (i.e., the construction of *game* as an exact concept) and then to provide the definition, with examples.

21.2. Graph Theoretic Formulation of the Tree Notion. A glance at Figures 21.1.1 and 21.1.2 shows that again the tree diagram provides an intuitively clear way of picturing the sequential logic of some phenomenon. In the case of the concept of a game, however, the tree becomes not merely a useful heuristic device but an important ingredient in the very definition of the concept. This is because a tree has a starting point and a series of possible paths to a variety of terminal points; thereby, it becomes a representation of one of the defining features of a game—namely, the existence of a beginning point and a variety of possible plays (in accordance with rules) leading to possible termination points of the play. Hence, in this context it becomes necessary to formalize the concept of a tree prior to defining the game notion.

Clearly, a tree is a special sort of graph. But even the notion of graph was treated only intuitively in earlier work, in connection with binary relational systems. To define a tree we first make exact the notion of an oriented or directed graph with no loops on its nodes. Harary, Cartwright, and Norman (1965) have called this sort of directed graph a digraph. The definition is as follows: Let (D, L) be a binary relational system (i.e., D is a set and L a relation on D). Then (D, L) is a digraph if and only if

(1) D is a finite set, and
(2) L is irreflexive on D [i.e., $(x, x) \notin L$, for all $x \in D$].

The elements of D are the nodes (or vertices or points), and the pairs in L are the (directed or oriented) lines.

To specify the distinguishing properties of a digraph that make it a tree, certain additional concepts are needed, as follows.

A semi path from point a to point b is a sequence of distinct points $x_1, x_2, \ldots,$ x_n, with $x_1 = a$, $x_n = b$, $x_i \in D$, such that for $i = 1, 2, \ldots, n-1$,

$$(x_i, x_{i+1}) \in L \text{ or } (x_{i+1}, x_i) \in L$$

Thus, for every adjacent pair (x_i, x_{i+1}) in the sequence a directed line exists in one direction or the other.

For example, the digraph in Figure 21.2.1

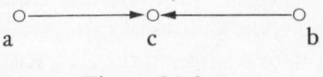

Figure 21.2.1

has a semipath from a to b, namely, $x_1 = a$, $x_2 = c$, $x_3 = b$. Note that we could not arrive at b from a tracing (through forward arcs) from a. This is the reason for the "semi-" prefix. A "path" would require $(x_i, x_{i+1}) \in L$ alone. Whenever a pair of points in a sequence satisfy the requirement for a path (from a to b), they satisfy the requirement for a semipath. Thus, any path is a semipath but not conversely. The above example is a semipath but not a path, because the pair $(x_2, x_3) = (c, b) \notin L$, but $(b, c) \in L$. The digraph in Figure 21.2.2

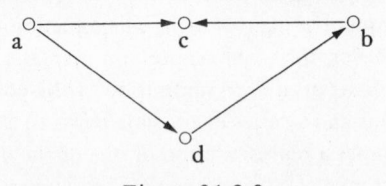

Figure 21.2.2

has a path from a to b, namely, a, d, b. Also, the sequence b, d, a is a semipath.

A digraph (D, L) is said to be (weakly) connected if and only if every pair of distinct points is joined by a semipath: take a and b arbitrarily from D, with $a \neq b$, and check for the existence of some semipath. If at least one such semipath exists from a to b, then (a, b) are joined—in that order. However, reversing the direction of the sequence preserves its character as a semipath; thus, only one direction needs to be checked. Formally, if (a, b) joined by a semipath, then (b, a) joined by a semipath. The symmetry here is not true of the path concept. For instance, in Figure 21.2.2, there is a path from a to b, but no path from b to a. To show that the graph of Figure 21.2.2 is connected, we consider each of the $\binom{4}{2} = 6$ couples of points and specify a semipath in each case:

Couple	Semipath
$\{a, c\}$	a, c
$\{a, b\}$	a, c, b
$\{a, d\}$	a, d
$\{b, c\}$	c, b
$\{b, d\}$	d, b
$\{c, d\}$	c, b, d

If the relation L on D is complete, by definition (see section 4.10) this means that for all x and y with $x \neq y$,

$$(x, y) \in L \text{ or } (y, x) \in L$$

Thus, taking any pair, a line joins them directly, in one direction or another (possibly both, because we satisfy sentence "p or q" when "p and q" holds). Thus, if a digraph is called complete when the relation L is complete on D, then we see that any complete digraph is connected, but the converse is not so, as exhibited by Figure 21.2.2, for which we concluded the graph is connected, but since $(c, d) \notin L$ and $(d, c) \notin L$, this graph is not complete.

The indegree of a point x in D is the number, $i(x)$, given by

$$i(x) = N\{y \in D: (y, x) \in L\}$$

where $N(X)$ means the number of elements in set X. The outdegree of a point x in D is the number, $o(x)$, given by

$$o(x) = N\{y \in D: (x, y) \in L\}$$

In Figure 21.2.2, we have

Point x	i(x)	o(x)
a	0	2
b	1	1
c	2	0
d	1	1

For example, $i(a) = 0$ because there is no line into a. In sociometry, the indegree is the number of choices received; the outdegree, the number of choices given.

The important type of digraph for the theory of games is called a tree. Intuitively, we have been using tree diagrams throughout this book and yet never stopped to produce a formal definition of when a graph formed a tree. It was not necessary. But for game theory, we want an exact concept of a tree.

(21.2.1) Definition. Let (D, L) be a digraph. Then (D, L) is a tree if and only if

(1) it is connected;

(2) there is exactly one point in D, say x_0, such that its indegree is zero: $i(x_0) = 0$;

(3) for every point $x \neq x_0$, we have indegree of one: $i(x) = 1$.

We can check that our intuitions are satisfied here by recalling a few probability tree diagrams, with the probability assignments removed. In Figure 21.2.3, on the left, we show the tree used for calculating transition probabilities in a certain learning model (recall section 10.7). Then, on the right, we eliminate the learning model notation and have an abstract digraph with

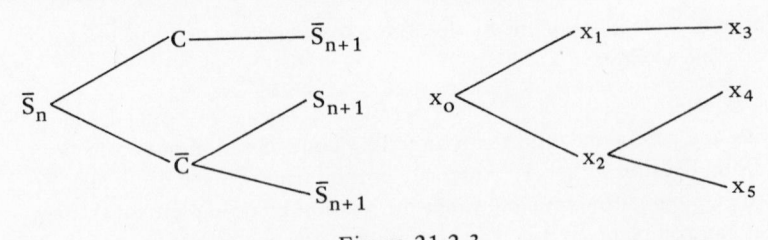

Figure 21.2.3

$$D = \{x_0, x_1, \ldots, x_5\}$$
$$L = \{(x_0, x_1), (x_0, x_2), (x_1, x_3), (x_2, x_4), (x_2, x_5)\}$$

Thus, $(x_i, x_j) \in L$ if and only if a line directed from left to right goes from x_i to x_j (in one step). Of course, in this case, we can see the connectedness property at a glance. This shows that condition (1) for a tree is satisfied. Conditions (2) and (3) are also true because no line is into x_0 and because for all points except x_0 only one line is into that point from the left.

Given a tree, there is a distinguished node, then, characterized by the unique property, so far as the collection of points is concerned, that $i(x_0) = 0$. Also, the tree diagrams make clear that there are points that terminate the digraph in the sense that no lines issue out from them: their outdegree is zero, $o(x) = 0$ for such x. In Figure 21.2.3, the set of terminal points is $\{x_3, x_4, x_5\}$. Put exactly, x is a terminal point of (D, L) if and only if $o(x) = 0$. Thus, a tree has a distinguished subset of points, its terminal set.

21.3. Game Trees. The examples of section 21.1 make it clear that a tree alone is not a sufficient characterization of a game. The tree notion formalizes the unfolding possibilities in the finite sequential process. But in rule-governed interactive choice situations a decision-point is associated with a particular role, whose encumbent is designated by the rules as the actor at that point: thus, a particular committee is required to make a choice from a set of possible alternatives. In games, this association is formalized by abstractly partitioning moves, each part associated with a player. We can show this in the tree diagram by using the integers $0, 1, \ldots, n$ (or any other set of n+1 distinct symbols), and the rule that a decision-point is assigned one and only one of these integers. The integer "0" is used to represent the moves where chance takes over. The shuffling of a deck of cards, for example, is represented as a move or decision-point in which the lines out from the point are the possible outcomes of the shuffling (namely, the possible orderings of the deck) and each line carries a probability assignment. To say that the rules include a proviso that the game begins by a dealer's shuffle means that an equal-probability assignment is made on the lines outward from the initial point of the tree. Complex games, such as poker, may have many chance moves. We may sum up this requirement that decision-points be associated with decision-makers by the stipulations:

(1) Consider the set of nonterminal points of the tree that will represent the game; then this set of points is partitioned into n+1 sets, called, say,

$$S_0, S_1, S_2, \ldots, S_n$$

where S_0 is the set of chance moves and may be empty:

(2) For each chance move in S_0, there is a probability assignment over the lines starting at that move.

There is an additional aspect of games specified by the rules: namely, the information available to the player concerning the results of other moves when it is his move. A minor change in this aspect can drastically alter the game. For instance, the proviso that each player not show his hand in five-card stud poker is a basic property of the game.

This aspect of the game requires representation in terms of the tree. A player makes moves within some S_j class of moves. If information is totally available to him, the moves in S_j are differentiated for the player, they are distinctive; but if some information is withheld from him, he may not be able to differentiate various moves in S_j. This suggests that information can be represented as a partition of S_j; the ideal limits are (1) all moves in S_j are differentiated, and so, player j has maximal information, and (2) all moves collapse into one class, so that player j is unable to "tell the difference" between moves available to him. On a tree diagram, one can enclose any pair of moves not differentiable for the player by a dotted closed curve. Such moves may be termed equivalent for the player. Some examples will help make this idea clear.

Example. (a) True-False Game. Two players 1 and 2, each secretly record one of the letters T, F on a slip of paper. They then compare their choices. If the choices agree, 1 wins; if not, 2 wins. One representation by a tree is shown in Figure 21.3.1.

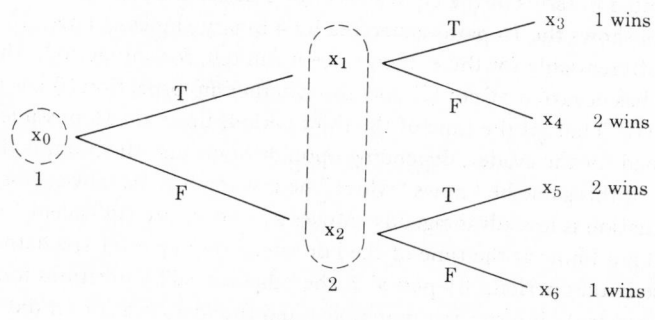

Figure 21.3.1

Note that we represent the secrecy by making moves x_1 and x_2 equivalent for 2: he cannot tell if 1 selected T or selected F. Here the "first move" was assigned to 1, but there is no real significance to this because of the information pattern, as we may call it, implied in 2's not knowing 1's selection. Nevertheless, Figure 21.3.2 shows a perfectly acceptable alternative representation.

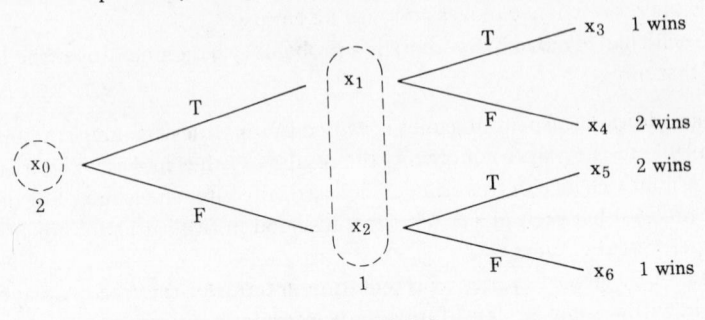

Figure 21.3.2

As long as the proper information pattern is represented, the assignment of "who goes first" is without import for this particular game. The information pattern is represented by the grouping of moves associated with players. Moves grouped together are the same for that player, in the sense that he lacks the information that would differentiate them.

Example. (b) Evasion Game. Rapoport (1966) treated an arms-control situation in terms of a game-theoretic model. A potential evader, player 1, may or may not choose to violate some arms agreement in a certain time period. There is an inspection agency, player 2, which must choose to inspect or not, in any time period. The information pattern is asymmetrical: 1's choice is hidden, but 2's is not, although we assume that if the inspector decides upon an inspection in a given period it will detect any violation if it occurred. Thus, in successive moves, the game develops not only out of the sequence of binary choices but the secrecy aspect represented in terms of the equivalence for 2 of certain possible moves.

Figure 21.3.3 shows the 16 paths generated by 4 binary choices. These 16 paths are all differentiable for the evader, since it can tell, for any period, whether an evasion (e) has occurred or not ($\bar{e}$) and also whether an inspection (i) has occurred or nor ($\bar{i}$). Thus, at the time of the third period, there are 16 possible "moves" defined for the evader, depending upon prior events: the point is that in any "play" of this game he knows "where" he is when it is his move. For the agency the situation is less advantageous. Moves x_1 and x_2 are equivalent for it, because it will not know at the time of the i or $\bar{i}$ decision in period 1, whether e or $\bar{e}$ is the case in that period. By period 2, the "theoretical" 8 positions it could be in are reduced to 3: it inspected in period 1 and found an evasion; it did not inspect in period 1; and it inspected in period 1 and did not find an evasion.

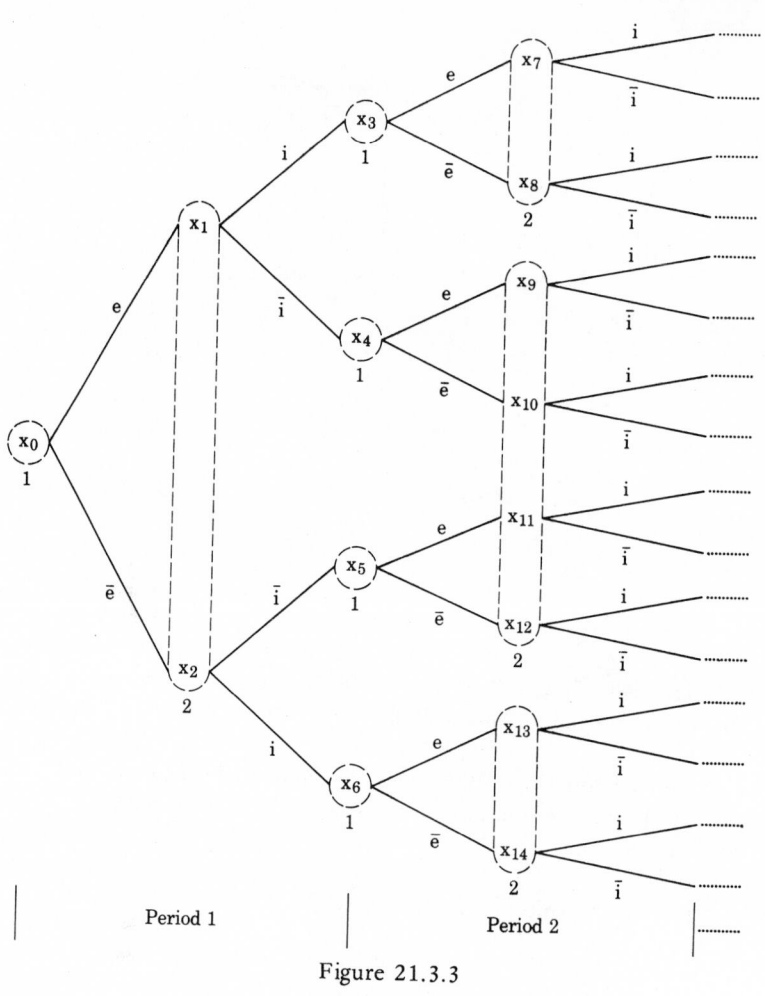

Figure 21.3.3

Example. (c) Abstract Game. A tree that might represent a game with both a chance move and nonchance moves is shown in Figure 21.3.4. The game begins with the chance move as shown, then player 1 moves, knowing the outcome of the chance move; then player 2 moves, and here the information pattern is selective: if 1's choice was made at x_1, then 2 cannot tell what 1 actually did, but if 1's choice was made at x_2, then 2 has information as to the alternative selected by 1. The game concludes after player 2 makes his choice. The (possible) terminal points are $x_7, x_8, \ldots, x_{16}$. This example shows that the logic of the game is completely prescribed by the tree, suitably labeled. The tree defines an "abstract game" that could be the structure of some real game or some real interactive choice situation modeled by an application of this abstract game.

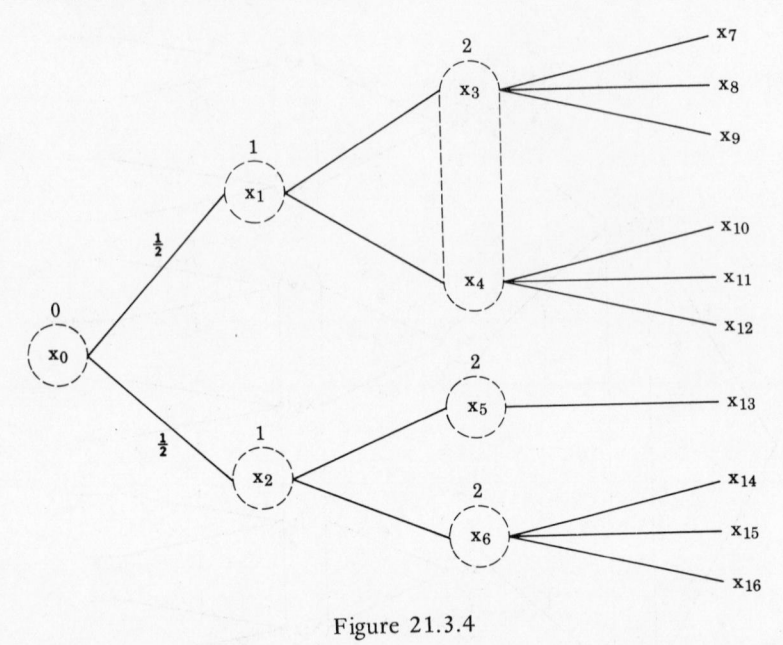

Figure 21.3.4

Perhaps the only missing item in Figure 21.3.4 is the set of outcomes and a mapping that associates an outcome with each terminal point of the game.

21.4. Axioms for a Finite Game; Examples. Having provided examples illustrating games or gamelike situations as trees, we move directly to the axiomatic definition of a game.

(21.4.1) Definition. A (finite) game is given by

$$G = (\Gamma, \mathcal{S}, \mathcal{P}, \pi, \Omega, \omega)$$

satisfying:

Axiom G1. Γ is a tree, (D, L).

Axiom G2. $\mathcal{S}$ is a partition of the set of nonterminal points of Γ, and we write

$$\mathcal{S} = \{S_0, S_1, \ldots, S_n\}$$

Axiom G3. $\mathcal{P}$ is a family of probability assignments, one for each point in S_0. If $v \in S_0$, and $o(v) = k$ (the outdegree of v is k), then we can let $\{p_v^{(1)}, p_v^{(2)}, \ldots, p_v^{(k)}\}$ denote the probability assignment associated with v.

Axiom G4. π is a family of sets, Π_i (i = 1, 2, ..., n) such that each Π_i is a partition of $S_i \in \mathcal{S}$, denoted

$$\Pi_i = \left\{ S_i^1, S_i^2, \ldots, S_i^j, \ldots, S_i^{m_i} \right\},$$

satisfying

(a) $v, v' \in S_i^j \quad \Rightarrow \quad o(v) = o(v')$

(b) $v, v' \in S_i^j \quad \Rightarrow \quad (v, v') \notin L \text{ and } (v', v) \notin L$

for $j = 1, 2, \ldots, m_i$. (Here m_i is the number of parts in Π_i.)

Axiom G5. Ω is a finite set and ω is a function with domain the terminal points of Γ and range in Ω.

Given this exact concept, we can stabilize the technical vocabulary of game theory. Γ is the game tree, of course. Nodes in D are the moves. They give rise to possible following moves via alternatives, represented as lines in L. The partition of the nonterminal moves, $\mathcal{S}$, is called the player partition. S_0 is the set of chance moves; and n is the number of persons or nonchance players. $\mathcal{P}$ gives the sets of probabilities of the alternatives at any chance move. π is called the information pattern. Π_i is the partition of player i's moves (S_i) into equivalence classes. The necessary conditions (a) and (b) of axiom G4 conform to this requirement: if two moves v, v' in a given class had different outdegrees, this would mean that a player might be able to discriminate them on this basis. Also, if one led to another they could be distinguished by this alone: hence, condition (b) requires that no line directly join them. Finally, Ω is termed the set of outcomes, and ω is a mapping that assigns an outcome to each terminal point of the game tree.

Example. (a) True-False Game. The game was represented by the tree diagram of Figure 21.3.1. The identification of the various objects of the game, applying definition (21.4.1) is as follows:

(1) $\Gamma = (D, L)$, where

$D = \left\{ x_0, x_1, \ldots, x_6 \right\}$

$L = \left\{ (x_0, x_1), (x_0, x_2), (x_1, x_3), (x_1, x_4), (x_2, x_5), (x_2, x_6) \right\}$

(2) $\mathcal{S} = \left\{ S_0, S_1, S_2 \right\}$, where

$S_0 = \varnothing \qquad$ (there are no chance moves)

$S_1 = \left\{ x_0 \right\}$

$S_2 = \left\{ x_1, x_2 \right\}$

(3) $\mathcal{P} = \varnothing \qquad$ (no chance moves, so no probability assignments)

(4) $\pi = \left\{ \Pi_1, \Pi_2 \right\}$, where

$\Pi_1 = \left\{ S_1 \right\}$

$\Pi_2 = \left\{ S_2 \right\}$

(5) $\Omega = \left\{ \text{"1 wins," "2 wins"} \right\} \equiv \left\{ I, II \right\}$

$\omega(x_3) = \omega(x_6) = I, \omega(x_4) = \omega(x_5) = II$

Example. (b) Evasion Game. Figure 21.3.3 does not completely specify a game, because the tree is tacitly incomplete, and of course, no outcomes are provided. Imagining the situation concluded at the end of period 2, then the tree has terminal points (say, $x_{15}, x_{16}, \ldots, x_{30}$) but no outcomes or outcome mapping. If this is supplied, a game model for this situation is specified. The other objects are as follows:

(1) $\Gamma = (D, L)$, where

$\quad D = \{x_0, x_1, \ldots, x_{30}\}$

$\quad L = \{(x_0, x_1), (x_0, x_2), \ldots, (x_{14}, x_{30})\}$

(2) $\mathcal{S} = \{S_0, S_1, S_2\}$

$\quad S_0 = \emptyset$ (no chance moves)

$\quad S_1 = \{x_0, x_3, x_4, x_5, x_6\}$ (evader's moves)

$\quad S_2 = \{x_1, x_2, x_7, x_8, \ldots, x_{14}\}$

(3) $\mathcal{P} = \emptyset$ (no chance moves)

(4) $\pi = \{\Pi_1, \Pi_2\}$

$\quad \Pi_1 = \{\{x_0\}, \{x_3\}, \{x_4\}, \{x_5\}, \{x_6\}\}$ (essentially S_1)

$\quad \Pi_2 = \{\{x_1, x_2\}, \{x_7, x_8\}, \{x_9, x_{10}, x_{11}, x_{12}\}, \{x_{13}, x_{14}\}\}$

or

$\quad S_2^1 = \{x_1, x_2\}$

$\quad S_2^2 = \{x_7, x_8\}$

$\quad S_2^3 = \{x_9, x_{10}, x_{11}, x_{12}\}$

$\quad S_2^4 = \{x_{13}, x_{14}\}$

Example. (c) Abstract Game. Figure 21.3.4 again did not completely specify an abstract game, because no set Ω or mapping ω exists. The remaining objects are as follows:

(1) $\Gamma = (D, L)$, where

$\quad D = \{x_0, x_1, \ldots, x_{16}\}$

$\quad L = \{(x_0, x_1), (x_0, x_2), (x_1, x_3), (x_1, x_4), (x_2, x_5), (x_2, x_6), (x_3, x_7), (x_3, x_8), (x_3, x_9), (x_4, x_{10}), (x_4, x_{11}), (x_4, x_{12}), (x_5, x_{13}), (x_6, x_{14}), (x_6, x_{15}), (x_6, x_{16})\}$

(2) $\mathcal{S} = \{S_0, S_1, S_2\}$, where

$\quad S_0 = \{x_0\}$

$\quad S_1 = \{x_1, x_2\}$

$\quad S_2 = \{x_3, x_4, x_5, x_6\}$

(3) $\mathcal{P} = \{\{p_{x_0}^{(1)}, p_{x_0}^{(2)}\}\}$, where $p_{x_0}^{(1)} = p_{x_0}^{(2)} = \frac{1}{2}$

(4) $\pi = \{\Pi_1, \Pi_2\}$, where

$\quad \Pi_1 = \{\{x_1\}, \{x_2\}\}$ (essentially S_1)

$\quad \Pi_2 = \{\{x_3, x_4\}, \{x_5\}, \{x_6\}\}$, and so $S_2^1 = \{x_3, x_4\}, S_2^2 = \{x_5\}, S_2^3 = \{x_6\}$

Example. (d) Appeals Game. We can represent the rule-governed interactive choice situation of Figure 21.1.1 as a game by labeling the decision-points and then using the tree diagram of Figure 21.4.1.

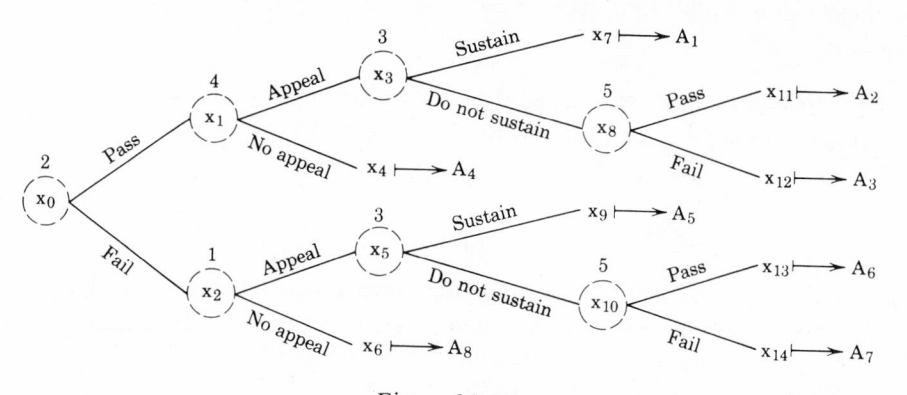

Figure 21.4.1

Here we have the game G given by the following:

(1) $\Gamma = (D, L)$, where

$D = \{x_0, x_1, \ldots, x_{14}\}$

$L = \{(x_0, x_1), (x_0, x_2), \ldots, (x_{10}, x_{14})\}$

(2) $\mathcal{S} = \{S_0, S_1, \ldots, S_5\}$

$S_0 = \emptyset$ (no chance factors included, "purely institutional")

$S_1 = \{x_2\}$

$S_2 = \{x_0\}$

$S_3 = \{x_3, x_5\}$

$S_4 = \{x_1\}$

$S_5 = \{x_8, x_{10}\}$

(3) $\mathcal{P} = \emptyset$

(4) $\boldsymbol{\pi} = \{\Pi_1, \Pi_2, \ldots, \Pi_5\} = \{\{S_1\}, \{S_2\}, \{S_3\}, \{S_4\}, \{S_5\}\}$ and each Π_i is essentially S_i because perfect information pattern.

(5) $\Omega = \{A_1, A_2, \ldots, A_8\}$, ω maps terminal points to Ω as shown in Figure 21.4.1.

Example. (e) In example (d), because of extensive communication, a possible stipulation that the second committee not know the result of the first examination is not likely to work. But at least theoretically it defines a new game, say G', given by all the objects of example (d) except the information pattern $\boldsymbol{\pi}$. This becomes

$$\boldsymbol{\pi}' = \{\Pi_1', \Pi_2', \ldots, \Pi_5'\}$$

where

$$\Pi_i' = \Pi_i, i = 1, 2, 3, 4$$
$$\Pi_5' = \{\{x_8, x_{10}\}\}$$

making it clear that player 5 has essentially one possible move, from its perspective.

21.5. The Game and a Play of the Game. The concept of game must be carefully distinguished from that of play. An appropriate analogy is

Game is to Tree

as

Play is to Path.

A play of a game, on some occasion and in accordance with the rules, will realize some definite path and, hence, via the terminal point, some definite outcome. This distinction is a crucial sociological distinction as well when applied to the class **RICS** of rule-governed interactive choice situations. Each such situation, as defined by rules, has an associated set of possible realizations: thus, the examination procedure treated in section 21.1 has eight possible paths for realization, depending upon the occasion. Essentially, an **RICS** is an institutional form. Any concrete occasion, on the other hand, requires definite human beings acting as individual units or as members of multi-individual groups acting in concert. Thus, in the concrete occasion, the institutional form is realized via one of its possible realizations, provided that the human beings involved act in accordance with the rules. Thus, the theory of **RICS** can proceed to study the logic of such situations by analytically relativizing to the hypothetical class of realized **RICS** in which the human beings "conform"—that is, indeed act as role-encumbents are required to act in the definition of the situation provided by the rules. If this theoretical strategy has value, it is in terms of the scientific methodology of initial idealization to simple cases, completely analyzed prior to the more cumbersome analysis of complex cases. The analysis of real role-encumbents then proceeds against the baseline model of the idealized role-encumbent, the "pure actor," so to speak.

21.6. Actors in Player Roles. Returning to the game concept, as given in definition (21.4.1), we see that the primitive notions do not include any set of players. But play requires players, not mere abstract classes of moves, just as interaction requires people and not just institutional forms. Our next step is to introduce players and, subsequently, to introduce preferences over outcomes, held by players.

Two new primitives are required and need to be identified in any application of a game model: a set A, whose members we term persons (for purely game-theoretic purposes) or actors (for more sociological interpretations); and also a notion of a mapping from A onto the set $\{S_1, S_2, \ldots, S_n\}$ of parts of the player partition of a game G. The mapping needs to be onto (surjective) because we want each part S_j to have its associated player. While many actors (individuals or groups) could be associated hypothetically with one class of moves, when they are, we treat them as forming a unitary entity. (For example, "France" vis-à-vis "England.") Thus, we can form the set A so that the mapping is injective (1-1): if players α_1 and α_2 are distinct, then the corresponding parts of the player partition are distinct. The axioms on the pair (A, r) where

$$r: A \rightarrow \{S_1, S_2, \ldots, S_n\}$$

are as follows:

Axiom P1. A is a finite set.

Axiom P2. r is bijective (1-1 correspondence).

If $r(\alpha_i) = S_j$ then actor, or "person," α_i is said to be the player of S_j. The symbol "r" is used to suggest the assignment of individuals or groups to roles in a social structure, so that we also can say that the role of α_i is to act at certain moves in G. An n-person game G with pair (A, r) is denoted (G, A, r) and called an n-person game with players.

Example. (a) True-False Game. Consider example (a) of section 21.3. The set and mapping

$$A = \{Tom, Mike\}$$
$$r(Tom) = \{x_0\} = S_1$$
$$r(Mike) = \{x_1, x_2\} = S_2$$

together make this a game with players, Tom and Mike, two definite human beings.

Example. (b) Evasion Game. Consider example (b) of section 21.3. The model is made definite with respect to a hypothetical interactive choice situation by identifying two actors, say $\{U.S.A., A.I.A.\}$, where A.I.A. is the "Arms Inspection Agency," and of course, the mapping r claims that the U.S.A. is playing the evader role vis-à-vis the agency. We then have a definite two-person game with players, as a model for some hypothetical arms-control situation.

Example. (c) Appeals Game. In Figure 21.4.1, the game model for the institutional form is made into a model with players by assigning named individuals or groups to each of the five sets $S_1, S_2, \ldots, S_5$ associated with roles in the situation. This "instantiates" the institutional form to these actors, in the specified set A of actors, assigned the specified roles. Note how we naturally identify a role with a class of (possible) moves.

21.7. Interpretation of the Game in Extensive Form.

In pure game theory, a game G satisfying definition (21.4.1) is usually treated with a set of numerical outcomes called payoffs, and since the interpretation is that each nonchance player obtains a payoff on termination of a play, the numerical outcomes are given as vectors with n components. In effect, then, the game theorist lets $\Omega = \Re^n$, where G is an n-person game. Then ω assigns a vector in $\Re^n$ to each terminal point. When this is done the usual assumption is that this numerical assignment is interval in scale-type, although "scale-type" does not figure in purely game-theoretical discourse. In any case, the name "game in extensive form" is given to a game G with $\Omega = \Re^n$.

Example. True-False Game. In example (a) of section 21.3 as formalized in example (a) of section 21.4, this is a two-person game, and so we need vectors in $\Re^2$. Thus, let $\Omega = \Re^2$ and

$$\omega(x_3) = \omega(x_6) = (1, -1)$$
$$\omega(x_4) = \omega(x_5) = (-1, 1)$$

with the interpretation, say, that (x, y) means player 1's payoff is x cents and player 2's payoff is y cents. The game is "zero-sum," which might mean that in fact the payoffs arise by a transfer of money from one to the other.

When game theory is considered in terms of the representational hypothesis (section 21.1) that it be used to construct games as models of arbitrary rule-governed interactive choice situations, a purely primitive payoff concept is inadequate.

To interpret the payoff concept, two methods may be used.

Method 1. Treat the numerical payoffs as meaningful objects for the players, as may be done when they are monetary quantities. Then treat the payoffs as the outcome set generating a space of alternatives over which a player has preferences. Postulate a model for this preference system. Postulate play principles such that, in some specified way, the player plays in accordance with the rules and in accordance with his preference system. One possibility is that the image of the terminal points under ω is taken as the set of pure alternatives for the NM system of Chapter 20. Then the player, say α_i, has a utility function u_i over the numerical payoffs. In playing, it is his utility for monetary gains and losses, and distribution of funds, that determines (in part) his choices. In this interpretation, the function ω taking terminal points into $\Re^n$ is not formally deprimitivised; rather, one assumes it means something as such to the actors in A, for otherwise, one could not very well claim that preferences over it exist.

Method 2. Deprimitivize the payoff function (i.e., construct it from an empirical basis). This assumes that the game has no real numerical payoffs associated with it, that the real outcomes are non-numerical, so that numbers enter the picture only by way of representation in the sense of measurement. This interpretation is the required method for using game models of many rule-governed interactive choice situations. These situations hardly ever come equipped with monetary payoffs, and when they do, method 1 may not apply, because the rule-governed outcomes may consist of more than monetary payoffs. For example, for the student taking an examination the pass-fail outcomes cannot be reasonably coded in monetary terms. We have to deal directly with the preference system over the defined outcomes. If we represent this system numerically, we construct the payoff function by composition:

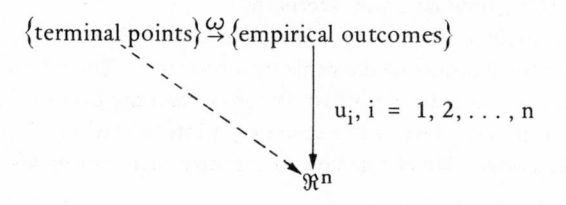

$$\{\text{terminal points}\} \overset{\omega}{\to} \{\text{empirical outcomes}\}$$

$$u_i, i = 1, 2, \ldots, n$$

$$\Re^n$$

where any terminal point of play maps into an outcome (in a qualitative set Ω, rather than $\Re^n$) which, via measurement theory for preferences of n players, leading to n numerical variables, we map into $\Re^n$. Thus, the mapping from the terminal points to $\Re^n$ is constructed. Given an abstract game in extensive form, the second method above abstractly deprimitivizes the payoff function. Thus, it interprets it in terms of player preferences.

Now let us note that under method 1, "players act to maximize payoff" is, in general, false. For example, father and son play checkers. Winner gives 25 cents to loser. It is hardly unusual to expect father to play occasionally to minimize his payoff, so that son can gain self-confidence. In method 2, on the other hand, "players act to maximize payoff" is a representational statement, standing for "each player plays to maximize utility." This statement means that each player plays such that whenever he forms a cognition of two forthcoming alternatives, one stemming from action a_1, the other from action a_2, then he chooses action a_i if the corresponding alternative is preferred, and he is indifferent between the two actions if he holds the relation of indifference for the two corresponding outcomes. For example, father faces on occasion the choice in which he prefers to see his son win, say,

$$\text{Action 1} \longrightarrow 25\cancel{c} \qquad u_2(25\cancel{c}) = 1$$
$$\text{Action 2} \longrightarrow -25\cancel{c} \qquad u_2(-25\cancel{c}) = 1.5$$

Then, he chooses action 2, of course. But this can be put, "he acts to maximize utility," because $1.5 > 1$. Thus, utility maximization is a consequence of the representational schema in which we locate the choice problem. It is a principle regulating the analysis of the situation: it cannot be false provided that we take into account the "correct" alternatives and the operative preferences are numerically represented. Put another way, when a person "fails to maximize utility," this means we did not map his alternatives correctly or we were mistaken about his preferences.

The argument given above applies as well to uncertain alternatives, assuming preferences exist over them. If we adopt a utility theory in which the utility of an uncertain combination is its expected utility, then we can say that the player maximizes expected utility. This is really the same as maximizing utility because "expected utility"—under the expected utility representation—is the utility of a particular uncertain combination of alternatives.

To sum up, a game in extensive form is a game G with outcome set $\Omega = \Re^n$. Thus, ω maps terminal points of the game into n-vectors. The ith component of any such n-vector is called the payoff to player i (when the play of the game has the given terminal point). Two major schemes of interpretation of the payoff function were described. We can name and organize them as follows:

Method 1: Monetary Payoff Interpretation. Defining condition: the payoff vectors are interpreted as meaningful outcomes for the players, and usually this means that the numbers are in monetary units. Options: (a) Players attempt to maximize payoff. This can be wrong. (b) Players have utility functions over the payoffs. This is the utility-of-money assumption, along with "players attempt to maximize utility." For most standard game-theoretical results to apply, this function will need to be a variable that is interval scale-type. A common assumption is that the utility of money is a "nice" function of money, say the log of money.

Method 2: Psychological Payoff Interpretation. Defining condition: the payoff vector is constructed by composition of two functions, the mapping ω into a qualitative outcome space Ω, and a utility fuction u_i defined over the alternatives based on Ω. Options: The main one is the choice of a theory of utility. The NM system (chapter 20) was intended to provide the general theoretical basis for the psychological payoff interpretation, in such a way as to support the empirical meaningfulness of numerical theorems within the theory of games. By and large, this requires that, if a game has any chance moves, the utility of an uncertain combination be its expected utility. Thus, a theory of utility, for general linkage to game theory, must deduce (or assume) the expected utility condition. The NM theory is not the only theory that does so, but it is the theory we will appeal to in our presentation of game theory with a psychological payoff interpretation.

21.8.   The Utility Postulate. Let (G, A, r) be an n-person game with players. By the deployment of the NM system, we can construct the game in extensive form, (i.e., supply it with a payoff function).

(21.8.1) Postulate. Suppose G has outcome set Ω. Let α be the space of alternatives (pure and uncertain combinations) based on Ω. Then for each actor α_i in A, there exists a binary relational system

$$(\alpha, \succeq_i)$$

that is an NM system.

As a consequence of postulate (21.8.1), we know that the representation and uniqueness theorems of NM theory can be applied to give us functions u_i ($i = 1, \ldots, n$) with domain α and range $\mathfrak{R}$. For terminal point τ (tau), if $\omega(\tau)$ is the outcome, then $u_i[\omega(\tau)]$ is the utility of outcome $\omega(\tau)$ for player i, ($i = 1, 2, \ldots, n$).

Thus the n-vector of compositions defines a mapping:

$$\tau \mapsto (u_1[\omega(\tau)], u_2[\omega(\tau)], \ldots, u_n[\omega(\tau)])$$

or, in other words, the function

$$u_i \circ \omega$$

maps terminal points into $\mathfrak{R}^n$. Hence, using this function, we can take the game G into a representation in which Ω is replaced by $\mathfrak{R}^n$. This representation is of the measurement type, the only new item being the n-vector representation arising from n distinct numerical variables.

As mentioned in section 21.7, the conclusion is that we have a game in extensive form in which "payoff" means "utility" for the given player—that is, actor in A playing role $r(\alpha_i)$ in the game G.

With this postulate we are affirming quite formally the remarks made earlier about idealization. An actor who satisfies an NM model is an idealized chooser and will be a bit too definite in his choices to suit psychologists (see Chapter 11) since the relation of preference-indifference ($\succsim_i$) is nonprobabilistic. But it has provided at least one psychological interpretation for game theoretical results that would otherwise be confined to applications dealing with monetary payoffs. Thus, we do not hesitate to refer to it to gain insight into interactive choice situations.

21.9. Setup and Strategic Analysis as Model-Building Phases. Intuitively, a strategy is given by a list of conditional statements of the form

If such-and-such is the situation, then I will act as follows. . . .

Here the antecedent conditions described by the conditional statements cover all possible contingencies. When a game is thought of in terms of a tree diagram, the conditionals are

If the play is at point v and it is my move, then I will choose. . . .

Here v must vary over all possible moves of the player if he is to have a strategy. Our aim now is to define the sets of possible strategies for each player in a game G, by appeal to definition (21.4.1).

Mathematically, a strategy of player i assigns to each possible move of player i (in set S_i) a definite alternative among those existing at that point. Since S_i is partitioned,

$$\Pi_i = \left\{ S_i^1, S_i^2, \ldots, S_i^{m_i} \right\}$$

such that within a class S_i^j the player cannot differentiate moves, we must assign the same choice to each move in the class. Another way of saying this is to regard any move in S_i^j as a representative move of the class and to define a strategy as follows:

(21.9.1) Definition. A mapping with domain Π_i that assigns to each S_i^j in Π_i one of the alternatives at a representative move in S_i^j is a strategy of player i.

We let Σ_i denote the set of all possible strategies of player i.

Example. (a) True-False Game. This continues examples (a) of sections 21.3 and 21.4. We have $\Pi_1 = \{S_1\}$, $\Pi_2 = \{S_2\}$, with $x_0 \in S_1$, and $x_1 \in S_2$ as representative moves (see Figure 21.3.1). The outdegrees $o(x_0) = 2$, $o(x_1) = 2$ show that each player has two possible alternatives. According to definition (21.9.1), the content of Σ_1 is given by

$$\Sigma_1 = \{\sigma_1^1, \sigma_1^2\}$$

where

$$\sigma_1^1(S_1) = T \qquad \text{(first possible strategy of first player)}$$
$$\sigma_1^2(S_1) = F \qquad \text{(second possible strategy of first player)}$$

Similarly, for the second player,

$$\Sigma_2 = \{\sigma_2^1, \sigma_2^2\}$$

where

$$\sigma_2^1(S_2) = T \qquad \text{(first possible strategy of second player)}$$
$$\sigma_2^2(S_2) = F \qquad \text{(second possible strategy of second player)}$$

In this example, strategies and choices are the same because each player has just one move in any play of the game.

Example. (b) Evasion Game. We continue examples (b) of sections 21.3 and 21.4. See Figure 21.3.3. We have, from section 21.4, example (b),

$$\Pi_1 = \{\{x_0\}, \{x_3\}, \{x_4\}, \{x_5\}, \{x_6\}\}$$
$$\Pi_2 = \{\{x_1, x_2\}, \{x_7, x_8\}, \{x_9, x_{10}, x_{11}, x_{12}\}, \{x_{13}, x_{14}\}\}$$

The set Σ_1 of all possible strategies of player 1 is given by considering that each strategy is an assignment of e (evade) or $\bar{e}$ (do not evade) to each of the five distinguishable moves in Π_1. Hence, one such strategy would be "$\bar{e}$ at x_0, but e otherwise," given formally in pointwise mapping notation by:

$$x_0 \mapsto \bar{e}, x_3 \mapsto e, x_4 \mapsto e, x_5 \mapsto e, x_6 \mapsto e$$

This is based on the idea of evading in the second period but not in the first. This strategy, by deciding e uniformly at all moves in the second period does not take account of information available to the evader. A strategy that does take this information into account is

$$x_0 \mapsto \bar{e}, \; x_3 \mapsto \bar{e}, \; x_4 \mapsto \bar{e}, \; x_5 \mapsto \bar{e}, \; x_6 \mapsto e$$

which applies the idea of (attempting a successful) evasion following an inspection that found compliant behavior (i.e., no evasion). In all, a binary choice is involved at each of 5 points, so there are $2^5 = 32$ possible strategies in Σ_1. The set Σ_2 can be investigated in a similar manner. Take a representative point in $\{x_1, x_2\}$, say x_1. There are two alternatives at x_1: i or ĩ. At $\{x_7, x_8\}$ we take x_7: again two alternatives i, ĩ; and so forth. There are 4 sets in Π_2, and for each we find the representative move has 2 alternatives. Thus, Σ_2 has $2^4 = 16$ possible strategies. One example* is

$$\bar{x}_1 \mapsto \tilde{i}, \; \bar{x}_7 \mapsto i, \; \bar{x}_9 \mapsto i, \; \bar{x}_{13} \mapsto i$$

which is the strategy of inspecting only in the second period.

Example. (c) Appeals Game. In the examination institution of example (d), section 21.4, Figures 21.4.1 and 21.1.1, we see that each Π_i is essentially S_i and that each S_i is rather small: Thus, for the most part the strategies are not complicated. Intuitively, we gain some insight into the distinctive game viewpoint if we look at the strategies of actor 3, the appeals committee, whose moves are x_3 and x_5. Here we have the following possibilities:

$$\left.\begin{array}{l} x_3 \mapsto \text{sustain} \\ x_5 \mapsto \text{sustain} \end{array}\right\} \text{first strategy}$$

$$\left.\begin{array}{l} x_3 \mapsto \text{sustain} \\ x_5 \mapsto \text{do not sustain} \end{array}\right\} \text{second strategy}$$

$$\left.\begin{array}{l} x_3 \mapsto \text{do not sustain} \\ x_5 \mapsto \text{sustain} \end{array}\right\} \text{third strategy}$$

$$\left.\begin{array}{l} x_3 \mapsto \text{do not sustain} \\ x_5 \mapsto \text{do not sustain} \end{array}\right\} \text{fourth strategy}$$

By reference to Figure 21.4.1 we can see what these strategies are. The first strategy is the formula "Sustain no matter what." The second strategy is "Sustain if appealed by a faculty man; do not sustain if appealed by student." The third strategy is "If appealed by a faculty man, do not sustain (i.e., accept the faculty challenge and call for new examination); if appealed by the student, sustain." Finally, the fourth strategy is "Do not sustain under any circumstances."

---

*Rather than $x_1 \to \tilde{i}$ it is convenient to write $\bar{x}_1 \to \tilde{i}$ to remind oneself that x_1 is only a representative move in a class $\bar{x}_1$. Wherever a class S_i^j contains more than one move, we will use this notation convention in the description of strategies as mappings.

Now such a committee doubtless would argue that it employs none of these strategies, for not one of them is based on the facts of the "case." We agree that this argument is valid. Therefore, the particular game model is inadequate to represent this situation. But this is far from showing the inadequacy of the game theory framework: it shows only that to meet this critique we should represent the situation by a different game model.

A new model might be based upon the idea that examination committees compare the performance (of the student) with standards. Then the appeals committee has a widened set of strategies, including "Sustain the appeal if the student was failed even though he satisfies standards." Figure 21.9.1 shows this new

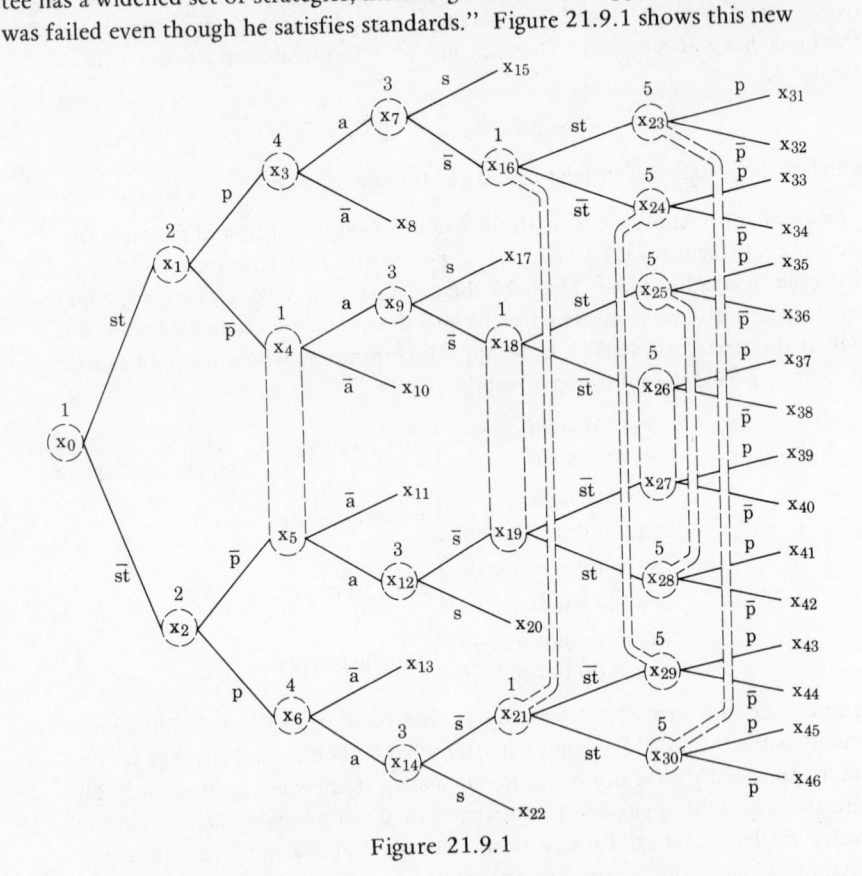

Figure 21.9.1

game representation of the situation. (In the figure, "st" means "satisfies standards"; "p" means "pass"; "a" means "appeal"; "s" means "sustain original decision"; and "x̄" means "not-x," for any symbol x.) We assume in this model that the candidate (actor 1) does not know whether he satisfied standards or not in his performance. This creates imperfect information for him. For example, moves x_{16} and x_{21} are indistinguishable because they are the two paths: $\overset{st}{\to} \overset{p}{\to} \overset{a}{\to} \overset{\bar{s}}{\to} x_{16}$ and $\overset{\overline{st}}{\to} \overset{p}{\to} \overset{a}{\to} \overset{\bar{s}}{\to} x_{21}$, where the only difference is in the unknown matter of satisfying standards. A second actor has imperfect information: the second examination committee (player 5), which also is not aware of the "real" performance in the first examination because it was not there (as was player 2, the first examination committee) and did not investigate to find out what occurred (as did the appeals committee, player 3). The player labeled 4 is the faculty apart from members of committees: we even assumed it functions as an audience for the first examination and, so, has information about the performance independent of the first examination committee's judgment $(p, \bar{p})$. Let us note now that after all the "fuss" created by the appeals committee, the following possibility (path to x_{32}) exists: a student satisfies standards (st), he passes (p), a faculty man appeals the verdict (a), the appeals committee does not sustain the original pass decision $(\bar{s})$, the student performs up to standards in the second examination (st), but he is failed $(\bar{p})$. The institutional form includes no right of appeal from the second examination.

In any case, the strategic possibilities are now enormous. The appeals committee (player 3) has 4 possible distinguishable moves in its set Π_3, namely,

$$\Pi_3 = \left\{\{x_7\}, \{x_9\}, \{x_{12}\}, \{x_{14}\}\right\}$$

These correspond to the partial paths:

$$\xrightarrow{\text{st}} \xrightarrow{\text{p}} \xrightarrow{\text{a}} x_7$$

$$\xrightarrow{\text{st}} \xrightarrow{\bar{\text{p}}} \xrightarrow{\text{a}} x_9$$

$$\xrightarrow{\overline{\text{st}}} \xrightarrow{\bar{\text{p}}} \xrightarrow{\text{a}} x_{12}$$

$$\xrightarrow{\overline{\text{st}}} \xrightarrow{\text{p}} \xrightarrow{\text{a}} x_{14}$$

There are two choices at any such "state of the game," and so, this committee has a total of $2^4 = 16$ possible strategies in its strategy set Σ_3. For example, one strategy is as follows:

$$\text{If} \xrightarrow{\text{st}} \xrightarrow{\text{p}} \xrightarrow{\text{a}} x_7 \text{ then s.} \qquad (x_7 \mapsto s)$$

$$\text{If} \xrightarrow{\text{st}} \xrightarrow{\bar{\text{p}}} \xrightarrow{\text{a}} x_9 \text{ then } \bar{s}. \qquad (x_9 \mapsto \bar{s})$$

$$\text{If} \xrightarrow{\overline{\text{st}}} \xrightarrow{\bar{\text{p}}} \xrightarrow{\text{a}} x_{12} \text{ then s.} \qquad (x_{12} \mapsto s)$$

$$\text{If} \xrightarrow{\overline{\text{st}}} \xrightarrow{\text{p}} \xrightarrow{\text{a}} x_{14} \text{ then } \bar{s}. \qquad (x_{14} \mapsto \bar{s})$$

Mathematically, this is the mapping defined parenthetically. It is the strategy that the committee would regard favorably because it leads to "fair" outcomes. In other words, the outcomes at terminal points obtained via paths based in part on this strategy are preferred by player 3 to those at other terminal points, such as, for example,

$$\xrightarrow{\text{st}} \xrightarrow{\bar{p}} \xrightarrow{\bar{a}} x_{10}$$

in which a satisfactory performance led to an uncontested failure. Note that the committee cannot strategically prevent this outcome: its ability to control the type of outcome is only partial, despite its presumed preference for "justice." However, since the first examination committee has information as to the st - $\overline{\text{st}}$ "choice" by the student, then assuming it prefers that $\bar{p}$ follow $\overline{\text{st}}$ only, the above "unfair" outcome will not be realized. Thus, this outcome can be avoided, assuming a "fair-minded" committee. In effect, a student who selects option a following the path $\overset{\text{st}}{\to} \overset{\bar{p}}{\underset{\to}{\mathbb{R}}}$ is claiming an unfair committee existed, but since he cannot distinguish the two paths

$$\xrightarrow{\text{st}} \xrightarrow{\bar{p}} x_4 \text{ and } \xrightarrow{\overline{\text{st}}} \xrightarrow{\bar{p}} x_5$$

his appeal is always made with uncertainty as to how he "really" performed.

This rather prolonged example illustrates that one can construct different game models for one and the same rule-governed interactive choice situation—Figures 21.4.1 and 21.9.1 present two models for one situation—so that the general idea is very concretely exemplified that the general logic of model-building applies as soon as one leaves the austere pure theory and its rigorously defined class of games, which function as possible models for social situations. In particular, the three basic phases of model-building (see Chapter 1) are realized here: (1) the setup phase, which means drawing an appropriately labeled tree diagram satisfying the G axioms, or, in more complex cases, bypassing the diagram but providing a formulation of the objects in the game model; (2) the analysis phase, which we illustrated in terms of strategic analysis: once the game model is rigorously specified in the setup phase, the definition of a strategy-set for a player applies immediately, and so, one can examine the strategic options of the players; and (3) the evaluation phase, which is the weak link in game models because it is not clear what we should demand of a game model: certainly not a prediction of the play, for this would confuse game with play (a mistake similar to confusing a grammar with an utterance, as is emphasized by Chomsky, 1965; see also Axten, 1972). But surely, a "proper" strategic analysis, one that fits our intuitions as to the possibilities of social action in the situation, is what we want.

CHAPTER TWENTY-TWO THE NORMAL FORM AND PLAYER ASSUMPTIONS

22.1. The Normal Form Concept in Mathematics. Given a game model, the entity constructed in the setup phase, the next step is strategic analysis. For the purpose of strategic analysis, the tree is a very unwieldy object: perhaps our discussion of the examination appeals procedure as a game convinced the reader that at some point the tree is too complex to permit cogent strategic analysis. Thus, NM were led to seek a simpler way of representing a game that nevertheless preserved the basic strategic structure. But this is a representation problem again, and the representation logic applies. Namely, the problem is to find a class of games (call it **M**) such that for any game G there exists a game in **M** to represent G (i.e., to use in place of G but preserving the structure of G). For the moment, imagine that it is clear when two games are "strategically isomorphic." Also, let the class of all games be denoted **G**. Then the appropriate picture is shown in Figure 22.1.1.

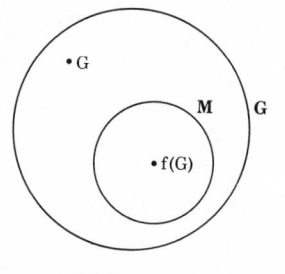

Figure 22.1.1

The representation theorem then needs to take the form: for any G in **G**, there exists a game, say f(G), in **M** such that f(G) is strategically equivalent to G. Then f(G) represents G for the strategic analysis. The uniqueness theorem will be as follows: Any two games in **M** representing G are themselves strategically equivalent.

631

The reader should recall that the representation problem of measurement theories is of the same generic type. Unlike the problem of representation of a phenomenon—as in the construction of a game-model of an interactive choice situation—this representation problem is purely mathematical (i.e., it involves abstract construction and deduction only).

In various parts of mathematics, the representation problem occurs. It has become conventional to say that the objects in the representing class are canonical or normal forms. Then the representation theorem simply claims that any object in the category of mathematically defined objects (e.g., any finite game) can be "put in normal form." Also, for a given definite object, it is desirable that there be an algorithm or set of rules to transform the object into normal form. When all this is accomplished, the pure theory—as of class **G**—can be confined to the smaller, and presumably simpler, class **M** and this "in all generality" because of the fact that **M** constitutes a class of normal forms.

We turn now to the specification of the class **M** of normal forms for strategic analysis.

22.2. The Class of Normal Forms for Game Theory.

Let Γ be a tree (D, L). Then we can define a level assignment over it as follows (see Harary, Norman, and Cartwright, 1965). The distinguished node x_0 is assigned 1, the class of moves following x_0 is assigned 2, the class of moves following any of those in class 2 is assigned 3, and so on until the class of moves such that all following moves are terminal receives the last integer, say m. We say that the tree has m levels. For example, in Figure 22.2.1 we show a 5-level tree.

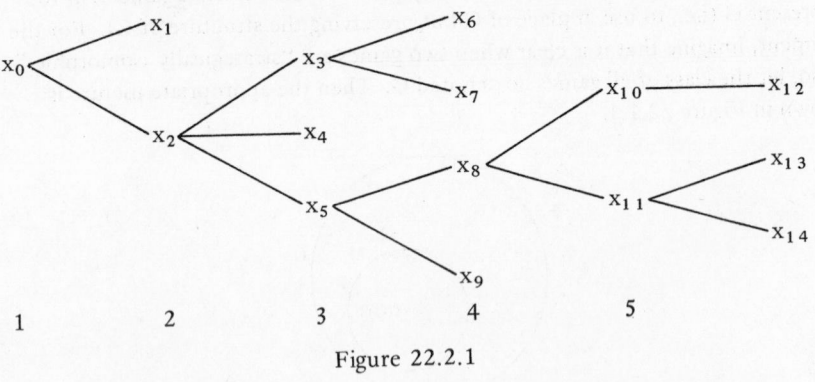

Figure 22.2.1

This level assignment differs from that of Harary, Norman, and Cartwright in detail (they begin with level 0 and assign m to the array of terminal points) but the modification is useful for a game tree. If Γ is part of a game, we would ordinarily say that "the game has m moves" if its tree has m levels. But we have let "move" refer to the sets of alternatives at points. Commonly, we also would say that

there are m turns in any play, and a player could ask "Is it my turn?" rather than "Is it my move?" Thus, we will technically designate the levels of a game tree turns.*
In Figure 22.2.1, the tree could be associated with a game in which there are 5 turns and 14 moves, of which only 7 are nonterminal moves possible in the 5 turns of the game. Recall a move v is terminal if $o(v) = 0$.

Now imagine a game with 2 players and no chance element. What is the simplest form of such a game strategically, assuming that each player has some moves assigned to him? Clearly, each player takes 1 turn and then the game is over. Thus, the tree would have m = 2, as in Figure 22.2.2, recapitulating the Evasion Game but taking only one period (see Figure 21.3.3) and not noting outcomes.

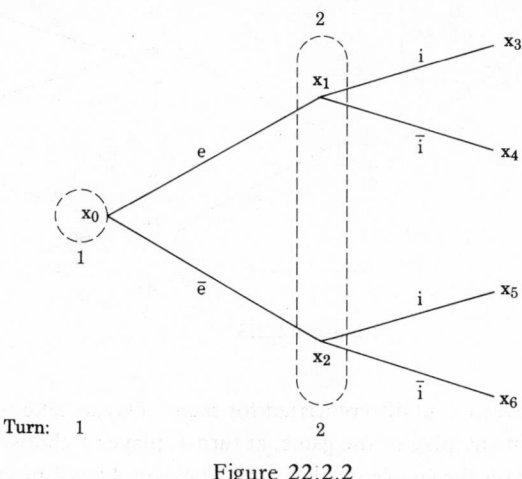

Turn: 1   2

Figure 22.2.2

Here we see a representation in which player 1 has the first turn, player 2 has the second turn, and then the game is over. Thus, there are m = 2 turns. Note that even if the information of player 2 permitted differentiation of moves x_1 and x_2, the number of turns would still be 2.

Now imagine a 3-person game. For example, suppose that we augment the game of Figure 22.2.2 by adding another nation, an onlooker, (player 3) whose options are "Support inspection schemes" or "Do not" ($s, \bar{s}$). Then we want this player to take the third turn: assume that by a spy-service, player 3 learns the decision of player 1 and, so, has perfect information. Figure 22.2.3 applies. There are 3 turns, with player i "taking" turn i, i = 1, 2, 3. Analysis of Figure 22.2.3 will help keep terminology clear: as defined, this game has 3 persons and no chance player. Players 1 and 3 have perfect information in the sense that their

*There appears to be no standard terminology in game theoretic discussions for what we call turns.

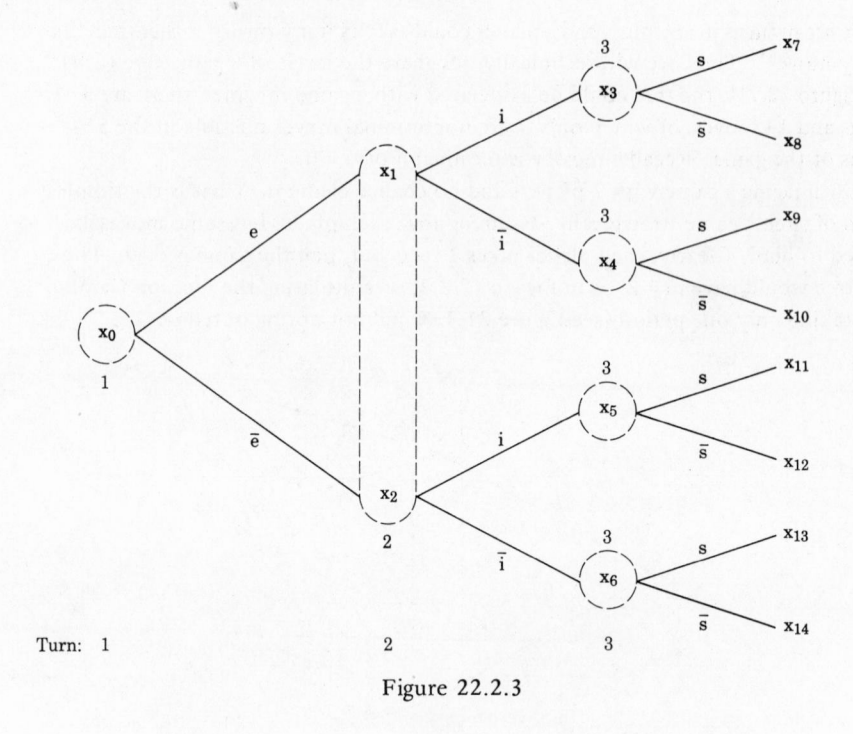

Turn: 1   2   3

Figure 22.2.3

assigned possible moves are all differentiated for them. Players take turns, of which there are 3. In any play of the game, at turn 1, player 1 chooses e or $\bar{e}$; at turn 2, depending on the choice of player 1, the player 2 has 2 possible moves, but according to the rules of the game, he cannot differentiate them and, so, essentially has only 1 possible move. Since we are in a play of the game, he chooses i or $\bar{i}$; finally, the third turn is taken by player 3, whose exact move depends upon the choices (paths) made in the earlier turns, but in any case he chooses s or $\bar{s}$ at his turn, using the information as to the past (earlier turns) in the game. This leads to a terminal point and "the game is over." More exactly, "the play is over."

The number of turns need not agree with the number of players: recall chess, checkers, poker, etc., or examine Figure 21.9.1. There are six turns, but only five players. Or, consider Figure 21.4.1: there are four turns and five players. Thus, a turn need not be associated with a unique player, nor need the number of turns be smaller than the number of players. But the most important types of games in extensive form have many more turns than players.

The intuition of NM was to seek to use the strategy notion in such a way that every game could be analyzed as a game in which each player took precisely one turn, no more, no less. Then an n-person game would be represented as an n-level tree: this would be true of chess, of ticktacktoe, of any game. If this could be done, then the construction of theory for strategic analysis could confine itself to the study of n-turn games. We provide details and examples in the forthcoming paragraphs.

22.3. Transformation to Normal Form.

Consider, again, the set Σ_i of all strategies of player i. Recall that by the Cartesian product of a series of sets $A_1, A_2, \ldots, A_n$ we mean

$$A_1 \times A_2 \ldots \times A_n = \left\{(a_1, a_2, \ldots, a_n): a_i \in A_i, i = 1, 2, \ldots, n\right\}$$

the set of all n-tuples where the ith component is drawn from A_i. Form the Cartesian product of strategy sets $\Sigma_1, \Sigma_2, \ldots, \Sigma_n$, and call it Σ:

$$\Sigma = \Sigma_1 \times \Sigma_2 \times \ldots \times \Sigma_n$$

If σ_i represents an arbitrary strategy of person i, as exactly defined by (21.9.1), then an arbitrary n-tuple in Σ can be indicated by

$$(\sigma_1, \sigma_2, \ldots, \sigma_n)$$

and we denote this arbitrary n-tuple by the symbol σ. Hence, σ is a list of n strategies, one per person in the game.

Example. (a) Evasion Game. Consider the two-period inspection-evasion game of example (b), section 21.9. Two players of the nonchance type exist, so two persons, in the technical sense. One strategy in Σ_1 for player 1, call it σ_1^0, is

$$\sigma_1^0 : x_0 \mapsto \bar{e}, x_3 \mapsto e, x_4 \mapsto e, x_5 \mapsto e, x_6 \mapsto e$$

and one definite strategy in Σ_2 for player 2, call it σ_2^0, is

$$\sigma_2^0 : \quad \bar{x}_1 \mapsto \bar{i}, \bar{x}_7 \mapsto i, \bar{x}_9 \mapsto i, \bar{x}_{13} \mapsto i$$

Intuitively, it seems that the pair of strategies (thus in $\Sigma = \Sigma_1 \times \Sigma_2$) should eventuate in some definite path of action, some play, if employed. Let us see (consult also Figure 21.3.3):

    Turn 1: Player 1's turn, at origin.
    Move: x_0
    Choice: $\bar{e}$ (by strategy σ_1^0)

    Turn 2: Player 2's turn, at move determined by choice at turn 1.
    Move: x_2
    Choice: $\bar{i}$ (by strategy σ_2^0 since x_2 is in $\bar{x}_1$)

    Turn 3: Player 1's turn, at move determined by choice at turn 2.
    Move: x_5
    Choice: e (by strategy σ_1^0)

Turn 4: Player 2's turn, at move determined by choice at turn 3.

Move: x_{11}

Choice: i (by strategy σ_2^0 since x_{11} is in $\bar{x}_9$)

End.

The sequence for this strategy pair, $\sigma^0 = (\sigma_1^0, \sigma_2^0)$, was

$$(\text{start}) \xmapsto{\sigma_1^0} \bar{e} \xmapsto{\sigma_2^0} \bar{i} \xmapsto{\sigma_1^0} e \xmapsto{\sigma_2^0} i \quad (\text{terminal})$$

From example (a), we can see that if each player is associated with a definite strategy, then (in a game with no chance moves) a unique play is specified. Thus, we could represent the two-person game in this manner:

Turn 1: Player 1 chooses a strategy from Σ_1.

Turn 2: Player 2 chooses a strategy from Σ_2.

End.

In other words, we transform the game into a two-turn game rather than a four-turn game.

In general, we have for n persons and with no chance moves the following:

Turn 1: Player 1 chooses a strategy from Σ_1.

Turn 2: Player 2 chooses a strategy from Σ_2.

.

.

.

Turn n-1: Player n-1 chooses a strategy from Σ_{n-1}.

Turn n: Player n chooses a strategy from Σ_n.

End.

The transformation to normal form is completed for G by noting the outcome assigned to the terminal point determined by $\sigma = (\sigma_1, \sigma_2, \ldots, \sigma_n)$.

This covers the case where there are no chance moves.

If there are chance moves, a definite terminal point does not exist even when we are given σ. Instead, if we let $\mathfrak{J}$ be the set of terminal points, with τ an arbitrary point in $\mathfrak{J}$, then the chance moves determine a probability assignment over $\mathfrak{J}$, given σ:

(22.3.1) $P_\sigma(\tau)$ = probability of termination at τ, given strategy n-tuple σ

so that

$$\sum_\tau P_\sigma(\tau) = 1$$

as τ varies over $\mathfrak{J}$.

Example. (b) Simplified Poker. This example is taken from Kemeny, Snell, and Thompson (1966) (see Figure 22.3.1). A description of the rules is as follows: An ordinary deck of cards is used. Cards are called "high" (H) if red and "low" (L) if black. Each of two person antes an amount a to the pot before the deal. Assume a dealer exists. He shuffles the cards well and deals one to each player. The person to the dealer's right is called player 1, the other is called player 2. Player 1 has the first turn after the deal. His option is to stay (s) or to raise (r). If he stays, cards are shown and the pot goes to the player with the higher card or, in the event of equivalence (HH or LL), they break even. If he raises, he puts amount b into the pot, and it becomes player 2's turn. Player 2 decides to fold (f) or to call (c). If he folds, player 1 gets the pot. If he calls, he puts amount b in the pot, and the cards are shown. Again, higher card takes the pot, or if the cards are the same, they break even. Figure 22.3.1 represents the game as a tree diagram; the dealer provides the chance move x_0, and the outcomes, shown in parentheses at terminal points, are the winnings or losses of player 1.

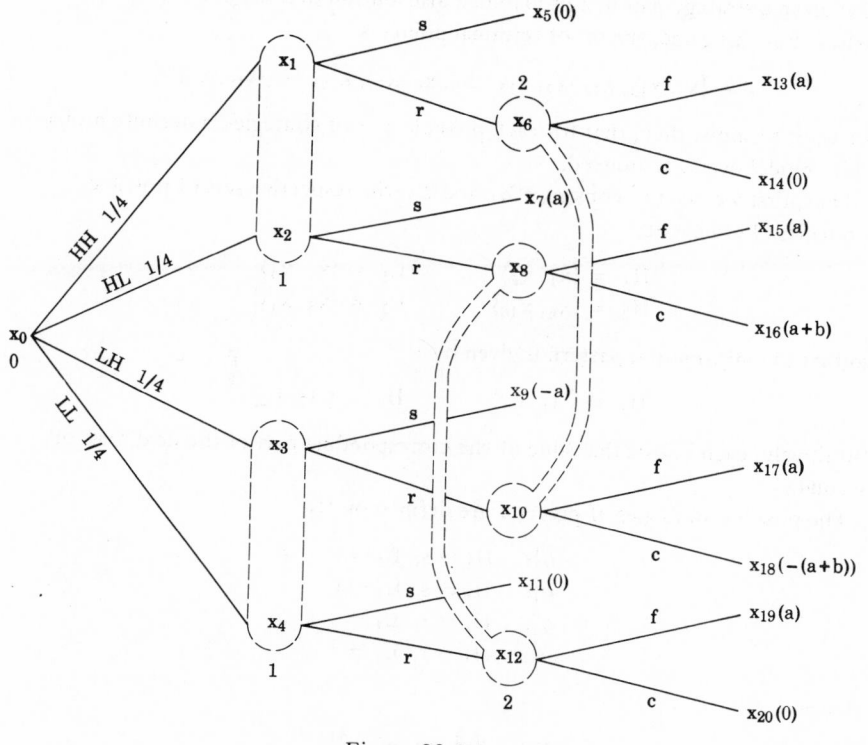

Figure 22.3.1

For example, the play

$$x_0 \xrightarrow{\text{HH}\left(\frac{1}{4}\right)} x_1 \xrightarrow{\quad r \quad} x_6 \xrightarrow{\quad f \quad} x_{13} \quad \text{(a)}$$

occurs if the deal gives both players a high card (chance 1/4). Player 1 raises by placing amount b in the pot, and then player 2 folds. Player 1 then wins amount a from player 2.

Note that the information sets are based on the idea that each player will know only the value of his own card after a definite deal, according to the rules. For example,

$$\left\{x_1, x_2\right\}$$

forms an information set for player 1 because he cannot tell the difference between the deals HH and HL at the time of his move. This game has three turns, the first belonging to "the chance player," (i.e., the dealer or the random device providing the cards to the players in a well-shuffled manner). Our immediate objective in this example is to provide a concrete example of expression (22.3.1): that given a strategy pair σ, a probability distribution over the terminal points exists. For this game, the set of terminal points is

$$\mathfrak{I} = \left\{x_5, x_{13}, x_{14}, x_7, x_{15}, x_{16}, x_9, x_{17}, x_{18}, x_{11}, x_{19}, x_{20}\right\}$$

We want to show, then, that for each possible pair of strategies, a definite probability model over $\mathfrak{I}$ is induced.

Thus, first we need to construct Σ_1 and Σ_2, the respective sets of possible strategies. Let us write

$$H_1 = \left\{x_1, x_2\right\} \qquad L_1 = \left\{x_3, x_4\right\}$$
$$H_2 = \left\{x_6, x_{10}\right\} \qquad L_2 = \left\{x_8, x_{12}\right\}$$

so that the information pattern is given by

$$\Pi_1 = \left\{H_1, L_1\right\} \qquad \Pi_2 = \left\{H_2, L_2\right\}$$

Intuitively, each knows the value of the corresponding part of the deal, first or second.

The possible strategies of player 1 are defined on Π_1:

$$\sigma_1^1: \quad H_1 \mapsto s, \ L_1 \mapsto s$$
$$\sigma_1^2: \quad H_1 \mapsto s, \ L_1 \mapsto r$$
$$\sigma_1^3: \quad H_1 \mapsto r, \ L_1 \mapsto s$$
$$\sigma_1^4: \quad H_1 \mapsto r, \ L_1 \mapsto r$$

Hence,

$$\Sigma_1 = \left\{\sigma_1^1, \sigma_1^2, \sigma_1^3, \sigma_1^4\right\}$$

For player 2 we have

$$\sigma_2^1: H_2 \mapsto f, \ L_2 \to f$$
$$\sigma_2^2: H_2 \mapsto f, \ L_2 \to c$$
$$\sigma_2^3: H_2 \mapsto c, \ L_2 \to f$$
$$\sigma_2^4: H_2 \mapsto c, \ L_2 \to c$$

so that,

$$\Sigma_2 = \left\{ \sigma_2^1, \sigma_2^2, \sigma_2^3, \sigma_2^4 \right\}$$

Hence,

$$\Sigma = \Sigma_1 \times \Sigma_2 = \left\{ (\sigma_1^1, \sigma_2^1), (\sigma_1^1, \sigma_2^2), (\sigma_1^1, \sigma_2^3), \dots , (\sigma_1^4, \sigma_2^4) \right\}$$

containing the 16 pairs of strategies.

Next we want the probability distributions, by (22.3.1), one per strategy-pair in Σ. For $\sigma \in \Sigma$, write $\sigma = ij$ if and only if $\sigma = (\sigma_1^i, \sigma_2^j)$. Thus, the typical distribution over $\mathfrak{I}$ of (22.3.1) has the typical term

$$P_{ij}(\tau) = \text{probability of termination at } \tau, \text{ given strategy 2-tuple } (\sigma_1^i, \sigma_2^j)$$

and

$$\sum_{\tau} P_{ij}(\tau) = 1 \qquad \text{(for fixed ij)}$$

The computation of the $P_{ij}(\tau)$ can be done by first determining the terminal point for each triple (deal, i, j) as shown in Table 22.3.1.

For a given column of Table 22.3.1 (thus fixed ij strategy pair) we have a probability of 1/4 for each entry. Thus, all other terminal points have chance 0 under these strategies, and we can regard our work as completed. For example, the probability distribution $\left\{ P_{11}(\tau) \right\}$ is

$$P_{11}(x_5) = P_{11}(x_7) = P_{11}(x_9) = P_{11}(x_{11}) = \tfrac{1}{4}$$
$$P_{11}(\tau) = 0 \qquad \text{(all other } \tau \text{ in } \mathfrak{I})$$

Thus, "Simplified Poker" illustrates the general principle that in a game with chance moves, the determination of a definite terminal point does not follow from a strategic choice by each player; instead, a definite probability of each terminal point follows from a strategic choice by each player.

Recall that an outcome is an object—for example, an amount of money or a label like "win"—assigned to a terminal point. This is true in any game G. When G has chance moves, these outcomes are not determined by the persons alone; their strategy choices determine only a probability distribution over terminal points. Now we can take this distribution and use it to define an uncertain combination of outcomes—namely, just by noting the outcomes and the probabilities. In "Simplified Poker," the tree diagram of Figure 22.3.1 gives the

Table 22.3.1 Data for Computation of Probability Distributions in Simplified Poker

Deal	Information Sets Applying		Termination Point by Strategy Pair															
			11	12	13	14	21	22	23	24	31	32	33	34	41	42	43	44
HH	$H_1,$	H_2	x_5	x_5	x_5	x_5	x_5	x_5	x_5	x_5	x_{13}	x_{13}	x_{14}	x_{14}	x_{13}	x_{13}	x_{14}	x_{14}
HL	$H_1,$	L_2	x_7	x_7	x_7	x_7	x_7	x_7	x_7	x_7	x_{15}	x_{16}	x_{15}	x_{16}	x_{15}	x_{16}	x_{15}	x_{16}
LH	$L_1,$	H_2	x_9	x_9	x_9	x_9	x_{17}	x_{17}	x_{18}	x_{18}	x_9	x_9	x_9	x_9	x_{17}	x_{17}	x_{18}	x_{18}
LL	$L_1,$	L_2	x_{11}	x_{11}	x_{11}	x_{11}	x_{19}	x_{20}	x_{19}	x_{20}	x_{11}	x_{11}	x_{11}	x_{11}	x_{19}	x_{20}	x_{19}	x_{20}

outcome for each terminal point and Table 22.3.1 presents the basis for the probabilities. Thus, suppose that $\sigma = (\sigma_1^1, \sigma_2^1)$ is the strategic choice of the players. Then, by Figure 22.3.1, we see the outcomes that are relevant:

$$\omega(x_5) = 0$$
$$\omega(x_7) = a$$
$$\omega(x_9) = -a$$
$$\omega(x_{11}) = 0$$

They are so because all other terminal points have probability 0 under this strategy. Then we find, by using the fact that x_5, x_7, x_9, and x_{11} each has probability 1/4, that a probability distribution over outcomes exists. Namely,

$$P_{11}(0) = \sum_{\omega(\tau)=0} P_{11}(\tau) = P_{11}(x_5) + P_{11}(x_{11}) = \tfrac{1}{2}$$

$$P_{11}(a) = \sum_{\omega(\tau)=a} P_{11}(\tau) = P_{11}(x_7) = \tfrac{1}{4}$$

$$P_{11}(-a) = \sum_{\omega(\tau)=-a} P_{11}(\tau) = P_{11}(x_9) = \tfrac{1}{4}$$

$$P_{11}(x) = 0 \qquad (\text{if } x \neq 0, a, -a)$$

In general, then, we have 16 probability distributions over outcomes in the outcome space Ω of the game G, each depending on a strategy-pair, which determines the $P_{ij}(\tau)$, and on the outcome mapping ω defined over $\tau \in \mathfrak{I}$, which determines $P_{ij}(x)$, x in the outcome space, given the $P_{ij}(\tau)$.

Let us now generalize to an arbitrary game. Let it be an n-person game with chance moves. Then we see that (1) from $\sigma \in \Sigma = \Sigma_1 \times \Sigma_2 \times \ldots \times \Sigma_n$, we obtain the distribution $\{P_\sigma(\tau), \tau \in \mathfrak{I}\}$, and (2) from ω defined over $\mathfrak{I}$ and with range the outcome space Ω, we obtain $\{P_\sigma(x), x \in \Omega\}$ by the rule

(22.3.2) $$P_\sigma(x) = \sum_{\omega(\tau)=x} P_\sigma(\tau) \qquad (\text{any } x \in \Omega)$$

that is, sum over the various mutually exclusive terminal points yielding the value x in Ω.

To put this in the terminology of the NM utility theory of Chapter 20, we have the pure alternatives in Ω (assigned to τ in game G), and then, via strategy n-tuples σ and outcome mapping ω, we have for each σ the uncertain combination in which the probabilities are $P_\sigma(x)$ for the pure alternatives x.

Let A_x mean the same as x but be used to coordinate to the notation of the NM system. There we had

$$(p_1 A_1, p_2 A_2, \ldots, p_r A_r)$$

for the arbitrary simple uncertain combination of pure alternatives A_i. This could be abbreviated

$$(p_i A_i)$$

and formally replacing i with x,

$$(p_x A_x)$$

is then the arbitrary uncertain combination of alternatives A_x. Obviously, for σ, we have $p_x = P_\sigma(x)$. That is, the players confront an alternative $(p_x A_x)$ of the uncertain type, given they each select a strategy. We can also write

$$(P_\sigma(x)A_x)$$

to make the strategic dependence obvious. In the case of the Simplified Poker game, when $\sigma = 11$, we had the alternative

$$(P_{11}(0)A_0, P_{11}(a)A_a, P_{11}(-a)A_{-a})$$

computed as

$$(\tfrac{1}{2}A_0, \tfrac{1}{4}A_a, \tfrac{1}{4}A_{-a})$$

Here the players would confront the (uncertain) alternative as shown, given they were to use strategy pair 11—that is, (σ_1^1, σ_2^1).

From a strategic analysis viewpoint players will be taken to be evaluating such entities as $(P_\sigma(x)A_x)$ under varying σ: the whole point of this construction is to have alternatives available for comparison (under the NM model, where any two alternatives, including two of the uncertain type, are evaluatively comparable).

To sum up, in representing G in normal form we can always meet the requirement of a finite game that a unique "outcome" be assigned to each terminal point by assigning an uncertain combination to the terminal point of the representing game determined by strategy n-tuple σ. Thus, schematically, a typical play of the normal form looks as follows:

That is, a path in the normal form, the game representing G, is labeled by the strategy n-tuple $(\sigma_1, \sigma_2, \ldots, \sigma_n) = \sigma$, and the outcome assigned to the terminal point is the simple uncertain alternative $(P_\sigma(x)A_x)$.

Example. (c) In Simplified Poker, with the strategy pair (σ_1^1, σ_2^1) we had

The tree diagram for the normal form representing simplified poker is shown in Figure 22.3.2. (Note that player 2 will not know player 1's choice at turn 1 when it is his move at turn 2.) Let us compare the outcomes in this game under two distinct paths determined, respectively, by (σ_1^1, σ_2^1) and (σ_1^2, σ_2^1). Note that in Figure 22.3.2 these two "outcomes" are denoted Θ_{11} and Θ_{21}, respectively,

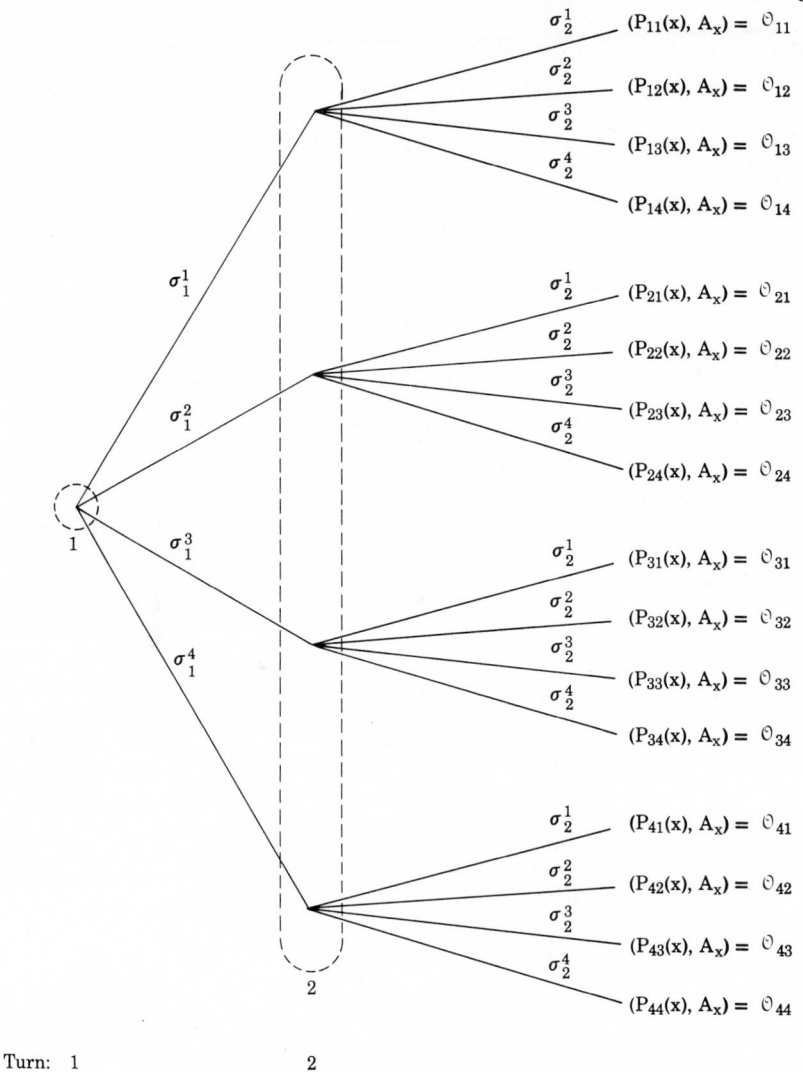

Turn: 1   2

Figure 22.3.2

each a simple uncertain combination of the basic outcomes A_x. To distinguish the outcomes A_x of the game in extensive form from the uncertain combinations serving as outcomes of the representing normal form game, we shall call the $\mathcal{O}_{ij}$ the normal form outcomes. We shall use this notation extensively in subsequent chapters on two-person theory. We have computed above

$$\mathcal{O}_{11} = (P_{11}(x)A_x) = (\tfrac{1}{2}A_0, \tfrac{1}{4}A_a, \tfrac{1}{4}A_{-a})$$

For (σ_1^2, σ_2^1), we obtain the outcome

$$\mathcal{O}_{21} = (P_{21}(x)A_x) = (\tfrac{1}{4}A_0, \tfrac{3}{4}A_a)$$

This arises as follows. From Table 22.3.1, we find that the possible terminal points under strategy pair (σ_1^2, σ_2^1) are x_5, x_7, x_{17}, and x_{19}, and under the chance move each has chance 1/4 of occurrence. Using mapping ω of Figure 22.3.1, we see that

$$x_5 \mapsto 0, x_7 \mapsto a, x_{17} \mapsto a, x_{19} \mapsto a$$

Hence,

$$P_{21}(0) = \sum_{\omega(\tau)=0} P_{21}(\tau) = P_{21}(x_5) = \tfrac{1}{4}$$

$$P_{21}(a) = \sum_{\omega(\tau)=a} P_{21}(\tau) = P_{21}(x_7) + P_{21}(x_{17}) + P_{21}(x_{19}) = \tfrac{3}{4}$$

By the NM utility model, player 1, for instance, can compare these two (uncertain) alternatives. Assume that player 1's preference for money outcomes is such that

$$(22.3.3) \qquad\qquad a \succ_1 0 \succ_1 -a$$

Next we note that

$$(22.3.4) \qquad\qquad (\tfrac{1}{4}A_0, \tfrac{3}{4}A_a) \succ_1 (\tfrac{1}{2}A_0, \tfrac{1}{4}A_a, \tfrac{1}{4}A_{-a})$$

if and only if, since u_1 is an order homomorphism,

$$u_1(\tfrac{1}{4}A_0, \tfrac{3}{4}A_a) > u_1(\tfrac{1}{2}A_0, \tfrac{1}{4}A_a, \tfrac{1}{4}A_{-a})$$

or by the expected utility condition,

$$\tfrac{1}{4}u_1(0) + \tfrac{3}{4}u_1(a) > \tfrac{1}{2}u_1(0) + \tfrac{1}{4}u_1(a) + \tfrac{1}{4}u_1(-a)$$

so that, equivalently after some numerical work,

$$u_1(a) > \frac{u_1(0) + u_1(-a)}{2}$$

which holds since (22.3.3) implies

$$u_1(a) > u_1(0) > u_1(-a)$$

From the condition (22.3.3), then, and the NM utility model, we conclude that (22.3.4) holds: the outcome of the game arising from strategy pair (σ_1^2, σ_2^1) is preferred by player 1 to the outcome arising from the pair (σ_1^1, σ_2^1); in normal form outcome notation: $\mathcal{O}_{21} \succ_1 \mathcal{O}_{11}$.

The generic point illustrated here is that strategic analysis operates under the guideline that players reason and will therefore compare strategies. In this instance, if player 1 regarded player 2 as committed to strategy σ_2^1 and if for some reason he was led to compare the outcomes of two strategies he could use, namely, σ_1^1 and σ_1^2, then he would prefer the strategy σ_1^2 to the strategy σ_1^1: assuming that he was cognizant of ω in G and that he knew the various chances of the various terminal points. This is a degree of knowledge no actual player is likely to have, of course.

22.4. Review of the Logic.

To review the logic of normal form: we want to represent G by a game that is strategically the same but in a class **M** of games we hope to be able to analyze in detail as to strategy. While to make G go into **M** in any particular case will be cumbersome, as Simplified Poker illustrates, the general theoretical exploration of problems of strategic analysis can safely confine itself to a class of games that at first sight appears to be unduly restrictive of the possibilities: namely, each player makes one move, he does this without rule-determined information as to other player's choices, and there are no chance moves. Yet, the point of the representation idea is that any game G can be put into this form, preserving strategic structure. Thus, theory construction can pass to the analysis of such a reduced set of possible games, assured that in principle all games are being studied.

Logically, then, in normal form game G appears as follows:

Turn 1: Selection from Σ_1.
Turn 2: Selection from Σ_2.

$\cdot$ $\quad$ $\cdot$

$\cdot$ $\quad$ $\cdot$

$\cdot$ $\quad\;$ $\cdot$

Turn_n: Selection from Σ_n.
End.

Outcome: Some $(P_\sigma(x)A_x)$ in α, the space of alternatives based on the Ω of G.

And all selections are made with no rule-determined information as to any other player's selection.

From definition (21.9.1) of the concept of strategy we can compute the number of strategies in Σ_i, $(i = 1, 2, \ldots, n)$. We know that

$$\Pi_i = \left\{ S_i^1, S_i^2, \ldots, S_i^{m_i} \right\} \qquad (i = 1, 2, \ldots, n)$$

is the partition of player i's moves. A strategy involves choosing an alternative at a representative move in each S_i^j. Let $o(v_i^j)$ be the number of alternatives (the outdegree) at node v_i^j in S_i^j. By axiom G4(a) this number is invariant in S_i^j. Hence, there are $o(v_i^1)$ ways to select an alternative in class S_i^1, $o(v_i^2)$ ways to select an alternative in $S_i^2, \ldots, o(v_i^{m_i})$ ways to select an alternative in class $S_i^{m_i}$. Hence, the number of strategies in Σ_i is the product of these terms:

$$(22.4.1) \qquad N(\Sigma_i) = o(v_i^1)\, o(v_i^2) \cdots o(v_i^{m_i}) \qquad (i = 1, 2, \ldots, n)$$

And, finally,

(22.4.2) $$N(\Sigma) = N(\Sigma_1) N(\Sigma_2) \cdots N(\Sigma_n)$$

gives the number of paths in the normal form.

Example. (a) In Simplified Poker,

$$\Pi_1 = \{S_1^1, S_1^2\}, \Pi_2 = \{S_2^1, S_2^2\}$$

where

$$S_1^1 = \{x_1, x_2\}, S_1^2 = \{x_3, x_4\}, S_2^1 = \{x_6, x_{10}\}, S_2^2 = \{x_8, x_{12}\}$$

The outdegree is 2 in each case. Hence, $N(\Sigma_i) = 2^2 = 4$ and $N(\Sigma) = 4^2 = 16$.

Example. (b) In Figure 21.9.1 we presented the more complex model in regard to the examination appeals institution. Let us compute the numbers $N(\Sigma_i)$ and $N(\Sigma)$, based upon the tree of the game model, first by tabulating the relevant outdegrees:

Player i	Π_i	$(o(v_i^1), o(v_i^2), \ldots, o(v_i^{m_i}))$
1	$\{\{x_0\}, \{x_4, x_5\}, \{x_{16}, x_{21}\}, \{x_{18}, x_{19}\}\}$	$(2, 2, 2, 2)$
2	$\{\{x_1\}, \{x_2\}\}$	$(2, 2)$
3	$\{\{x_7\}, \{x_9\} \{x_{12}\}, \{x_{14}\}\}$	$(2, 2, 2, 2)$
4	$\{\{x_3\}, \{x_6\}\}$	$(2, 2)$
5	$\{\{x_{23}\}, \{x_{24}, x_{29}\}, \{x_{25}, x_{28}\}, \{x_{26}, x_{27}\}\}$	$(2, 2, 2, 2)$

Hence,

$$N(\Sigma_1) = 2^4 = 16$$
$$N(\Sigma_2) = 2^2 = 4$$
$$N(\Sigma_3) = 2^4 = 16$$
$$N(\Sigma_4) = 2^2 = 4$$
$$N(\Sigma_5) = 2^4 = 16$$

and

$$N(\Sigma) = 2^{16}$$

Note that the binary choices at each nonterminal move are a special case of the general model where each $o(v_i^j)$ in the tabulation may be distinct.

22.5. Interpretations of the Payoffs. Consider now a game in extensive form (see section 21.7): here the outcome space, for an n-person game, is $\Re^n$. The outcome mapping is called the payoff function.

Recall that $\mathfrak{I}$ is the set of terminal points of G. Let G have chance moves. Then for each strategy n-tuple σ, we recall that a probability distribution $\{P_\sigma(\tau)\}$ was defined on $\mathfrak{I}$. Thus, $(\mathfrak{I}, \mathfrak{B}, P_\sigma)$ may be regarded as a finite probability model (see Chapter 9). The payoff function for player i has domain $\mathfrak{I}$ and goes into $\mathfrak{R}$, the real numbers. It follows that this payoff for player i is a random variable. Call it X_i. We see the structure:

(1) G has a finite number of terminal points, $\mathfrak{I}$;

(2) G has payoff function ω: $\mathfrak{I} \to \mathfrak{R}^n$;

(3) G has a chance player; and hence,

(4) G has associated distributions of the form P_σ, $\sigma \in \Sigma$, defined on $\mathfrak{I}$,

(5) X_i can be defined on $\mathfrak{I}$ as the ith component of $\omega(\tau)$,

$$X_i(\tau) = x_i \quad \text{iff} \quad \omega(\tau) = (x_1, x_2, \ldots, x_i, \ldots, x_n)$$

so that

(6) X_i is a random variable, the probability distribution of which depends on σ.

In fact,

(22.5.1)
$$P_\sigma(X_i = x_i) = \sum_{x_i \text{ fixed}} P_\sigma(\tau)$$

$i = 1, 2, \ldots, n$. The payoff functions are now random variables with expectations, say

(22.5.2)
$$E_\sigma(X_i) = \sum_{x_i} x_i P_\sigma(X_i = x_i)$$

The discussion in section 21.7 of the two methods of interpretation of the concept of payoff function applies now to the consideration of the meaning to be ascribed to the expectations $E_\sigma(X_i)$, $i = 1, 2, \ldots, n$, $\sigma \in \Sigma$. The basic query is, apart from the law of large numbers, and so repeated independent plays (with identical strategy n-tuple σ) does $E_\sigma(X_i)$ have any meaning? We give brief answers under each method.

Method of Monetary Payoffs. X_i is the gain or loss, in money, of player i. $E_\sigma(X_i)$ is the expected money gain or loss if the players employ strategy n-tuple σ. X_i has meaning for the actor, but $E(X_i)$ may not. $\overline{X}_i$ has meaning for the actor: the average across some set of plays, of his earnings. But then he has to consider that strategy vectors (the σ) may vary from play to play and so one cannot simply relate $\overline{X}_i$ to $E_\sigma(X_i)$. Under a highly "rationalistic" rendition of this interpretation, one simply informs the player i that $E_\sigma(X_i)$ is approximately $\overline{X}_i$ in which the average is over repeated, independent plays all of which involve the identical strategy n-tuple σ. Neither the analyst nor the player need assume such repetitions will ever exist. Rather, the analyst or the player may use this sort of information in making preferential judgments over comparisons involving gambles (i.e., uncertain combinations of monetary payoffs). But this essentially means passing to the interpretative scheme under which the player has some utility function over the space of gambles (and pure money outcomes). In this case, the meaning of $E_\sigma(X_i)$ depends on the functional relation between X_i and utility. Discussions of this problem may be found in Bartos (1967), Luce and Raiffa (1957), and Chernoff and Moses (1959).

Method of Psychological Payoffs. X_i is constructed, theoretically and perhaps to some degree in actual measurements, via the composition of the outcome function of some given game and the utility function defined over the space of alternatives for which these outcomes serve as the pure alternatives. Under postulate (21.8.1) this utility satisfies the expected utility condition. It follows that $E_\sigma(X_i)$ stands for the utility of an uncertain alternative in which the outcomes have probability distribution based on $P_\sigma(\tau)$, τ a terminal point of the game. Thus, in this case, the expected payoff may be used in the strategic analysis to calculate preferential relations, via the representation theorem relating utilities to preferences.

Example. In some way, utility mappings have been defined over outcomes $\{A, B, C\}$ such that for player 1 of the game,

$$u_1(A) = 1$$
$$u_1(B) = .6$$
$$u_1(C) = 0$$

Then suppose under strategy n-tuple σ,

$$P_\sigma(A) = .5, P_\sigma(B) = .5, P_\sigma(C) = 0$$

and under σ',

$$P_{\sigma'}(A) = .1, P_{\sigma'}(B) = .8, P_{\sigma'}(C) = .1$$

There is the question of which strategy n-tuple is ideally better for player 1 in the sense that if he knew P_σ and $P_{\sigma'}$, then he himself would regard the strategy n-tuple leading to a preferred alternative as better. (For the present, we neglect the question of which strategy components of σ or σ' can be controlled by player 1.) Applying the expected utility condition to the σ case

$$u_1(.5A, .5B) = .5u_1(A) + .5u_1(B)$$
$$= .5 + .3$$
$$= .8$$

whereas, in the σ' case,

$$u_1(.1A, .8B, .1C) = .1u_1(A) + .8u_1(B) + .1u_1(C)$$
$$= .10 + .48$$
$$= .58$$

Thus, because

$$u_1(.5A, 5B) > u_1(.1A, .8B, .1C),$$

we conclude, by order-homomorphism,

(22.5.3) $\qquad$ (.5A, .5B) $\succ_1$ (.1A, .8B, .1C)

and so, under the given assumptions, player 1 will regard the strategy n-tuple σ as better than the strategy n-tuple σ'. In this example, we have terminal point τ ranging over $\{A, B, C\}$. Hence,

$$X_1(\tau) = u_1(\tau)$$

and so,

$$E_\sigma(X_1) = .5u_1(A) + .5u_1(B) = .8 = u_1(.5A, .5B)$$
$$E_{\sigma'}(X_1) = .1u_1(A) + .8u_1(B) + .1u_1(C) = .58 = u_1(.1A, .8B, .1C)$$

Thus, we have computed two expectations, compared them, and concluded that (22.5.3) holds. Starting from (22.5.3), we show in the same way that the expectations are unequal. Hence,

(22.5.4) $\qquad E_\sigma(X_1) > E_{\sigma'}(X_1) \Leftrightarrow (P_\sigma(x)A_x) \succ_1 (P_{\sigma'}(x)A_x)$

(where $A_x = A, B, C$), and under the "ideally better" notion,

(22.5.5) $(P_\sigma(x)A_x) \succ_1 (P_{\sigma'}(x)A_x)$ implies σ ideally better than σ'.
So, by transitivity of the conditional,

(22.5.6) $E_\sigma(X_1) > E_{\sigma'}(X_1)$ implies σ ideally better than σ' for player 1.

Thus, (22.5.4) presents a valid interpretation of the expected payoff relevant to any one play: one uncertainty combination is preferred to another. Then (22.5.5) goes a step further in proposing a strategic significance for this preference, and so, (22.5.6) gives the strategic meaning for actor 1, of the expected payoff comparison. The actor knows nothing of $E_\sigma(X_1)$ and $E_{\sigma'}(X_1)$—they are in the analyst's representation of the game in extensive form—but by the analyst's assumptions about the actor, the analyst's computational comparison reflects or represents a potential real strategic comparison made by an idealized actor, assigned to the game as player 1. Finally, we remind the reader that the actor ordinarily controls only σ_i of σ. Hence, the induced preference over Σ does not yield a definite course of action for the actor to realize the preferred outcome. We return to this problem in section 22.7.

22.6. Normal Form and Strategic Analysis. When we "normalize" a game in extensive form, then the interpretation of the normal form depends upon the meaning assigned to the payoff function. In the purely mathematical sense, the normal form now has the following appearance:

Turn 1: Selection from Σ_1.

Turn 2: Selection from Σ_2.

• •

• •

• •

Turn n: Selection from Σ_n.
End.
Outcome.
Payoffs.

Here the information pattern makes the order of turns irrelevant. A joint selection by n players yields a termination point τ, which maps to $\omega(\tau)$, the outcome; this in turn maps to $\Re^n$, a payoff vector. Under the psychological construction, the outcome in this form is $(P_\sigma(x)A_x)$ and the vector of payoffs is

(22.6.1) $[u_1(P_\sigma(x)A_x), u_2(P_\sigma(x)A_x), \ldots, u_n(P_\sigma(x)A_x)]$

which by the expected utility condition may be represented as

(22.6.2) $[E_\sigma(X_1), E_\sigma(X_2), \ldots, E_\sigma(X_n)]$

Now all the conceptual background has led to the simple mapping

$$\sigma = (\sigma_1, \sigma_2, \ldots, \sigma_n) \mapsto (x_1, x_2, \ldots, x_n) \in \Re^n$$

in which we say that each player chooses a strategy and then the payoffs are the x_i. But x_i here is purely numerical. "Payoff" is open to interpretation. The most solid interpretation is that this payoff vector is (22.6.2), interpreted by (22.6.1). Then we can reason with the payoffs in the conceptual light cast by their evaluative meaning for the players, because (22.6.1) is itself shorthand for a set of n binary relational systems of preference. For at least idealized actors, then, it is reasonable to suppose that the evaluation of the two strategy vectors σ, σ' itself relates to this collection of relational systems of preferences over certain and uncertain outcomes of the strategic choices.

The numerical shortcut is thus: from $\sigma \mapsto x_i$, $\sigma' \mapsto x_i'$ with $x > x_i'$ conclude that σ is ideally better (for player i) than σ'. But the entire history of game theory shows that it is painfully unclear to know exactly what this payoff calculus really assumes. Perhaps the heavy conceptual mathematics of this section has helped the reader to see the difficulties inherent in the detailed setup of an interactive choice structure. We have tried to do this without confusing the issue of the interpretation of the expected payoff with a related problem: that the player cannot "choose" σ over σ' in the general case because σ and σ' are vectors of strategies of distinct players. The conceptual problem that presents itself—the objects of individual preference are outcomes determined interactively via σ, while each individual chooses only one component of σ—is the core problem for the theory of games, as we shall see at the end of the next section.

22.7. Player Assumptions and the Fundamental Problem of Game Theory. In their analysis of the conceptual foundations of game theory, Luce and Raiffa (1957) located three standard assumptions made about the players of games. In this section, the aim is to describe these assumptions.

We can regard these assumptions as concerning an n-person game with players (G, A, r): G is a game, A a set of actors, and r assigns each actor to a unique player role in the game G.

The first assumption we have treated in the previous chapter: it is the idea that each actor has a preference system over the space of alternatives generated by the outcomes of G and that this system can be represented as a utility function satisfying the expected utility condition. Neumann and Morgenstern explicitly constructed the theory behind this assumption, as noted in Chapter 20. Postulate (21.8.1) explicitly introduces this theory as a basis for the interpretation of numerical results in game theory. Once the game is in normal form, we simply say that the players have preferences over outcomes.

The second major assumption located by Luce and Raiffa is that each actor, in taking his role in G, knows the game G, knows the roles of other actors, and, finally, is cognizant of their preference systems over outcomes of G. As we frame this assumption, then, it has three parts:

(1) knowing the game G,
(2) knowing the roles assigned to other actors in A,
(3) knowing the preferences over outcomes in G held by each actor in his assigned role.

The first two components of this assumption seem reasonable idealizations, corresponding fairly well to the symbolic interactionist concept (derived from G. H. Mead, 1934) that each role-player takes the role of the other. In fact, the game of baseball is often cited to illustrate the operation of this idea (e.g., Bierstedt, 1970).

The third component of this knowledge assumption is more difficult to justify. One interpretation is based on the role idea. A role, as understood by actors, includes not only certain modes of actions within a structure—as in the rule-governed interactive choice situations—but tacit preferential biases. In other words, built into role conceptions are outcome preferences. In example (d), section 21.4, the examination appeals procedure, for instance, the role of "student taking examination"—defined, as in the G sense, in abstraction from a definite set of actors—would be interpreted by actors taking this and other roles in the inter-active system as including a preference for the outcome of ultimate "pass." We can say that the student will prefer to pass without first failing and winning an appeal. An actor α in A—associated via r with "student" (i.e., r(α) = student)—who prefers ultimate failure is an oddity, a deviant case exemplifying the tacit assumption by all players that certain roles have certain associated evaluative "directions."

Thus, on closer examination, the third component of the knowledge assumption may be intimately linked to the others: knowing an actor is assigned to a role (second component) and knowing the game (first component) often means knowing the preferences. Strictly speaking, however, this is a special social structural assumption of a set of "adequately motivated" actors who, via a socialization process, have internalized the role "definitions" in G. Thus, the set A is a set of idealized actors, who conform to the theoretical picture of the scene provided by their social system.

The argument that (3) coordinates to some notion of sociocultural "binding-in" of the actors, so that their preferences in relation to G can be inferred by other actors, has been in abstraction from the first player assumption. The cultural argument would provide perhaps a useful scope condition for (3) if we assumed only an ordinal preference over pure alternatives. A check of the above paragraphs reveals a tacit assumption that G has no chance moves. Introducing chance moves into G immediately activates the conceptual requirement for the extension of preferences to uncertain outcomes. The first player assumption is applicable in this case. If so, the system $(\alpha, \succ_\alpha)$ held by actor α over the (pure and uncertain) alternatives based on the outcomes of G is highly unlikely to be known by actor β; this is so even if they both know $\succ_\alpha$ and $\succ_\beta$, as restricted to the pure alternatives, the outcomes of G. Thus, in the general case of an arbitrary game G including elements of uncertainty, it appears that part (3) of the knowledge assumption is not satisfied by actors.

The conclusion is that part (3) is perhaps the strongest idealization of the game-theoretic player assumptions and that strategic analysis of a game model using (3) is likely to be mainly of theoretical interest in the sense that the conclusions take the form "If the actors in A, in assignments to roles in G, were to know each other's complete preferences, then. . . ." This is not to say that additional assumptions will not be made; the mainline of any argument might be cast in the form "If X were the case, then if Y were the case, then it would follow that. . . ." A nesting of such conditionals is not unusual in purely theoretical arguments (e.g., about hypothetical experiments, hypothetical particles, hypothetical chemical reactions). A difference exists, however. The usual "contrary-to-fact" nestings of theoretical arguments are all within the framework of some abstract analytical system (e.g., Newtonian mechanics, thermodynamics) whose main principles are functioning as rules of inference, so deep is the accord reached on their formulation. Thus, they fail to appear in the line of reasoning explicitly, yet anyone following the argument is basing his understanding on these principles. In game theory, our presentation to this point does not include any such basic principle. The explicit introduction of an acceptable general principle would make the hypothetical arguments of game-theoretic analysis more important. The conditionals would then refer to potential matters of fact, while the principle would generate consequences under those hypothetical conditions.

The final assumption located by Luce and Raiffa resembles such a principle.

This third and final assumption about players is called a "law of behavior" by Luce and Raiffa. But as they point out, it is at once more and less than a principle of choice behavior. Methodologically, it is less because it appears to be inherently nonfalsifiable and more because it is a regulative principle guiding the analysis of game models.

The general principle may be illustrated by reference to the NM utility theory. Let A_1 and A_2 be two alternatives. Let r_1 and r_2 be two possible responses or courses of action such that, from the actor's point of view, there are certain chances of A_1 and A_2 contingent upon the selected course of action; this may be shown in terms of the tree diagram in Figure 22.7.1. $P_i(j)$ is the chance of

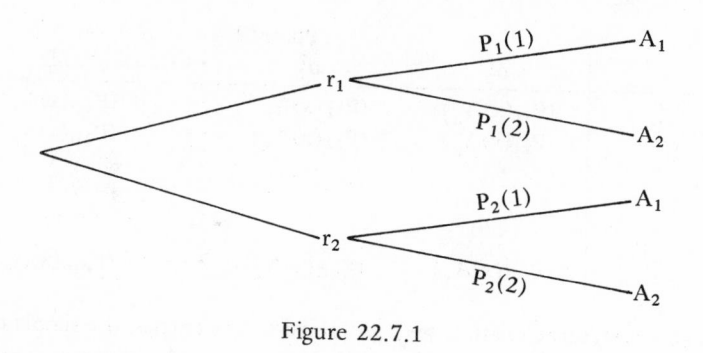

Figure 22.7.1

alternative A_j if action r_i is taken. Using the notation of Chapter 20, the choice situation may be denoted as in Figure 22.7.2.

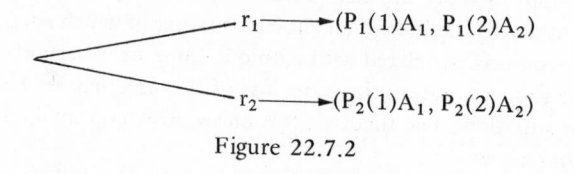

Figure 22.7.2

What we shall call the interpretation-of-action principle says:

If $(P_1(1)A_1, P_1(2)A_2) \succ_\alpha (P_2(1)A_1, P_2(2)A_2)$, then α will choose r_1.

That is, the actor acts so as to make an optimal choice. The nonfalsifiability arises from the fact that should α choose r_2, we would interpret this as an indication that the outcome preferences or likelihoods were not expressed correctly for his choice situation.

Applying this principle to game theory, we have a first identification:

let $r_1 = \sigma$, $r_2 = \sigma'$ and then

$$\text{if } (P_\sigma(x)A_x) \succ_\alpha (P_{\sigma'}(x)A_x)$$
then α will choose σ.

But, of course, this is nonsense: actor α chooses component strategy σ_i in strategy vector σ when the role assignment of α is that of player i. Of course, the principle has a certain force: ideally, as in (22.5.5), σ is better than σ' for α, but α ordinarily does not have a choice between σ and σ' unless these differ only in the ith component.

What we see, then, is that there is no straightforward passage from a preference scheme over alternatives to optimal action in an interactive choice situation. Instead, for instance, the two-person situation has the form of a matrix with typical entry the normal form outcome $o_{ij} = (P_{ij}(x)A_x)$:

(22.7.1)

	Player 2			
	σ_2^1	σ_2^2	$\cdots$	σ_2^n
σ_1^1	$(P_{11}(x)A_x)$	$(P_{12}(x)A_x)$	$\cdots$	$(P_{1n}(x)A_x)$
σ_1^2	$(P_{21}(x)A_x)$	$(P_{22}(x)A_x)$	$\cdots$	$(P_{2n}(x)A_x)$
Player 1 $\vdots$	$\vdots$	$\vdots$	$\vdots$	$\vdots$
σ_1^m	$(P_{m1}(x)A_x)$	$(P_{m2}(x)A_x)$	$\cdots$	$(P_{mn}(x)A_x)$

Even though a preference relation $\succ_\alpha$ is defined on the entries, the simple uncertain alternatives based on outcomes $\{A_x\}$, actor α assigned a player role will choose not a row and a column (which together determine a strategy vector σ) but only a row or only a column depending on his player role.

Hence, the attempt to apply the interpretation-of-action principle seems to break down: contrary to the individual choice situation in which each of an actor's possible actions is associated with a unique (pure or uncertain) alternative, in the interactive choice situation there is a loss of this uniqueness. Each actor knows this is the situation. The theorists NM know it. An optimal choice for each player is not defined.

Worse yet, consider the meaning of "rational action." The definition provided by Parsons (1937, p. 58) will suit our purposes:

Action is rational in so far as it pursues ends possible within the conditions of the situation, and by the means which, among those available to the actor, are intrinsically best adapted to the end for reasons understandable and verifiable by positive empirical science.

Operating in a tradition of economic theory that focused on the choice situation as a problem for rational action NM (in effect) interpreted the breakdown of the interpretation-of-action principle as a problem in the extension of the rational action notion to interactive choice situations (see NM, 1947, Chap. 1).

Here the "ends" are given by player preferences, whatever they may be, over alternatives based on outcomes of interaction, and the "means" are the strategies. But "intrinsically best" is not yet even a sensible statement. The players and NM realize the optimal choice is not clearly given. While the players may arrive at their own "solutions" to this problem, the rational action definition suggests a problem for science: namely, define for the actors "the intrinsically best adapted" actions (i.e., optimal strategies). Thus, NM were led to the following:

(22.7.2)   Fundamental Problem of the Theory of Games. Define "rational action" (i.e., optimal strategy) for actors in n-person games, for any game, for any n.

Thus, far from assuming rationality, the whole apparatus of game theory emerges as a solution—or, rather, a series of theoretical answers—to the fundamental problem. That this problem is a problem for science and yet is "normative" is one of the sources of difficulty for sociologists approaching game theory and, even more so, for psychologists. From a psychological standpoint, the whole NM approach is suspect precisely because it is not "positive empirical science." Restle foundations and Luce's theory (see Chapter 11) both lead to the view that the actors in the capacity of choosers will in their own limitations place "conditions" on the situation. Certified by empirical science, these psychological conditions (one would think) provide part of the basis for the "intrinsically best-adapted" actions verifiable by science. But NM psychology was limited to the player assumptions given above. On the other hand, classic economic theory operates in the same manner. That game theory should end up not describing actually chosen actions in interactive choice situations is no problem: this was never its aim. But that it should define rational action in such situations in abstraction from positive psychological science is a relevant criticism.

One final point on the interpretation-of-action principle, thought of in terms of Figure 22.7.2: we may regard the principle as an active rule that players use in interpreting and forecasting their behaviors. If α sees that β has two action possibilities, to predict β's action, α will "model" the situation of β, seeking possible consequences evaluated in terms of what α knows of β's preferences.* In game theory, α knows all there is to know in this regard; also, he knows the possible strategies of β. Hence, even though the interpretation-of-action principle in general breaks down, α still can infer that of any two strategies that β

---

*In fact, because of external power differences, actor α may have helped determine the preferences of β. Game theory takes the preferences as givens, so they could have been formed under some external threat. (See, however, section 25.14.)

might choose for a fixed strategy choice by α, that β will choose the one that leads to the preferred alternative. This inference uses all three basic player assumptions: (1) β and α have preference systems over alternatives, (2) α knows the role β plays in the game and, so, the strategy choices of β, and he knows β's preferences, and (3) α uses the interpretation-of-action principle, so far as he can, to reason about the strategic situation of β, using his knowledge of the game and of β's role and preferences.

We may also say that there is a symbolic aspect to such a situation: actions are interpreted by the actors in terms of alternatives and preferences. A given actor "encodes" the results of his own reasoning process into an overt action, and another actor "decodes" that action as having a certain significance in terms of what he knows about the game and about the other player.

22.8. Summary of the Concepts for the Two-Person Case. We summarize the conceptual apparatus leading from an arbitrary two-person game to the normal form of such a game. We begin with a finite game

$$G = (\Gamma, \mathcal{S}, \mathcal{O}, \boldsymbol{\pi}, \Omega, \omega)$$

where $\mathcal{S} = \{S_0, S_1, S_2\}$, so that S_0 is the set of chances moves in Γ, S_1 is the set of moves of player 1, and S_2 is the set of moves of player 2. The set Ω is the outcome set of G, and ω assigns an outcome in Ω to each point in $\mathfrak{I}$, the set of terminal points on the tree Γ. We define Σ_1 as the set of all strategies of player 1, where

$$N(\Sigma_1) = o(v_1^1) \, o(v_1^2) \cdots o(v_1^{m_1})$$

and $o(v_1^j)$ is the outdegree of any move in part S_1^j ($j = 1, 2, \ldots, m_1$) given by the information partition Π_1 in Π. Similarly, we define Σ_2. Then we let $\Sigma = \Sigma_1 \times \Sigma_2$ be the set of all strategy vectors (σ_1^i, σ_2^j) in Σ. We also write $\sigma = ij$ for a typical strategy vector. A strategy vector determines a unique probability distribution over $\mathfrak{I}$, the terminal points. Since $\omega(\tau)$ is an outcome in Ω, a distribution also exists on outcomes in Ω, obtained by adding the probabilities of all τ that yield the same outcome x in Ω,

$$P_{ij}(x) = \sum_{\omega(\tau)=x} P_{ij}(\tau)$$

Then a unique uncertain combination of outcomes is associated with strategy vector $\sigma = ij$—namely, in NM notation

$$(P_{ij}(x)A_x)$$

where A_x is the same as x. We write

$$\mathfrak{O}_{ij} = (P_{ij}(x)A_x)$$

to denote this typical normal form outcome [see Figure 22.3.2 and matrix (22.7.1).]

By NM utility theory the game G can be put in extensive form by the composition of outcome mapping ω with the utility functions, u_i (i = 1, 2). The latter are defined over the space of alternatives generated by outcomes in Ω. We have denoted these by

$$u_1(P_{ij}(x)A_x), \; u_2(P_{ij}(x)A_x)$$

which makes sense even if the A_x ($\equiv$ x) in Ω are non-numerical. Hence, we have $u_1(o_{ij})$, $u_2(o_{ij})$ when we go to normal form. In the normal form, we call $u_1(o_{ij})$ the payoff to player 1 and u_2 (oij) the payoff to player 2. The interpretation is given by the NM utility theory. In particular types of games, this notation simplifies considerably, as we shall see in the next chapter.

One particular special case deserves mention here. Suppose G has $S_0 = \emptyset$, so that there are no chance moves. Then a strategy vector σ determines a terminal point in $\mathfrak{J}$ and, so, a definite outcome in Ω. Then if σ = ij, the expression $(P_{ij}(x)A_x)$, which stands for a full probability distribution over the outcomes $x \equiv A_x$ in Ω, becomes zero everywhere except at one point. Hence, o_{ij} becomes a single outcome in Ω. That is, in this case the concept of normal form outcome reduces to the concept of outcome in Ω. If there is no uncertainty element in G, then the outcomes of the normal form are the actual outcomes of G, otherwise they are uncertain combinations of such actual outcomes.

CHAPTER TWENTY-THREE THEORY OF TWO-PERSON ZERO-SUM GAMES

23.1. The Meaning of Zero-Sum. We introduce the basic techniques of strategic analysis with the simplest class of two-person games, the so-called zero-sum games. Intuitively, opposite interests exist between a pair of actors. The game is in normal form. In the numerical representation, a payoff matrix is defined

$$(a_{ij})$$

where

$$a_{ij} = \text{payoff to player 1 if player 1 chooses his } i^{th} \text{ strategy, player 2} \\ \text{chooses his } j^{th} \text{ strategy.}$$

And, $-a_{ij}$ is the payoff to player 2.

It is important to see the payoffs as numerical representations of preference systems held by (arbitrary) actors, say α and β, taking roles in the game. Thus, we need to consider the associated object for each such ij pair, the normal form outcome,

$$o_{ij} = \text{the pure or uncertain alternative that arises if player 1 } (\alpha) \\ \text{chooses his } i^{th} \text{ strategy, player 2 } (\beta) \text{ chooses his } j^{th} \text{ strategy.}$$

And by analogy,

$$(o_{ij})$$

is the collection of all such entities.

We need notation for strategies. Previously, we used

$$\sigma_i \in \Sigma_i$$

for strategy σ_i of player i, in set of strategies possible for i, Σ_i. To avoid cumbersome notation, we shift now, with two players, to the notation

$$\Sigma_1 = \{\alpha_1, \alpha_2, \ldots, \alpha_m\}$$
$$\Sigma_2 = \{\beta_1, \beta_2, \ldots, \beta_n\}$$

to suggest that actor α has typical strategy α_i ($i = 1, 2, \ldots, m$) and actor β has typical strategy β_j ($j = 1, 2, \ldots, n$). (Note that hereafter n refers to the number of strategies of β and not, as in earlier chapters on general game theory, to the number of players.) Thus, a joint strategy pair is (α_i, β_j), an element of $\Sigma_1 \times \Sigma_2$. We have

$$(\alpha_i, \beta_j) \mapsto o_{ij} \mapsto a_{ij}$$

where the first mapping arises because (α_i, β_j) determines a terminal point (of the game in normal form) and so a normal form outcome o_{ij}. The second mapping is the numerical representation of player 1's preferences (i.e., of actor α's orientation, given he has the role of player 1). Thus,

$$a_{ij} = u_1(o_{ij})$$

and this is under the postulate (21.8.1), which makes a utility of a normal form outcome o_{ij} an expected utility of actual outcomes of the game in extensive form.

Example. (a) Suppose we have a game which originally, in extensive form, had no chance moves. As we know from the discussion at the end of section 22.8, the normal form outcomes are then the same as the original outcomes. For example, let (o_{ij}) be given by

$$
\begin{array}{cc}
 & \begin{array}{cc} \text{Player 2} \\ \beta_1 \qquad\qquad \beta_2 \end{array} \\
\text{Player 1} \quad
\begin{array}{c} \alpha_1 \\ \\ \alpha_2 \end{array}
&
\begin{pmatrix}
\begin{array}{l}\text{1 wins}\\ \text{2 loses}\end{array} & \begin{array}{l}\text{1 loses}\\ \text{2 wins}\end{array} \\
\begin{array}{l}\text{1 loses}\\ \text{2 wins}\end{array} & \begin{array}{l}\text{1 wins}\\ \text{2 loses}\end{array}
\end{pmatrix}
\end{array}
$$

Thus, if α chooses α_1 and β chooses β_1, then (o_{11}) is the outcome: 1 wins, 2 loses. Suppose

$$\begin{array}{l}\text{1 wins}\\ \text{2 loses}\end{array} \succ_\alpha \begin{array}{l}\text{1 loses}\\ \text{2 wins}\end{array}$$

which says α prefers to win; similarly, β prefers to win,

$$\begin{array}{l}\text{1 loses}\\ \text{2 wins}\end{array} \succ_\beta \begin{array}{l}\text{1 wins}\\ \text{2 loses}\end{array}.$$

A simple order-preserving map, say u_1, will represent $\succ_\alpha$,

$$u_1 \begin{pmatrix}\text{1 wins}\\ \text{2 loses}\end{pmatrix} = 1$$

$$u_1 \begin{pmatrix} 1 \text{ loses} \\ 2 \text{ wins} \end{pmatrix} = -1$$

Then, rather than use numbers like 12 and -12, to represent β's preference $\succ_\beta$ we let u_2 be given by

$$u_2 \begin{pmatrix} 1 \text{ wins} \\ 2 \text{ loses} \end{pmatrix} = -1$$

$$u_2 \begin{pmatrix} 1 \text{ loses} \\ 2 \text{ wins} \end{pmatrix} = 1$$

then we have

$$
\begin{aligned}
(a_{ij}) &= \begin{pmatrix} u_1(\mathcal{O}_{11}) & u_1(\mathcal{O}_{12}) \\ u_1(\mathcal{O}_{21}) & u_1(\mathcal{O}_{22}) \end{pmatrix} \\
&= \begin{pmatrix} u_1\begin{pmatrix} 1 \text{ wins} \\ 2 \text{ loses} \end{pmatrix} & u_1\begin{pmatrix} 1 \text{ loses} \\ 2 \text{ wins} \end{pmatrix} \\ u_1\begin{pmatrix} 1 \text{ loses} \\ 2 \text{ wins} \end{pmatrix} & u_1\begin{pmatrix} 1 \text{ wins} \\ 2 \text{ loses} \end{pmatrix} \end{pmatrix} \\
&= \begin{pmatrix} 1 & -1 \\ -1 & 1 \end{pmatrix}
\end{aligned}
$$

Also, we see that, since we were judicious in our choice of u_2,

$$u_2(\mathcal{O}_{ij}) = -u_1(\mathcal{O}_{ij}) \qquad \text{(all i, j)}$$

and so,

$$u_1(\mathcal{O}_{ij}) + u_2(\mathcal{O}_{ij}) = 0 \qquad \text{(all i, j)}$$

and so as function,

$$u_1 + u_2 = 0$$

This sum of exactly zero is meaningful in a pragmatic sense, not empirically; that is, we could just as well represent $\succ_\alpha$ and $\succ_\beta$ such that the sum is nonzero. The zero-sum condition is merely a convenient numerical representation of opposite preferences. Nothing in game theory will ever depend upon the sum being zero, per se, because all the thinking is NM utility thinking: the u_i are representations of empirical systems of the form $(\alpha, \succsim)$. We are actually "adding pounds and feet" when we add u_1 and u_2 because distinct empirical systems are represented. Yet, this (empirical) nonsense is good form for people embedded in a numerically-oriented culture. (Game theory could readily be developed with no numbers at all, for that matter.) But if we are to assure ourselves that at no point do we draw conclusions that are empirically interpreted in contravention to the ideas guiding game-theoretic work, we must revert back to $\succ_\alpha$ and $\succ_\beta$, as "abstract" and formidable as such non-numerical notation seems. Our first application of the benefit of this reversion to the domains of the utility functions occurs in obtaining a clear empirical meaning for "zero-sum": strictly inverse preference orderings.

Example. (b) Suppose that in some game there is considerable room for chance to play a role. Then in the normal form, Θ_{ij} is not a definite actual outcome of the original game—for example, winning a poker hand—but a probability distribution over such outcomes [see matrix (22.7.1)]. However, the NM utility theory supplies a meaning for $u_1(\Theta_{ij})$. Substantively, it supplies the meaning that if Θ_{ij} and $\Theta_{i'j'}$ are any two such uncertain alternatives

$$u_1(\Theta_{ij}) \geqslant u_1(\Theta_{i'j'})$$

means (represents, stands for)

$$\Theta_{ij} \underset{1}{\succcurlyeq} \Theta_{i'j'}$$

and in turn, this latter preference arises by the extension process of the NM theory. In numerical terms, it then provides the special meaning that $u_1(\Theta_{ij})$ is given by the expected utility of the actual outcomes of the original game. But once we are in normal form, all we need is the fact that a_{ij} means something for strategic reasoning in one play of the game. According to section 22.5, we can regard the a_{ij}, even when they refer to expected payoffs, as psychologically referent to the evaluative process in one play, and in fact, in no way connected with the concept of repeated play. Thus, given a matrix

$$\begin{pmatrix} 1 & -1 \\ -1 & 1 \end{pmatrix}$$

of expected payoffs, under the NM interpretation this object is a representation that can be analyzed in exactly the same terms as the previous matrix, representing the game with no chance components: thus, we see how remarkable the normal form turns out to be in its representational role.

23.2. Equilibrium. Players will be examining the outcome array (Θ_{ij}) to decide upon strategies. One way to think about what a player does is that he engages in "virtual" choices, tests to see what the other player would do, modifies his "choice," tests again, and so forth until he arrives at some "equilibrium," if it exists for the game. Let us make this equilibrium concept exact. (Although the theory here is static, there is an implicit "mental dynamics" for which this notion of equilibrium exemplifies the generic equilibrium idea; see Chapter 8.) We first define the notion and then provide some intuitive basis for the definition.

(23.2.1) Definition. A pair $(\alpha_{i_0}, \beta_{j_0})$ of strategies is an equilibrium pair if and only if

(1) no Θ_{ij_0} is preferred to $\Theta_{i_0 j_0}$ by α (i.e., player 1):

$$\Theta_{i_0 j_0} \succeq_\alpha \Theta_{ij_0} \qquad \text{(all } i = 1, 2, \ldots, m)$$

and,

(2) no $\Theta_{i_0 j}$ is preferred to $\Theta_{i_0 j_0}$ by β (i.e., player 2):

$$\Theta_{i_0 j_0} \succeq_\beta \Theta_{i_0 j} \qquad \text{(all } j = 1, 2, \ldots, n)$$

For (1) think of (Θ_{ij}) as follows:

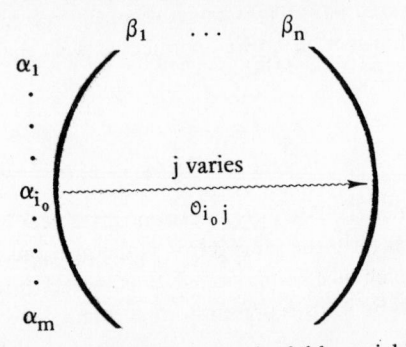

so that $\Theta_{i_0 j_0}$ is such that if player 1 were to consider his manipulable variable (row i) under the condition that the column variable is fixed at j_0, he would do no better. $\Theta_{i_0 j_0}$ is at least tied for "best" in column j_0.

For part (2) of definition (23.2.1) think of the following:

$$
\begin{array}{c}
\quad\quad \beta_1 \quad \cdots \quad\quad \beta_n \\
\begin{array}{c} \alpha_1 \\ \cdot \\ \cdot \\ \cdot \\ \alpha_{i_0} \\ \cdot \\ \cdot \\ \cdot \\ \alpha_m \end{array}
\left(
\begin{array}{c}
\text{j varies} \\ \xrightarrow{\hspace{3cm}} \\ \Theta_{i_0 j}
\end{array}
\right)
\end{array}
$$

Here $\Theta_{i_0 j_0}$ is such that if β considers his manipulable variable, column j, he could do no better in row i_0 than at outcome $\Theta_{i_0 j_0}$.

The intuitive meaning of an equilibrium pair of strategies must be explored. Recall the knowledge that α has about β: (1) β has certain possible actions Σ_2, (2) β has certain possible outcomes (Θ_{ij}), and (3) β has a preference system $\succeq_\beta$ with respect to these outcomes. By the interpretation-of-action principle (IAP), α can "derive" the virtual, or not overt, action choice of β for any given α_i strategy.

For example, suppose α considers α_1. Then IAP as applied by α instantiates to:

(1) possible actions of β: $\Sigma_2 = \{\beta_1, \beta_2, \ldots, \beta_n\}$
(2) possible outcomes for β: (Θ_{1j}), $(j = 1, 2, \ldots, n)$
(3) (relevant) preferences of β: $\succeq_\beta$ restricted to (Θ_{1j})

Immediately, α obtains for the virtual choice by β, using IAP, that class of strategies in Σ_2 that produce outcomes highest in β's preference order. Suppose, for simplicity of exposition, some one choice is thereby determined, say β_{j_1}. Then, the outcome $\mathcal{O}_{1j_1}$ can be evaluated by α by his own preferences, restricting comparisons to column j_1. [The mention of "column" is intended to suggest the matrix $(\mathcal{O}_{ij})$ but not to impute to the actors reasoning about columns or rows—this is our representational form.] Thus, at some "time" in this process of reasoning, say t_1, we would have for the state of α's reasoning as to choices

$$t_1 : (\alpha_1, \beta_{j_1})$$

where

$$\mathcal{O}_{1j_1} \gtrsim_\beta \mathcal{O}_{1j} \qquad \text{(all } j\text{)}$$

and a "transition" to

$$t_2 : (\alpha_{i_2}, \beta_{j_1})$$

where

$$\mathcal{O}_{i_2 j_1} \gtrsim_\alpha \mathcal{O}_{ij_1} \qquad \text{(all } i\text{)}$$

Reasoning about β again, using IAP, α evaluates $\mathcal{O}_{i_2 j}$ with $\gtrsim_\beta$, and thus, we have a new transition:

$$t_3 : (\alpha_{i_2}, \beta_{j_3})$$

where

$$\mathcal{O}_{i_2 j_3} \gtrsim_\beta \mathcal{O}_{i_2 j} \qquad \text{(all } j\text{)}$$

Here at t_3 the virtual outcome is one of the optimal choices for β, as deduced by α, given that α reasons IAP-wise as to what β would arrive at in outcome comparisons holding α_{i_2} fixed.

Thus, the "reasoning dynamics" of α generates a trajectory of virtual strategic choices by both actors, preceding any actual choice. This trajectory is produced by (1) the "initial condition" given by some tentative choice by α to see what would happen; (2) the "dynamic law" applied to produce "state-transitions," namely, IAP; and (3) the fixed boundary condition that the knowledge assumptions are satisfied.

But β satisfies the same generic principles. Thus, for a pair of actors, what we have is an initial condition for each and a trajectory of reasoning for each.

Suppose, now, that the initial virtual choice by α is α_{i_0} and for β the choice is β_{j_0}, where $(\alpha_{i_0}, \beta_{j_0})$ is an equilibrium pair. Then the meaning of definition 23.2.1 is, in these dynamic reasoning terms, that the reasoning process does not lead to a transition out of "state" $(\alpha_{i_0}, \beta_{j_0})$ for either player. If the system (G, A, r)—under game-theoretic player assumptions—begins at $(\alpha_{i_0}, \beta_{j_0})$ as the virtual strategic choices of the actors, then it stays there. Thus, the usual concept of equilibrium is employed.

Thus, for α we have the following:

Time	State	α-Reasoning
t_1:	$(\alpha_{i_0}, \beta_{j_0})$	Considering α_{i_0} as my choice, I see that $\theta_{i_0 j_0} \succsim_\beta \theta_{i_0 j}$, for all j, so β has no reason to do anything else than β_{j_0} if he reasons that I will select α_{i_0}.
t_2:	$(\alpha_{i_0}, \beta_{j_0})$	Considering β_{j_0} as his choice, I see that $\theta_{i_0 j_0} \succsim_\alpha \theta_{i j_0}$, for all i, so I really cannot do better. There is no reason for me to consider some other choice if β will choose β_{j_0}.

Therefore, for the α reasoning process, $(\alpha_{i_0}, \beta_{j_0})$ is a state with no reason on either component for shifting. For β, we have the following:

Time	State	β-Reasoning
t_1:	$(\alpha_{i_0}, \beta_{j_0})$	Considering β_{j_0} as my choice, I see that $\theta_{i_0 j_0} \succsim_\alpha \theta_{i j_0}$, for all i, so α would have no reason to change his mind about α_{i_0} if he reasons that I will select β_{j_0}.
t_2:	$(\alpha_{i_0}, \beta_{j_0})$	Considering α_{i_0} as α's possible choice, I see $\theta_{i_0 j_0} \succsim_\beta \theta_{i_0 j}$, this for all j, so I really have no reason to change from β_{j_0} if α will choose α_{i_0}.

Therefore, the β-reasoning process preserves $(\alpha_{i_0}, \beta_{j_0})$ as the state of the process (i.e., the pair of reasoning processes). This means that if the pair $A = \{\alpha, \beta\}$ starts the reasoning process in state $(\alpha_{i_0}, \beta_{j_0})$, it will have no reason to depart from it. At this point, then, each actor "encodes" his virtual choice into an overt choice. That is, α chooses α_{i_0} and β chooses β_{j_0}.

Example. (a) A game with an equilibrium pair. Consider:

$$\begin{array}{cc} & \begin{array}{cc} \beta_1 & \beta_2 \end{array} \\ \begin{array}{c} \alpha_1 \\ \alpha_2 \end{array} & \begin{pmatrix} \theta_{11} & \theta_{12} \\ \theta_{21} & \theta_{22} \end{pmatrix} \end{array}$$

with corresponding payoff matrix (with the psychological interpretation):

$$(a_{ij}) = \begin{array}{c} \alpha_1 \\ \alpha_2 \end{array} \begin{pmatrix} 4 & 2 \\ 6 & -1 \end{pmatrix}$$

We look for equilibria:

(1) (α_1, β_1): No equilibrium, because $a_{21} > a_{11}$ means $\theta_{21} \succ_\alpha \theta_{11}$, violating condition (1) of definition (23.2.1).

(2) (α_2, β_1): No equilibrium, because $a_{21} > a_{22}$ implies $-a_{22} > -a_{21}$, and this means $\theta_{22} \succ_\beta \theta_{21}$, violating condition (2) of definition (23.2.1).

(3) (α_2, β_2): No equilibrium, because $a_{12} > a_{22}$ means $O_{12} \blacktriangleright_\alpha O_{22}$, violating condition (1) of definition (23.2.1).

(4) (α_1, β_2): Equilibrium here because $O_{12} \sim O_{12}$ and,

(a) $a_{12} > a_{22}$ means $O_{12} \blacktriangleright_\alpha O_{22}$, satisfying (1) of definition (23.2.1).

(b) $a_{12} < a_{11}$ implies $-a_{12} > -a_{11}$, meaning $O_{12} \blacktriangleright_\beta O_{11}$, satisfying (2) of definition (23.2.1), as above for $O_{12} \blacktriangleright_\beta$ 12.

Thus, the pair (α_1, β_2) is an equilibrium pair. Then for this game (1) an equilibrium exists and (2) it is unique. There are zero-sum two-person games in which (1) fails and those in which (1) obtains but (2) fails.

Example. (b) A game with no equilibrium pair. Given the matrix

$$\begin{pmatrix} 5 & 3 \\ 2 & 4 \end{pmatrix}$$

we embed it in the game context by interpreting it as (a_{ij}) under the psychological interpretation and then apply the definition (23.2.1) to check for the existence of equilibrium pairs. The reader can verify by using all pairs of possible strategies that none satisfies both conditions of the definition. The following cycle characterizes this game:

That is, no matter what the initial virtual choices, each player reasons in a circle back to his starting point, which he left because of advantage elsewhere. Thus, this game is not "strictly determined" in the implications for action by players satisfying the three assumptions (i.e., "standard players," as we may call them).

Representing the strategy pairs as nodes and reasoning transitions by directed arrows, with vertical arrows for α since α "controls the row choice" and horizontal arrows for β since β "controls the column choice," we have the two graphs of Figure 23.2.1. Graph (a) refers to the game of example (a), with equilibrium pair (α_1, β_2). Graph (b) represents the present example. The rule is to draw the lines in the direction of gain since, by IAP, each actor chooses optimally. In Figure 23.2.1 we see in (a) that the point labeled "12" has no arrows leaving it: if the reasoning process as at "12," then it stays there. This is the equilibrium of the reasoning process. But every node of graph (b) has an arrow leaving it; hence, there is no equilibrium.

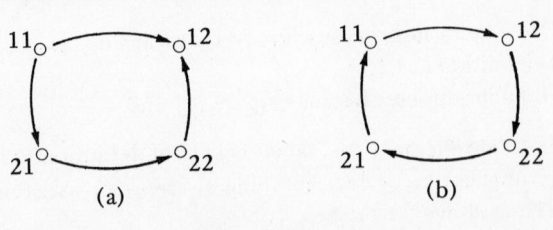

Figure 23.2.1

The matrix of example (b) demonstrates the first of the following propositions, while the second is demonstrated in example (c) below.

(23.2.2) Proposition.   An equilibrium pair for a zero-sum two-person game may not exist.

(23.2.3) Proposition.   If an equilibrium pair exists, it may not be unique.

Example. (c) (Luce and Raiffa, 1957) Let the payoff matrix be as follows:

$$\begin{array}{c} \\ \alpha_1 \\ \alpha_2 \end{array} \begin{array}{ccc} \beta_1 & \beta_2 & \beta_3 \\ \left(\begin{array}{ccc} 4 & 5 & 4 \\ 3 & 0 & 1 \end{array} \right. & & \left. \vphantom{\begin{array}{c}4\\3\end{array}} \right) \end{array}$$

Then (α_1, β_1) is an equilibrium pair: α cannot gain with β_1 fixed, and β cannot gain with α_1 fixed. But, also, (α_1, β_3) is an equilibrium pair: hold β_3 fixed, note no gain for α in switching; hold α_1 fixed, note no gain for β in switching. (Note how we begin to abbreviate the description of the reasoning process and begin relying more and more upon the numerical representation for quick deductions; also, see Figure 23.2.2.)

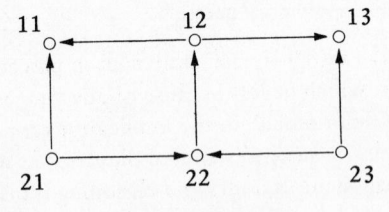

Figure 23.2.2

(23.2.4) Proposition.   If a zero-sum two-person game has more than one equilibrium pair, all these pairs are a matter of indifference to the players.

To continue example (c), (α_1, β_1) and (α_1, β_3) are equilibrium pairs. We have to compare $\mathfrak{o}_{11}$ and $\mathfrak{o}_{13}$ in terms of the relations $\succsim_\alpha$ and $\succsim_\beta$. But,

$$a_{11} = a_{13} \text{ means } u_1(\mathfrak{o}_{11}) = u_1(\mathfrak{o}_{13})$$
$$\text{means } \mathfrak{o}_{11} \sim_\alpha \mathfrak{o}_{13}$$

and it also implies $-a_{11} = -a_{13}$, which means $o_{11} \sim_\beta o_{13}$.

Propositions (23.2.2) and (23.2.3) are proved by counterexample and example, respectively, which have been given. Proposition (23.2.4) is proved in Luce and Raiffa (1957, p. 66).

The equilibrium concept leads us to the next definitions.

(23.2.5) Definition. Let G be a two-person zero-sum game with an equilibrium pair, say $(\alpha_{i_0}, \beta_{j_0})$. Then α_{i_0} is called an equilibrium strategy of α, and β_{j_0} is called an equilibrium strategy of β.

From proposition (23.2.4), we see that:

(23.2.6) Proposition. Any two equilibrium strategies of a player lead to outcomes of indifference to the player, provided that the other player also uses an equilibrium strategy.

(23.2.7) Definition. Let G be a two-person zero-sum game. Then if G has an equilibrium pair it is called strictly determined.

23.3. **Security.** In our attempt to provide a sound intuitive meaning for the concept of an equilibrium pair, we introduced two hypothetical conditions, one explicitly and the other implicitly. First, we asked what would "happen" in the reasoning processes if the initial condition were an equilibrium pair, in the technical sense of definition (23.2.1), and found that the state of the system would remain the same, in good agreement with the generic meaning of equilibrium. Second, however, we tacitly assumed that our players conducted their reasoning processes without benefit of any information not provided by (1) the rules of the game, (2) the roles assigned to the two actors, or (3) the known preferences.[*] In real plays of games, on the other hand, actors often provide clues as to their forthcoming actions, unintentionally on the part of the "encoder" of the clue but perhaps (via "spying," say) intentionally on the part of the decoder. The question, then, is this: If α knows that β may be able to decode his (α's) choice before he intentionally encodes it into action, how can α take account of this possibility in his strategic reasoning?

Example. Consider the game matrix of example (a), section 23.2:

$$\begin{pmatrix} 4 & 2 \\ 6 & -1 \end{pmatrix}$$

The possibility exists that α becomes enticed by his best outcome o_{21} (with payoff 6) and virtually chooses α_2. That is, he chooses α_2 but does not yet encode

[*]The three categories correspond to the three player assumptions of section 22.7.

it into action. Now if β discovers (decodes) this choice, β will notice (by knowledge assumptions) that he can gain by choosing β_2. Thus, although α encodes α_2 anticipating (for whatever reason) that β will choose β_1 when the choices are manifested, the outcome is o_{22}: β's best outcome.

Thus, this example indicates that within the confines of the game and the player assumptions (especially IAP) some definite "dangers" exist for the players. When interests are inverse—preferences opposite, utilities zero-sum—this suggests a notion of security. How can a player secure his choice from "premature" decoding by the other, with possible disastrous consequences?

On this basis, Luce and Raiffa (1957) developed the following interpretative theory for the equilibrium concept.

Suppose that β were to discover α intends to play strategy α_i. Then, from the point of view of α—forewarned that this could occur by "positive empirical science"—the choice of β should be anticipated. When α examines this problem, via IAP, he concludes that β would choose that strategy which is his best, given α_i. Reasoning in this way, for each α_i actor α finds the anticipated strategy of β upon a hypothetical "security leak" of α_i to β. Each such choice α_i may be compared, by α, to every other in terms of which security leak will create the least damage—that is, which of the n anticipated β choices, one per α choice disclosure, is best for α (and so worst for β). This defines the best security level of α. Similarly, β has a best security level. The basic theorem of zero-sum two-person games is that if G has an equilibrium pair, then the best security levels for the players are precisely the equilibrium strategies.

To develop this theory in mathematical form and to provide a proof of this theorem, we pause to explicate the mathematics of maximum and minimum operators on vectors: vectors, of course, because these are the rows (columns) of a game matrix and maximum-minimum operators to represent calculations related to extremal points on preference orderings. For example, the anticipated choice of β, when α supposes β discovers α will play α_i, yields a payoff for β that is the minimum of ith row of (a_{ij}).

23.4. Mathematics of the Max and Min Operators. Let $(b_1, b_2, \ldots, b_n)$ be an n-tuple (vector) of numbers. There exists a number x such that

$$x \geqslant b_i \qquad (i = 1, 2, \ldots, n)$$

and x is actually equal to one or more of the b_i.

Example. (a)

If $(b_1, b_2, b_3) = (3, 4, 2)$, then $x = 4 = b_2 \geqslant b_i$ (all i)

If $(b_1, b_2, b_3) = (3, 2, 3)$, then $x = 3 = b_1 = b_3 \geqslant b_i$ (all i)

This number x will be denoted $\max_i b_i$

Example. (b) Let $(b_1, b_2, b_3) = (7, e, \pi)$. Then $\max_i b_i = 7$. For an explicit list of numbers we simply write "max." Thus,

$$\max(7, e, \pi) = 7.$$

Similarly, there will exist a number y such that

$$y \leqslant b_i \qquad \text{(all i)}$$

and we write $y = \min_i b_i$.

Example. (c)

If $(b_1, b_2, b_3, b_4) = (7, e, \pi, -1)$, then $\min_i b_i = -1$.

If $(b_1, b_2, \ldots, b_{10}) = (7, 7, 7, \ldots, 7)$, then $\min_i b_i = 7 = \max_i b_i$.

For an explicit list of numbers we write "min." Thus, $\min(7, e, \pi, -1) = -1$.

Let (a_{ij}) be a matrix of order m $\times$ n.

Holding i fixed, we can consider the row maximum and the row minimum. Here we have

$$(b_1, b_2, \ldots, b_n) = (a_{i1}, a_{i2}, \ldots, a_{ij}, \ldots, a_{in})$$

and we write

$$\max_j(a_{ij}) = \max_j b_j = \max(a_{i1}, a_{i2}, \ldots, a_{ij}, \ldots, a_{in})$$

Similarly,

$$\min_j(a_{ij}) = \min_j b_j = \min(a_{i1}, a_{i2}, \ldots, a_{ij}, \ldots, a_{in})$$

Example. (d) Let

$$(a_{ij}) = \begin{pmatrix} a_{11} & a_{12} \\ a_{21} & a_{22} \end{pmatrix} = \begin{pmatrix} 10 & 6 \\ 4 & 8 \end{pmatrix}$$

Then

$$\max_j(a_{1j}) = \max(a_{11}, a_{12}) = \max(10, 6) = 10$$

$$\max_j(a_{2j}) = \max(a_{21}, a_{22}) = \max(4, 8) = 8$$

$$\min_j(a_{1j}) = \min(a_{11}, a_{12}) = \min(10, 6) = 6$$

$$\min_j(a_{2j}) = \min(a_{21}, a_{22}) = \min(4, 8) = 4$$

Similar operations can be performed on column vectors of a matrix (a_{ij}):

$$\max_i(a_{ij}) = \max \begin{pmatrix} a_{1j} \\ a_{2j} \\ \cdot \\ \cdot \\ a_{nj} \end{pmatrix}$$

$$\min_i(a_{ij}) = \min \begin{pmatrix} a_{1j} \\ a_{2j} \\ \cdot \\ \cdot \\ \cdot \\ a_{nj} \end{pmatrix}$$

Example. (e) Again,

$$(a_{ij}) = \begin{pmatrix} 10 & 6 \\ 4 & 8 \end{pmatrix}$$

$$\max_i(a_{i1}) = \max \begin{pmatrix} 10 \\ 4 \end{pmatrix} = 10$$

$$\min_i(a_{i1}) = \min \begin{pmatrix} 10 \\ 4 \end{pmatrix} = 4$$

$$\max_i(a_{i2}) = \max \begin{pmatrix} 6 \\ 8 \end{pmatrix} = 8$$

$$\min_i(a_{i2}) = \min \begin{pmatrix} 6 \\ 8 \end{pmatrix} = 6$$

(23.4.1) Convention. It will be useful to consider an agreement to record the row minima at the right of the matrix, the column maxima at the bottom.

Hence, for the above matrix, we write

$$\begin{pmatrix} 10 & 6 \\ 4 & 8 \end{pmatrix} \begin{matrix} 6 \\ 4 \end{matrix}$$

$$\begin{matrix} 10 & 8 \end{matrix}$$

or, more explicitly, to help fix ideas,

10	6
4	8

$6 = \min_j(a_{1j})$

$4 = \min_j(a_{2j})$

$$\begin{matrix} 10 & 8 \\ \| & \| \\ \max_i(a_{i1}) & \max_i(a_{i2}) \end{matrix}$$

Note that it makes sense to say, in this case,

$$\max_i(a_{i1}) \neq \min_j(a_{2j})$$

because

$$\max_i(a_{i1}) = 10 \neq 4 = \min_j(a_{2j})$$

On the other hand, in the matrix

$$(a_{ij}) = \begin{pmatrix} 3 & 0 & -1 \\ 3 & -1 & -3 \\ 1 & 4 & -10 \end{pmatrix}$$

we have,

$$\max_i(a_{i3}) = \min_j(a_{1j})$$

which can be verified by first recording all the extrema by convention (23.4.1):

$$\begin{pmatrix} 3 & 0 & -1 \\ 3 & -1 & -3 \\ 1 & 4 & -10 \end{pmatrix} \quad \begin{array}{l} -1 \longleftarrow \min_j(a_{1j}) \\ -3 \\ -10 \end{array}$$

$$\begin{array}{ccc} 3 & 4 & -1 \longleftarrow \max_i(a_{i3}) \end{array}$$

Next let us note that the extrema recorded on the sides of the matrix form two vectors. For the last matrix, the row minima form the vector

$$(-1, -3, -10).$$

This can be operated upon, for example, to produce

$$\max(-1, -3, -10) = -1$$

Similarly, the column maxima form the vector

$$(3, 4, -1)$$

which yields under the min operator

$$\min(3, 4, -1) = -1$$

Thus, in this case we have the computations:

$$-1 = \min_j(a_{1j})$$
$$-3 = \min_j(a_{2j})$$
$$-10 = \min_j(a_{3j})$$

so that

$$-1 = \max(-1, -3, -10) = \max(\min_j a_{1j}, \min_j a_{2j}, \min_j a_{3j})$$
$$= \max_i(\min_j a_{ij})$$

This suggests the operator

$$\max_i \min_j a_{ij}$$

to mean "Take the maximum of the row minima" if read from left-to-right. From inside-to-outside it means "Take the minimum for row i, each i, then take maximum of these minima."

We can picture this operation as follows:

That is, min is applied to each row i, and then max is applied on the margin. Thus,
$$\min_j \qquad\qquad \max_i$$

$$\max_i \min_j \begin{pmatrix} 7 & 6 \\ 4 & 1 \end{pmatrix}$$

is found in two steps:

Now consider the operator that first forms the column maxima (for each j, $\max_i a_{ij}$) and then finds the minimum of these column maxima: $\min_j \max_i a_{ij}$.

The picture is as follows:

Thus,

$$\min_j \max_i \begin{pmatrix} 7 & 6 \\ 4 & 1 \end{pmatrix}$$

is found in two steps:

Thus, for the matrix

$$(a_{ij}) = \begin{pmatrix} 7 & 6 \\ 4 & 1 \end{pmatrix}$$

we have obtained the equality

(23.4.2)
$$\min_{j} \max_{i} a_{ij} = \max_{i} \min_{j} a_{ij}$$

Here are two sample computations, based on the query: For the given matrix (a_{ij}), does (23.4.2) hold?

(1) Matrix:

$$(a_{ij}) = \begin{pmatrix} 6 & 1 \\ 3 & 2 \end{pmatrix}$$

Work:

Answer: (23.4.2) holds.

(2) Matrix:

$$(a_{ij}) = \begin{pmatrix} 10 & 4 \\ 3 & 8 \end{pmatrix}$$

Work:

Answer: (23.4.2) does *not* hold.

Thus, putting both pictures together:

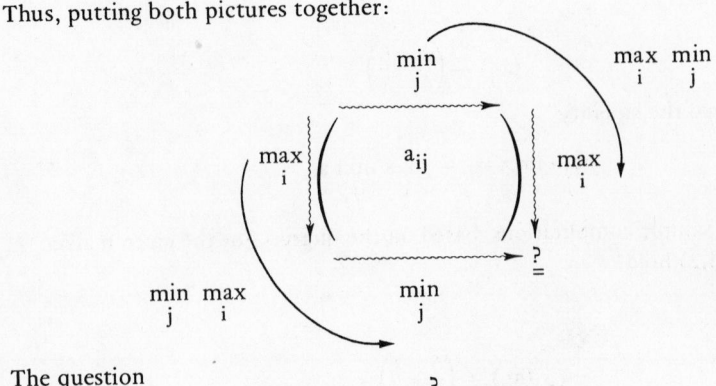

The question

$$\max_i \min_j a_{ij} \overset{?}{=} \min_j \max_i a_{ij}$$

can be thought of as a question of "commutativity." When the identity holds we say the operators $\max_i$ and $\min_j$ commute: order of application to (a_{ij}) is irrelevant.

Subsequently, these operators play a role in general abstract reasoning. To see how this works let us prove a proposition.

(23.4.3) Proposition. For any matrix (a_{ij}),

$$\max_i \min_j a_{ij} \leqslant \min_j \max_i a_{ij}$$

Proof. We first prove a different property, and then use it to obtain (23.4.3). Thus, we require to prove an intermediate result, and conventionally this type of statement is called a lemma.

Lemma. Each column max $\geqslant$ each row min.

Proof of lemma: The max in column j is $\max_i a_{ij}$. It satisfies

$$\max_i a_{ij} \geqslant a_{kj} \qquad \text{(any k)}$$

But a_{kj} (any k) is in row k and so satisfies

$$a_{kj} \geqslant \min_{j'} a_{kj'} \qquad \text{(j fixed, j' variable)}$$

Therefore, by transitivity,

$$\max_i a_{ij} \geqslant \min_{j'} a_{kj'} \qquad \text{(any k)}$$

and this holds for any j.

Returning now to the main proof, we note that if B_1 and B_2 are two lists of numbers such that

$$x \geqslant y \qquad \text{(for any x in } B_1, \text{ y in } B_2)$$

then even the minimum of B_1 is at least as large as the maximum of B_2:

$$\min B_1 \geqslant \max B_2$$

Letting

$$B_1 = (\max_i a_{i1}, \max_i a_{i2}, \ldots, \max_i a_{in})$$
$$B_2 = (\min_j a_{1j}, \min_j a_{2j}, \ldots, \min_j a_{mj})$$

we see that

$$\min B_1 = \min_j \max_i a_{ij}$$
$$\max B_2 = \max_i \min_j a_{ij}$$

whereby we conclude from the lemma, that since

$$\max_i a_{ij} \geqslant \min_{j'} a_{kj'} \qquad \text{(all j, all k)}$$

it follows that

$$\min_j \max_i a_{ij} \geqslant \max_i \min_j a_{ij}$$

which is the same as (23.4.3).

23.5. Security Levels, Saddle-Points, and the Max-Min Operators.

Consider once again definition (23.2.1). Translating this into utility terms, we obtain: $(\alpha_{i_0}, \beta_{j_0})$ is an equilibrium pair iff

$$\text{(1)} \quad u_1(\varnothing_{i_0 j_0}) \geqslant u_1(\varnothing_{i j_0}) \qquad \text{(all i)}$$
$$\text{(2)} \quad u_2(\varnothing_{i_0 j_0}) \geqslant u_2(\varnothing_{i_0 j}) \qquad \text{(all j)}$$

But $u_2 = -u_1$, so that (2) becomes, after substitution and multiplying by -1,

$$\text{(2a)} \quad u_1(\varnothing_{i_0 j_0}) \leqslant u_1(\varnothing_{i_0 j}) \qquad \text{(all j)}$$

But $(a_{ij}) = (u_1(\varnothing_{ij}))$, so that (1) and (2a) become

$$\text{(1a)} \quad a_{i_0 j_0} \geqslant a_{i j_0} \qquad \text{(all i)}$$
$$\text{(2b)} \quad a_{i_0 j_0} \leqslant a_{i_0 j} \qquad \text{(all j)}$$

But (1a) means $a_{i_0 j_0} = \max_i a_{i j_0}$; and (2b) means $a_{i_0 j_0} = \min_j a_{i_0 j}$.

Hence, $(\alpha_{i_0}, \beta_{j_0})$ is an equilibrium pair iff

(23.5.1)
$$a_{i_0 j_0} = \max_i a_{i j_0} = \min_j a_{i_0 j}.$$

Couched in terms of payoffs, (23.5.1) says that $(\alpha_{i_0}, \beta_{j_0})$ is an equilibrium if and only if the payoff simultaneously yields the maximum of column j_0 and the minimum of row i_0. A matrix entry $a_{i_0 j_0}$ satisfying (23.5.1) is called a saddle-point of the matrix (a_{ij}). Thus, an equilibrium exists if and only if the matrix has a saddle-point.

We return now to security levels, as anticipated in section 23.3. We define

$$(23.5.2) \qquad\qquad \text{S.L.}\,\alpha_i = \min_j a_{ij}$$

as the security level of α in "row i." (This represents α's anticipation that upon a security leak of α_i, β would choose his best outcome under that condition, which, by opposite preferences, is α's worst.) In payoff language, S.L. α_i is α's anticipated payoff under a security leak of α_i.

We write

$$\text{B.S.L.}\,\alpha \equiv \text{Best S.L.}\,\alpha_i$$

to mean the best (for α) of the security levels, as anticipated by varying the virtual choice α_i. Thus,

$$(23.5.3) \qquad\qquad \text{B.S.L.}\,\alpha = \max_i \min_j a_{ij}$$

Example. (a)

$$
(a_{ij}) = \begin{array}{c} \overset{\underset{\displaystyle \min}{j}}{\xrightarrow{\hspace{2cm}}} \\ \begin{pmatrix} 4 & 2 & 1 \\ 6 & 0 & 3 \\ -1 & 1 & 2 \end{pmatrix} \end{array} \begin{array}{l} 1 \\ 0 \\ -1 \end{array} \Big\} \begin{array}{c} \max \\ i \end{array}
$$

①

Hence, B.S.L. $\alpha = 1$.

For β, we have to consider

$$(23.5.4) \qquad\qquad \text{S.L.}\,\beta_j = \max_i a_{ij}$$

because a disclosure of β_j by β could be anticipated by him to lead to α's choosing the best outcome under those conditions: in payoffs, the highest in column j.
Also,

$$\text{B.S.L.}\,\beta \equiv \text{Best} \max_i a_{ij}$$

where "Best" is from the β point of view, which means the maxima with respect to $(-a_{ij})$ entries but minima with respect to (a_{ij}). Thus,

$$(23.5.5) \qquad\qquad \text{B.S.L.}\,\beta = \min_j \max_i a_{ij}$$

In the matrix given above in example (a), we have

$$\max_{i}\left\{\begin{pmatrix} 4 & 2 & 1 \\ 6 & 0 & 3 \\ -1 & 1 & 2 \end{pmatrix}\right.$$

$$6 \quad 2 \quad 3$$

$$\xrightarrow{\hspace{2cm}} \textcircled{2}$$

$$\min_{j}$$

Hence, B.S.L. $\beta = 2$.

Putting both players' security level thinking together in matrix terms:

$$\min_{j}$$

$$\max_{i}\left\{\begin{pmatrix} 4 & 2 & 1 \\ 6 & 0 & 3 \\ -1 & 1 & 2 \end{pmatrix}\begin{matrix} 1 \\ 0 \\ -1 \end{matrix}\right\} \max_{i}$$

$$\textcircled{1}$$

$$6 \quad 2 \quad 3$$

$$\xrightarrow{\hspace{2cm}} \textcircled{2}$$

$$\min_{j}$$

The meaning of the inequality of the operators, then, is

$$\text{B.S.L.} \ \alpha \neq \text{B.S.L.} \ \beta,$$

for this game.

Security-level reasoning, then, would lead α to virtually choose α_1, which is associated with B.S.L. α. It would lead β to virtually choose β_2, which is associated with B.S.L. β. Thus, the initial condition established is (α_1, β_2). Now reasoning from this point, the players will be led to adopt these strategies if they form an equilibrium pair. But in this case, at least, they do not: for β would make the transition from (α_1, β_2) to (α_1, β_3); then α would carry it to $(\alpha_2, \beta_3), \ldots$, with each player mirror-reasoning the other's virtual choices because of his knowledge.

The reasoning processes form a cyclical trajectory shown in Figure 23.5.1.

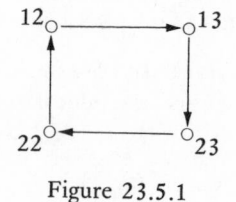

Figure 23.5.1

Since this is a circular reasoning process, the security level strategic ideas do not "lead anywhere," in this case.

To summarize the results thus far on zero-sum two-person games:

(1) An equilibrium pair may not exist.

(2) There may be more than one equilibrium pair, but this presents no problem, because all lead to outcomes that are among themselves evaluatively equivalent for the actors.

(3) Each player has a best security level; this always exists.

(4) But these levels may not specify strategies that are equilibrium strategies.

Let us coordinate at this point the security level analysis to the max-min operator version of the equilibrium notion, namely (23.5.1): $(\alpha_{i_0}, \beta_{j_0})$ is an equilibrium pair if and only if $a_{i_0 j_0}$ is a saddle-point; that is,

$$(23.5.6) \qquad\qquad a_{i_0 j_0} = \max_i a_{i j_0} = \min_j a_{i_0 j}$$

Since

$$(23.5.7) \qquad \text{B.S.L. } \alpha = \max_i \min_j a_{ij}, \text{ B.S.L. } \beta = \min_j \max_i a_{ij}$$

and these may not agree, there is a tempting conjecture at hand, suggested by the operator versions of the ideas: namely, could the value $a_{i_0 j_0}$ required to satisfy the equilibrium condition (23.5.6) be such that the two levels of (23.5.7) are equal? Conversely, might equal best security levels yield equilibrium strategies? The basic answer is affirmative, as follows: If, and only if, the max and min operators commute does the game have an equilibrium pair of strategies, and then by (23.5.6), we can locate a value $v = a_{i_0 j_0}$ that is a saddle-point, simultaneously column maximum and row minimum. Since all equilibrium pairs yield outcomes related by indifference, they must have this same value. In the next section, we provide the proof of this connection between equilibrium and the commuting operator condition. For the present, we note that this means that when (and only when) equilibrium strategies exist for the players (of a zero-sum, two-person game) will there be an outcome of the game that simultaneously realizes the best security levels for both players. This is the value of the game. Thus, when equilibrium strategies exist, the players can use them to achieve this value:

$$(23.5.8) \qquad\qquad v = \text{B.S.L. } \alpha = \text{B.S.L. } \beta$$

In measurement terms $(a_{ij}) = (u_1(\mathcal{O}_{ij}))$, so v is a measure on the scale of u_1. Since u_1 is interval-level as to scale-type, the value is given "up to positive linear transformations." Also, with

$$v = u_1(\mathcal{O}_{i_0 j_0})$$

where $(\alpha_{i_0}, \beta_{j_0})$ is an equilibrium pair, we need to consider the meaning of the outcome $o_{i_0 j_0}$. In the general case, $o_{i_0 j_0}$ is an uncertain alternative, a probability mixture of pure alternatives (given on the terminal points of the game prior to normal-form transformation). Then $u_1(o_{i_0 j_0})$ is an expected utility, and so v is the expected utility of (pure) outcomes arising from α's choice of α_{i_0}, β's choice of β_{j_0}. The real substantive meaning of v is in reference to the systems

$$(\alpha, \succ_\alpha), (\alpha, \succ_\beta)$$

which locate $o_{i_0 j_0}$ in two preference orders. The value is a compact numerical representation of this complex relational location existing in two psychological systems.

We learn from the above work that to discover if G has an equilibrium all we need to do is to check to see if the max min and min max operators commute or, equivalently, if the matrix has a saddle-point. Thus, the method of discovering whether a conceptual property of a game holds is made strictly computational by the mathematical theory providing a link between concepts (equilibrium, security levels) and formal operators.

Example. (b) Given

$$\begin{pmatrix} 6 & 0 & 3 & 1 \\ 6 & 0 & 1 & 3 \\ 4 & -1 & 0 & 1 \\ 2 & 1 & 1 & 2 \end{pmatrix}$$

assumed to represent a zero-sum two-person game in normal form, find the value if it exists.

Computation:

Computational conclusion: v exists and equals 1.

Interpretation: This game has an equilibrium pair (α_4, β_2), so α_4 is an equilibrium strategy for α, and β_2 is an equilibrium strategy for β.

Ramifications: If α chooses α_4 and β chooses β_2, then they jointly achieve their best security levels. If α chooses α_4 but β departs from β_2, α will get at least as good an outcome. If β chooses β_2 but α departs from α_4, then β will get at least as good an outcome.

Let us conclude this section by noting the conceptual difference between the various items discussed: (1) the concept of equilibrium pair of strategies; (2) the representation in numerical form of this concept, yielding a numerical rendition of the definition of the concept in terms of the existence of a saddle-point; (3) a conceptualization of the security problem in zero-sum games, leading to the notion of a best security level for each player; (4) a representation in numerical form of this conceptualization, yielding an interest in the two operators $\max_i \min_j$ and $\min_j \max_i$; and (5) a theorem relating the two numerical representations—a saddle-point exists iff the operators commute—that is interpreted in terms of the concepts represented: a criterion of equilibrium is that the equilibrium strategies also be such as to yield each player his best security level.

23.6. Proof of the Basic Theorem. We now provide the proof in detail of the theorem mentioned in the preceding paragraph.

(23.6.1) Theorem. Let (a_{ij}) be an m $\times$ n matrix of real numbers. Then (a) if there exists an entry, say $a_{i_0 j_0}$, such that $a_{i_0 j_0}$ is a saddle-point of the matrix, then the operators $\max_i \min_j$ and $\min_j \max_i$ commute, and (b) if the operators commute, then there exists an entry that is a saddle-point.

Proof. (a) By hypothesis $\exists a_{i_0 j_0}$ such that

$$(1) \quad a_{i_0 j_0} = \max_i a_{i j_0} = \min_j a_{i_0 j}$$

Since by the notion of $\max_i$

$$\max_i \min_j a_{ij} \geqslant \min_j a_{i'j} \qquad \text{(for any } i')$$

putting $i' = i_0$ yields

$$(2) \quad \max_i \min_j a_{ij} \geqslant \min_j a_{i_0 j}$$

Similarly, because

$$\min_j \max_i a_{ij} \leqslant \max_i a_{ij'} \qquad \text{(for any } j')$$

putting $j' = j_0$ yields

$$(3) \quad \min_j \max_i a_{ij} \leqslant \max_i a_{i j_0}$$

Now using $a_{i_0 j_0}$ from (1) in (2) and (3), we conclude

$$(4) \quad \begin{aligned} \max_i \min_j a_{ij} &\geqslant a_{i_0 j_0} \\ \min_j \max_i a_{ij} &\leqslant a_{i_0 j_0} \end{aligned}$$

But recall the proposition (23.4.3), proved earlier, that for any (a_{ij}),

$$(5) \quad \max_i \min_j a_{ij} \leqslant \min_j \max_i a_{ij}$$

Combining (4) and (5),

$$a_{i_0 j_0} \leqslant \max_i \min_j a_{ij} \leqslant \min_j \max_i a_{ij} \leqslant a_{i_0 j_0}$$

Now if $\max_i \min_j a_{ij} \neq \min_j \max_i a_{ij}$, then $a_{i_0 j_0} < a_{i_0 j_0}$, an absurdity. Hence,

$$(6) \quad \max_i \min_j a_{ij} = \min_j \max_i a_{ij}$$

which was to be shown.

Proof. (b) By hypothesis, (6) holds and we want to conclude

$$(7) \quad \exists a_{i_0 j_0} \text{ such that (1) above holds.}$$

Now because the vector of row minima,

$$(\min_j a_{ij})$$

has a maximum, occurring at some (not necessarily unique) row index i_0, we have

$$\exists i_0 \text{ such that } \min_j a_{i_0 j} = \max_i \min_j a_{ij}$$

Similarly, with respect to the maxima under the columns, there exists a minimal column,

$$\exists j_0 \text{ such that } \max_i a_{ij_0} = \min_j \max_i a_{ij}$$

But since the operators commute—that is, (6) holds by hypothesis—we can conclude that there exists a pair (i_0, j_0) such that

$$(8) \quad \min_j a_{i_0 j} = \max_i a_{ij_0}$$

Now since $a_{i_0 j_0}$ is in column j_0,

$$a_{i_0 j_0} \leqslant \max_i a_{ij_0}$$

and since $a_{i_0 j_0}$ is in row i_0,

$$a_{i_0 j_0} \geqslant \min_j a_{i_0 j}$$

But then (8) implies

$$a_{i_0 j_0} = \max_i a_{ij_0} = \min_j a_{i_0 j}$$

and so, (7) holds.

This concludes the proof.

23.7. Rational Action: The Problem for Theory. The conceptual problem posed by theorem (23.6.1) is that it cannot provide "positive empirical science" with a general solution to the problem of defining rational action for the players in zero-sum two-person games, because saddle-points may not exist. In other words, only if the game is strictly determined does this theorem provide a concept of rational action: each player chooses an equilibrium strategy, which guarantees him his best security level. If the other player—according to this definition of rational action in this class of games—should act nonrationally (i.e., use some nonequilibrium strategy) by strict adherence to an equilibrium strategy, the player may even do better than his best security level. But what if no saddle-point exists? In other words, what if there is no equilibrium pair?

The objective of conceptualizing rational action in any arbitrary game must be kept in mind. The equilibrium strategies, not existing when the game matrix has no saddle-point, are akin to the absence of solutions to certain equations (a point made in the monograph by Rapoport, 1970). Namely,

(1) G has no equilibrium

is analogous to

(2) $x^2 + 1 = 0$ has no solution.

Now (1) is true because of the way we define "equilibrium", and (2) is true because of the way we define "solution." Implicitly, the domain is the set of real numbers $\Re$. If we now want $x^2 + 1 = 0$ to have a solution, we have a problem of extension: namely, redefine "solution" such that x has a domain in which the equation is satisfied. This was accomplished by introducing the complex numbers. Instead of (2), because $i^2 = -1$, we have

(2a) $x^2 + 1 = 0$ has a solution.

The predicate "is a solution" has been redefined to include the earlier concept as a special case; moreover the goal of the redefinition is to give solutions to a whole class of equations that in the earlier theory did not possess solutions.

Turning now to (1), the game theoreticians NM had already committed themselves to developing a concept of rational action for the entire class of games and, so, in particular, for the zero-sum two-person games. Therefore, equating "equilibrium strategy" and "rational action," as above, NM find (1) intolerable. The concept of equilibrium requires extension to produce a concept of rational action valid for the entire class of zero-sum two-person games. In the following sections, we study the mathematics that accomplishes this extension.

23.8. Concept of Mixed Strategy. Let G be a zero-sum two-person game such that, according to definition (23.2.1), no equilibrium pair exists. As we know from (23.5.1), this means the matrix (a_{ij}) of the game in normal form has no saddle-point. We call G and (a_{ij}) nonstrictly determined. For example,

$$\begin{pmatrix} -3 & -1 \\ 0 & -2 \end{pmatrix}$$

is nonstrictly determined, as may be seen by

$$\begin{array}{c} \text{min} \\ j \end{array}$$

$$\underset{i}{\text{max}} \left\{ \begin{pmatrix} -3 & -1 \\ 0 & -2 \end{pmatrix} \begin{array}{c} -3 \\ -2 \end{array} \right\} \underset{i}{\text{max}}$$

$$\begin{array}{cc} 0 & -1 \end{array} \bigcirc{-1} \quad \bigcirc{-2}$$

$$\begin{array}{c} \text{min} \\ j \end{array}$$

using theorem (23.6.1): of course, for a small matrix in concrete form the commuting operator condition is not really needed, since one can check for a saddle-point directly.

Our immediate problem is to extend the equilibrium concept to nonstrictly determined games (with two players and zero-sum). The key to doing so is the idea of a mixed strategy.

(23.8.1) Definition. Let G be any finite game. Then any probability distribution over a strategy set Σ_i is called a mixed strategy (of player i).

Since the game is finite, so is Σ_i. Thus, Σ_i becomes the outcome space of a finite probability model. For two-person games, we have

$$\Sigma_1 = \{\alpha_1, \alpha_2, \ldots \alpha_m\}$$
$$\Sigma_2 = \{\beta_1, \beta_2, \ldots \beta_n\}$$

An arbitrary mixed strategy of player 1 may be denoted

$$x = (x_1, x_2, \ldots, x_m)$$

and for player 2,

$$y = (y_1, y_2, \ldots, y_n)$$

Of course,

$$\sum_{i=1}^{m} x_i = 1, \qquad x_i \geqslant 0 \text{ (all i)}$$

$$\sum_{j=1}^{n} y_j = 1, \qquad y_j \geqslant 0 \text{ (all j)}$$

The set of all possible mixed strategies of player i (i = 1, 2) is called his strategy space. In particular, for player 1 the strategies $\alpha_1, \alpha_2, \ldots, \alpha_m$ are each in the strategy space because the mixed strategies

$$(1, 0, 0, \ldots, 0)$$
$$(0, 1, 0, \ldots, 0)$$
.

.

°

$$(0, 0, 0, \ldots, 1)$$

are tantamount to these strategies. The terminology of game theory is that of calling α_i a pure strategy of player 1. Then we have the embedding of the α_i in the strategy space as special cases. Similar remarks apply to player 2.

The interpretation of a mixed strategy is that the player employs some sort of random scheme to select a pure strategy. For instance, player 1 might use a table of random numbers. The analogy with the solution of equations applies here:

mixed strategy is to (pure) strategy

as

complex number is to (real) number.

The goal of the enlargement of the strategy set is to provide an equilibrium definition—and, so, a conceptualization of rational action—for nonstrictly determined zero-sum two-person games.

In the situation of joint choice of (pure) strategies by random schemes the opposition of interests assures the reasonableness of the following:

(23.8.2)   Assumption. Mixed strategy x and mixed strategy y, if applied by the players, are applied independently.

Note that

$$x_i = P(\alpha_i) \qquad (i = 1, 2, \ldots, m)$$
$$y_j = P(\beta_j) \qquad (j = 1, 2, \ldots, n)$$

if the players choose x, y, respectively. Hence, assumption (23.8.2) means

(23.8.3) $P(\alpha_i \text{ and } \beta_j) = P(\alpha_i)P(\beta_j) = x_i y_j \qquad (\text{all } i, j)$

if x and y are chosen.

The payoff to player 1 if the event "α_i and β_j" occurs is a_{ij}. Thus, the payoff is a random variable with expectation,

$$\sum_{(i,j)} a_{ij}P(a_{ij}) = \sum_{(i,j)} a_{ij}P(\alpha_i \text{ and } \beta_j)$$

where the sum is over the entire matrix (a_{ij}). Because of (23.8.3) we have

$$\sum_{(i,j)} a_{ij}x_iy_j$$

as the expectation of the payoff (random variable) to player 1. For player 2, of course $(-a_{ij})$ gives the possible values of his (random variable) payoff and

$$\sum_{(i,j)} (-a_{ij})x_iy_j$$

his expectation, under mixed strategies x and y. Thus, the expected payoff to player 2 is the negative of the expected payoff to player 1. The most important mathematical entity in the extended strategy space formulation is now defined.

(23.8.4) Definition. The mapping

$$(x, y) \mapsto \sum_{(i,j)} a_{ij}x_iy_j$$

which associates to each pair of possible mixed strategies the expected payoff to player 1 is called the expectation function. We denote it by

$$E(x, y) = \sum_{(i,j)} a_{ij}x_iy_j$$

where

$$x = (x_1, x_2, \ldots, x_i, \ldots, x_m), y = (y_1, y_2, \ldots, y_j, \ldots, y_n)$$

Example. Given (a_{ij}) as follows:

$$\begin{pmatrix} 0 & 1 & 3 \\ 4 & 10 & 1 \\ -1 & 2 & 2 \end{pmatrix}$$

we find that no saddle-point exists because

Thus, suppose that the players utilize random devices to yield the mixed strategies

$$x = (.5, .3, .2)$$
$$y = (.5, .2, .3)$$

The terms

$$P(\alpha_i \text{ and } \beta_j) = P(\alpha_i)P(\beta_j) = x_i y_j \qquad (\text{all } i, j)$$

can be regarded in matrix multiplication form, treating x as a column vector:

$$xy = \begin{pmatrix} x_1 \\ x_2 \\ x_3 \end{pmatrix} (y_1 \ y_2 \ y_3) = \begin{pmatrix} x_1 y_1 & x_1 y_2 & x_1 y_3 \\ x_2 y_1 & x_2 y_2 & x_2 y_3 \\ x_3 y_1 & x_3 y_2 & x_3 y_3 \end{pmatrix} = (x_i y_j)$$

which yields

$$\begin{pmatrix} .5 \\ .3 \\ .2 \end{pmatrix} (.5 \ .2 \ .3) = \begin{pmatrix} .25 & .10 & .15 \\ .15 & .06 & .09 \\ .10 & .04 & .06 \end{pmatrix}$$

Then we require that each a_{ij} be multiplied by the corresponding $x_i y_j$ and the results summed. If we let X mean term-by-term multiplication of matrices (as in section 6.4), we can arrange the computations of the terms $a_{ij}x_i y_j$ as follows: Abstractly:

$$(a_{ij}) \ X \ (x_i y_j) = (a_{ij} x_i y_j)$$

Concretely,

$$\begin{pmatrix} 0 & 1 & 3 \\ 4 & 10 & 1 \\ -1 & 2 & 2 \end{pmatrix} X \begin{pmatrix} .25 & .10 & .15 \\ .15 & .06 & .09 \\ .10 & .04 & .06 \end{pmatrix} = \begin{pmatrix} 0 & .10 & .45 \\ .60 & .60 & .09 \\ -.10 & .08 & .12 \end{pmatrix}$$

Then all the numbers in the matrix on the right are summed to yield the expected value 1.94. Thus, the value of the expectation function $E(x, y)$ at $x = (.5, .3, .2)$, $y = (.5, .2, .3)$ is 1.94.

The expectation here is interpretable in the probability sense of section 10.5 because we interpret the probabilities as objective elements determined by a random device. Also, because $(a_{ij}) = (u_1(\mathfrak{o}_{ij}))$ and u_1 is interval in scale-type, the expectation is empirically meaningful. But, of course, we want to know what it means to the actor.

To interpret the expectation function, consider that we have assumed our players to satisfy the NM utility system, so that collections of uncertain outcomes, with specified probabilities, have evaluative significance per se in the sense that as between two such uncertain combinations the relation $\succsim$ holds in one or the other direction. Hence, if we can interpret $E(x, y)$ in terms of an uncertain combination of alternatives, we do not need to rely upon actual repetitions of play to provide the expectation with meaning. Rather, $E(x, y)$ will have meaning for a choice in one play of the game.

Note that since we have the correspondence

$$(\alpha_i, \beta_j) \mapsto \mathfrak{o}_{ij}$$

we have under mixed strategies

$$P(\Theta_{ij}) = P(\alpha_i \text{ and } \beta_j) = P(\alpha_i)P(\beta_j) = x_i y_j$$

The uncertain alternative

(23.8.5) $(x_1 y_1 \Theta_{11}, x_1 y_2 \Theta_{12}, \ldots, x_i y_j \Theta_{ij}, \ldots, x_m y_n \Theta_{mn})$

is well defined, for every x and y. We abbreviate it by

$$(x_i y_j \Theta_{ij})$$

Now, consider the meaning of Θ_{ij}. Either (a) Θ_{ij} is an actual concrete outcome of a game in extensive form (e.g., a tie score) or (b) Θ_{ij} is an uncertain combination of such pure alternatives. In case (a), $(x_i y_j \Theta_{ij})$ is a simple uncertain combination and, so, is in the space α of alternatives generated by the set of pure alternatives. In case (b), $(x_i y_j \Theta_{ij})$ is a compound uncertain combination. But in either case, according to basic NM utility theory (Chapter 20), the object $(x_i y_j \Theta_{ij})$ enters into the relation $\succsim$ with other alternatives, and in particular, as we vary x, y over the strategy spaces, the relation $\succsim$ orders these associated uncertain alternatives.

Thus, the point is that for all x and y

$$(x_i y_j \Theta_{ij})$$

is in the space α of the preference systems $(\alpha, \succsim_\alpha)$, $(\alpha, \succsim_\beta)$ of the two actors taking the roles of player 1 and player 2, respectively.

Hence, the associated utility functions u_1, u_2 with

$$u_2 = {}^- u_1$$

to represent the inverse relationship between $\succsim_\alpha$ and $\succsim_\beta$ are defined on the $(x_i y_j \Theta_{ij})$. Using u_1, we see that because of the expected utility condition,

$$u_1(x_i y_j \Theta_{ij}) = \sum_{(i,j)} u_1(\Theta_{ij}) x_i y_j$$
$$= \sum_{(i,j)} a_{ij} x_i y_j$$

Noting the expectation at the right in the last identity, we conclude

(23.8.6) $E(x, y) = u_1(x_i y_j \Theta_{ij})$

Thus, we interpret the expectation function as providing, at each pair of possible mixed strategies x, y, the associated utility to α of the uncertain combination $(x_i y_j \Theta_{ij})$. Of course, just as in the case of pure strategies, the opposition of interests assures that the player α cannot count on knowing or controlling the mixed strategy y. So, he prefers some particular uncertain alternatives to others but has only partial control as to which uncertain alternative will arise.

Finally, because of the entire idea behind the NM system we note the meaning-fulness of comparing, ordinally, various values of $E(x, y)$ at different domain arguments:

$$E(x, y) > E(x', y') \quad \text{iff} \quad u_1(x_i y_j \circ_{ij}) > u_1(x_i' y_j' \circ_{ij})$$
$$\text{iff} \quad (x_i y_j \circ_{ij}) \succ_\alpha (x_i' y_j' \circ_{ij})$$

where

$$x' = (x_1', x_2', \ldots, x_i', \ldots, x_m'), \, y' = (y_1', y_2', \ldots, y_j', \ldots, y_n')$$

The conclusion is that once we accept the NM system as holding for the players, we can interpret mixed strategies as having meaning for the actors in a one-shot play of the game: we need not imagine that they actually repeat the play to empirically interpret $E(x, y)$. Of course, because the probabilities x, y are now intended to be identified by using some random device, $E(x, y)$ also has the usual relationship to averages over potential repetitions. But these averages may never actually be realized, because we do not contemplate two actors in a standoff repetition of a zero-sum game N times with identical and independent mixed strategies in each play. The real force of the mixed strategy concept must come from the same source that, for the player, gives meaning to any uncertain combination of pure alternatives. If he found intuitive meaning in comparing such entities in the first place—which he did, in our view, when we imputed an expected utility function to him—then he will find intuitive meaning in mixed strategies. Thus, the theory that proposes to extend the rationality concept from strictly determined to nonstrictly determined games is entirely consistent with its psychological basis in its concept of mixed strategies.

23.9. Generalization of the Equilibrium Notion. We now extend the definition of an equilibrium pair. Recall that $(x_i y_j \circ_{ij})$ abbreviates (23.8.5).

(23.9.1) Definition. A pair (x^0, y^0) of mixed strategies is an equilibrium pair if and only if

$$(1) \quad (x_i^0 y_j^0 \circ_{ij}) \succeq_\alpha (x_i y_j^0 \circ_{ij}) \qquad \text{(all x)}$$

and

$$(2) \quad (x_i^0 y_j^0 \circ_{ij}) \succeq_\beta (x_i^0 y_j^0 \circ_{ij}) \qquad \text{(all y)}$$

Note the analogue to definition (23.2.1).

From definition (23.9.1), condition (1), we obtain

$$(x_i^0 y_j^0 \circ_{ij}) \succeq_\alpha (x_i y_j^0 \circ_{ij}) \quad \text{iff} \quad u_1(x_i^0 y_j^0 \circ_{ij}) \geq u_1(x_i y_j^0 \circ_{ij})$$
$$\text{iff} \quad E(x^0, y^0) \geq E(x, y^0)$$

using (23.8.6) in the second step.

Similarly, from condition (2) we obtain

$$(x_i^0 y_j^0 \circ_{ij}) \succsim_\beta (x_i^0 y_j \circ_{ij}) \quad \text{iff} \quad u_2(x_i^0 y_j^0 \circ_{ij}) \geqslant u_2(x_i^0 y_j \circ_{ij})$$
$$\text{iff} \quad -u_2(x_i^0 y_j^0 \circ_{ij}) \leqslant -u_2(x_i^0 y_j \circ_{ij})$$
$$\text{iff} \quad u_1(x_i^0 y_j^0 \circ_{ij}) \leqslant u_1(x_i^0 y_j \circ_{ij})$$
$$\text{iff} \quad E(x, y) \leqslant E(x^0, y)$$

Combining these results:

(23.9.2) Definition of Equilibrium in Numerical Representation. A pair (x^0, y^0) of mixed strategies is an equilibrium pair iff

$$(1) \quad E(x^0, y^0) \geqslant E(x, y^0) \qquad \text{(for all x)}$$

and

$$(2) \quad E(x^0, y^0) \leqslant E(x^0, y) \qquad \text{(for all y)}$$

The interpretation of the reasoning process is identical to that given earlier, because the objects of reasoning are meaningful entities comparable with other such entities for the two actors.

23.10. The Minimax Theorem. Consider now the generalization of the concept of security level to mixed strategies. If α provides β with a clue that he will use x, then α anticipates that β would employ his mixed strategy leading to the best outcome for β. Thus, for each x we have a mixed strategy security level (MSSL),

$$(23.10.1) \quad \text{MSSL } x = \min_y E(x, y) \qquad \text{[compare (23.5.2)]}$$

The notation then makes clear, for the best mixed strategy security level,

$$\text{BMSSL } \alpha = \max_x \text{MSSL } x.$$

Thus,

$$(23.10.2) \quad \text{BMSSL } \alpha = \max_x \min_y E(x, y) \qquad \text{[compare (23.5.3)]}$$

Similarly, for each y, taking β's point of view toward the security problem,

$$(23.10.3) \quad \text{MSSL } y = \max_x E(x, y) \qquad \text{[compare (23.5.4)]}$$

Hence,

$$(23.10.4) \quad \text{BMSSL } \beta = \min_y \max_x E(x, y) \qquad \text{[compare (23.5.5)]}$$

Hence, for a best mixed strategy security level, "α applies the operator" $\max_x \min_y$ and "β applies the operator" $\min_y \max_x$.

Then theorem (23.6.1), interpretable as asserting that an equilibrium pair exists if and only if the operators commute, suggests a generalization. But recall the goal: to provide a concept of rational action for the entire class of games of zero-sum two-person type. Thus, we do not want a mere criterion of existence: we want a satisfied existence criterion. Therefore, we are led to two conjectures. First, the relationship of commuting operators to the numerical form of the equilibrium definition holds for the extension and, second, the operators do commute. The first conjecture may be validated by using the properties of the inequality relation and the definitions. The proposition is stated here without proof.

(23.10.5) Proposition. An equilibrium pair of mixed strategies exists if and only if the operators commute:

$$\max_x \min_y E(x, y) = \min_y \max_x E(x, y)$$

The second conjecture is the fundamental conjecture, and it is valid, as originally demonstrated by NM in their famous "minimax theorem." The theorem, called by Owen (1968) "the most important of game theory," is not elementary (i.e., its proof is difficult). Thus, we merely refer the reader to the treatment by Owen.

(23.10.6) Minimax Theorem. $\max_x \min_y E(x, y) = \min_y \max_x E(x, y)$

(23.10.7) Corollary. Every zero-sum two-person game has an equilibrium pair of mixed strategies.

Given these propositions, we can summarize the equilibrium properties of any two-person zero-sum game. By the minimax theorem we first extend the concept of value of a game by noting that the theorem implies that the best mixed strategy security level of α [BMSSL α as given in (23.10.2)] is the same as the best mixed strategy security level of β [BMSSL β as given in (23.10.4)]. Hence, generalizing definition (23.5.8), we define the value v by

(23.10.8) $\qquad\qquad v = \text{BMSSL } \alpha = \text{BMSSL } \beta$

Given v, then the basic results about equilibrium are as follows:

(23.10.9) Properties of Equilibrium.

(1) $E(x^0, y) \geqslant v$ (any y)
(2) $E(x, y^0) \leqslant v$ (any x)
(3) $E(x^0, y^0) = v$

These properties interpretatively assert the following:

(1) By playing an equilibrium or maximin strategy x^0—guaranteed to exist by

(23.10.7)—player 1, α, can assure himself his best mixed strategy security level (which is v) and, in fact, may do better if, under this concept of rational action, the other player chooses nonrationally in the sense of using some y not given as an equilibrium strategy.

(2) Similarly, playing equilibrium or minimax strategy y^0, player 2, β, can assure himself his best mixed strategy security level. [The inequality appears in the reverse of that in (1) only because of the representation of u_2 as $-u_1$.]

(3) If both players behave rationally, in the specific sense of using equilibrium strategies, they each hold the other to best mixed strategy security level, which, of course, depends on the structure of the game.

We note that $v = u_1(x_i^0 y_j^0 \theta_{ij})$ for actor α and $-v = u_2(x_i^0 y_j^0 \theta_{ij})$ for actor β, so the value is the utility of the equilibrium outcome, for each player.

There is one aspect of the security level concept deserving mention. That is, BSL α of (23.5.3) always exists (unlike the saddle-point) and BMSSL α of (23.10.2) always exists. But the latter supercedes the former because, in being the maximum over all mixed strategies x, it is the maximum over all the embedded pure strategies and, therefore,

$$\text{BMSSL } \alpha \geqslant \text{BSL } \alpha$$

(with a similar result for β). In passing to mixed strategies, therefore, α cannot possibly do worse than in the pure strategy case and may do better if he chooses rationally in this specific sense.

Finally, we note that the work of this section is a thorough generalization of the results for the strictly determined games. The value as defined by (23.10.8) specializes to the computed value in the case of a saddle point. The equilibrium pair (x^0, y^0) becomes the equivalent to the equilibrium pair in the case of a saddle point; for instance, if (α_2, β_1) is the equilibrium pair for a strictly determined game, then in terms of mixed strategies $\alpha_2 = (0, 1), \beta_1 = (1, 0)$ and the mixed strategy equilibrium pair is $(x^0, y^0) = ((0, 1), (1, 0))$. In other words, the conceptual and computational results for strictly determined games form a special case of the more general results subsumed under mixed strategy analysis.

23.11. Computations and Interpretation Problems.

In this section, we outline a practical technique for actually locating the equilibrium (mixed) strategies in the nonstrictly determined case where the game matrix is 2 × 2. Further details may be found in Kemeny, Snell, and Thompson (1966) and for higher-order matrices by the same authors in their book addressed especially to students of business (Kemeny, Schleifer, Snell, and Thompson, 1962).

Let

$$\begin{pmatrix} a_{11} & a_{12} \\ a_{21} & a_{22} \end{pmatrix}$$

be such that no saddle-point exists. Hence, it can represent an arbitrary non-strictly determined zero-sum two-person game where each player has two pure strategies. We want to find (x^0, y^0), an equilibrium pair of strategies. By definition, a mixed strategy for the row player is a pair (x_1, x_2) such that

$$x_1 + x_2 = 1, x_1 \geqslant 0, x_2 \geqslant 0$$

A mixed strategy for the column player is (y_1, y_2) with

$$y_1 + y_2 = 1, y_1 \geqslant 0, y_2 \geqslant 0$$

Recalling the method of computation of $E(x, y)$ from the example of section 23.8, we have the elementwise product

$$(a_{ij}x_iy_j) = (a_{ij}) \times (x_iy_j)$$
$$= \begin{pmatrix} a_{11} & a_{12} \\ a_{21} & a_{22} \end{pmatrix} \times \left(\begin{pmatrix} x_1 \\ x_2 \end{pmatrix} (y_1, y_2) \right)$$

where the second product is ordinary matrix multiplication.

Now, since we want to find x^0, we set $x = x^0$. At first, y is made pure, $y = (1, 0)$. This yields the computation of $E(x^0, (1, 0))$:

$$\begin{pmatrix} a_{11} & a_{12} \\ a_{21} & a_{22} \end{pmatrix} \times \left(\begin{pmatrix} x_1^0 \\ x_2^0 \end{pmatrix} (1 \ 0) \right) = \begin{pmatrix} a_{11} & a_{12} \\ a_{21} & a_{22} \end{pmatrix} \times \begin{pmatrix} x_1^0 & 0 \\ x_2^0 & 0 \end{pmatrix} = \begin{pmatrix} a_{11}x_1^0 & 0 \\ a_{21}x_2^0 & 0 \end{pmatrix}$$

Adding all terms of the latter matrix gives

$$E(x^0, (1,0)) = a_{11}x_1^0 + a_{21}x_2^0$$

Next set $y = (0, 1)$ to compute $E(x^0, (0, 1))$:

$$\begin{pmatrix} a_{11} & a_{12} \\ a_{21} & a_{22} \end{pmatrix} \times \left(\begin{pmatrix} x_1^0 \\ x_2^0 \end{pmatrix} (0 \ 1) \right) = \begin{pmatrix} a_{11} & a_{12} \\ a_{21} & a_{22} \end{pmatrix} \times \begin{pmatrix} 0 & x_1^0 \\ 0 & x_2^0 \end{pmatrix} = \begin{pmatrix} 0 & a_{12}x_1^0 \\ 0 & a_{22}x_2^0 \end{pmatrix}$$

A sum over the final matrix gives

$$E(x^0, (0, 1)) = a_{12}x_1^0 + a_{22}x_2^0$$

We obtain two equations in y^0 in a similar way. Thus, we have the system

$$(23.11.1) \quad \begin{aligned} E(x^0, (1, 0)) &= a_{11}x_1^0 + a_{21}x_2^0 \\ E(x^0, (0, 1)) &= a_{12}x_1^0 + a_{22}x_2^0 \\ E((1, 0), y^0) &= a_{11}y_1^0 + a_{12}y_2^0 \\ E((0, 1), y^0) &= a_{21}y_1^0 + a_{22}y_2^0 \end{aligned}$$

We appeal now to the properties of equilibrium, given in (23.10.9), (1) and (2). By (1),

$$(23.11.2) \quad \begin{aligned} E(x^0, (1, 0)) &\geqslant v \\ E(x^0, (0, 1)) &\geqslant v \end{aligned}$$

and by (2),

(23.11.3)
$$E((1, 0), y^0) \leqslant v$$
$$E((0, 1), y^0) \leqslant v$$

Using the identities of (23.11.1) to substitute for the corresponding terms in the inequalities, (23.11.2) and (23.11.3), we obtain the system of four inequalities

(23.11.4)
$$a_{11}x_1^0 + a_{21}x_2^0 \geqslant v$$
$$a_{12}x_1^0 + a_{22}x_2^0 \geqslant v$$
$$a_{11}y_1^0 + a_{12}y_2^0 \leqslant v$$
$$a_{21}y_1^0 + a_{22}y_2^0 \leqslant v$$

Since $x_2^0 = 1 - x_1^0$, $y_2^0 = 1 - y_1^0$, we have to solve for three unknowns: x_1^0, y_1^0, and v, in terms of the parameters of the problem, (a_{ij}). In the system (23.11.4) a strict inequality would lead to a contradiction (see Kemeny, Snell, and Thompson, 1966); although this is not true for higher-order systems, which then become more laborious to solve. Using this fact, (23.11.4) becomes a system of four ordinary algebraic equations in three unknowns. The first pair can be equated (because both are equal to v), and we solve for the only unknown x_1^0. The second pair yield y_1^0 in the same way. Then we can obtain an expression for v by returning to (23.11.4) with x_1^0 or y_1^0 in terms of the parameters. The general results are as follows:

(23.11.5)
$$x_1^0 = \frac{a_{22} - a_{21}}{(a_{11} + a_{22}) - (a_{12} + a_{21})}$$

(23.11.6)
$$y_1^0 = \frac{a_{22} - a_{12}}{(a_{11} + a_{22}) - (a_{12} + a_{21})}$$

(23.11.7)
$$v = \frac{a_{11}a_{22} - a_{12}a_{21}}{(a_{11} + a_{22}) - (a_{12} + a_{21})}$$

Note that this solution (x_1^0, y_1^0, v) requires that

$$a_{11} + a_{22} \neq a_{12} + a_{21}$$

But in fact, Kemeny, Snell, and Thompson (1966) indicate that the following proposition holds: a 2 × 2 matrix (a_{ij}) has no saddle-point if and only if the entries in one diagonal both exceed the entries in the other diagonal. Thus, once one has determined by inspection that a 2 × 2 matrix (a_{ij}) has no saddle-point, the proposition yields either (23.11.8a) or (23.11.8b):

(23.11.8) (a) $\begin{array}{l} a_{11} > a_{12}, a_{11} > a_{21} \\ a_{22} > a_{12}, a_{22} > a_{21} \end{array}$ (b) $\begin{array}{l} a_{12} > a_{11}, a_{12} > a_{22} \\ a_{21} > a_{11}, a_{21} > a_{22} \end{array}$

Recalling that for any numbers x, y, z, w, if $x > y$ and $z > w$, then

$$x + z > y + w$$

we obtain, from (23.11.8),

(23.11.9)   (a) $a_{11} + a_{22} > a_{12} + a_{21}$ (b) $a_{12} + a_{21} > a_{11} + a_{22}$

so that in either case the denominator of (23.11.5), (23.11.6), and (23.11.7) is nonzero. Hence, a solution exists.

Example. (a) To provide a concrete illustration of the interpretative problem in regard to a formal solution, let (a_{ij}) be given by the matrix with no saddle-point:

$$\begin{pmatrix} 1 & 0 \\ -1 & 2 \end{pmatrix}$$

Then, using (23.11.5), (23.11.6), and (23.11.7), we see that

$$x_1^0 = \tfrac{3}{4}, y_1^0 = \tfrac{1}{2}, v = \tfrac{1}{2}$$

Thus, an actor α taking the role of row player has available the equilibrium strategy in which he chooses row 1 with probability 3/4, row 2 with probability 1/4. Column player using an equilibrium strategy chooses with equal probabilities. The expected payoffs are $v = E(x^0, y^0) = 1/2$. What this all means however, depends upon the game (or interactive choice situation) leading to this normal form.

One very special case of importance to experimentalists involves presenting the game in terms of row, column, and monetary or "score" payoffs as listed in the matrix. This game is, from our standpoint, only given in the sense of the definition of a game in extensive form with "monetary interpretation." That is (see section 22.5), the outcome space contains numbers that have meaning for the actors in some everyday sense: for instance, so many pennies, or perhaps so many "points." But from a psychological standpoint, the actors assigned row and column roles have preference systems over the entries as so interpreted. Our rapid formal solution for x_1^0, y_1^0 and v in terms of pennies or points may have little to do with the actual preference systems of these actors: empirically, it will depend strongly upon their "attitudes toward risk-taking," which is a mode of talking about the relations $\succcurlyeq_\alpha, \succcurlyeq_\beta$ over uncertain alternatives.

To apply the theory properly—from an interpretative standpoint—would require knowing the utility of the numerical outcomes or making a reasonable assumption. Suppose the outcomes are in money terms. This is the (θ_{ij}) matrix. Applying u_1 yields $(u_1(\theta_{ij}))$, and it is the latter whose interpretation is coherent with respect to the conceptual foundations of the numerical notions of equilibrium strategies and value of the game. If the θ_{ij} amounts are small, it is not unreasonable to assume that for $m > 0$,

$$u_1(\theta_{ij}) = m\theta_{ij} + b \qquad \text{(all i, j)}$$

Also, $u_2 = -u_1$ is assumed. This linearity assumption yields the following results. Taking the given θ_{ij} matrix of money outcomes as the a_{ij} terms in (23.11.5),

(23.11.6), and (23.11.7) and applying the transformation

$$a_{ij} \mapsto ma_{ij} + b$$

yields the transformed solution

$$x_0^1 \mapsto x_0^1$$
$$y_0^1 \mapsto y_0^1$$
$$v \mapsto mv + b$$

in which we see that the equilibrium strategy pair remains invariant and the value v shifts to mv + b, the utility for player 1 of the money value v. Then since u_1 is an interval scale variable, we can admissibly transform it under the linear function,

$$u_1 \mapsto \frac{1}{m}(u_1 - b)$$

Then, since mv + b is in u_1 terms,

$$mv + b \mapsto \frac{1}{m}(mv + b - b) = v$$

and so, we can interpret v (in utility units) as the utility of the computed value v (in money units) in some utility scale representing the money preferences of player 1. From this it follows that from the two assumptions,

(1) for small amounts of money, utility is a linear function of the amount, $u_1(\theta_{ij}) = m\,\theta_{ij} + b$

(2) $u_2 = -u_1$

we can apply the entire formal solution procedure to the matrix of money outcomes (θ_{ij}) and interpret the solution in terms of the concepts developed in this chapter (equilibrium, security level, etc.).

The above argument was for a payoff matrix of order 2 × 2. However, in Kemeny, Snell, and Thompson (1966) it is proved that if (a_{ij}) is an arbitrary m × n payoff matrix with value v, then under the linear mapping $a_{ij} \mapsto ma_{ij} + b$ the new matrix yields a solution that has the same equilibrium strategy pair as (a_{ij}) and whose value is just mv + b.

Example. (b) In Section 23.1, example (a), we were led to the matrix:

$$\begin{pmatrix} 1 & -1 \\ -1 & 1 \end{pmatrix}$$

There is no saddle-point. Three "methods" show this: (1) direct inspection, appealing to the definition of saddle-point: there is no row minimum that is a column maximum; (2) application of the operators min max and max min,
$$ j \quad i \qquad i \quad j$$
which do not yield the same result; and (3) the proposition mentioned prior to (23.11.8), noting that in one diagonal the terms are positive, in the other negative.

In particular the difference of the diagonal sums, required for the denominator of (23.11.5) is

$$(a_{11} + a_{22}) - (a_{12} + a_{21}) = 4$$

Hence, $x_1^0 = 2/4 = 1/2$. Using (23.11.6), $y_1^0 = 1/2$. Then (23.11.7) yields $v = 0$. The solution is the triple

$$(\tfrac{1}{2}, \tfrac{1}{2}, 0)$$

According to the derivation of this game matrix, given in section 23.1, the solution may be interpreted for player 1 as yielding a strategy—say, toss a fair coin to select α_1 or α_2—such that the worst he can do on his utility scale is to arrive at the average of the utilities assigned "1 wins, 2 loses" and "1 loses, 2 wins."

We note that the best security level of α in this example is $\min_{j} \max_{i} (a_{ij})$:

$$\begin{matrix} & \overset{\displaystyle \min_{j}}{\longrightarrow} & \\ \begin{pmatrix} 1 & -1 \\ -1 & 1 \end{pmatrix} & \begin{matrix} -1 \\ -1 \end{matrix} & \Big\downarrow \; \underset{i}{\max} \\ & \boxed{-1} & \end{matrix}$$

This compares unfavorably with $v = 0$, the best mixed strategy security level, as claimed in section 23.10. Similarly, BSL $\beta = \min_{j} \max_{i} (a_{ij}) = 1$, which is -1 on β's scale, comparing unfavorably with the BMSSL $\beta = v = 0$.

The question of interpreting this solution now arises. Two persons are playing, or rather are about to play. We know that

$$A_1 \succ_\alpha A_2$$
$$A_2 \succ_\beta A_1$$

where in A_1 "α wins, β loses," in A_2 "α loses, β wins." Assuming the NM approach, an uncertain combination,

$$(pA_1, (1-p)A_2) \qquad (0 \leq p \leq 1)$$

has psychological meaning to the actors. For a binary set

$$\Omega = \{A_1, A_2\}$$

the free assignment, a choice of zero and unit,

$$A_1 \mapsto 1$$
$$A_2 \mapsto 0$$

always yields the expected utility:

$$(pA_1, (1-p)A_2) \mapsto p \qquad \text{(for player 1).}$$

The prior assignment for α,

$$A_1 \mapsto 1$$
$$A_2 \mapsto -1$$

is the scale of the value v. But this scale u_1, relates to the $A_1 \mapsto 1$, $A_2 \mapsto 0$ scale (say u_1') by

$$u_1' = \tfrac{1}{2}u_1 + \tfrac{1}{2}$$

which is readily seen by drawing a (u_1, u_1') coordinate system and plotting the two points $(1, 1)$ and $(-1, 0)$. Thus,

$$v' = \tfrac{1}{2} \qquad (\text{because } v = 0)$$

The transformed value is just the probability that α gets A_1 when he employs a device to realize the maximin strategy $x^0 = (1/2, 1/2)$. Similarly, on the β-scale, we have under u_2:

$$A_1 \mapsto -1$$
$$A_2 \mapsto 1$$

and we want to put this u_2 scale into u_2' form:

$$A_1 \mapsto 0$$
$$A_2 \mapsto 1$$
$$(pA_1, (1-p)A_2) \mapsto 1-p \qquad (\text{for player 2}).$$

Thus, the identical transformation will do the job,

$$u_2' = \tfrac{1}{2}u_2 + \tfrac{1}{2}$$

and we conclude that for β, the value becomes, as expected,

$$v' = \tfrac{1}{2} \qquad (\text{because } v = 0)$$

and is interpretable for β as the probability that β gets A_2, his favored outcome. Hence, each player in this game has the same chance of obtaining his favored outcome. Considerations of this kind prompt Kemeny, Snell, and Thompson (1966) to call a game with value $v = 0$ *fair*.

23.12. Conclusions on Rationality. In the second half of this chapter, we have seen how the theory developed in the first half was extended to provide a concept of rational action in a whole class of games not covered by the preceding concept and yet how this was done via the rigorous methodology of logical extension. The concept of rational action is now as follows: each player chooses an equilibrium (mixed) strategy that guarantees him his best mixed strategy security level, which in turn is at least as good as his pure-strategy best security level. Moreover, if the other player, according to this definition of rationality in this class of games, should act nonrationally (i.e., use some nonequilibrium strategy), then by strict adherence to an equilibrium strategy the player may even do better than his best mixed strategy security level. In principle, there is no zero-sum two-person game that is excluded from this rationality principle. Of course, as in the solution of equations, knowing that a solution exists—that a situation has an associated rational choice—is not at all the same as actually solving for that choice. We have illustrated computational procedures only for the simplest class, namely, the 2 X 2 games, referring the reader to Kemeny, Schleifer, Snell, and Thompson (1962) for solution procedures for larger matrices.

CHAPTER TWENTY-FOUR GENERAL TWO-PERSON GAMES: BACKGROUND

24.1. Meaning of General Two-Person Game. By a "general" two-person game, we mean a two-person game in which the actors assigned to the player roles do not necessarily have opposite preferences. Thus, these games have the potential for cooperation. The solution theory for such games is quite involved. The objective of developing a concept of rationality for these games has been subjected to a variety of approaches. We first outline the basic ideas taken in all of these approaches as points of departure.

The expectation function has to be refined for representation of two partly-coincident, partly-opposed preference systems. Even if we assume independence [recall assumption (23.8.2)], we have

$$(24.1.1) \quad \begin{array}{l} \text{(a)} \;\; E_1(\mathbf{x}, \mathbf{y}) = \sum_{(i,j)} a_{ij}x_iy_j = \text{expectation function for player 1,} \\[2mm] \text{(b)} \;\; E_2(\mathbf{x}, \mathbf{y}) = \sum_{(i,j)} b_{ij}x_iy_j = \text{expectation function for player 2.} \end{array}$$

Recall that in section 23.8, we had $(-a_{ij})$ as the payoff matrix for player 2; thus, $(b_{ij}) = (-a_{ij})$ for this special case of strictly opposing preferences. Then we could treat only one expectation function, namely, that called $E_1(\mathbf{x}, \mathbf{y})$ here. But for the general case, we require (b_{ij}), given interpretatively as

$$(b_{ij}) = (u_2(\sigma_{ij}))$$

That is, the utility to the second player of the outcome arising when the pure strategies used are α_i and β_j is denoted b_{ij}.

24.2. An Example of the Setup Phase in a General Two-Person Game Model. Let us consider a situation giving rise to a general two-person game. In the decision-problems of a married couple, one often sees the so-called battle-of-the sexes game, in which there is a choice to be made between two activities, say a and b, in which they engage individually or together. (For example, a = baseball game, b = shopping; a = golf, b= tennis; a = film 1, b = film 2.) We assume a and b are such that the husband (H) has the preference a $\succ_H$ b, the wife (W) has the preference b $\succ_W$ a. On the other hand, they share a desire to be together. Let t mean "together," s mean "separate," as choice elements. That is, we see each actor facing the dual problem,

$$\Omega_1 = \{t, s\} \qquad t \succ_H s, t \succ_W s$$
$$\Omega_2 = \{a, b\} \qquad a \succ_H b, b \succ_W a$$

in which a choice is required from Ω_1 as well as Ω_2, with consensus on Ω_1, but opposite preferences as to the activities. The actors individually may be assumed to give more or less preference to the Ω_1 choice dimension vis-à-vis Ω_2 as a choice dimension. In other words, Ω_1 and Ω_2 are such that they require evaluative ordering. Suppose for husband and wife, it is "more important" to be active together than to engage in some particular activity.
Then the product set

$$\Omega = \Omega_1 \times \Omega_2 = \{(t, a), (t, b), (s, a), (s, b)\}$$

can be ordered by each actor: namely,

$$(t, a) \succ_H (t, b) \succ_H (s, a) \succ_H (s, b)$$
$$(t, b) \succ_W (t, a) \succ_W (s, b) \succ_W (s, a)$$

using a "lexicographic ordering" idea: first, together; second, "my" activity. (This ordering technique was discussed earlier, in Chapter 12.)

Clearly the systems $(\Omega, \succ_H)$ and $(\Omega, \succ_W)$ are not opposite, although they do not exhibit complete accord. There is motivation for cooperation in t $\succ_H$ s, t $\succ_W$ s — but the activities' opposition has introduced a conflict element. This exemplifies a "general" game: neither identity nor strict opposition of interests. Now a numerical representation involves, for the husband, using the NM utility theory of Chapter 20:

$$u_1(t, a) = 1$$
$$u_1(s, b) = 0$$
$$u_1(t, b) = p_1 \quad \text{iff} \quad (t, b) \sim_H (p_1(t, a), (1-p_1)(s, b))$$
$$u_1(s, a) = p_2 \quad \text{iff} \quad (s, a) \sim_H (p_2(t, a), (1-p_2)(s, b))$$

For the wife,

$$u_2(t, b) = 1$$
$$u_2(s, a) = 0$$
$$u_2(t, a) = q_1 \quad \text{iff} \quad (t, a) \sim_W (q_1(t, b), (1-q_1)(s, a))$$
$$u_2(s, b) = q_2 \quad \text{iff} \quad (s, b) \sim_W (q_2(t, b), (1-q_2)(s, a))$$

Now to return to the game element in this example, who has which moves? It is customary (see Rapoport, 1966) to represent the situation as follows: each makes one move, namely, a choice from $\{a, b\}$, but the rules of the game do not forbid communication to achieve realization of choice t, not itself formally represented as a choice at a move. In particular, each is identified as a player with two pure strategies, namely, a and b. Then in this customary interpretation, we have

$$
\begin{array}{c}
W \\
\begin{array}{cc}
a & b
\end{array} \\
H \begin{array}{c} a \\ b \end{array} \begin{pmatrix} 0_{11} & 0_{12} \\ 0_{21} & 0_{22} \end{pmatrix}
\end{array}
$$

where

0_{11}: they do a together
0_{12}: H does a, W does b
0_{21}: H does b, W does a
0_{22}: they do b together.

From the prior analysis we have two viewpoints toward these outcomes, as follows:

$$
\begin{array}{cc}
\underline{\text{H viewpoint}} & \underline{\text{W viewpoint}} \\
W & W \\
\begin{array}{cc} a & b \end{array} & \begin{array}{cc} a & b \end{array} \\
H \begin{array}{c} a \\ b \end{array} \begin{pmatrix} (t, a) & (s, a) \\ (s, b) & (t, b) \end{pmatrix} & H \begin{array}{c} a \\ b \end{array} \begin{pmatrix} (t, a) & (s, b) \\ (s, a) & (t, b) \end{pmatrix}
\end{array}
$$

These map into payoffs, using the two utility functions, as follows:

$$
\frac{\text{H viewpoint}}{\begin{pmatrix} (t, a) & (s, a) \\ (s, b) & (t, b) \end{pmatrix}} \xrightarrow{u_1} \frac{\text{H payoffs } (a_{ij})}{\begin{pmatrix} 1 & p_2 \\ 0 & p_1 \end{pmatrix}} \qquad (0 < p_2 < p_1 < 1)
$$

$$
\frac{\text{W viewpoint}}{\begin{pmatrix} (t, a) & (s, b) \\ (s, a) & (t, b) \end{pmatrix}} \xrightarrow{u_2} \frac{\text{W payoffs } (b_{ij})}{\begin{pmatrix} q_1 & q_2 \\ 0 & 1 \end{pmatrix}} \qquad (0 < q_2 < q_1 < 1)
$$

Hence, forming a matrix of pairs of payoffs,

$$(24.2.1) \quad ((a_{ij}, b_{ij})) = \begin{pmatrix} (1, q_1) & (p_2, q_2) \\ (0, 0) & (p_1, 1) \end{pmatrix} \quad \begin{pmatrix} 0 < p_2 < p_1 < 1 \\ 0 < q_2 < q_1 < 1 \end{pmatrix}$$

The two expectation functions are

$$
E_1(x, y) = \sum_{(i,j)} a_{ij}x_iy_j = a_{11}x_1y_1 + a_{12}x_1y_2 + a_{21}x_2y_1 + a_{22}x_2y_2
$$
$$(24.2.2) \qquad = x_1y_1 + p_2x_1y_2 + p_1x_2y_2$$

and

$$E_2(x, y) = \sum_{(i,j)} b_{ij}x_iy_j = b_{11}x_1y_1 + b_{12}x_1y_2 + b_{21}x_2y_1 + b_{22}x_2y_2$$

(24.2.3) $$= q_1x_1y_1 + q_2x_1y_2 + x_2y_2$$

24.3. The Mapping into Payoff Space. The above example has illustrated, by construction from an illustrative interactive choice situation, that general games have as many expectation functions as there are players with distinct preference systems. Thus, for two players the mixed strategy pair (x, y) goes into a point in two-space: $(E_1(x, y), E_2(x, y))$, the first coordinate of the point interpretable as the expected payoff to player 1, the second component interpretable as the expected payoff to player 2.

A visualization of the mapping

$$(x, y) \mapsto (E_1(x, y), E_2(x, y))$$

is essential for comprehending the theory. Note that in the case of two pure strategies for each player,

$$x = (x_1, 1-x_1)$$
$$y = (y_1, 1-y_1)$$

because x and y are probability distributions. Thus, we can think of the above mapping as follows (see Figure 24.3.1):

$$(x_1, y_1) \mapsto (E_1(x, y), E_2(x, y))$$

where (x_1, y_1) varies over a square of points (domain of this mapping) and the images form some figure in two-space. This map may be denoted (E_1, E_2).

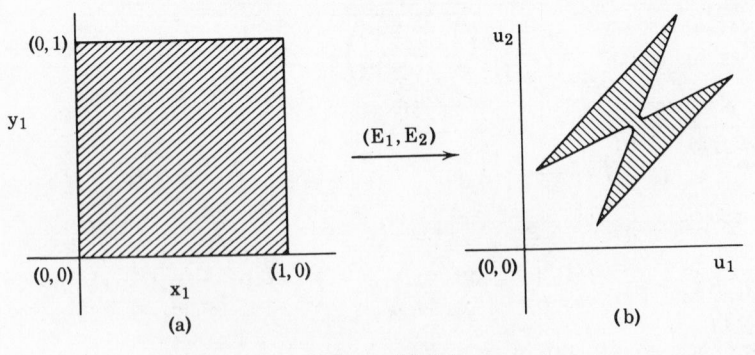

Figure 24.3.1

A concrete instance of the point (x, y) going into $(E_1(x, y), E_2(x, y))$ may be useful. For this purpose, let the matrix (24.2.1) be

$$\left(\left(a_{ij}, b_{ij}\right)\right) = \begin{pmatrix} (1, .8) & (.6, .5) \\ (0, 0) & (.7, 1) \end{pmatrix}$$

Also, let us find the image of mixed strategy pair

$$x = (.6, .4)$$
$$y = (.3, .7)$$

Substitution into (24.2.2) yields

$$\begin{aligned} E_1(x, y) &= (.6)(.3) + (.6)(.6)(.7) + (.7)(.4)(.7) \\ &= .180 + .252 + .196 \\ &= .628 \end{aligned}$$

Substitution into (24.2.3) yields

$$\begin{aligned} E_2(x, y) &= (.8)(.6)(.3) + (.5)(.6)(.7) + (.4)(.7) \\ &= .144 + .210 + .280 \\ &= .634 \end{aligned}$$

Hence, the image point is (.628, .634).

In general, the players are not restricted to

$$x = (x_1, x_2)$$
$$y = (y_1, y_2)$$

with two independent parameters,

$$(x_1, y_1)$$

Instead they have n and m pure strategies, respectively, so that,

$$x = (x_1, x_2, \ldots, x_m)$$
$$y = (y_1, y_2, \ldots, y_n)$$

with independent parameters

$$(x_1, x_2, \ldots, x_{m-1}, y_1, y_2, \ldots, y_{n-1})$$

However, although this makes the domain of (E_1, E_2) large, the range remains the payoff space of two dimensions (u_1, u_2). In any case, the image—in the mapping sense—of the domain is called the payoff region. The theory depends strongly on the analysis of the properties of the payoff region and, hence, upon general geometric notions to be outlined in the next sections. We return to the payoff region, then, after a preparatory study of the geometric concepts themselves.

24.4. Line Segments in Vector Terms. The notion of line segment in two-space is familiar. In a coordinate plane two points, the end points, are sufficient to specify the line segment L as shown in part (a) of Figure 24.4.1.

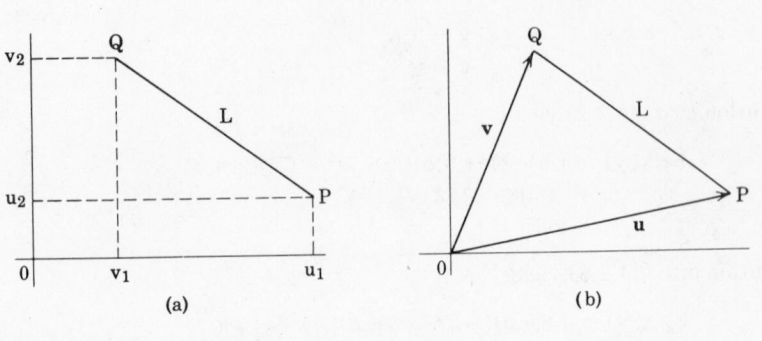

(a)   (b)

Figure 24.4.1   (a) Line segment L in a coordinate plane. (b) Line segment in relation to the vectors associated with the endpoints.

Points P and Q are only the endpoints of L. We want to specify L as a set of points. But points P and Q are in correspondence with number-pairs and so with vectors $u = (u_1, u_2)$, $v = (v_1, v_2)$, as shown in part (b) of Figure 24.4.1. This suggests thinking of L as a set of tips of vectors from the origin, where the vectors are all between vectors u and v. Thus, let x be an arbitrary such vector as in Figure 24.4.2 part (a).

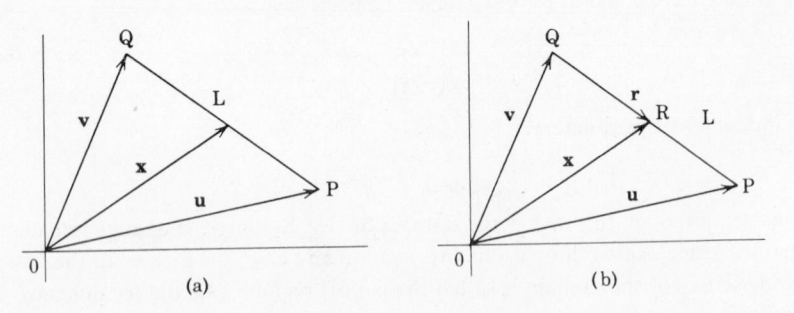

(a)   (b)

Figure 24.4.2   (a) A point in L corresponds to the tip of a vector. (b) The representation of point R in L as $x = v + r$.

Now let us formulate our objective: to arrive at a vector representation of L. This means that the intuitive idea of "tips of vectors between u and v" must be formulated in purely vector language.

Thus, we must specify x in terms of u and v as the vector correspondent of the idea that P and Q determine L. The "tip" idea seems sound. The point R, which is the tip of the vector x, is also the tip of the vector r starting at Q and terminating at R, as shown in Figure 24.4.2 (b). Then the geometric interpretation of vector addition shows that x = v + r, where we think of "arriving" at point R by two successive displacements: origin to Q and then to R. The conception of letting R vary along the line segment L now corresponds to letting r vary, but preserving its starting point at Q and not letting it go beyond P. Thus, r is always some shrinking of the vector from Q to P. As we know from section 6.1, this is represented as a scalar multiple of the vector from Q to P, in which the scalar cannot go beyond the bounds of 0 and 1 [to "shrink" but not to reverse direction (<0) or expand (>1)]. Hence, if we know the vector from Q to P, then r is represented easily.

But the vector from Q to P must satisfy the addition rule that when added to v, it yields u. Thus, the schema is

$$v + (\text{vector from Q to P}) = u$$

and, so, solving by subtracting v from both sides,

$$\text{vector from Q to P} = u - v$$

Thus, with p such that $0 \leqslant p \leqslant 1$, we have

$$r = p(u - v)$$

by our requirement that r be a shrinking of the vector from Q to P. Hence, the representation of R as

$$x = v + r$$

becomes

(24.4.1) $x = v + p(u - v)$ $(0 \leqslant p \leqslant 1)$

Example. We consider the two points in the plane

$$P(3, 1), Q(1, 3)$$

Determine the analytical formula, in the form (24.4.1), specifying the points x on the line segment joining P and Q.

Here we are given

$$u = (3, 1), v = (1, 3)$$

so that

$$u - v = (3, 1) - (1, 3) = (2, -2)$$

Hence,

$$x = v + p(u - v) = (1, 3) + p(2, -2)   (\text{where } 0 \leqslant p \leqslant 1)$$

Thus, we can think of vector x—and, so, point R—as "traced out" as p varies from 0 (x = v) to 1 (x = u). This is a pseudodynamic interpretation (as if p were time) of great value in thinking about static geometric entities. Thus, L is now conceived as a tracing out of a moving point, described by the tip of x.

Rearrangement of (24.4.1), using vector algebra, can be done:

$$
\begin{aligned}
x &= v + p(u - v) &&\text{(given)} \\
&= v + (pu - pv) &&\text{(distributive law)} \\
&= (v + pu) - pv &&\text{(associativity of vector addition)} \\
&= (pu + v) - pv &&\text{(commutativity of vector addition)} \\
&= pu + (v - pv) &&\text{(associativity)}
\end{aligned}
$$

Hence, using the distributive law once again,

(24.4.2) $x = pu + (1 - p)v$

To trace out L we require all the p values between 0 and 1, inclusive. Thus, in two-space, we are led to represent line segment L by the set of all x satisfying (24.4.1) or, equivalently, (24.4.2).

In vector terms, then, the line determined in two-space by two given vectors u and v is

(24.4.3) $L(u, v) = \left\{ x: x = pu + (1-p)v, \text{where } 0 \leqslant p \leqslant 1 \right\}$

Our objective is accomplished: L has been represented in purely vector terms. We mention in passing that by imagining x, u, and v to be vectors in n-space (i.e., n-tuples), we can take formula (24.4.3) as defining the concept of line segment in any vector space. The same idea will apply to all of the ideas of this chapter.

24.5. Convex Sets. For the purposes of the theory of general two-person games the most important concept based on line segments is convexity. Hereafter, we denote the plane, or set of all 2-tuples, by $\mathfrak{R}^2 = \mathfrak{R} \times \mathfrak{R}$.

(24.5.1) Definition. Let $S \subseteq \mathfrak{R}^2$, so that S is some set of points in the plane. Then S is convex if and only if for any points u and v in S, the line segment between u and v is contained in S. That is,

if u, v in S, then $L(u, v) \subseteq S$.

Examples. Consider the sets of points shown in Figure 24.5.1. If S is the set of points (a), then S is convex: picture picking any two points on the boundary or in the figure, then trace a line segment between them; if this segment does not go outside of the figure then the figure is convex. Consider (b) as a set S of points, then S is convex. Consider (c). Then it is not convex. It requires only a counterexample to prove nonconvexity, so choose the right endpoints of the upper and lower boundary line segments, respectively. A line segment is determined, of course, but it falls outside of the set itself. For the

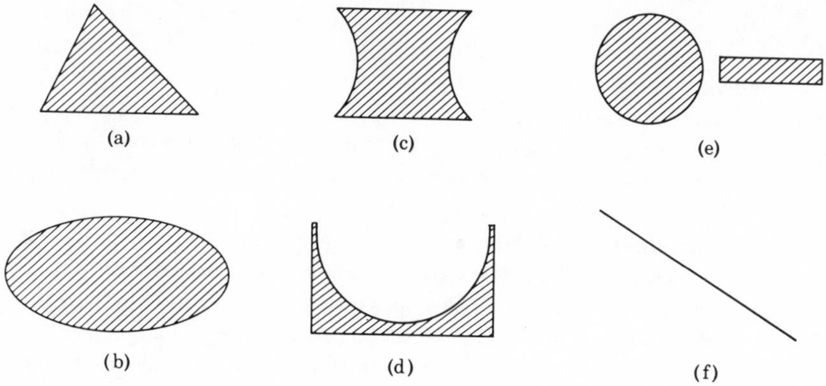

Figure 24.5.1

same reason (d) is not convex, choosing the upper endpoints of the two vertical boundary line segments as the basis of the counterexample. In (e), we have two figures. But define S to be the union of these two figures (sets of points), in the usual set-theoretic sense. Then any point in the circular figure and any point in the rectangular figure determine a line segment that at some point falls outside of the union S. Thus, S is not convex. The line segment (f) is convex because it is such as to include any sub-line segment. Consider, finally, the set S to be the boundary only of figure (a). Then this S is not convex: we have to go through the interior with any line segment joining points on different bounding line segments. Consider (b) in the sense of boundary only; then, again, the set is not convex. Thus, a set of points must be rather precisely specified before we can determine if it forms a convex set.

Some ideas of convex set theory can now be developed and in two parallel ways: by appeal to the geometric intuition and by rigorous definitions and proofs based on the exact concept of a convex set.

The first example of this parallel development is given by noting that if (a) and (b), in Figure 24.5.1, thought of as interior-plus-boundary in both cases, were to overlap then the overlap would also be convex (see Figure 24.5.2). If it were not,

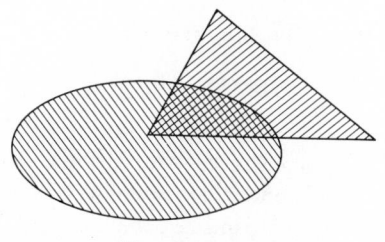

Figure 24.5.2

there would be two points lying in the interior of the triangle that were not joined by a line segment in the triangle. In other words, the test "pick two points in the doubly-ruled region and check the line segment" necessarily is a realization of the test "pick two points in the triangle and check," which already "passed" the test. Hence, the overlap was included in the "pass" decision. This leads to the conjecture that for any collection of convex sets in the plane, their intersection is convex.

To prove this conjecture formally, we have to represent the notion "any collection of convex sets." Let I be any set such that its members are used to keep track of, or index, the convex sets. Thus, S_i (with $i \in I$) is a set and $\{S_i, i \in I\}$ is the whole collection. Since I can be the set of first ten integers, all integers, or even all real numbers, "any collection" is specified in that no particular number of such sets is specified. The intersection can be denoted (see section 3.6) $\underset{i \in I}{\cap} S_i$, or, simply, $\cap S_i$.

Thus, the theorem to be proved is:

(24.5.2) *Theorem.* Let $\{S_i, i \in I\}$ be any collection of convex sets. Then

$$S = \cap S_i$$

is convex.

Proof. Suppose that u and v are any points in S. Then, they determine $L(u, v)$ and we want to conclude,

$$L(u, v) \subseteq S$$

Since $S = \cap S_i$, it follows that u and v are in S_i, for each i. Since each S_i is convex,

$$L(u, v) \subseteq S_i \qquad \text{(each i)}$$

which means

$$\text{if } x \in L(u, v), \text{ then } x \in S_i \qquad \text{(each i)}$$

and so,

$$\text{if } x \in L(u, v), \text{ then } x \in \cap S_i$$

by the meaning of intersection. But this means

$$L(u, v) \subseteq \cap S_i$$

which was to be shown.

For a set of points that is not convex it is necessary for certain purposes to consider the smallest convex set containing the points. For example, in Figure 24.5.3, a figure and its corresponding smallest convex set are paired, the latter denoted H(S) if S is the original set of points.

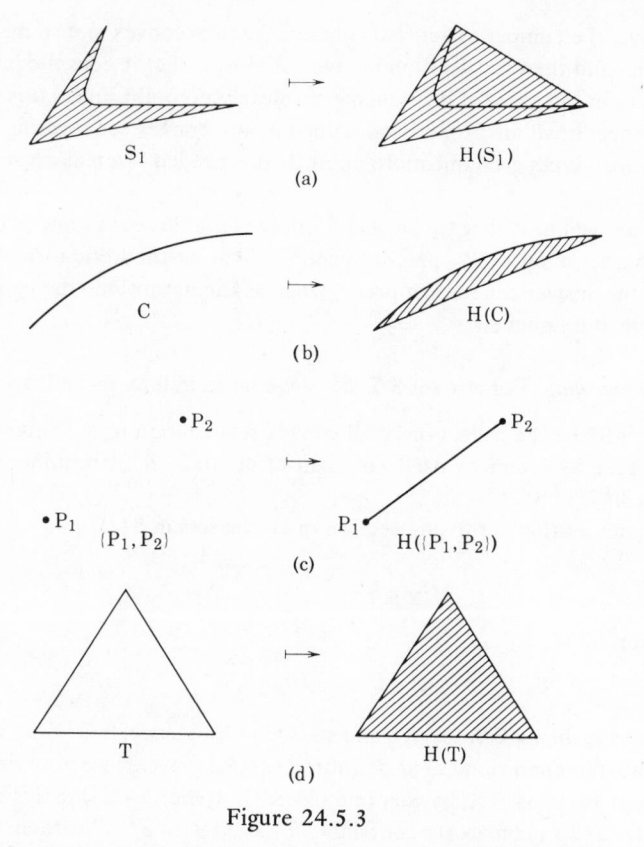

Figure 24.5.3

Now that we know what we want—the smallest convex set containing set S—
we can formally define the concept.

(24.5.3) Definition. Let S be any set of points in the plane. Then by the convex hull of S—denoted H(S)—we mean the smallest convex set containing S.
Exactly, H(S) satisfies three conditions:

   (1) $S \subseteq H(S)$,
   (2) H(S) is convex,
   (3) If S' is convex and $S \subseteq S'$, (so that S' is a convex set containing S), then
$H(S) \subseteq S'$.

Condition (1) assures that H(S) contains S; for example, in Figure 24.5.3 part
(b), H(C) contains C: every point on curve C is also in H(C).

Condition (2) demands that H(S) be convex.

Condition (3) systematizes the idea of "smallest convex set containing S."
This is done by placing H(S) in "competition" with any other set satisfying (1)

and (2)—thus, the competing set also contains S and is convex and so might do the job we want—and then the decision in favor of H(S) is that it does the job less superfluously in the sense of introducing absolutely no point unless it is really needed. Hence, it will be properly contained in any convex set covering S, if that set, say S', introduces even one more point than is needed "to make S into a convex set."

But how do we know that for any set S in the plane there is a convex hull? And is such a set unique? We practice proof-style in set-theoretic formulations by making the answer a matter of proof, whereas the definition—the convex hull— tacitly assumed uniqueness.

(24.5.4) Theorem. For any set $S \subseteq \Re^2$, the convex hull exists and is unique.

Proof. Let S^* be the collection of all convex sets containing S. This set is not empty, because $\Re^2$ is convex itself—any pair of points in $\Re^2$ determines a line in $\Re^2$—and so $\Re^2$ is in S^*.

Consider the set that is the intersection of all the sets in S^*:

$$\underset{X \in S^*}{\cap} X \quad (\equiv \cap X)$$

By definition of S^*,

$$S \subseteq X \qquad (\text{all } X)$$

so that if $x \in S$, then $x \in X$ (all X), and so, $x \in \cap X$. Hence, $S \subseteq \cap X$. This shows that $\cap X$ satisfies condition (1) of definition (24.5.3). Next, we note that since each X is convex, so is $\cap X$, by theorem (24.5.2). Hence, $\cap X$ satisfies condition (2). Now let S' be a convex set containing S. Then S' in S^*. But then, since the intersection of sets is always contained in the sets, $\cap X \subseteq S'$. Thus, condition (3) is satisfied. This shows that $\cap X$ is a convex hull of set S (i.e., it is a set satisfying the definition). But the logic of a uniqueness proof is required (see section 2.14).

To show that $\cap X$ is *the* convex hull of S, we introduce the assumption that a set, say K, is a convex hull of S, perhaps not the same as $\cap X$ and we aim to show that in fact K is just $\cap X$. Thus, if K is such a set,

$$K \subseteq X \qquad (\text{all } X \text{ in } S^*)$$

by condition (3). Thus,

$$\text{if } x \in K, \text{ then } x \in X \qquad (\text{all } X)$$

so that

$$\text{if } x \in K, \text{ then } x \in \cap X$$

or, $K \subseteq \cap X$. If we can show that, conversely, $\cap X \subseteq K$, then we have the desired conclusion that $K = \cap X$. But, since K is a convex set containing S and since $\cap$ X satisfies condition (3) we conclude (taking $S' = K$) that $\cap X \subseteq K$. Hence, $K = \cap X$.

By this proof we see that the convex hull exists and is uniquely given by

$$H(S) = \underset{X \in S^*}{\cap} X$$

where S^* is the family of all convex sets containing S.

(24.5.5) Corollary. If S is convex, then $H(S) = S$.

Proof. S convex implies S is in S^*, and so the intersection $\cap X = H(S)$ is contained in it: $H(S) \subseteq S$. Also, however, $S \subseteq H(S)$. Hence, having shown containment mutually, $H(S) = S$.

Of course, we easily see the obvious character of the corollary. For example, begin with a line segment, and since it is already convex, it is carried into itself by "the operation" H. On the other hand, Figure 24.5.3 shows H carrying nonconvex sets into their convex hulls.

Thus, in general, theorem (24.5.4) shows that

$$S \mapsto H(S) = \underset{X \in S^*}{\cap} X$$

specifies a mapping or transformation from the subsets of $\Re^2$ into the convex subsets of $\Re^2$. Thus, operator H carries sets into their smallest convex "covers" (i.e., their convex hulls).

24.6. Elementary Point-Set Concepts.

The next notions are all rather intuitive and elementary geometric ideas. Hence, they are presented here in a rather formal series of definitions as a preliminary to our second major geometric concept, compactness. This section is also of interest in showing a connection between concepts from set theory, geometry, and vector algebra. It is conventional to speak of "point-sets" in this context: any subset of the plane.

To define the elementary point-set notions, we use vector notation and two additional concepts about vectors: length and distance.

(24.6.1) Definition. By the length of a vector x, denoted $|x|$, we mean

$$|x| = \sqrt{x_1^2 + x_2^2}$$

where $x = (x_1, x_2)$. (And in n-space, $|x| = \sqrt{x_1^2 + x_2^2 + \ldots + x_n^2}$).

Example. If $x = (4, 3)$, then

$$|x| = |(4, 3)| = \sqrt{4^2 + 3^2} = \sqrt{25} = 5$$

(24.6.2) Definition. For any vectors x, y in $\Re^2$ we define the Euclidean distance from x to y by

$$d(x, y) = |x - y|$$

(For the intuitive idea, see Figure 6.1.6.)

Example. Let $x = (3, 1)$, $y = (1, 3)$. Then

$$d(x, y) = |(3, 1) - (1, 3)| = |(2, -2)| = \sqrt{2^2 + (-2)^2} = \sqrt{8} = 2\sqrt{2}$$

(24.6.3) Proposition. The distance $d(x, y)$ is a metric in $\Re^2$, in the sense of definition (11.3.2).

Not only does $x = y$ imply $d(x, y) = 0$, but also $d(x, y) = 0$ implies $x = y$. Also, $d(x, y) \geqslant 0$ since $|x - y| \geqslant 0$, and $d(x, y) = |x - y| = |y - x| = d(y, x)$, showing that Euclidean distance is symmetric.

Hereafter, we use "point" and "vector" interchangeably, thinking geometrically: the tip of the vector is the point corresponding to it.

(24.6.4) Definition. By a circle of radius r with center at the origin we mean the point-set

$$\big\{x \colon d(0, x) = r\big\} \qquad (r > 0)$$

(24.6.5) Definition. By a circle with center x_0 and radius r we mean the point-set

$$\big\{x \colon d(x_0, x) = r\big\} \qquad (r > 0)$$

(24.6.6) Definition. By a neighborhood of point x_0 we mean the point-set

$$\big\{x \colon d(x_0, x) < r\big\}$$

for any $r > 0$ and we write $N(x_0, r)$.

Hence, a neighborhood of a point is just any circular region with center at the point. Note, however, the boundary where $d(x_0, x) = r$ is excluded from the neighborhood, by definition.

Example. The point-sets

$$S_1 = \big\{x \colon d(0, x) = 1\big\}$$
$$S_2 = \big\{x \colon d(0, x) < 1\big\}$$

are, respectively, the "unit circle" (of radius 1) with center at the origin and the interior of this circle: all points x such that the distance from the origin to the point x is less than 1. Hence, S_2 is a neighborhood of point 0, the origin. The sets

$$S_3 = \big\{x \colon d((1, 1), x) = 1\big\}$$
$$S_4 = \big\{x \colon d((1, 1), x) < 1\big\}$$

are, respectively the unit circle centered at $(1, 1)$ and the interior of the unit circle centered at $(1, 1)$. Hence, S_4 is a neighborhood of point $(1, 1)$.

(24.6.7) Definition. By a boundary point of a point-set S we mean a point x_0 such that for any $r > 0$,

$$N(x_0, r) \cap S \neq \emptyset \text{ and } N(x_0, r) \cap \bar{S} \neq \emptyset$$

In other words: every neighborhood of x_0, no matter how small, contains some point in S and some point in the complement of S. Note that a boundary point of S is a boundary point of $\bar{S}$, the complement of S, and conversely.

Example. In Figure (24.6.1) the point x_0 is a boundary point of the set

$$S = \{x: d(0, x) \leqslant r\}$$

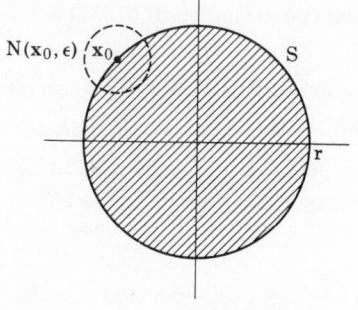

Figure 24.6.1

Here no matter how small we make the radius, calling it ϵ to think of "small" by mathematical convention, we will find a point in $N(x_0, \epsilon)$ that is in S,

$$N(x_0, \epsilon) \cap S \neq \emptyset$$

and a point in $N(x_0, \epsilon)$ that is in $\bar{S}$,

$$N(x_0, \epsilon) \cap \bar{S} \neq \emptyset$$

Since this is true for all $r = \epsilon > 0$, it follows that x_0 is a boundary point by definition (24.6.7). Note that we are merely formalizing geometric intuition. (Reminder: the concepts apply in n-space as well, where pictures can no longer be drawn; the point-sets are n-dimensional.)

(24.6.8) Definition. By the boundary b(S) of a point-set S we mean the set of all its boundary points.

Note that definition (24.6.7) implies that $b(S) = b(\bar{S})$: a set and its complement have the same boundary.

Example. (a) In Figure 24.6.1, let r = 1, so the set b(S) is given by

$$b(S) = \{x: d(0, x) = 1\}$$

In other words, the boundary is the unit circle.

Example. (b) Define

$$S' = \{x: d(0, x) < 1\}$$

Then the boundary of S' is still the unit circle. The difference is that S' and $b(S')$ are disjoint, in this example, whereas in example (a), $b(S) \subseteq S$. The distinction is reflected in the following two definitions.

(24.6.9) Definition. A point-set S is closed iff $b(S) \subseteq S$.

(24.6.10) Definition. A point-set S is open iff $b(S) \cap S = \emptyset$.

(24.6.11) Proposition. A point-set S is closed iff its complement $\overline{S}$ is open.

Proof. If S is closed, then $b(S) \subseteq S$. Hence, $b(S) \cap \overline{S} = \emptyset$. But by definition (24.6.7), $b(\overline{S}) = b(S)$. Hence, $b(\overline{S}) \cap \overline{S} = \emptyset$ and $\overline{S}$ is open. Conversely, if $\overline{S}$ is open, then $b(\overline{S}) \cap \overline{S} = \emptyset$. Hence, $b(S) \cap \overline{S} = \emptyset$ and so $b(S) \subseteq S$. Hence, S is closed.

(24.6.12) Corollary. $\Re^2$ and $\emptyset$ are both open and closed.

Proof. Since $N(x_0, r) \cap \emptyset = \emptyset$, all x_0, definition (24.6.7) implies that $b(\emptyset) = \emptyset$. Hence, $b(\Re^2) = \emptyset$ since $\Re^2 = \overline{\emptyset}$. Hence, $b(\emptyset) \cap \emptyset = \emptyset$ and $b(\Re^2) \cap \Re^2 = \emptyset$ shows that $\emptyset$ and $\Re$ are open. But then proposition (24.6.11) implies both are closed.

(24.6.13) Definition. By the interior i(S) of a point-set S we mean all its non-boundary points:

$$i(S) = S - b(S)$$

(24.6.14) Definition. A point-set S is bounded if and only if it is contained in some neighborhood of the origin:

$$S \subseteq N(0, r) \qquad \text{(some } r > 0)$$

Example. The unit-square

$$\{(x_1, x_2): 0 \leqslant x_1 \leqslant 1, 0 \leqslant x_2 \leqslant 1\}$$

is bounded, for we can take r = 2, for instance, to enclose the square in the interior of a circle centered at the origin.

(24.6.15) Definition. A point-set S is unbounded if and only if it is not bounded.

Example. The point set

$$\{(x_1, x_2): x_1 > 0\}$$

is the entire right half plane. Taking a circle N(0, r) we can still find points x such that $|x| > r$. Hence, no neighborhood of the origin can ever enclose this set.

We illustrate these various elementary concepts of point-set theory in the following.

Example. Consider the various point-sets closely associated with the circle of radius 1, centered at the origin, as in Figure 24.6.2. Both S_1 and S_2 are bounded. S_1 is its own boundary set. S_2 is open. S_3 has a boundary set S_1 and contains it; thus, it is closed. However, it is unbounded because it cannot be enclosed in a circle, no matter how high we might let the radius r be. Intuitively, we would say that S_3 has an inner, but not an outer, rim. We could make S_3 into an open set by deleting S_1 from it, calling the result S_4. But S_4 would still be unbounded. Hence, an unbounded set can be either open or closed.

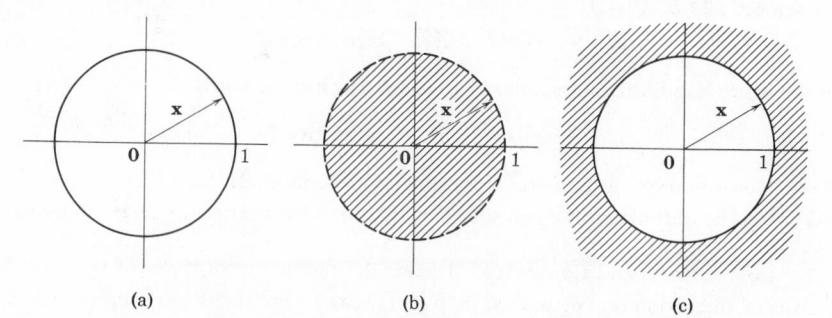

(a)   (b)   (c)

Figure 24.6.2   (a) $S_1 = \{x: |x| = 1\}$; $b(S_1) = S_1$;   (b) $S_2 = \{x: |x| < 1\}$; $b(S_2) = S_1$;
(c) $S_3 = \{x: |x| \geqslant 1\}$; $b(S_3) = S_1$.

Finally, note that S_1 is not convex: lines joining its points cross into S_2. But the latter set is convex. On the other hand, S_3 is not convex because one can locate pairs of points that determine lines passing through S_2 and, so, outside S_3.

The concepts of open and closed sets relate to the neighborhood concept. A characterization of an open set can be given in terms of neighborhoods: that is, a biconditional can be proved of the form, "S is open iff . . ." where the right side refers to neighborhoods.

(24.6.16) Proposition. A point-set S is open if and only if it contains a neighborhood of each of its points.

Proof. Suppose S is open. Then $b(S) \cap S = \emptyset$. Hence, by definition (24.6.13), $i(S) = S$. Hence, each point of S is such that it has a neighborhood contained in S, by negating the condition in definition (24.6.7). Conversely, if S contains a neighborhood of each of its points, each such point is not a boundary point of S. Hence, $S \cap b(S) = \emptyset$. Thus, S is open.

In any theory, when a characterization

$$\text{property A} \qquad \text{iff} \qquad \text{property B}$$

is proved, it can function in other treatments of the theory as a definition of property A. This simple point bears noting because when we pass from intuitive bundles of properties—as in systematizing any bundle of as yet not precise ideas—we cannot expect to include all the conceptual interconnections in one statement, "the definition." Rather we expect to take some property we believe to relate equivalently to the concept-to-be-defined as a contextually framed "definition." Then we can try to arrive at the other associated properties via proofs based on definitional connections and, in a less purely mathematical context, empirically falsifiable assumptions.

To review "characterization-logic" in a concrete case, look at the definition of open set, (24.6.10):

$$(1) \quad \text{S is open} \qquad \text{iff} \qquad b(S) \cap S = \emptyset$$

On the other hand, the characterization (24.6.16) has the form

$$(2) \quad \text{S is open} \qquad \text{iff} \qquad \text{condition N}$$

where "condition N" abbreviates the property described. In Bartle (1964), we find (2) as the definition of open set. But then he deduces readily that (1) holds.

24.7. Compactness and Convexity. Having defined many of the intuitive notions of the geometry of two-space, we may now pass to the next major concept needed for game theory. The first was the idea of a convex set. The present concept is that of a point-set that is both closed and bounded.

(24.7.1) Definition. A point-set S is compact if and only if it is both closed and bounded.

Thus, a compact set satisfies two conditions:

$$(1) \quad b(S) \subseteq S$$
$$(2) \quad S \subseteq N(0, r) \text{ for some } r > 0$$

Example. (a) Consider again the sets in Figure 24.6.2. From our earlier discussion,

S_1: closed and bounded; hence, compact

S_2: open and bounded; hence, not compact

S_3: closed and unbounded; hence, not compact.

Example. (b) Consider the sets of points in the plane, given in Figure 24.5.3, in terms of a set and its convex hull. In (a), S_1 is closed (we always assume the solid boundary line means that the boundary is in the set) and it is bounded. Thus, S_1 is compact. Also, $H(S_1)$ is both compact and convex. In (b), curve C is closed because it is its own boundary, $b(C) = C$, and clearly it is bounded. Thus, C is compact, and $H(C)$ is both compact and convex. In (c) the two points are being considered as a point-set, $\{P_1, P_2\} = S$, say. We will return to this example momentarily. In (d), triangle T—boundary only—is obviously its boundary $b(T) = T$, and so, it is closed. Also, since it is bounded, it is compact. The convex hull $H(T)$ is convex and compact. Returning to (c), with $S = \{P_1, P_2\}$, this is the first time we have asked about "neighborhood" properties of isolated points. Let us see what the definitions imply. First, we know S is not convex. Second, we have to imagine a choice of an origin, say P_1, to study S via analytic method. Thus, S is given the coordinate system embedding: $P_1(0, 0)$, $P_2(1, 0)$. Hence, equivalently, S is $\{0, u\}$ where $u = (1, 0)$. Obviously, $d(0, u) = 1$, applying definition (24.6.2). Third, we investigate the boundary concept for $S = \{0, u\}$, appealing to definition (24.6.7), which specializes to the following: x_0 is a boundary point of $\{0, u\}$ iff, for every $r > 0$,

$$(1) \quad N(x_0, r) \cap \{0, u\} \neq \emptyset$$
$$(2) \quad N(x_0, r) \cap \overline{\{0, u\}} \neq \emptyset$$

Clearly if x_0 has distances from the points in the set,

$$\epsilon_1 = d(0, x_0)$$
$$\epsilon_2 = d(u, x_0)$$

then by choosing $r < \min(\epsilon_1, \epsilon_2)$, the neighborhood of radius r about x_0 will fail to include both 0 and u. The picture to have in mind is shown in Figure 24.7.1.

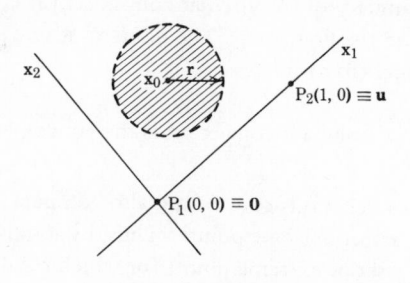

Figure 24.7.1

It arises by superimposing a coordinate system on the original point-set $\{P_1, P_2\}$. Here a neighborhood of arbitrary point x_0 has radius r chosen to exhibit the counterexample to x_0 being a boundary point. This can be done for any x_0 not in $\langle 0, u \rangle$. On the other hand, for any x_0 the point x_0 is in its own neighborhood $N(x_0, r)$. Hence, if $x_0 = 0$ or $x_0 = u$, then it satisfies both conditions (1) and (2) for a boundary point. Hence if $S = \langle 0, u \rangle$, then $b(S) = S$. It follows from (24.6.9) that the set is closed. Of course, we see immediately that S is bounded. Thus, S is compact, by definition (24.7.1). Thus, the properties of the set of two isolated points in part (c) of Figure 24.5.3 are compactness but not convexity.

We have been providing precise formulations of intuitive space concepts with the aim of providing a basis for two exact concepts that are central to the geometry of two-space, convexity and compactness, in its usage in game theory. A convex set contains the line segment joining each pair of its points, and a compact set is closed and bounded. In the next few pages we relate these two concepts, but without proofs in order not to get overly involved in the inner problems of this no longer elementary mathematical background.

(24.7.2) Definition. Let C be a convex set. An extreme point of C is a point x in C such that there do not exist distinct points u and v in C such that

$$x = \tfrac{1}{2} u + \tfrac{1}{2} v$$

Two point-sets in Figure 24.7.2 indicate the meaning of this definition.

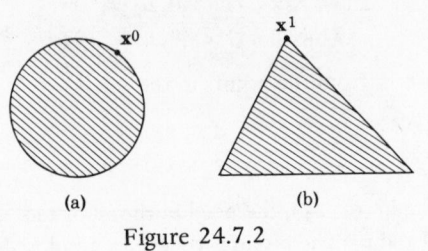

(a) (b)

Figure 24.7.2

The point x^0 on the boundary of the circular convex set (a) is an extreme point: so is every other point on the boundary. The point x^1 is one of three extreme points in the triangular set (b) of points.

(24.7.3) Proposition. A compact convex set is the convex hull of its extreme points.

Example. In Figure 24.7.2, (a) is convex and also compact. We saw above that its boundary forms the set of extreme points. Thus, by looking at the circular boundary only, we see a set of extreme points for which region (a) is the smallest convex set containing this circle. In (b), the three extreme points can be imagined as given first: then the triangular set is the convex hull of these points.

(24.7.4) Definition. Let $x^1, x^2, \ldots, x^m$ be m points. A convex linear combination of these points is given by

$$p_1 x^1 + p_2 x^2 + \ldots + p_m x^m$$

where

$$p_i \geqslant 0 \qquad \text{(all i)}$$

$$\sum_{i=1}^{m} p_i = 1$$

Note that with m = 2, we have a point on line segment as characterized in (24.4.2), with $u = x^1$, $v = x^2$.

(24.7.5) Proposition. Let C be any compact convex set. Then any point x in C can be represented as a convex linear combination of at most three of the extreme points of C.

Example. (a) The assertion is that three extreme points suffice to specify a point in a given compact convex set, via convex linear combinations. For instance, consider the compact convex set in Figure 24.7.3.

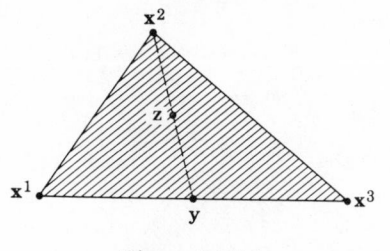

Figure 24.7.3

Then we note that there are only three extreme points here, and so, these three should suffice to represent any point in the set. In particular, using definition (24.7.4), we have the convex linear combinations

$$x^1 = 1x^1$$
$$x^2 = 1x^2$$
$$x^3 = 1x^3$$
$$y = \tfrac{1}{2} x^1 + \tfrac{1}{2} x^3$$

To find a representation for z in terms of the x^i, we note that z is the midpoint of the line segment determined by y and x^2, and so,

$$z = \tfrac{1}{2} y + \tfrac{1}{2} x^2$$
$$= \tfrac{1}{2} (\tfrac{1}{2} x^1 + \tfrac{1}{2} x^3) + \tfrac{1}{2} x^2$$
$$= \tfrac{1}{4} x^1 + \tfrac{1}{4} x^3 + \tfrac{1}{2} x^2$$

and hence,

$$z = .25x^1 + .50x^2 + .25x^3$$

Example. (b) The figure in two-space, given in Figure 24.7.4,

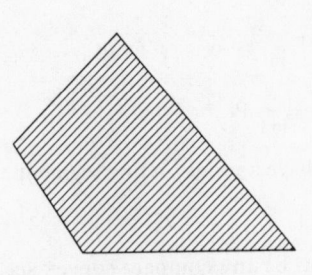

Figure 24.7.4

has four extreme points, but since it is convex and compact proposition (24.7.5) assures us only three are required to represent any point in it by a convex linear combination of extreme points. This can be seen by "triangulation," as then we see that the figure is a union of two compact convex sets A and B as shown in Figure 24.7.5.

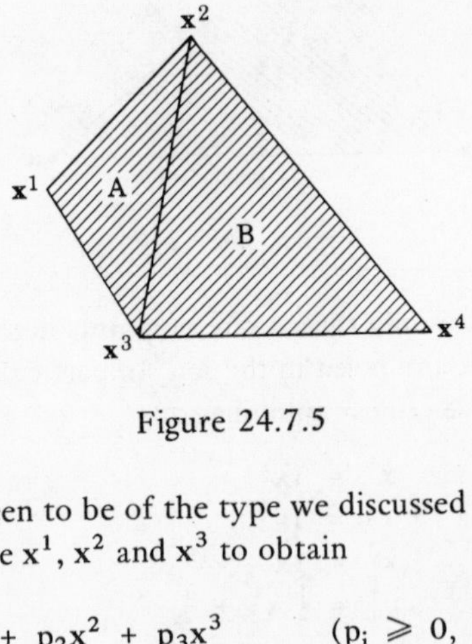

Figure 24.7.5

But both A and B are seen to be of the type we discussed in the previous example: hence, if x is in A we use x^1, x^2 and x^3 to obtain

$$x = p_1x^1 + p_2x^2 + p_3x^3 \qquad (p_i \geqslant 0, \sum_{i=1}^{3} p_i = 1)$$

However, if x' is in B, we use

$$x' = q_2x^2 + q_3x^3 + q_4x^4 \qquad (q_i \geqslant 0, \sum_{i=2}^{4} q_i = 1)$$

In any case, we represent the point using only three of the extreme points. Note the analogy to representation of points in the plane via coordinates.

Example. (c) If C is the circular region of Figure 24.7.2(a), we know that the entire boundary consists of extreme points. Thus, each boundary point is represented with the triviality of x = 1x, and interior points are represented as on line segments joining two boundary points, so two extreme points are sufficient. In any case proposition (24.7.5) applies, even though the set is not in any way triangular as given.

Finally, we cite a useful theorem proved in Owen (1968).

(24.7.6) *Theorem.* The convex hull of any point-set is precisely the set of all points that are convex linear combinations of its elements.

For example, a set of three isolated points in the plane such that they form the vertices of a triangle is transformed into its convex hull in two steps, as it were: the line segment (a set of convex linear combinations of two vertices) joining any pair of vertices is introduced, then any interior point of this triangle (as we know from our study of Figure 24.7.3) is a convex linear combination of the three vertices we start with. Hence, the triangular region that is the union of the triangle formed by the three points and its interior is the convex hull of the three points: it is the smallest convex set that contains these three points.

Recall, that in section 24.3, we had the mapping specified by

$$(x_1, y_1) \mapsto (E_1(x, y), E_2(x, y))$$

with domain in $\Re^2$ and range $\Re^2$. Figure 24.3.1 can be seen to specify that this mapping carries the shaded square, a compact convex set, into the payoff region. This latter set is apparently not convex: note that it has points such that line segments joining them are not in the set. Also, the payoff region is closed and bounded, as shown, and, so, compact. Thus, under this mapping one property of the domain point-set is preserved, compactness, and another lost, convexity.

The mathematics of the problem of characterizing the properties preserved in space mappings is difficult. It leads deeper and deeper into topology. Generally, the mappings we deal with in science are of the type termed continuous. What does this mean? Avoiding a precise and general characterization, we can say that the context for a continuity problem is "topological"—that is, it refers to properties based on the idea of neighborhoods. A concept that is definable purely in terms of neighborhoods is topological. For example, the notions of open set, closed set, boundary, and boundedness are all topological. Hence, compactness is a topological property. However, the concept of convex linear combination is not. A mapping that preserves any and all topological properties is a continuous mapping. Hence, it is not surprising that in mapping into the payoff region, assuming the mapping is continuous, convexity is lost while compactness is preserved.

CHAPTER TWENTY-FIVE STRATEGIC ANALYSIS OF GENERAL TWO-PERSON GAMES

25.1. Review of the Mapping into Payoff Space. Recall from sections 24.1-24.3 that if the mixed strategies are used independently (an assumption to be removed in section 25.3),

$$E_1(x, y) = \sum_{(i,j)} a_{ij}x_iy_j$$

$$E_2(x, y) = \sum_{(i,j)} b_{ij}x_iy_j$$

which yields the point in the plane

$$(E_1(x, y), E_2(x, y)) = (\Sigma a_{ij}x_iy_j, \Sigma b_{ij}x_iy_j)$$

$$= \Sigma(a_{ij}x_iy_j, b_{ij}x_iy_j) \qquad \text{(a vector of sums is the sum of vectors)}$$

$$(25.1.1) \qquad\qquad = \Sigma x_iy_j(a_{ij}, b_{ij})$$

and in this latter form we see that since

$$x_iy_j \geqslant 0 \text{ and } \sum_{(i,j)} x_iy_j = 1$$

we have a convex linear combination [recall definition (24.7.4)] of the set $\langle(a_{ij}, b_{ij})\rangle$ of payoff-pairs, which are points in two-space. For a 2 X 2 game, the first component of (25.1.1) was given by formula (24.2.2) and the second, by formula (24.2.3), and a concrete example was given. Another example follows, exhibiting the vector form (25.1.1).

Example. If

$$(a_{ij}) = \begin{pmatrix} 1 & 1 \\ 2 & -1 \end{pmatrix}, (b_{ij}) = \begin{pmatrix} 2 & -2 \\ -1 & 2 \end{pmatrix}$$

or

$$\begin{pmatrix} (1,2) & (1,-2) \\ (2,-1) & (-1,2) \end{pmatrix} = \left((a_{ij}, b_{ij}) \right)$$

and

$$x = (x_1, x_2), y = (y_1, y_2)$$

then, using (25.1.1),

$$(E_1(x, y), E_2(x, y)) = \Sigma x_i y_j (a_{ij}, b_{ij})$$
$$= x_1 y_1 (1,2) + x_1 y_2 (1,-2) + x_2 y_1 (2,-1) + x_2 y_2 (-1,2)$$

a point in two-space.

25.2. The Shape of the Payoff Region.

We now specialize to the case in which the matrices are 2 × 2 in order to answer the question, in a relatively "simple" context, as to the shape of the payoff region. Thus, we have

$$\left((a_{ij}, b_{ij}) \right) = \begin{pmatrix} (a_{11}, b_{11}) & (a_{12}, b_{12}) \\ (a_{21}, b_{21}) & (a_{22}, b_{22}) \end{pmatrix}$$

(25.2.1) $(E_1(x, y), E_2(x, y)) = x_1 y_1 (a_{11}, b_{11}) + x_1 y_2 (a_{12}, b_{12})$
$$+ x_2 y_1 (a_{21}, b_{21}) + x_2 y_2 (a_{22}, b_{22})$$

When (x, y) is varied, the set of points described by (25.2.1) is generated. Although each particular combination of the four points (a_{ij}, b_{ij}) is a convex linear combination of them (the four coefficients $x_i y_j$ are non-negative and sum to unity) there are not three independent coefficients in (25.2.1) but two, namely, x_1 and y_1. Thus, we can expect that not every convex linear combination of the four payoff points will be generated. This is already revealed in Figure 24.3.1.

Hence, we imagine the four payoff points (a_{ij}, b_{ij}) plotted in the plane, and we ask which of the six line segments connecting the four points may be generated by a convex linear combination of the particular pair of points, using independent mixed strategies. This question having been answered for each pair we will then go on to seek additional information yielding a visualization of the shape of the payoff region.

In the following the phrase "independently generated" in reference to a line segment means "traced out by using independent mixed strategies."

Property 1. The interior of the line segment from (a_{11}, b_{11}) to (a_{22}, b_{22}) cannot be independently generated from this pair of points.

Proof. If a point is in the interior of the line segment, according to section 24.4 it must have the form

$$p(a_{11}, b_{11}) + (1-p)(a_{22}, b_{22}) \qquad (0 < p < 1)$$

which means that in (25.2.1) we must have, to independently generate the segment,

$$x_1 y_1 = p > 0$$
$$x_2 y_2 = 1{-}p > 0$$
$$x_1 y_2 = 0$$
$$x_2 y_1 = 0$$

But the first two conditions yield

$$x_1 > 0, y_1 > 0, x_2 > 0, y_2 > 0$$

which contradicts the second two conditions.

Property 2. The interior of the line segment from (a_{12}, b_{12}) to (a_{21}, b_{21}) cannot be independently generated from this pair of points.

Proof. If so, then, the point in the interior has the form

$$p(a_{12}, b_{12}) + (1{-}p)(a_{21}, b_{21}) \qquad (0 < p < 1)$$

and so, in (25.2.1) we have

$$x_1 y_1 = 0$$
$$x_1 y_2 = p > 0$$
$$x_2 y_1 = 1{-}p > 0$$
$$x_2 y_2 = 0$$

so that the two nonzero conditions imply

$$x_1 > 0, y_2 > 0, x_2 > 0, y_1 > 0$$

contradicting the two zero conditions.

Property 3. The interior of the line segments from (a_{11}, b_{11}) to (a_{12}, b_{12}) and from (a_{21}, b_{21}) to (a_{22}, b_{22}) are independently generated.

Proof. These points take the form, in the first case, of

$$p(a_{11}, b_{11}) + (1{-}p)(a_{12}, b_{12}) \qquad (0 < p < 1)$$

so that in (25.2.1)

$$x_1 y_1 = p$$
$$x_1 y_2 = 1{-}p$$
$$x_2 y_1 = 0$$
$$x_2 y_2 = 0$$

If $x_1 = 1$, then $y_1 = p$ and $x_2 = 0$. Then all four conditions are satisfied by this pair of values of x_1 and y_1. Hence, the values

$$x_1 = 1, \quad 0 < y_1 < 1$$

generate the interior of the line segment. Similarly, in the case of points (a_{21}, b_{21}) and (a_{22}, b_{22}) the interior of the line segment is generated by

$$x_2 = 1, \quad 0 < y_1 < 1$$

in which player 1 uses pure strategy α_2.

The analogy suggested here is: let player 2 choose a pure strategy ($y_1 = 1$ or $y_2 = 1$), and let x_1 vary. This yields the following.

Property 4. The interior of the line segments from (a_{11}, b_{11}) to (a_{21}, b_{21}) and from (a_{12}, b_{12}) to (a_{22}, b_{22}) are independently generated.

Proof. As in the case of property 3.

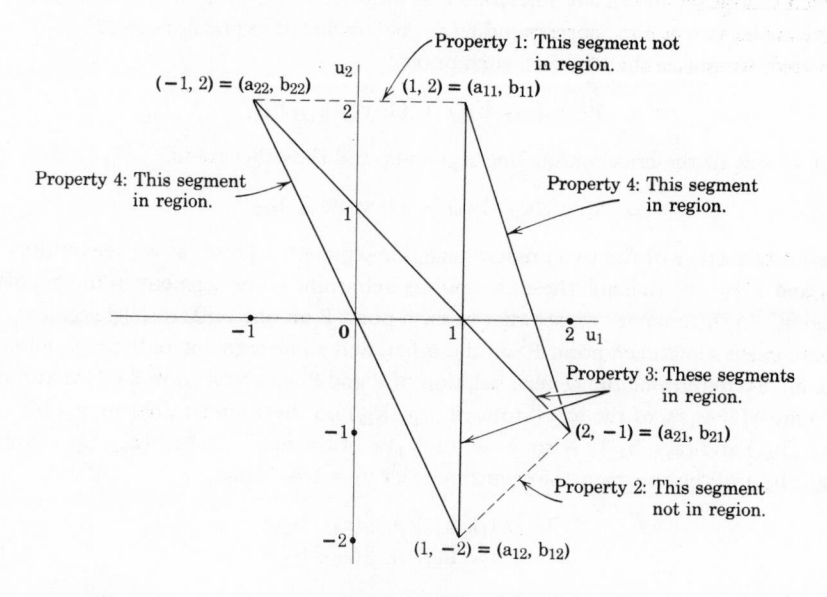

Figure 25.2.1

When we plot the special configuration of four points of the example of section 25.1 and then systematically draw and label line segments in accordance with properties 1–4 we arrive at a partial determination of the payoff region as shown in Figure 25.2.1.

For example, the line segment connecting (a_{11}, b_{11}) and (a_{22}, b_{22}) is not in the payoff region, in this case. This says a little more than property 1 in general asserts, since not only is it the case that no point of this line segment may be generated by independent mixed strategies over the two endpoints (a_{11}, b_{11}) and (a_{22}, b_{22}) but also in this case no point on this segment may be generated by any other independent mixed strategy. The same is true, in this case, of the points of the line segment joining (a_{12}, b_{12}) and (a_{21}, b_{21}). For an example where properties 1 and 2 do not yield this result, look ahead to Figure 25.6.1, in which $(a_{11}, b_{11}) = (.7, .8)$ and $(a_{22}, b_{22}) = (.3, .3)$. In this latter case, the points on the line segment between this pair of points are independently generated through convex linear combinations of other points. How this comes about will be clearer after the following discussion.

Further determination of the shape of the payoff region can be made by writing (25.2.1) in the form

$$(25.2.2) \quad x_1[y_1(a_{11}, b_{11}) + (1-y_1)(a_{12}, b_{12})] + (1-x_1)[y_1(a_{21}, b_{21}) + (1-y_1)(a_{22}, b_{22})]$$

which can be geometrically interpreted as follows: First, y_1 is chosen and this determines two points, corresponding to the bracketed expressions in (25.2.2). Second, we notice that the first such point,

$$P: \quad y_1(a_{11}, b_{11}) + (1-y_1)(a_{12}, b_{12})$$

lies on one of the crisscrossing line segments, and the other point,

$$P': \quad y_1(a_{21}, b_{21}) + (1-y_1)(a_{22}, b_{22})$$

lies on the other of the two crisscrossing line segments. Third, as we see by the x_1 and $1-x_1$ coefficients, these two points determine a line segment in the payoff region. In this manner we see that to each point P on one crisscrossing segment there exists a matched point P' on the other, and a line segment in the region joins them. To determine the generic relation of P and P' recall (section 24.4) that P is a point "$(1-y_1)\%$ of the way" toward (a_{12}, b_{12}) on the segment determined by (a_{11}, b_{11}) and (a_{12}, b_{12}). Also, P' is "$(1-y_1)\%$ of the way" toward (a_{22}, b_{22}) from (a_{21}, b_{21}) in their segment. For instance, let $y_1 = 1/4$. Then,

$$P: \quad \tfrac{1}{4}(a_{11}, b_{11}) + \tfrac{3}{4}(a_{12}, b_{12})$$
$$P': \quad \tfrac{1}{4}(a_{21}, b_{21}) + \tfrac{3}{4}(a_{22}, b_{22})$$

and we have to join P and P' with a line segment.

A rough sketch in preparation for a more careful figure would be as in Figure 25.2.2. These considerations lead us to conjecture that the entirety of the two triangular subregions determined by the crisscross is part of the payoff region. But a similar analysis based on a rearrangement of (25.2.1) in the form with y_1 and $1-y_1$ as the "outer coefficients" leads to sketches of the kind shown in Figure 25.2.3.

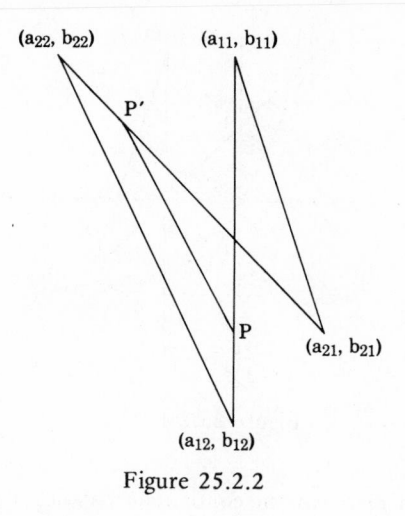

Figure 25.2.2

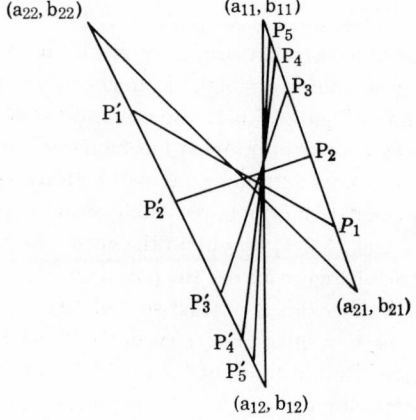

Figure 25.2.3

For instance, the pair of points

$$P_5: \tfrac{15}{16}(a_{11}, b_{11}) + \tfrac{1}{16}(a_{21}, b_{21})$$
$$P_5': \tfrac{15}{16}(a_{12}, b_{12}) + \tfrac{1}{16}(a_{22}, b_{22})$$

as weighted by y_1 and $1-y_1$ generates a line segment from P_5 to P_5' in the region.

The crossing of these segments around, but not necessarily through, the point of the original crisscross creates a rounding-off at this region and the result is that the region has the shape shown in Figure 25.2.4.

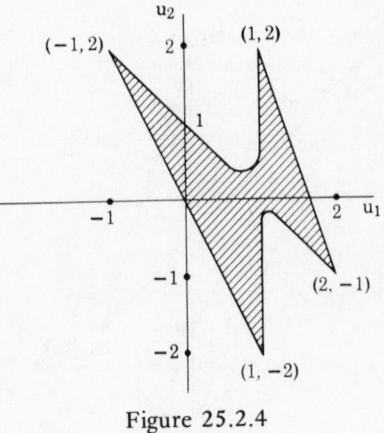

Figure 25.2.4

Now we see that mathematically the continuous mapping under (E_1, E_2) has taken the unit-square of mixed strategy probabilities into a set of points in the plane. The mapping is continuous, so that any and all neighborhood properties are preserved. Note the compact—closed and bounded—character of the unit-square is preserved. But note also that convexity is not preserved, so the payoff region under independently chosen strategies is, in general, not convex.

However, if one examines Figure 25.6.1, also obtained under independent mixed strategies, one sees a convex set. What has happened in this latter case is that the four pure payoff points determine a closed boundary given by the four segments of properties 3 and 4, and then, reasoning as in the process of constructing Figures 25.2.2 and 25.2.3, we obtain the entire interior of this region. Hence, in the special case of Figure 25.6.1 the payoff region is convex under independent strategies. However this is a rather special case that contributes to the peculiar interest we have in this game, reserving it for analysis in section 25.6. For the present, the reader should think of Figure 25.2.4 as the typical payoff region for a general two-person game in which the players use independent mixed strategies.

25.3. Joint Mixed Strategies. Suppose that the players are not only in a not necessarily zero-sum situation but that the rules do not forbid cooperation. Then it is no longer necessary to hold to the restriction of independence,

$$P(\alpha_i \text{ and } \beta_j) = P(\alpha_i)P(\beta_j) = x_i y_j$$

as in assumption (23.8.2). Instead they can select $P(\alpha_i \text{ and } \beta_j)$ together. Thus, abstractly, we have the parameters

$$z_{ij} = P(\alpha_i \text{ and } \beta_j) \qquad (i=1, 2, \ldots, m; j=1, 2, \ldots, n)$$

In the general case, then, the uncertain alternative becomes

$$(z_{11}\,0_{11},\ z_{12}\,0_{12},\ \ldots,\ z_{mn}\,0_{mn})$$

or

$$(z_{ij}\,0_{ij})$$

Then,

$$u_1(z_{ij}\,0_{ij}) = \sum_{(i,j)} z_{ij}u_1(0_{ij}) = \Sigma z_{ij}a_{ij}$$
$$u_2(z_{ij}\,0_{ij}) = \sum_{(i,j)} z_{ij}u_2(0_{ij}) = \Sigma z_{ij}b_{ij}$$

which defines a mapping,

$$z \mapsto (E_1(z),\ E_2(z))$$

where

$$E_1(z) = \Sigma z_{ij}a_{ij}$$
$$E_2(z) = \Sigma z_{ij}b_{ij}$$

Hence, a joint mixed strategy $z = (z_{ij})$, as it is called, gives rise to a point in two-space whose first coordinate is the expected payoff to player 1 and whose second coordinate is the expected payoff to player 2. We may also say that now correlated mixed strategies are allowable; this locution reminds us that the correlation might arise in more ways than one—for example, perhaps by tacit rather than consciously planned cooperation.

A special case is that of independent mixed strategies,

$$z_{ij} = x_i y_j \qquad \text{(all i, j)}$$

but we now have a wider generality.

The idea of a convex linear combination is again applicable:

$$z \mapsto (E_1(z),\ E_2(z)) = (\Sigma z_{ij}a_{ij},\ \Sigma z_{ij}b_{ij})$$
$$= \Sigma(z_{ij}a_{ij},\ z_{ij}b_{ij})$$
$$= \Sigma z_{ij}(a_{ij},\ b_{ij})$$

In particular, in the case that (a_{ij}) and (b_{ij}) are 2×2, we have

$$\left((a_{ij},\ b_{ij})\right) = \begin{pmatrix} (a_{11},\ b_{11}) & (a_{12},\ b_{12}) \\ (a_{21},\ b_{21}) & (a_{22},\ b_{22}) \end{pmatrix}$$

and

$$z = (z_{11},\ z_{12},\ z_{21},\ z_{22})$$

or, if one prefers,

$$z = \begin{pmatrix} z_{11} & z_{12} \\ z_{21} & z_{22} \end{pmatrix}$$

with $z_{ij} \geqslant 0$, $\Sigma z_{ij} = 1$. Then,

$$(25.3.1) \quad (E_1(z), E_2(z)) = z_{11}(a_{11}, b_{11}) + z_{12}(a_{12}, b_{12}) + z_{21}(a_{21}, b_{21}) + z_{22}(a_{22}, b_{22})$$

is the explicit expression for the points in the payoff region, represented by convex linear combinations of the pure strategy payoff points.

The basic distinction that now arises between this more general payoff region and that of section 25.2 is that properties 1 and 2 describe a restriction "artificially" induced on the z-possibilities by the idea that $z_{ij} = x_i y_j$. In allowing the players to choose a z of the correlated type, we open up new strategic and payoff possibilities. This can be seen by noticing that in property 1 the two points were (a_{11}, b_{11}) and (a_{22}, b_{22}). We can generate the line segment here by

$$z_{11} + z_{22} = 1, z_{12} = 0, z_{21} = 0$$

so that (25.3.1) becomes on the right

$$z_{11}(a_{11}, b_{11}) + z_{22}(a_{22}, b_{22})$$

with $z_{22} = 1 - z_{11}$ and $z_{11}, z_{22} \geqslant 0$. Analogously, in property 2 we let

$$z_{12} + z_{21} = 1, z_{11} = 0, z_{22} = 0$$

and we obtain

$$z_{12}(a_{12}, b_{12}) + z_{21}(a_{21}, b_{21})$$

with $z_{21} = 1 - z_{12}$, and $z_{12}, z_{21} \geqslant 0$.

In Figure 25.2.4, the payoff region would contain these lines and, by convex linear combinations of points "opposite" each other on these lines, all points interior to the polygon determined by the original four payoff points (a_{ij}, b_{ij}). The general result can be expressed as:

(25.3.2) Proposition. Under joint mixed strategies z, the payoff region of any 2 X 2 general two-person game is the convex hull of the four pure strategy payoff points.

These points (a_{ij}, b_{ij}) are the extreme points of the compact and convex set forming the payoff region [see proposition (24.7.3)], as shown in Figure 25.3.1.

(25.3.3) Proposition [(m X n generalization of (25.3.2)]. Under joint mixed strategies, the payoff region of an m X n two-person game is the convex hull of the mn points in two-space representing the pure strategy payoffs to the two players. That is, calling the payoff region R*,

$$(25.3.4) \qquad R^* = H\langle\!\langle (a_{ij}, b_{ij}) \rangle\!\rangle$$

in the convex hull notation of section 24.5.

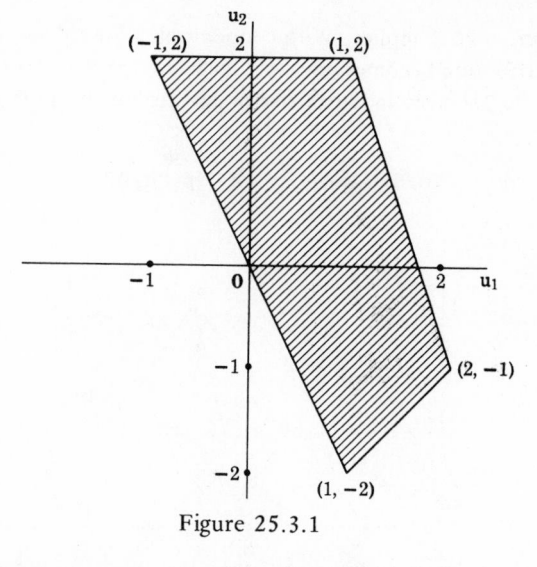

Figure 25.3.1

Both propositions (25.3.2) and (25.3.3) follow immediately from theorem (24.7.6). The set of points

$$\left\{ (a_{ij}, b_{ij}) \right\}$$

in two-space is given. Then because z varies over all possible probability distributions on these points, and each such z provides a convex linear combination of the (a_{ij}, b_{ij}), we conclude that the set of all such convex linear combinations—for us, the pairs of expected payoffs—is the convex hull of the payoff points (a_{ij}, b_{ij}).

Example. We return to the battle-of-the-sexes game of section 24.2, where we arrived at the form (24.2.1) reproduced here:

$$
\text{(25.3.5)} \qquad
\begin{array}{c}
 & \qquad\qquad \mathbf{W} \\
 & \quad \mathbf{A} \qquad\quad \mathbf{B} \\
\mathbf{H} \begin{array}{c} a \\ b \end{array}
\left(
\begin{array}{cc}
(1, q_1) & (p_2, q_2) \\
(0, 0) & (p_1, 1)
\end{array}
\right)
\end{array}
$$

where

$$0 < p_2 < p_1 < 1 \text{ and } 0 < q_2 < q_1 < 1$$

With $p_1 > p_2$ and $1 > q_2$, there is motivation to jointly favor bb over ab. Similarly, there is motivation to jointly favor aa over ab. Thus, the problem is then reduced to some solution of the conflict of interest between aa and bb. If some probabilistic solution were proposed it would involve some point on the line segment from $(1, q_1)$ to $(p_1, 1)$. This means nonindependence, for otherwise property 1 of section 25.2 applies. With cooperation allowable—as in this in-stance—the z distribution becomes relevant: correlated mixed strategies may be selected. Figure 25.3.2 gives an illustrative payoff region: we plot first the four payoff points

$$\langle (0, 0), (1, q_1), (p_1, 1), (p_2, q_2) \rangle$$

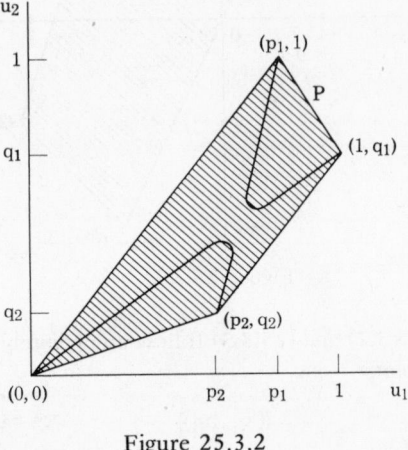

Figure 25.3.2

Then proposition (25.3.2) applies because we assume that they may jointly select the probabilities. But then we simply draw the quadrangle determined by the four points. The payoff region is the set of all points interior to, or on the boundary of, this figure. Figure 25.3.2 also shows the more restricted region that would obtain if the players use independent mixed strategies. Now we can see the geo-metric interpretation of why H and W might want to arrive at some joint (mixed) strategy between the two possibilities: aa and bb. The corresponding pure payoffs are the extreme points $(p_1, 1)$ and $(1, q_1)$. Thus, a joint mixed strategy between these players generates an expected payoff point,

$$z_{11}(1, q_1) + z_{22}(p_1, 1)$$

where

$$z_{11} + z_{22} = 1, z_{11} \geqslant 0, z_{22} \geqslant 0$$

so that it lies on line segment called P in Figure 25.3.2, determined by $(1, q_1)$ and $(p_1, 1)$. (The reason for the P notation will become apparent in the next section.)

25.4. The Pareto Optimal Set. The exact notion required to specify the special character of boundary segment P in Figure 25.3.2 is the relation over the points in two-space, given by the following:

(25.4.1) Definition.   Let (u, v) and (u', v') be points in two-space. Then (u, v) is jointly dominated by (u', v') if and only if

$$(1) (u, v) \neq (u', v')$$
$$(2) u' \geqslant u, v' \geqslant v$$

Of course, we are interested in this numerical relation as a representation of certain facts about the two players. In particular, in the payoff region R^* [recall definition (25.3.4)] we note that under definition (25.4.1) (1), since a single (generally uncertain) alternative gets mapped into a point in R^*, for both players (u, v) and (u', v') are two points representing a pair of uncertain alternatives that are not a matter of indifference to both players. By (2), one alternative, mapped into (u', v'), is such that neither player strictly prefers the other alternative.

Geometrically, a sketch of the set of all points in the plane jointly dominating point (u, v) is given in Figure 25.4.1.

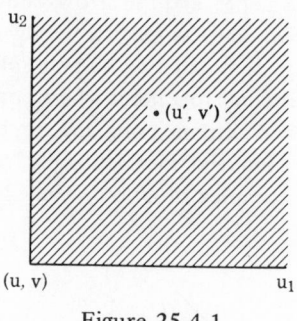

Figure 25.4.1

This means that all points in the unbounded set to the right and above (u, v) jointly dominate (u, v), and also all points on the vertical and horizontal boundaries of this set jointly dominate (u, v). The exception is that (u, v) does not jointly dominate itself.

Now consider the relation as it is restricted to points in R^*: the unbounded set becomes bounded. Within the convex hull—say in Figure 25.3.2—we can trace upward and to the right to locate points jointly dominating any point from which we trace. In other words, consider any positively sloped line in R^*; then "positive slope" means that both u_1 and u_2 are increasing (both players consider the alternatives represented by the tracing upward along the line as getting "better and better"). Hence, our interest shifts immediately to the boundaries. A boundary line of R^* is either (1) positively sloped, (2) negatively sloped, (3) zero-sloped, or (4) "infinite-sloped" (i.e., vertical). Type (1) boundaries are traced up to an extreme point that jointly dominates all other points on that boundary. Type (3) boundaries are traced horizontally to an extreme point: player 2 is indifferent, but player 1 is moving up along his utility function. Type

(4) boundaries are traced up to an extreme point: player 1 is indifferent, but player 2 is gaining.

Finally, there is type (2), negatively sloped boundary lines. In this case, if the boundary line is such that there are points jointly dominating it, we proceed upward to those points, as in Figure 25.4.2.

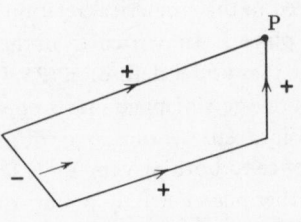

Figure 25.4.2

Here the signs are those of the slopes, and the arrows show us "tracing" upward as described. In the sketch the point P is special: we always trace up to it and we cannot leave it. It is jointly dominated by no point of the convex hull shown. For the players, point P corresponds to a point where both prefer the alternative to any other. Thus, since it is an extreme point and, so, a pair of payoffs in the matrix (a_{ij}, b_{ij}) the players are induced, motivationally, to jointly choose the pure strategies yielding P. A more complex situation is shown in Figure 25.4.3.

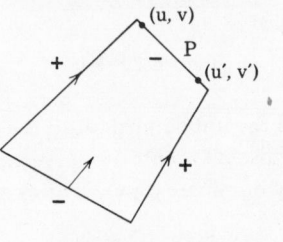

Figure 25.4.3

Here we trace up to the extreme points of line segment (not point) P by the positively sloped boundaries, and from the lower negatively sloped boundary we trace up toward P also. Once in P, however, points like (u, v) and (u', v') do not jointly dominate each other: thus, the idea of tracing up and to the right terminates on a set forming an upper negatively sloped boundary of the payoff region. Another such set is sketched in Figure 25.4.4. Here P consists of two negatively sloped boundary line segments.

(25.4.2) Definition. Let R* be the payoff region—that is, the convex hull of the extreme points obtained from the matrix (a_{ij}, b_{ij}). Then the set of all points

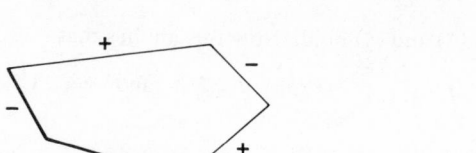

Figure 25.4.4

in R* such that a point in this set is not jointly dominated by any point in R* is termed the Pareto optimal set. It is denoted P.

In the battle-of-the-sexes game, Figure 25.3.2, the line segment labeled P is the Pareto optimal set.

The prior discussion made clear that P must be a subset of the boundary points of R* and that this excludes boundary lines with positive slope. Since, by definition, points in P do not jointly dominate each other it might be thought that this is all there is to say about P. However, P does have a fundamental property: geometrically, we can conjecture what this is by the negative slope. One player's gain is the other player's loss. In fact, we can frame the situation in P as follows.

(25.4.3) Proposition. If two pure or uncertain alternatives are mapped into the Pareto optimal set P, then the two players have strictly opposing preferences for these two alternatives.

(Note: The converse is false because any other negatively sloped line segment in R* is the image of a set of such "zero-sum" alternatives.)

Proof of (25.4.3). (Practice in applied logic) A strict proof from definitions (25.4.1) and (25.4.2) is as follows: Let (u, v) and (u', v') be in P, and assume $(u, v) \neq (u', v')$. Then, because of definition (25.4.2) of P, it is not the case that

$$(1) \quad u' \geqslant u \text{ and } v' \geqslant v$$

because (u, v) is not jointly dominated by (u', v'), a point in R*. Also, it is not the case that

$$(2) \quad u \geqslant u' \text{ and } v \geqslant v'$$

because (u', v') is not jointly dominated by (u, v), a point in R*. The negation of (1) by tautological equivalence yields

$$\text{not-}(u' \geqslant u) \quad \text{or} \quad \text{not-}(v' \geqslant v)$$

and this means

$$(3) \quad u' < u \quad \text{or} \quad v' < v$$

Similarly, (2) yields

$$(4) \quad u < u' \quad \text{or} \quad v < v'$$

Thus, both (3) and (4) hold. Now this implies that

$$(5a)\quad u' < u \quad \text{and} \quad v < v'$$

or

$$(5b)\quad u < u' \quad \text{and} \quad v' < v$$

because other possible combinations are contradictory of the asymmetry of the "less than" relation.

If (5a) holds this means that any alternatives mapped into (u, v) and (u', v') are characterized by opposite preference: player 1 prefers the alternative represented by (u, v); player 2 prefers the alternative represented by (u', v'). If (5b) holds, an analogous situation exists of opposite preference. Since (5a) or (5b) holds and since (u, v) and (u', v') are arbitrary distinct points in P, proposition (25.4.3) follows.

25.5. The Negotiation Set and Rationality.   Security levels are defined for the players, in terms of their use of mixed strategies $\mathbf{x}$ and $\mathbf{y}$, respectively. It is instructive to compare their best mixed strategy security levels with points in the Pareto optimal set. Recall that

$$\text{BMSSL } \alpha = \max_{\mathbf{x}} \min_{\mathbf{y}} E(\mathbf{x}, \mathbf{y}) \qquad \text{[definition (23.10.2)]}$$

in the case that the game was zero-sum. Since $E(\mathbf{x}, \mathbf{y})$ was required to be considered in two forms for the general two-person game,

$$E_1(\mathbf{x}, \mathbf{y}) = \text{expected payoff to player 1, } \alpha$$
$$E_2(\mathbf{x}, \mathbf{y}) = \text{expected payoff to player 2, } \beta$$

the concepts of best mixed strategy security levels, as in numerical form, both become "max-min" because E_1 is in terms of u_1 and E_2 is in terms of u_2. Hence, under the assumption of independent choices of mixed strategies, we can define:

(25.5.1)   Best Mixed Strategy Security Levels.

$$v_1 = \text{BMSSL } \alpha = \max_{\mathbf{x}} \min_{\mathbf{y}} E_1(\mathbf{x}, \mathbf{y})$$
$$v_2 = \text{BMSSL } \beta = \max_{\mathbf{x}} \min_{\mathbf{y}} E_2(\mathbf{x}, \mathbf{y})$$

The point (v_1, v_2) in two-space has the usual best security level meaning for each player. A typical location of the point, within the convex hull R^*, is shown by the sketch in Figure 25.5.1. [Note that (v_1, v_2) has to be within the payoff region for independent mixed strategies, which is not itself indicated in the figure.]   The set of all Pareto optimal points equal to, or jointly dominating, the BMSSL's of the players defines a special subset of P.

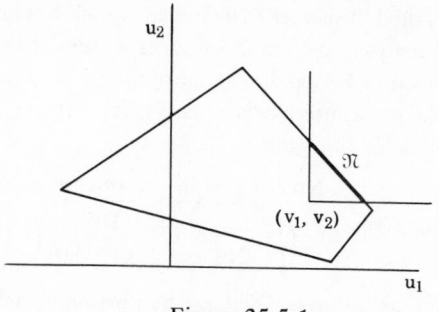

Figure 25.5.1

(25.5.2) Definition. By the negotiation set of a general two-person game, we mean the set $\mathfrak{N}$ given by

$$\mathfrak{N} = \left\{(u_1, u_2): \; (u_1, u_2) \in P \text{ and } u_1 \geqslant v_1, u_2 \geqslant v_2\right\}$$

where (u_1, u_2) ranges over the payoff region.

Thus, $\mathfrak{N}$ is a subset of the Pareto optimal set, but in it the players have the additional advantage of preserving security.

The set $\mathfrak{N}$ is associated with a set of joint strategies that map into it under the expectations (E_1, E_2). Let us denote this set $\mathfrak{N}^-$. Thus,

(25.5.3) $$\mathfrak{N}^- = \left\{z: \; (E_1(z), E_2(z)) \in \mathfrak{N}\right\}$$

A joint strategy not achievable by independent mixed strategies may be called a cooperative action. The cooperative actions in $\mathfrak{N}^-$ are as a whole better than those not in $\mathfrak{N}^-$. This is true in terms of the preference patterns of the players and in terms of security considerations. The cooperative actions in $\mathfrak{N}^-$ are rational for this reason: (1) they maintain security and, while doing so, capitalize on any patterns of shared preference by (2) guaranteeing some expected payoff in the Pareto optimal set. Thus, rational actions in a general two-person game are given by $\mathfrak{N}^-$ and yield the expected payoffs in $\mathfrak{N}$.

From this point of view, we can say that any condition imposed by or upon the players that prohibits a choice from $\mathfrak{N}^-$ forces the actors into nonrational action.

Note that we maintain our view that NM were trying to evolve by mathematical means a concept of rational action of wider and wider applicability. Thus, $\mathfrak{N}^-$ is to us a specification of the rational actions, and so a concept of rationality is given for the general two-person game. In the case that actions in $\mathfrak{N}^-$ are not chosen, we do not say that the players are irrational or that the concept is not general enough, we say instead that conditions induce nonrational action.

25.6. Prisoner's Dilemma: Setup and Strategic Analysis. The point of the present section is to illustrate the basic ideas of previous sections in the context of a special 2 × 2 game, called "Prisoner's Dilemma," which has become the object of much discussion, analysis, and empirical investigation. (See, for example, Luce and Raiffa, 1957; section 5.4; and, Rapoport and Chammah, 1965.)

In normal form with the usual interpretation a typical matrix (o_{ij}) of outcomes (not psychological payoffs) for this game is

$$
\begin{array}{c}
 & \text{No confession} & \text{Confession} \\
\begin{array}{c} \text{No confession} \\ \text{Confession} \end{array} &
\begin{pmatrix} (3, 3) & (20, 1) \\ (1, 20) & (10, 10) \end{pmatrix}
\end{array}
$$

where the interpretation is as follows: There are two prisoners held in custody by the district attorney. Each is suspected to be an accomplice of the other in a crime for which sentences range in years to as high as 20, but the D.A. will not be able to prove it unless he can get them to confess. To do so he separates them and offers them a lure in the familiar manner: a promise of a lesser sentence, say 10 years. But if only one of them confesses, the differential impact of the confession on the D.A., the judge, and the jurors is such as to make a comparatively light sentence the outcome for the confessor. These remarks provide the basis for outcomes $o_{22} = (10, 10)$ when they both confess and $o_{12} = (20, 1)$ or $o_{21} = (1, 20)$ when only one confesses. Finally, if the D.A. can get neither to confess, he can assure a conviction on a lesser charge, yielding 3 years in prison for each. The two persons are placed in isolation from each other to preclude their collusion to outwit the D.A.

Preferences of prisoners α and β may be assumed to be inverse with respect to the number of years. Thus, for each player,

$$1 \blacktriangleright 3 \blacktriangleright 10 \blacktriangleright 20$$

and we can set the endpoints to 1 and 0 respectively to obtain

$$u_1(1) = 1, u_1(20) = 0$$
$$u_2(1) = 1, u_2(20) = 0$$

For concreteness, assume also

$$u_1(3) = .7, u_1(10) = .3$$
$$u_2(3) = .8, u_2(10) = .3$$

Then we obtain the payoff matrix

$$
\begin{array}{c}
 & \beta_1 & \beta_2 \\
\big((a_{ij}, b_{ij})\big) = \begin{array}{c} \alpha_1 \\ \alpha_2 \end{array} &
\begin{pmatrix} (.7, .8) & (0, 1) \\ (1, 0) & (.3, .3) \end{pmatrix}
\end{array}
$$

We wish to analyze formally the game model given by this matrix, in the sense of strategic analysis. First, we compute each player's BMSSL. This can be done by treating each of the matrices (a_{ij}), (b_{ij}) separately.

For (a_{ij}) we have the following:

$$
\begin{array}{c}
\underset{j}{\min} \\
\max_{i} \left\{ \begin{pmatrix} .7 & 0 \\ 1 & .3 \end{pmatrix} \begin{array}{c} 0 \\ .3 \end{array} \right\} \max_{i} \\
\begin{array}{cc} 1 & .3 \end{array} \underset{\underset{j}{\min}}{\,} \overset{\textstyle .3}{\overset{\textstyle }{}} = \textcircled{.3} \\
\end{array}
$$

Here we have the opportunity to review the fact that a saddle-point exists if and only if the operators commute [see theorem (23.6.1)]. They do commute, and the value of BSL α is .3. By choosing strategy α_2 ("confess") player 1 (α) can assure himself an outcome worth .3 to him on his utility scale. Since the game is strictly determined,

$$v_1 = \text{BMSSL } \alpha = \text{BSL } \alpha = .3$$

where v_1 is the first of the two best security levels we want to compute.

For (b_{ij}) we temporarily transpose to take the usual row viewpoint:

$$
\begin{array}{c}
\underset{j}{\min} \\
\max_{i} \left\{ \begin{pmatrix} .8 & 0 \\ 1 & .3 \end{pmatrix} \begin{array}{c} 0 \\ .3 \end{array} \right\} \max_{i} \\
\begin{array}{cc} 1 & .3 \end{array} \underset{\underset{j}{\min}}{\,} \overset{\textstyle .3}{\overset{\textstyle }{}} = \textcircled{.3} \\
\end{array}
$$

Again the saddle-point exists, the player β has an equilibrium pure strategy ($\beta_2 \equiv$ "confess") yielding him .3 in his utility units. Thus, as above,

$$v_2 = \text{BMSSL } \beta = \text{BSL } \beta = .3$$

so that

$$(v_1, v_2) = (.3, .3)$$

for this game. Also, the strictly determined pure strategies, each of which guarantees the player at least his v_i, are both "confess." In computing the point (v_1, v_2), we have not assumed anything apart from empirical meaningfulness. Any strategic meaning it has is, so to speak, saddled with the original zero-sum formulation: but the present game is not zero-sum. This point will be taken up again in our subsequent discussion.

Our second formal task is to determine the negotiation set because in our view this is the set of rational actions, apart from the communication restrictions.

Figure 25.6.1 plots the payoff region (note that it forms a convex set even under independent mixed strategies, because of the special configuration of the four payoffs, as discussed earlier in section 25.2). Also, this figure shows the point $(v_1, v_2) = (.3, .3)$, the Pareto optimal set (the boundary line segments shown with negative signs), and the induced negotiation set $\mathfrak{N}$ (the thick boundary line segments).

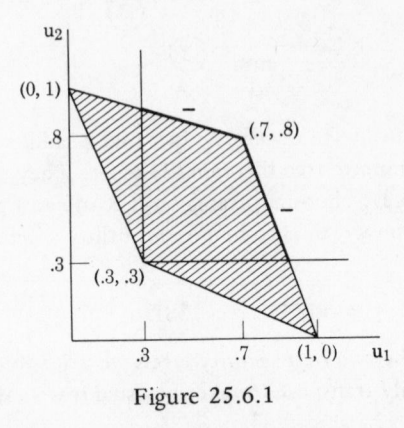

Figure 25.6.1

With these formal tasks accomplished, we turn now to the interpretation of this analysis for the situation modeled by the game.

By rationality here one means a joint action from $\mathfrak{N}^-$ thus leading to $\mathfrak{N}$. The choice (α_1, β_1)—both do not confess—is elegantly simple in the regard: it lies in $\mathfrak{N}^-$ because it has the payoff pair $(.7, .8)$ in $\mathfrak{N}$ and it is intuitively obvious to the players (prisoners) as a stategic option (see Schelling, 1960, in regard to "prominence"). Unfortunately, the isolation of the prisoners forbids communication. Hence, to act rationally they would implicity have to choose jointly. Thus, there would be a trust aspect to their strategic thought. Such trust would be rational.

Then a lack of trust in this situation is nonrational: since the situation, as they see it, is not zero-sum—as standard players, they each know that each prefers the lighter sentence—a reversion to zero-sum security-level thought is nonrational. Since it, in fact, occurs—otherwise there would be no dilemma to speak of—this means that the D.A. has induced nonrational action on the part of the prisoners.

But what does "in fact occurs" really mean? Extant discussions of the Prisoner's Dilemma game suffer from a certain cultural provinciality in which it is tacitly assumed that the (v_1, v_2) pair is the "natural solution." Then this natural solution is associated with the equilibrium concept, and an interpretation is found for it: it is "rational action for each individual." For example, in Luce and Raiffa (1957, p. 96) we find this statement: "Since the players each want to maximize utility, α_2 and β_2 are their 'rational' choices. Of course, it is slightly uncomfortable that two so-called irrational players will both fare much better than two so-called rational ones. Nonetheless, it remains true that a rational player (an α_2 or β_2 conformist) is always better off than an irrational player." This statement employs the term "rational player" in the sense of the zero-sum concept of rationality and leads to the designation of the "no-confession" strategy as, by contrast, "irrational." The "dilemma" is in the tacit assumption of cultural boundary conditions in which otherwise standard idealized players—satisfying the knowledge, preference, and interpretation-of-action assumptions—experience the game in the manner that makes the "individualistic" strategic choice "natural." But we must emphasize the game is not zero-sum and that, far from being "irrational," the implicit cooperative action of nonconfession is the most prominent rational action. In appropriate cultural boundary conditions, it is simple to imagine actors assigned players roles in a Prisoner's Dilemma situation who immediately adopt the no-confession strategy choice by implicit cooperative action, based on their rationality in the negotiation set sense. In fact, the alleged "natural" and incorrectly designated "rational" choice of "confession" is a datum about a culture harnessed to a tacit concept of rationality that at this point contravenes the theory of games.

Our vision of social life, as it exists in contemporary "civilized" culture, distorts this basic theoretical result of game theory into a "dilemma" in which players are "rational" (the quotes signifying the tacit reversion to zero-sum assumptions) but they both get less than they would if they were "irrational" (the quotes signifying the nomenclature wrongly applied to a rational choice in this situation). It appears that from this standpoint the trust basis of social life (see, for example, Conviser, 1970) has a definite relation to the meaning of rationality in general social situations. In other words, social organization is the tacit and imperfect working out of rationality in the natural situation of man, in which there are partially opposing and partially coinciding interests.

25.7. Prelude to a Solution Theory. We remind the reader that this chapter has provided but the foundation for the general two-person theory. Many, if not most, theorists have not been satisfied with the rationality concept outlined above. In particular, they have not been satisfied with a "solution" $\mathfrak{N}$ that is ordinarily a boundary segment. The urge has been to specify a unique point as the solution. Also, the fact that the pairs of points in $\mathfrak{N}$ are in an opposite-preference relation, means that there is a new theoretical problem: how to specify that action in $\mathfrak{N}^-$ that in some sense both players can accept. This notion has led to the many "solution theories" presented in Luce and Raiffa (1957) and in Rapoport (1966). In the next paragraphs, we provide only one of the theories that leads to a definite point in $\mathfrak{N}$.

25.8. Requirements for an Arbitration Scheme. The theory to be presented is a conjunction of the "Nash solution theory" (Nash, 1950) and the Shapley "status quo concept" (Shapley, 1953). The two together yield an axiomatic system with definite implications for strategic analysis.

The Nash solution theory aims to specify a unique point as "the arbitrated solution" of the game. Nash (1950) framed the problem in terms of an "arbiter," an actor (or device that represents an actor) who must formulate a principle that leads for any game of a given class of games to a unique expected payoff point. Hence, there is a need to set up axioms specifying a mapping from a specified class of games into two-space. This mapping, apart from changes of scale, should be unique. The mapping is called an arbitration scheme, with image point the arbitrated solution. Detailed discussion of the conceptual basis appears in Luce and Raiffa (1957; section 6.4).

The a priori conditions demanded of the scheme are as follows:

(1) The arbitrated solution must be in $\mathfrak{R}$.

(2) The arbitration scheme must be compatible with the NM theory of utility or, more generally, with any theory extending preference to uncertain alternatives that yields an interval-scale utility function satisfying the expected utility condition.

(3) The arbitrated solution has a certain symmetry character, to be discussed subsequently along with:

(4) The arbitrated solution should reflect the threat capabilities of the players.

25.9. Axioms for the Shapley-Nash Scheme. The objects needed for the axiomatic structure are as follows:

G: an arbitrary two-person game (m × n) with payoffs (a_{ij}, b_{ij}).

Z: joint mixed strategy space, the set of all z yielding probabilities z_{ij} of (α_i, β_j), so that $z_{ij} = P(\circ_{ij})$.

$E_1(z) = \sum_{(i,j)} a_{ij} z_{ij}$: expected payoff to player 1.

$E_2(z) = \sum_{(i,j)} b_{ij} z_{ij}$: expected payoff to player 2.

Thus (E_1, E_2) together give

$$Z \to \mathfrak{R}^2$$

as in the preceding paragraphs; that is

$$z \mapsto (E_1(z), E_2(z))$$

As before, we let R* be the payoff region, the image of Z under (E_1, E_2). It is the convex hull of the points (a_{ij}, b_{ij}).

(25.9.1) Definition. Let $(u^*, v^*) \in R^*$. A pair consisting of R* with point (u^*, v^*) is called a bargaining game; we write

$$[R^*, (u^*, v^*)]$$

The point (u^*, v^*) is called the status quo point: interpretatively, it is where the players "are" or "will be" if the arbitration scheme "fails."

If an arbitration scheme fails in a particular play, it will be interpreted here as a departure from rationality: that is, we take the Nash arbitration scheme to define rational action within the rationality of the set of possible rational actions denoted $\mathfrak{N}^-$ earlier. To have a name for this adoption of the arbitrated solution, we say that the players act with inner rationality. We make the following interpretative assumption, which is not part of the Shapley-Nash theory.

(25.9.2) Assumption. If two players do not act with inner rationality, then they do not act rationally.

This means that if they fail to arbitrate, they will revert to some nonrational pattern of action (for example, to zero-sum security-level thought). This assumption we regard as the justification for the next assumption.

(25.9.3) Assumption of the Shapley Solution. The status quo point (u^*, v^*) is the pair (v_1, v_2) given by best mixed strategy security levels (i.e., the maximin values).

Under (25.9.3), the class of bargaining games for our presentation is given by

$$\{[R^*, (v_1, v_2)]\}$$

where (v_1, v_2) are, in the usual way,

$$v_1 = \max_{\mathbf{x}} \min_{\mathbf{y}} E_1(\mathbf{x}, \mathbf{y})$$
$$v_2 = \max_{\mathbf{x}} \min_{\mathbf{y}} E_2(\mathbf{x}, \mathbf{y})$$

(25.9.4) Definition. A mapping F with domain all bargaining games of the form $[R^*, (v_1, v_2)]$ and range in $\mathfrak{R}^2$ is called a Shapley arbitration scheme. The point

$$(u_0, v_0) = F([R^*, (v_1, v_2)])$$

is called a Shapley solution.

The Nash theory deals with any arbitration scheme. Here we specialize it to the Shapley arbitration schemes and arrive at the desired theory using the Nash axioms:

(25.9.5) Definition. A Shapley arbitration scheme F,

$$[R^*, (v_1, v_2)] \mapsto (u_0, v_0) = F([R^*, (v_1, v_2)])$$

is a Shapley-Nash scheme if and only if:

Axiom N1. $(u_0, v_0) \in \mathfrak{N}$.

Axiom N2. If $T \subseteq R^*$ and $(u_0, v_0) \in T$, then

$$(T, (v_1, v_2)) \mapsto (u_0, v_0) = F([T, (v_1, v_2)])$$

Axiom N3. If

$$u_1 \mapsto u_1' = a_1 u_1 + b_1 \qquad (a_1 > 0)$$
$$u_2 \mapsto u_2' = a_2 u_2 + b_2 \qquad (a_2 > 0)$$

so that

$$R^* \to R^{*'} = \left\{ (u_1', u_2') \in \Re^2 \right\}$$
$$(v_1, v_2) \mapsto (v_1', v_2') = (a_1 v_1 + b_1, a_2 v_2 + b_2)$$
$$(u_0, v_0) \mapsto (u_0', v_0') = (a_1 u_0 + b_1, a_2 v_0 + b_2)$$

then

$$F([R^*, (v_1', v_2')]) = (u_0', v_0')$$

Axiom N4. If R^* is such that

$$(u, v) \in R^* \qquad \text{iff} \qquad (v, u) \in R^*$$

then

$$\text{if } v_1 = v_2, \text{ then } u_0 = v_0.$$

25.10. Interpretation of the Axioms. Our next step is to briefly indicate the interpretation of each axiom.

(1) Axiom N1 means that an arbitrated solution must lie in $\Re$. Thus, it satisfies all the properties of such expected payoff points: (a) $u_0 \geqslant v_1, v_0 \geqslant v_2$, so both players cannot do worse than under reversion to the "status quo," which is for us (using the Shapley scheme) no worse than their best security levels; in turn, assumption (25.9.2) makes these prominent in the theory because we argue that a failure to arbitrate will mean a failure of rationality in the cooperative sense and, so, a reversion to the strategic thought carried over from zero-sum conditions; (b) the point is realizable by some strategic choices (i.e., it is in R^*); and, (c) it is not jointly dominated by any other point in R^*, and so, it is in the Pareto optimal set.

(2) Axiom N2 says that if the game were restricted to a smaller subset of strategic possibilities, producing a payoff region $T \subseteq R^*$, then, if nevertheless the solution (u_0, v_0) can still be reached by a strategy in T, the arbitration scheme must be coherent in the sense of designating this same point as the arbitrated solution.

(3) Axiom N3 is a requirement for coherence with respect to the scale-type of the two utility variables u_1, u_2. As such it is a requirement of the metatheory of measurement emphasized in Luce (1959a): the "form" of a scientific theory must have certain invariance aspects induced by the scale-types of its fundamental variables. In detail, the axiom starts with any admissible transformation—see section 7.4—of each variable; this induces a new, but empirically equivalent,

payoff region, status quo point, and arbitrated solution. Hence, any scheme F operating on this empirically equivalent bargaining game must produce an empirically equivalent solution: the admissibly transformed original solution.

(4) Axiom N4 is the symmetry axiom. To see how this axiom makes sense, recall that players are assigned numbers "1" and "2" and we so map into R* with axes (u_1, u_2). But we could just as well have called the "2" player by name "1" and "1" by "2." Thus, to every bargaining game of the form

$$[R^*, (v_1, v_2)]$$

there is an associated equivalent game

$$[R^{*S}, (v_2, v_1)]$$

arrived at by flipping R* about the 45° line. Figure 25.10.1 gives an example. The basic flip mapping is $(x, y) \mapsto (y, x)$.

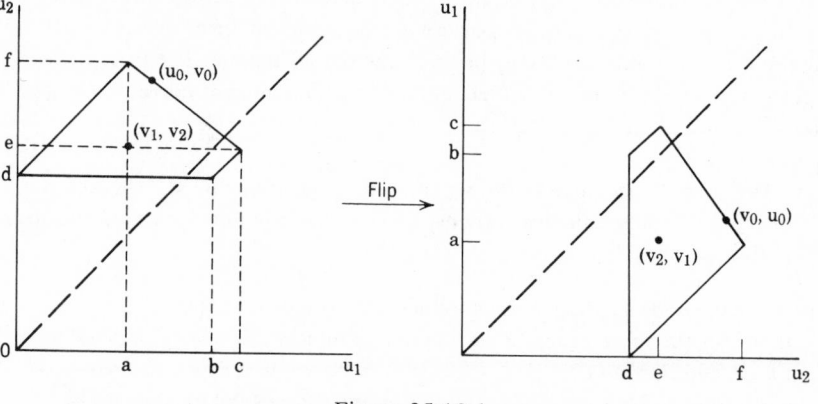

Figure 25.10.1

In another way of saying the same thing, whatever the numerical names may be, actors α and β are such that on the left in Figure 25.10.1, the NM utility model for α maps $(\alpha, \succsim_\alpha)$ into the x-axis, and the model for β maps $(\alpha, \succsim_\beta)$ into the y-axis. On the right side, $(\alpha, \succsim_\alpha)$ is mapped into the y-axis, while $(\alpha, \succsim_\beta)$ goes into the x-axis. Since no matter which representation is chosen we have the identical empirical systems, no difference should occur in arbitration as represented numerically. Thus, if we let R^{*S} be the flipped, symmetric image of R*, we demand

(25.10.1)   If $[R^*, (v_1, v_2)] \mapsto (u_0, v_0)$ under F
then $[R^{*S}, (v_2, v_1)] \mapsto (v_0, u_0)$ under F.

This seems to be dictated by the whole idea of equivalence of various different numerical representations of the same empirical systems. Now consider the special case in which the scales are such that flipping leaves the figure invariant—that is, R^* and its image under $(x, y) \mapsto (y, x)$ are the same. But then $R^{*S} = R^*$, and for the moment call this region A. Now suppose, in addition, that $v_1 = v_2 = v^*$ (temporary notation again). Then substituting into (25.10.1),

$$\text{If} \quad [A, (v^*, v^*)] \mapsto (u_0, v_0) \text{ under F}$$
$$\text{then} \quad [A, (v^*, v^*)] \mapsto (v_0, u_0) \text{ under F}$$

which, since F must be a mapping (i.e., have a unique image point for its argument), implies that $(u_0, v_0) = (v_0, u_0)$. But then $u_0 = v_0$. This is the conclusion of axiom N4. Thus, in our interpretation, N4 is again imposed on the arbitration scheme by the metatheoretical demand that numerical forms of theory conform to the scaling aspects of the fundamental variables. (It seems that most game theorists think of this axiom in terms of not including into arbitration properties extrinsic to the strategic structure—for example, the number of employees who implement a corporate strategy—although, of course, this structure can reflect properties like resources and other power-base capacities in terms of the model leading to the pure strategies "lying behind" the payoff region. The interpretation given here seems to be a rigorous metatheoretical argument corresponding to the intuitive idea of excluding "extrinsic properties," but see also p. 753.)

25.11. The Basic Consequence: Product Formula with Examples. There is one basic conceptual result of the Nash theory and a corresponding numerical formula for computation.

(25.11.1) Theorem. A Shapley-Nash scheme exists and is unique.

The proof for the more general case may be found in Nash's work, and a discussion, in Luce and Raiffa (1957).

For our purposes the proof is significant not only because of existence and uniqueness but because it leads to an explicit formula for the solution.

(25.11.2) Shapley-Nash Solution Formula. The point $(u_0, v_0) = F([R^*, (v_1, v_2)])$ is the unique point at which the following function has a maximum:

$$g(u_1, u_2) = (u_1 - v_1)(u_2 - v_2)$$

where $u_1 \geqslant v_1$, $u_2 \geqslant v_2$, and $(u_1, u_2) \in R^*$. This is also called the product formula.

Example. (a) (Requires calculus) Suppose that $(v_1, v_2) = (0, 0)$. Then (u_0, v_0) is the point at which

$$g(u_1, u_2) = u_1 u_2 \qquad [u_1 \geqslant 0, u_2 \geqslant 0, (u_1, u_2) \in R^*]$$

Imagine that R^* has the form sketched in Figure (25.11.1). Then we know that the point lies in P, the Pareto optimal set, where the points (u_1, u_2) must satisfy the equation of this line:

$$u_2 = 1 - u_1$$

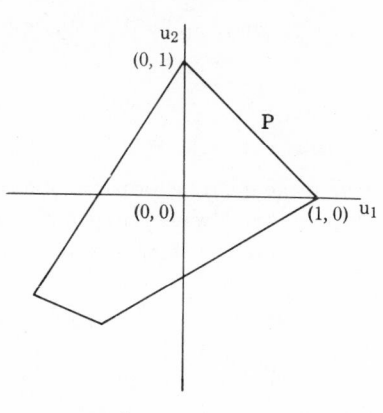

Figure 25.11.1

Thus, the product formula becomes

$$g(u_1, u_2) = u_1(1 - u_1)$$

a function of one variable. Hence, since the maximum must occur where the slope of the function $u_1(1 - u_1)$ is parallel to the horizontal axis, its slope (and, so, derivative) must be 0 there. Differentiating,

$$\frac{d}{du_1}[u_1(1-u_1)] = \frac{d}{du_1}(u_1 - u_1^2)$$
$$= \frac{d}{du_1}u_1 - \frac{d}{du_1}u_1^2$$
$$= 1 - 2u_1$$

Then setting this to 0 yields the solution, which is the value of u_0:

$$u_0 = \tfrac{1}{2}$$

Because the point (u_0, v_0) lies in P, $v_0 = 1 - u_0 = 1/2$. Thus, the Shapley-Nash arbitrated solution is

$$(\tfrac{1}{2}, \tfrac{1}{2})$$

Example. (b) The battle-of-the-sexes game (25.3.5) when plotted (see Figure 25.3.2) does not present us with the simplicity of the equation

$$u_2 = 1 - u_1$$

for the Pareto optimal set, as in the previous example. Thus, since the variables are interval in scale-type, we are led to find a pair of transformations

$$u_1 \mapsto a_1 u_1 + b_1$$
$$u_2 \mapsto a_2 u_2 + b_2$$

that will put the game in "easy" form to apply the technique of example (a). Thus, we want:

$$\begin{pmatrix} (1, q_1) & (p_2, q_2) \\ (0, 0) & (p_1, 1) \end{pmatrix} \to \begin{pmatrix} (1, 0) & (-a, -b) \\ (-c, -d) & (0, 1) \end{pmatrix}$$

Concrete matrices will help make the transformation easier to follow and lead readily to the conviction that one can always put such a game into the simple form for strategic analysis in terms of the Shapley-Nash product formula.

Suppose, then, we are given

(25.11.3)
$$\begin{pmatrix} (1, .5) & (.5, .2) \\ (0, 0) & (.8, 1) \end{pmatrix}$$

This means we want for the (a_{ij})

$$\begin{pmatrix} 1 & .5 \\ 0 & .8 \end{pmatrix} \to \begin{pmatrix} 1 & -a \\ -c & 0 \end{pmatrix}$$

Hence, we want

$$1 \mapsto 1$$
$$.8 \mapsto 0$$

providing the sufficient pair of points $(1, 1)$, $(.8, 0)$ to determine a line and, so, a linear transformation of the u_1-scale, which defines (a_{ij}), namely,

$$u_1 \mapsto 5u_1 - 4$$

Applying this transformation to the remaining two entries,

$$.5 \mapsto 5(.5) - 4 = 2.5 - 4 = -1.5$$
$$0 \mapsto 5(0) - 4 = -4$$

Thus,

$$a = 1.5, c = 4$$

Next we transform (b_{ij}):

$$\begin{pmatrix} .5 & .2 \\ 0 & 1 \end{pmatrix} \to \begin{pmatrix} 0 & -b \\ -d & 1 \end{pmatrix}$$

The points $(1, 1)$ and $(.5, 0)$ provide the scale transformation for u_2:

$$u_2 \mapsto 2u_2 - 1$$

and so we map the other two entries:

$$.2 \mapsto 2(.2) - 1 = .4 - 1 = -.6$$
$$0 \mapsto 2(0) - 1 = -1$$

Thus,

$$b = .6, d = 1$$

The transformation yields the resulting (a_{ij}, b_{ij}):

(25.11.4)
$$\begin{pmatrix} (1, \ 0) & (-1.5, \ -.6) \\ (-4, \ -1) & (0, \ 1) \end{pmatrix}$$

which if plotted determines the four extreme points of a convex hull such that the Pareto optimal set has equation

$$u_2' = 1 - u_1' \qquad (0 \leqslant u_1' \leqslant 1)$$

(see Figure 25.11.2). However, the security levels are not $(0, 0)$, so we cannot routinely conclude that $u_0' = 1/2$. The next section provides the solution for this more general case.

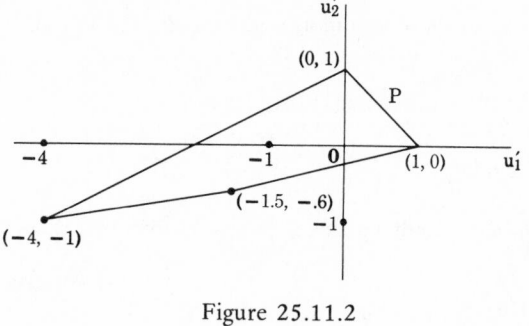

Figure 25.11.2

Although we have labeled the variables in Figure 25.11.2 with primes because in this context we obtained them from an original pair of scales, in the subsequence these primes play no role, and they are therefore to be omitted. The point is that we can obtain a "canonical" form for the species of two-person games with original form (25.3.5)—emerging from the analysis of a mixture of opposition and agreement in preferences—such that this latter "easier" form is the object of analysis. To coordinate to Chapter 1, the form (25.3.5) is obtained in the setup phase of game-modeling, and we then put it into a (different) canonical form with preservation of all empirical meaning, in preparation for the analysis phase of game-modeling: strategic analysis.

25.12. Strategic Analysis of an Abstract General Game. We now perform calculations and interpretations providing a strategic analysis of the game, with $a < c$, $b < d$, and $a, b, c, d > 0$,

(25.12.1)
$$\begin{pmatrix} (1, \ 0) & (-a, -b) \\ (-c, \ -d) & (0, \ 1) \end{pmatrix}$$

the canonical form of the abstract battle-of-the-sexes game. We do not assume that $(v_1, v_2) = (0, 0)$.

We have

$$\begin{aligned} g(u_1, u_2) &= (u_1 - v_1)(u_2 - v_2) \\ &= (u_1 - v_1)(1 - u_1 - v_2) \end{aligned}$$

giving us the function, say $g(u_1)$,

$$g(u_1) = (1 + v_1 - v_2)u_1 - u_1^2 - v_1(1 - v_2)$$

with the derivative set equal to 0,

$$0 = \frac{d}{du_1} g(u_1) = 1 + v_1 - v_2 - 2u_1$$

Solving for u_1 as a function of v_1 and v_2 under maximization and calling the value u_1^*, we have

(25.12.2) $u_1^* = \frac{1}{2} [1 - (v_2 - v_1)]$

Thus, the following holds:

(25.12.3) Proposition. For any battle-of-the-sexes game, that is, game in form (25.12.1),

(a) $u_0 - v_1 = v_0 - v_2$	if	$0 \leqslant u_1^* \leqslant 1$
(b) $u_0 = 0, v_0 = 1$	if	$u_1^* < 0$
(c) $u_0 = 1, v_0 = 0$	if	$u_1^* > 1$

Proof. Examining Figure 25.11.1, the typical form of the payoff region corresponding to matrix (25.12.1), we see that points in the Pareto optimal set P are between 0 and 1 in both components. Since the arbitrated solution (u_0, v_0) must be in P it follows that the domain of u_0 is the interval from 0 to 1. If $0 \leqslant u_1^* \leqslant 1$, then $u_1^* = u_0$ because then the point in the domain of the function $g(u_1)$ that maximizes the function lies in the required domain. Then,

$$u_0 = \frac{1}{2} [1 - (v_2 - v_1)]$$
$$v_0 = 1 - u_0$$

and so $u_0 - v_0 = v_1 - v_2$, implying (a). If $u_1^* < 0$, a point at which $g(u_1)$ has a maximum is to the left of the interval from 0 to 1 in which we must have point u_0. Since g will be monotone decreasing over this interval [to see this, sketch the function $g(u_1)$], the point $u_0 = 0$ will yield the maximum of $g(u_1)$ over the 0-1 interval. Similarly, if $u_1^* > 1$, then a maximum of $g(u_1)$ occurs to the right of

$u_1 = 1$ and the function must be monotone increasing over the 0-1 interval, yielding a maximum at 1. Hence, $u_0 = 1$.

The equality (a) provides a clear way to visualize the arbitrated solution, as indicated in Figure 25.12.1. Because of (25.12.3)(a), the points (v_1, v_2) and (u_0, v_0) are the coordinates of two diagonally opposite corners of a square with sides of equal length, $u_0 - v_1$ and $v_0 - v_2$. Also, this means to find (u_0, v_0) all we need to do is draw the 45° line up from (v_1, v_2): the point of intersection with the line $u_2 = 1 - u_1$ is the arbitrated solution, (u_0, v_0). This geometric visualization is persuasive of the fact that the arbitration scheme maps the status quo into an arbitrated solution.

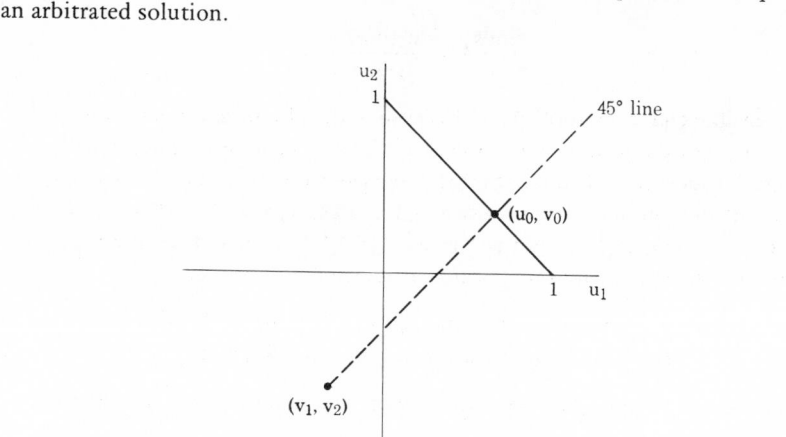

Figure 25.12.1

We can interpret (25.12.3)(a) as follows. The difference $u_0 - v_1$ is on the u_1-scale. Thus, it is a (non-negative) distance measured on this scale. Hence, since distances on such an interval scale are ratio-level in scale-type, we can interpret $u_0 - v_1$ as the distance-gained by H (row player) and note it has an arbitrary scale unit. Similarly, $v_0 - v_2$ is a distance measured on the u_2-scale and, so, a ratio-level measure, the distance-gained by W (column player). Then (a) seems to equate two distances measured in those particular units such that a proportionality constant, say k, is unity. To see this transform $u_0 - v_1$, $v_0 - v_2$ by admissible transformations,

$$u_0 - v_1 \mapsto a_1(u_0 - v_1) \qquad (a_1 > 0)$$
$$v_0 - v_2 \mapsto a_2(v_0 - v_2) \qquad (a_2 > 0)$$

In the new units,

$$u_0' - v_1' = (a_1 u_0 + b_1) - (a_1 v_1 + b_1) = a_1(u_0 - v_1)$$
$$v_0' - v_2' = (a_2 v_0 + b_2) - (a_2 v_2 + b_2) = a_2(v_0 - v_2)$$

and the identity (a) of proposition (25.12.3) implies, then,

$$u_0' - v_1' = \frac{a_1}{a_2}(v_0' - v_2') \qquad \left(\frac{a_1}{a_2} > 0\right)$$

This may be written

(25.12.4) $u_0' - v_1' = k(v_0' - v_2')$ $(k > 0)$

As we maintained above, the real meaning of proposition (25.12.3)(a) involves an interscale constant whose dimension,

$$\frac{\dim a_1}{\dim a_2} = \frac{u_1\text{-unit}}{u_2\text{-unit}}$$

yields a sensible equation in (25.12.4) because both sides have the same dimension. Then (a) of the proposition is the case of $k = 1$, yielding a dimensionally meaningful proposition. The fact that the two gains by arbitration are equal is an artifact of the canonical form for strategic analysis (25.12.1). What is not an artifact is the proportionality of these gains: (25.12.4) with k an arbitrary positive constant. To analyze the meaning of this proportionality, we let

$$g_1 = u_0 - v_1 = \text{distance gained by player 1}$$
$$g_2 = v_0 - v_2 = \text{distance gained by player 2}$$

Then the general result (25.12.4) can be expressed in purely gain notation as

(25.12.5) $g_1 = kg_2$ $(k > 0)$

and proposition (25.12.3)(a) expresses this generalization in terms of the canonical form in which $k = 1$.

Let H and W be two actors such that they play three distinct battle-of-the-sexes games successively. Suppose they satisfy player assumptions and also agree to abide by the arbitrated solution in each game. Then calling the games G_1, G_2, G_3, we have three gain-points, each with non-negative components:

$$\left(g_1^{(1)}, g_2^{(1)}\right), \left(g_1^{(2)}, g_2^{(2)}\right), \left(g_1^{(3)}, g_2^{(3)}\right)$$

where

$$g_j^{(i)} = \text{distance gained in game } G_i \text{ by player } j \qquad (i = 1, 2, 3; j = 1, 2)$$

Suppose these $g_j^{(i)}$ are all distinct, and imagine them plotted in the plane: because of proportionality they lie on a line through the origin, as shown in Figure 25.12.2. A linear transformation through the origin—the meaning of the proportionality in geometric terms—preserves the ratio of any two values on the domain in mapping into the range. (Recall that this is a property usually used in explicating ratio scales, as in section 7.4, but this is true more generally and here relates two empirically distinct ratio scales, the gains.) Thus, let us see what this means to H,

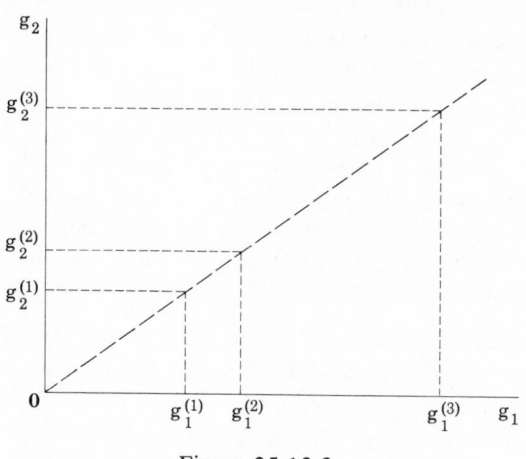

Figure 25.12.2

the actor identified as player 1 with gains on the g_1-axis. We can see that from H's point of view the arbitration scheme defines gains for him such that whenever, in different situations (games), it defines H's gain as c times as much as in some other game, it does the same for W, where c here is meaningful because g_1 and g_2 are ratio-type variables. For example, for H, in Figure 25.12.2,

$$g_1^{(3)} = 2g_1^{(2)}$$

a gain in G_3 twice that of G_2, and for W, the same is true as a result of arbitrations:

$$g_2^{(3)} = 2g_2^{(2)}$$

Consider now that if we had

$$g_2^{(3)} = 4g_2^{(2)}$$

then H would say, "How is it that a so-called arbitration scheme yields W a gain in G_3 that is four times as great as her gain in G_2 but for me the G_3 gain is only twice as great? I am willing to grant that the gains from arbitration depend on one's own preferences, and I am not really able to compare $g_1^{(3)}$ and $g_2^{(3)}$, but if the scheme is to make any sense then I want to be assured that in using it when I experience what amounts to a doubling of gain, then W also should experience what amounts to a doubling of her gain and not a quadrupling of it. This would mean that the scheme is attending not so much to the strategic situation as to some particularistic aspect that W carries along with her into these situations. But then if we reversed roles—call me 'W' and her 'H'—I would be favored. But this is ridiculous because I don't even share those particularistic traits of W. The point is the scheme has to be symmetric or it doesn't make any sense."

In a similar way, W would see the disadvantage of a quadrupling versus a doubling. In the absence of proportionality, the gains (g_1, g_2) would fluctuate radically over the plane in various games. True, W would get the better of the current comparison, but she might reason that if proportionality does not hold, there is no telling what will happen in the next game in terms of the arbitrated solution. "This time maybe it was because I am a female, but next time the ruling may depend on eye-color or hair-length or drawing ability and, in that case, maybe H is favored. The whole thing is ridiculous, anyway, because these other properties are irrelevant. The scheme has to be symmetric or it doesn't make any sense." This is the meaning of the linear function given in formula (25.12.5).

Example. In continuing example (b) of section 25.11, after transformation to canonical form, we have matrix (25.11.4). The best mixed strategy security levels in terms of this matrix are needed; that is, we require (v_1, v_2). To compute v_1 we use equation (23.11.7) on the (a_{ij}) matrix,

$$\begin{pmatrix} 1 & -1.5 \\ -4 & 0 \end{pmatrix}$$

This yields $v_1 = -.77$. Similarly, to compute v_2 we use equation (23.11.7) with the (b_{ij}) matrix,

$$\begin{pmatrix} 0 & -.6 \\ -1 & 1 \end{pmatrix}$$

This yields $v_2 = .15$. The reader may wish to visually plot $(v_1, v_2) = (-.77, .15)$ in Figure 25.11.2 to verify by the $45°$-line method of Figure 25.12.1 that the arbitrated solution will yield a point very near to $(0, 1)$. Substituting into formula (25.12.2), we find

$$u_1^* = .04$$

Hence, $(u_0, v_0) = (.04, .96)$. The gain of H, player 1, is $g_1 = u_0 - v_1 = .81$; the gain of W, player 2, is $g_2 = v_0 - v_2 = .81$, which shows that, in accordance with formula (25.12.5), with $k = 1$ here,

$$g_1 = g_2$$

This equality expresses the symmetry of the arbitration scheme, as indicated in our analysis above, using arbitrary constant k in the general case (i.e., where all possible scales representing preferences are allowed and not only the special scales for the canonical form). In the case at hand, then, both H and W exceed their security-level points. Since the arbitrated solution (.04, .96) lies in the negotiation set, strategic action to attain it is rational. Apart from this, it provides a unique solution to the problem of choice within the class of rational actions.

25.13. Rationality Revisited. The most important interpretation, for our purposes, of the solution concept given by the Shapley-Nash scheme is that it can be used to refine the concept of rational action in a general two-person game. The game begins with the partial opposition and partial coincidence of interests. The first step is to define rational action for the players as any action leading to payoffs in $\mathfrak{N}$. Then the next step is to recognize the opposition of preferences within $\mathfrak{N}$ by formulating an arbitrated solution that yields a definite rational action. If the players accept rationality in the first step (i.e., they are determined to behave rationally), then the only new conditions are those of axioms N1-N4 of definition (25.9.5). But, of these, only N2 is at all problematic, because N1 is natural once we pass the first stage, while N3 and N4 are not referent to the players so much as they are to the empirical meaningfulness of the arbitrated solution. (On the other hand, see Luce and Raiffa (1957) for a discussion of N2). Hence, the two-stage rationality concept is given by the following:

(1) rationality: explicit or tacit commitment to realize a payoff point in the negotiation set, followed by,

(2) inner rationality: application of the Shapley-Nash arbitration scheme, yielding a unique payoff point (u_0, v_0) in the negotiation set.

The reader should understand that this concept is not given as some dogma but as one intellectually interesting and defensible way to interpret some of the achievements of game theory in regard to the problem of rationality in interactive choice situations.

25.14. Threat Capabilities and the Solution. We have not yet shown that the arbitration scheme takes into account the threat capabilities of the players. First, we have to make clear what the "threat value" of a strategy is. The battle-of-the-sexes game has two forms we have used. One form emerges from measurement, as in matrix (24.3.1) reproduced here with a minor notation change:

$$
\begin{array}{c}
 & \hspace{2.5em} W \\
 & \hspace{1em} A \hspace{2em} B \\
(25.14.1) \hspace{3em} H \begin{array}{c} A \\ B \end{array} \begin{pmatrix} (1, q_1) & (p_2, q_2) \\ (0, 0) & (p_1, 1) \end{pmatrix}
\end{array}
$$

Recall the parameters satisfy:

$$0 < p_2 < p_1 < 1$$
$$0 < q_2 < q_1 < 1$$

The second form emerges after scale transformations to obtain a canonical form for strategic analysis, as in matrix (25.12.1), reproduced here:

$$W$$

$$A \qquad B$$

(25.14.2) $\qquad H \begin{array}{c} A \\ B \end{array} \begin{pmatrix} (1, 0) & (-a, -b) \\ (-c, -d) & (0, 1) \end{pmatrix}$

Here the parameters satisfy:

$$a < c, b < d;$$
$$a, b, c, d > 0$$

In the first form we think of $(0, 0)$ as the worst that can occur from both players' point of view: separate activities and each engaging in the nonpreferred activity. W, player 2, prefers B to A, we may recall from section 24.2, while H prefers A to B. W can use activity B to threaten H in the sense that if H fails to go along with her (BB), the payoff for H is the term $p_2 < p_1$. Similarly, player 1, H, can use activity A to threaten W in the sense that if W fails to go along with him (AA), the payoff for W is $q_2 < q_1$. The analytical question is, To what extent are these threat capabilities realistically reflected in the inner rationality of the Shapley-Nash solution? We note that p_2 corresponds to $-a$, and q_2 corresponds to $-b$. Thus,

(25.14.3) Definition. In the canonical form (25.14.2) for strategic analysis of the battle-of-the-sexes game, the threat value of player 2 is $-a$ and the threat value of player 1 is $-b$.

Our problem is to study formula (25.12.2)

$$u_0 = \tfrac{1}{2} (1 - (v_2 - v_1))$$

simplifying to the case (a) of proposition (25.12.3). The solution u_0 needs to be expressed in terms of the threat values. Thus, we need to write v_1 and v_2 in terms of the parameters of matrix (25.14.2), using the general formula derived earlier (23.11.7), first for (a_{ij}) then for (b_{ij}):

$$v_1 = \frac{a_{11} a_{22} - a_{12} a_{21}}{(a_{11} + a_{22}) - (a_{12} + a_{21})} = \frac{0 - ac}{1 - [-(a+c)]} = \frac{-ac}{1 + a + c}$$

$$v_2 = \frac{b_{11} b_{22} - b_{12} b_{21}}{(b_{11} + b_{22}) - (b_{12} + b_{21})} = \frac{0 - bd}{1 - [-(b+d)]} = \frac{-bd}{1 + b + d}$$

Hence,

(25.14.4) $\qquad u_0 = \dfrac{1}{2} + \dfrac{1}{2} \left\{ \dfrac{bd}{1 + b + d} - \dfrac{ac}{1 + a + c} \right\}$

Thus, to study how u_0 behaves as a function of a and of b, the absolute magnitudes of the threat values, we use partial differentiation to obtain

$$\frac{\partial u_0}{\partial a} = -\frac{1}{2} \frac{c(1+c)}{(1+a+c)^2}$$

$$\frac{\partial u_0}{\partial b} = \frac{1}{2} \frac{d(1+d)}{(1+b+d)^2}$$

Since $c > 0$ and $d > 0$, we conclude that

(25.14.5) $$\frac{\partial u_0}{\partial a} < 0, \frac{\partial u_0}{\partial b} > 0$$

Thus, the conclusions are as follows:

(1) u_0 is a monotone decreasing function of a: the greater the absolute magnitude of the threat value of player 2 (vis-à-vis player 1), the smaller the u_0, the "share" of the arbitrated payoffs given to player 1.

(2) u_0 is a monotone increasing function of b: the greater the absolute magnitude of the threat value of player 1 (vis-à-vis player 2), the larger the u_0.

If we say "threat potential" for such absolute magnitudes of the threat values, the intuitive idea we have here is that the greater the husband's threat potential, the more likely they are to go together to his favored activity. This interpretation is valid because the (u_0, v_0) solution is a (u_1, u_2) point. Hence, it is the utility, in each partner's scale, of some uncertain alternative in which—since we are in the negotiation set for this game—with some probability z_{11} they engage in activity a together and with the complementary probability $z_{22} = 1 - z_{11}$ they engage in activity B together. The threat potential (b) of the husband is the absolute magnitude of the utility that his wife has for her favored activity in case he sticks with activity A and she refuses to go along.

Example. In the game (25.11.4) we studied again in the example of the preceding section, we found that $(u_0, v_0) = (.04, .96)$, whereas the threat potential of H is .6 and that of W is 1.5. Now consider a second game, altering only the threat potentials:

$$\begin{pmatrix} (1, 0) & (-1, -.9) \\ (-4, -1) & (0, 1) \end{pmatrix}$$

Here we have decreased the threat potential of W from 1.5 to 1 and increased the threat potential of H from .6 to .9. Calculating the solution u_0 from formula (25.14.4), we find that $u_0 = .32$, $v_0 = .68$. Hence, H has moved up from .04 to .32 and W has moved down from .96 to .68 in the arbitrated solution, corresponding to an increase in threat potential for H and a decrease in threat potential for W. This result is in accordance with expressions (25.14.5).

25.15. Summary of the Extension Process. Let us summarize the concept of rationality for two-person general games and then relate it to the concept developed in the zero-sum context. "Rationality" means:

(1) Cooperative rationality: The players first use any coincidence of interests to narrow the options of action, by eliminating options in which both can do better or one can do better without injury to the other's interests. This determines the Pareto optimal set for the game. Moreover, since in this set the interests are opposed, the players then commit themselves to action that at least preserves the security of each. This determines the negotiation set for the game. Cooperative rationality means the use of strategic options that map into payoffs in the negotiation set.

(2) Inner rationality: Once committed to the negotiation set, the players each agree to abide by the arbitration scheme of Shapley-Nash. This scheme will pinpoint a definite realizable payoff point in the negotiation set as the arbitrated solution of the remaining conflict of interests. Moreover, it will do this in such a way as to realistically reflect the threat potentials determined by the strategic analysis of the game. However, by employing a principle of symmetry, it does this in a truly universalistic way; that is, no properties of the players can affect the arbitrated solution except insofar as they are reflected in the structure of the game itself and in their preferences.

The thesis is that this concept applies to any finite two-person game. Thus, in some sense, the concept applies even to the zero-sum situation. Let us see what it implies for rational action in a zero-sum game.

First, we note that a zero-sum game is the special case in which

$$(a_{ij}, b_{ij}) = (a_{ij}, -a_{ij})$$

Thus, if we plot these pairs of payoffs, we arrive at a series of points on the line

$$u_2 = -u_1$$

in the (u_1, u_2) plane.

From this it follows that the convex linear combinations that are the independent mixed strategies also define points on this line and that the convex hull R^* is just the line segment joining the two most extreme points plotted: one in the upper left and the other in the lower right (see Figure 25.15.1 for an example: the solid line segment is the payoff region R^*).

Given R^*, our first step is to define the Pareto optimal set. By the negative slope and the absence of any points in R^* not on the segment, we conclude at once that

$$P = R^*$$

Next we want the negotiation set. This means first finding the security levels (v_1, v_2). However, $v_2 = -v_1$, and the fact that this is the image point of some pair of (equilibrium) strategies assures us that this point is on the payoff line segment. Thereby the geometric idea of finding points in R^* satisfying $u_1 \geqslant v_1$ and $u_2 \geqslant v_2$ leads to nothing but the point (v_1, v_2) itself.

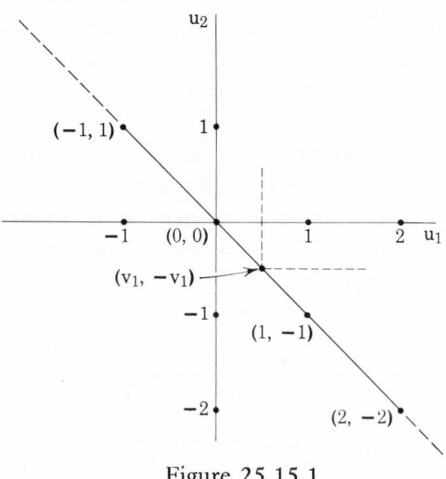

Figure 25.15.1

Hence, the intersection of this latter set with the Pareto optimal set $P = R^*$ is the negotiation set in its strict sense, now containing only one point:

$$\mathfrak{N} = \left\{ (v_1, -v_1) \right\}$$

If we now apply the Shapley-Nash formula in a purely formal way, without thinking about the strategic aspect, we have

$$\max g(u_1, u_2) = (u_1 - v_1)(u_2 - v_2)$$

where

$$u_2 = -u_1$$
$$v_2 = -v_1$$

so that

$$g(u_1, u_2) \equiv g(u_1) = (u_1 - v_1)(v_1 - u_1)$$
$$= -(u_1 - v_1)^2$$

a function whose maximum is 0 because elsewhere it is strictly negative, and the 0 is achieved with solution v_1; that is,

$$u_0 = v_1$$
$$v_0 = -u_0 = -v_1$$

gives the Shapley-Nash arbitrated solution. It is merely the pair of best mixed strategy security levels, which means that in the general concept of rationality the rational action for the players in a zero-sum game is to behave in accordance with strict security considerations (i.e., to use equilibrium strategies). But this is the very concept of rational action defined in section 23.12: in "the concept of rational action . . . each player chooses an equilibrium (mixed) strategy that guarantees him his best mixed strategy security level." Let us call this concept, developed for a zero-sum game, zero-sum rationality.

(25.15.1) Proposition. If a general two-person game is zero-sum, then it is rational to act in accordance with zero-sum rationality.

The phrase "it is rational" refers to the more general concept of two-stage rationality outlined above. In other words, zero-sum rationality receives a certain justification from a higher-order notion of rational action based on the potentialities for cooperation. Methodologically, proposition (25.15.1) shows that the rationality concept has been extended, in the strict sense, from the class of all two-person zero-sum games to the class of all two-person general games.

The logical structure of the extension is summarized in Table 25.15.1.

Table 25.15.1   Extension of Rationality Concept

Type of 2-Person Game	Rationality Concept	Security Guarantee
Strictly determined zero-sum	Equilibrium strategy	BSL
Zero-sum	Equilibrium mixed strategy	BMSSL
General	Arbitrated solution	At least as good as BMSSL

If we denote the strictly determined zero-sum type as G_1, the zero-sum type by G_2 and the general type by G_3, then

$$G_1 \subset G_2 \subset G_3$$

so that each type of game includes the lower type as a special subtype. Correspondingly, the most general rationality concept involves the cooperative and inner rationality leading to the arbitrated solution. But this applies for all games in G_3. When we specify the game is zero-sum, yielding class G_2, then the cooperative rationality specializes to the zero-sum rationality of equilibrium mixed strategy. Finally, if we specify that the game is strictly determined as well, then the subclass G_1 is selected and the rationality involves the choice of an equilibrium pure strategy, outlined in section 23.2. In each case, as we move up in extension, security is preserved: the security of cooperative rationality is at least as good as zero-sum security.

The following may be noted. In games in G_3 but not in G_2—general, not zero-sum—the failure to use cooperative and inner rationality, leading to an arbitrated solution, is nonrational. In other words, in games not zero-sum to act with zero-sum rationality is not at a sufficiently "high" level of rationality.

Philosophically, the humanistic appeal of cooperative action as rational per se is the most significant aspect of the extension. A purely theoretical investigation supports the idea that security-level strategic thought is a one-sided and often insufficient aspect of rationality. Only when the rules of the game, coupled with the player's preferences, generate a strictly opposing set of interests, can one admit the adequacy of the rationality of the zero-sum way of thought. As applied to almost any situation outside of parlor games with clear win-lose outcomes, the theory of games defines rationality in the situation as intrinsically cooperative. The word "optimal" applies, then, to strategies explicitly or implicitly (via trust) correlated with each other to produce outcomes, or uncertain combinations of outcomes, that for both parties are ordinarily an improvement over outcomes arising from the inappropriate (i.e., nonrational) use of the principles of "optimal behavior" for a zero-sum situation.

SUGGESTIONS FOR FURTHER READING (PART FOUR)

Chapter 20. Our treatment of utility theory is based upon Luce and Raiffa (1957). Another exposition is found in Chernoff and Moses (1959). A comprehensive survey of utility and subjective probability is given by Luce and Suppes (1965).

Chapter 21. The axiomatic concept of a game is found in von Neumann and Morgenstern (1947) and in Owen (1968).

Chapter 22. A good discussion of strategies and passage to normal form is given by Rapoport (1960) as well as Luce and Raiffa (1957), who discuss the player assumptions in great detail. On the fundamental problem of game theory, see the opening chapter of von Neumann and Morgenstern (1947).

Chapter 23. For zero-sum two-person games, we follow closely the conceptualization of Luce and Raiffa (1957). A brief treatment with proofs is given in Owen (1968). Rapoport (1966) gives an exposition of the theory, focusing on conceptual analysis, while for the pure mathematics of the game matrix a good source is Kemeny, Snell, and Thompson (1966).

Chapter 24. For more background on general two-person games, see Rapoport (1966). For the mathematics of point-sets in n-space, see Bartle (1964) or Royden (1963).

Chapter 25. For general two-person game theory, the discussion by Luce and Raiffa (1957) is recommended. Another treatment of the Nash scheme is given by Owen (1968). Rapoport (1966) considers not only the Shapley but other solution theories. For criticisms of the Nash scheme, see Harsanyi (1956). For a discussion of the rationality concept in game theory, see Rapoport (1960, 1966, 1970). We have not treated n-person theory, which is found in Rapoport (1970). Theories of the play of games appear in Rapoport (1966), as well as Suppes and

Atkinson (1960), Shubik (1964), and Bartos (1967). For a treatment of game theory as a "tool for the moral philosopher," see Braithwaite (1955). The potential for variation on the "classic" themes of game theory, to include a wide variety of topics of interest to social scientists (e.g., dynamic games, stochastic games, market games), is exhibited in the series of volumes "Contributions to the Theory of Games." (See Shapley, 1953, for a precise reference.)

A GUIDE TO THE LITERATURE IN MATHEMATICAL SOCIOLOGY

Contributions to the field are made primarily in research papers published in sociological journals and anthologies. Some particular professional publications regularly publish articles in mathematical sociology:

(1) *Journal of Mathematical Sociology*
(2) *American Sociological Review*
(3) *American Journal of Sociology*
(4) *Sociometry*
(5) *Behavioral Science*
(6) *Sociological Methodology* (a yearbook published by the American Sociological Association)
(7) *General Systems* (a yearbook published by the Society for General Systems Research).

Also relevant are the *Journal of Mathematical Psychology*, and several European-based journals: *Human Relations, Acta Sociologica, Sociology,* and *Quality and Quantity.* In addition, a useful *Mathematics and Computing Newsletter,* containing reviews, book lists, and research notes, is published by the British Sociological Association, under the editorial direction of A.P.M. Coxon of the Department of Sociology, University of Edinburgh, Scotland.

The literature of the field can be categorized into various types:

Bibliographies. Two comprehensive bibliographies are Paris-Steffens (1966) and Holland and Steuer (1970). The latter is annotated. Both, however, are already a bit dated (the latter bibliography covers only material published by the spring of 1968). More recent bibliographies are best found in reviews of particular subjects in the literature listed under the next category.

Reviews and Expository Essays. Two reviews of the related field of mathematical social psychology have appeared recently: Abelson (1967) and Rosenberg

764

(1968). A review of mathematical anthropology is provided by White (1972). Review papers in mathematical sociology are harder to locate. Coxon (1970) gives a useful overview, while Fararo (1969) and Mayer (1971) are expository at a relatively nontechnical level. More technical, but a bit dated, are the review expositions by Coleman (1960) and Karlsson (1958). A thorough review is given by Rapoport (1963). More specialized reviews (of social mobility) are given by Pullum (1970) and Mayer (1972). Kaplan (1952) and Rapoport (1960) give lucid critical discussions of the field. A highly critical statement from an expert in the history of sociological theory is given by Martindale (1963). An older but still relevant statement is given by Arrow (1951).

Readers. By far the most outstanding example is the collection of papers edited by Lazarsfeld and Henry (1968). A classical set was edited by Lazarsfeld (1954). Other collections of papers are Arrow, Karlin, and Suppes (1960), Criswell, Solomon and Suppes (1962), Massarik and Ratoosh (1965), and, in more limited areas, the game theory anthology edited by Shubik (1964), the causal-models book edited by Blalock (1971), the mathematical anthropology collection edited by Kay (1971), and the measurement papers in Churchman and Ratoosh (1959). One should also mention a two-volume set of readings in mathematical psychology that contains many papers of interest to sociologists, edited by Luce, Bush, and Galanter (1963) to supplement their 3-volume *Handbook of Mathematical Psychology.* Also relevant is the systems theory anthology edited by Buckley (1968), and the two volumes of *Sociological Theories in Progress* edited by Berger et al. (1966, 1972).

Texts. A very elementary, brief, and readable text is that by Beauchamp (1970). More extensive is the elementary text by Bartos (1967), which covers Markov models and game theory. An elementary text which emphasizes graph-theoretic techniques and is written from a modern mathematical viewpoint is that by Doreian (1970). At a more advanced level, the set of studies in Kemeny and Snell (1962) forms a useful text, covering such topics as dynamic systems, Markov chains, graphs, and dynamic programming. By far the most useful general text in the logic of probability model-building, with the bonus that one learns some psychology simultaneously, is that by Atkinson, Bower, and Crothers (1965), which is accompanied by a separately bound set of problems, with solutions, developed by Batchelder, Bjork, and Yellott (1966). Mathematical textbooks with an emphasis on applications in social science form a related type of text: in graph theory there is Harary, Norman, and Cartwright (1964); in difference equations there is Goldberg (1961); in general finite mathematics there is Kemeny, Snell, and Thompson (1966); and in general mathematics, including calculus, there is Bishir and Drews (1970) and the more statistically oriented survey of probability, calculus, and mathematical statistics by Gelbaum and March (1969). In the area of stochastic processes, most of the texts require a strong background in mathematics. A recent text with an applied emphasis, including models of social mobility, is that by Bhat (1972). Of course, in Markov chain theory there is Kemeny and Snell (1960) for finite chains, and for more general theory, one can study Bartholomew (1967) or Karlin (1966), where the level of prerequisite mathematics is quite high however. In abstract algebra, a text that more than most tries to emphasize intuitive meanings is Fraleigh (1967), but there is nothing like an applied emphasis even here. Texts in theory construction in which formal, but not necessarily quantitative, techniques are used are worth mentioning. Foremost from a case study viewpoint are the two volumes edited by Berger, Zelditch, and Anderson (1966, 1972). Other books in theory construction that are instructive range from the now almost classical statement by Zetterberg (1964) to more recent studies by Dubin (1969), Blalock (1969), Stinchcombe (1968), Abell (1971), Mullins (1971), and Reynolds (1971). In a class by itself as an expository masterpiece is the work by Berger et al. (1962), which includes an annotated bibliography by B. Finnie and T. Mayer.

Original Monographs. In this category one should place three classical books: Dodd (1942), Zipf (1949), and Rashevsky (1951). They were each subjected to critiques that may be found in some of the papers in the first category above. A "second generation" of original monographs is roughly demarcated from these three "early" works and comprises Arrow (1963), Blumen, Kogan, and McCarthy (1955), Cohen (1963), Coleman (1964, a, b),McPhee (1963), Simon (1957), and White (1963). Other research monographs include Fararo and Sunshine (1964), Lazarsfeld and Henry (1968), Rapoport and Chammah (1965), Richardson (1960), Suppes and Atkinson (1960), and White (1970).

Research Papers. By far the largest and fastest growing category of literature in the field is that of research papers published in journals, yearbooks, and anthologies. The growing edge of the field is best represented here, especially among some highly creative younger men who have not put their ideas into monograph form. It would be pointless to list citations in this category: hardly a month goes by that the list is not expanded to include new papers by previously unknown persons. The present bibliography does not do full justice to all the extant contributions in this category, simply because it is confined to those papers explicitly mentioned at some point in the main body of the book. Particularly significant is the number of people contributing to a modern sociometry: a good, but already dated, bibliography in this area is to be found in Doreian (1970).

BIBLIOGRAPHY

Abell, P. (1968). "Measurement in Sociology: I. Measurement and Systems." Sociology, 2, 1–20.

Abell, P. (1969). "Measurement in Sociology: II. Measurement Structure and Sociological Theory." Sociology, 3, 397-411.

Abell, P. (1971). Model Building in Sociology. London: Weidenfeld and Nicolson.

Abelson, R. P. and Rosenberg, M. J. (1958). "Symbolic Psycho-logic: A Model of Attitudinal Cognition." Beh. Sci., 3, 1-13.

Abelson, R. P. (1967). "Mathematical Models in Social Psychology," in L. Berkowitz, ed., Advances in Experimental Social Psychology. Vol. 3. New York: Academic Press.

Anderson, A. R. and Moore, O. K. (1962). "Toward a Formal Analysis of Cultural Objects." Synthese, 14, 144-170.

Anderson, T. W. and Goodman, L. A. (1957). "Statistical Inference About Markov Chains." Ann. Math. Stat., 28, 89-110.

Arrow, K. J. (1951). "Mathematical Models in the Social Sciences," in D. Lerner and H. D. Lasswell, eds., The Policy Sciences. Stanford: Stanford University Press, Chap. 8.

Arrow, K. J., Karlin, S., and Suppes, P., eds. (1960). Mathematical Methods in the Social Sciences. Stanford: Stanford University Press.

Arrow, K. J. (1963). Social Choice and Individual Values. 2nd ed. New York: Wiley.

Asch, S. (1951). "Effects of Group Pressure upon the Modification and Distortion of Judgment," in H. Guetzkow, ed., Groups, Leadership and Men. Pittsburgh: Carnegie Press, pp. 177-190.

Atkinson, R. C., Bower, G. H., and Crothers, E. J. (1965). An Introduction to Mathematical Learning Theory. New York: Wiley.

Axten, N. (1972). "Generative Theories in the Human Sciences," in Proceedings: Third Annual Pittsburgh Conference on Modeling and Simulation. Department of Electrical Engineering, University of Pittsburgh.

Bailey, N. T. J. (1964). The Elements of Stochastic Processes with Applications to the Natural Sciences. New York: Wiley.

Bales, R. F. (1950). A Set of Categories for the Analysis of Small Group Interaction. Am. Soc. Rev., 15, 257-263.

Bartholomew, D. J. (1967). Stochastic Models for Social Processes. New York: Wiley.

Bartle, R. G. (1964). The Elements of Real Analysis. New York: Wiley.

Bartos, O. (1967). Simple Models of Group Behavior. New York: Columbia University Press.

Batchelder, W. H., Bjork, R. A., and Yellott, J. I. Jr. (1966). Problems in Mathematical Learning Theory with Solutions. New York: Wiley.

Beauchamp, M. A. (1970). Elements of Mathematical Sociology. New York: Random House.

Bellman, R. (1961). Adaptive Control Processes: A Guided Tour. Princeton: Princeton University Press.

Berger, J., Cohen, B. P., Connor, T. L., and Zelditch, M., Jr. (1966). "Status Characteristics and Expectation States: A Process Model," in J. Berger, M. Zelditch, Jr. and B. Anderson, eds., Sociological Theories in Progress. Vol. I. Boston: Houghton Mifflin, Chap. 3.

Berger, J. Cohen, B. P., Snell, J. L., and Zelditch, M., Jr. (1962). Types of Formalization in Small-Group Research. Boston: Houghton Mifflin.

Berger, J., Cohen, B. P., and Zelditch, M., Jr. (1966). "Status Characteristics and Expectation States," in J. Berger, M. Zelditch, Jr., and B. Anderson, eds., Sociological Theories in Progress. Vol. I. Boston: Houghton Mifflin, Chap. 2.

Berger, J., Connor, T. L., and Fisek, M. H., eds. (in press). Expectation-States Theory: A Theoretical Research Program. Cambridge: Winthrop Press.

Berger, J., Connor, T. L., and McKeown, W. (1969). "Evaluations and the Formation and Maintenance of Performance Expectations." Hum. Rels., 22, 186-198.

Berger, J. and Snell, J. L. (1957). "On the Concept of Equal Exchange." Beh. Sci., 2, 111-118.

Berger, J., Zelditch, M., Jr., and Anderson, B. (1966, 1972). Sociological Theories in Progress. Vols. I and II. Boston: Houghton Mifflin.

Bhat, U. N. (1972). Elements of Applied Stochastic Processes. New York: Wiley.

Bierstedt, R. (1970). The Social Order. 3rd ed. New York: McGraw-Hill.

Bishir, J. W. and Drewes, D. W. (1970). Mathematics in the Behavioral and Social Sciences. New York: Harcourt, Brace and World.

Blalock, H. M., Jr. (1969). Theory Construction. Englewood Cliffs: Prentice-Hall.

Blalock, H. M., Jr., ed. (1971). Causal Models in the Social Sciences. Chicago: Aldine.

Blau, P. M. (1964). Exchange and Power in Social Life. New York: Wiley.

Blau, P. M. and Duncan, O. D. (1967). The American Occupational Structure. New York: Wiley.

Blumen, I. M., Kogan, M., and McCarthy, P. J. (1955). The Industrial Mobility of Labor as a Probability Process. Ithaca: Cornell University Press.

Boudon, R. (1968). "A New Look at Correlation Analysis," in H. M. Blalock, Jr. and A. B. Blalock, eds., Methodology in Social Research. New York: McGraw-Hill, pp. 199-235.

Boyd, J. P. (1969). "The Algebra of Group Kinship." J. of Math. Psych., 6, 139-167.

Boyle, R. P. (1969). "Algebraic Systems for Normal and Hierarchial Sociograms." Sociometry, 32, 99-119.

Bradley, R. A. (1954a). "Rank Analysis of Incomplete Block Designs. II. Additional Tables for the Method of Paired Comparisons." Biometrika, 41, 502-537.

Bradley, R. A. (1954b). "Incomplete Block Rank Analysis: On the Appropriateness of the Model for a Method of Paired Comparisons." Biometrics, 10, 375-390.

Bradley, R. A. (1955). "Rank Analysis of Incomplete Block Designs. III. Some Large-Sample

Results on Estimation and Power for a Method of Paired Comparisons." Biometrika, **42**, 450-470.

Bradley, R. A. and Terry, M. E. (1952). "Rank Analysis of Incomplete Block Designs: I. The Method of Paired Comparisons." Biometrika, **39**, 324-345.

Braithwaite, R. B. (1955). Theory of Games as a Tool for the Moral Philosopher. Cambridge: Cambridge University Press.

Braithwaite, R. B. (1956). Scientific Explanation. Cambridge: Cambridge University Press.

Brauer, F. and Nohel, J. A. (1967). Ordinary Differential Equations. New York: Benjamin.

Buckley, W., ed. (1968). Modern Systems Research for the Behavioral Scientist. Chicago: Aldine.

Bush, R. R. (1963). "Estimation and Evaluation," in R. D. Luce, R. R. Bush, and E. Galanter, eds., Handbook of Mathematical Psychology. Vol. I. New York: Wiley, pp. 429-469.

Bush, R. R. and Mosteller, F. (1955). Stochastic Models for Learning. New York: Wiley.

Carnap, R. (1956). Meaning and Necessity: A Study in Semantics and Modal Logic. Chicago: University of Chicago Press.

Carnap, R. (1958). An Introduction to Symbolic Logic and Its Applications. New York: Dover.

Carnap, R. (1969). (1) The Logical Structure of the World, and (2) Pseudoproblems in Philosophy. Berkeley: University of California Press.

Cartwright, D. and Harary, F. (1956). "Structural Balance: A Generalization of Heider's Theory." Psych. Rev., **63**, 277-294.

Cartwright, D. and Harary, F. (1970). "Ambivalence and Indifference in Generalizations of Structural Balance." Beh. Sci., **15**, 506-511.

Chernoff, H. and Moses, L. E. (1959). Elementary Decision Theory. New York: Wiley.

Chomsky, N. (1965). Aspects of the Theory of Syntax. Cambridge, Mass.: M.I.T. Press.

Chomsky, N. and Miller, G. A. (1963). "Introduction to the Formal Analysis of Natural Languages," in R. D. Luce, R. R. Bush, and E. Galanter, eds., Handbook of Mathematical Psychology. Vol. II. New York: Wiley.

Churchman, C. W. and Ratoosh, P., eds. (1959). Measurement: Definitions and Theories. New York: Wiley.

Cohen, B. P. (1963). Conflict and Conformity: A Probability Model and Its Application. Cambridge, Mass.: M.I.T. Press.

Cohn, P. M. (1965). Universal Algebra. New York: Harper and Row.

Coleman, J. S. (1960). "The Mathematical Study of Small Groups," in H. Solomon, ed., Mathematical Thinking in the Measurement of Behavior. New York: The Free Press, part I.

Coleman, J. S. (1964a). An Introduction to Mathematical Sociology. New York: The Free Press.

Coleman, J. S. (1964b). Models of Change and Response Uncertainty. Englewood Cliffs: Prentice-Hall.

Coleman, J. S. (1968). "The Mathematical Study of Change," in H. M. Blalock, Jr. and A. B. Blalock, eds., Methodology in Social Research. New York: McGraw-Hill, Chap. 11.

Conviser, R. (1970). A Theory of Interpersonal Trust. Ph.D. thesis. Department of Social Relations, Johns Hopkins University.

Cox, D. R. and Lewis, P. A. W. (1966). The Statistical Analysis of Series of Events. London: Methuen.

Coxeter, H. S. M. and Moser, W. O. (1965). Generators and Their Relations for Discrete Groups. Berlin: Springer.

Coxon, A. P. M. (1970). "Mathematical Applications in Sociology: Measurement and Relations." Int'nl. J. of Math. Educ. in Sci. and Tech., 1, 159-174.

Criswell, J., Solomon, H., and Suppes, P., eds. (1962). Mathematical Methods in Small Group Processes. Stanford: Stanford University Press.

Davies, A. F. (1967). Images of Class. Sydney: Sydney University Press.

Davis, A., Gardner, B. B. and Gardner, M. R. (1941). Deep South: A Social Anthropological Study of Caste and Class. Chicago: University of Chicago Press.

Davis, J. A. (1967). "Clustering and Structural Balance in Graphs." Hum. Rels., 20, 181-188.

Davis, R. L. (1954). "Structure of Dominance Relations." Bull. Math. Biophysics, 16, 131-140.

Dodd, S. C. (1942). Dimensions of Society. New York: Macmillan.

Doreian, P. (1969). "A Note on the Detection of Cliques in Valued Graphs." Sociometry, 32, 237-242.

Doreian, P. (1970). Mathematics and the Study of Social Relations. London: Weidenfeld and Nicolson.

Doreian, P. and Hummon, N. P. (1972). "Sociological Modeling: Structural Versus Processual Approaches," in Proceedings: Third Annual Pittsburgh Conference on Modeling and Simulation. Department of Electrical Engineering, University of Pittsburgh.

Dubin, R. (1969). Theory Building. New York: The Free Press.

Dumont, L. (1970). Homo Hierarchicus. London: Weidenfeld and Nicolson.

Duncan, O. D. (1961). "A Socio-economic Index for All Occupations," in A. J. Reiss et al. Occupations and Social Status. New York: The Free Press, Chap. 6.

Duncan, O. D. (1966). "Methodological Issues in the Analysis of Social Mobility," in N. J. Smelser and S. J. Lipset, eds., Social Structure and Mobility in Economic Development. Chicago: Aldine, Chap. 2.

Fararo, T. J. (1968). "Theory of Status." Gen. Sys., 13, 177-188.

Fararo, T. J. (1969a). "Nature of Mathematical Sociology." Soc. Res., 36, 75-92.

Fararo, T. J. (1969b). "Stochastic Processes," in E. Borgatta, ed., Sociological Methodology 1969. San Francisco: Jossey-Bass, Chap. 8.

Fararo, T. J. (1970a). "Status Dynamics," in E. Borgatta and G. Bohrnstedt, eds., Sociological Methodology 1970. San Francisco: Jossey-Bass, Chap. 16.

Fararo, T. J. (1970b). "Strictly Stratified Systems." Sociology, 4, 85-104.

Fararo, T. J. (1970c). "Theoretical Studies in Status and Stratification." Gen. Sys., 15, 71-101.

Fararo, T. J. (1972a). "Dynamics of Status Equilibration," in J. Berger, M. Zelditch, Jr. , and B. Anderson, eds., Sociological Theories in Progress. Vol. II. Boston: Houghton Mifflin, Chap. 9.

Fararo, T. J. (1972b). "Status, Expectations and Situation: A Formulation of the Structure Theory of Status Characteristics and Expectation States." Qual. and Quant., 6, 37-98.

Fararo, T. J. and Sunshine, M. (1964). A Study of a Biased Friendship Net. Syracuse: Syracuse University Youth Development Center.

Feller, W. (1966). An Introduction to Probability Theory and Its Applications. Vol. II. New York: Wiley.

Flament, C. (1963). Applications of Graph Theory to Group Structure. Englewood Cliffs: Prentice-Hall.

Foster, C. C., Rapoport, A., and Orwant, C. (1963). "A Study of a Large Sociogram II." Beh. Sci., 8, 56-65.

Fraleigh, J. B. (1967). A First Course in Abstract Algebra. Reading, Mass.: Addison-Wesley.

Friedell, M. (1967). "Organizations as Semilattices." Am. Soc. Rev., 32, 46-54.

Friedell, M. (1969). "On the Structure of Shared Awareness." Beh. Sci., 14, 28-39.

Galanter, E. (1966). Textbook of Elementary Psychology. San Francisco: Holden-Day.

Galtung, J. (1966). "Rank and Social Integration: A Multidimensional Approach," in J. Berger, M. Zelditch, Jr., and B. Anderson, eds., Sociological Theories in Progress. Vol. I. Boston: Houghton Mifflin, Chap. 7.

Gelbaum, B. R. and March, J. G. (1969). Mathematics for the Social and Behavioral Sciences. Philadelphia: Saunders.

Ginsberg, R. B. (1971). "Semi-Markov Processes and Mobility." J. of Math. Soc., 1, 233-262.

Glass, D. V. and Hall, J. R. (1954). "Social Mobility in Great Britain: A Study of Intergenerational Changes in Status," in D. V. Glass, ed., Social Mobility in Great Britain. London: Routledge and Kegan Paul.

Goldberg, S. (1960). Probability: An Introduction. Englewood Cliffs: Prentice-Hall.

Goldberg, S. (1961). Introduction to Difference Equations. New York: Wiley.

Goodman, L. A. (1961). "Statistical Methods for the 'Mover-Stayer' Model." J. Amer. Stat. Assoc., 56, 841-868.

Grossman, I. and Magnus, W. (1964). Groups and Their Graphs. New York: Random House.

Hall, A. D. and Fagen, R. E. (1968). "Definition of System," in W. Buckley, ed., Modern Systems Research for the Behavioral Scientist. Chicago: Aldine, Chap. 10.

Harary, F., Norman, R. Z., and Cartwright, D. (1965). Structural Models: An Introduction to the Theory of Directed Graphs. New York: Wiley.

Harsanyi, J. C. (1956). "Approaches to the Bargaining Problem Before and After the Theory of Games: a Critical Discussion of Zeuthen's, Hick's, and Nash's Theories." Econometrica, 24, 144-157.

Hedburg, M. (1961). "The Turnover of Labor in Industry, an Actuarial Study." Acta Sociologica, 5, 129-143.

Heider, F. (1958). The Psychology of Interpersonal Relations. New York: Wiley.

Hempel, C. G. (1952). Fundamentals of Concept Formation in Empirical Science. Chicago: University of Chicago Press.

Henry, N. W., McGinnis, R., and Tegtmeyer, H. W. (1971). "A Finite Model of Mobility." J. of Math. Soc., 1, 107-118.

Herbst, P. G. (1963). "Organizational Commitment: A Decision Process Model." Acta Sociologica, 7, 34-35.

Higman, B. (1964). Applied Group Theoretic and Matrix Methods. New York: Dover.

Hodge, R. W., Siegel, P. M., and Rossi, P. H. (1966). "Occupational Prestige in the United States: 1925-1963," in R. Bendix and S. M. Lipset, eds., Class, Status and Power. 2nd ed. New York: The Free Press, pp. 322-334.

Holland, J. and Steuer, M. D. (1970). Mathematical Sociology: A Selective Annotated Bibliography. New York: Schocken.

Holland, P. and Leinhardt, S. (1971). "Transitivity in Structural Models of Small Groups." Comp. Group Studies, 2, 107-124.

Homans, G. C. (1961). Social Behavior: Its Elementary Forms. New York: Harcourt, Brace and World.

Howard, R. A. (1971). Dynamic Probabilistic Systems. Vol. I. New York: Wiley.

Hu, S. T. (1965). Elements of Modern Algebra. San Francisco: Holden-Day.

Jacobson, N. (1951). Lectures in Abstract Algebra. Vol. I. Princeton: Van Nostrand.

Jaeckel, M. (1971). "Coleman's Process Approach," in H. Costner, ed., Sociological Methodology 1971. San Francisco: Jossey-Bass, Chap. 9.

Kaplan, A. (1952). "Sociology Learns the Language of Mathematics." Commentary, 14, 274-284.

Karlin, S. (1966). A First Course in Stochastic Processes. New York: Academic Press.

Karlsson, G. (1958). Social Mechanisms. New York: The Free Press.

Kay, P., ed. (1971). Explorations in Mathematical Anthropology. Cambridge, Mass.: M.I.T. Press.

Kemeny, J. G. and Snell, J. L. (1960). Finite Markov Chains. Princeton: Van Nostrand.

Kemeny, J. G. and Snell, J. L. (1962). Mathematical Models in the Social Sciences. Boston: Ginn.

Kemeny, J. G., Schleifer, A., Jr., Snell, J. L., and Thompson, G. L. (1962). Finite Mathematics with Business Applications. Englewood Cliffs: Prentice-Hall.

Kemeny, J. G., Snell, J. L., and Thompson, G. L. (1966). Introduction to Finite Mathematics. 2nd ed. Englewood Cliffs: Prentice-Hall.

Kolmogorov, A. (1933). Foundations of Probability. English edition, 1950. New York: Chelsea.

Kurosh, A. G. (1963). Lectures on General Algebra. New York: Chelsea.

Land, K. C. (1969). "Principles of Path Analysis," in E. Borgatta, ed., Sociological Methodology 1969. San Francisco: Jossey-Bass, Chap. 1.

Land, K. C. (1970). "Mathematical Formalization of Durkheim's Theory of Division of Labor," in E. Borgatta and G. Bohrnstedt, eds., Sociological Methodology 1970. San Francisco: Jossey-Bass, Chap. 15.

Land, K. C. (1971). "Some Exhaustible Poisson Process Models of Divorce by Marriage Cohort." J. of Math. Soc., 1, 213-232.

Landau, H. G. (1951). "On Dominance Relations and the Structure of Animal Societies: I. Effect of Inherent Characteristics." Bull. Math. Biophysics, 13, 1-19.

Lang, S. (1965). Algebra. Reading, Mass.: Addison-Wesley.

Langer, R., Krohn, K., and Rhodes, J. (1968). "Transformations, Semigroups, and Metabolism," in M. D. Mesarovic, ed., Systems Theory and Biology. New York: Springer, pp. 130-140.

Lazarsfeld, P. F. (1954). "A Conceptual Introduction to Latent Structure Analysis," in Lazarsfeld, ed., Mathematical Thinking in the Social Sciences. New York: The Free Press.

Lazarsfeld, P. F. and Henry, N. W. (1968). Latent Structure Analysis. New York: Houghton Mifflin.

Lazarsfeld, P. F. and Henry, N. W., eds. (1968). Readings in Mathematical Social Science. Cambridge, Mass.: M.I.T. Press.

Levine, J. H. (1972). "A Two-Parameter Model of Interaction in Father-Son Status Mobility." Beh. Sci., 17, 455-465.

Lewis, G. (1971). "The Accuracy of Alternative Stochastic Models of Participation in Group Discussion." J. Math. Soc., 1, 263-276.

Lorge, I. and Solomon, H. (1960). "Group and Individual Performance in Problem Solving Related to Previous Exposure to Problem, Level of Aspiration, and Group Size." Beh. Sci. 5, 28-38.

Lorrain, F. and White, H. C. (1971). "Structural Equivalence of Individuals in Social Networks." J. Math. Soc., 1, 49-80.

Luce, R. D. (1959). Individual Choice Behavior. New York: Wiley.

Luce, R. D. (1959p). "On the Possible Psychophysical Laws." Psych. Rev., 66, 81-95.

Luce, R. D., Bush, R. R., and Galanter, E., eds. (1963, 1965). Handbook of Mathematical Psychology. Volumes I, II, and III. New York: Wiley.

Luce, R. D., Bush, R. R., and Galanter, E., eds. (1963). Readings in Mathematical Psychology. Vols. I and II. New York: Wiley.

Luce, R. D. and Raiffa, H. (1957). Games and Decisions. New York: Wiley.

Luce, R. D. and Suppes, P. (1965). "Preference, Utility, and Subjective Probability," in R. D. Luce, R. R. Bush, and E. Galanter, eds., Handbook of Mathematical Psychology. Vol. III. New York: Wiley, Chap. 19.

McFarland, D. D. (1970). "Intragenerational Social Mobility as a Markov Process: Including a Time-Stationary Markovian Model That Explains Observed Declines in Mobility Rates." Am. Soc. Rev., 35, 463-476.

McGinnis, R. (1968). "A Stochastic Model of Social Mobility." Am. Soc. Rev., 33, 712-722.

McGinnis, R., Myers, G. C., and Pilger, J. (1963). "Internal Migration as a Stochastic Process." Paper presented at the 34th session, International Statistical Institute, Ottawa, Canada.

MacLane, S., and Birkhoff, G. (1967). Algebra. New York: Macmillan.

McPhee, W. N. (1963). Formal Theories of Mass Behavior. New York: The Free Press.

McPhee, W. N. (1966). "When Culture Becomes a Business," in J. Berger, M. Zelditch, Jr., and B. Anderson, eds., Sociological Theories in Progress. Vol. I. Boston: Houghton Mifflin, Chap. 10.

Martindale, D. (1963). "Limits to the Uses of Mathematics in the Study of Sociology," in J. C. Charlesworth, ed., Mathematics and the Social Sciences. Philadelphia: American Academy of Political and Social Sciences.

Massarik, F. and Ratoosh, P., eds. (1965). Mathematical Explorations in Behavioral Science. Homewood, Ill.: Irwin-Dorsey.

Mayer, T. (1971). "Mathematical Sociology: Some Educational and Organizational Problems of an Emergent Sub-discipline." Int'l. J. of Math. Educ. in Sci. and Tech., 2, 217-232.

Mayer, T. (1972). "Continuous Time Models of Intragenerational Mobility," in J. Berger, M. Zelditch, Jr., and B. Anderson, eds., Sociological Theories in Progress. Vol. II. Boston: Houghton Mifflin.

Mead, G. H. (1934). Mind, Self, and Society. Chicago: University of Chicago Press (paperback, edited by C. W. Morris, 1962).

Mood, A. M. and Graybill, F. A. (1963). Introduction to the Theory of Statistics. 2nd ed. New York: McGraw-Hill.

Morrison, P. A. (1967). "Duration of Residence and Prospective Migration: The Evaluation of a Stochastic Model." Demography, 4, 559-560.

Mullins, N. C. (1971). The Art of Theory: Construction and Use. New York: Harper and Row.

Nash, J. F. (1950). "The Bargaining Problem." Econometrica, 18, 155-162.

Newell, A. and Simon, H. A. (1972). Human Problem Solving. Englewood Cliffs: Prentice-Hall.

Oeser, O. A. and Harary, F. (1962). "A Mathematical Model for Structural Role Theory I." Hum. Rels., 15, 89-110.

Oeser, O. A. and Harary, F. (1964). "A Mathematical Model for Structural Role Theory II." Hum. Rels., 17, 3-18.

Owen, G. (1968). Game Theory. Philadelphia: Saunders.

Paris-Steffens, J. R., Kurtz, N. R., McPhee, W. N., and Rose, E. (1966). Formal Theory in the Behavioral Sciences: A Bibliography. Institute of Behavioral Sciences, University of Colorado, Boulder, Colorado.

Parsons, T. (1937). Structure of Social Action. New York: McGraw-Hill.

Parsons, T., Bales, R. F., and Shils, E. A. (1953). Working Papers in the Theory of Action. New York: Free Press.

Parzen, E. (1960). Modern Probability Theory and Its Applications. New York: Wiley.

Parzen, E. (1962). Stochastic Processes. San Francisco: Holden-Day.

Pfanzagl, J. (1968). Theory of Measurement. New York: Wiley.

Popper, K. (1959). The Logic of Scientific Discovery. New York: Basic Books.

Prais, S. J. (1955). "The Formal Theory of Social Mobility." Population Studies, 9, 72-81.

Pullum, T. W. (1970). "What Can Mathematical Models Tell Us About Occupational Mobility?" Soc. Inquiry, 40, 258-280.

Pyke, R. (1961). "Markov Renewal Processes: Definitions and Preliminary Properties." Ann. of Math. Stat., 32, 1231-1242.

Rapoport, A. (1959). "Uses and Limitations of Mathematical Models in Social Science," in L. Gross, ed., Symposium on Sociological Theory. New York: Row-Peterson, Chap. 11.

Rapoport, A. (1960). Fights, Games, and Debates. Ann Arbor: University of Michigan Press.

Rapoport, A. (1963). "Mathematical Models of Social Interaction," in R. D. Luce, R. R. Bush, and E. Galanter, eds., Handbook of Mathematical Psychology. Vol. II. New York: Wiley.

Rapoport, A. (1966). Two-Person Game Theory. Ann Arbor: University of Michigan Press.

Rapoport, A. (1970). N-Person Game Theory. Ann Arbor: University of Michigan Press.

Rapoport, A. and Chammah, A. M. (1965). Prisoner's Dilemma: A Study of Conflict and Cooperation. Ann Arbor: University of Michigan Press.

Rapoport, A. and Horvath, W. J. (1961). "A Study of a Large Sociogram I." Beh. Sci., 6, 279-291.

Rashevsky, N. (1951). Mathematical Biology of Social Behavior. Chicago: University of Chicago Press.

Restle, F. (1961). Psychology of Judgment and Choice. New York: Wiley.

Reynolds, P. D. (1971). A Primer in Theory Construction. Indianapolis: Bobbs-Merrill.

Richardson, L. F. (1960). Arms and Insecurity. Pittsburgh: Boxwood Press.

Rogoff, N. (1953). Recent Trends in Occupational Mobility. New York: The Free Press.

Rosenberg, S. (1968). "Mathematical Models of Social Behavior," in G. Lindzey and E. Aronson, eds., The Handbook of Social Psychology. Vol. I. Reading, Mass.: Addison-Wesley, Chap. 3.

Royden, H. L. (1963). Real Analysis. New York: Macmillan.

Rumelhart, D. L. and Greeno, J. G. (1971). "Similarity Between Stimuli: An Experimental Test of the Luce and Restle Choice Models." J. of Math. Psych., 8, 370-381.

Saaty, T. (1968). Mathematical Models of Arms Control and Disarmament. New York: Wiley.

Sanchez, D. A. (1968). Ordinary Differential Equations and Stability Theory: An Introduction. San Francisco: Freeman.

Schelling, T. C. (1960). The Strategy of Conflict. Cambridge, Mass.: Harvard University Press.

Schwarz, R. J. and Friedland, B. (1965). Linear Systems. New York: McGraw-Hill.

Scott, D. and Suppes, P. (1958). "Foundational Aspects of Theories of Measurement." J. of Symbolic Logic, 23, 113-128.

Shapley, L. S. (1953). "A Value for n-Person Games," in H. W. Kuhn and A. W. Tucker, eds., Contributions to the Theory of Games, II. (Annals of Mathematical Studies, 28). Princeton: Princeton University Press.

Shubik, M., ed. (1964). Game Theory and Related Approaches to Social Behavior. New York: Wiley.

Simon, H. (1957). Models of Man. New York: Wiley, Chap. 6.

Skvoretz, J. (1971). "Occupational Prestige: A Study in the Logic of Measurement." Unpublished paper, Department of Sociology, University of Pittsburgh.

Skvoretz, J., Windell, P., and Fararo, T. J. (1972). "Luce's Axiom and Occupational Prestige: Test of a Measurement Model." Unpublished paper, Department of Sociology, University of Pittsburgh.

Spilerman, S. (1970). "The Causes of Racial Disturbance: A Comparison of Some Alternative Explanations." Am. Soc. Rev., 35, 627-649.

Stinchcombe, A. L. (1968). Constructing Social Theories. New York: Harcourt, Brace and World.

Stouffer, S. and others. (1949). The American Soldier: Adjustment During Army Life. Vols. I and II. Princeton: Princeton University Press.

Strodtbeck, F. L., James, R. M., and Hawkins, C., (1957) "Social Status in Jury Deliberations." Am. Soc. Rev., 22, 713-719.

Strodtbeck, F. L. and Mann, R. D. (1956). "Sex Role Differentiation in Jury Deliberations." Sociometry, 19, 3-11.

Suppes, P. (1957). Introduction to Logic. Princeton: Van Nostrand.

Suppes, P. and Atkinson, R. C. (1960). Markov Learning Models for Multiperson Interactions. Stanford: Stanford University Press.

Suppes, P. and Zinnes, J. L. (1963). "Basic Measurement Theory," in R. D. Luce, R. R. Bush, and E. Galanter, eds., Handbook of Mathematical Psychology. Vol. I. New York: Wiley, Chap. 1.

Tarksi, A. (1946). Introduction to Logic and to the Methodology of Deductive Sciences. 2nd ed., rev. New York: Oxford University Press.

Teuter, K. (1969). "Towards a Formal Concept of Class Consciousness." Unpublished paper, Department of Sociology, University of Pittsburgh.

Tinter, G. (1968). Methodology of Mathematical Economics and Econometrics. Chicago: University of Chicago Press.

Torgerson, W. S. (1958). Theory and Methods of Scaling. New York: Wiley.

Toulmin, S. (1960). The Philosophy of Science. New York: Harper Torchbooks (first published by Hutchinson, London, 1953).

von Neumann, J. and Morgenstern, O. (1947). Theory of Games and Economic Behavior. 2nd ed. Princeton: Princeton University Press.

White, D. R. (1972). "Mathematical Anthropology," in J. J. Honigmann, ed., Handbook of Social and Cultural Anthropology. Chicago: Rand-McNally.

White, H. C. (1963). An Anatomy of Kinship. Englewood Cliffs: Prentice-Hall.

White, H. C. (1970). Chains of Opportunity. Cambridge, Mass.: Harvard University Press.

Whitehead, A. N. (1929). Process and Reality. New York: Macmillan.

Yale, P. (1968). Geometry and Symmetry. San Francisco: Holden-Day.

Zehna, P. W. and Johnson, R. L. (1962). Elements of Set Theory. Boston: Allyn and Bacon.

Zelditch, M., Jr. (1964). "Family, Marriage, and Kinship," in R. E. L. Faris, ed., Handbook of Modern Sociology. Chicago: Rand-McNally.

Zelditch, M., Jr., Berger, J., and Cohen, B. P. (1966). "Stability of Organizational Status Structures," in J. Berger, M. Zelditch, Jr., and B. Anderson, eds., Sociological Theories in Progress. Vol. I. Boston: Houghton Mifflin.

Zetterberg, H. (1964). On Theory and Verification in Sociology. Rev. ed. Totowa, N.J.: Bedminster Press.

Zipf, G. K. (1949). Human Behavior and the Principle of Least Effort. Reading, Mass.: Addison-Wesley.

AUTHOR INDEX

SUBJECT INDEX